55

Springer Series in Chemical Physics

Edited by Fritz P. Schäfer

Springer Series in Chemical Physics

Editors: Vitalii I. Goldanskii   Fritz P. Schäfer   J. Peter Toennies

Managing Editor: H. K. V. Lotsch

Volume 40 **High-Resolution Spectroscopy of Transient Molecules**
By E. Hirota

Volume 41 **High Resolution Spectral Atlas of Nitrogen Dioxide 559–597 nm**
By K. Uehara and H. Sasada

Volume 42 **Antennas and Reaction Centers of Photosynthetic Bacteria**
Structure, Interactions, and Dynamics
Editor: M. E. Michel-Beyerle

Volume 43 **The Atom-Atom Potential Method.** Applications to Organic Molecular Solids
By A. J. Pertsin and A. I. Kitaigorodsky

Volume 44 **Secondary Ion Mass Spectrometry SIMS V**
Editors: A. Benninghoven, R. J. Colton, D. S. Simons, and H. W. Werner

Volume 45 **Thermotropic Liquid Crystals, Fundamentals**
By G. Vertogen and W. H. de Jeu

Volume 46 **Ultrafast Phenomena V**
Editors: G. R. Fleming and A. E. Siegman

Volume 47 **Complex Chemical Reaction Systems**
Mathematical Modelling and Simulation
Editors: J. Warnatz and W. Jäger

Volume 48 **Ultrafast Phenomena VI**
Editors: T. Yajima, K. Yoshihara, C. B. Harris, and S. Shionoya

Volume 49 **Vibronic Interactions in Molecules and Crystals**
By I. B. Bersuker and V. Z. Polinger

Volume 50 **Molecular and Laser Spectroscopy**
By Zu-Geng Wang and Hui-Rong Xia

Volume 51 **Space-Time Organization in Macromolecular Fluids**
Editors: F. Tanaka, M. Doi, and T. Ohta

Volume 52 **Multiple-Photon Laser Chemistry**
By. R. V. Ambartzumian, C. D. Cantrell, and A. Puretzky

Volume 53 **Ultrafast Phenomena VII**
Editors: C. B. Harris, E. P. Ippen, G. A. Mourou, and A. H. Zewail

Volume 54 **Physics of Ion Impact Phenomena**
Editor: D. Mathur

Volume 55 **Ultrafast Phenomena VIII**
Editors: J.-L. Martin, A. Migus, G. A. Mourou, and A. H. Zewail

Volume 1–39 are listed on the back inside cover

J.-L. Martin · A. Migus
G. A. Mourou · A. H. Zewail (Eds.)

Ultrafast Phenomena VIII

Proceedings
of the 8th International Conference,
Antibes Juan-Les-Pins, France, June 8–12, 1992

With 523 Figures

Springer-Verlag
Berlin Heidelberg New York
London Paris Tokyo
Hong Kong Barcelona
Budapest

Professor Jean-Louis Martin
Professor Arnold Migus

Laboratoire d'Optique Appliquée, Ecole Polytechnique
ENS Techniques Avancées, F-91120 Palaiseau, France

Professor Gerald A. Mourou

Electrical Engineering and Computer Science Department
University of Michigan, Ann Arbor, MI 48109, USA

Professor Ahmed H. Zewail

Department of Chemistry, California Institute of Technology
Pasadena, CA 91125, USA

ISBN-13: 978-3-642-84912-1 e-ISBN-13: 978-3-642-84910-7

DOI: 10.1007/978-3-642-84910-7

Softcover reprint of the hardcover 1st edition 1993

Typesetting: Camera ready by authors/editors
54/3140 - 5 4 3 2 1 0 - Printed on acid-free paper

Preface

This volume contains papers presented at the Eighth International Conference on Ultrafast Phenomena held at Antibes Juan-les-Pins, France, from June 8 to 12, 1992. The conference was the eighth in the series initiated in 1978. The first Picosecond Phenomena meeting held in 1978 brought together about hundred scientists from various disciplines who were exploring the then emerging technologies of picosecond pulses and applications to the fields of physics, chemistry, biology, and electronics. Fourteen years later the Antibes meeting demonstrated the vitality and expansion of this field. The over 400 registered participants from 25 countries, including 85 students, represented a wide range of scientific disciplines sharing common interest in ultrafast phenomena. The particular success of this meeting continues to lie in its truly interdisciplinary character which facilitates the diffusion of concepts across different areas of physics, chemistry and biology. Progress was reported in studies of primary and fundamental processes in matter. These include molecular vibrations or phonons, collisions in condensed or dense media, charge transfer or conductivity, elementary atomic motion, molecular conformational changes, and others. Several papers in the volume are from overview talks, which outline progress and perspectives in the different disciplines. This book provides the most up-to-date record of current research and references in the field of ultrafast phenomena.

Acknowledgements

These interdisciplinary exchanges would have not been possible without grants from the following organizations: Centre National de la Recherche Scientifique (Départements Sciences Physiques et Mathématiques, Chimie, Sciences pour l'Ingénieur and Sciences de la Vie), Centre National d'Etudes des Télécommunications, Commissariat à l'Energie Atomique (Direction des Sciences de la Matière), Direction des Recherches et Etudes Doctorales (MEN), Direction des Recherches et Etudes Techniques, Ecole Nationale Supérieure des Techniques Avancées, Institut National de la Santé et de la Recherche Médicale, Ministère de la Recherche et de l'Espace, Programme Ultimatech (CNRS), and Secrétariat Général à la Défense Nationale.

We are indebted to the Société Française d'Optique, Société Française de Physique, European Optical Society, and Optical Society of America for their cooperation.

Much appreciation is owed to those who contributed to the success of this meeting, members of the Committee who agreed to spend week-ends selecting papers, the staff directed by Marie-Jo Lécuyer who helped us so efficiently in organizing and managing on site, Guy Hamoniaux and co-workers for taking care of companies in the organization of the exhibition, and Mr. Bouvignies with the Service d'Edition de l'ENSTA who did such a great job with the Programs and the Digests. The involvement of industry contributed also to the success of this meeting. We thank the following companies for their support and active commitment: E.G.G. (Division Instruments), Quantel, Spectra Physics, Coherent, Optilas (representing Continuum and Hadland), Hamamatsu Photonics (France), B.M. Industries, ARP (representing also Photek), Quantronix-PSI, Instruments S.A (Division Jobin-Yvon), Opton (representing Clark Instruments, Medox Research and Opton), Quartz et Silice, Medox Elect. Opt., Melles Griot, IOP Publishing, and Sopra (representing Oxford Laser). We also acknowledge support from Klinger (Newport/Micro-Controle), Groupe Saint-Gobain, and Thompson S.A.

The American Committee gratefully acknowledges the assistance of Barbara Hicks and of the Optical Society of America, and supports from the following governmental agencies: National Science Foundation, Air Force Office of Scientific Research and Defense Advanced Research Program Agency (DARPA), as well as additional contributions from the following companies: Schwartz Electro-Optics, New Focus, Spectra-Physics Lasers, Newport/Klinger Products, Clark Instrumentation, Milton Roy, Cleveland Crystals, Coherent, Lumonics, Hamamatsu Photonics, and Thorlabs.

Palaiseau, France	*Jean-Louis Martin*
Palaiseau, France	*Arnold Migus*
Ann Arbor, MI	*Gérard A. Mourou,*
Pasadena, CA	*Ahmed H. Zewail*
August 1992	

Contents

Part III Spectroscopy and Advances in Measurements

Part IV Tools: Sources and Amplifiers

Part VIII **Biology: Primary Dynamics, Electron and Energy Transfer**

Part I

Overview and General Prospects

Additive Pulse Modelocking and Kerr-Lens Modelocking

H.A. Haus

Department of Electrical Engineering and Computer Science, Research Laboratory of Electronics, Massachusetts Institute of Technology, Cambridge, MA 02139, USA

Abstract. Both Additive Pulse Modelocking and Kerr Lens Modelocking permit a fully analytic description allowing comparisons between the two schemes. Advantages and disadvantages of each are discussed. A figure of merit is proposed.

1. Introduction

Additive Pulse Modelocking (APM)[1–4] and Kerr Lens Modelocking (KLM)[5] are fairly new methods of generating short pulses from solid state lasers. Both produce an equivalent fast saturable absorber action (response time shorter than the pulse-width) that can be tailored to the system at hand. For this reason it is possible to modelock solid state lasers with long gain relaxation times. Normally, lasers with long gain relaxation times Q-switch when modelocking with available "natural" saturable absorbers is attempted. Further, saturable absorbers are wavelength specific, whereas both APM and KLM can be made to operate over a wide wavelength range.

APM started with coupled cavities that require active interferometric feedback stabilization. The laser beam is injected into an auxiliary cavity containing loss ($L < 1$ is the amplitude multiplier), a Kerr medium, and a fiber as shown in Fig. 1. The mode

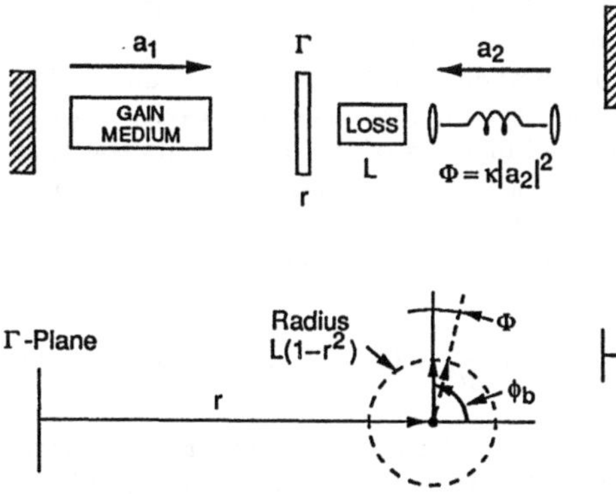

Fig. 1. Schematic of coupled cavity APM system; Reflection coefficient of auxiliary cavity in Γ-plane for narrowband pulse train.

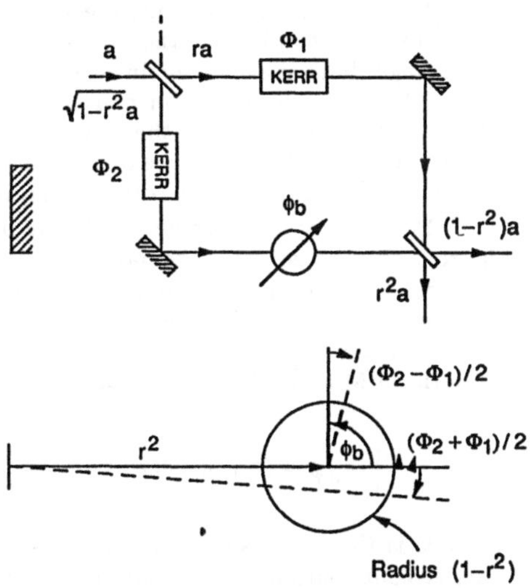

Fig. 2. Nonlinear Mach-Zehnder interferometer as transformer from nonlinear phase to amplitude modulation.

Springer Series in Chemical Physics, Vol. 55 Ultrafast Phenomena VIII
Editors: J.-L. Martin · A. Migus · G.A. Mourou · A.H. Zewail

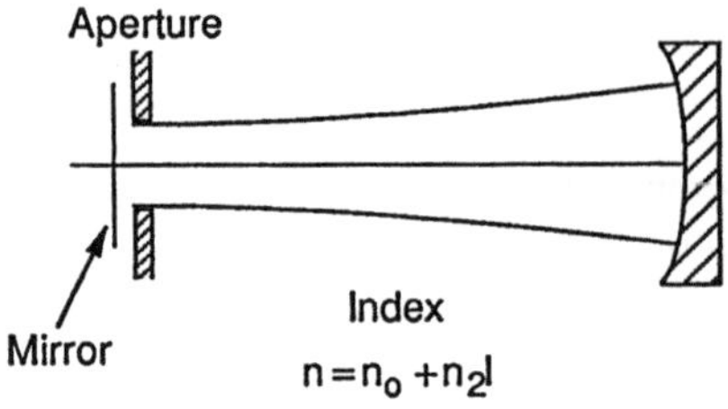

Fig. 3. Monolithic cavity Kerr Lens Resonator

amplitudes in the two cavities are a_1 and a_2. The Kerr medium shifts the phase of the returning pulse, producing a larger phase shift at the peak of the pulse, smaller phase shifts in the wings of the pulse. As this returning pulse is made to interfere at the mirror separating the two cavities, a shorter pulse is sent back into the main cavity, provided that the auxiliary cavity has the proper bias phase. This is shown schematically in Fig. 1 by the reflection coefficient presented to the main cavity by the auxilary cavity as viewed through the mirror. The nonlinear phase shift Φ changes the reflection coefficient. It may appear puzzling to treat the reflection coefficient experienced by the pulse as frequency independent, and only a function of intensity. In general, the reflection coefficient is a periodic function of frequency. However, for a pulse train periodic in time with a round trip time corresponding to the pulse repetition time all returning "echoes" from the auxiliary cavity superimpose. Further, when the bandwidth of the pulse is narrow enough to consider the bias phase ϕ_b as constant over the pulse spectrum, the treatment in terms of a nonlinear reflection coefficient is legitimate.

The bias phase must be maintained by feedback stabilization. The need for stabilization can be obviated, if no auxiliary cavity is used and a self-stabilized nonlinear interferometric element is introduced in the main cavity. Figure 2 shows a nonlinear Mach-Zehnder (M-Z) interferometer with Kerr media in the two arms. If the intensity in the two arms is different, the Kerr phase shifts Φ_1 and Φ_2 are different. Upon recombining the two beams, the difference in phase shifts is transformed into an amplitude change. This is shown again schematically in the complex transmission coefficient plane. The M-Z interferometer can be realized in a self-stabilizing form by using two polarizations propagating in the same (nonlinear) isotropic fiber. An elliptic polarization rotates under the influence of the Kerr self phase shift. With proper placement of polarizers and analyzers, the rotation can be transformed into amplitude modulation.

KLM produces a saturable amplitude modulation by self-focusing and aperturing. Figure 3 shows a simple schematic of a KLM system, consisting of a cavity with an aperture. Increases in the intensity produce self-focusing of the beam at its waist, an increase of the beam diameter at the mirror. An aperture placed at the waist causes intensity dependent loss, the loss decreasing with increasing intensity.

Both APM and KLM can be described by the master equation[7]

$$\left[-j\psi-(\ell+jx)+g\left(1+\frac{1}{\Omega_g^2}\frac{d^2}{dt^2}\right)+jD\frac{d^2}{dt^2}+(\gamma-j\delta)|a|^2\right]a=0 \tag{1}$$

where ψ is the phase shift per pass, ℓ is the loss, g the gain per pass, x is the linear reactive contribution of the media in the resonator, Ω_g is the gain bandwidth; D is the group velocity dispersion (GVD) parameter; $\gamma|a|^2$ and $\delta|a|^2$ are the self-amplitude and self-phase modulations.

Equation (1) is solved by the ansatz:

$$a_o(t)=A_o[\text{sech}(t/\tau)]^{(1+j\beta)} \tag{2}$$

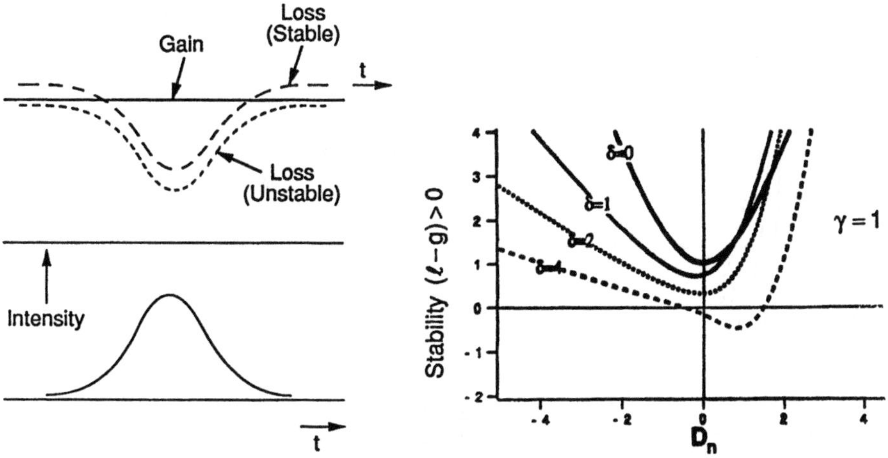

Fig. 4. Stability against growth of narrowband noise preceding and following the pulse. For stability $\ell - g > 0$.

where β is the chirp parameter. When the ansatz (2) is introduced into (1), $a_o(t)$ acquires multipliers that are sech^2 and constants. Setting these multipliers equal to zero results in four real equations for $\psi, g - \ell, \tau$, and β[7].

Not the entire range of solutions is physically realizable. The gain must be less than the loss preceding and following the pulse in order to prevent the growth of noise. This amounts to the stability criterion shown in Fig. 4. The parameter $\ell - g$ must be positive preceding and following the pulse. This stability criterion is necessary, but not sufficient. It has been shown that the pulses obtained from (1) with $\delta = D = 0$ are stable[8]. Computer simulations and analytical studies[9] show that in the limit of strong soliton behavior (large δ, small γ and large Ω_g) self-amplitude modulation represented by γ tends to destabilize the pulses.

The evaluation of the coefficients in the master equation (1) for any particular system can be done analytically both for APM and KLM when some approximations are made[10]. In the APM case, one expands the expressions to first order in the nonlinear phase shifts Φ_i. In the KLM case, one assumes a gaussian beam profile and approximates the Kerr self focusing by replacing the gaussian index profile with a parabolic one of the same radius of curvature at the peak[11]. When this is done, a simple analytic transformation of the q-parameter of the beam is obtained that requires no integration[12]. This analytic treatment, which we do not repeat here, permits a comparison of the two methods of modelocking. Figure 5 shows plots of the self-amplitude modulation parameter (γ) divided by the loss parameter ℓ in MW^{-1} for the APM system of Fig. 2 utilizing bulk components, and the KLM system using composite cavities of lengths L_1 and L_2($L_1 = L_2$ for plot) and an aperture at one of the mirrors backed against the Kerr medium of length L as shown in the inset. The monolithic KLM system has a value of $\gamma/\ell = 0.13$ MW^{-1}. One can see that the bulk component APM system has values comparable to the monolithic KLM. The composite cavity KLM shown in the inset has much larger values "bought" by operating near the mode stability boundary ($1 - L_1/f \to 0$). If a fiber of 1 meter length is used for the Kerr medium, with a mode diameter of 9 μ, the values of Fig. 5a are multiplied by 3,750.

The analytic treatment also arrives at values of a figure of merit, the parameter $M = \gamma/\delta\ell$, where ℓ is the loss parameter, γ is the self-amplitude modulation parameter

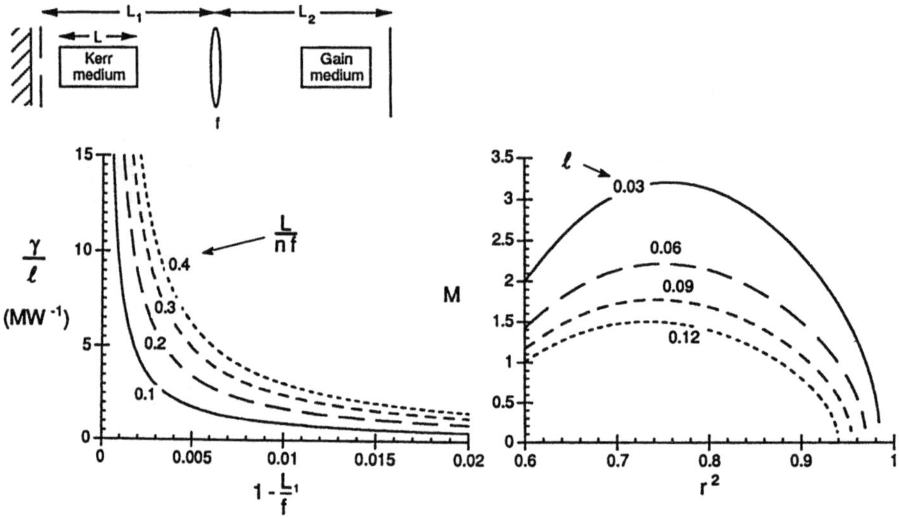

Fig. 5. The self-amplitude modulation parameter for APM (bulk components) and for KLM (composite cavity).

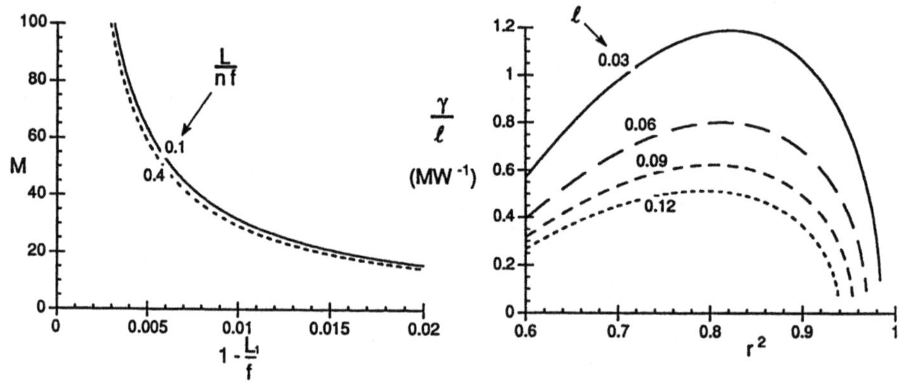

Fig. 6. The figure of merit for APM and KLM (composite cavity).

divided by the self-phase shift parameter δ. The rationale for the use of the parameter M as an indicator of quality is the following. The greater the value of γ, the more one may overcome bandwidth limitation. Further, the best performance is obtained when the chirp is eliminated by negative group velocity dispersion. The larger δ, the larger is the required negative GVD, generally produced by the introduction of prism pairs. However, prism pairs introduce not only GVD, but also third order dispersion, which increases the pulse width[13,14]. Further, we have mentioned above that the near soliton-like operation tends to be destabilized by self-amplitude modulation. This is another reason why one wants small δ and large γ.

The figure of merit for self stabilized APM using a nonlinear M-Z is shown in Fig. 6. The best value is of the order of unity. The figure of merit for the monolithic KLM system of Fig. 3 is also of the order of unity. However, in the composite cavity, the figure of merit can be made much larger by operating near the stability boundary of the cavity mode. It should be mentioned, that M can also be made large in the coupled cavity APM systems. Thus, large figures of merit are achievable at the cost of mode or interferometer stability respectively. Decreased stability leads to greater amplitude fluctuations and timing fluctuations of the modelocked pulses. The decision as to which systems will achieve both low noise operation and short pulses will have to await further work.

This work was supported in part by the National Science Foundation Grant 9012787-ECS and Air Force Office of Scientific Research Contract F49620-91-C-0091.

References

[1] L.F. Mollenauer and R. H. Stolen, Opt. Lett. **9**, 13-15 (1984)

[2] K.J. Blow and B.P. Nelson, Opt. Lett. **13**, 1026-1028 (1988).

[3] P.N. Kean, X. Zhu, D.W. Crust, R.S. Grant, N. Langford, W. Sibbett, Opt. Lett. **14**, 39-41 (1989).

[4] E.P. Ippen, H.A. Haus, and L.Y. Liu, J. Opt. Soc. Am. B **6**, 1736-1745 (1989).

[5] D.K. Negus, L. Spinelli, N. Goldblatt, and G. Feuget, Digest of Topical Meeting on Advanced Solid State Lasers, OSA, Washington DC 1991 (postdeadline paper).

[6] D.E. Spence, P.N. Kean, and W. Sibbett, Opt. Lett. **16**, 42-44 (1991).

[7] H.A. Haus, J.G. Fujimoto, and E.P. Ippen, J. Opt. Soc. Am. B **8**, 2068-2076 (1991).

[8] P.L. Hagelstein, IEEE J.. Quantum Electron. **QE-14**, 396-398 (1978).

[9] H. A. Haus and A. Mecozzi, "Noise in modelocked lasers," submitted to IEEE J. Quantum Electron.

[10] H.A. Haus, J. Fujimoto, and E.P. Ippen, "Analytic theory of additive pulse modelocking and Kerr lens modelocking," accepted by IEEE Journ. Quantum Electron.

[11] J.H. Marburger, Prog. Quantum Electronics, **4**, 35 (1975).

[12] D. Huang, M. Ulman, L.H. Acioli, H.A. Haus, and J.G. Fujimoto, Opt. Lett. **17**, 511 (1992).

[13] C.-P. Huang, H.C. Kapteyn, J. McIntosh, and M. Murnane, Opt. Lett. **17**, 139 (1992).

[14] J.M. Jacobson, K. Naganuma, H.A. Haus, J. G. Fujimoto, CLEO/QELS'92, Anaheim, CA May 10-15, 1992, Paper CTuU2.

Molecular Control Spectrometer

Y. Yan[1], B.E. Kohler[2], R.E. Gillilan[1], R.M. Whitnell[1], K.R. Wilson[1], and S. Mukamel[2]

[1]Department of Chemistry, University of California San Diego, La Jolla, CA 92093-0339, USA
[2]Department of Chemistry, University of Rochester, Rochester, NY 14627, USA

Abstract: It is now possible to move beyond the observation of ultrafast molecular dynamics to the control of those dynamics. We discuss here some of the theoretical and experimental issues involved in performing molecular dynamic control experiments. The resolution of these issues can provide the basis for a "molecular control spectrometer" which we are now in the process of building, which tightly couples time domain theory and experiment to not only control molecular processes, but also to discover fundamental properties of molecular systems.

There has been a remarkable growth in the application of ultrafast laser technology to the observation and interpretation of ultrafast molecular dynamics. Dynamics in the subpicosecond regime, which in the not too distant past would have seemed to be inaccessible to observation, can now be probed in many cases with commercial off-the-shelf lasers. The result has been, as this series of conferences demonstrates, a deeper and fuller understanding of how chemical processes, in gas phase and in solution, happen on the microscopic level.

While there is still much to be learned by probing and observing molecular dynamics on the femtosecond time scale, there are also new challenges to be met. A considerable body of theoretical work has been created over the past few years in the area of the control of molecular dynamics by light (by, among others, Tannor, Rice and Kosloff[1], Brumer and Shapiro[2], Rice[3], Rabitz[4], Warren[5], and Bandrauk[6]). The fundamental idea behind these studies is that, by driving a molecular system with an appropriately tailored light field, a desired outcome can be obtained, such as molecular rearrangement, an increased probability of dissociation, the preferential breaking of one molecular bond over another, or the preferential formation of a particular product or product state. However, the experimental tests of theoretical control predictions are only beginning to appear. For example, Gordon and co-workers[7] have recently reported experimentally modulating the ionization rate of HCl using the coherent control methodology of Brumer and Shapiro, and Crim and co-workers[8] and Zare and co-workers[9] have used vibrational excitation of a bond to control the products of both photodissociation reactions[8] and bimolecular gas phase reactions[8,9].

The goal now is to meld the existing considerable theoretical knowledge with solutions to the real world problems confronting molecular control experiments. Among the questions central to this undertaking are:

i) What if the molecular systems are not in a pure state? It is often difficult experimentally to create a pure eigenstate of the molecular system. This is especially true if one wishes to examine the possibility of control at finite temperatures or in a liquid solution or other condensed phase system. Pure state molecular dynamics can be described by a single wavefunction ψ, and the corresponding density matrix is given by

Springer Series in Chemical Physics, Vol. 55 Ultrafast Phenomena VIII
Editors: J.-L. Martin · A. Migus · G.A. Mourou · A.H. Zewail

$$\rho = |\psi\rangle\langle\psi|. \tag{1}$$

However, the ensemble behavior of dynamic processes (such as dephasing and energy relaxation) can only be transparently implemented in the (Liouville space) density matrix formulation[10-13]. A specific representation of Liouville space (e.g. the Wigner representation) allows a clear semiclassical picture in phase space.

This phase space picture of control can be expressed as follows[14]. Suppose one has an outcome $A(t_f)$ to optimize that can be expressed as an expectation value over an operator $\hat{A}$:

$$A(t_f) = \mathrm{Tr}\left[\hat{A}\rho(t_f)\right]. \tag{2}$$

The outcome depends on the radiation electric field *E(t)* through the dipole coupling term *–DE(t)*, where *D* is the transition dipole moment. Now consider a variation, *δE(t)*, away from the true radiation field *E(t)* that optimizes this outcome. From perturbation theory, the first-order change in the outcome is found to be[14]

$$\delta A^{(1)}(t_f) = -\int_{t_0}^{t_f} d\tau\, \delta E(\tau) K(\tau), \tag{3}$$

where the kernel *K(τ)* is given by

$$K(\tau) = 2\,\mathrm{Im}\left\{\mathrm{Tr}\left[\tilde{A}(\tau) D \rho(\tau)\right]\right\}. \tag{4}$$

The quantity $\rho(\tau)$ is calculated by propagating the initial phase space distribution forward from time t_0 while $\tilde{A}(\tau)$ is calculated via backward propagation of $\hat{A}$ from time t_f. In the pure state case, Equation (4) reduces to the control equations used by Rabitz[4] and co-workers and by Rice, Tannor, Kosloff and co-workers[1,15] for optimal molecular control. However, the phase space formulation of the control problem facilitates the treatment of complex systems and allows the theory to be implemented in a unified way using quantum, semiclassical or classical methods.

ii) What if the ensemble of light pulses is also not a pure state? In most experiments, the light pulses created are not in a specified phase relationship with respect to one another. However, most of the theoretical work on molecular control has assumed a fixed phase relationship. While the work of Scherer, Fleming and co-workers[16] and of Fayer and co-workers[17] has demonstrated the ability to create pulses in a fixed phase relationship, such techniques add to the complexity of these experiments. Is such a fixed phase relationship necessary for control? By performing control calculations using an ensemble of pulses[14] with random phase relationships, one can examine the robustness of the control dynamics with respect to pulse phase.

iii) What light pulses are experimentally realizable? The theoretically optimum light to meet an objective does not necessarily correspond to those fields that can be achieved in the laboratory. However, a wider range of fields is becoming available, for example, from very bright higher repetition rate continuum femtosecond sources[18], such as the one we are using, thereby allowing in principle the synthesis of ultrafast light fields containing a wide range of frequencies. In addition, the types and shapes of ultrashort fields that can be created are becoming more varied. One example is the Fourier plane manipulation of Weiner, Heritage,

Nelson and co-workers[19]. However, even though the universe of experimental light fields is becoming larger, theory must still find the best field within that universe. Many examples of molecular control theory consider unrestricted fields $E(t)$. The optimal field is proportional to the optimal control kernel $K(t)$, in the case that the total energy input to the molecular system from the field $E(t)$ is constrained to be constant. The resulting converged field is the optimal field $E(t)$. Such fields are not guaranteed to be experimentally realizable. In order to mesh with experiment, it is convenient to consider a parameterized electric field, $E\left(\tau;\{\alpha_j\}\right)$ which corresponds to the realizable range of light fields. The parameters, $\{\alpha_j\}$, of such a field may include such quantities as the sub-pulse carrier frequencies, widths and amplitudes as well as the time delays and relative phases between pulses. One then optimizes the parameters of the field rather than the unrestricted field. An example of optimization of parameters can be found in the work of Rice and co-workers on the control of I_2 and HgAr photodissociation[20].

The optimization of parameters turns the problem into one of calculus rather than control. One wishes to find the set of parameters that optimizes the outcome $A(t_f)$. However, the outcome also now depends on the parameters $\{\alpha_j\}$. For any given set of field parameters, Equation (2) can be used to calculate the outcome while Equation (3) can be used to find the first derivative of the outcome with respect to changes in the parameters[14]:

$$\frac{\partial A\left(t_f;\{\alpha_j\}\right)}{\partial \alpha_i} = -\int_{t_0}^{t_f} d\tau \frac{\partial E(\tau;\{\alpha_j\})}{\partial \alpha_i} K(\tau). \tag{5}$$

Given the values of A and the first derivatives of A with respect to each of the $\{\alpha_j\}$, standard minimization techniques can then be used to calculate an optimized set of parameters that give the best realization of the objective, given the experimental limitations.

iv) How long are we willing to let our computers run? The molecular control calculations described above require a lengthy optimization process in order to find the optimal electric field. As the systems we wish to control become more complex, or as we have to deal with ensembles of light or molecules, the required computational effort will increase. It has been shown that both the spectral and the temporal profiles of the fields are important in the analysis of nonlinear spectroscopy[11,13,21]. These profiles couple in general to different types of motion such as electronic populations and coherences. It may be possible for example under certain conditions to introduce an independent "snapshot spectrum" in which the field can be considered both infinitely short and monochromatic. This limit is physically realizable without violating the Heisenberg energy-time uncertainty[11,13,21]. In this case, the interpretation of pump-probe measurements simplifies considerably. Yan[22] has recently developed a Liouville-space formalism for weak field pump-probe spectroscopy involving the use of a mixed time-frequency, causality preserving representation in which the light-matter coherent interference is transparent. In the case of pump and probe pulses that do not overlap in time, Yan[22] has shown that the probe absorption $S(t_d)$, where t_d is the delay time, is given by

$$S(t_d) = \int_{-\infty}^{\infty} d\omega' \int_{-\infty}^{\infty} d\omega'' \int_{0}^{\infty} d\tau\, F(\tau - t_d, \omega', \omega'') M(\tau, \omega', \omega''). \tag{6}$$

In Equation (6), F depends only on light field properties and M depends only on molecular ensemble properties. Thus, this formulation of pump-probe spectroscopy allows the convenient splitting of the molecular calculations and the light field calculations into separate parts. In principle, one would only have to calculate the molecular properties once. The effect on the molecules of a variety of light fields could then be calculated rapidly. Results such as these, as well as the continuing increase in computational speed and algorithmic efficiency, can make increasingly larger systems accessible to molecular control theory and, eventually, to experiment.

What we have described here are some of the issues relevant to the construction of a "molecular control spectrometer," as discussed by Rabitz and coworkers[4,23], which we are now in the process of building. Such a spectrometer will closely couple theory and experiment in the following iterative manner. First, for a given molecular system, the theoretical side of the spectrometer (a computer) will take the relevant molecular description (potential energy and transition dipole moment surfaces, etc.) and experimental description (the nature of the available universe of light fields) and calculate the light field expected to best achieve a particular objective, for example, the control of the outcome of a chemical reaction. The parameters for this field are then transferred from the theoretical computer to the computer controlling the experimental side of the spectrometer which adjusts the optical system to produce the desired field. The experimental response of the sample (which will not exactly match that predicted by the theoretical computer because the molecular description is only approximate) is fed back into the theoretical computer, which improves the molecular description to reflect this new information, and computes a new test light field designed to minimize the uncertainty in the molecular description. This new light field is sent to the experimental computer, which measures a new molecular response, and so on, iterating to convergence between experiment and theory. Note that such a spectrometer achieves two goals: not only does it provide a way of controlling molecular processes, but in producing that control, it also discovers new information about molecular systems. In particular, the potential energy and transition dipole moment surfaces become increasingly refined so that control can be achieved. This refinement can be done in an efficient manner: The parts of the potential energy surface that are derived most accurately are precisely those that are important in obtaining the desired outcome.

The implementation of ultrafast molecular control in the "molecular control spectrometer" mode requires a close link between theory and experiment that is perhaps unparalleled in chemical physics. The iterative convergence between theory and experiment should in principle then result in both the discovery of the basic nature of molecular systems and the ability to control the dynamics of such systems.

References

[1] D. J. Tannor and S. A. Rice, J. Chem. Phys. **83**, 5013 (1985); D. J. Tannor, R. Kosloff and S. A. Rice, J. Chem. Phys. **85**, 5805 (1986).

[2] P. Brumer and M. Shapiro, Acc. Chem. Res. **22**, 407 (1989); in *Mode Selective Chemistry*, J. Jortner, R. D. Levine, B. Pullman, eds. (Kluwer, Dordrecht, 1991).

[3] S. A. Rice, in *Mode Selective Chemistry*, J. Jortner, R. D. Levine, B. Pullman, eds. (Kluwer, Dordrecht, 1991); M. Zhao and S. A. Rice, J. Chem. Phys. **95**, 2465 (1991).

[4] H. Rabitz and S. Shi, *Advances in Molecular Vibrations and Collision Dynamics* **1A**, 187 (1991).
[5] W. S. Warren, Science **242**, 878 (1988).
[6] S. Chelkowski, A. D. Bandrauk and P. B. Corkum, Phys. Rev. Lett. **65**, 2355 (1990).
[7] S.-P. Lu, S. M. Park, Y. Xie and R. J. Gordon, J. Chem. Phys. **96**, 6613 (1992).
[8] F. F. Crim, M. C. Hsiao, J. L. Scott, A. Sinha and R. L. Vander Wal, Phil. Trans. R. Soc. Lond. A **332**, 259 (1990); F. F. Crim, Science **249**, 1387 (1990); A. Sinha, M. Hsiao and F. F. Crim, J. Chem. Phys. **94**, 4928 (1991).
[9] M. J. Bronikowski, W. R. Simpson, B. Girard and R. N. Zare, J. Chem. Phys. **95**, 8647 (1991).
[10] S. Mukamel, Phys. Rep. **93**, 1 (1982); Adv. Chem. Phys. **70**, 165 (1988).
[11] S. Mukamel, Annu. Rev. Phys. Chem. **41**, 647 (1990).
[12] Y. J. Yan and S. Mukamel, J. Chem. Phys. **89**, 5160 (1988); Phys. Rev. A. **41**, 6485 (1990).
[13] Y. J. Yan, L. E. Fried and S. Mukamel, J. Phys. Chem. **93**, 8149 (1989).
[14] Y. J. Yan, R. E. Gillilan, R. M. Whitnell, K. R. Wilson and S. Mukamel, to be submitted.
[15] R. Kosloff, S. A. Rice, P. Gaspard, S. Tersigni and D. J. Tannor, Chem. Phys. **139**, 201 (1990).
[16] N. F. Scherer, A. Ruggiero, M. Du and G. R. Fleming, J. Chem. Phys. **93**, 856 (1990); N. F. Scherer, R. J. Carlson, A. Matro, M. Du, A. J. Ruggiero, V. Romero-Rochin, J. A. Cina, G. R. Fleming, and S. A. Rice, J. Chem. Phys. **95**, 1487 (1991).
[17] J. T. Fourkas, W. L. Wilson, G. Wackerle, A. E. Frost and M. D. Fayer, J. Opt. Soc. Am. B **9**, 1905 (1989).
[18] F. Salin, J. Squier, G. Mourou and G. Vaillancourt, Opt. Lett. **16**, 1964 (1991). Such a source is commercially available from MXR, Inc. and Clark Instrumentation.
[19] A. M. Weiner, J. P. Heritage and R. N. Thurston, Opt. Lett. **11**, 153 (1986); A. M. Weiner and J. P. Heritage, Rev. Phys. Appl. **22**, 1619 (1987); A. M. Weiner, D. E. Leaird, G. P. Wiederecht and K. A. Nelson, Science, **247**, 1317 (1990).
[20] B. Armstrup, R. J. Carlson, A. Matro and S. A. Rice, J. Phys. Chem. **95**, 8019 (1991).
[21] L. E. Fried and S. Mukamel, J. Chem. Phys. **93**, 3063 (1990).
[22] Y. J. Yan, Chem. Phys. Lett. (submitted).
[23] R. S. Judson and H. Rabitz, Phys. Rev. Lett. **68**, 1500 (1992).

Internal Motions of Proteins

M. Karplus

Department of Chemistry, Harvard University,
Cambridge, MA 02138, USA, and
Institut de Biologie Moléculaire et Cellulaire, Université Louis Pasteur,
F-67084 Strasbourg, France

The technological advances in ultrafast spectroscopy described at this meeting -- that 20 fs measurements in the visible region are now "routine" and that 200 fs time resolution in the infra red is possible, if not routine -- suggest that it would be useful to review the types of protein motions that might be of interest to study with such techniques. In so doing it is important to realize that ultrafast measurements may be necessary but are not sufficient to examine the full dynamic range of protein motions, particularly those of biological importance. First, the time scale of the significant protein motions extends from fs to s. Thus, it is important to have techniques that make possible uniform measurements over a very wide time range. Second, the most useful results will come from measurements that give the most detailed structural information. From that point of view, optical measurement that determine the change in the electronic spectrum with time, although easiest to do, are the least useful. Time resolved infra-red and Raman measurements are more intimately related to structural change and so are better suited for obtaining information on the atomic scale. Of even greater interest might be x-ray measurements on the fs to ps time scale. If the progress in the x-ray area described at this meeting by Shank and coworkers comes to fruition it may well open up the possibility of time-resolved structural studies for proteins. Such a development would provide results that could be compared directly with simulations. However, it is important to realize that when an event is triggered in a population of protein molecules (e.g. photodissociation of myoglobin, which has been discussed at this meeting), the coordinates of each molecule change differently as a function of time from those of every other molecule. The dispersion of the indvididual responses observed in a series of molecular dynamics simulation of myoglobin photodissociation suggests that a careful analysis is needed to determine the effect of dynamic disorder on the crystal spectrum [1]. Preliminary estimates indicate that the effect will be significant in "smearing out" the x-ray spectrum but not so great as to eliminate all the information. (Detailed simulations and analysis using a time dependent analog of the work described in Kuriyan *et al.* [2] are planned.) Third, to study protein dynamics in a more general sense, triggers other than those in present use will have to be invented. So far, all the systems that have been studied (and reported on at this conference) have natural triggers in the optical range. Three of them (bacterial photosynthetic reaction center, bacteriorhodopsin, and rhodopsin) have the absorption of optical photons as the natural initiator of the biological event of interest (electron transfer in the first case and chromophore isomerization in the two others). In myoglobin, by contrast, photodissociation of the ligand does not occur in nature but is used to trigger events that can be followed by ultrafast spectroscopy. The heme chromophore absorbs the photon that leads to dissociation and its optical spectrum is monitored to follow the reaction. Recently femtosecond infra-red studies of a heme band and of the ligand have been described (see the reports of Hochstrasser and Anfinrud at this meeting). However, there are many dynamic phenomena in proteins where no such obvious trigger exists. The invention of suitable methods to trigger interesting events is a challenge to the imagination and skill of the experimentalists at this meeting. In principle, the introduction of chromophoric groups, infra-red excitation, and rapid heating might serve to initiate events. In practice, special methods suited to specific problems will have to be introduced. It is developments in this area that will determine the range of future applications of ultrafast techniques to the time dependence of processes of functional importance in proteins and other macromolecules of biological interest.

Springer Series in Chemical Physics, Vol. 55 **Ultrafast Phenomena VIII**
Editors: J.-L. Martin · A. Migus · G.A. Mourou · A.H. Zewail

There are a number of general references on protein dynamics and simulation methods. Two useful books are by McCammon and Harvey [3] and by Brooks *et al.* [4]. The two systems that have been studied both by dynamic simulations and ultrafast techniques are the bacterial photosynthetic reaction center [5] and myoglobin [1]. Other dynamic events in proteins that are of interest, have been simulated, but have not been studied by ultrafast techniques primarily because of the lack of an obvious trigger include the "lid" conformational transition in triosephosphate isomerase [6], the activated events in enzyme catalysis [7], ion migration through channels [8] and protein folding [9].

References

1. Ligand Binding and Protein Relaxation in Heme Proteins: A Room Temperature Analysis of NO Geminate Recombination, *Biochemistry* **30**, 3975-3987 (1991), by J. W. Petrich, J.-C. Lambry, K. Kuczera, M. Karplus, C. Poyart, and J.-L. Martin.
2. Effect of Anisotropy and Anharmonicity on Protein Crystallographic Refinement: An Evaluation by Molecular Dynamics, *J. Mol. Biol.* **190**, 227-254 (1986), by J. Kuriyan, G. A. Petsko, R. M. Levy, and M. Karplus.
3. J. A. McCammon and S. Harvey, Dynamics of Proteins and Nucleic Acids (Cambridge University Press, Cambridge, 1987).
4. C. L. Brooks III, M. Karplus, and B. M. Pettitt, *Proteins: A Theoretical Perspective of Dynamics, Structure, & Thermodynamics,* Adv. Chem. Phys. **LXXI** (John Wiley & Sons, 1988).
5. Chromophore-Protein Interactions and the Function of the Photosynthetic Reaction Center: A Molecular Dynamics Study, *Proc. Natl. Acad. Sci. USA* **89**, 75-79 (1992), by H. Treutlein, K. Schulten, A. Brünger, M. Karplus, J. Deisenhofer, and H. Michel
6. Anatomy of a Protein Conformational Change: Hinged "Lid" Motion of the Triosephosphate Isomerase Loop, *Science* **249**, 1425-1428 (1990), by D. Joseph, G. A. Petsko, and M. Karplus.
7. Computer Simulation and Analysis of the Reaction Pathway of Triosephosphate Isomerase, *Biochemistry* **30** 5826-5832 (1991), by P. A. Bash, M. J. Field, R. C. Davenport, G. A. Petsko, D. Ringe, and M. Karplus.
8. Ion Transport in a Gramicidin-like Channel: Dynamics and Mobility, *J. Phys. Chem.* **95**, 4856-4868 (1991), by B. Roux and M. Karplus.
9. Protein Folding Bottlenecks: A Lattice Monte Carlo Simulation, *Phys. Rev. Lett.* **67**, 1665-1668), by E. Shakhnovich, G. Farztdinov, A. M. Gutin, and M. Karplus.

Some Theoretical Aspects of Electron Transfer in Supermolecules

J. Jortner and M. Bixon

School of Chemistry, Tel Aviv University, Ramat Aviv, Tel Aviv 69978, Israel

Abstract. We consider structural, intramolecular, medium energetic and medium dynamic control of electron transfer in synthetic and natural supermolecules. Optimization principles are formulated for ultrafast, highly efficient and stable charge separation.

1. Introduction

Electron transfer (ET) reactions are ubiquitous in chemistry, physics and biology [1]. ET in supermolecules falls into two general categories:
(1) Man-made synthetic supermolecules [2-5] consisting of an electron donor (D), and an electron acceptor (A), linked by a nonrigid or a rigid molecular bridge (B).
(2) Nature-made supermolecules such as the photosynthetic reaction centers (RC) of bacteria and plants [6], where the primary process of the conversion of solar energy into chemical energy proceeds via a sequence of well organized, highly efficient, directional and specific ET processes between prosthetic groups embedded in the protein medium.

Photophysical and photobiological ET from an electronically excited donor proceeds via superexchange mediated or direct processes: $DBA \xrightarrow{h\nu} D^*BA \xrightarrow{ET} D^+BA^-$ or $DA \xrightarrow{h\nu} D^*A \xrightarrow{ET} D^+A^-$.

The control of ET in DBA or DA systems in solution [1], in a solid [1], in a protein medium [1,6] or within an "isolated" solvent-free supermolecule [7] can be accomplished by:
(1) Structural control. "Molecular engineering" of the D, A and B subunits determines the molecular energetics and the direct D-A or superexchange D-B-A electronic coupling.
(2) Intramolecular dynamic control of the nuclear equilibrium configurational changes (i.e., nuclear distortions) accompanying ET [1].
(3) Medium control of "conventional" ET in a solvent or a cluster. The function of the medium on the $DBA \rightarrow D^+BA^-$ ET is [1-7]: (i) The energetic stabilization of the ionic states. (ii) The coupling of the electronic states with the medium nuclear motion, which originates from short-range and long-range interactions in polar solvents, short-range interactions with C-H group dipoles in nonpolar hydrocarbons and with polar amino acid residues in proteins.
(4) Dynamic medium control of ET [8] involving: (i) The medium acting as a heat bath. The relaxation of the medium polar modes is often fast on the time scale of the electronic ET processes, which then constitute the rate determining step. (ii) Dynamic solvation effects of DBA or/and D^+BA^- determine the ET dynamics when condition (i) is violated and solvent-controlled ET may be exhibited.

The structural, intramolecular, solvent and dynamic control of ET will allow for the formulation of an optimization principle allowing for the design of molecular systems where ET is: (a) Ultrafast (in the time scale of ~ 1 psec to ~ 100 fsec) overwhelming any energy waste processes. (b) Highly efficient, eliminating any back reactions. (c) Stable with respect to the predictable variation of molecular and medium properties. (d) Practically invariant with respect to temperature changes.

2. Theoretical Foundations

(2.A) Basic ET theory. The nonadiabatic ET rate is given by [1-6,9] $k = (2\pi/\hbar)V^2F$, where V is the electronic coupling and F is the thermally averaged nuclear vibrational Franck-Condon factor. This microscopic description rests on the following description and conditions:

Springer Series in Chemical Physics, Vol. 55   Ultrafast Phenomena VIII
Editors: J.-L. Martin · A. Migus · G.A. Mourou · A.H. Zewail

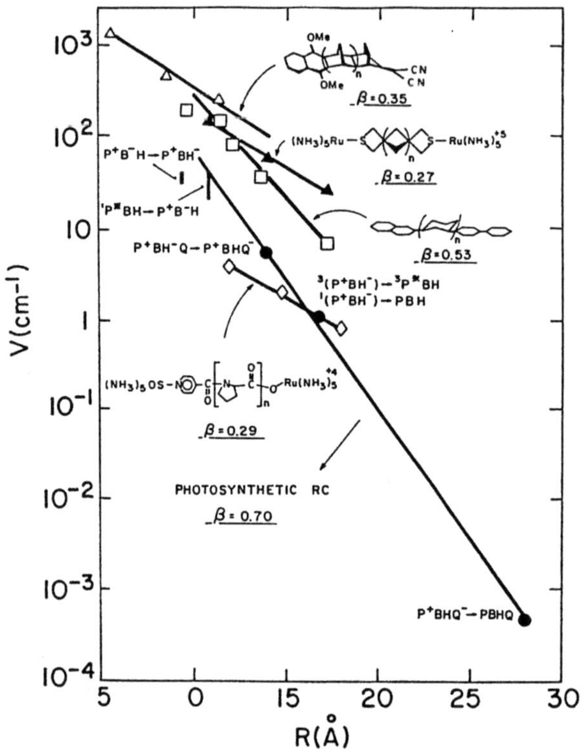

Figure 1. The distance dependence of V. The electronic coupling for ET in the RC (direct exchange and terminated superexchange) and for synthetic supermolecules (intramolecular superexchange), as obtained from kinetic data (K) and spectroscopic data (S), are displayed vs the center-to-center D-A separation. Data from references 2-5, 6 and 13. The β values are presented in the figure.

(A) ET is described as a radiationless transition.
(B) Applicability of the Born-Oppenheimer separability of electronic and nuclear motion, allowing for the description of the system in terms of diabatic potential surfaces.
(C) The electronic coupling is sufficiently weak to warrant the description of the radiationless transition in the nonadiabatic limit.
(D) Insensitivity of microscopic ET rates to medium dynamics. This state of affairs is realized when one of the rollowing conditions is realized: (i) The common situation of fast medium vibrational dynamics, which allows for the separation of time scales with the microscopic ET rate constants constituting the rate determining step [8]. (ii) The microscopic ET rates weakly depend on the distribution in the initial D^*BA vibronic manifold [10].

(2.B) Electronic coupling. The electronic coupling in the DBA system $V = V_{DA} + V_{super}$ consists of a sum of direct D-A exchange contributions [1,9] $V_{DA} = \langle \phi^0_{DA} | H | \phi^0_{D^+A^-} \rangle$ between the electronic states of DA and D^+A^-, and a superexchange [1,9] $V_{super} = V_{DB}V_{BA}/\delta E_B$ where δE_B is the vertical energy difference between the potential energy surfaces of DBA and DBA^-. For superexchange between closed shell D, B and A systems both direct and superexchange interactions require the evaluation of individual pair V_{DA}, V_{DB} or V_{BA} coupling [6,13], while for intramolecular superexchange in D-B-A supermolecules quantum mechanical calculations were utilized [11]. In both cases many-electron computations, transcedenting the naive one-electron picture, have to be invoked. The accumulated information concerning the distance dependence of both direct and superexchange interactions is that both interactions are expected to exhibit an approximate exponential distance dependence [1-6,9]

$$V = a \exp(-\beta R_{DA}) \tag{2.1}$$

where R_{DA} is the (either edge-to-edge or center-to-center) D-A distance. The distance dependence of intramolecular interactions in synthetic supermolecules and the RC (Fig. 1) was inferred from the analysis of ET rates or optical charge transfer spectra [2-6,13]. Such analysis involved hidden assumptions, e.g., the proper description of Franck-Condon factors for both thermal and optical ET, and the neglect of the contribution of D^*BA electronic excitations to the oscillator strength of CT optical transitions. The general trend confirms Eq. (2.1).

(2.C) Nuclear Franck-Condon factors. The nuclear Franck-Condon factor F contains the thermal average of the overlap integrals between the initial and final vibrational states of the system, under the restriction of energy conservation. For two harmonic potential energy surfaces characterized by identical reduced masses $\{\mu_i\}$, frequencies $\{\omega_i\}$, normal modes $\{Q_i\}$ and nuclear displacement $\{\Delta Q_i^0\}$ of the equilibrium coordinates, F takes the form of a Fourier transform [9]

$$F = (2\pi)^{-1} \exp(-G) \int_{-\infty}^{\infty} dt \exp(-i\Delta Et) \exp[G_+(t) + G_-(t)] \tag{2.2}$$

where ΔE is the energy gap,

$$G_{\pm}(t) = \sum_i (\Delta_i^2/2) \begin{Bmatrix} \bar{v}_i+1 \\ \bar{v}_i \end{Bmatrix} \exp(\pm i\omega_i t) \tag{2.3}$$

$$G = G_+(0) + G_-(0) \tag{2.4}$$

$$\bar{v}_i = [\exp(\hbar\omega_i/k_BT)-1]^{-1} \tag{2.5}$$

$$\Delta_i = (\mu_i\omega_i/\hbar)^{1/2} \Delta Q_i^0 \; . \tag{2.6}$$

The vibrational modes in Eqs. (2.2)-(2.6) incorporate:

(i) Low-frequency medium modes, which are usually approximated by a single frequency ω_m with an effective Δ_m. "Intelligent guesses" result for glasses in $\omega_m \simeq 10\text{-}100\ cm^{-1}$. It seems that for a polar solvent a phonon picture is not applicable, while for a protein medium $\omega_m \approx 100\ cm^{-1}$ [12]. The medium reorganization energy is $\lambda_m = S_m\hbar\omega_m$, where $S_m = \Delta_m^2/2$. For polar solvents $\lambda_m = 4000\text{-}8000\ cm^{-1}$ [1], while for the protein RC $\lambda_m = 800\text{-}3000\ cm^{-1}$ [13].

(ii) Intramolecular modes in the range $\omega \simeq 100\text{-}3000\ cm^{-1}$. Often, but not exclusively, the medium modes can be approximated by a single molecular frequency $\omega_c \simeq 1500\ cm^{-1}$ and a dimensionless shift $S_c = \Delta_c^2/2 \simeq 1$ [12].

The quantum nature of the intramolecular vibrational modes is usually prevalent at all the relevant temperatures, as $k_BT << \hbar\omega_c$. When the medium modes can be handled in terms of the high temperature limit ($k_BT > \hbar\omega_m$), Eq. (2.2) reduces to [9]

$$F = (4\pi\lambda_m k_BT)^{-1/2}\exp(-S_c) \sum_{n=0}^{\infty} \frac{S_c^n}{n!} \exp\left[-\frac{(\Delta E+\lambda_m+n\hbar\omega_c)^2}{4\lambda_m k_BT}\right] , \tag{2.7}$$

which is useful for the calculation of nuclear contributions to the ET rate over a broad temperature domain. In the limit $S_c = 0$, Eq. (2.7) reduces to the classical Marcus relation [1], $F_{cl}(S_c=0) = (4\pi\lambda_m k_BT)^{-1/2} \exp[-(\Delta E+\lambda_m)^2/4\lambda_m k_BT]$, with a Gaussian activation energy. The (free) energy dependence of F (Fig. 2) provides a demonstration of the classical Marcus relation [1] in the normal region ($\Delta E \geq -\lambda_m$) and the activationless domain ($\Delta E = -\lambda_m$), while in the inverted region ($\Delta E \leq -\lambda_m$) marked deviations from the classical relation are exhibited (i.e., $F > F(S_c=0)$) due to the vibrational excitation of the quantum modes of D and A, accompanying ET.

(3.D) Optimizing F. Activationless ET. When the nuclear potential across the medium coorinate surface intersect at the minimum of the initial DA state, i.e., $-\Delta E = \lambda_m$, the dominating contribution to the sum in F, Eq. (2.7) (for typical values $S_c \simeq 1$) originates from the

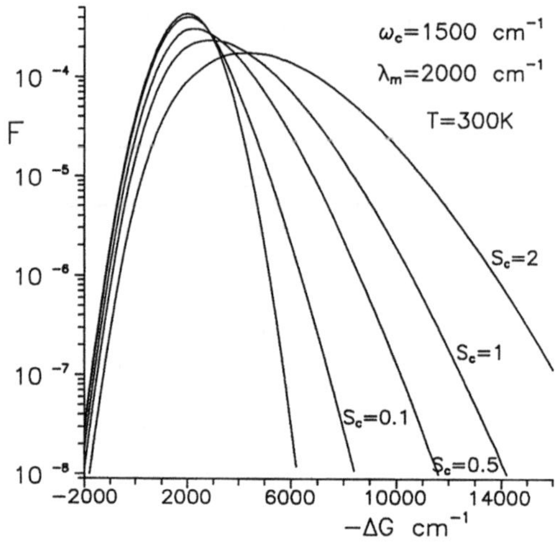

Figure 2. The energy gap dependence of the Franck-Condon factor, Eq. (2.7).

n=0 term, so that $F(\Delta E=-\lambda_m) \simeq (4\pi\lambda_m k_B T)^{-1/2} \exp(-S_c)$. This situation corresponds to activationless ET, with a weak temperature dependent $k \propto T^{-1/2}$, nonarrhenius type rate. Here the Franck-Condon factor is maximized, involving the optimization of the nuclear contribution to the ET rate.

In the nonadiabatic limit the optimal activationless rate is

$$k = (2\pi V^2/\hbar)\, F(\Delta E=-\lambda_m) \exp(-S_c) \quad . \tag{2.8}$$

For characteristic values λ_m = 4000 cm^{-1} for ET in polar solvents or $\lambda_m \simeq$ 1000 cm^{-1} in protein RC together with $S_c \simeq 1$, we estimate (at room temperature) the activationless ET rate $(k/\text{sec}^{-1}) \simeq (1.4\text{-}2.7) \times 10^8 (V/\text{cm}^{-1})^2$. The upper limit of the maximized activationless ET rate is determined by the breakdown of the nonadiabatic formalism, which requires the extension of the nonadiabatic theory.

3. Extensions of Conventional ET Theory.

(3.A) Adiabatic ET. In section (2.A) we have spelled out the conditions for the applicability of the nonadiabatic ET theory. Condition (C) for weak electronic coupling, formulated within the framework of the theory of radiationless transitions, implies that the widths of the resonances (i.e., the decay rates k_i of individual initial states) are smaller than their separations (determined by an effective medium vibrational frequency $\sim \omega_m$) i.e., $\gamma = k_i/\omega_m < 1$. The coupling parameter is $\gamma = 2V^2/\hbar\omega_m(\lambda_m\hbar\omega_m)^{1/2} < 1$. This parameter is similar to the Landau-Zenner parameter. Taking characteristic values of $\hbar\omega_m$ = 100 cm^{-1} both for a protein medium and for a polar medium, nonadiabatic ET prevails for $V \lesssim 200$ cm^{-1} in a polar solvent ($\lambda_m \simeq$ 4000 cm^{-1}) and $V \lesssim 100$ cm^{-1} for a protein medium (λ_m = 1000 cm^{-1}). In Fig. 3 we have marked the upper limit for the V values, which still correspond to nonadiabatic ET. When the electronic coupling is sufficiently strong, i.e., $\gamma > 1$, the adiabatic limit for ET applies. The adiabatic ET rate is then given by the Holstein formula $k = (\omega_m/2\pi)\exp(-E_a/k_B T)$. For activationless ET the adiabatic ET rate is $k \approx \omega_m/2\pi$.

(3.B) Solvent-controlled ET. The breakdown of assumption (D) implies that solvent relaxation, rather than the microscopic electronic processes constitute the rate determining step for ET. For ET in a system solely characterized by coupling to the medium (S_c = 0), which corresponds to the normal Marcus region ($\Delta E \geq -\lambda_m$) the realization of solvent-controlled ET is determined by the magnitude of the solvent adiabaticity parameter [8] $\kappa = 4\pi V^2 \langle\tau\rangle / \hbar\lambda_m$, where $\langle\tau\rangle$ is the medium relaxation time, induced by a constant charge distribution. In polar solvents $\langle\tau\rangle$ is the longitudinal dielectric relaxation time τ_L, e.g., τ_L = 200 fsec for water, $\tau_L \simeq$

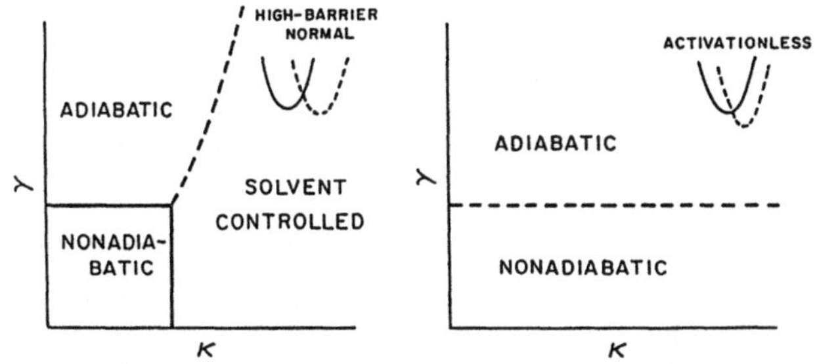

Figure 3. Domains for ET for the normal limit and the activationless case.

190 fsec for acetonitrile and τ_L = 1.2 psec for methyl acetate [8]. For protein medium of the RC molecular dynamics simulations give $\langle\tau\rangle$ = 100 fsec over the temperature domain 10K-300K [14]. The medium controlled ET rate is given by $k = k^{NA}/(1+\kappa)$, where k^{NA} is the nonadiabatic rate. In the limit $\kappa >> 1$ $k \propto k^{NA}/\kappa \propto \langle\tau\rangle^{-1}$, being independent of V.

Is this formalism relevant for activationless ET? Medium controlled ET will be manifested only provided that the microscopic rates are sensitive to the details of the distribution of the initial states. For activationless ET model calculations reveal that microscopic ET rates k_i are quite insensitive to the initial vibrational state [10]. This weak excess energy dependence of k_i is compatible with the weak temperature dependence of activationless ET. Accordingly, activationless ET is invariant with respect to medium relaxation dynamics, with the rate being independent of κ (Fig. 3). This analysis provides a possible explanation for the recent experimental observations of some ET rates in synthetic molecules, which substantially exceed the τ_L^{-1} limit predicted for solvent-controlled ET. Yoshihara et al. [15] reported ultrafast ET rates in (Nile bule$^+$)(TMPD) with $k \simeq (100\ \text{fsec})^{-1}$ which correspond to $k \simeq 50/\tau_L$, while Michel-Beyerle, Staab et al. [16] observed for ET in porphyrine-quionone cyclophanes the independence of k ($\approx (1\ \text{psec})^{-1}$) on κ (= 0.1-400) over a broad region. ET in both systems seems to correspond to activationless ET. The primary ET reactions in the photosynthetic RC are nearly activationless [12,13], whereupon the protein medium dynamics (with $\langle\tau\rangle \simeq$ 100 psec [14]) does not manifest explicitly in the rates. Thus activationless ET can be appreciably faster than previously expected, not being limited by medium dynamics.

(3.C) An upper limit for activationless ET. For activationless ET the nonadiabatic limit is expected to prevail, being characterized by $k \propto V^2$ with increasing V, until the adiabatic limit for ET will be achieved constituting an upper limit for the rate. An approximate estimate for the upper limit for the activationless ET rate can be inferred from the adiabatic limit, which for a typical value of $\omega_m \simeq 100\ \text{cm}^{-1}$ results in $k^{-1} \gtrsim 300$ fsec. The fastest room temperature ET rates recorded up to date in synthetic supermolecules [15,16] and in the RC [17,18] are either lower or comparable to the theoretical upper limit corresponding to adiabatic activationless ET.

References

[1]R. A. Marcus and N. Sutin, Biochim. Biophys. Acta **811**, 265 (1985).

[2]G. L. Closs, L. T. Calcaterra, N. J. Green, K. W. Penfield, and J. R. Miller, J. Phys. Chem. **90**, 3673 (1986).

[3]C. A. Stein, N. A. Lewis, and G. J. Seitz, J. Am. Chem. Soc. **104**, 2596 (1982).

[4]S. S. Isied, A. Vassilian, J. F. Wishart, C. Creutz, H. A. Schwarz, and N. Sutin, J. Am. Chem. Soc. **110**, 635 (1988).

[5]K. W. Penfield, J. R. Miller, M. N. Paddon-Row, E. Cotsaris, A. M. Oliver, and N. S. Hush, J. Am. Chem. Soc. **109**, 5061 (1987).

[6]J. Jortner and B. Pullman, eds., Perspectives in Photosynthesis (Kluwer Academic Publishers, Dordrecht, 1990).

[7]J. Jortner, M. Bixon, H. Heitele, and M. E. Michel-Beyerle, Chem. Phys. Lett. (in press).

[8]I. Rips and J. Jortner, J. Chem. Phys. **87**, 6513 (1987).

[9]J. Jortner, J. Chem. Phys. **64**, 4860 (1976).
[10]M. Bixon and J. Jortner (to be published).
[11]P. Siddarth and R. A. Marcus, J. Phys. Chem. **94**, 2985 (1990).
[12]M. Bixon and J. Jortner, J. Phys. Chem. **95**, 1941 (1991).
[13]M. Bixon, J. Jortner, and M. E. Michel-Beyerle, Biophys. Biochim. Acta **1056**, 30 (1991); and to be published.
[14]K. Schulten and M. Tesch, Chem. Phys. **158**, 421 (1991).
[15]T. Kobayashi, Y. Takagi, H. Kandori, K. Kemnitz, and K. Yoshihara, Chem. Phys. Lett. **180**, 416 (1991).
[16]H. Heitele, F. Pöllinger, K. Kremer, M. E. Michel-Beyerle, M. Futscher, G. Voit, J. Weiser, and H. Staab, Chem. Phys. Lett. (in press).
[17]W. Holzapfel, U. Finkele, W. Kaiser, D. Oesterhelt, H. Scheer, H. U. Stilz, and W. Zinth, Proc. Natl. Acad. Sci. USA **87**, 5168 (1990).
[18]C. K. Chan, T. J. DiMango, K. X. Q. Chen, J. Norris, and G. R. Fleming, Proc. Natl. Acad. Sci. USA **88**, 11202 (1991)

Femtosecond Time-Resolved Spectroscopy of Magneto-Excitons

D.S. Chemla[1], *J.B. Stark*[2], *and W.H. Knox*[3]

[1]Physics Department, University of California at Berkeley and Materials Sciences Division, Lawerence Berkeley Laboratory, 1 Cyclotron Road, Berkeley, CA 94720, USA
[2]AT&T Bell Laboratories, Murray Hill, NJ 07974, USA
[3]AT&T Bell Laboratories, Holmdel, NJ 07733, USA

Abstract We present femtosecond time resolved investigations of the evolution of the nonlinear optical response of semiconductor quantum wells as their quasi-two dimensional electronic states are further confined into quasi-zero dimensions by a strong magnetic field perpendicular to the quantum wells.

1. Introduction

In semiconductor quantum wells (QW), excitons (X) are confined in one direction by the crystal potential and thus acquire a reduced (quasi-2D) dimensionality when the QW thickness become smaller than the X-Bohr radius, (a_o). The wealth of new physical properties and applications that has resulted from this effect [1] has triggered numerous attempts to make quasi-1D and quasi-0D structures. These efforts have been plagued by the difficulty of obtaining defect free samples and of eliminating size fluctuations. Further confinement within the QW plane, however, can be obtained by applying a large magnetic field, H, perpendicular to that plane [2-6]. The relative electron-hole (e-h) motion then becomes limited by the cyclotron radius, $r_c = (c/eH)^{1/2}$. Thus, by varying H, the dimensionality of the QW-electronic states can be tuned continuously from quasi-2D, when $a_o << r_c$, to quasi-0D, when $a_o >> r_c$, in materials with excellent quality and uniformity. Coulomb correlated e-h states in a magnetic field are called magneto-excitons, (MX). Further interest in MX systems stems from exact many-body results, according to which, at large H and thermodynamic equilibrium, an ensemble of spin polarized MX behave like a gas of non-interacting and non-polarizable point-like Bosons which should therefore exhibit collective behavior [7,8]. In this lecture we present femtosecond time resolved investigations of the evolution of the nonlinear optical response of GaAs QWs as their quasi-2D electronic states are further confined into quasi-0D by a strong magnetic field perpendicular to the plane of the QWs [9-12].

2. Nonlinear Optical Response of Quasi-0D Magneto-excitons

When a magnetic field is applied perpendicular to a QW, the quasi-2D e-h pairs experience the total potential $V(r) = [(\lambda r/2)^2 - 2/r]R_y$, which consists of the sum of the quadratic-potential imposed by H and the Coulomb potential. The dimensionless parameter $\lambda = (a_o/r_c)^2 = \omega_c/2R_y$ measures the ratio of the Magnetic and Coulomb confinement lengths and zero-point energies, (ω_c is the cyclotron frequency and R_y is the X-Rydberg). For $\lambda = 0$ the e-h pairs form usual quasi-2D Xs; for $\lambda \rightarrow \infty$ they tend toward e and h in Landau Levels (LL); and for the intermediate values, they form MXs.

The experimental linear absorption spectra of a high quality GaAs/AlGaAs-QW structure ($L_z = 84Å$) at low temperature, T = 4K, and in a perpendicular magnetic field, is shown in Fig. 1. The spectra were measured with σ_- polarized light as H was stepwise tuned from, $\lambda = 0 \rightarrow \approx 4$ ($H = 0 \rightarrow 12T$). The transition from 2D behavior at low fields to 0D behavior at high fields and the way the LLs originate from the bound 2D-X states are nicely displayed.

Fig. 2 compares the 12T-linear absorption for σ_- (dashed line) and σ_+ (solid line). They are Zeeman-split in agreement with the QW selection rules at the Γ-point of the Brillouin zone. At such a high field the confinement in all dimensions is so strong that the absorption strength is almost zero between the 1s and 2s MX, indicating that the MX are almost diagonal in the LL basis. These spectra are in good agreement with the theoretical calculations of Ref.[6].

Springer Series in Chemical Physics, Vol. 55 **Ultrafast Phenomena VIII**
Editors: J.-L. Martin · A. Migus · G.A. Mourou · A.H. Zewail

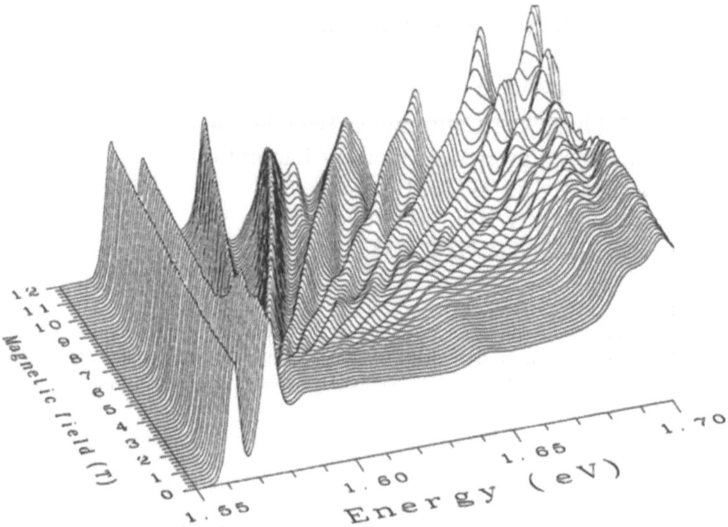

Figure 1: Linear absorption of a $L_z = 84Å$ GaAs/AlGaAs quantum well structure for σ_- polarized light at various magnetic fields $H = 0 \rightarrow 12$ Tesla.

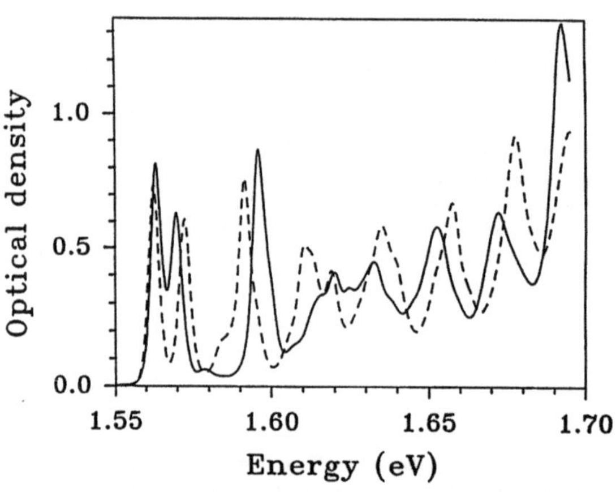

Figure 2: Comparison of the $H = 12$Tesla absorption spectra of the $L_z = 84Å$ GaAs/AlGaAs quantum well structure for σ_- (dashed line) and σ_+ (solid line) polarized light.

Experiments comparing the nonlinear optical responses at $H = 0$ and $H = 12T$ were performed at 4K using the pump/probe technique with circular polarized light and a $\approx 100fs$ time resolution [9-11]. The most direct way of interpreting such experiments is to consider that the pump creates populations of MXs (i.e. a linear superposition of excited states), and the probe measures the change of absorption induced by the presence of these populations.

Fig. 3a&b presents the differential absorption spectra (DAS), $\Delta\alpha L_z$, seen by a σ_--probe for respectively σ_--pump and σ_+-pump, of the same intensity, resonant with the lowest 1s-hh MX at $h\nu \approx 1.56eV$. Very strong responses are observed up to photon energies as high as $h\nu = 1.67eV$. To separate the spin-dependent nonlinearities from the spin-independent ones we subtract the two sets of spectra in Fig. 3c. It is clearly seen that all the response above $h\nu \approx 1.59eV$ disappears in these

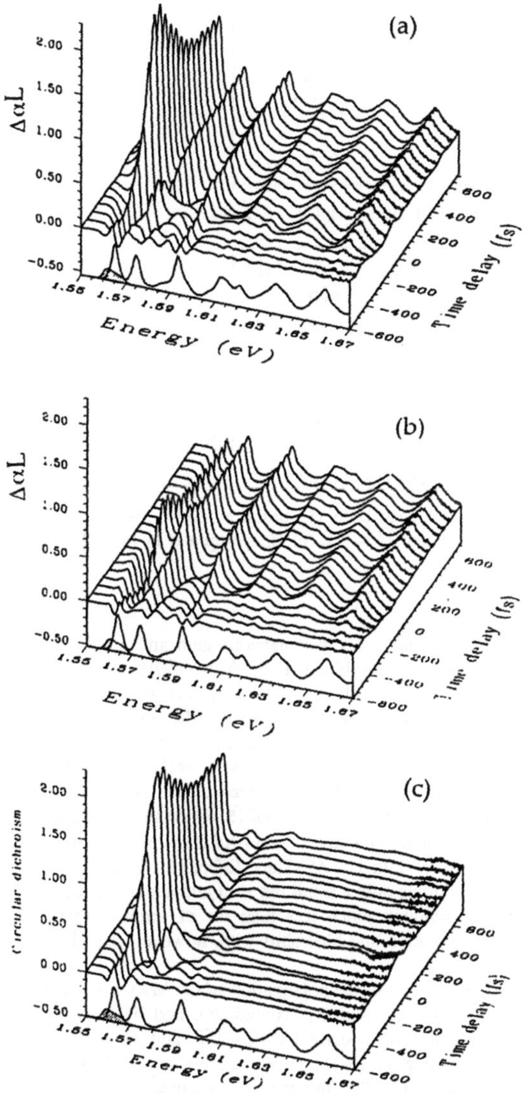

Figure 3: Differential absorption spectra, $-\Delta\alpha L_z$, vs photon energy and time delay for a σ_--probe, and (a) σ_--pump, (b) σ_+-pump. The circular dichroism spectra, (c) = (a) - (b), isolate the spin dependent interactions.

differential circular dichroism (DCD) spectra. This demonstrates that the high energy MXs (2s and above) are only sensitive to the charge density associated with the resonantly excited 1shh–σ_- or 1shhσ_+-MXs. These charge density effects correspond to collisional broadening and dielectric screening. Conversely the 1s-MXs ($h\nu < 1.59$eV) are sensitive to the spin of the MXs created by the pump as seen from the magnitude and the lineshape of the DAS and DCD. In order to display more clearly this behavior we compare, in Fig. 4, a blowup of the lower portion of the absorption spectra seen by a σ_--probe at $H = 0$T and $H = 12$T and for three time delays, $\Delta t = -660$fs, 0fs and +660fs after excitation by a σ_--pump and a σ_+-pump. The $\Delta t = -660$fs the spectra are essentially that of the unexcited sample. For

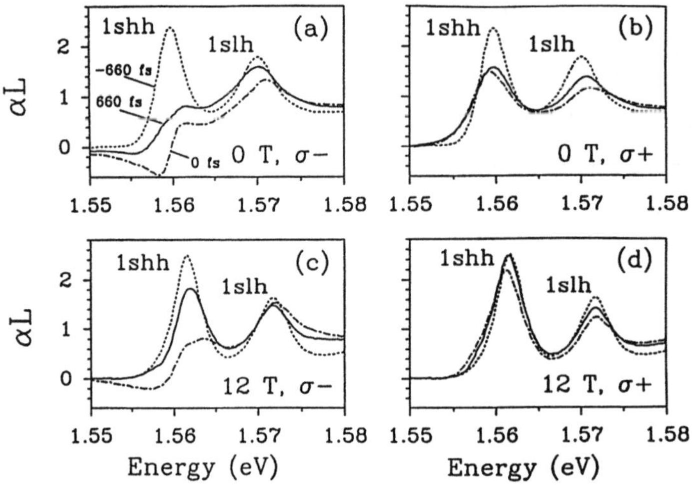

Figure 4: Absorption around the 1s exciton resonances seen by a σ_{-}-probe at $\Delta t = -660fs$, $0fs$ and $+660fs$ for: (a) σ_{-}-pump and $H = 0T$, (b) σ_{+}-pump and $H = 0T$, (c) σ_{-}-pump and $H = 12T$, and (d) σ_{+}-pump and $H = 12T$.

pump and probe both σ_{-} polarized, the 1sh-MX response is an instantaneous gain and a blue shift at $\Delta t = 0$, which evolves toward a strong saturation and a smaller blue shift at $\Delta t = 660fs$. This is due to phase space filling (PSF) and exchange (EXCH) by the coherent ($\Delta t = 0$) and then the relaxed ($\Delta t = 660fs$) $1shh{-}\sigma_{-}$-MXs created by the pump. The difference between the $H = 0T$ and the $H = 12T$ cases is only qualitative. The high field effects are similar but significantly attenuated as compared to the low field ones (for example the blue shift changes from $1.9meV \rightarrow 0.3meV$). For both fields the 1slh experiences an instantaneous blue shift around $\Delta t = 0$, as seen on the DCD of Fig. 3c, which evolves to a small red shift and collisional broadening for $\Delta t = 660fs$. Since the 1shh and 1slh originate from distinct e and h states, the 1slh response in not due to PSF and EXCH by the real 1shh MXs. It comes from the pump induced virtual-populations of 1slh i.e. AC-Stark Effect [1]. When the pump polarization is reversed to σ_{+}, the $1shh{-}\sigma_{-}$-MX exhibits only a small red shift and almost no saturation at $H = 12T$, whereas at $H = 0$ it still saturates although much less than for σ_{-}-pump. As for the $1slh{-}\sigma_{-}$-MX it experiences a small but distinct saturation and blue shift with, however, the $H = 12T$ response attenuated as compared to that at $H = 0T$.

Two sets of effects have to be distinguished, i) the effects of the polarization of the MXs created by the pump (i.e. the difference between the DAS spectra for σ_{-} and σ_{+} excitations), and ii) the effects of magnetic confinement (i.e. the difference between the DAS spectra for $H = 0T$ and $H = 12T$). The polarization effects can be explained intuitively by considering [9-11]: i) Pauli exclusion and the symmetry of the e and h states out of which the MXs are built, and ii) the "molecular" exciton-exciton interaction potential which is attractive in the singlet state and repulsive in the triplet state. PSF and EXCH are strongly active when the pump and probe MXs involve the same e and/or h states. The sign of the shift experienced by the probe MXs is explained by the attractive or repulsive character of the MX-MX interaction. Considering the evolution of the nonlinear response vs H, we note that for pure parabolic bands (neglecting Coulomb interaction), the e and h wavefunctions depend only on r_c and are therefore mass independent. As H is increased the magnetic confinement starts to dominate the Coulomb interaction, and the e and h are forced into almost identical and overlapping wavefunctions. Hence the quasi-0D MX at high field occupy much smaller volume and, at the same density, show much less PSF. Furthermore, as the magnetic potential, $(\lambda r/2)^2$, increases it confines more and more e and h on the top of one another making the MXs more rigid and locally neutral and therefore much less polarizable. This has two consequences, i) The MX-MX interaction is reduced at high field and the shifts of the resonances, although with the same sign as for $H = 0T$, become much smaller. ii) The inter-MX dielectric screening (similar to the band gap renormalization induced by e-h plasma), itself is strongly attenuated. This last remark explains why the significant saturation of the $1shh{-}\sigma_{-}$-MX seen at $H = 0T$ for σ_{+}-pump, which originates from this dielectric screening, almost disappears at at $H = 12T$. This is the essence of the aforementioned exact thermodynamic equilibrium many-body results [7,8] which are therefore supported by our experiments. To go beyond these qualitative arguments and to explain the ultrafast dynamics that we investigate a time dependent many-body theory of MX-nonlinearities was developed [12,10]. It follows the unrestricted Hartree-Fock theory, recently introduced by Schmitt-Rink et al. [13,14], which has been very successful in describing the near band gap nonlinear optical effects in semiconductors [15,19]. The Hartree-Fock theories are more appropriate in the case of MXs since they become exact in the extreme magnetic limit, $\lambda \rightarrow \infty$ [20,21]. In this approach, the MX wavefunctions are expanded on the LL orbitals, ϕ_v. The optically connected conduction and valence band LLs form a set of two-level systems labeled by v. The density matrix of the semiconductor breaks into 2×2 blocks;

$$\hat{n}_v(t) = \begin{bmatrix} n_{c,v}(t) & \psi_v(t) \\ \psi_v^*(t) & n_{v,v}(t) \end{bmatrix} \tag{1}$$

where $n_{(c,v),v}(t)$ are the components of the populations on the conduction and the valence band LL and $\psi_v(t)$ are the components of the pair amplitude. The density matrix obeys the Liouville equation. The difference between this model and a collection of independent two-LL systems is due to the coupling between the LL by the Coulomb interaction, $V_{v,v'}$. It modifies the physics in two ways. Firstly the conduction and valence band energies are renormalized by the excited populations,

$$\varepsilon_{j,v}(t) \rightarrow \varepsilon_{j,v} - \sum_{v'} V_{v,v'}\, n_{j,v}(t) \tag{2a}$$

with $j = e$ or h. Secondly the coupling with the electromagnetic field, which is expressed by the Rabi frequency is modified according to:

$$\mu E(t) \rightarrow \Delta_v(t) = \mu E(t) + \sum_{v'} V_{v,v'} \psi_{v'}(t). \tag{2b}$$

This expresses the fact that the optically connected LLs at v do not experience the applied field, $\mu E(t)$, rather they see the selfconsistent "local field", $\Delta_v(t)$, which is the sum of the applied field and the "molecular" field, due to all the other e-h LLs [10,12]. The interaction between MXs appears in the Liouville equation as an "exchange" term; $\sum_{v'} V_{v,v'} [\psi_{v'} n_v - \psi_v n_{v'}]$. This term vanishes indentically on the diagonal, $v = v'$, giving the clue for explaining the observed difference in the interaction between MXs in the same state and in different states at low and high magnetic field. For small fields, MX are composed of many LL and, therefore, there is a strong interaction between all the MXs whenever one MX-state is photoexcited. At high field the MXs become more and more diagonal in the LL basis, the interaction within a MX state vanishes, yet it persists between different MX states. The residual interaction is mostly due to the self-energy corrections, which explains the high field disappearance of the 1s-MX blue-shift found experimentally. The attractive inter-MX self-energy correction explains the experimental red shift of the 2s-MX induced by a population of photoexcited 1s-MS. At high field MXs behave like two-LL systems and their nonlinear optical response become dominated by Pauli exclusion. Numerical solutions of the Liouville equation using a development on a basis of 1000-LL, shows that indeed Coulomb correlation is completely quenched in the extreme magnetic limit and reproduced fairly well the experimental results [10,12].

3. Effects of Reduced Density of States on Carrier Relaxation

An important aspect of the $2D \rightarrow 0D$ transition is that it affects directly the density of states. This restricts the number of available initial and final states for any transition, and therefore strongly influences relaxation. We have investigated this effect by exciting nonthermal carrier populations about $25\,meV$ above the lowest 1shh-resonance at $H = 0T$ and $H = 12T$. The DCD spectra at $\Delta t = 0fs$ and $\Delta t = 200fs$ are shown in Fig.5, for the cases without magnetic field and in Fig. 6, for the case with magnetic field.

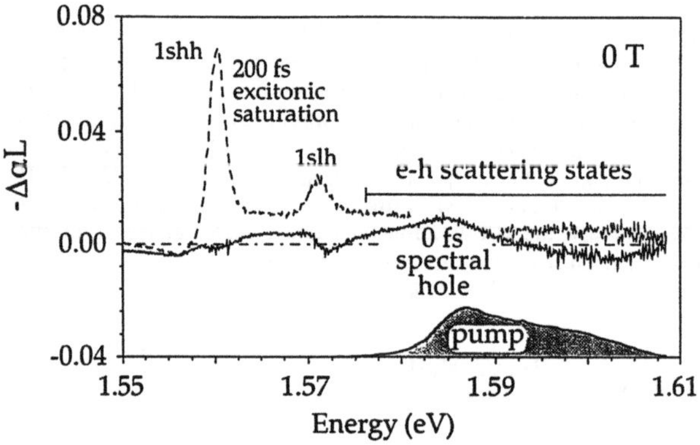

Figure 5: Circular dichroism spectra without magnetic field, for excitation 25meV above the 1shh resonance and at $\Delta t = 0fs$ (solid line) and $\Delta t = 200fs$ (dashed line).

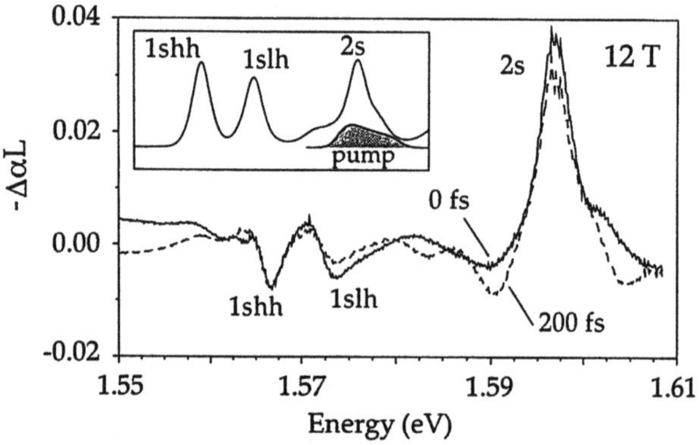

Figure 6: Circular dichroism spectra with a $H = 12T$ magnetic field, for excitation 25meV above the 1shh resonance and at $\Delta t = 0fs$ (solid line) and $\Delta t = 200fs$ (dashed line).

The DCD spectra separate the PSF effects due to the excited populations from charge density effects such as screening [11].

When $H = 0T$ one sees, in agreement with the previous observations [22,23], an instantaneous, $\Delta t = 0$, spectral hole burning in the continuum at an energy slightly lower than the pump photon central energy and no response at the 1shh.

Hole burning in the continuum is the signature of the PSF induced by the nonthermal populations. At $\Delta t = 200fs$ the carriers have thermalized among themselves by carrier-carrier scattering and occupy the states at the bottom of the band out of which the 1shh and 1slh excitons are made, blocking the transitions to these excitons. Hence the DAS reproduces the profile of these two resonances and the spectral hole has disappeared [22,23]. We note that because the excess energy is smaller than that of the optical phonons, the thermalization is internal to the plasma i.e. the average energy remains constant but the e-h populations establish very quickly a thermal distribution among themselves, although at a temperature different from that of the lattice. When the magnetic field is applied, the same pump excites the 2s-MX as shown in the inset of Fig. 4b. There is immediately a strong PSF signal at the 2s-MX and, as discussed above, a small signal due to dielectric screening at the 1shh-MX and 1slh-MX (note the different scales of the Fig. 4a and 4b). The important result is that the spectra at $\Delta t = 0fs$ and $\Delta t = 200fs$ are almost identical. The nonthermal populations are blocked at 25meV above the lowest states. They cannot emit optical phonons and the carrier-carrier scatterings are quenched because of the very limited number of final states available in the quasi-0D density of states. The confinement in quasi 0D therefore results in a dramatic reduction of the thermalization rates [11].

4.Conclusion We have explored the femtosecond dynamics of the nonlinear optical response of Magneto-excitons, as the quasi-2D quantum well electronic states are further confined in quasi-0D by a strong magnetic field. It was found, in agreement with theory, that at high field Coulomb correlation between magneto-excitons in the same state is quenched. It, however, persists between magneto-excitons in different states giving strong nonlinear responses. These results also show that a gas of magneto-excitons is a unique two-component manybody system that behaves very differently from the one-component systems such as electron gas in the fractional quantum Hall effect or Wigner crystal. Furthermore, the magnetic field induces a restriction of the number of states available for transition which almost completely quenches the relaxation of nonthermal populations. This produces qualitatively different carrier dynamics that must be accounted for in any quasi-0D system or device.

Acknowledgment: We have benefited from the theoretical support of S. Schmitt-Rink, and W. Schäfer. The work of D.S. Chemla is supported by the Director, Office of Energy Research, Office of Basic Energy Research, Division of Materials Sciences of the US Department of Energy under Contract No. DE-AC03-76SF00098. J.B. Stark would like to acknowledge support through NSF grant EET-871841.

REFERENCES

1. S. Schmitt-Rink, D.S. Chemla, and D.A.B. Miller, Phys. Rev. B32, 6601 (1985)
2. O. Akimoto, and H. Hasegawa, J. Phys. Soc. Jnp., 22, 181 (1967)
3. M. Shinada, and K. Tanaka, J. Phys. Soc. Jnp., 29,1258 (1970)
4. S.R.E. Yang, and L.J. Sham, Phys. Rev. Lett. 58, 2598 (1987)
5. G.E.W. Bauer, and T. Ando, Phys. Rev. B38, 6015 (1988)
6. H. Chu, and Y.C. Chang, Phys. Rev. B40, 5497 (1989)
7. I.V. Lerner and Yu. E. Lozovik, Zh. Eksp. Teor. Fiz. 80, 1488 (1981) [Sov. Phys. JEPT 53, 763 (1981)]
8. D. Paquet, T.M. Rice, and K. Ueda, Phys. Rev. B32, 5208 (1985)
9. J.B. Stark, W.H. Knox, D.S. Chemla, W. Schäfer, S. Schmitt-Rink and C. Stafford, Phys. Rev. Lett. 65, 3033 (1990)
10. S. Schmitt-Rink, J.B. Stark, W.H. Knox, D.S. Chemla, W. Schäfer, Appl. Phys. A 53, 491 (1991)
11. J.B. Stark, W.H. Knox, D.S. Chemla, Phys. Rev. Lett. 68, 3080 (1992) and to be published in Phys. Rev. B
12. C. Stafford, S. Schmitt-Rink, W. Schäfer, Phys. Rev. B41, 10,000 (1990-I)
13. S. Schmitt-Rink, D.S. Chemla, Phys. Rev. Lett. 57, 2752 (1986).
14. S. Schmitt-Rink, D.S. Chemla, H. Haug, Phys. Rev. Lett. B37, 941 (1988).
15. W. Schäfer, Adv. Solid State Phys. 28, 63, (1988), W Schäfer, K.H. Schuldt, and R. Binder, Phys. Stat. Sol. B150, 407 (1988)
16. C. Ell, J.F. Muller, K. El Sayed, L. Banyai, and H. Haug, Phys. Stat. Sol. B150, 393 (1988) and Phys. Rev. Lett. 62, 304 (1989)
17. R. Zimmermann, Phys. Stat. Sol. B146, 545 (1988) and R. Zimmermann and M. Hartmann Phys. Stat. Sol. B150, 365 (1989)

18. I. Balslev, R. Zimmermann, and A. Stahl, Phys. Rev B40, 4095 (1989)

19. R. Binder, S.W. Koch, M. Lindberg, N. Peyghambarian, and, W. Schäfer, Phys. Rev. Lett. 65, 899 (1990)

20. C. Kallin, and B.I. Halperin, Phys. rev. B30, 5655 (1984)

21. A.B. Dzyubenko, and, Yu. E. Lozovik, J. Phys. A24, 415 (1991)

22. W.H. Knox, C. Hirlimann, D.A.B. Miller, J. Shah, D.S. Chemla, and C.V. Shank, Phys. Rev. Lett. 56, 1191 (1986).

23. W.H. Knox, D.S. Chemla, G. Livescu, J.E. Cunningham, and J.E. Henry, Phys. Rev. Lett. 61, 1290 (1988), W.H. Knox, in "Hot Carriers in Semiconductor Nanostructures", J. Shah ed. Academic Press (1992).

High-Order Harmonic Generation in Strong Laser Fields

A. L'Huillier and P. Balcou

Service des Photons, Atomes et Molécules, Centre d'Etudes de Saclay,
F-91191 Gif-sur-Yvette, France

Abstract. The recent advances concerning high-order harmonic generation processes in strong laser fields are discussed. We show how the width of the plateau inherent to the production of the high harmonics varies with the pulse width, the laser wavelength and the atomic medium. Phase matching effects are discussed in terms of the coherence length owing to the laser focusing. Finally, we show that the idea of generating attosecond light pulses by adding harmonics with constant intensities will need rephasing of the high harmonics.

1. Introduction

As is well known, the interaction of a laser field with an atomic gas leads to generation of odd harmonics of the fundamental frequency. When the incident laser is intense and of short pulse duration, high order harmonics can be created [1-3]. Instead of rapidly decreasing as expected within lowest-order perturbation theory, the intensities of these harmonics in general are found to be approximately constant, from the 5th or 7th order up to a very high order. This behavior is characteristic of a strong (non-perturbative) laser-atom interaction.

This paper is organized as follows. First, we discuss the microscopic properties of these high-order harmonic generation processes. We examine how the length of the plateau varies with the parameters that affect the response of the individual atoms to the radiation field. Then, we consider the influence of coherence, phase matching effects on the conversion efficiency. In conclusion, we briefly discuss the possibility for these processes to lead to the generation of ultrashort pulses, below 0.1 fs.

2. Microscopic properties of harmonic generation processes.

Studies of harmonic intensity dependences show that the signal saturates when the medium gets ionized. Before the saturation intensity, it increases rapidly, but not as rapidly as within lowest-order perturbation theory. It even shows structures, resonances in the case of low-order harmonics. To increase the width of the plateau, one must then make the atomic medium more difficult to ionize. This can be done by reducing the pulse width. Indeed, we found that the number of harmonics produced in xenon at about the same intensity and at 1 μm increases from 21 at 36 ps to 29 at 1.2 ps. One can also use lighter atomic systems which ionize at a higher laser intensity. Thus, much higher harmonics can be observed in neon. We detected the 53rd harmonic, but we were limited by our detection system for the observation of shorter wavelengths.

Figure 1 illustrates the influence of the wavelength on harmonic generation. Here, we compare results obtained at 1053 nm (full squares), 616 nm (open circles) and 308 nm (open squares) using short pulse lasers (1.2, 2 and 1.4 ps respectively). All of the

Springer Series in Chemical Physics, Vol. 55 Ultrafast Phenomena VIII
Editors: J.-L. Martin · A. Migus · G.A. Mourou · A.H. Zewail

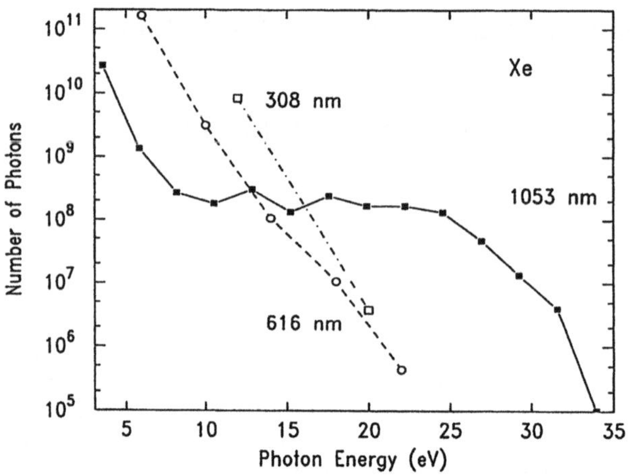

Fig. 1. Comparison between results obtained in xenon at 1053 nm, 616 nm and at 308 nm.

results have been obtained at intensities above saturation. The comparison between the three results shows that (i) the plateau behaviour, which is remarkable at 1053 nm, disappears at shorter excitation wavelengths (the harmonic intensities decrease more rapidly with increasing order for shorter wavelengths) (ii) the conversion efficiency is higher for low energy photons and lower for high energy photons when using short incident wavelengths. The laser-atom interaction regime departs more strongly from the perturbative limit for longer incident wavelengths. The plateau is more prononced and high harmonics can be produced more efficiently.

This conclusion seems to contradict the result recently obtained by Sarukura et al. [2]. They observed up to the 25th harmonic (125 eV) of a 248 nm excimer laser using a very high laser intensity (above 10^{17} W/cm^2) and a pulse width of 300 fs. In spite of the short incident wavelength, they obtained an extremely high generated photon energy, the second highest reported to date, the first one being the 170 eV observed with a 800 nm Ti:Sapphire laser [4]. However, recent calculations show that the harmonic emission in their experiment might be due to ions and not neutral atoms. Ions which are more difficult to ionize than neutrals can produce much more harmonics. Thus, Krause et al. [5] predict generation of photons of more than 400 eV energy in He^+ irradiated by the second harmonic of a Nd-Glass laser at an intensity of 5 $\times 10^{15}$ W/cm^2. These calculations do not include propagation effects which may alter this prediction, reducing the conversion efficiency for ions. We now discuss the influence of these effects for neutral media.

3. Coherence properties of harmonic generation processes

Phase matching of the high harmonics can be conveniently discussed by introducing the notion of an effective coherence length, defined as the length over which the qth harmonic field E_q can be built up coherently [6]. This coherence length L_{coh} can be due to dispersion in the medium ($L_{coh}^{disp} = \pi/\Delta k$, where Δk is the phase mismatch) and/or to focusing of the incident radiation ($L_{coh}^{foc} = b\tan[\pi/(q-1)]/2$), the latter effect being in general dominant. In most of the experimental situations when the ionization of the

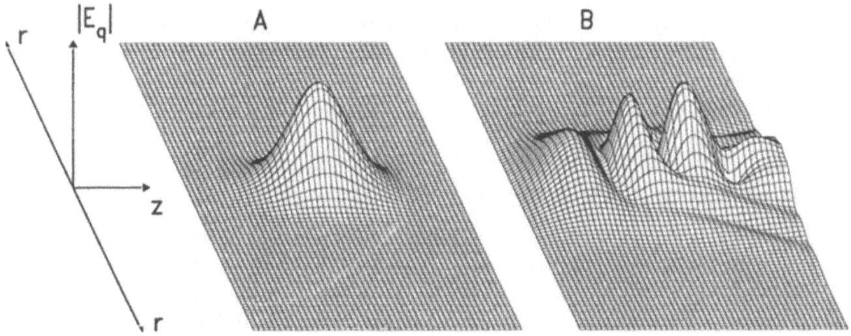

Fig. 2. Graphical representation of $|E_q(r,z)|$ ($q = 13$) in the nonlinear medium. The laser propagates from the left to the right. The polarization is assumed to follow a I^{13} power law (weak field limit) in (a) and a I^5 power law in (b) (simulating a strong field regime).

medium is not important, $L_{coh}^{disp} > L > L_{coh}^{foc}$, so that only the effect of focusing needs to be taken into account. We show in Figure 2 how the 13th harmonic is generated in a 2 mm long uniform nonlinear medium.

The key result is that phase matching of the high harmonics depends dramatically on the variation of the *amplitude* of the polarization field in the medium. In a weak field limit, for a tight focused geometry, the harmonic field created in the first half of the medium gets almost completely cancelled out by destructive interference effects in the second half of the medium (Fig. 2(a)). In strong laser fields, the atomic response is non perturbative and the variation of the harmonic rate with intensity is in average much less rapid than in the weak field limit. Consequently, the harmonic field gets defocused and phase matching oscillations can develop over a longer length (Fig. 2(b)), so that a significant harmonic intensity gets out of the medium.

These phase matching oscillations were observed experimentally by varying the position of the focus in the gas jet [7]. Indeed, in a "semi-infinite medium approximation", i.e. such that the polarization amplitude is negligible at the entrance of the medium, one can easily show that varying the focus position in the medium is equivalent to displacing the exit edge of the medium thus directly probing the oscillations shown in Fig. 2(b). The experiments were performed with a Nd-YAG laser at intensities of the order of 10^{13} W/cm^2. We observed oscillations in the variations of the 7th, 9th and 13th harmonics as function of the position of the focus. The period of these oscillations are close to twice the coherence lengths owing to focusing, thus confirming our interpretation.

These arguments allow us to give a clear interpretation of previous experimental results consisting in studying the number of photons for several interaction geometries [1]. The harmonic signal was found to follow a simple b^3 power law, independently of the process order. Indeed, the strength of the harmonic field outside the medium depends on the length over which it can be built up coherently, so that the intensity of the harmonic field leaving the medium should be approximately proportional to b^2. This yields a b^3 scaling for the integrated intensity, since the focal section of the harmonic beam depends on b. This should be valid as long as $L_{coh} \leq L$. In the case of a collimated geometry where $L_{coh} \geq L$, the harmonic field does not depend on b, and the integrated intensity (the number of harmonic photons) should vary linearly with the confocal parameter b.

This provides also a simple means of increasing the conversion efficiency. If the intensity at best focus is higher than the minimum required for producing a given harmonic, one can simply defocus the beam relative to the gas jet. The coherence length increases away from the focus and the signal increases. This allowed us to optimize the conversion efficiency in xenon up to 5×10^{-7}, with numbers of photons per pulse as high as 4×10^{8} at 20 eV.

4. Towards attosecond light pulses ?

Finally, we briefly discuss the idea recently mentioned by several groups [8] that the plateau harmonics might lead to ultrashort pulses. Suppose that one is able to get rid of the first and last few harmonics, keeping only those of the same amplitude, for example the seventh to the 21st harmonics in xenon (Fig. 1). If these harmonics have the same phase, then the addition of the electric fields oscillating at frequencies 7ω, 9ω, etc.. 21ω leads to ultrashort pulses (whose width at half maximum is smaller than 0.1 fs) separated by the the fundamental half period of 1 fs, as shown in Fig. 3(a).

However, this attracting idea lies on the assumption that all the harmonics have the same phase. Although they have a well defined phase relationship with the fundamental, in general they do not have the same phase. Fig. 3(b) shows the result of a calculation in xenon adding the harmonic field complex amplitudes at 5×10^{13} W/cm^2 for a single xenon atom. The regularity of the structures shown in Fig. 3(a) has disappeared. If one includes also the effect of propagation in a thin nonlinear medium, which introduces an additional dephasing between the harmonics, the sharp structures broaden, as illustrated in Fig. 3(c). The plateau does not behave as a mode-locked laser, but rather as a chaotic multimode laser, with sharp and indeed very short substructures.

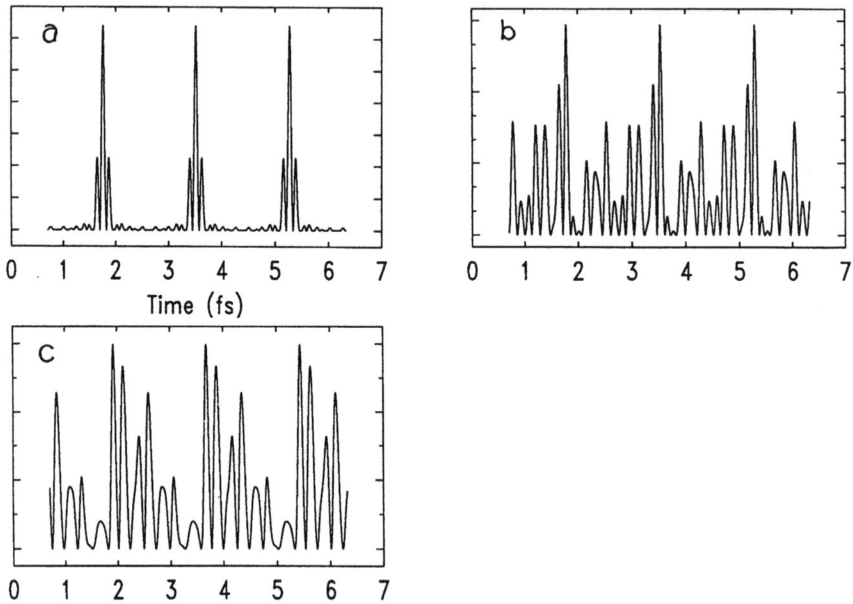

Fig. 3. Intensity of the electric field corresponding to harmonics # 7 to # 21 in xenon. (a) assuming constant phases and amplitudes; (b) taking the complex dipole moments obtained from the solution of the time-dependent Schrödinger equation; (c) including propagation

5. Conclusion

In conclusion, we would like to emphasize that these processes might lead to useful XUV light sources. The generated radiation is coherent and bright. Its pulse width follows that of the pump field and can therefore be very short (100 fs), enabling the possibility to implement pump-probe techniques. The investigation of these processes is worth persuing, not only for its basic interest as a probe of the behavior of atoms in strong laser fields but also because it might lead to short pulse, bright, compact and possibly tunable sources of coherent radiation in the XUV range.

References

[1] X. F. Li, A. L'Huillier, M. Ferray, L. A. Lompré, G. Mainfray and C. Manus, Phys. Rev. A **39**, 5751 (1989) ; A. L'Huillier, L. A. Lompré, G. Mainfray and C. Manus, in "Atoms in Intense Laser Fields", Ed. M. Gavrila, Academic Press, Orlando, Florida, in press.

[2] N. Sarakura, K. Hata, T. Adachi, R. Nodomi, M. Watanabe and S. Watanabe, Phys. Rev. A **43** 1669 (1991).

[3] K. Miyazaki and H. Sakai, J. Phys. B **25**, L83 (1992)

[4] J. J. Macklin, J. D. Kmetec and C. L. Gordon III, to be published.

[5] J. L. Krause, K. J. Schafer and K. C. Kulander, Phys. Rev. Lett. **68**, 3535 (1992).

[6] A. L'Huillier, K. J. Schafer and K. C. Kulander, Phys. Rev. Lett. **66**, 2200 (1991) ; J. Phys. B, **24** 3315 (1991).

[7] A. L'Huillier, P. Balcou and L. A. Lompré, Phys. Rev. Lett. **68**, 166 (1992).

[8] S. E. Harris, J. J. Macklin and T. W. Hänsch, preprint ; G. Farkas and C. Toth, preprint ; G. Mourou and G. Mainfray, private communication.

QED at 10^{20} W/cm^2

A.C. Melissinos

Department of Physics and Astronomy, University of Rochester, Rochester, NY 14627, USA

We report on an experiment [1] in preparation at the Stanford Linear Accelerator Center (SLAC) to study the interaction of electrons and photons in an electric field exceeding the critical value. The critical field can be equivalently defined as the field in which an electron gains in one Compton wavelength energy equal to its rest mass [2]

$$E_c = \frac{m^2c^3}{e\hbar} = 1.3 \times 10^{16} \text{ V/cm} \quad . \tag{1}$$

Or as the magnetic field in which the critical synchrotron radiation frequency equals the electron energy [3]

$$\gamma B_c = \frac{m^2c^2}{e\hbar} = 4.4 \times 10^{13} \text{ Gauss} \quad , \tag{2}$$

where $\gamma = E_e/mc^2$ is the ratio of the electron's total energy to its rest mass. Or as the average field seen by an electron bound to a nucleus with $\alpha Z = 1$

$$E_c = \frac{eZ}{a_0^2} = (\alpha Z)^3 \, \frac{m^2c^3}{e\hbar} \quad , \tag{3}$$

where $a_0 = \hbar^2/mZe^2$ is the Bohr radius.

We define the parameter Υ as the Lorentz invariant ratio of the electric field $E^* = 2\gamma E$ seen by an electron in a wave field E to the critical field

$$\Upsilon \equiv 2\gamma \, \frac{eE\hbar}{m^2c^3} \quad . \tag{4}$$

We also define a second Lorentz invariant parameter which is proportional to the classical vector potential of a wave field of frequency ω, through

$$\eta = \frac{eE}{\omega mc} \quad . \tag{5}$$

When η approaches or exceeds unity, multiphoton effects become dominant. Furthermore, an electron moving in this wave field, executes transverse motion as a result of which it acquires an effective mass [4]

$$\overline{m}^2 = m^2\gamma_\perp^2 = m^2(1+\eta^2) \, . \tag{6}$$

While QED is well established because of the great accuracy obtained from the perturbative calculation of higher order graphs as in the case of the Lamb shift, muonium and anomalous magnetic moment, it has not been tested near the critical field regime. In principle, perturbative calculations are not valid at such field strengths and

Springer Series in Chemical Physics, Vol. 55 Ultrafast Phenomena VIII
Editors: J.-L. Martin · A. Migus · G.A. Mourou · A.H. Zewail

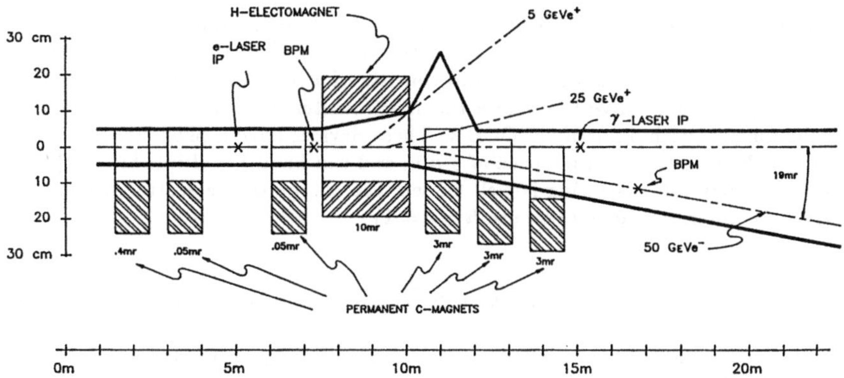

Figure 1. The experimental arrangement.

new phenomena such as spontaneous e^+e^- pair production set in. It has even been speculated that at the critical field QED may undergo a transition to a strongly coupled phase [5].

The electric field at the (diffraction limited) focus of a laser beam of power P is given by

$$E_{\rm rms}[\mathrm{V/cm}] = \sqrt{120\pi I} = \frac{8\pi}{(f/d)\lambda[\mathrm{cm}]}\ \sqrt{P(\mathrm{Watts})} \quad . \tag{7}$$

For an intensity $I = 10^{20}$ W/cm^2, the field reaches $E \simeq 2 \times 10^{11}$ V/cm which is still five orders of magnitude away from the critical field. To reach this field it is necessary to scatter high energy electrons from the focused laser spot. At SLAC the 50 GeV electron beam provides $\gamma \sim 10^5$ and thus according to Eq. (4) $\Upsilon = 1$ is reached when $I = 10^{19}$ W/cm^2. We have approached this intensity, using a chirped pulse [6] YLF/glass laser system ($\lambda = 1054$ nm) delivering 1.5 Joules in a 1.5 ps pulse, focussed to an area $A = 20\ \mu\mathrm{m}^2$.

Our intention is to investigate:

(a) Nonlinear, i.e. multiphoton, Compton scattering by measuring the differential cross section for the high energy γ-rays [7]

$$n\omega_0 + e \longrightarrow \omega + e' \quad . \tag{8}$$

(b) Pair production in the laser-electron interaction, a process also referred to as "beamstrahlung" [8]

$$n\omega_0 + e \longrightarrow e^+e^- + e' \quad . \tag{9}$$

We detect the high energy positrons in the range 5 GeV $< p_+ <$ 25 GeV.

(c) Pair production in "light by light scattering" with real photons [9]. The high energy back-scattered photons from the first ω-e collision are brought to a second interaction point where they are scattered from the laser beam

$$n\omega_0 + \omega \longrightarrow e^+e^- \quad . \tag{10}$$

(d) The effective mass spectrum of the produced e^+e^- pairs. Structure in that spectrum would indicate that the QED interaction is modified in the strong field [10].

The experiment will operate at a rate of 1 Hz, and the arrangement of the electron-laser interaction region is shown schematically in Fig. 1. Of great concern is the reduction of backgrounds, in particular synchrotron radiation, from the high energy electron beam since they can saturate the detector. The detector is located 100 m downstream from the interaction point and consists of a magnetic pair spectrometer. High energy γ-rays are converted before the spectrometer and the e^+e^- are tracked behind the magnet with an array of CCD's. These devices have a pixel size of 22.5 × 22.5 μm^2 yielding a resolution in space of 6 μm, and can handle multiple hits [11].

Nonlinear Compton scattering is well known, the energy of the scattered photons being given for intense fields by (we use $\hbar = c = 1$)

$$\omega = n\omega_0 \frac{4\gamma^2}{1 + 2\gamma^2(1 - \cos\theta) + [2n\gamma\omega_0/m + \eta^2/2](1 + \cos\theta)} , \tag{11}$$

where $E_e = \gamma m$ is the electron energy and θ is the laboratory scattering angle, which is of order $1/\gamma$. It is convenient to introduce the dimensionless parameters

$$x = 4\omega_0 E_e/m^2, \qquad y = \omega/E_e \quad .$$

The differential cross section can be given in closed form when the incident photons are circularly polarized [12]

$$\frac{d\sigma_n}{dy} = \frac{2\pi r_0^2}{x} \left\{ -\frac{4}{\eta^2} J_n^2(z) + \left(2 + \frac{u^2}{1+u}\right) \left[J_{n-1}^2(z) + J_{n+1}^2(z) - 2J_n^2(z)\right] \right\} . \tag{12}$$

In the above equations the index n labels the number of photons absorbed from the field of the laser and the parameters u, z are defined through

$$u \simeq \frac{y}{1-y}, \qquad y_{\max} = \frac{nx}{1 + \eta^2 + nx}, \qquad z = \eta\sqrt{1+\eta^2}\, \frac{2}{x} \sqrt{u\left(\frac{nx}{1+\eta^2} - u\right)} . \tag{13}$$

In Fig. 2 we show the differential cross section for

$$\begin{aligned} E_e &= \gamma m = 50 \text{ GeV} \;, \\ \omega_0 &= 1.17 \text{ eV } (\lambda = 1{,}054 \text{ nm}) \;, \\ \eta &= 0.4 \;. \end{aligned}$$

In Fig. 3 we show the spectrum of the high energy photons at the second interaction point. The incident laser pulse was assumed to be in the UV ($\lambda = 350$ nm) which can be achieved by tripling the IR pulse. We have obtained an efficiency of 40% for this process.

Pair creation by n laser photons is the cross-channel process from nonlinear Compton scattering, and for circular polarization of both the laser photons and the high-energy photon is given [12] in analogy to Eq. (12) by

$$\frac{d\sigma_n}{dy} = \frac{2\pi r_0^2}{x} \left\{ \frac{4}{\eta^2} J_n(z) + (u - 2) \left[J_{n-1}^2(z) + J_{n+1}^2(z) - 2J_n^2(z)\right] \right\} , \tag{14}$$

where (for $n\omega_0 \ll \omega$)

$$x = \frac{4\omega_0\omega}{m^2}, \qquad y = \frac{E_e}{\omega}, \qquad y_{\max,\min} = \frac{1}{2} \pm \sqrt{\frac{1}{4} - \frac{1+\eta^2}{nx}}, \qquad u = \frac{1}{y(1-y)} , \tag{15}$$

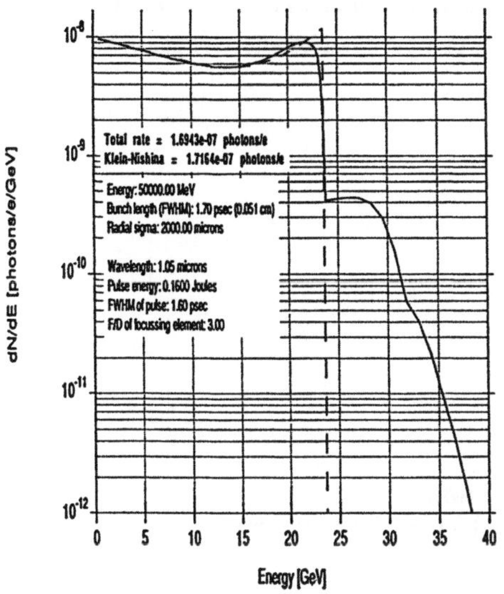

Figure 2. Multiphoton Compton scattering.

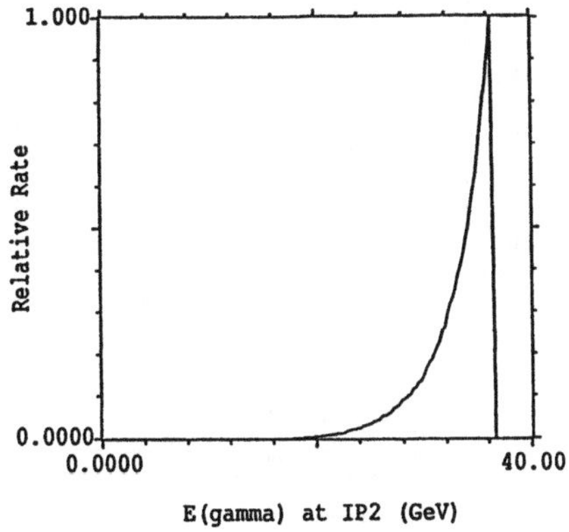

Figure 3. Energy spectrum of the backscattered high energy γ rays reaching the secong IP.

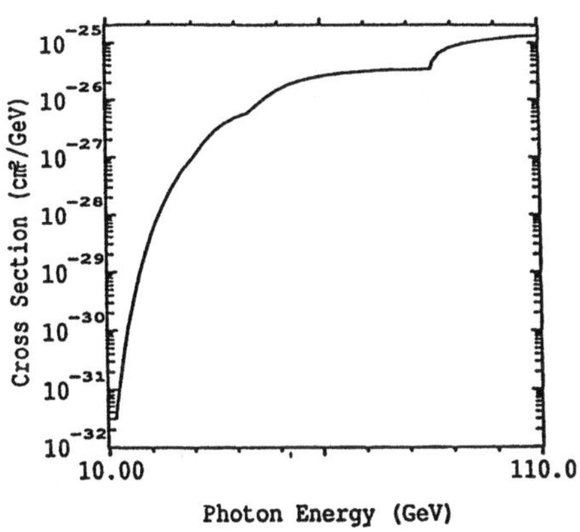

Figure 4. Total cross section for pair pro– duction as a function of electron energy.

and z is defined as before. The cross section exists only for those values of n such that y_{max} is real.

The cross section for $\eta = 0.4$ and $\omega_0 = 3.5$ eV, is shown in Fig. 4 as a function of the energy of the γ. The thresholds for pair production via n laser photons are given by

$$E_n = m^2(1+\eta^2)/n\omega_0 \tag{16}$$

and occur for $n = 1$ at 86.2 GeV, $n = 2$ at 43.1 GeV, and $n = 3$ at 28.7 GeV as can be seen in the Figure. For γ's from Compton backscatter from a 50-GeV electron beam, the maximum energy is 36.5 GeV at which energy three laser photons are required for pair production.

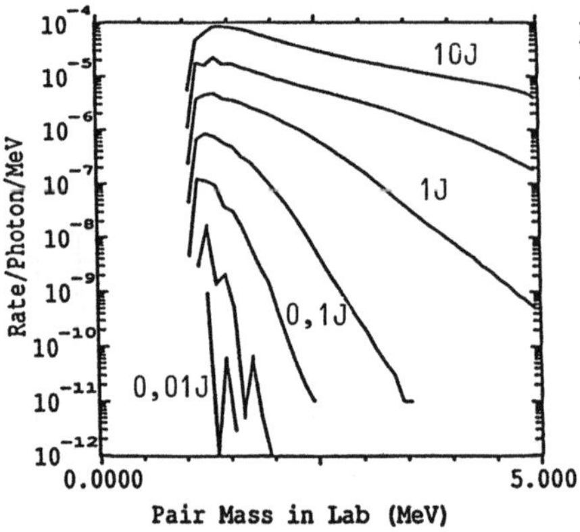

Figure 5. Differential rate for production of e^+e^- pair of effective mass M.

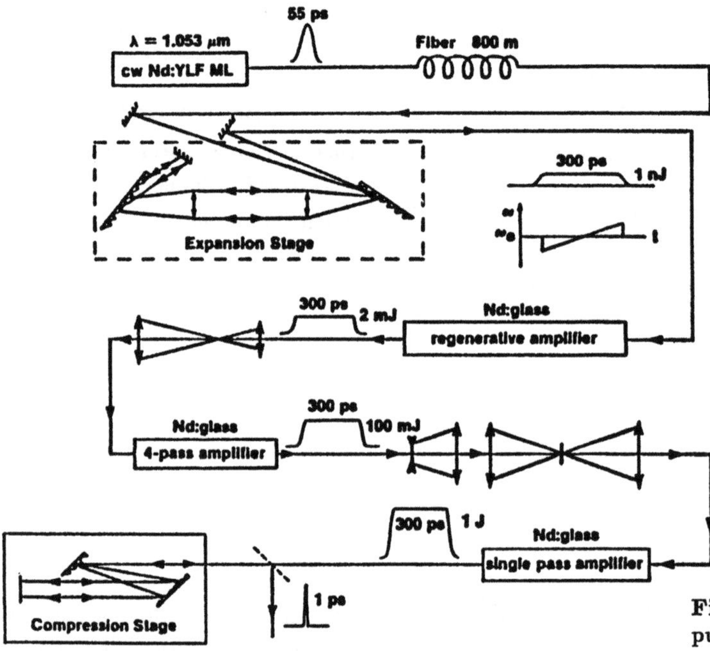

Figure 6. Chirped pulse amplification.

Of particular interest is the differential rate for the production of an e^+e^- pair of effective mass M, shown in Fig. 5 for different laser energies. It is assumed that a 36.5 GeV photon beam collides with a laser beam of 3.5 eV photons. The laser is focused in a lens of $f/D = 3$, and the high-energy photon beam has a radial σ of 60 μm. The seven curves are, from bottom to top, for 0.01, 0.3, 0.1, 0.3, 1, 3, and 10 Joule laser-pulse energies. Corresponding to these are $\eta_{\max} = 0.21$, 0.37, 0.67, 1.16, 2.12, 3.68, 6.71, and $\Upsilon_{\max} = 0.11$, 0.18, 0.33, 0.58, 1.05, 1.82, and 3.32, respectively.

The laser system is shown schematically in Fig. 6. At present it operates at a low repetition rate, which is not adequate for the contemplated experiments. To achieve the 1 Hz rate we will use a slab amplifier developed at the Laboratory for Laser Energetics. This amplifier has been in operation for over a year delivering nanosecond pulses with superior beam quality, suitable for efficient tripling [13].

References

[1] **Study of QED at Critical Field Strength in Intense Laser-High Energy Electron Collisions at the Stanford Linear Accelerator:** J. G. Heinrich, C. Lu, K. T. McDonald, Joseph Henry Laboratories, Princeton University, Princeton, NJ 08544; C. Bamber, T. Blalock, S. Boege, R. Hirosky, A. C. Melissinos, D. Meyerhofer and Th. Kotseroglou, Department of Physics and Department of Mechanical Engineering, University of Rochester, Rochester, NY 14627; T. Barklow, D. L. Burke, P. Chen, R. Field, A. C. Odian and J. E. Spencer, Stanford Linear Accelerator Center, Stanford University, Stanford, CA 94309; R. B. Palmer, Stanford Linear Accelerator Center, Stanford, CA 94309 and Brookhaven National Laboratory, Upton, NY 11973.

[2] For a review see K.T. McDonald, *Proposal for Experimental Studies of Nonlinear Quantum Electrodynamics*, Princeton U. Report DOE/ER/3072-32 (Sept. 1986).

[3] J. Schwinger, *The Quantum Correction in the Radiation by Energetic Accelerated Electrons*, Proc. Nat. Acad. Sci. **40** (1954) 132.

[4] See for instance Sec. 40 of V.R. Berestetskii, E.M. Lifshitz and L.P. Pitaevskii, *Quantum Electrodynamics*, 2nd ed., (Pergamon Press, 1982).

[5] S.L. Adler, *A New Embedding of Quantum Electrodynamics in a Non-Abelian Gauge Structure*, Phys. Lett. **B221** (1989) 39; D.G. Caldi, , Comments Nuc. Part. Phys. **19** (1899) 137.

[6] P. Maine, *et al.*, *Generation of Ultrahigh Peak Power Pulses by Chirped Pulse Amplification*, IEEE J. Quantum Electron. **QE-24**, 398 (1988). D. Strickland, and G. Mourou, Optics Commun. **56**, 219 (1985).

[7] R.H. Milburn, *Electron Scattering by an Intense Polarized Photon Field*, Phys. Rev. Lett. **10** (1963) 75; H.R. Reiss, *Absorption of Light by Light*, J. Math. Phys. **3** (1962) 59; *Production of Electron Pairs from a Zero-Mass State*, Phys. Rev. Lett. **26** (1971) 1072.

[8] See for instance M. Bell and J.S. Bell, *End Effects in Quantum Beamstrahlung*, Part. Acc. **24** (1988) 1; R. Blankenbecler and S.D. Drell, *Pair Production from Photon-Photon Collisions*, Phys. Rev. Lett. **61** (1988) 2324; M. Jacob and T.T. Wu, *Quantum Calculation of Beamstrahlung*, Nucl. Phys. **B303** (1989) 373, 389; P. Chen and K. Yokoya, *Field-Gradient Effect in Quantum Beamstrahlung*, Phys. Rev. Lett. **61** (1988) 1101; V. N. Baier, V. M. Katkov and V. M. Strakhovenko, Nucl. Phys. **B328**, 387 (1989).

[9] G. Breit and J.A. Wheeler, *Collision of Two Light Quanta*, Phys. Rev. **46** (1934) 1087.

[10] See for instance T. Cowan and J. Greenberg, in *Physics of Strong Fields*, ed. W. Greiner (Plenum, N.Y. 1987); the original papers are J. Schweppe *et al.*, *Observation of a Peak Structure in Positron Spectra from U+Cm Collisions*, Phys. Rev. Lett. **51** (1983) 2261; T. Cowan *et al.*, *Observation of Correlated Narrow-Peak Structures in Positron and Electron Spectra from Superheavy Collision Systems*, Phys. Rev. Lett. **56** (1986) 444; P. Salabura *et al.*, Phys. Lett. **B245** (1990) 153; W. Koenig *et al.*, *On the Momentum Correlation of (e^+e^-) Pairs Observed in U+U and U+Pb Collisions*, Phys. Lett. **B218** (1989) 12.

[11] C.J.S. Damerell *et al.*, *A Vertex Detector for SLD*, Nucl. Instr. and Meth. **A275** (1989) 484; *A CCD-Based Vertex Detector for SLD*, Nucl. Instr. and Meth. **A288** (1990) 236.

[12] N.B. Narozhnyi, A.I. Nikishov and V.I. Ritus, *Quantum Processes in the Field of a Circularly Polarized Electromagnetic Wave*, Soviet Physics JETP **20** (1963) 622; see also Sec. 101 of Ref. 4.

[13] Y. Wang, B. Lether-Davis, Y.-H. Chuang, R. S. Craxton, D. D. Meyerhofer, Optics Letters, **16** (1991) 1862.

Part II

Elementary Dynamics: Chemistry, Biology and Physics

Femtochemistry

A.H. Zewail

Arthur Amos Noyes Laboratory of Chemical Physics,
California Institute of Technology, Pasadena, CA 91125, USA

Abstract. This paper addresses some recent developments in femtosecond chemistry. Different applications are outlined for probing, in real-time, the elementary steps - bond breaking and bond forming - of chemical reactions.

Since 1985, progress has been made in the use of femtosecond lasers to directly observe and study the femtochemistry of chemical reactions in gas phase/molecular beams (for reviews, see Refs. 1-5). The processes of bond breaking and bond forming involve elementary nuclear motions, and these motions are generally classified into two categories:

Half-collision Reactions: (1)

$$A \overset{*}{-} BC \rightarrow [A\cdots BC]^{\ddagger *} \rightarrow A + BC$$

(bond breakage)

Full-collision Reactions: (2)

$$A + BC \rightarrow [A\cdot\cdot B\cdot\cdot C]^{\ddagger} \rightarrow AB + C$$

(bond breakage/bond formation).

Both types of motions have been observed in a variety of systems[1-5] and, here, we only highlight the recent advances made and some new directions.

1. Transition States and Coherent Evolution of Products

Two selected examples serve to illustrate the concept. In the "one-dimensional" reaction, involving two atoms:

$$Na{*}I \rightarrow [Na\cdot\cdot I]^{\ddagger *} \rightarrow Na + I, \quad (3)$$

the activated complexes were observed, together with their resonance oscillations. This is due to the avoided crossing caused by the fundamental nature of the chemical bond--covalent [$\Psi(Na, I)$] and ionic [$\psi(Na^+, I^-)$] contributions. Moreover, the product, e.g., Na atoms, are formed by a *coherent* mechanism (quantized steps) and not by an incoherent kinetic mode. This represented the first example where coherent motion of nucleii can be observed from reactants to products.

The second example involves the coherent evolution of a reaction from the transition state to final products. The phases are well-defined and we termed the motion as yielding coherent trajectories:

Springer Series in Chemical Physics, Vol. 55 Ultrafast Phenomena VIII
Editors: J.-L. Martin · A. Migus · G.A. Mourou · A.H. Zewail

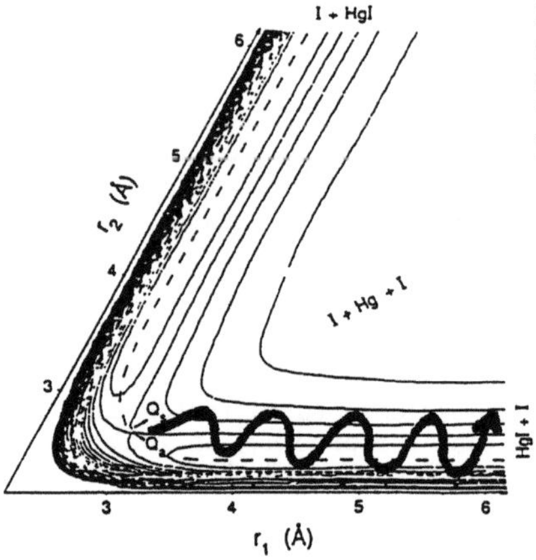

Fig. 1. The 2-D potential which describes the observed dynamics of coherent evolution from the transition state [IHgI]‡* to product HgI + I.

$$I^{*}HgI \rightarrow [I\cdot\cdot HgI]^{\ddagger *} \rightarrow I + HgI. \quad (4)$$

In this case, the potential is two-dimensional (two coordinates; Fig. 1) and there is a well-defined saddle point transition state. The product, HgI, was observed in vibrationally coherent states and the temporal evolution of both the rotational and vibrational motions was monitored. Such behavior has recently been found in the predissociation of complexes[6] and in the photodissociation of I_3^-.[7] Manz[8] has also mapped out the evolution theoretically.

2. Coherent Wave Packets and Dynamics in Solutions and Clusters

The wave packet of I_2 has become a standard for calibrating the behavior of bound state coherent nuclear motion. Both the vibrational and rotational motions have been observed and directly used to obtain the anharmonic potential of the motion. The phase evolution of the packet and its recurrences have also been reported (Fig. 2). Knowing the nature of the

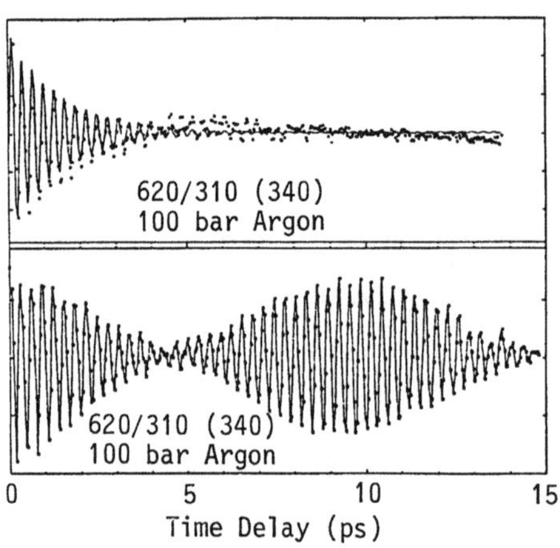

Fig. 2. The wave packet motion observed in I_2 under collisionless condition (bottom) and at 100 bar (top).

motion in time, we have recently used such packets as a probe of *solute - solvent* interactions.[5] By studying I_2 in rare gases at high pressures, we were able to explore the phenomena of caging of two iodine atoms by Ar, the wave packet dephasing and vibrational relaxation, all on the time scale in which they occur. The Wilson's group[9] has elegantly shown that the dynamics of these processes can be obtained theoretically by considering the interaction potential between the solute and solvent atoms and the nature of the motion (coherence,...etc.).

The other direction of the problem was also considered by forming clusters of I_2 with rare gas atoms (of certain composition). We were able to observe, in real-time, the evaporation of the different rare gas atoms and relate these observations to the theory of VP and IVR.[10]

3. Ground State Bimolecular Reactions

Some time ago, we, in collaboration with the late Professor R. Bernstein, used molecular beam techniques to initiate bimolecular reactions on ground state surfaces, e.g., $H + CO_2 \rightarrow HO + CO$. Recently, we have developed a method for probing the collision complex of a bimolecular encounter ($Br + I_2$) and measured, in real-time, the lifetime of such species.[11] The translational energy can be varied, and from these experiments one has a direct view of the dynamics in the transition state region:

$$Br + I - I \rightarrow [Br \cdot\cdot I \cdot\cdot I]^{\ddagger} \rightarrow BrI + I. \qquad (5)$$

4. Femtosecond Control of a Chemical Reaction

Our first attempt to control, in addition to probing, a chemical reaction was successful this year.[12] In a bimolecular encounter between I_2 and Xe, it was shown that the product yield XeI can be switched "on" or "off" depending on the time delay between an initiation pulse and a control pulse. The results (Fig. 3) are very promising and we hope to extend the approach to other systems. We note that this switching has defined a "zero-of-time" for a bimolecular reaction under a normal full collision encounter with a full impact parameter distribution.

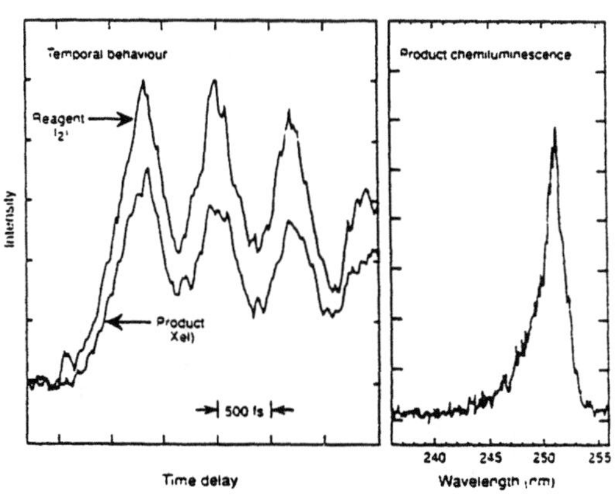

Fig. 3. The fs control of the reaction Xe + I_2 (12).

5. Ultrafast Electron Diffraction and Molecular Structure

Some time ago, an idea for observing isolated molecular structures on the *ultrafast* time scale was proposed.[4] The analysis[13] showed that such experiments were feasible, and only recently have we been successful in recording structures with ultrashort time resolution (molecules: I_2, CCl_4).[14] We have also been able to observe chemical changes in the bond breaking process of CF_3I. We believe that this is an exciting advance that has the potential of revolutionizing our ability to study chemical and structural changes on the ultrashort time scale (Fig. 4, 5).

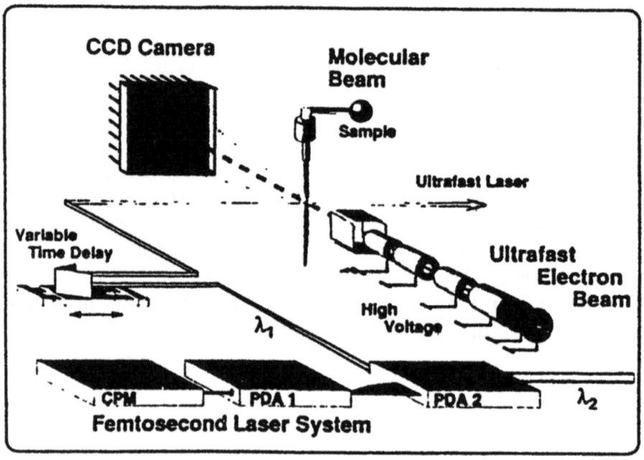

Fig. 4. Schematic of the ultrafast electron diffraction apparatus built at Caltech.

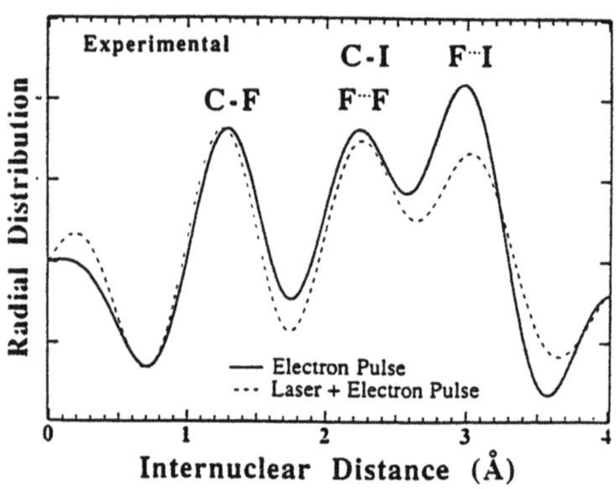

Fig. 5. Electron diffraction obtained for CF_3I, with ultrashort electron pulse and in the presence of a laser field (radical distribution function).

6. Bimolecular, General Reactions

The full A + BC collision has an ill-defined zero-of-time. Different molecules at low pressures and in molecular beams have different mean free times (typically µs) and this "averages out" the ps to fs dynamics of interest. We are currently involved in using stimulated emission (or Raman) techniques to prepare [A••B••C]‡ collision complexes at different internuclear separations (on the ground state potential) and with

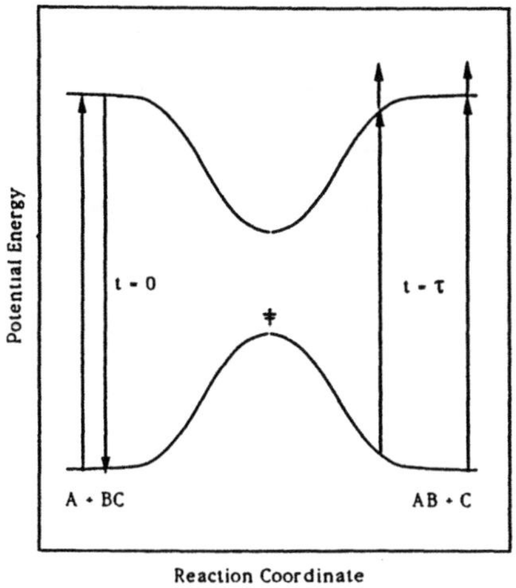

Fig. 6. The concept for probing bimolecular reactions A + BC, on the fs time scale. The probing is by LIF or MPI.

different kinetic energies. Work is now in progress to consider the theoretical and experimental aspect of the approach as applied to A + BC and ABA type reactions (Fig 6).

Conclusions

With fs time resolution, the above developments are providing the real-time methodology for observing and studying the elementary and fundamental steps of chemistry - the breaking and making of chemical bonds, which occur on the fs time scale. Clearly more applications are continuing, and with the electron diffraction techniques, a great number of exciting studies, especially in large systems, are promised.

This work was supported by the AFOSR and by the NSF. During this Conference, I enjoyed discussions of the theoretical analysis of experiments (6) with B. Girard.

References

1. A. H. Zewail, Science **242**, 1645 (1988).
2. A. H. Zewail and R. B. Bernstein, Chem. & Eng. News, **24** (November 7, 1988).
3. L. R. Khundkar and A. H. Zewail, Ann. Rev. Phys. Chem. **41**, 15 (1990).
4. A. H. Zewail, Faraday Discuss. Chem. Soc. **91**, 207 (1991).
5. A. H. Zewail, M. Dantus, R. M. Bowman, and A. Mokhtari, J. Photochem. Photobiol. A: Chem. **62/3**, 301 (1992).
6. A. J. Kaziska and M. R. Topp, Chem. Phys. Lett. **180**, 423 (1991).
7. S. Ruhman, in this proceeding.
8. J. Manz, to be published.
9. Y. Yan, R. M. Whitnell, K. R. Wilson, and A. H. Zewail, Chem. Phys. Lett. **193**, 402 (1992).
10. D. M. Willberg, M. Gutmann, J. J. Breen, and A. H. Zewail, J. Chem. Phys. **96**, 198 (1992); M. Gutmann, D. M. Willberg, and A. H. Zewail **ibid**, submitted.

11. M. Gruebele, I. Sims, E. D. Potter, and A. H. Zewail, J. Chem. Phys. 95, 7763 (1991).
12. E. D. Potter, J. L. Herek, S. Pedersen, Q. Liu, and A. H. Zewail, Nature 355, 66 (1992).
13. J. C. Williamson and A. H. Zewail, Proc. Natl. Acad. Sci. 88, 5021 (1991).
14. J. C. Williamson, M. Dantus, S. Kim, and A. H. Zewail, submitted.

Transient Dichroism Studies of I_2 Predissociation in Solution

N.F. Scherer[1,3], *L.D. Ziegler*[2], *D. Jonas*[1], and *G.R. Fleming*[1]

[1]Department of Chemistry and The James Franck Institute, The University of Chicago, 5735 S. Ellis Avenue, Chicago, IL 60637, USA
[2]Department of Chemistry, Boston University, Boston, MA 02215, USA
[3]Current address: Department of Chemistry, University of Pennsylvania, Philadelphia, PA 19104-6323, USA

One and two-color time-domain measurements of the resonant dichroic response of iodine in n-hexane are described. The appearance time of the nuclear response increases with blue probe tuning reflecting excited state wavepacket motion along the reaction coordinate.

1. Introduction

The utility of direct time-domain measurements of rates of chemical processes has long been appreciated [1]. However, the last few years have witnessed the application of femtosecond pump-probe measurements for elucidation of spectroscopic information concerning processes occurring in condensed phases [2]. Provided that vibrational coherence persists and that the linewidths are inhomogeneously broadened [3], the results of femtosecond nonlinear measurements may be Fourier transformed to extract "sub-linewidth" vibrational dynamics of solutes. The study of reaction dynamics in solution is expected to be similarly accessible.

Here we report the results of measurements of the femtosecond optically-induced dichroism of 5-10 mM I_2-alkane solutions. The separate measurement of the real and imaginary contributions to the time dependent response function [3a] of the reactive moiety facilitates the study of weakly absorbing solutes in small concentration. Furthermore, in the case of iodine the electronic state origin of the vibrational beats observed in the third-order polarization response is readily established, unlike the wavepacket motions observed in dye molecules in solution [2].

The dynamics of iodine dissociation, geminate recombination and vibrational relaxation in a variety of solvents has been reviewed by Harris and co-workers [4], who estimated curve crossing and recombination to occur in less than 2 ps. The solution dynamics reported here and in [5] are clearly distinguished from isolated molecule gas phase dynamics [6] by the rapid decay of the excited state coherence and prominence of the ground state coherence. Resonance Raman studies by Sension and Strauss [7] have established that the inverse optical dephasing rate (T_2) of iodine in simple non-interacting liquids is 300 fs.

2. Experimental Method

The experimental approach used here, as described elsewhere [5], is a two-pulse pump-probe polarization measurement performed with 25-30fs duration pulses. The crossed-polarizer heterodyne detection scheme allows separate measurement of the resonant dichroism and birefringence. The measurements are preformed at 100kHz repetition rate [8] with pulses obtained from a spectrally selected and amplified portion of a white light continuum. The two-color measurements also use the amplified continuum for the pump (5nJ @ 580 nm) and the unamplified continuum for the probe (<200pJ). The beams double-pass through separate F-2 prism-pairs for chirp compensation and are brought to a common focus in the sample. In both cases the pump beam is amplitude-modulated and the change in transmission of the probe beam through the analyzer is detected and processed in a lock-in amplifier.

Springer Series in Chemical Physics, Vol. 55 Ultrafast Phenomena VIII
Editors: J.-L. Martin · A. Migus · G.A. Mourou · A.H. Zewail

3. Results and Discussion

Figure 1 shows the optically induced dichroism of iodine in n-hexane for 580 nm pump and probe (denoted 580/580). The two oscillatory contributions to the signal are evident from 40fs delay. Simple exponential decay components are also present and correspond to excited state population relaxation and I_2 rotational reorientation; hence all three of the vibronic contributions to the third-order polarization are detected and assigned. The pi-phase shift of the excited state dynamics indicates an absorptive probe process to a higher lying electronic state [9].

Figure 2 shows the dichroism signal for 580/475 nm pump and probe. The fit to the data is decomposed into three constituent components showing the prominence of the excited state contribution as compared with the ground state. In contrast with Fig. 1, the magnitude of the nuclear contribution to the signal is larger than the instantaneous electronic contribution. The enhanced absorption of the excited state vibrational coherence signal reflects a stronger optical transition for the subsequent nuclear dynamics than for t=0 probing. The delay in the appearance of the excited state signal corresponds to >1/2 the B-state period.

Figure 3 shows the time delay for the appearance of the excited state vibrational coherence contribution to the dichroism signal. The 580/580 and 615/615 data show only a sign change reflecting absorption from the B <- X Franck Condon (FC) region of the B-state coordinate. The delay increases with shorter wavelength probing. The two-color measurements reveal delays that correspond to absorptive probing, possibly in the right turning point region of the B-state and beyond. The bluer probing follows the time course of the reaction past the and reflects the flux of the excited state wavepacket passing through the "probe window."

The combination of the large delay time, greater than half the B-state period, and the prominence of the nuclear dynamics at 580/475 and 580/460 argues for three conclusions; (i) that these dichroism signals arise from iodine molecules that are further along the reaction coordinate than the B-state allows, (ii) that the electronic character of the reaction has changed, and (iii) vibrational dephasing results from de-population via curve crossing as opposed to simple vibrational population relaxation and is the principle cause of optical dephasing. The one and two-color studies can be understood to sample the reaction coordinate from the bound B-state onto a dissociative, ^1a_g [10], potential curve. However, we have also studied the solvent dependence of the B-state predissociation. The rate of vibrational dephasing correlates with solvent polarizability, which indicates that the O^-_u potential may be involved [11]. The direct observation of caging effects may be possible with a shorter wavelength probe!

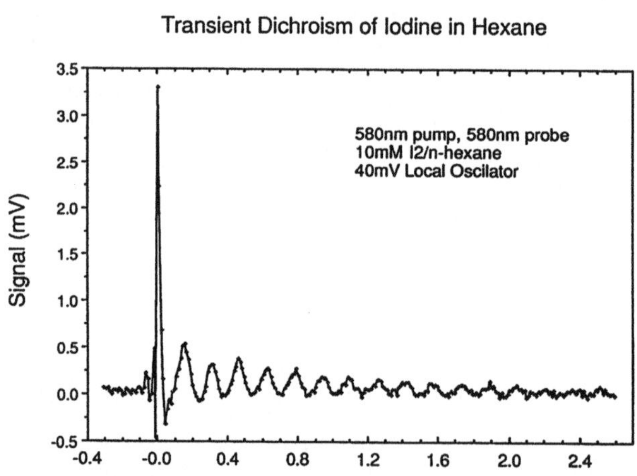

Fig. 1 Transient dichroism of iodine in hexane. Singular value decomposition analysis [5] yields $110 cm^{-1}$ and $210 cm^{-1}$ cosine components with equal amplitudes, 200fs and 900fs exponential damping times and with π and 0 phases, respectively.

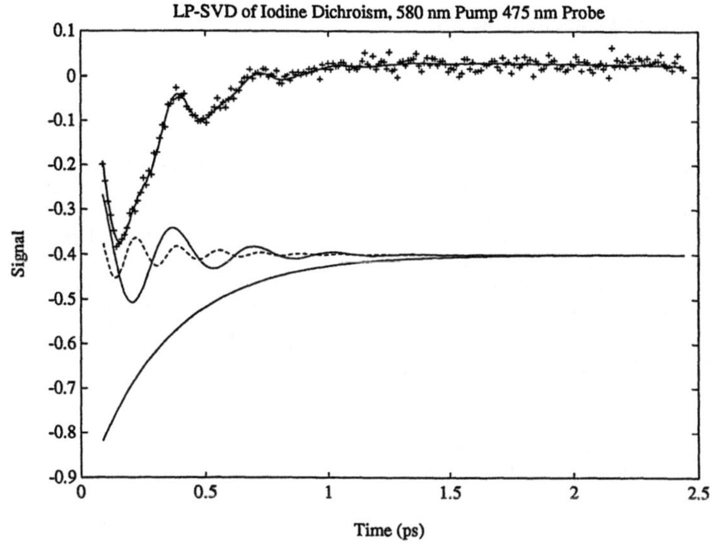

Fig. 2 Singular value decomposition for 580/475nm transient dichroism. The ground and excited state vibrational coherence components are shown along with an exponential build-up. Excited state appears with >2π phase shift.

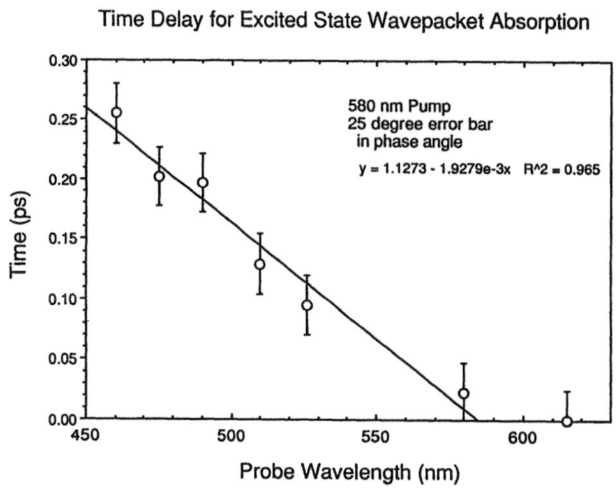

Fig. 3 Time of appearance of excited state wavepacket absorption contribution to transient dichroism signal. Zero time implies absorption $O^+_g(^1\Sigma)$ <- B from the left turning point region. 150fs corresponds to 1/2 period on the B-state surface for 580nm excitation.

This work was supported by a grant from the NSF.

References

1. G.R. Fleming, " Chemical Applications of Ultrafast Spectroscopy", Oxford, 1986
2. M.J. Rosker, F.W. Wise and C.L. Tang, Phys. Rev. Lett. **57**, 321 (1986); I.A. Walmsley, M. Mitsunaga and C.L. Tang, Phys. Rev. A **38**, 4681 (1988).
3. (a) S. Mukamel, Ann. Rev. Phys. Chem. **41**, 647 (1990); (b) W.T. Pollard, S.-Y. Lee and R.A. Mathies, J. Chem. Phys. **92**, 4012 (1990).
4. A.L. Harris, J.K. Brown and C.B. Harris, Ann. Rev. Phys. Chem., **39**, 341 (1988).
5. N.F. Scherer, L.D. Ziegler and G.R. Fleming, J. Chem. Phys., **96**, 5544 (1992).
6. R.M. Bowman, M. Dantus and A.H. Zewail, Chem. Phys. Lett., **161**, 297 (1989).
7. R.J. Sension and H.L. Strauss, J. Chem. Phys. **85**, 3791 (1986).
8. A.J. Ruggiero, N.F. Scherer, G.M. Mitchell, G.R. Fleming and J.N. Hogan, J. Opt. Soc. Am. B., **B8**, 2061 (1991).
9. R.S. Mullikan, J. Am. Chem. Soc., **74**, 811 (1952).
10. J. Tellinghuisen, J. Chem. Phys., **82**, 4012 (1985).
11. K. Nakagawa, et al., Chem. Phys. Lett., **106** 259 (1986).

Investigation of the Primary Event in Vision Using 10 fs Blue-Green Optical Pulses

R.W. Schoenlein[1], *L.A. Peteanu*[2], *Q.W. Wang*[2], *R.A. Mathies*[2], *and C.V. Shank*[1]

[1]Lawrence Berkeley Laboratory, MS 70-193A,
1 Cyclotron Rd., Berkeley, CA 94720, USA
[2]Department of Chemistry, University of California, Berkeley, CA 94720, USA

Abstract. The primary event in vision, the *cis-trans* isomerization of rhodopsin, is resolved with the use of 10 fs blue-green optical pulses. Rhodopsin 11-*cis* is excited with a 35 fs pump pulse at 500 nm, and the transient changes in absorption are measured from 450 to 620 nm using a 10 fs probe pulse at 500 nm, and a 10 fs probe pulse at 620 nm. Within 200 fs, an increased absorption is observed between 550 and 620 nm, indicating the formation of photoproduct on this time scale. These measurements demonstrate that the first step in vision is essentially complete in only 200 fs.

Recent developments in the generation and amplification of femtosecond optical pulses in the blue-green spectral regime[1] now enable the investigation of ultrafast processes in a variety of biological systems with a time resolution of 10 fs. One of the most important ultrafast processes occurring in nature is the detection of light by the visual system. This process is initiated by the absorption of a visible photon by the retinal pigment rhodopsin and culminates in the stimulation of the optic nerve. Nearly thirty years ago, Yoshizawa and Wald[2] determined that the initial chemical reaction in vision is the photoisomerization of the retinal chromophore in rhodopsin. Subsequent studies have shown that this first step in vision occurs within a few picoseconds[3]. However, measurements of the initial reaction dynamics have been limited by laser system resolution, and a complete understanding of this process remains an important problem in photochemistry and biology. We report here the room-temperature investigation of the *cis-trans* isomerization in rhodopsin using 10 fs blue-green pulses. Our results demonstrate that the first step in vision occurs on a time scale of 200 fs[4].

The photoreactive component of the rhodopsin molecule is 11-*cis* retinal, a vitamin-A derivative, which is bound within the protein opsin by a Shiff base linkage. The absorption maximum of rhodopsin in the 11-*cis* form is ~500 nm. The absorption of a photon causes the retinal prosthetic group to isomerize about the C_{11}-C_{12} double bond, forming the photoproduct bathorhodopsin with an absorption maximum at ~535 nm. Following this initial isomerization, thermally mediated conformational changes occur which result in the release of opsin and all-*trans* retinal. Previous measurements of the initial isomerization using transient absorption spectroscopy[3] have shown the formation of a photoproduct within the 6 ps laser pulsewidth, but were unable to resolve the initial isomerization process. In addition, resonance Raman spectroscopy[6] and time-resolved Raman studies[7] have confirmed that an initial rhodopsin photoproduct with a twisted all-*trans* structure is formed in less that 30 ps. Raman intensity analysis have suggested that the isomerization dynamics occur on a subpicosecond time-scale[8], and theoretical simulations of the isomerization dynamics are qualitatively consistent with this picture[5,9]. However, the time scale for the formation of the primary photoproduct has never been experimentally determined.

The recent development of a blue-green femtosecond laser system[1], now makes it possible to directly study the time-course of rhodopsin isomerization. The laser system consists of a colliding-pulse modelocked (CPM) dye laser producing 50 fs pulses at 620 nm. These pulses are amplified to microjoules energies in a dye amplifier pumped at 400 Hz by a XeF excimer laser. The amplified pulses are used to generate a white-light continuum in a 1-mm jet of ethylene glycol, and the blue-green portion of the continuum is re-amplified to the microjoule level in a second amplifier pumped by the same excimer laser. We use a standard pump-probe technique in which a femtosecond pump pulse excites the rhodopsin sample, and the resulting changes in absorption are measured with a probe pulse, which is delayed in time with respect to the pump. The 35 fs pump pulse at 500 nm comes directly from the amplifier. The 10 fs probe pulse at 500 nm is created using an optical fiber with a sequence of gratings and prisms

Springer Series in Chemical Physics, Vol. 55 **Ultrafast Phenomena VIII**
Editors: J.-L. Martin · A. Migus · G.A. Mourou · A.H. Zewail

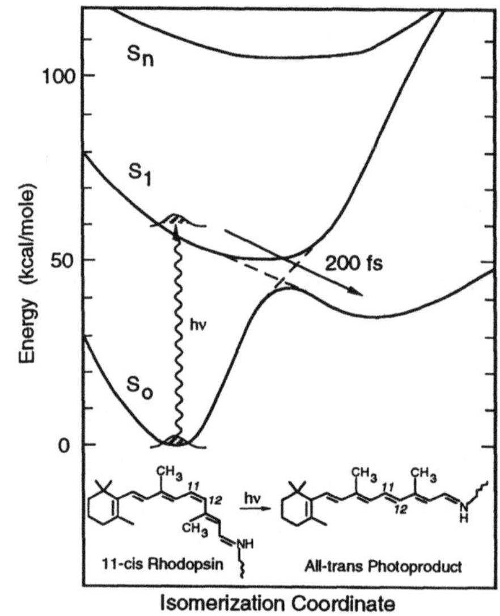

Fig. 1. Schematic ground-state and excited-state potential-energy surfaces for the 11-*cis*→11-*trans* isomerization in rhodopsin adapted from reference [5]. The reaction path of the photoisomerization is indicated by the nonadiabatic potential surfaces (broken lines).

for phase compensation. A 10 fs probe pulse at 620 nm is generated in the same way using part of the amplified CPM pulses before the continuum jet. The large bandwidths of the 10 fs red and green pulses allows us to resolve the spectral dynamics of the rhodopsin molecule (from 450 to 650 nm) following excitation by the narrow-band (~15 nm) pump pulses.

The rhodopsin sample is prepared from outer-segments of bovine retinas which are purified by ultracentrifugation in a continuous sucrose gradient, lysed with cold water, and pelleted by additional centrifugation. Rhodopsin from 400 retinas is solubilized (5% Ammonyx-LO, 20 mM MOPS, pH 7.4) and then concentrated to an optical density of 15 OD/cm (at 500 nm). A 3-ml sample flows through a 300-μm wire-guided jet at a sufficient velocity to ensure complete replacement of the sample between each pair of pump-probe pulses. The pump pulses are focused on the rhodopsin jet to an energy density of ~400 $\mu J/cm^2$. At these pump fluences, only ~10% of the exposed sample is bleached by the pump. The weaker probe pulses are similarly focused on the sample (~30 $\mu J/cm^2$). Transient changes in absorption ($\Delta T/T$) are measured with two different techniques. Time-resolved measurements at specific wavelengths are obtained by spectrally filtering the probe pulse (after passing through the sample) and combining differential detection with lock-in amplification. Differential spectral measurements over the entire bandwidth of the 10 fs probe pulse are made with a spectrometer and a dual diode array detector. In all measurements, the maximum signal ($\Delta T/T$) is a few percent, and the linearity is verified in order to avoid saturation effects.

The transient change in absorption is measured at 500, 535, 550, and 570 nm, following excitation of rhodopsin by a 35 fs pulse at 500 nm (Fig. 2a). Measurements at 500 nm probe the initial bleach and partial recovery of the ground state absorption of 11-*cis* rhodopsin. At early times, we observe a transient excited-state absorption as evidenced by the negative differential signal ($\Delta T/T<0$) near zero delay. The arrival of the pump pulse induces an absorption at ~500 nm, assigned to the $S_1 \rightarrow S_n$ transition (Fig. 1), which interferes with the ground-state bleach signal. As the wavepacket created in the first excited state moves away from the Franck-Condon region, the excited-state absorption disappears and the full bleach of the rhodopsin absorption is revealed by 125 fs. A rapid partial recovery of the initial bleach at 500 nm is observed, which has a time constant of ~250 fs. The subsequent long-time recovery of the bleach occurs with a time constant of ~8 ps.

Transient absorption changes at 550 nm and 570 nm reveal the kinetics of photoproduct formation. At 570 nm, near the peak absorption of the photoproduct, we observe a rapidly developing absorption ($\Delta T/T<0$) which reaches a maximum by 200 fs. Beyond 200 fs there is very little change in the absorption, indicating that the photoproduct is formed on this time scale. This conclusion is supported by the appearance of absorption at 550 nm, on the

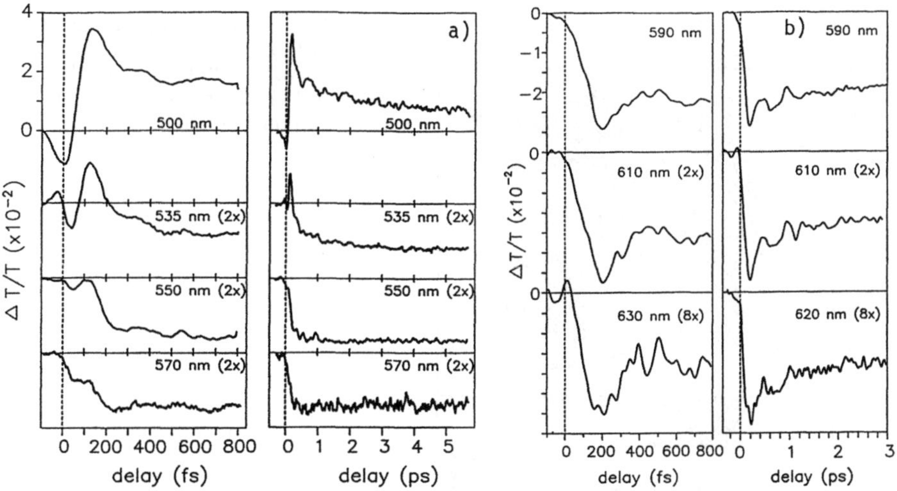

Fig. 2. Transient absorption measurements of 11-*cis* rhodopsin after excitation with a 35 fs pump pulse at 500 nm (a) 10 fs probe pulse at 500 nm, (b) 10 fs probe pulse at 620 nm.

blue side of the photoproduct absorption band, following a delay of 200 fs. Between 200 fs and 1 ps, the absorption at 550 nm gradually increases, and then remains unchanged out to 6 ps delay.

Measurements at 535 nm, on the red edge of the 11-*cis* rhodopsin absorption band and on the blue edge of the photoproduct absorption, are more complicated. The excited-state ($S_1 \rightarrow S_n$) absorption induced by the pump ($\Delta T/T<0$) is apparent at short times between 0 and 100 fs. This induced absorption does not appear instantaneously as is the case at 500 nm, but is delayed by ~50 fs. The delay is attributed to the dynamics of the excited-state absorption resulting from spreading and/or motion of the wavepacket excited on the S_1 surface as it leaves the Franck-Condon region. By 100 fs we observe the appearance of the ground-state bleach, and by 200 fs the photoproduct absorption begins to dominate and the signal changes sign. The long time (1 to 5 ps) behavior at 535 nm results from a slow recovery of the ground-state absorption (as observed at 500 nm) as well as vibrational cooling of the photoproduct which causes the absorption spectrum to shift to shorter wavelengths. In addition, we observe oscillatory behavior at all probe wavelengths, though this is particularly evident at 500 nm and at 570 nm between 0 and 200 fs.

Using the 10 fs probe pulse at 620 nm, we can investigate the spectral dynamics at the red edge of the photoproduct absorption. Figure 2b shows the transient change in absorption ($\Delta T/T<0$) at 590, 610 and 630 nm, following excitation of rhodopsin by a 35 fs pump pulse at 500 nm. Measurements at all three wavelengths show a rapid developing absorption ($\Delta T/T<0$) which reaches a maximum by 200 fs. The appearance of this absorption corresponds to the formation of photoproduct on this time scale consistent with the measurements at 570 nm shown in Fig. 2a. On a picosecond time scale we observe a partial recovery of the absorption which results from vibrational cooling and conformational relaxation of the rapidly formed photoproduct. Furthermore we observe oscillatory behavior at all probe wavelengths which is a result of coherent vibrational excitation by the femtosecond pump pulse.

Complementary information about the isomerization kinetics is provided by the differential spectral measurements shown in Fig. 3. Measurements at shorter wavelengths are obtained using the 10 fs blue-green probe pulse, and measurements at longer wavelengths are obtained using the 10 fs probe pulse in the red. The break in the curves marks the separation between the two measurements. At 33 fs delay, the increase in absorption between 490 and 540 nm results from the excited-state ($S_1 \rightarrow S_n$) transition which dominates the bleach signal at short times. Between 33 fs and 100 fs, the initial rhodopsin bleach between 470 and 540 nm becomes evident. The filling-in of the bleach signal is observed at longer delays, consistent with the time-resolved measurements shown in Fig. 2a. Although we cannot exclude the

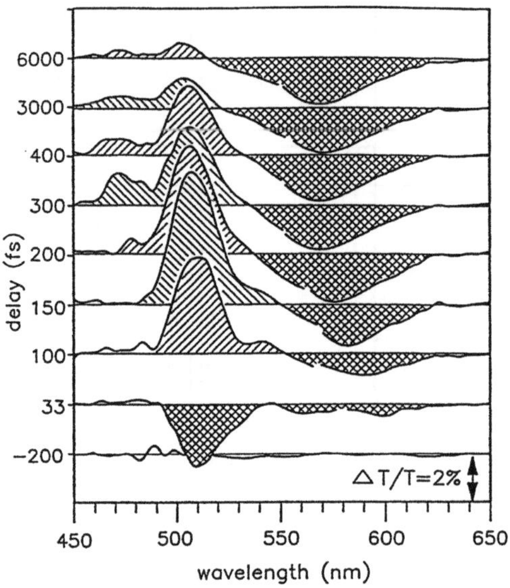

Fig. 3. Difference spectra measurements of *11-cis* rhodopsin at various delays following a 35 fs pump pulse at 500 nm (~10 fs probe at 500 nm and ~10 fs probe at 620 nm).

possibility that some residual excited-state population contributes to the recovery of the bleach at long times, our interpretation is supported by the fact that vibrational cooling and conformational relaxation are known to occur in bacteriorhodopsin on a similar time scale (~3 ps)[10].

The initial appearance of photoproduct is indicated by the differential absorption observed between 550 and 620 nm at 33 fs delay (Fig. 3). Between 33 fs and 200 fs, the photoproduct absorption increases and shifts to shorter wavelengths. After 200 fs, the overall amplitude of the photoproduct absorption (540 to 620 nm) remains unchanged, indicating that the isomerization is complete on this time scale. In addition, the blue shift of the photoproduct absorption over a time scale of several hundred femtoseconds is clear evidence of vibrational cooling and conformational relaxation occurring in the ground state photoproduct. As a result of this process, time resolved measurements on the blue edge of the photoproduct absorption (Fig. 2a, 535 nm) show an increase in absorption, and measurements at on the red edge of the absorption (Fig. 2b) show a decrease in absorption over the time scale of the cooling.

The difference spectra at long delays are in reasonable agreement with previous measurements made with substantially longer pulses[11,12,13]. Difference spectra at 6 ps delay (Fig. 3) indicate an isosbestic (~515 nm) which is within the range of values previously reported. Although our measurements of the differential absorption maxima (~500 nm and ~550 nm) are slightly shifted from what others have observed, this may be due to measurement uncertainty or the shorter wavelength (500 nm) of the excitation pulses used here, or both. In addition, the signal ratio $\Delta\alpha_{batho}/\Delta\alpha_{rhodopsin} \approx 3$ measured at 40-ps delay (not shown), is consistent with single-photon excitation[13]. Since we use pulses that are shorter than the ground-state-recovery and photoproduct-formation times, and fluences (~400 $\mu J/cm^2$ or ~0.1 photons absorbed per molecule) that are much lower than previous studies, the possibility of multiphoton photolysis is minimized. Finally, because the time scale of our measurements is faster than the photo→bathorhodopsin transition, we assume that the initial photoproduct is the intermediate, photorhodopsin, identified by Shichida et al.[11].

These femtosecond measurements have temporally resolved the isomerization of the 11-*cis* retinal prosthetic group in rhodopsin. The rapid increase in absorption between 540 and 580 nm shows that the all-*trans* photoproduct is formed in only 200 fs. This confirms earlier suggestions that a rapid non-radiative isomerization of the retinal chromophore is necessary in order to explain the low fluorescence yield of rhodopsin as well as the photosensitivity and high quantum efficiency of the isomerization[8]. Furthermore, the non-exponential kinetics observed between 550 and 620 nm, (Fig. 2) indicate that the isomerization can not be described by a first-order process. The appearance of the excited-state absorption at ~500 nm at short times is consistent with theoretical models which indicate the presence of an excited-state absorption[5].

This absorption disappears as a result of the rapid isomerization and corresponding changes in oscillator strengths as the chromophore becomes non-planar. The lack of any obvious evidence for stimulated emission from the excited state also indicates that the torsional wavepacket on the excited-state surface moves rapidly out of the Franck-Condon region. In addition, the 100-200 fs oscillatory behavior in the time-resolved measurements indicates that non-stationary vibrational states are excited by the short pulses. The vibrational frequency of these oscillations ($\sim$100 cm^{-1}) is consistent with known low-frequency torsional modes of rhodopsin[8], suggesting that excited-state torsional oscillations[5,9] may modulate the appearance of photoproduct. However, impulsive Raman excitation of the rhodopsin ground state can also contribute to coherent vibrational oscillations in these measurements.

In conclusion, we have time-resolved the spectral dynamics of the primary step in vision. Our results indicate that the *cis-trans* isomerization of rhodopsin is essentially complete in only 200 fs and is one of the fastest chemical reactions ever studied. This observation has important implications for the photochemistry of vision. First, 200 fs is faster than typical vibrational dephasing and relaxation times, suggesting that the photochemistry occurs from a vibrationally coherent system. Indeed, we see indications of coherent vibrational oscillations which are rapidly damped as the photoproduct is formed. The significance of such vibrational coherence in the photochemistry of vision has been discussed in several theoretical studies[5,9]. Second, the speed of the isomerization process contradicts the traditional picture of photochemistry which assumes vibrational relaxation in the excited state followed by partitioning to photoproduct and to reactant. Our results indicate an essentially barrierless transition in the formation of photoproduct, suggesting that the isomerization follows a nonadiabatic potential surface (broken lines in Fig. 1) which reflects the strong coupling between the rhodopsin excited-state and the ground-state of the photoproduct. This presents experimental evidence for a new paradigm for visual photochemistry that may be relevant for a variety of photochemical and photobiological processes.

This work supported by NIH grant EY 02051 (RM) and NSF grant CHE 86-15093 (RM) and by DOE contract DE-AC03-76SF00098 (CS). LAP acknowledges support from NIH Postdoctoral Training Grant T32EY07043.

References

1. R.W. Schoenlein, J.-Y. Bigot, M.T. Portella, and C.V. Shank, *Appl. Phys. Lett.*, **58**, 801 (1991).
2. T. Yoshizawa and G. Wald, *Nature*, **197**, 1279 (1963).
   G. Wald, *Science,* **162**, 230 (1968).
3. G.E. Busch, M.L. Applebury, A.A. Lamola, and P.M. Rentzepis, *Proc. Natl. Acad. Sci. USA,* **69**, 2802 (1972).
4. R.W. Schoenlein, L. A. Peteanu , R.A. Mathies and C.V. Shank, *Science* **254**, 412 (1991).
5. R.R. Birge and L.M. Hubbard *J. Am. Chem. Soc.* **102** 2195 (1980).
   R.R. Birge *Annu. Rev. Phys. Chem.* **41** 683 (1990).
6. G. Eyring et al., *Biochemistry*, **19**, 2410 (1980).
7. G. Hayward, W. Carlsen, A. Siegman, and L. Stryer, *Science*, **211**, 942 (1981).
8. G.R. Loppnow and R.A. Mathies, *Biophys. J.,* **54**, 35 (1988).
9. R.M. Weiss and A. Warshel *J. Am. Chem. Soc.* **101** 6131 (1979).
10. S.J. Doig, P.J. Reid and R.A. Mathies *J. Phys. Chem.* **95** 6372 (1991).
11. Y. Shichida, S. Matuoka, and T. Yoshizawa, *Photobiochem. Photobiophys*, **7**, 221 (1984).
12. T.G. Monger, R.R. Alfano, and R.H. Callender, *Biophys J.*, **27**, 105 (1979).
13. J.D. Spalink, A.H. Reynolds, P.M. Rentzepis, W. Sperling, and M.L. Applebury, *Proc. Natl. Acad. Sci.,* **80,** 1887 (1983).

Mechanisms of Charge Separation in Bacterial Reaction Centers

M.H. Vos[1;2], *F. Rappaport*[1], *J.-C. Lambry*[1], *C. Rischel*[1], *J. Breton*[2], and *J.-L. Martin*[1]

[1]Laboratoire d'Optique Appliquée, INSERM U275, Ecole Polytechnique-ENSTA, F-91120 Palaiseau, France
[2]Service de Biophysique, Département de Biologie, Centre d'Etudes Nucleaires de Saclay, F-91191 Gif-sur-Yvette Cedex, France

Abstract. In photosynthetic reaction centers coherence phenomena occur on the same time scale as the primary electron transfer reaction. The origin of these phenomena is discussed and a model is presented for near-adiabatic electron transfer coupled to a low frequency vibrational mode which conserves its phase during the reaction.

1. Introduction

The key processes of photosynthetic energy conversion occur in the socalled reaction center. The very fast (~17Å in ~3 ps) and highly efficient (quantum yield near unity) primary electron transport step has been studied with a variety of direct and indirect techniques and has, in addition, gained considerable theoretical attention [1].

The reaction center is a membrane bound pigment-protein complex. The structure of reaction centers of purple bacteria has been resolved by X-ray crystallography. Four bacteriochlorophylls (P_L, P_M, B_L and B_M), two bacteriopheophytins (H_L and H_M) and one or two quinones (Q) are bound to two of its subunits L and M. The center displays approximate C_2 symmetry, with the C_2-axis running through the center of two closely lying bacteriochlorophylls at the periplasmic side of the complex and a non-heme iron atom at the cytoplasmic side of the complex. The six bacteriochlorin pigments have their Q_Y absorption bands in the near-infrared. It is generally agreed that in the ground state P_L and P_M have a strong excitonic interaction; they are usually referred to as P. The lowest energy electronic transition band of the reaction center (at 870 nm in *Rb. sphaeroides* R-26) is attributed to the lower exciton band of P.

After excitation P is oxidized (in about 3 ps at room temperature and about 1 ps at cryogenic temperatures), the bacteriopheophytin acceptor H_L, located 17Å from P, is reduced on the same time scale. The reaction has been mostly analyzed in terms of linear kinetic models between clearly distinct states. The speed of the reaction is to high to be explained in terms of a simple direct reaction. Therefore mediation of the intermediate B_L chromophore has been proposed, in the form of a virtual (super-exchange, [2]) or of a real $P^+B_L^-$ state [3]. In the latter scheme, which was proposed after observation of transients faster than P^* decay, B_L^- is reoxidized at a higher rate than it is formed.

These schemes all assume that the reactions follow exponential kinetics. In homogeneous systems, this is the case if nuclear motion along the reaction coordinate is thermally equilibrated on the time scale of the reaction and if the reaction is non-adiabatic. A hint on whether these assumptions are valid can be obtained from studies on the temperature dependance of the electron transfer rate. The temperature dependance of the decay of P^* has been modelled in terms of the thermally averaged Franck-Condon factor, thus assuming coupling to vibrationally relaxed modes and non-adiabatic electron transfer (weak coupling), using a single mode representation [4]. This approach yielded a good description for *Rb. sphaeroides* using a 80 cm^{-1} mode, but showed deviations for *Rps. viridis*. Vibrations of 80 cm^{-1} correspond to a period of about 400 fs. Comparing this value with the observed overall rate of ~1 ps at low temperature, it is clear that, for this reaction, the limit of applicability of the weak coupling approximation approaches. This is particularly true for the analysis of the presumed second step in a two-step reaction scheme, which yields a 30 cm^{-1} mode (1 ps period) with a reaction time of ~0.3 ps at 25K [5].

Springer Series in Chemical Physics, Vol. 55 Ultrafast Phenomena VIII
Editors: J.-L. Martin · A. Migus · G.A. Mourou · A.H. Zewail

Furthermore, in our lab we have recently observed that the optical transients of reaction centers at low temperature are modulated by oscillatory features in the range of 10-100 cm^{-1} [6], persisting for a few picoseconds. If these features reflect nuclear vibrational modes, this implies that those modes are not completely dephased (and hence certainly not relaxed) on the time scale of primary electron transport. In this paper we further discuss the origin of the oscillatory features and describe a model for electron transport coupled to modes in which the phase is preserved on the time scale of electron transport.

2. Experimental

Our apparatus for femtosecond pump-probe experiments with pulses of 45 fs (in some cases the pump pulse was longer due to spectral selection) has been described before [6].

Reaction centers of *Rb. sphaeroides* R-26 were prepared as before. It was found that the amplitude of oscillations (~2 ps period) in the stimulated emission of these reaction centers (as in Figure 1B) depended on the preparation. The optical absorbtion spectra of the preparations were not different. Apparently, the low frequency modes are influenced by subtle changes in preparation and detergent concentration and this may lead to a dispersive (heterogeneous) frequency distribution. The data shown are taken from the same sample in which the oscillatory features (under conditions of a narrow pump bandwidth) were reproducible.

3. Results and Discussion

As reported before [6], oscillatory features were observed in the D_{LL} mutant of *Rb. capsulatus*, in which H_L is absent. In this mutant no overall change in shape of the transient spectrum is observed, indicating that only P^* is formed and no other chromophore is reduced [7]. The oscillatory features, with clearly distinct periods of 500 fs and 2 ps were observed both in the kinetics of stimulated emission and in transient absorption. As in this mutant no charge separation takes place, these features must reflect modulation within the P^* state or the ground state.

In functional reaction centers of *Rb. sphaeroides* R-26, oscillatory features with a period of ~700 fs, were observed in the absorption region and with a period of ~2 ps in the stimulated emission region. The latter results are shown in Figure 1. In this case, interestingly, the oscillations were not observed when a broad band (i.e. shortest) pump pulse was used (25 nm, i.e. exciting essentially the entire P_- band), but were present when exciting with a (longer) 10 nm broad pulse. In the case of reaction centers which do perform electron transport, these features may reflect, in addition to spectral modulation, modulated changes in the population of the states.

The dephasing of the oscillatory features takes place with a time constant of about 1 ps, i.e. on the timescale of electron transfer, in both mutant and functioning reaction centers. With the present data we are unable to determine whether dephasing is faster in functioning reaction centers, due to the facts that only a few periods are observed and that the initial kinetics are not well defined in *Rb. sphaeroides* R-26, because of kinetic congestion.

The origin of the oscillations may be electronic or vibronic. In the case of electronic coherence, an electron tunnels back and forth between two closely spaced electronic states after non-stationary impulsive excitation [8]. It can be imagined that such states exists in reaction centers, i.e. neutral and charge transfer excited states may be closely spaced. However, the fact that different frequencies in one reaction center complex are observed would imply that several of such states are involved. Also, the disappearance of the 2 ps oscillation in *Rb. sphaeroides* R-26 with the shortest (i.e spectrally broadest) pump pulses seems hard to reconcile with such a mechanism. This possibility will be further investigated in an extensive study of effects of pump and probe wavelength.

Hence the features most probably reflect nuclear vibrational motion. In the case of the mutant this must be modulation of the absorption and stimulated emission spectrum by the nuclear coordinates. The excited state absorption is probed in a region where the ground state also absorbs, so contributions of motion of a wave packet in the ground state, populated by pump pulse impulsive stimulated Raman processes cannot be excluded. However, due to the large Stoke shift, the efficiency of such processes is very low in reaction centers, in contrast to bacteriorhodopsin [9]. Furthermore, similar frequencies are observed in the stimulated emission

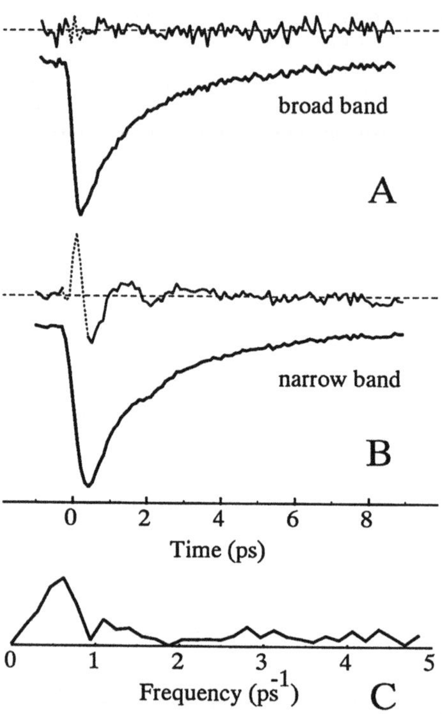

Fig. 1 Kinetics of stimulated emission of reaction centers of *Rb. sphaeroides* R-26 at 100 K probed at 930 nm. The pump pulse was centered at 870 nm and had a spectral width (fwhm) of 25 nm (A) or 10 nm (B). Residuals were obtained by subtracting a fit of a biexponential decay convoluted with the instrument response function from the data. (C) Fourier transform of the residuals in B.

($S_1 \rightarrow S_0$) and the excited state absorption ($S_1 \rightarrow S_n$), indicating that wavepacket motion on the excited state potential energy surface is probed.

Regardless of the exact nature of the vibrations, an important point is that the phase of the nuclear motions is not lost on a time scale of 1 ps after synchronisation. This is especially important in view of the low frequency of the oscillations. Vibrational modes in this range are global protein modes [10] and thus are likely to govern relative motion of the different chromophores. Collective motion of chromophores may bring about a high, but temporary restricted, probability of population of the transition zone, which is lost after dephasing in the charge separated state, thus providing a high 'barrier' for back reaction.

Recently several authors have theoretically adressed the question of coherence in electron transfer reactions [11,12]. To explore the possibility of coherent electron transfer we constructed a simple model in terms of wave packet motion on the excited state potential energy surface. The model takes into account the experimental conditions and essentially has as free parameters the vibrational frequency of a single coupling mode, the probability of transition to the charge separated state upon passing the transition zone and a parameter describing the dephasing. The experimentally observed damping of the oscillations may be due to anharmonicity, or elastic (pure dephasing) or inelastic (thermalization) energy exchange with the surrounding protein. We previously explored the possibility of damping of the oscillations by anharmonicity [1,6]. We now have extended the model to allow for pure dephasing by introducing diffusive broadening of the wave packet. The model is presented in Figure 2. Assuming that the stimulated emission is directly proportional to the entire P^* population, that the transition zone is located at the bottom of the P^* well ('activationless') and allowing only diffusive broadening as a dephasing mechanism, the experimental features are well reproduced (cf. Figure 1, narrow band). It should be noted that although pure dephasing probably occurs on the ps timescale, other decay

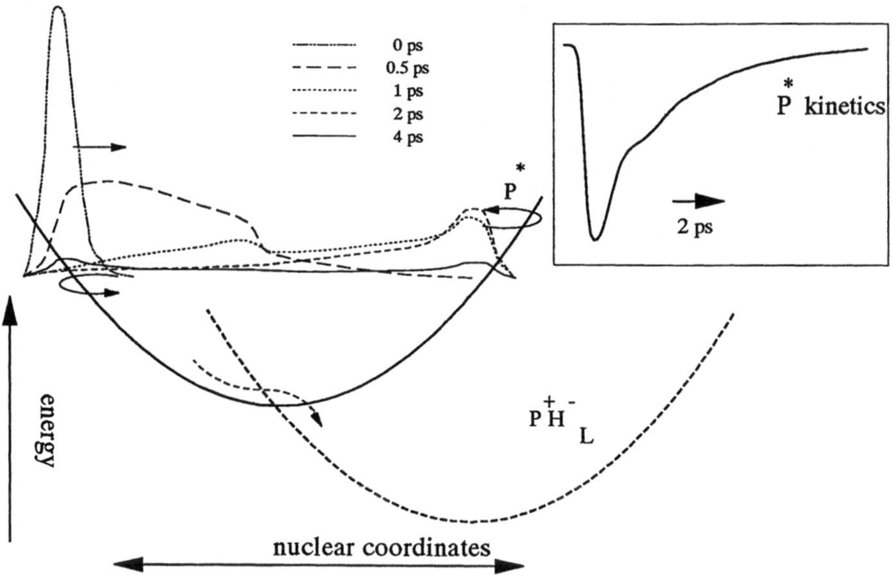

Fig. 2 Wave packet motion on the P* surface along the reaction coordinate of primary electron transport. The distribution of the population of the eigenstates by the pump pulse (here the narrow band conditions are taken) are calculated from the spectra of the pump pulse and the ground state. Excitation at 870 nm corresponds to an average excess energy of 500 cm^{-1}. Electron transfer is modelled by a probability of leak when passing through the bottom of the well. Pure dephasing is modelled by diffusive broadening of the wave packet. Wave packets are calculated at various times assuming a coupling mode of 10 cm^{-1}. (Inset) Simulated P* population kinetics, plotted negative to facilitate comparison with the experimental kinetics.

mechanisms presumably also play a role. Especially, some anharmonicity has to be introduced to model the effect of the spectrum of the pump probe on the oscillations (Figure 1).

Our model is necessarily very simple due to the lack of knowledge of the potential energy surfaces shape and dephasing mechanisms. This may be overcome by more elaborate studies on coherence in proteins. Also, direct evidence on the coupling of electron transfer to a low frequency, non-dephased mode must await experiments selectively probing the product state. Nevertheless the demonstrated consistency of our experiments with such a mechanism allows us to regard coherent electron transfer as an attractive option.

4. Acknowledgment

M.H.V. was supported by a grant of the EEC Science Program during the course of this work.

5. References

1. J.-L. Martin and M.H. Vos, Ann. Rev. Biophys. Biomol. Struct. **21**, 199 (1992)
2. M.B. Michel-Beyerle, M. Plato, J. Deisenhofer, H. Michel, M. Bixon and J. Jortner, Biochim. Biophys. Acta **932**, 52 (1988)
3. W. Holzapfel, U. Finkele, W. Kaiser, D. Oesterhelt, H. Scheer, H.U Stilz and W. Zinth, Chem. Phys. Lett. **160**, 1 (1989)
4. G.R. Fleming, J.-L. Martin and J. Breton, Nature **333**, 190 (1988)

5. C. Lauterwasser, U. Finkele, H. Scheer and W. Zinth, Chem. Phys. Lett. **183**, 471 (1991)
6. M.H. Vos, J.-C. Lambry, S.J. Robles, D.C. Youvan, J. Breton and J.-L. Martin, Proc. Natl. Acad. Sci. USA **88**, 8885 (1991)
7. J. Breton, J.-L. Martin, J.-C. Lambry, S.J. Robles and D.C. Youvan, in *Reaction Centers of Photosynthetic Bacteria*, ed. M.E. Michel-Beyerle (Springer, Berlin), p. 293 (1990)
8. K. Leo, J. Shah, E.O. Gobel, T.C. Damen, S. Schmitt-Rink, W. Schafer and K. Kohler, Phys. Rev. Lett. **66**, 201 (1991)
9. S.L. Dexheimer, Q. Wang, L.A. Peteanu, W.T. Pollard, R.A. Mathies and C.V. Shank, Chem. Phys. Lett. **188**, 61 (1992)
10. N. Go, T. Noguti and T. Nishigawa, Proc. Natl. Acad. Sci. USA **80**, 3696 (1983)
11. S.S. Skourtis, W. Bialek, A.J.R. da Silva and J.N. Onuchic, to be published
12. J.M. Jean, R.A. Friesner and G.R. Fleming, J. Chem. Phys. **96**, 5827 (1992)

Coherent Phonons in Superconducting Materials

W. Albrecht, Th. Kruse, and H. Kurz

Institute of Semiconductor Electronics, Technical University of Aachen, Sommerfeldstr. 24, W-5100 Aachen, Fed. Rep. of Germany

Abstract. We report the first time-resolved observation of optically excited coherent phonon modes in superconducting $Y_1Ba_2Cu_3O_{7-\delta}$ ($\delta<0.1$). A significant change in amplitude and decay time of coherently vibrating Ba and Cu(2) ions is observed in the superconducting state. A close correlation between the amplitude of the Ba-mode and the optical excited quasiparticles is found below T_c.

Transient pump/probe experiments on fully c-axis oriented $YBa_2Cu_3O_{7-\delta}$ films on $SrTiO_3$ are performed using 80 fs pulses at 2 eV and a highly sensitive detection scheme[1]. The samples are mounted within a cryostat for measurements in a temperature range from 4 - 330 K. Laser-ablated und DC-sputtered films (T_c = 90 K) are used with a thickness of 300 nm to suppress multiple interference effects.

Figure 1 compares the normalized reflectivity signal of superconducting $YBa_2Cu_3O_{7-\delta}$ at 300 K and 40 K. At 300 K the thermomodulation response in reflectivity is known from investigations of carrier heating and cooling in metallic systems [2].

At T = 40 K (Fig. 1), however, the negative reflectivity due to the creation of quasiparticles [3] dominates the thermomodulation signature completely.

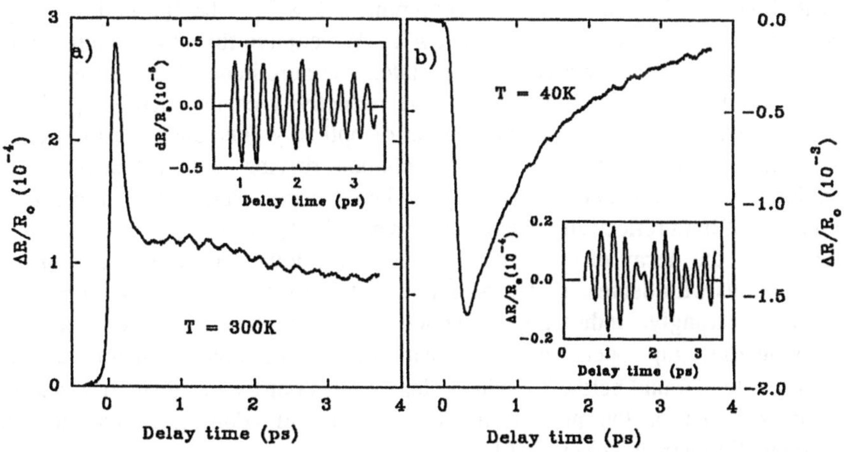

Fig.1: Time resolved reflectivity change of a 300 nm thick $Y_1Ba_2Cu_3O_{7-\delta}$ film on $SrTiO_3$ at 300 K (a) and 40 K (b). The insets show the oscillations, when the slow decaying background is subtracted.

Springer Series in Chemical Physics, Vol. 55 Ultrafast Phenomena VIII
Editors: J.-L. Martin · A. Migus · G.A. Mourou · A.H. Zewail

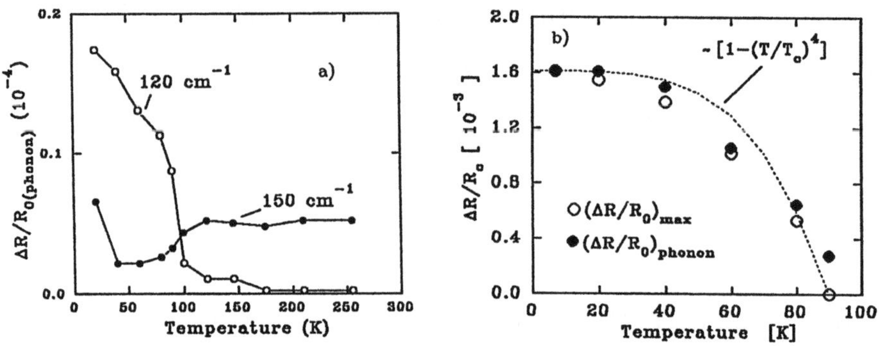

Fig. 2: a) Temperature dependence of the reflectivity change $(\Delta R/R_o)_{phonon}$ induced by coherent modes in superconducting $YBa_2Cu_3O_{7-\delta}$.
b) Comparison between temperature dependence of the maximum negative reflectivity change $(\Delta R/R_o)_{max}$ and of the Ba-phonon induced reflectivity change, normalized to a value at 10 K.

The slow decaying components of reflectivity signatures are modulated periodically as shown in fig. 1. In the insets of fig. 1, the oscillations recovered from the background, are displayed.

Fast Fourier transformation spectra exhibit pronounced A_{1g} phonon modes at 120 cm^{-1} and 150 cm^{-1} which have been assigned to vibrations of Ba (120cm^{-1}) and of Cu(2) (150cm^{-1}) in Raman experiments [4,5].

The amplitude of reflectivity modulation, $(\Delta R/R_o)_{phonon}$, induced by the coherent vibration of metallic ions varies strongly at the transition temperature. As shown in Fig 2a, the 150 cm^{-1} Cu(2) mode remains fairly constant upon cooling down to 120 K and passes through a minimum below T_c. The Ba-mode is very weak above T_c, but increases sharply below T_c. We explain this strong increase as a specific feature of the superconducting state. The amplitude of the Ba-mode is strictly related to the maximum of negative reflectivity change $(\Delta R/R_o)_{max}$ (see fig. 2b), which is determined by the amount of optically generated quasiparticles. This close correlation implies that the phonon driving source below T_c is determined by the number of optically broken Cooper pairs. The temperature dependence of both reflectivity changes obeys a $[1-(T/T_c)^4]$-law.

Above T_c the decay of coherent phonons slows down due to a freeze-out of decay channels with decreasing temperature. At T_c, however, the decay times start to increase strongly, indicating a reduced electron-phonon coupling as the superconducting gap opens up. A linear relation between temperature and dephasing rate is found below T_c, which clearly rules out the existence of a gap in the energy range of the phonon modes observed in addition to the normal superconducting gap at 320 cm^{-1} [6].

For nonsuperconducting $YBa_2Cu_3O_{6.3}$ only one phonon mode at 143 cm^{-1} is detected, whose decay remains fairly constant (1 ps) in a wide range of temperature.

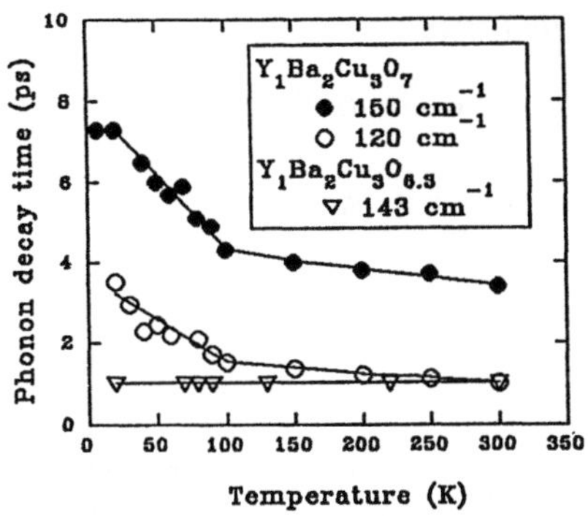

Fig. 3: Decay times of phonon modes in superconducting $YBa_2Cu_3O_{7-\delta}$ and semiconducting $YBa_2Cu_3O_{6.3}$ as a function of lattice temperature. The lines are guides for the eye.

[1] M. Strahnen, W. Kütt, H. Kurz, VME-Research, eds. C. Eck and C. Parkman, (Elsevier Science Publishers (North Holland) 1988), p. 69.

[2] S.D. Brorson, A. Kazeroonian, D.W. Face, T.K. Cheng, G.L. Doll, M.S. Dresselhaus, G. Dresselhaus, E.P. Ippen, T. Venkatesan, and A. Inam, Solid State Commun. **74**, 1305 (1990)

[3] S.G. Han, Z.V. Vardeny, K.S. Wong, O.G. Symko, G. Koren, Phys. Rev. Lett. **65**(21), 2708 (1990).

[4] A. Mascarenhas, H. Katayama-Yoshida, J. Pankove, S.K. deb, Phys. Rev. B. **39**(7), 4699 (1989)

[5] B. Friedl, C. Thomsen, H.U. Habermaier, M. Cardona, Solid State Com.,**81**, 989 (1992)

[6] R. Zeyher and G. Zwicknagel, Z. Phys. B **78**, 175 (1990)

Displacive Excitation of Coherent Phonons

T.K. Cheng, J. Vidal, H.J. Zeiger, E.P. Ippen, G. Dresselhaus, and M.S. Dresselhaus

Massachusetts Institute of Technology, 77 Massachusetts Ave., Cambridge, MA 02139, USA

Recently, we have shown that coherent lattice vibrations can be induced and subsequently detected in certain semimetals and semiconductors (e.g. Bi, Sb, Te, Ti_2O_3) via time-resolved optical pump-probe measurements using a CPM oscillator.[1] The experimental data show in each case that only totally symmetric lattice modes (i.e., A_1 symmetry) are coherently excited, even though other symmetry modes of comparable Raman cross-section exist. Furthermore, careful measurement of the coherent phonon phase reveals that the excitation mechanism for coherent phonons in these materials appears to be qualitatively different from that responsible for impulsive stimulated Raman scattering previously seen in organic dyes and crystals.

A phenomenological model has been constructed for coherent phonon generation in the materials under study and is referred to as displacive excitation of coherent phonons (DECP).[2] In the DECP model, we argue that the quasi-equilibrium coordinates of only the totally symmetric phonon modes are "instantaneously" shifted due to electronic excitation by the optical pump pulse. The lattice, finding itself displaced from its excited state quasi-equilibrium, undergoes coherent cosinusiodal oscillation about its new equilibrium position until the oscillation is damped out via electron-phonon and/or phonon-phonon scattering events. The fit of the experimental data to the DECP model is excellent.

Here, this technique for generating coherent phonons is used to investigate the temperature dependent behavior of Ti_2O_3. It is well-known that between 300K to 570K dramatic changes take place in Ti_2O_3 with respect to both lattice and electronic properties, turning it from a semiconductor to a metal. Temperature dependent X-ray measurements reveal a significant deformation of its corundum structure along the A_{1g} symmetry lattice coordinate, but with no change in lattice symmetry. Transport measurements indicate a smoothly varying drop in its resistivity by over a factor of 30. Spontaneous Raman scattering measurements show a softening of the optical phonon frequencies by up to 10%.

In the figure we show the temperature dependent pump-probe data for Ti_2O_3. The pronounced oscillatory components in the data are the signatures of the lower frequency Raman-active A_{1g} phonon mode. We observe that this mode frequency decreases (7.0 THz -> 6.2 THz) as a function of increasing temperature consistent with conventional Raman results.

Because the reflectivity changes are so large, it is of interest to estimate the magnitude of the actual phonon vibrational amplitude in Ti_2O_3. By simply assuming that there exists a mapping between static equilibrium properties and the time-resolved properties in Ti_2O_3, we can estimate the magnitude of the vibrational amplitude. This calibration would be provided by temperature dependent X-ray and optical reflectivity measurements. Temperature dependent X-ray data can accurately measure the anomalous changes in the Ti-Ti and Ti-O separations as a function of increasing temperaure. Rice and Robinson have found that through the metal-insulator transition, the ions move along a vibrational coordinate of A_{1g} symmetry, the same which we coherently excite.[3] Accompanying this change in lattice parameters is also a marked change in the optical reflectivity of the material which has been measured as a function of temperature. By relating the reflectivity and X-ray data through their temperature dependences, a mapping can be established between the static reflectivity of the material and the magnitude of the Ti-Ti separation. From 400K to 500K, we have measured in a separate steady-state experiment that a 10% decrease in the optical reflectivity at 2 eV corresponds to an increase in the nearest neighbor Ti-Ti separation by a distance of ~ 0.06Å. If the phonon dynamics of the system can be assumed to appear quasistatic to the electrons on the timescale of the A_{1g} oscillation period, then a 5% change in transient reflectivity due to the coherent phonon (obtained by deconvolving the raw data with the pulse autocorrelation) would correspond to a

Springer Series in Chemical Physics, Vol. 55 Ultrafast Phenomena VIII
Editors: J.-L. Martin · A. Migus · G.A. Mourou · A.H. Zewail

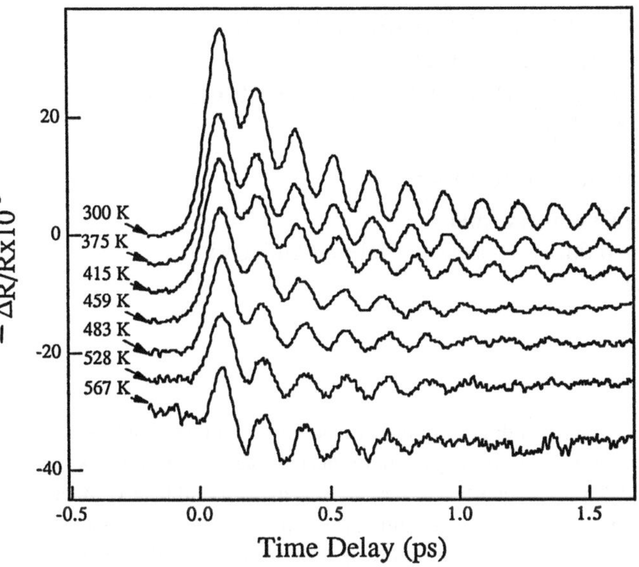

Figure : temperature consistent with conventional Raman results.

vibrational amplitude of 0.03 Å (1.2 % change in Ti-Ti separation). It must be noted that the accuracy of this mapping relies on the assumption that temperature dependent changes in refractive index in the equilibrium measurements arise primarily from deformations along A_{1g} phonon coordinates and not other phonon coordinates.

Vibrational amplitudes of this size suggest that modulating an insulator-metal transition at THz frequencies is possible. The band model parameters for semiconductors are dependent upon the lattice parameters of the crystal via the deformation potential. This can be verified experimentally by pressure dependent energy gap measurements. In a coherent phonon generation experiment, it is possible to induce a coherent lattice vibration of large amplitude as estimated above. We propose that this modulation of the lattice parameter can in turn modulate the energy bands of a material (and thus the energy gap). For an appropriate bias temperature (so that the energy gap is very small) and a coherent lattice vibration of sufficient amplitude, the conduction and valence bands in the material can be made to overlap at a frequency of ~ 7 THz. Whether or not transport properties are modulated at ~ 7 THz depends on the dynamical equilibration time for intervalley carrier scattering between the valence and conduction band. This simple picture, of course, assumes that the rigid band approximation is valid.

This work was supported in part at MIT by JSEP DAAL03-92-C-0001, AFOSR F49620-91-C-0091, and ONR 76804N0001492J1439. TKC ackowledges support of a doctoral fellowship by the IBM Corporation.

[1]T. K. Cheng, S. D. Brorson, A. K. Kazeroonian, J. S. Moodera, G. Dresselhaus, M. S. Dresselhaus, E. P. Ippen, Appl. Phys. Lett. **57**, 1004 (1990).

[2]T. K. Cheng, J . Vidal, H. J. Zeiger, G. Dresselhaus, M. S. Dresselhaus and E. P. Ippen, Appl. Phys. Lett. **59**, 1923 (1991); H. J. Zeiger, J . Vidal, T. K. Cheng, E. P. Ippen, G. Dresselhaus and M. S. Dresselhaus, Phys. Rev. **B45**, 768 (1992).

[3]C. E. Rice and W. R. Robinson, Acta Cryst **B33**, 1342 (1977).

Femtosecond Time-Resolved Photodissociation of Triiodide Ions in Alcohol Solution: Directly Observed Photoinduced Vibrational Coherence of Reactants and Products

U. Banin, A. Waldman, and S. Ruhman

Department of Physical Chemistry, and
Farkas Center for Light-Induced Processes, Hebrew University,
Jerusalem 91904, Israel

In this letter we report a study of triiodide ion photodissociation in ethanol solution using femtosecond time resolved absorption spectroscopy. Our results indicate that direct photocleavage of I_3^- using 60 femtosecond UV pulses leads to a coherently vibrating ensemble of I_2^- fragments within 300 fsec. The impulsive nature of the excitation also imposes vibrational coherence in the triiodide ions remaining in the ground state following absorption. The relaxation both of phase coherence and population is followed directly.

The triiodide ion dissociates upon excitation in two intense near UV bands, centered at 350 and 290 nm, to diiodide and atomic iodine radicals [(1)]. In solution caging of the fragments limits the extent of product formation to ~ 10% [(2)]. The I_2^- fragment absorbs in two broad bands which lead to dissociation, one which overlaps the parent spectrum and peaks at 360 nm and a second band of moderate intensity is centered at 740 nm, well separated from the reactant absorptions. Transition dipoles of the absorption bands are polarized along the molecular axis for both the diiodide and the triiodide ions. The normal modes of vibration for the parent ion are below 150 cm^{-1}, and both the symmetrical stretch of I_3^- and the vibrational of relaxed I_2^- have frequencies of ~ 110 cm^{-1}. *All the molecular motions of the reactants and products are directly resolvable using 60 fsec pulses.*

In our experiments photolysis of 3 mM solutions of I_3^- in ethanol is induced at 308 nm and the sample is probed either with portions of a white light continuum or a secondary doubled fundamental pulse at 308 nm. Pulses are generated with an antiresonant ring dye laser which is amplified in three dye stages at 1 kHz. Nominal pulse widths are 60 fsec. The system is based on Nd;YLF pumping lasers, and a novel servo stabilization scheme which allows continuous operation for hours. It is based on a double compensation setup with piezo-electric cavity length control for low frequency fluctuations or drift (0-1 Hz), and slight modulation of the RF frequency feeding the mode locker for correction of high frequency variations (1-250 Hz). The error signal is derived from feeding a portion of the dispersed output from the laser onto a PIN photodiode.

The transient transmission kinetics at early time delays are outlined in figure 1 for two probe frequencies within the visible absorption band of the diiodide fragment. Throughout this range excitation brings about a rise in absorption which is system response limited. This initial rise is most intense at the λ_{max} of I_2^-, and gives way to a rapid decrease in absorption which is followed after ~ 300 fsec with an even slower rise in the absorbance upon which damped oscillations are superimposed. The phase of the oscillations at 620 and 880 nm which are to the blue and red of λ_{max} (I_2^-) respectively, differ by precisely π radians. This is more clearly

Springer Series in Chemical Physics, Vol. 55 Ultrafast Phenomena VIII
Editors: J.-L. Martin · A. Migus · G.A. Mourou · A.H. Zewail

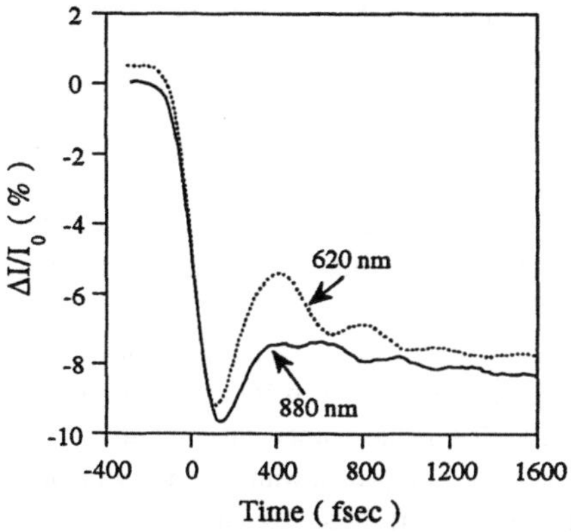

Fig.1] Transient transmission data for triiodide at two probing frequencies, 620 and 880 nm.

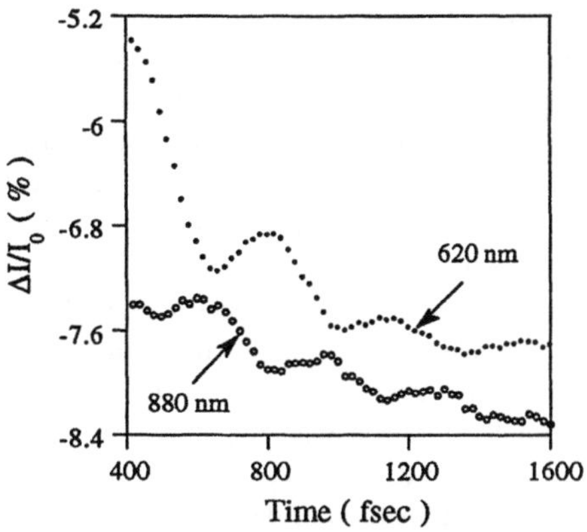

Fig.2] Data from Fig.1 on expanded scale accentuating opposed phase of oscillations at the two wavelengths.

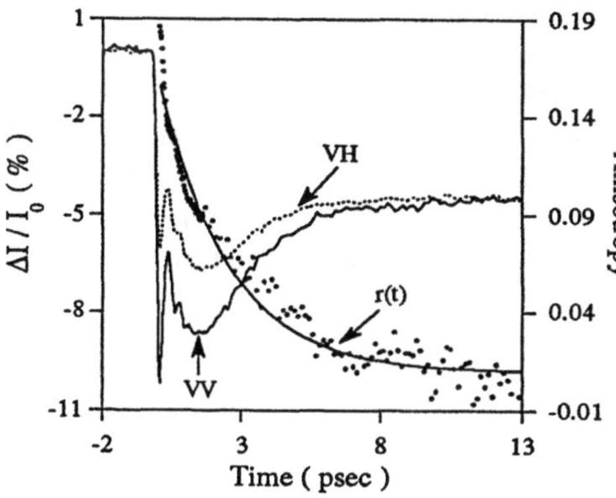

Fig.3] Anisotropy and transmission data at 620 nm. The decay of the anisotropy has been coarsly fit to a single exponential.

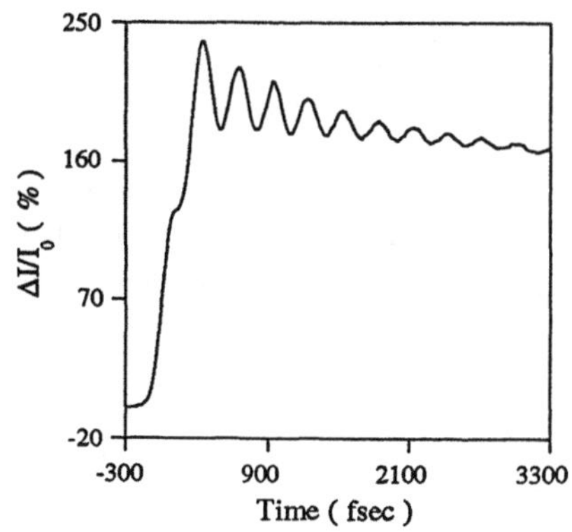

Fig.4] Transient transmission for probe pulses at 308 nm following excitation of triiodide at this frequency.

demonstrated in the expanded presentation of the oscillatory phase of the data in figure 2. The frequency of the oscillation is nearly 95 cm^{-1}, and decays with a time constant of 400 fsec. Scans which extend to longer times conducted with parallel and perpendicular relative polarizations of the pump and probe show that the product reorients on a 4 psec timescale as demonstrated in figure 3, and the absorption spectrum narrows with a 4 psec timescale following the initial 3 psec after excitation. Finally results of probing the transient absorption at 308 nm, the wavelength of the pump pulses, are depicted in figure 4. An initial instantaneous bleach is superimposed with a 110 cm^{-1} oscillation which decays within 1.4 psec. The anisotropy of this bleach has been measured, and decays with a time constant of nearly 15 psec.

We associate the initial absorption throughout the visible with excited state-excited state transitions of the I_3^- in the dissociative state, and the rapid decay which precedes oscillations with the stage of dissociation. We interpret the oscillations to indicate that nascent I_2^- fragments are formed in a coherent vibrational state. The opposed phase of the modulations to the red or blue of the band center add credence to this contention. In view of the induction time for appearance of the oscillations, and the large discrepancy of the anisotropy in the visible from that of the ground state triiodide, it is unlikely that the coherent motion is taking place on an excited surface of I_3^-. Accordingly the decay of the oscillation in the absorption measures dissipation of fragment vibrational coherence. The narrowing of the absorption band is either due to vibrational population relaxation, or solvent reorganization. Similar T_1 times have recently been measured for ions in polar solvents [3].

We relate the modulations in the transients at 308 nm to coherent vibrations in the symmetric stretch mode of I_3^- remaining in the ground state due to Resonant Impulsive Stimulated Raman Scattering [4]. Essentially the rapid excitation has burnt a coherent hole in the ground state vibrational population, the filling of which is accompanied with coherent vibration. A fourier transform of the oscillations in the data reproduces recorded Raman spectra of this mode.

The status of the observed diiodide ions with respect to caging / recombination is unclear. CW experiments predict strong caging[2] though most of the bleach in the UV absorption is accountable as transient I_2^- absorption, and replenishment of triiodide absorption is slow. This may result from the fact that the photoproducts do not correlate with the ground potential, and require an electron transfer for regaining fragments that will.

These questions, and the dependence of our reported results on the solvent used will be addressed experimentally in the near future. In view of the presented results we are convinced that this will prove an exciting new molecular system for study of ionic chemistry in solution. In particular the multitude of directly measurable dynamic variables, including the *absolute phase of excess vibrational energy deposited in products* will allow us to study this reaction in unprecedented detail.

REFERENCES

1] P.Fournier de Violet, R.Bonneau & J.Joussot-Dubien, *Chem. Phys. Lett.*, **28**, 569 (1974).

2] J.C.Roy, W.H.Hamill & R.R.Williams Jr., *JACS*, **77**, 2953 (1955).

3] J.C.Owrutsky, Y.R.Kim, M.Li, M.J.Sarisky & R.M.Hochstrasser, *Chem. Phys. Lett.* **184**, 368 (1991).

4] J.Chesnoy & A.Mokhtari, *Phys. Rev. A*, **38**, 3566 (1988).

Vibrational Coherence in Charge Transfer

K. Wynne, C. Galli, P.J.F. De Rege, M.J. Therien, and R.M. Hochstrasser

Department of Chemistry, University of Pennsylvania,
Philadelphia, PA 19104, USA

1. Introduction

The role of the solvent on chemical reactions is one of the most important questions in physical chemistry. Although a large body of knowledge has accumulated over the years through gas phase studies, theoretical calculations of potential surfaces, resonance Raman scattering, time resolved infrared absorption and transient absorption studies, some fundamental questions still persist. For example, in adiabatic reactions the standard Kramers theory predicts a reaction rate that is always less than or equal to the relaxation rate of the solvent. Yet a number of examples exist where the reaction rate is equal to or even greatly exceeds the solvent relaxation time.

Recently there was a great deal of interest in the electron transfer (ET) reaction. In the conventional theories for ET, a vibrational coordinate plays an essential role: in the Sumi-Marcus theory [1] this mode facilitates ET from the reactant state whereas in the Jortner-Bixon theory [2] it provides for additional decay channels in the inverted regime. In these theories it is a priori assumed that these vibrational motions are thermally equilibrated and motion along the classical reaction coordinate is limited by the longitudinal relaxation time of the solvent. Recently, however, ultrafast ET has been observed in which the transfer rate greatly exceeds the relaxation time of the solvent [3]. This has raised questions as to whether vibrational coherence in promoting modes might facilitate conversion of reactants to products.

Vibrational coherence during an ET reaction has been observed by J.-L. Martin and coworkers in the reaction centers of the purple bacteria of Rhodobacter sphaeroides R-26 [4]. Although the solvent (i.e. the protein in which the reaction center is embedded) is in principle well defined, the complexity of the reactant molecule makes a detailed analysis quite difficult. Studies of ET in two model systems having less complexity will be described in the present paper.

Initial experiments were focused on studying direct ET in electron donor-acceptor (EDA) complexes [5]. Complexes of tetracyanoethylene (TCNE) with pyrene or perylene are of interest for the following reasons: i) although the individual molecules absorb light in the UV, the complexes show an absorption band in the visible region (at 650nm and 800nm respectively with a FWHM of ~6000 cm^{-1}) that facilitates **direct** ET upon absorption of a photon. Charge recombination is thus free from complications introduced by electronic states intermediate to the ground and charge transfer (CT) states. ii) The reported charge recombination (CR) rates in acetonitrile are faster than the longitudinal relaxation time of the solvent (~500fs and ~300fs respectively [6] in acetonitrile which has a 0.56ps relaxation time). iii) A complete analysis utilizing resonance Raman scattering is available for a similar EDA complex, TCNE with hexamethylbenzene (HMB) [7]. In addition to the high frequency C-C and C-N stretches in TCNE, a low frequency dimer oscillation (169cm^{-1}) is a key promoting mode. iv) The ET rate in a series of similar EDA complexes as a function of driving force is not in accord with either the Sumi-Marcus or the Jortner-Bixon theory and is thus essentially unexplained [6].

The porphyrin-quinones are also of great interest since this class of molecules shows very fast ET and can be considered a model system for part of the photosynthetic process. Most studies on porphyrin-quinones have been performed with picosecond resolution and consequently the fastest rate, less than 150fs in phenyl-Zn-porphyrin-quinone [8], could only be estimated. Below we will present the first ultrafast studies on free-base triphenyl-porphyrin-quinone (5-(1',-4'-benzoquinonyl)-10,15,20-triphenylporphyrin or PQ).

Springer Series in Chemical Physics, Vol. 55 **Ultrafast Phenomena VIII**
Editors: J.-L. Martin · A. Migus · G.A. Mourou · A.H. Zewail

2. Experimental

The laser system which is used for these experiments is based on a colliding pulse mode-locked (CPM) laser producing 50-70fs pulses at 620nm. These pulses are amplified in a copper-vapor laser pumped bow-tie amplifier yielding pulses with an energy of ~3μJ at a repetition rate of 8kHz. About half of this light is used to pump the samples and the other half is focussed in a 2mm thick flow cell with ethylene glycol, yielding a continuum which can be used for probing from ~450nm to ~700nm. 10nm wide slices from the continuum can be compressed in a double pass prism compressor to 70-100fs. Overlap between the pump and probe pulses is determined by crosscorrelation in a KDP crystal or by pump-probe in a nile-blue solution. The EDA complexes were prepared by dissolving commercially available TCNE, HMB and pyrene in various solvents. The PQ complex was synthesized, dissolved in toluene, glycerol-triacetate (GTA) or chloroform (~1mM concentration), degassed and stored in a sealed quartz cell under nitrogen.

3. Results and Discussion

Pump-probe experiments were performed on TCNE-HMB complexes in cyclohexane and TCNE-pyrene complexes in acetonitrile and benzonitrile. In the TCNE-HMB complex, a transient bleach signal is observed with a recovery time of 16.6±0.2 ps reflecting the charge recombination. Superimposed on this signal an oscillation can be discerned with a frequency of 162±0.4 cm^{-1} and a decay time of 1.4±0.2 ps. The frequency and decay time of this beat signal correspond well with the resonance Raman data. The beat has zero phase shift at both 620 and 650 nm probe wavelengths. These results strongly indicate an oscillation in the ground state of the complex. The zero phase shift corresponds to a cos(ωt) oscillation on a positive transmission change signal, corresponding to a bleach of the ground state absorption.

In the TCNE-pyrene complex in acetonitrile a 1.43±0.03 ps recovery without oscillations is observed. The pyrene cation radical absorption at 460nm was found to decay with 270±3 fs in acetonitrile and 1.65±0.01 ps in benzonitrile. These decay times are remarkably short compared with the solvent relaxation times of 0.56 and 5.9 ps respectively and furthermore the decay traces are markedly nonexponential. However, well defined oscillations attributable to the excited state were not yet observed, which may in part be due to the long probe pulses employed.

The porphyrin-quinone complex, PQ, used in the studies of ET, is the first of a series of porphyrin-quinone complexes we intend to study in the near future. The pump pulse at 620nm is bracketed by the $Q_x(1,0)$ and $Q_x(0,0)$ band of the porphyrin. A metallated porphyrin-quinone complex with appropriate meso-substituents is now being synthesized which has its Q(0,0) band at 620nm. After pumping PQ at 620 nm a transient absorption, that arises from the charge transfer (CT) state P^+Q^-, extending over the range 480-680 nm has been probed. It was found to decay due to charge recombination with time constants of 2.5±0.9 ps in chloroform, 12.2±0.2 ps in glycerol-triacetate (GTA) and 19.5±0.4 ps in toluene. Note that the slower decay in toluene relative to that in GTA is perhaps unexpected since the solvent relaxation time of toluene is 5.7 ps and that of GTA 125 ps (at 293K). One possible explanation for this phenomenon is a change in the CT state potential in GTA relative to that in toluene.

Absorption from the CT state is not the only feature of the pump-probe signal of PQ however: further studies at 680 nm reveal a transient stimulated emission signal from the $Q_x(0,0)$ band. Assuming that the stimulated emission originates from the locally excited (LE) state which decays to the CT state, a fit to this data yields an ET time constant of 59±1.4 fs (in toluene at room temperature). In some of the data taken at 680 nm a small magnitude oscillation is visible just after the stimulated emission signal is gone.

Interesting results are obtained around 500 nm. In toluene at 500 and 510 nm a transient bleach signal is observed that evolves into an absorption signal with a large overshoot. The bleach signal can be interpreted as a replica hole on the $Q_y(1,0)$ band that disappears due to the growing LE or CT state absorption. Interestingly enough, about 200 fs (at 500 nm) and 300 fs (at 510 nm) after the overshoot one observes a small recurrence suggesting that coherence plays a role in the ET

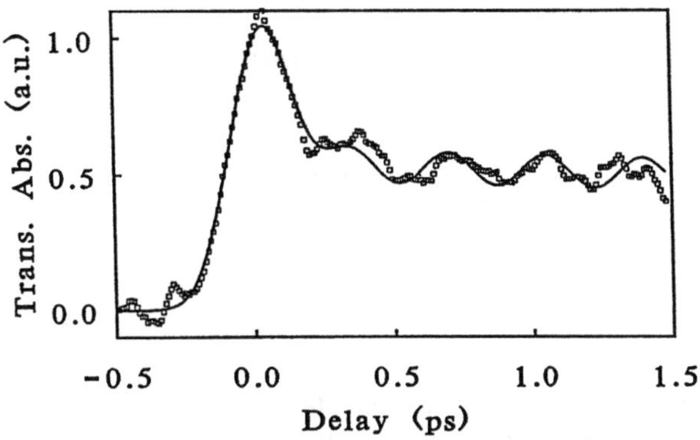

Figure 1. Transient absorption from free-base triphenyl-porphyrin-quinone in glycerol-triacetate, pumped at 620 nm and probed at 520 nm.

reaction observed here. Longer lasting coherence is observed at 520 nm in GTA (see the Fig.1): a large amplitude beat with a frequency of approximately 95 cm^{-1} continues long after the ET is concluded (i.e. for much longer than 59 fs). The phase of the beat is approximately π (see above) which would indicate that this is an excited state oscillation. The fact that this beat is observed around 500 nm and not at 620 nm where the ground state absorbs and the signal to noise ratio is much higher, is a further indication that the oscillation does not originate in the ground state. Elimination of ground state oscillations leaves open two possible interpretations of these beats: either nuclear coherence is induced in a spectator mode which survives the ET reaction, or coherence is **induced by** the reaction.

In conclusion there are strong indications that coherence of nuclear or electronic origin plays a role in the ET transfer reaction in PQ complexes. Detailed studies of a variety of PQ complexes possessing Q(0,0) bands centered at 620 nm and varied ET and CR driving forces is in progress.

Acknowledgement

This research was supported by grants from NSF, NIH and the Searle Scholars Program. KW acknowledges financial support from the Netherlands Organization for Scientific Research (NWO).

References

1 H. Sumi, R.A. Marcus, J. Chem. Phys. **84**, 4894 (1986).
2 a) J. Jortner, M. Bixon, J. Chem. Phys. **88**, 167 (1988), b) J. Jortner, ibid. **64**, 4860 (1976).
3 G.C. Walker, E. Åkeson, A.E. Johnson, N.E. Levinger P.F. Barbara, J. Phys. Chem. **96**, 3728 (1992).
4 M.H. Vos, J.-C. Lambry, S.J. Robles, D.C. Youvan, J. Breton, J.-L. Martin, Proc. Natl. Acad. Sci. USA **88**, 8885 (1991).
5 a) R.S. Mulliken, W.B. Person, *Molecular Complexes*, Wiley, New York, 1969, b) G. Briegleb, *Elektronen-Donator-Acceptor-Komplexe*, Springer, Berlin, 1961.
6 T. Asahi, N. Mataga, J. Phys. Chem. **95**, 1956 (1991).
7 F. Markel, N.S. Ferris, I.R. Gould, A.B. Myers, J. Am. Chem. Soc. in press.
8 a) J. Rodriguez, C. Kirmaier, M.R. Johnson, R.A. Friesner, D. Holten, J.L. Sessler, J. Am. Chem. Soc. **113**, 1652 (1991), b) M.A. Bergkamp, J. Dalton, T.L. Netzel, J. Am. Chem. Soc. **104**, 253 (1982).

Ultrafast Dynamics in Solution: Wavepacket Motion and the Cage Effect in Iodine

Y. Yan[1], *R.M. Whitnell*[1], *K.R. Wilson*[1], *and A.H. Zewail*[2]

[1]Department of Chemistry, University of California San Diego, La Jolla, CA 92093-0339, USA

[2]Arthus Amos Noyes Laboratory of Chemical Physics, California Institute of Technology, Pasadena, CA 91125, USA

The photodissociation of I_2 in solution has been studied for many years by many groups[1-3]. In a recent series of experiments at Caltech[2], ≈60-90 fs tunable laser pulses are used to pump I_2, both in the gas phase and in compressed Ar solution, from the ground state to the electronically excited B state. The dynamics of I_2 in the B state is then monitored by a time-delayed probe pulse which excites the I_2 molecule further, to the f or E state. Absorption or fluorescence resulting from the probe pulse is measured as a function of delay time between the two pulses. The initial pumped packet is influenced by the coherent nature of the pump pulse and the Franck-Condon overlaps between the initial thermal distribution of the ground state and the B state vibrational manifold. The dynamics of the wavepacket on the B state is governed by the shape of the potential and the nature of the solvent environment.

The Caltech experiments[2] used a variety of pump energies to excite the I_2 molecule to the B state. We have performed quantum and classical simulations[3,4] in order to elucidate the underlying dynamics on going from the gas phase toward the solution phase limit, considering three pump wavelengths used in the Caltech experiments: 520 nm, for which the resulting wavepacket on the B state is bound; 505 nm, where the wavepacket has both bound and dissociative components and 490 nm, where the wavepacket is entirely dissociative.

The presence of Ar gas at 100 bar has a significant influence on the time evolution of the I_2 wavepacket, including an increase in the asymptotic signal as compared to that from the gas phase, as shown in Figure 1. The Ar experiments display strongly damped recurrences in the transient spectra at all pump wavelengths at which these oscillations occur. Collisions with the solvent influence this damping of the recurrences, causing an increase in the rate of dephasing. In the gas phase, Zewail and co-workers also observe a rephasing of the spectrum for both 520 nm and 620 nm pump pulses, but in the 100 bar Ar solution, this rephasing is no longer evident. The recurrence peaks in the solution phase spectra are broadened as compared to the gas phase spectrum.

In this work, we carry out classical and quantum simulations with the goal of understanding the underlying physics leading to the experimental spectra. We calculate quantum and classical dynamics of the I_2 molecule in the gas phase and compute classical molecular dynamics in 100 bar Ar solution. We begin with the I_2 molecule on the X state, then excite it to the B state and compute the dynamics on the B state. From the resulting B state dynamics, we compute the absorption (or equivalently the laser induced fluorescence) as a function of delay time between the pump and probe pulse. We refer the reader to Ref. [3] for a detailed description of the computational techniques and relevant references. The result of these simulations is shown in Figure 1.

Springer Series in Chemical Physics, Vol. 55 Ultrafast Phenomena VIII
Editors: J.-L. Martin · A. Migus · G.A. Mourou · A.H. Zewail

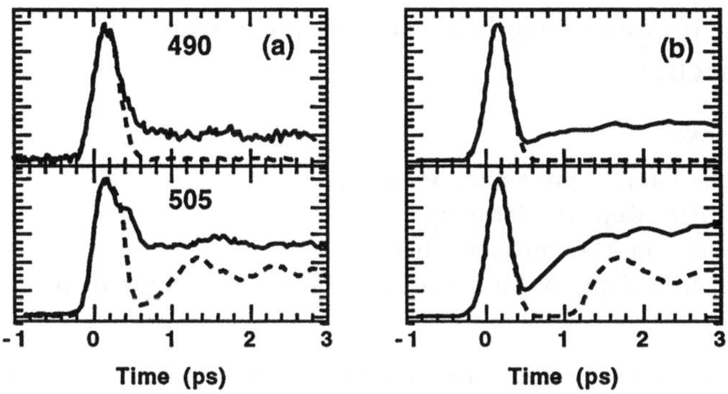

Figure 1. (a) Experimental and (b) calculated I_2 transient spectra for 490 and 505 nm pump pulses. Dashed curve, gas phase; solid curve, 100 bar Ar solution.

The results demonstrate that simulation is capable of reproducing the experimental results. We are therefore able to analyze the simulation results more deeply with confidence that they do in fact correspond to physical reality. Analysis of the quantum and classical simulations and comparison with the experimental results at the three pump wavelengths below, at, and above the B state dissociation limit shows that the experimental signals result from several processes. These processes include the "unphasing" of the coherent vibrational wavepacket as it evolves in time, the dephasing effect of Ar solvent interactions on the vibrational evolution of the ensemble of I_2 molecules, the vibrational energy loss of the I_2 to the solvent (which has the effect of increasing the signal as it moves the vibrational coordinate probability into a region more amenable to excitation by the probe pulse), solvent caging (which can result from interactions similar to those leading to vibrational decay) and recombination of the I atoms to reform I_2 molecules. The current results show that it is now possible to observe experimentally such processes as vibrational dephasing and energy relaxation, caging and recombination on the timescales on which they occur.

References

[1] A. L. Harris, J. K. Brown and C. B. Harris, Annu. Rev. Phys. Chem. **39**, 341 (1988) and references therein and in Ref. [3].

[2] R. M. Bowman, M. Dantus and A. H. Zewail, Chem. Phys. Lett. **161**, 297 (1988); A. H. Zewail, M. Dantus, R. M. Bowman and A. Mokhtari, J. Photochem, Photobiol. A: Chem **62**, 301 (1992).

[3] Y. J. Yan, R. M. Whitnell, K. R. Wilson and A. H. Zewail, Chem. Phys. Lett. (in press).

[4] Y. J. Yan, Chem. Phys. Lett. (submitted).

Femtosecond Time-Resolved Ionization Spectroscopy of Polyatomic Molecules

M. Seel[1] *and W. Domcke*[2]

[1]Institut für Medizinische Optik, Universität München, W-8000 München 40, Fed. Rep. of Germany
[2]Institut für Physikalische und Theoretische Chemie, Technische Universität München, W-8046 Garching, Fed. Rep. of Germany

Abstract. Computer simulations of two-pulse time-resolved ionization spectroscopy are presented. The computations are performed for a model of the pyrazine molecule which is based on ab initio calculations. The calculated time-resolved photoelectron spectra reflect ultrafast population dynamics, coherent vibrational motion and vibrational dephasing in the excited states.

It is generally appreciated that femtosecond pump-probe techniques have the potential to provide novel and important information on elementary photophysical and photochemical processes. It has been demonstrated by several applications that time-resolved two-pulse ionization, that is, optical excitation and subsequent ionization of a molecular system with two short laser pulses, is a promising tool to observe excited state dynamics.[1] This work aims at the simulation of femtosecond pump-probe ionization spectroscopy of polyatomic molecules from first principles.

The material system is described quantum mechanically and represented as an appropriate set of electronic states which is coupled to an appropriate set of vibrational modes. The radiation is described as a classical time-dependent field. The interaction of the material system with the radiation field is treated non-perturbatively by solving the time-dependent Schrödinger equation with time-dependent Hamiltonian. An efficient discretization of the ionization continua allows the numerically exact treatment of the ionization process even for complex molecular systems.

As an example of nontrivial and ultrafast intramolecular dynamics we consider the internal-conversion dynamics of the strongly vibronically coupled $S_1(n\pi^*)$ and $S_2(\pi\pi^*)$ excited states of pyrazine. The model includes the four spectroscopically most relevant vibrational modes and accounts, on a microscopic level, for ultrafast and non-Markovian electronic decay as well as fast vibrational dephasing processes.[2] The electronic ground state S_0 of the molecule as well as the low-lying $I_0(n^{-1})$ and $I_1(\pi^{-1})$ states of the pyrazine cation with associated one-electron Coulombic continua have been included in order to model the two-pulse ionization process. In extension of earlier work [3], all electronic potential-energy surfaces are now characterized by ab initio electronic structure calculations [4]. The model has furthermore been generalized to take into account the vibronic coupling of the two cation states. Figure 1a gives a schematic view of the model system.

The results emerging from the computer simulations of pump-probe ionization for this model of a polyatomic molecule can be briefly summarized as follows. If frequency and duration of the probe pulse are suitably chosen for one-photon ionization out of the coupled S_1–S_2 manifold, the total ion yield as a function of the pulse delay time reflects the essential structures of the population probability of the diabatic S_1 state. It is interesting to note that previous investigations have shown that the population probability of the diabatic S_2 state can be monitored via detection of stimulated emission.[5]

Figure 1b shows the calculated time- and energy-resolved photoelectron spectrum. The shift of the maxima of the photoelectron spectrum as a function of the pulse delay

Springer Series in Chemical Physics, Vol. 55 Ultrafast Phenomena VIII
Editors: J.-L. Martin · A. Migus · G.A. Mourou · A.H. Zewail

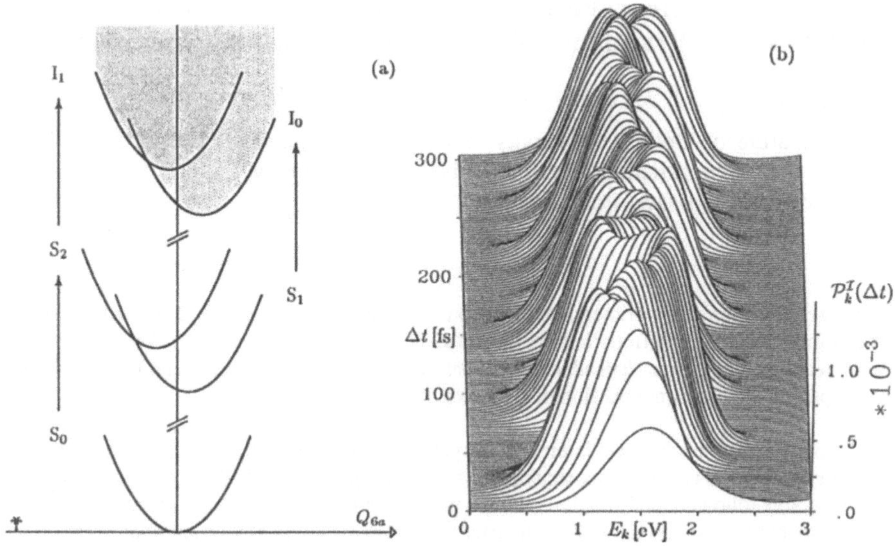

Figure 1. (a) A cut through the potential-energy surfaces along the normal coordinate Q_{6a}. The shaded areas symbolize the ionization continua. The arrows indicate possible transitions induced by the laser pulses. (b) The photoelectron energy distribution is given as a function of the pulse delay time. The probe pulse duration is 6 fs.

time reflects the initially coherent vibrational motion of the nonstationary state prepared by the pump pulse in the S_1–S_2 manifold. The damping of the shift with increasing delay time reflects the dephasing of the coherence of the vibrational motion.

To summarize, it has been shown that a fully microscopic description of multiphoton ionization with ultrashort laser pulses is possible even for complex molecular model systems featuring multidimensional vibrational motion and ultrafast internal conversion. The insight gained from such computer simulations should be helpful for the interpretation of time-resolved ionization experiments on polyatomic molecules.

Computation time on a Cray Y-MP has generously been provided by the Leibnitz Rechenzentrum der Bayerischen Akademie der Wissenschaften.

[1] J. M. Wiesenfeld and B. I. Greene, Phys. Rev. Lett. **51**, 1745 (1983);
T. Baumert, M. Grosser, R. Thalweisser and G. Gerber, Phys. Rev. Lett. **67**, 3753 (1991);
M. Dantus, M. H. M. Janssen and A. H. Zewail, Chem. Phys. Lett. **181**, 281 (1991).

[2] R. Schneider, W. Domcke and H. Köppel, J. Chem. Phys. **92**, 1045 (1990).

[3] M. Seel and W. Domcke, J. Chem. Phys. **95**, 7806 (1991).

[4] L. Seidner, G. Stock, A. L. Sobolewski and W. Domcke, J. Chem. Phys. **96**, 5298 (1992).

[5] G. Stock, R. Schneider and W. Domcke, J. Chem. Phys. **90**, 7184 (1989).

A Study of Nuclear Vibrational Wave Packets in Na_2 by Time- and Frequency-Resolved Fluorescence Upconversion

I.A. Walmsley[1], *T.J. Dunn*[1], *J. Sweetser*[1], *and C. Radzewicz*[2]

[1]The Institute of Optics, University of Rochester, Rochester, NY 14620, USA
[2]Institute of Experimental Physics, University of Warsaw, ul. Hoza 69, Warsaw, Poland

Abstract. We present experimental results from measurements of the time-dependent energy spectrum of fluorescence from Sodium dimers excited by a short optical pulse. We have observed directly the time-dependent Stokes shift of the emission, the decay and revival of the wave packet and coherent emission near the exciting wavelength.

There has been much recent interest in the physics of the generation and detection of wave packets in atoms[1] and small molecules[2] using short optical pulses. The dynamics of the wave packets following photoexcitation often provide new insights into physical and chemical processes, even if they do not always improve the determination of atomic and molecular constants. In particular, several of these studies have been concerned with some of the contrasts between the classical and non-classical aspects of the wave packet dynamics, such as revivals and fractional revivals[3], and quantum interference[4] between two wave packets.

A characteristic of molecular nuclear vibrational wave packet is that the mean position of the packet follows a classical trajectory, whereas it is the position dispersion that exhibits the non-classical effects. For example, it has been predicted that quadrature squeezing of the wave packet's position can occur in an harmonic molecule when excited by a short, but finite duration pulse[5]. In such a situation the positional quantum noise of the oscillator can be substantially less than that of the ground vibrational state at certain points in the nuclear motion, and the location of these points can be chosen to some degree by imposing phase structure on the pulse. Figure 1 shows the results of a calculation of the squeezing for a molecule with harmonic X- and A-states whose minima are displaced with respect to one another. For parameters which correspond roughly to those of our experiments, the wave packet may be squeezed to less than one-half of the ground-state position dispersion, and this occurs at about 10 degrees from the classical turning points. The nuclear coordinate may be squeezed further if the nuclei in the A-state are less tightly bound than in the X-state.

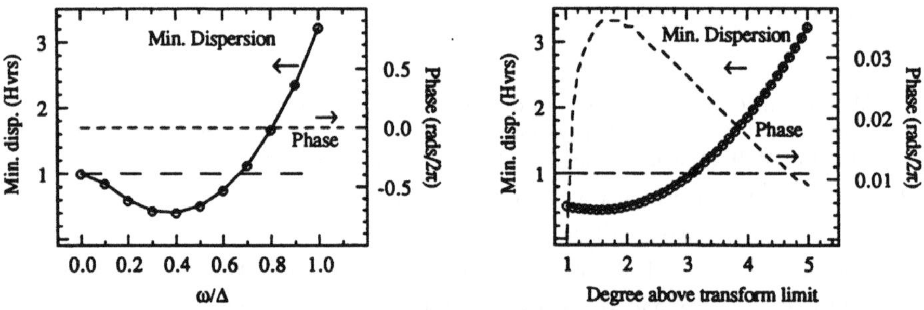

Figure 1. The position dispersion of a wave packet in the excited state of an harmonic molecule with vibrational frequency ω, as a function of (left) the pulse duration $(\sqrt{2}\Delta)^{-1}$, for a transform-limited exciting pulse, and (right) the linear chirp of the pulse with ω=0.3Δ. The potential displacement is twice the width of the ground state.

Springer Series in Chemical Physics, Vol. 55 Ultrafast Phenomena VIII
Editors: J.-L. Martin · A. Migus · G.A. Mourou · A.H. Zewail

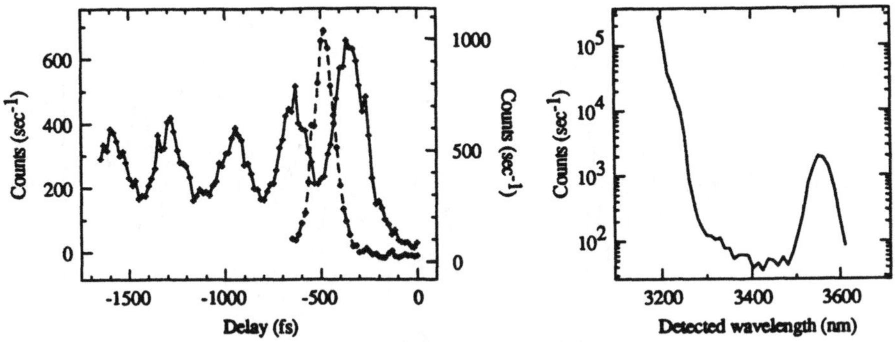

Figure 2. (Right) Time-resolved fluorescence at 810nm from Na_2 at 720K excited by a 60fs pulse at 630nm. The dashed line is the temporal response of the detection system. (Left) Spectrum of emission at a maximum of the fluorescence modulation, showing localization of the wave packet at the classical outer turning point.

Evidently an experiment designed to detect such effects must be able to measure the dispersion as well as the mean position of the nuclear coordinate. A possible technique for such a measurement is time- and frequency-resolved fluorescence upconversion[6]. The principal idea behind this method is that as the molecule vibrates (while in an excited electronic state) it may spontaneously emit a photon whose frequency depends upon the relative position of the nuclei at the time of emission. Thus a measurement of the energy spectrum of the fluorescence emitted within a given time window delayed with respect to the excitation pulse reveals not only the mean position of the wave packet, but also its shape[7].

Figure 2 show the time dependent fluorescence at 810nm emitted from an extended sample of sodium dimers at 720K when excited by a 60fs pulse at 630nm. The period of the modulation is 308fs, which corresponds to a wave packet centered at v=10 in the $A^1\Sigma_u$ state. The modulation depth is quite large, indicating not only that the signal arises predominantly from this state, but also that the packet is quite localized at the classical outer turning point of its trajectory. Note also that the maxima occur at (n+1/2) periods following the excitation pulse, since it takes one-half of a vibrational period for the packet to propagate to the outer turning point. As the wave packet passes the minimum of the A-state potential it has a rather large extent. Thus the fluorescence is broadband and not so deeply modulated. Also shown is the fluorescence spectrum taken at one of the maxima, illustrating again that the wave packet is well-localized at this position. The narrowest width of the spectral peak at 810nm that we have seen is about 8nm, and is determined by the resolution of our detection system. This corresponds to a wave packet extent of about 0.4Å, approximately the width of the ground vibrational state in this electronic state.

The anharmonicity of the nuclear potential is evident from the decay of the sinusoidal beat pattern to a more complicated signal at longer delays, as shown in Figure 3. Harmonic motion is evident out to about 8ps (some 20 periods), but has completely disappeared at 18ps (lower trace in right figure). A revival of the beats at the fundamental period, but with greatly reduced modulation, occurs at 26ps (upper trace in right figure). A calculation based on known Na_2 potentials suggests that a fractional revival, characterized by a beat pattern with a 151fs period, rather than a full revival, should occur at this delay. The experimental sensitivity may preclude observation of the complete, and fairly complex, revival signal. A full revival is expected near 50ps[8]. We have observed a small revival at this delay, but with much reduced modulation, probably because of the large range of rotational quantum numbers that are excited.

Since the wave packet is also quite well-localized at the inner turning point, it might be expected that the emission at the exciting pulse wave length would also be quite strong and have

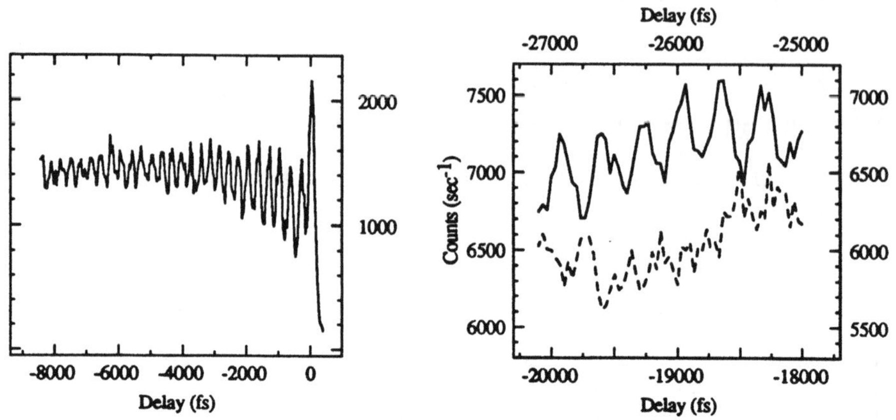

Figure 3. (Left) The long-time dynamics of the fluorescence, showing the decay of the modulation. (Right) The disappearence of the modulation at 16ps (lower trace), and a revival at 26ps (upper trace).

relatively narrow bandwidth. In fact, we observe that the radiation at 630nm is about 10^4 times as intense as that at 810nm, and is very directional, occurring only within one-half degree of the exciting beam direction. There are a number of reasons why this might be so. First, the polarization at this frequency contains a "coherent" component, due to the interference of the excited state wave packet and the ground state "hole" wave packet that is absent at the longer wavelengths. Second, the emission is phase matched and group velocity matched at 630nm whereas the different dispersion near 810nm prevents such a coherent effect from arising at that wavelength. The temporal structure of the emission at 630nm is quite complicated and is still not completely understood, but possibly arises from the interference of vibrational and rotational wave packets in several different electronic states. This is still under investigation.

References

[1] J. Yeazell and C. R. Stroud, Phys. Rev. Lett., **60**, 1494 (1988)
[2] P. Cong, A. Mokhtari and A. H. Zewail, Chem. Phys. Lett., **172**, 190 (1990)
[3] J. Yeazell, M. Malalieu and C. R. Stroud, Phys. Rev. Lett., **64**, 2077 (1990)
[4] N. F. Scherer ,A. Matro, L. D. Ziegler, M. Du, R. J. Carlson, J. A. Cina and G. R. Fleming, Jnl. Chem. Phys., **95**, 1487 (1991)
[5] J. Janszky and A. N. Vinogradov, Phys. Rev. Lett., **64**, 2771 (1990)
[6] J. Shah, IEEE Jnl. Quant. Elec., **QE-24**, 276 (1988)
[7] P. Kowalczyk, C. Radzewicz, J. Mostowski and I. A. Walmsley, Phys. Rev. A, **42**, 5622 (1990)
[8] T. Baumert, V. Engel, C. Röttgermann, W. T. Strunz, and G. Gerber, Chem. Phys. Lett., **191**, 639 (1992)

Ultrafast Dynamics of Solid C_{60}*

S.L. Dexheimer[1,2], *D.M. Mittleman*[1,2], *R.W. Schoenlein*[1], *W. Vareka*[2], *X.-D. Xiang*[2], *A. Zettl*[1,2], *and C.V. Shank*[1,2]

[1]Lawrence Berkeley Laboratory, MS: 70-193A,
1 Cyclotron Road, Berkeley, CA 94720, USA
[2]Department of Physics, University of California, Berkeley, CA 94720, USA

There has been a great deal of recent interest in C_{60} as a possible new material for electronic and nonlinear optical applications. The work presented here focuses on the fast dynamics of solid C_{60} following photoexcitation, and shows that this system has complex relaxation dynamics, including components on a femtosecond time scale. C_{60} crystallizes in an fcc lattice, with relatively weak interactions between the covalently bound molecules. Thin film samples are prepared by thermal sublimation of purified C_{60} powder, and are studied with optical pulses as short as 12 fs in duration. For excitation into the lowest optical band, pulses centered at 620 nm are generated using the femtosecond laser system described previously[1], with the addition of a second amplification stage prior to compression of the chirped pulse. Optical pulses at blue-green wavelengths[2] (450 – 550 nm) are used for excitation into higher-lying electronic states. Time-resolved measurements of the differential transmittance are made using a pump-probe technique, and wavelength resolution is achieved by dispersing or filtering the probe beam after transmission through the sample.

Time-resolved measurements in both the red and blue-green spectral regions show nonexponential relaxation dynamics. The response is modulated by oscillations at characteristic molecular vibrational frequencies, indicating that impulsive photoexcitation of this molecular solid generates coherent intramolecular vibrational motion. Figure 1 shows the negative time-resolved differential transmittance detected at 580 nm using pulses of 12 fs duration centered at 620 nm. The signal corresponds to an induced absorption, indicating that the response is dominated by the excited electronic state. The trace shows a fast oscillatory modulation that persists on a picosecond time scale. The frequency content of the oscillatory part of the response can be resolved by Fourier analysis, and the Fourier power spectrum is shown in the inset. Vibrational modulation is evident at frequencies in good agreement with the fully symmetric vibrational modes of C_{60} determined by Raman spectroscopy[3]. The 495 cm^{-1} component (67 fs period) corresponds to the A_g radial breathing mode, and the 1467 cm^{-1} component (23 fs period) corresponds to the A_g pentagonal pinching mode.

For excitation into the lowest optical band, the time-resolved response gives an excellent fit to the stretched exponential decay function, $\exp[-(t/\tau)^{\beta}]$. This function, which was initially applied to the decay dynamics of photoexcited C_{60} by Cheville and Halas[4], is characterized by a distribution of time scales and has been observed for a wide range of dynamical processes in a variety of disordered condensed matter systems. In our work, we find that the function fits the data even at short times, indicating that the distribution of relaxation times includes components as short as tens of femtoseconds. The response is unchanged between room temperature and liquid nitrogen temperature, which argues against lattice dynamics or thermally activated carrier processes as the dominant relaxation mechanism on this time scale. However, both the time constant τ and stretching parameter β show a dramatic dependence on excitation density, as seen in Figure 2. The highest power levels correspond to photoexcitation of 10 – 20% of the molecules, suggesting that interactions between photoexcited molecules may significantly contribute to the non-exponential dynamics. Excitation into the next highest optical band using pulses centered at 500 nm creates an induced transmittance around 500 nm, evolving into an induced absorption

* This work was supported by the U.S. Department of Energy under contract DE-AC03-76SF00098.

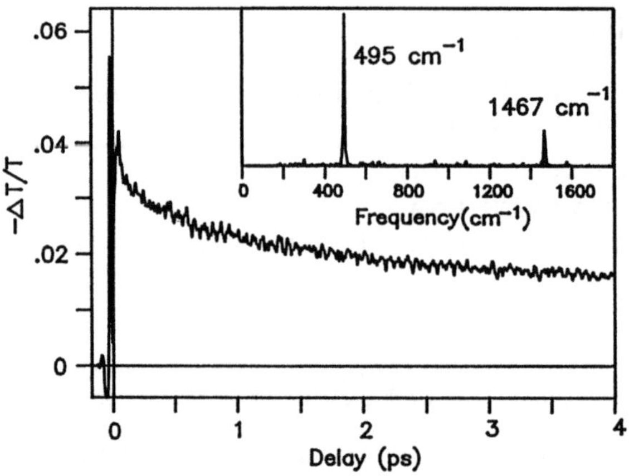

Figure 1
Light-induced negative differential transmittance ($-\Delta T/T$) of a C_{60} thin film detected at 580 nm as a function of time delay between 12 fs pump and probe pulses centered at 620 nm. The Fourier power spectrum of the oscillatory part of the response is shown in the inset.

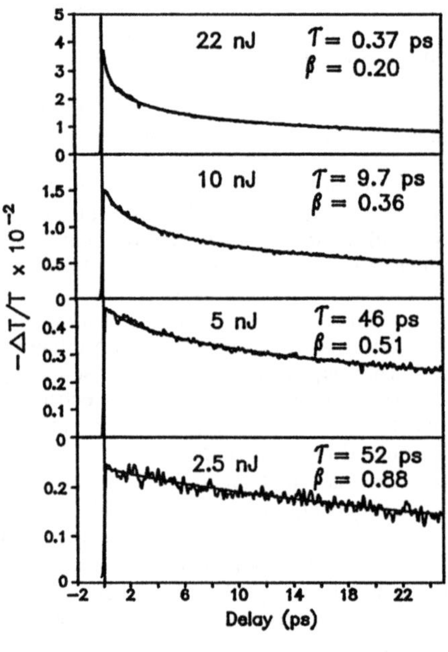

Figure 2 Light-induced negative differential transmittance as a function of pulse energy measured using amplified CPM pulses at 620 nm focused to a spot ~30 μm in diameter. The smooth curves through the data traces represent fits to the stretched exponential decay function with the displayed parameters.

at longer wavelengths. The spectral response, with a fast decay from the higher electı state, is suggestive of fast intramolecular energy relaxation on a time scale of ~100 fs.

References

1. R.L. Fork, C. H. Brito Cruz, P. C. Becker, and C. V. Shank, *Opt. Lett.* **12** (1987) 483.
2. R.W. Schoenlein, J.-Y. Bigot, M.T. Portella, and C.V. Shank, *Appl. Phys. Lett.* **58** (1 801.
3. D.S. Bethune, G. Meijer, W.C. Tang, H.J. Rosen, W.G. Golden, H. Seki, C.A. Brown M.S. de Vries, *Chem. Phys. Lett.* **179** (1991) 181.
4. R.A. Cheville and N.J. Halas, *Phys. Rev.* B **45** (1992) 4548.

Femtosecond Dynamics of Molecular and Cluster Ionization and Fragmentation

T. Baumert, R. Thalweiser, V. Weiß, and G. Gerber

Fakultät für Physik, Universität Freiburg,
W-7800 Freiburg, Fed. Rep. of Germany

Abstract. The real-time dynamics of molecular (Na_2 , Na_3) and cluster Na_n (n=4-21) multiphoton ionization and -fragmentation has been studied in beam experiments applying femtosecond pump-probe techniques in combination with ion and electron spectroscopy. Wave packet motion in the dimer Na_2 reveals two independent multiphoton ionization processes while the higher dimensional motion in the trimer Na_3 reflects the chaotic vibrational motion in this floppy system. The first studies of cluster properties (energy, bandwidth and lifetime of intermediate resonances Na_n^*) with femtosecond laser pulses give a striking illustration of the transition from "molecule-like" excitations to "surface-plasma"-like resonances for increasing cluster sizes. Time-resolved fragmentation of cluster ions Na_n^+ indicate that direct photo-induced fragmentation processes are more important at short times than the statistical unimolecular decay.

1. Introduction

Femtosecond techniques have been advanced in recent years to directly probe molecular motions in real time. The time evolution of the induced coherences has been studied either by measuring the total fluorescence emitted from excited electronic states [1] or by ionizing the molecule and measuring the transient ionization spectrum as a function of the pump-probe time delay [2]. Molecular vibrational wave packet motion has been observed for a number of systems. The spreading and full recurrence of the wave packet is seen in the I_2-fluorescence signal [3] and in the Na_2-ionization signal [4].
Clusters and in particular metal-clusters have been the subject of many experimental and theoretical studies. They form the link between surface chemistry and molecular physics. Clusters show for different sizes very distinct features ranging from molecule-like properties for small particles to solid-like properties seen in large aggregates. Until now there have been no studies of cluster size-dependent phenomena and the dynamics of metal-cluster ionization and fragmentation employing femtosecond laser pulses. Two major issues in cluster physics which are adressed in this contribution are i) energy,width and lifetime of Na_n^* (molecular or surface plasmon-like) resonances and ii) stability and fragmentation dynamics of metal cluster ions Na_n^+ .

2. Experimental

Femtosecond laser pulses of 50-100fs time duration and of 0.1-50 μJ energy in the wavelength range 400nm to 800nm are generated in a home-built colliding-pulse mode-locked (CPM) ring dye laser. The pulses are amplified in a bow-tie amplifier, which is pumped by an excimer laser at 308nm, pulse compressed, selected from a white light continuum, amplified again and compressed before entering the molecular/cluster beam-laser interaction region. A Michelson arrangement was used to delay the probe laser relative to the pump laser. The tunable ultrashort pulses are employed to induce and to probe molecular transitions and cluster plasmon resonances in order to study the time evolution of the different ionization and fragmentation processes. Sodium molecules and clusters are produced by

Springer Series in Chemical Physics, Vol. 55 Ultrafast Phenomena VIII
Editors: J.-L. Martin · A. Migus · G.A. Mourou · A.H. Zewail

a seeded beam expansion in which sodium vapor is co-expanded with the argon carrier gas. This technique provides efficiently cooled clusters. The final molecular and cluster continuum states are analyzed by time-of-flight (TOF) photoelectron and ion spectroscopy, and by measuring the kinetic energy of the formed ionic fragments. Details of the experimental arrangement are given elsewhere [5].

3. Results and Discussion

Femtosecond time-resolved multiphoton ionization of Na_2 reveals unexpected features of the dynamics of the absorption of many photons by a diatomic molecule [2]. The wave packet motions in different bound molecular potentials show that incoherent contributions from direct photoionization of a singly excited electronic state and from excitation and autoionization of a bound doubly excited molecular state determine the time evolution of the multiphoton ionization. Electronic autoionization of Na_2^{**} (n'l',nl) leads to ground state molecular ions Na_2^+ and to ionic and neutral fragments $Na^+ + Na(3s) + e^-$. Figure 1 shows the transient Na_2-photoionization signal as well as the transient Na^+ photofragmentation spectrum. The transient Na_3-photoionization spectrum, shown in Fig.2, is more complex due to the more complicated three-dimensional wave packet motions in the excited B-state and in the electronic ground state . The observed chaotic vibrational motion is a signature of the floppy Na_3 system. Surprisingly enough, spectroscopic data like vibrational eigenfrequencies and pseudorotation energies can still be obtained from the Fourier Transformation of the time domain measurement.
The photoabsorption spectra of larger clusters Na_n with n≥4 are yet not observed in two-

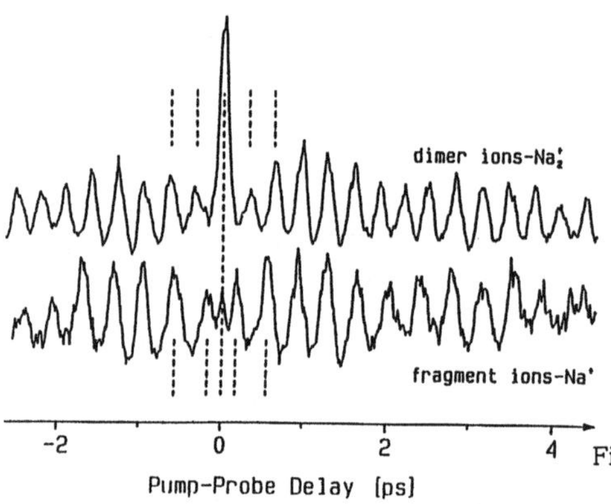

Fig.1 Transient ionization -Na_2^+- and photo-fragmentation -Na^+- spectra

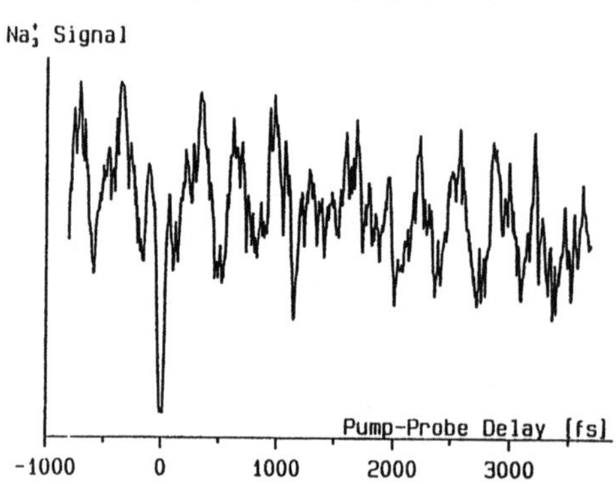

Fig.2 Transient Na_3^+ spectrum

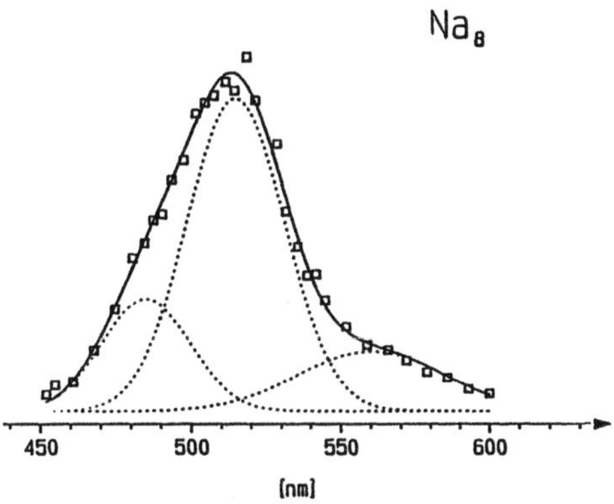

Fig.3 fs two-photon ionization spectrum of Na_8

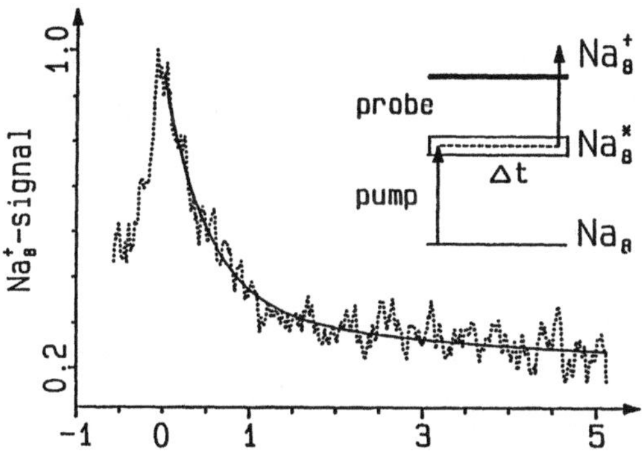

Fig.4 fs time-resolved decay of the Na_8^* resonance at 515nm

photon ionization using nanosecond laser pulses because of the fast decay of the intermediate Na_n^* states. These states are of primary interest in view of the size-dependent transition from molecule-like absorption to collective excitation of valence electrons. We have measured the energy and bandwidth of these resonances for Na_3 to Na_{21} by femtosecond two-photon ionization in the range 450nm to 750nm. The most striking result is obtained for the cluster Na_8 . On the basis of the Mie-Drude model theory predicts for this spherically symmetric metallic cluster a single intense band corresponding to a classical surface-plasma oscillation. Figure 3 shows the two-photon ionization spectrum of Na_8 measured with a femtosecond pulse. The spectrum clearly shows three resonances at 485nm, 515nm and 560nm with different widths. No direct measurements of the intermediate resonance lifetime(s) have been reported so far. Employing femtosecond pump-probe techniques we have measured the decay time(s) for the observed Na_8^* resonances . A pump laser excites the resonance and a time delayed probe laser probes the population by ionizing the excited neutral cluster. The transient Na_8^+ spectrum obtained at the center of the 515nm transition is shown in Fig. 4 . The observed fast decay (τ=0.45ps) of the resonance does not correspond to the measured bandwidth. For Na_8 therefore, the observation of three absorption bands with different widths and different decay dynamics is difficult to interpret within the classical picture of a single "surface-plasmon" like resonance. The transient Na_n^+ spectra obtained for the cluster sizes n=3-21 show a variety of different decay patterns and lifetimes ranging from 0.3ps to several ps. The absorption spectra and the lifetimes measured for the clusters Na_3 to Na_{21} give a striking illustration of the transition from molecular excited states to collective electronic oscillations in larger clusters.

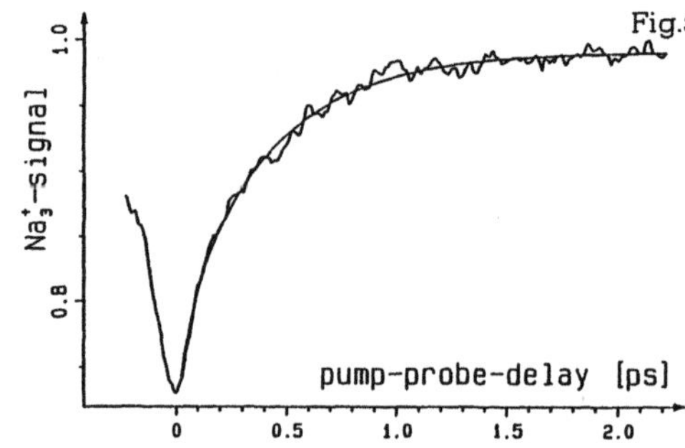

Fig.5 Transient cluster fragmentation spectrum of ejected neutral Na_3 ; τ=0.4ps

Time-resolved fragmentation of small cluster ions Na_n^+ show that ejection of neutral dimer Na_2 and trimer Na_3 photofragments occur on ultrashort time scales of 2.5ps and 0.4ps, respectively. This and the absence of cluster heating reveals that photo-induced fragmentation processes are important at short times rather than the statistical unimolecular decay.

This work has been supported by the Deutsche Forschungsgemeinschaft through the Sonderforschungsbereich 276

4. References

1. L. Khundkar, A. Zewail ; Annu. Rev. Phys. Chem. 41, 15 (1990),and references therein
2. T. Baumert, B. Bühler, R. Thalweiser, G. Gerber ; Phys. Rev. Lett. 67, 3573 (1991)
3. R. M. Bowman, M. Dantus, A. Zewail ; Chem. Phys. Lett. 161, 297 (1989)
4. T. Baumert,V. Engel,C. Röttgerman,W. Strunz,G. Gerber ; Chem.Phys.Lett. 191, 639 (1992)
5. T. Baumert et al. J. Phys. Chem. 95, 8103 (1991)

Dephasing and Beats of Excitonic-Enhanced Transitions of J-Aggregates Measured by Femtosecond Time-Resolved Resonance CARS

V.F. Kamalov[1,2], *R. Inaba*[1,3], *and K. Yoshihara*[1]

[1]Institute for Molecular Science, Myodaiji, Okazaki 444, Japan
[2]Photochemistry Department, Institute of Chemical Physics, Russian Academy of Sciences, Kosygin str. 4, Moscow 119977, Russia
[3]Department of Chemistry, Faculty of Science, The University of Tokyo, Bunkyo-ku, Tokyo 113, Japan

Abstract: J-aggregate of carbocyanine dyes in solutions were studied by femtosecond resonance coherent anti-Stokes Raman scattering (CARS). Subpicosecond dephasing of excitonic-enhanced Raman transitions of J-aggregates is observed at room temperature, which is accompanied by femtosecond beats of Raman transitions.

Introduction: J-aggregates, firstly observed by Jelley [1] and Scheibe [2] in 1930's, are the subject of reviving interest due to the high values of optical susceptibilities of excitonic transition, originated from delocalization of electronic excitation in this one-dimentional molecular aggregate [3]. Resonant enhancement of third-order susceptibility $\chi^{(3)}$ of J-aggregates of pseudoisocyanine solutions was observed and the peak value of excitonic resonance was estimated to be up to 10^{-5} cm^3/erg [4].

Experimental: BIC J-aggregates formed in water/ethylene glycol mixture (9/1 volume ratio) had sharp (300 cm^{-1})absorption band with the maximum at 595 nm. The intensity of CARS signal at the anti-Stokes frequency $\omega_a=2\omega_1-\omega_2$ was measured as a function of delay time Δt between femtosecond pulses (85 fs FWHM) at 592 nm, in the resonance with the excitonic transition. Picosecond pulses (7 ps FWHM) of second dye laser were tunable in 610-690 nm region. Two beams ω_1 and ω_2, with the peak power 2 kW and 0.1 kW, respectively, and probe beam ω_1 with 2 kW, were focused by lens (F=4.0 cm) in jet stream of J-aggregate solution with a thickness of 300 µm. CARS beams were linearly polarized, all polarization vectors were parallel to one another, which corresponded to measurement of CARS signal determined by χ_{1111} component of $\chi^{(3)}(\omega_a,\omega_1,\omega_1-\omega_2)$ tensor.

Results and Discussion: The CARS signal intensity plotted against Δt at the fixed frequency difference 2245 cm^{-1} is shown in fig.1. The CARS signal is determined by electronic susceptibility of J-aggregate at this frequency difference, no Raman contributions should be expected at this frequency. CARS signal at this $\Delta\omega$ coincides practically with the response function of spectrometer, showing that dephasing time of excitonic transition of J-aggregate is shorter than 100 fs.

CARS intensity at frequency difference 1000 cm^{-1} presented in Fig.1 shows a fast response at short Δt followed by a longer monoexponential decay with τ = 0.43 ps. Two components in CARS decay profile can be assigned to excitonic and Raman susceptibilities, correspondingly. The absolute value of signal increased significantly when ω_2 tuned closer to

Editors: J.-L. Martin · A. Migus · G.A. Mourou · A.H. Zewail

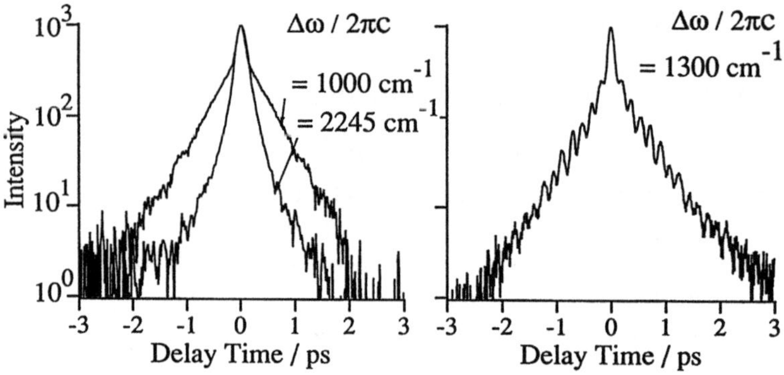

Figure 1. Time-resolved CARS signal of J-aggregate at $\Delta\omega/2\pi c$=2245 cm^{-1} (determined by excitonic transition), 1000 cm^{-1} (excitonic and single Raman mode), 1300 cm^{-1} (excitonic and two Raman modes).

excitonic transition (from 685 nm for 2245 cm^{-1} to 630 nm for 1000 cm^{-1}) due to resonant enhancement.

Assignment of subpicosecond component to Raman transition is also based on the CARS decay measurements, which gives different behavior of decay at different ω_2. CARS decay measured at frequency difference 1300 cm^{-1} was shown in fig.1. Here CARS signal decay accompanying peaks with period of about 180 fs. This 180 fs period corresponds to high frequency, about 5.5 THz (185 cm^{-1}), beats. This beats are due to coherent excitation of two different vibrational modes of J-aggregate by broad spectrum of femtosecond pulse (85 fs duration corresponds 400 cm^{-1}). The decay time at this $\Delta\omega$ (1300 cm^{-1}) is 0.44 ps. We cannot determine exact values of frequencies of vibrational modes beatting (which is possible in principle by Fourier transformation).

Conclusion: Dephasing of excitonic transition of BIC J-aggregates is estimated as shorter than 100 fs at room temperature. Excitonically-enhanced Raman transitions of J-aggregate are observed, with the dephasing time on the subpicosecond scale. Ultrahigh frequency beats of J-aggregate vibrational modes are observed.

References:

1. E.E.Jelley, Nature **138**, 1009 (1936).
2. G.Schreibe, Angew.Chem. **49**, 563 (1936).
3. S.de Boer, D.A.Wiersma, Chem.Phys.Lett. **165**, 45 (1990).
4. V.L.Bogdanov et.al., JETP Lett. **53**, 105 (1991).

Excited States Dynamics of the Special Pair Dimer

P.O.J. Scherer and S.F. Fischer

Physikdepartment T38, Technische Universität München, W-8046 Garching, Fed. Rep. of Germany

Abstract. Quantum calculations of the MRINDO/CI type are performed on a representative model of the reaction center including four bacteriochlorophylls, two pheophytins and some protein residues. A low lying Rydberg state of the dimer is found which has a large permanent dipole and couples only weakly to P*.

Experimental results [1] show the participation of an intermediate state for the transfer of the electron from the special pair dimer P to the pheophytin H_l. There are different models which assume either a second electronic state of the dimer or a state which involves the dimer and the accessory monomer B_l and is spatially extended. In this paper we present a new type of dimer excitation which extends over the whole reaction center.

The MRINDO program treats valence orbitals plus outer atomic orbitals (2s,2p for hydrogen, 3s,3p for second row elements and 4s,4p for magnesium) in a semiempirical way. A detailed description is found in [2]. The valence orbitals are treated in the INDO model with Zerner's parametrization, the outer orbitals in the CNDO approach. The resonance integrals involving outer orbitals were reduced by a factor 0.05 as compared to the valence orbital parameters. Excited states were calculated from the lowest 100 single excited configurations (CI). For the hexamer we included 1416 valence plus 2124 outer orbitals.

The lowest Rydberg orbital of a single bacteriochlorophyll is at –2.2 eV , about 1 eV above the lowest unoccupied valence orbital. The energy is lowered with increasing size of the system. For the dimer it is –2.6 eV and if we include also the accessory monomers and the pheophytines the energy is dropped to –3eV and comes close to the lowest unoccupied valence orbitals. Fig.1 shows the distribution of electron density over the reaction center.

The CI calculation gives the lower dimer band at 1.38 eV and the Q_y transitions of the accessory monomers and pheophytines between 1.6 and 1.8 eV. The lowest transition into Rydberg orbitals is found at 1.7 eV. It carries no oscillator strength,

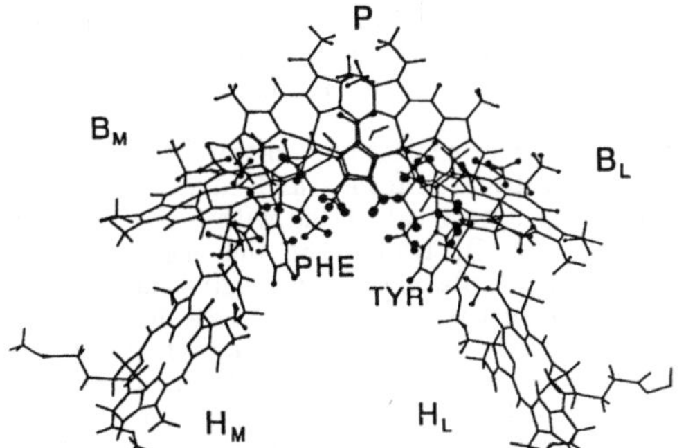

HM:0.01 BM:0.04 PL:0.39 PM:0.35
BL:0.09 HL:0.01 PHE:0.04 TYR:0.07

Fig.1 electron densities of the lowest Rydberg orbital

Springer Series in Chemical Physics, Vol. 55 Ultrafast Phenomena VIII
Editors: J.-L. Martin · A. Migus · G.A. Mourou · A.H. Zewail

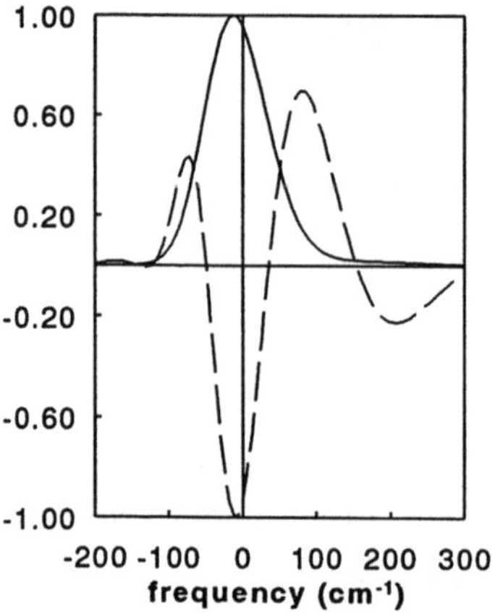

Fig.2 absorption - Stark effect ...

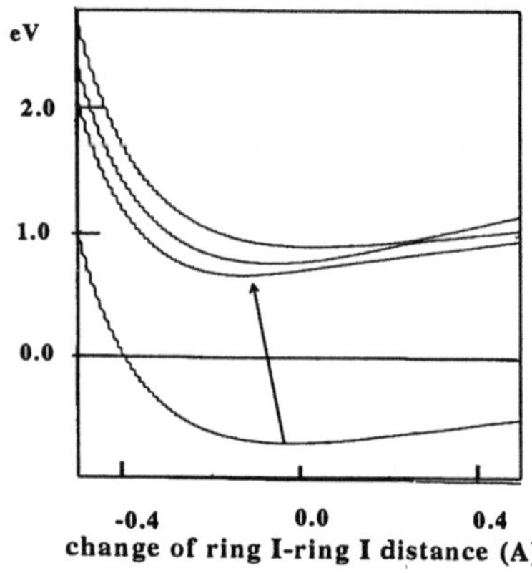

Fig. 3 energies of the lowest dimer states

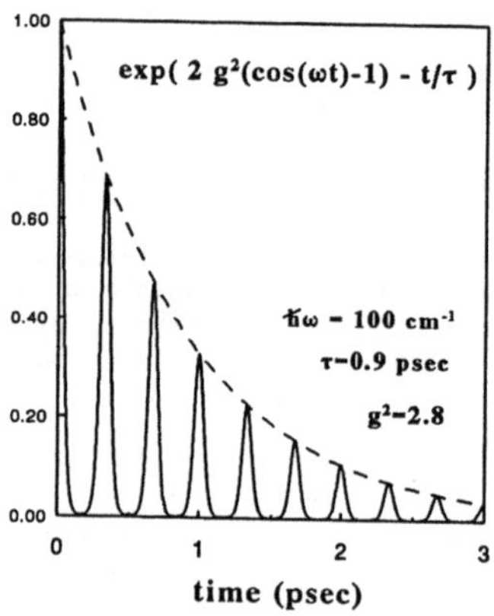

Fig. 4 stimulated emission

couples only weakly to P* but has a rather large change of permanent dipole of ≈ 20 Debyes along the symmetry axis.

In principle weak coupling of such a state to the dimer band could explain the Stark effect (a simulation is shown in fig.2). It could also induce the charge separation process. However, the biexponential decay characteristics observed by Holzapfel et al [1] would have to be reinterprated such that the faster rate must preceed the slower in the order $P^* \rightarrow P_r$ followed by $P_r \rightarrow P^+H_1^-$.

We also calculated the dependence of the dimer excitations on the distance of the two halves (fig. 3). The groundstate energy was estimated from standard van der Waals and Coulombic interactions. The excitation energies were calculated within the extended exciton model [3] where the coupling between the Q_y transitions and the intradimer charge transfer states are taken from the INDO calculations. The results suggest that upon excitation of the dimer to P* the distance of the two dimer halves decreases by about 0.05 A. Such a coupling of the dimer excitation to the intradimer vibration would be consistent with the observed Stokes shift $g^2\hbar\omega \approx 280cm^{-1}$ and also with the oscillatory behaviour [4] of the stimulated emission (fig. 4 shows a simulation for coherent broadband excitation without taking vibrational relaxation into account).

References:

1 Holzapfel,W., Finkele U.,Kaiser W., Oesterhelt D., Scheer H., Stilz H.U. and Zinth W., Proc.Natl.Acad.Sci. USA (1990) 87, 5168–5172

2 Prasad R.S. and Rai B.N., Theor. Chim. Acta (1990) 77, 343

3 P.O.J.Scherer and S.F.Fischer in: chlorophylls, ed. H.Scheer,CRC press 1991,1079–1093

4 Vos M.H., Lambry J.C., Robles St.J., Youvan D.C., Breton J. and Martin J.L., Proc.Natl.Acad.Sci.USA(1991) 88,8885–8889

Creation of an Anti-Wavepacket in a Rydberg Atom

L.D. Noordam[1,2], *H. Stapelfeldt*[2,3], *D.I. Duncan*[2], *and T.F. Gallagher*[2]

[1]FOM-Institute AMOLF, Kruislaan 407,
NL-1098 SJ Amsterdam, The Netherlands
[2]University of Virginia, Charlottesville, VA 22901, USA
[3]Institute of Physics, University of Aarhus, DK-8000 Aarhus C, Denmark

Abstract. Starting with Ba atoms in a single Rydberg state we have observed the population of adjacent Rydberg states when the atoms are exposed to an intense 1.5-ps laser pulse. The initial Rydberg state population is redistributed by a Raman two-photon process.This superposition of states describes a wavefunction with a hole. The hole or anti wavepacket oscillates in the atomic potential. During the short pulse the hole is near the core and further ionisation of the wave function is precluded. The atom is stable against photoionisation by the intense short pulse.

1. Introduction

At the 1990 UFP conference the observation of several kinds of wavepackets was reported [1]. All of these wavepackets have the common characteristic that the wavefunction is localised due to a coherent superposition of eigenstates. The localised 'particle' evolves in course of time. The oscillation time of the wavepacket is determined by the inverse spacing of the states out of which the wavepacket is formed. We reported on the localisation of an atomic electron wavefunction by a coherent superposition of Rydberg states [1a]. The observation of exciton wavepackets was also reported [1b], as well as the observations of optic phonons in the time domain [1c]. At an earlier conference the CalTech group reported the observation of the oscillation of vibrational wavepackets in diatomic molecules [2]. Recently, we have shown that a coherent superposition can also lead to a hole in a wavefunction [3]. This anti wavepacket fills the stationary state space where a small fraction of the wavefunction is absent. This hole or anti wavepacket shows similar time behaviour as the wavepacket. We focus here on the creation of an anti wavepacket in a Rydberg atom. However, the concept of an anti wavepacket is as general as the wavepacket concept.

2. Experiment

We exposed barium atoms in a single Rydberg state to an intense laser pulse of duration short compared to the roundtrip time of the electron in the Rydberg state. During the short pulse the wavefunction can be seen as being frozen. Ionisation of the extended Rydberg wavefunction takes place predominantly near the ionic core.The laser pulse burns a hole at the ionic core in the wavefunction of the initial Rydberg state. The resulting wavefunction can no longer be described as a single stationary state, but is instead a coherent superposition of many states. This superposition is formed by a Raman process via the continuum. Both the frequencies of the absorbed photon and stimulated emitted photon fit within the bandwidth. In other words, the difference frequency of the Raman process fits within the bandwidth of the short laser pulse. As a manifestation of the creation of an anti wavepacket, we observed the population of adjacent Rydberg states. In the actual experiment the initial barium Rydberg state was prepared by two ns-lasers (6s6s -> 5d6p -> 6snd, see also figure 1). The small bandwidth of the ns-lasers ensures that only one Rydberg state is excited. This Rydberg state was then exposed to a 1.5 ps pulse with a peak intensity of 4×10^{12} W/cm^2. The population of the individual Rydberg states after the intense laser pulse was probed by pulsed electric field ionisation. In figure 2 we present the experimental observation of the transfer of population from the initial Rydberg state to neighbouring states by the intense laser pulse. In principle the population of adjacent Rydberg sates can also occur by a two photon V-transition via a bound state. However

Springer Series in Chemical Physics, Vol. 55 Ultrafast Phenomena VIII
Editors: J.-L. Martin · A. Migus · G.A. Mourou · A.H. Zewail

Fig. 1

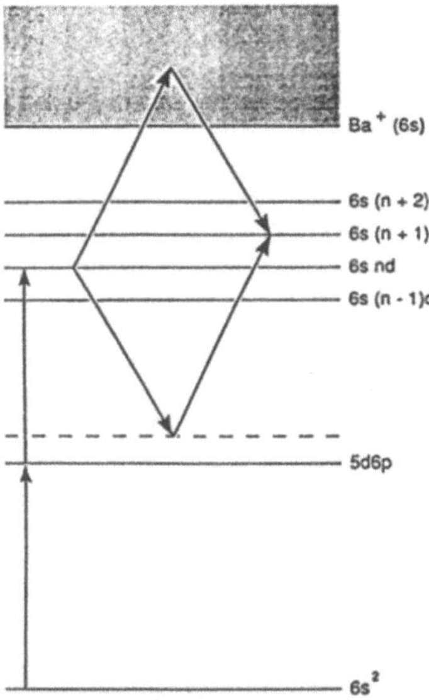

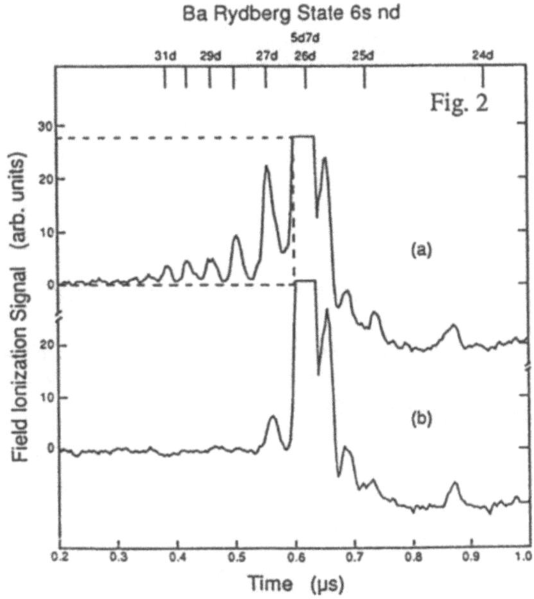

Fig. 1 Schematic energy diagram of the two ways of populating adjacent states starting from one Rydberg state. The Λ-path corresponds to coupling via the continuum. As shown, the initial Rydberg state is created by two step resonant ns-laser excitation from the ground state via the 5d6p state prior to exposure to the intense ps laser.

Fig. 2. Field-ionisation spectrum of Ba Rydberg atoms when the initial state is the 26d Rydberg state. Trace (a) is with the ps laser while trace (b) is without the ps laser. The peaks to the left of the saturated 26d signal correspond to Raman excitation via the continuum of the 27d-31d states.

we tuned the frequency of the intense laser over the (26d -> 5d6p) bound-bound transition and found no enhancement of the population redistribution. Therefore we conclude that the redistribution takes place via the continuum.

3. Anti wavepacket

We also calculated the time evolution of the wavefunction created by the redistribution of the population. As the initial state we chose the 42s hydrogen Rydberg state. This state has a classical orbiting time of 11 ps. Starting from the formalism given by Parker and Stroud [4] we calculated the amplitudes, a_n, of the adjacent Rydberg s-states. Population of d-states is rather unlikely. Using the calculated amplitudes we constructed the time-dependent radial Rydberg wavefunction:

$$\Psi(r,t) = \Sigma_n a_n \exp(i\omega_n t) R_n(r) \quad ,$$

where $R_n(r)$ is the radial part of the Rydberg state wavefunction. In figure 3 we plot the difference between the wavefunction, actually $\delta(r\Psi)^2$, before the pulse (a pure 42s-state) and after the 2-ps pulse (a superposition of states around 42s). The pulse ionises some of the wavefunction close to the core, hence the large difference between the wavefunctions near the origin.This anti wavepacket oscillates radially back and forth between the core and the outer turning point of the potential at 3500 atomic units. In figure 3b it is demonstrated that the anti wavepacket is near the outer turning point after half an oscillation (5.5 ps).

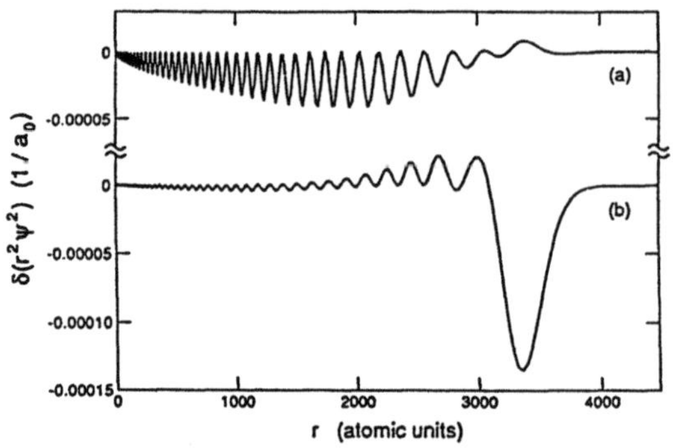

Fig. 3. Calculation of the anti wavepacket immediately after the intense pulse (a), and after 5.5 ps (b). The difference between the wavefunction with and without irradiation by the intense pulse is plotted: $\delta(r\Psi)^2$.

4. Stabilization against photoionisation

These Rydberg anti wavepackets can also be seen in the perspective of stabilization against photoionisation. Since the pulse duration is very short compared to the characteristic evolution time of the wavefunction, the wavefunction can be seen as being frozen during the pulse. Once the intense short pulse has depleted the core, region further ionisation is precluded since the major fraction of the wavefunction is far away from the core and will not penetrate the core during the short pulse. The atoms is stable against ionisation [4] in a transient way. A more intense short pulse will hardly enhance the ionisation, whereas a longer pulse will cause further ionisation of the Rydberg wavefunction. In the frequency domain this stabilization can be seen as follows. The initial Rydberg state and the adjacent populated states form together with the continuum a Λ-system [5] in which the continuum is not populated (a so-called black resonance).

References

[1] Ultrafast Phenomena VII, New York, Springer-Verlag (1990), [a] L. D. Noordam et al p. 139, [b] K. Leo et al p 250, and T. Tokizaki et al p 253, [c] Weiner et al p 331, and Cho. et al p 337.

[2] Ultrafast Phenomena VI, Heidelberg, Springer-Verlag (1988), A. H. Zewail p. 500.

[3] L. D. Noordam, H. Stapelfeldt, D. I. Duncan and T. F. Gallagher, Phys. Rev. Lett. **68**, 1496 (1992).

[4] J. Parker and C. R. Stroud, Jr. Phys. Rev. A **41**, 1602 (1990).

[5] P. M. Radmore and P. L. Knight, Phys. Lett. **102a**, 180 (1984)

Squeezing of the Molecular Vibrations by Femtosecond Laser Pulses

A.V. Vinogradov[1] *and J. Janszky*[2]

[1]P.N. Lebedev Physics Institute, 117924 Moscow, Russia
[2]Research Laboratory for Crystal Physics, P.O. Box 132, H-1502 Budapest, Hungary

In the last few years a considerable interest was manifested to the stimulation of Franck-Condon transitions in molecules by ultrashort (femtosecond-range) laser pulses [1]. Since the spectrum of such a pulse definitely overlaps several vibrational levels, the molecule is put into a nonstationary vibrational state by the electron transition. This state represents a wave packet: a coherent superposition of vibrational eigenstates. In a number of papers [2-6] we studied how the properties of this wave packet relate to the characteristics of the laser pulse, i.e. to its duration and phase modulation, and how they can be seen in pump-probe spectroscopic experiments. It was demonstrated that when the spectral width of the pulse is smaller than the width of the absorption band, but much larger than the frequency of vibrations, the vibrational ground state of the molecule transforms during the excitation into a squeezed state with reduced value of the quantum uncertainty of the coordinate. All our previuous results were obtained for Franck-Condon transition resulting exlusively from a spatial shift of harmonic nuclear potentials.

Now we want to present new results for a more general model of Franck-Condon transition, which also includes the change of the vibrational frequency. In this case the vibrational wave function of electronically-excited molecule can be written in the form:

$$|v,t\rangle=\int_{-\infty}^{+\infty}dt_1\exp\left\{i\Delta t_1-\frac{u^2}{2}t_1^2\right\}D(t_1-t)S(t_1-t)\,|0\rangle. \tag{1}$$

Here Δ is the detuning of the electronical resonance; u, the reciprocal duration of the light pulse of Gaussian form; D and S are the well known displacemen and squeezing operators:

$$D=\exp\left\{-\kappa(b^+-b)\right\},\quad S=\exp\left\{\frac{1}{2}r(b^2-b^{+2})\right\}.$$

Squeezing is coming from the frequency change during Franck-Condon transition,

$r=\frac{1}{2}\ln\frac{\omega_i}{\omega_e}$, where ω_i, ω_e, are the vibrational frequencies in the initial and excited electronic states of a molecule. κ and is equal to the ratio of spatial shift of the nuclear potentials to the amplitude of zero vibration in the excited state. In the limit of instanteneous excitation, when $u\to\infty$, the wave function $|v,t\rangle\approx D(-t)S(-t)|0\rangle$ already possess the squeezing property, which, for $\omega_i<\omega_e$ at time t=o (just after the excitation) corresponds to the enhancement of the coordinate uncertainty with respect to its value in the ground vibrational state of the electronically excited molecule. We have studied the properties of the wave function (1) for large, but finite values of u. Using mathematical methods, developed in [3-6], we found that for $u \gg \omega_e$ and $\kappa \gg 1$ the

Springer Series in Chemical Physics, Vol. 55 Ultrafast Phenomena VIII
Editors: J.-L. Martin · A. Migus · G.A. Mourou · A.H. Zewail

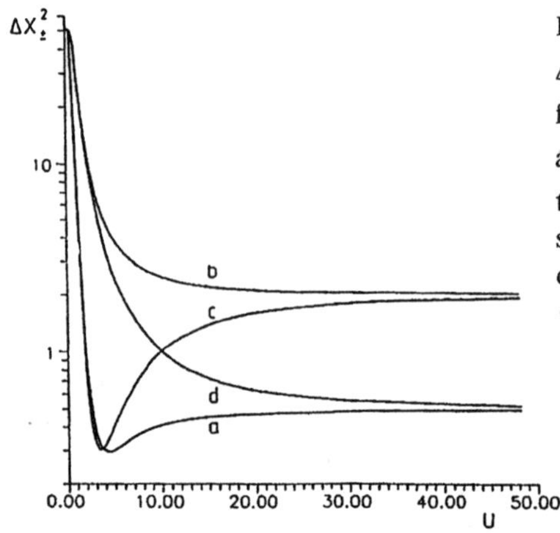

Fig.1 The uncertainties of Δx_+^2 and Δx_-^2 as a function of the ratio u $/\omega_e$ for $\kappa = 5$. For curves a, b $2\omega_e = \omega_i$, and c, d : $\omega_e = 2\omega_i$. We can see that the squeezing may have opposite sign for the same frequency change depending on the pulse duration (curve C).

wave function (1) corresponds to the minimal uncertainty state, in which the uncertainties of the conjugate variables $x_+=(b^+ +b)$, $x_- = i(b^+ - b)$ obey the equations

$$\Delta x_+^2 = m\frac{u^2}{2m\kappa^2\omega_e^2+u^2}, \quad m=\frac{\omega_e}{\omega_i}, \quad \Delta x_+\Delta x_- = 1.$$

Let us note that $(2\ m\ \kappa^2\omega_e^2)^{1/2}$ has the physical meaning of the absorption spectrum linewidth. So, in the case $u^2 >> 2\ m\ \kappa^2\omega_e^2$ the limit of instanteneous excitation is realized and $\Delta x_+^2 \approx m > 1$.When $u^2 >> 2\ m\ \kappa^2\omega_e^2$ a molecule appears in the squeezed state of the opposite type with the reduced uncertainty of the coordinate $\Delta x_+^2 = \frac{u^2}{2\kappa^2\omega_e^2} < 1$. It is surprizing that this reduced value of Δx_+^2 does not depend on the vibrational frequency in the initial state. This indicates that the squeezing mechanism proposed in [3-6], which we explain as quantum interference effect due to the nuclei motion during the excitation, is of general nature and does not depend on the explicit form of the initial vibrational state. This basic statement is also proved by numerical calculations shown in Fig.1, and by consideration of the excitation of a molecule not from the ground vibrational state, but from the thermal equilibrium state [6].

References

1. M.Grueble, A.H.Zewail. Phys. Today, 24 (May 1990).
2. An.V.Vinogradov, J.Janszky.Sov.Phys.Solid State, 27, 598 (1985).
3. An.V.Vinogradov, J.Janszky. Kratk.Soobshch.Fiz. N 4, 45, (1989).
4. J.Janszky, An.V.Vinogradov, Phys.Rev.Lett. 64, 2771 (1990).
5. J.Janszky et al. T. Kobayashi, An.V. Vinogradov. Opt.Commun. 76, 30 (1990).
6. An.V.Vinogradov, J.Janszky. Sov.Phys. JETP, 73, 211 (1991).

Part III

Spectroscopy and Advances in Measurements

Spectroscopic Applications of Phase-Locked Femtosecond Pulses

N.F. Scherer[1], M. Cho[1], L.D. Ziegler[2], M. Du[1], A. Matro[1], J. Cina[1], and G.R. Fleming[1]

[1]Department of Chemistry and
The James Franck Institute, The University of Chicago,
5735 S. Ellis Avenue, Chicago, IL 60637, USA
[2]Department of Chemistry, Boston University, Boston, MA 02215, USA

Linear and nonlinear spectroscopies with phase locked pulses are described and the relationship to ultrafast solvation dynamics discussed.

1. Introduction

The ability to generate pulse sequences whose members are short with respect to typical molecular vibrational periods and whose relative phase can be controlled opens up new possibilities for the study and manipulation of molecules. Optimal control theory [1] has been applied to obtain pulse sequences designed to break specific bonds in molecules. In general it is found that both phase and amplitude must be controlled to produce the desired degree of control. In conventional ultrafast spectroscopy the precision of the delay time setting is much lower than that corresponding to a few tenths of the laser wavelength. Consequently experiments, particularly those using repetitive scans of the delay, average over the optical phase and do not directly observe quantum interference effects. The "quantum beats" observed in many such experiments arise from the ensemble so that, for example, a correctly timed second excitation pulse will excite new members of the ensemble such that they oscillate in phase with the initially excited molecules. By contrast, if the delay stage can be moved in, for example, integer multiples of the laser carrier frequency it becomes possible to observe contributions of the kind shown in diagram (III) of Figure 1. Here the two interactions with the excitation fields occur with separate pulses and the diagram is drawn to show that fluorescence is detected. Diagrams (I) and (II), on the other hand, represent contributions resulting from two interactions with the same excitation pulse and preclude observation of interference effects resulting from the relative phases of E_1 and E_2.

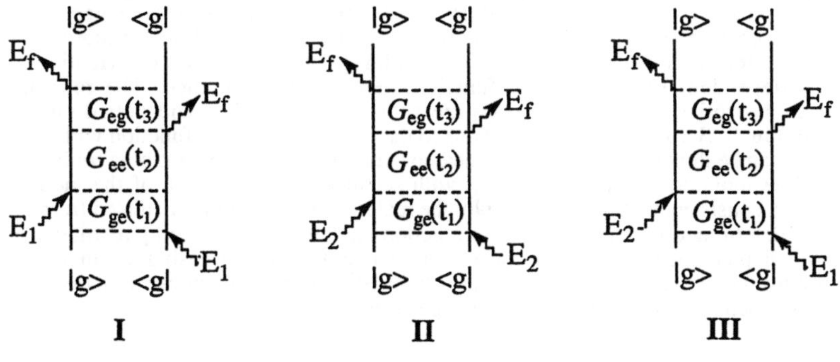

Fig.1 Double-sided Feynman diagrams for the fluorescence. One out of six Liouville pathways for the spontaneous light emission is shown here. The initial ground state density matrix interacts with the first field (E_1 in I and III, and E_2 in II) resulting in an optical coherence state. The second interaction with the field creates the population in the excited state. Finally, two interactions with spontaneous outgoing fields produce a fluorescence photon. $G_{ab}(t_i)$ represents the propagation of |a><b| for t_i.

Editors: J.-L. Martin · A. Migus · G.A. Mourou · A.H. Zewail

Note that the system is in a coherence during the t_1 period. If the total electric field is written as

$$E(r,t) = E_1(t+\tau_1) \exp(ik_1r - i\omega_1 t) + E_2(t) \exp(ik_1r - i\omega_1 t - i\phi) \quad (1)$$

then interferences can be observed if the relative phase ϕ is controlled, and the contributions from Diagrams (I) and (II) are suppressed. For example, in a two level system if $\phi=0$ for some frequency in the pulse, Ω_L, then we expect a beat frequency at $\omega_{ab} - \Omega_L$, where $\omega_{ab} \equiv (E_a - E_b/h)$. In the case of a vibronic system we will observe recurrences due to the motion of vibrational wavepackets and constructive ($\phi=0$) or destructive ($\phi=\pi$) interference can be observed between vibronic coherences prepared by the first pulse and those prepared by the second pulse [2-4]. However, the possibilities are by no means limited to in-phase or out-of-phase fluorescence interferograms and in this paper, after describing the experimental method, we will discuss a number of other applications of the phase locking technique. In particular, we will describe how $\phi = 0$, $\phi = \pi/2$ combinations can be used to separately obtain the real and imaginary parts of molecular response functions and susceptibilities.

2. Experimental Method

Phase locking of pulses that are separated by many times their duration is achieved by monitoring their electric field interference after the pulses are "stretched" in a monochromator. It is straightforward to observe this interference for times at least thre orders of magnitude greater than the pulse duration (25-50 fs). A lock-in amplifier is provided with a reference from the sinusoidal modulation applied to a piezo electric transducer holding the end mirror of one arm of a Michelson interferometer. The amplitude of the mirror displacement is about $\lambda/10$. The interference signal is detected by the lock-in and the output is used as an error signal to adjust the relative lengths of the two arms of the interferometer. The relative phase angle of the common locked frequency component of the two pulses is fixed at $2n\pi$ $(2n+1)\pi$ or $(2n\pm 1/2)\pi$ for in-phase, out-of-phase, and in-quadrature pulses respectively. Here n is an integer.

3. Applications to Linear Spectroscopy

Use of the phase locking capability to separately determine both the real and imaginary contributions to the linear response has been illustrated in experiments on iodine vapor. By combining the in-phase and in-quadrature fluorescence interferograms of Fig. 2 according to the prescription of reference 4 the real (dispersive) and imaginary (absorptive) contributions to the frequency dependent linear response function were directly obtained. The resulting dispersive and absorptive response functions are shown as the dashed and dotted curves, respectively, in Fig. 3. Because both contributions to the complex response function within the laser bandwidth can be directly measured, the linear experiment with phase control is equivalent to an optical free induction decay experiment with sensitivity to the phase of the induced polarization. As a consistency check, we also plot the real response function calculated from the experimentally determined imaginary response function via the Kramers-Kronig relation. The slight vertical displacement of the Kramers-Kronig result with increasing frequency is accounted for by the sensitivity of the directly obtained dispersive response function to a pronounced asymmetry of the experimental pulse spectrum.

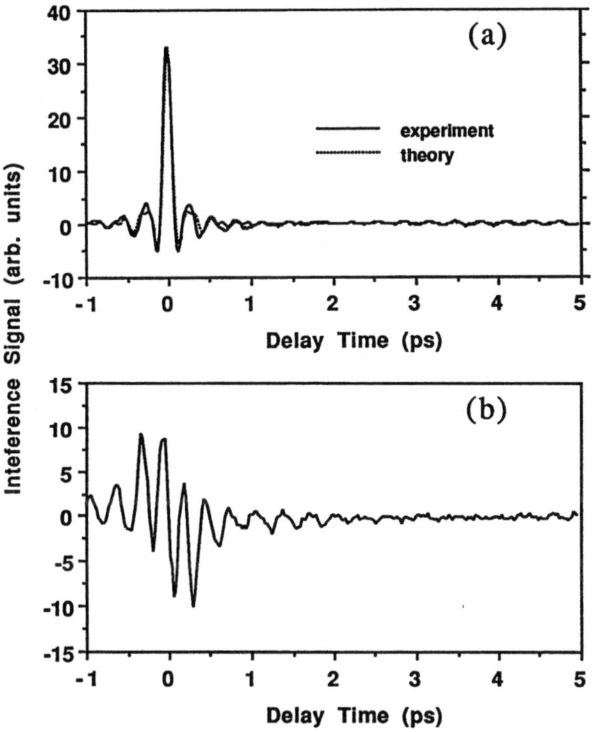

Fig. 2 (a) Experimental interferogram compared to a calculated interferogram locked in phase at 607.15 nm. (b) In-quadrature experimental interferogram phase-locked at 607.06 nm.

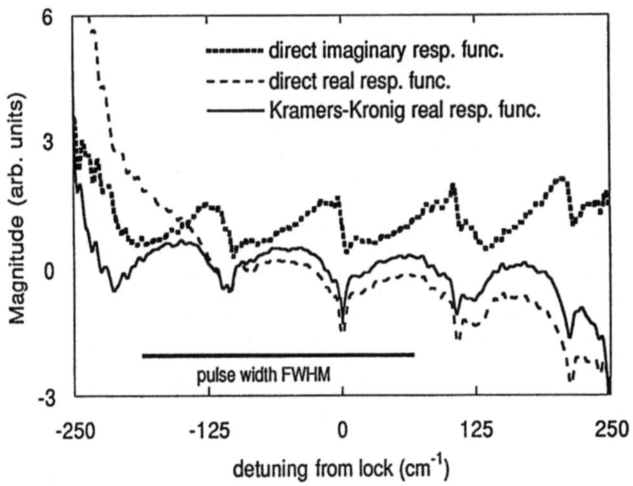

Fig. 3 Real and imaginary parts of the experimentally determined linear response function with the real part of the linear response function obtained via the Kramers-Kronig transformation of the experimentally determined imaginary part. The horizontal bar shows the FWHM of the laser pulse in relation to the locked frequency.

4. Applications to Nonlinear Spectroscopy

The fluorescence studies described above are a form of linear spectroscopy. In this case the molecular dynamics is often obscured by dephasing resulting from inhomogeneity: the rapid initial drop in the fluorescence interferogram of Figure 2a resulting from the rotational energy level distribution is an example of such an effect. In nonlinear spectroscopy experimental arrangements, such as the photon-echo method, can be devised to defeat the inhomogeneous broadening and reveal the underlying dynamics. Nonlinear optical measurements have been described by Mukamel [5] in terms of a nonlinear response function. In general the calculation of a specific nonlinear measurement requires only a few of the individual Liouville space pathways which are complex (i.e. have real and imaginary parts) rather than the entire response function which is real. By analogy with the previous section it may be imagined that the real and imaginary contributions to the response function can be measured separately using phase-locking techniques. This is indeed the case [6] and by controlling the phase factor between the two pulses to be in-phase or in-quadrative one obtains the imaginary and real parts of the response functions. In the limit of large inhomogeneous broadening two response functions, R_2 and R_3 (and their complex conjugates), contribute to the response function. Figure 4 shows the pulse sequence for a new technique, heterodyne detected stimulated photon echo (HSPE) spectroscopy.

The HSPE signal in the impulsive and large inhomogeneous broadening limits with $\omega_1 = \omega_2 = \omega_{LO}$ is

$$\begin{aligned} S_{HSPE}(\tau',\tau,\psi) &= 2\mathrm{Re}\,\{[R_2(\tau',\tau,\tau') + R_3(\tau',\tau,\tau')]\exp(i\chi)\} \\ &= 2\mathrm{Re}[R_2(\tau',\tau,\tau') + R_3(\tau',\tau,\tau')] \text{ for } \chi = 0 \\ &\text{or } -2\mathrm{Im}[R_2(\tau',\tau,\tau') + R_3(\tau',\tau,\tau')] \text{ for } \chi = \pi/2 \end{aligned} \tag{2}$$

Thus, a complete record of the complex nonlinear response functions can be obtained. We can express all the other photon echoes in terms of the HSPE signal [6], for example, the stimulated photon echo signal is given by

$$S_{SPE}(\tau',\tau) = [S_{HSPE}(\tau',\tau,\chi = 0)]^2 + [S_{HSPE}(\tau',\tau,\chi = \pi/2)]^2 \tag{3}$$

Related expressions for two pulse echoes and accumulated photon echoes are given in [6]. Similarly pump-probe measurements, with and without phase locking can be related to the HSPE signal.

Recently two groups have calculated photon echo signals from (classical) dynamics simulations [7-8]. In the work of Fried et al. [7] the simulated response function is used to calculate both two and three-pulse photon echoes and pump-probe signals. However, as noted above anyone of these signals, if measured experimentally would not suffice to completely determine the response function. The HSPE experiment provides this capability. Both simulations and experiment point to the major importance of inertial solvent dynamics in determining the short time behavior of the echo signal. The role of quantum effects on dynamics involving high frequency motions and on the propagation of the solvent dynamics in systems in which electronic coherence is present is poorly understood. Thus experimental measurements of the response function will be valuable. The same short-time solvent motions mentioned above also appear in pump-probe signals, time resolved Stokes shift measurements and in the Optical Kerr effect of the pure solvent. We have recently been exploring a description in terms of the underlying molecular motions. In this approach the solvent is represented by a realistic spectral density for the intermolecular modes and discrete frequencies for the intramolecular modes. This approach has enabled us to derive a simple approximate relation between Optical Kerr studies and fluorescence Stokes shift measure-ments [9]. On a more formal level it is possible to relate the coupling constants which enter into the total Hamiltonian to the pair distribution function for the liquid.

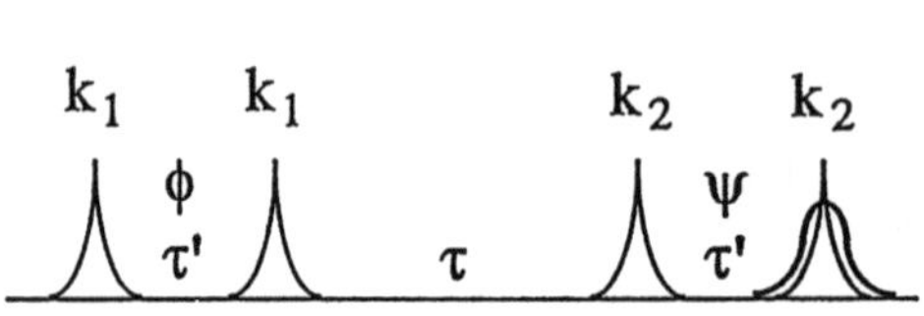

Fig. 4. Pulse sequences for the heterodyne-detected stimulated photon echo (HSPE) measurement. The phase relations between the two consecutive pulses are adjusted as ϕ and ψ. χ in the context is defined as $\chi = \psi - \phi$. Thick solid curve denotes the echo signal.

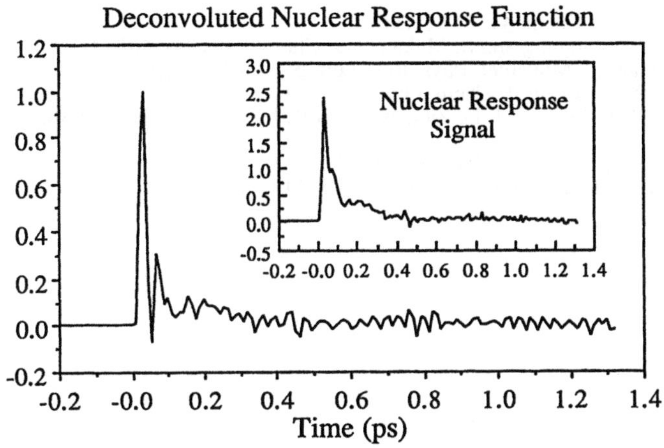

Fig. 5 The nuclear response function is drawn after deconvoluting the laser intensity correlation function from the nuclear response signal shown in the inset. The electronic coherent artifact arising from the higher-order nonlinearity of the material is eliminated by using a Fourier transform technique.

A primary conclusion from this approach is that the short time dynamics in polar liquids is dominated by underdamped motion of many oscillators: the broad distribution of frequencies gives rise to the rapid decay of the solvation correlation function, or the rise and decay of the Optical Kerr signal. The inset in Figure 5 shows experimental data for the Optical Kerr response in pure water. The curve in Figure 5 shows the nuclear response. Note the ultrafast (~25 fs) decay and the obvious oscillation at 50 fs.

In order to elucidate the origin of the ultrafast initial decay and the oscillatory features we have modeled the Kerr signal using a spectral density for the hindered rotations (librations) of water obtained via computer simulation [10]. The librational spectral density has a mean frequency of about 600 cm^{-1}, but extends from 100-1100 cm^{-1} with significant amplitude. The oscillators are modeled as Langevin oscillators with a damping constant which is small compared to the mean frequency. Both the rapid initial drop and the oscillations are present. They result respectively from the ensemble dephasing of the whole distribution and from a frequency representing the mean of the distribution. These results are completely consistent with the ultrafast solvation response found by Maroncelli and Fleming in their molecular dynamics simulation of water [11]. We wish to emphasize that none of the oscillators are overdamped, the rapid decay simply arises from the width of the distribution. If one wished to model the decay by one or a small number of oscillators a large damping constant would be required (i.e. the concept of friction starts to appear as the description is coarsened), but the underlying elementary motions are free (inertial). This inertial behavior dominates the solvation response in all systems we have studied and is discussed in more detail in the paper of Rosenthal et al. in this volume.

This work was supported by a grant from NSF. We thank Shaul Mukamel for much valuable input.

5. References

1. S.A. Rice in Mode Selective Chemistry, eds. J. Jortner et al., Kluwer (Dordrecht) p. 485 (1991).
2. N.F. Scherer, A.J. Ruggiero, M. Du and G.R. Fleming, J Chem. Phys. 93, 856 (1990).
3. N.F. Scherer, R.J. Carlson, A. Matro, M. Du, A.J. Ruggiero, V. R.-Rochin, J.A. Cina, G.R. Fleming and S.A. Rice, J. Chem. Phys. 95, 1487 (1991).
4. N.F. Scherer, A. Matro, L.D. Ziegler, M. Du, R.J. Carlson, J.A. Cina and G.R. Fleming, J. Chem. Phys. 96, 4180 (1992).

5. S. Mukamel, Ann. Rev. Phys. Chem. 41, 647 (1990).
6. M. Cho, N.F. Scherer, G.R. Fleming and S. Mukamel, J. Chem. Phys. 96, 5618 (1992).
7. L.E. Fried, N. Bernstein and S. Mukamel, Phys. Rev. Lett. 68, 1842 (1992).
8. N.E. Shemetulskis and R. F. Loring, J. Chem. Phys., in press.
9. M. Cho, S.J. Rosenthal, N.F. Scherer, L.D. Ziegler and G.R. Fleming, J. Chem. Phys. 96, 5037 (1992).
10. G.I. Szasz and K. Heinzinger, J. Chem. Phys. 79, 3467 (1983).
11. M.P. Maroncelli and G.R. Fleming, J. Chem. Phys. 89, 5044 (1988).

Use of Piecewise Phase-Swept Pulses to Counteract Inhomogeneous Decay in Wave Packet Interferometry

L.W. Ungar, A. Matro, and J.A. Cina

Department of Chemistry and
The James Franck Institute, University of Chicago,
5735 S. Ellis Avenue, Chicago, IL 60637, USA

Abstract. Calculations show that application of a piecewise linear phase sweep to one of two phase-locked pulses, creating a piecewise constant phase pulse, can counteract the effects of rotational and other inhomogeneities.

In the recently developed technique of wave packet interferometry [1-2] two ultrashort laser pulses with a fixed optical phase relationship coherently superpose nuclear wave packets on an excited electronic potential energy surface. The composite nuclear wave function prepared in the excited electronic state from initial state $|v\,J\,M\rangle$ is: [1b]

$$\langle e|\Psi(t)\rangle = i\sqrt{2\pi}\,\mu \sum_{v',\,J'=J\pm1,\,M'} |v'\rangle\, e^{-i\left(\varepsilon'_{el,\,v',J'}\right)t} \langle v'|v\rangle\, E_1\!\left(\varepsilon'_{el,\,v',J'} - \varepsilon_{v,J}\right)$$

$$\times\left[1 + e^{i\left(\varepsilon'_{el} + \varepsilon'_{v'} - \varepsilon_v + \varepsilon'_{J'} - \varepsilon_J - \Omega_L\right)t_d - i\phi}\right] |J'\,M'\rangle \langle J'\,M'|\cos\theta|J\,M\rangle.$$

Subscripts el, v, and J label the electronic, vibrational, and rotational energies respectively, primed for the excited electronic state, unprimed for the ground state. The pulses $E_1(\omega)$ and $E_2(\omega) = E_1(\omega)\exp[i(\omega - \Omega_L)t_d - i\phi]$ are phase-locked at frequency Ω_L with relative phase ϕ and time delay t_d. μ is the transition dipole moment and θ is the angle between the field polarization and the molecular axis. The field amplitude E_1 changes only slightly over the range of $\varepsilon'_{J'} - \varepsilon_J$. But the dependence of the optical phase shift on the rotational transition frequency produces a J-dependent phase factor, $\exp[i(\varepsilon'_{J'} - \varepsilon_J)t_d]$, which multiplies the vibrational wave packet prepared by the delayed pulse. Thus, in a rotationally inhomogeneous sample, with thermal population of a range of initial J values, a variety of composite wave functions is generated. [1b]

We propose to prepare nearly identical vibrational waveforms throughout an inhomogeneous sample by phase modulating one of the two pulses. Adding a frequency dependent phase to $E_2(\omega)$ [3] opposite to that due to the detuning from the bare vibronic transition will counteract the J-dependent phase for a chosen time delay $\underline{t}_d$:

$$\underline{E}_2(\omega) = E_2(\omega)\, e^{i\left(\varepsilon'_{el} + \varepsilon'_{v'} - \varepsilon_v - \omega\right)\underline{t}_d}, \quad \text{for } \varepsilon'_{v'-1} < \omega - \varepsilon'_{el} + \varepsilon_v < \varepsilon'_{v'}.$$

The applied phase is piecewise linear: it varies linearly over small intervals of frequency, between discontinuous jumps coinciding with the vibronic transitions of the system. The resulting pulse $\underline{E}_2(\omega)$ has an unchanged power spectrum but, for time delay $t_d = \underline{t}_d$, has a piecewise constant phase that matches its unswept counterpart at each vibronic transition frequency, then remains constant until the next vibronic transition.

This phase structure would eliminate the J dependence of the relative phase between the wave packets and hence most of the inhomogeneity of the composite waveforms. As shown

Springer Series in Chemical Physics, Vol. 55 Ultrafast Phenomena VIII
Editors: J.-L. Martin · A. Migus · G.A. Mourou · A.H. Zewail

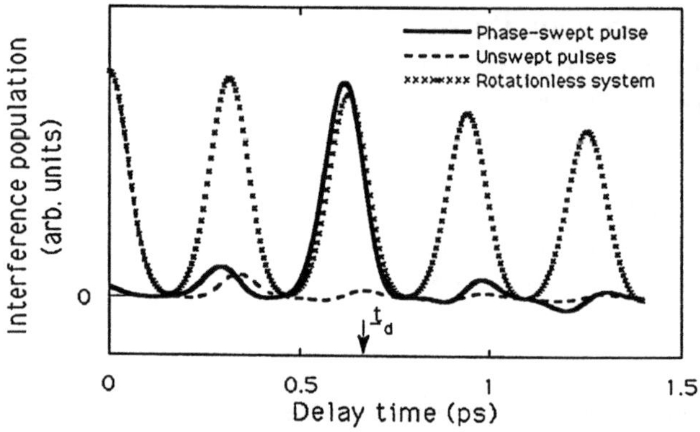

Figure 1. Interference contribution to the excited state population of I_2 as a function of t_d for phase-swept (——) and unswept (----) pulses. Also included is a calculation for unswept pulses applied to a system without rotations (xxx). The temperature is 298 K. Locked frequency and carrier frequency are 16470 cm^{-1}, resonant with the v=2 to v'=10 transition. The unswept pulse envelope is Gaussian with FWHM of 50 fs centered at t_d = 0.66 ps. The applied phase is chosen for v = 2 and $\underline{t}_d$ = 0.66 ps. I_2 parameters are literature values cited in Refs. 1a and 1b.

in the calculation for gaseous I_2 in Fig. 1, when $t_d = \underline{t}_d$ the excited state population due to interference between the wavepackets shows only slight inhomogeneous decay.

The technique proposed here would counteract the decay from any source of inhomogeneity, not just rotations as in I_2. However, inhomogeneity greater than the vibrational spacing (i.e. $\varepsilon'_{J'} - \varepsilon_J > \varepsilon'_{v'} - \varepsilon'_{v'-1}$) will not be fully compensated. Thermal population of multiple vibrational energy levels also will still cause decay. The ability of a piecewise linear phase sweep to superpose wave packets with identical relative phase throughout real samples with moderate inhomogeneity could find use in coherent control of chemical reaction dynamics. [4]

We thank Graham Fleming and Norbert Scherer for helpful discussions that stimulated this work. This work was supported by grants from the National Science and Camille and Henry Dreyfus Foundations and by an NSF Graduate Fellowship to LWU.

References

[1] a) N.F. Scherer, R.J. Carlson, A. Matro, M. Du, A.J. Ruggiero, V. Romero-Rochin, J.A. Cina, G.R. Fleming, and S.A. Rice, J. Chem. Phys. **95**, 1487 (1991). b) N.F. Scherer, A. Matro, L.D. Ziegler, M. Du, R.J. Carlson, J.A. Cina, and G.R. Fleming, J. Chem. Phys. **96**, 4180 (1992).

[2] R. Bavli, V. Engel, and H. Metiu, J. Chem. Phys. **96**, 2600 (1992). J.T. Fourkas, W.L. Wilson, G Wäckerle, A.E. Frost and M.D. Fayer, JOSA B **6**, 1905 (1989).

[3] A.M. Weiner, J.P. Heritage, and E.M. Kirschner, JOSA B **5**, 1563 (1988).

[4] D.J. Tannor and S.A. Rice, J. Chem. Phys. **83**, 5013 (1985).

Ultrafast Nonlinear Spectroscopy with Chirped Optical Pulses

E.T.J. Nibbering, F. de Haan, D.A. Wiersma, and K. Duppen

Ultrafast Laser and Spectroscopy Laboratory, University of Groningen, Nijenborgh 4, NL-9749 AG Groningen, The Netherlands

Abstract. Chirped optical pulses can be used in nonlinear scattering experiments to study dynamics at ultrafast time scales. By varying the chirp rate the character of the matter–field interaction can be changed from impulsive to steady-state. This allows for a detailed assessment of all time scales on which dynamics takes place in a system. Impulsive and chirped scattering experiments are compared.

The generation of optical pulses with an enhanced linear frequency chirp has attracted a lot of attention in the last ten years as a means to compress ultrashort pulses to even shorter durations [1]. The purpose of this operation generally is to increase the time resolution of an optical experiment to the point where meaningful dynamic information about a system can be obtained. The matter–field interaction then preferably occurs in the impulsive limit, i.e. the excitation process must be faster than the relevant relaxation times.

Dynamical information on ultrashort time scales can be obtained in other ways as well. Tokizaki et al. were the first to employ chirped pulses in time resolved nonlinear experiments instead of pulses that were compressed to their transform limited width [2]. They observed that the result of two beam scattering of the type shown in the insert of Fig. 1 was identical for picosecond chirped pulses and femtosecond compressed pulses. Recently, we used chirped and compressed pulses to study the dynamics of molecular electronic states in liquids on a femtosecond time scale [3]. As shown in Fig. 1, the trace obtained with chirped pulses of picosecond duration is distinctly broader and more symmetric than the trace obtained with compressed pulses of 9 femtoseconds duration. In this paper we will briefly discuss the reason for this difference, and indicate how it can be exploited to identify dynamical processes.

With compressed pulses the experiment of Fig. 1 can be interpreted as the generation of photon echoes. The dephasing/rephasing mechanism, that is essential to the formation of an echo, is in this liquid state system due to a finite correlation time in the system–bath interaction (non-Markovian dynamics) instead of the presence of inhomogeneous width [4]. With chirped pulses the interaction between system and optical field is fundamentally different. The excitation process does not necessarily occur in the impulsive limit, and all fields are present at the same time instead of being properly time ordered. A consequence is that now in principle more permutations in time ordering of the optical fields have to be considered than with compressed pulses. In time domain perturbation theory this leads to additional terms in the nonlinear response that can be displayed as additional double sided Feynman diagrams.

The importance of the extra terms in the nonlinear response function depends mainly on two things: the chirp rate of the optical pulses and the time scale(s) on which dynamics takes place in the system. In the limit of very fast chirp rate the field sweeps through the material resonances in a time short compared to the shortest relaxation time. In those circumstances the excitation occurs impulsively, and the result of the experiment should be identical to that with compressed optical pulses. The typical asymmetric signal trace of a photon echo experiment is then observed. When the chirp rate is made slower, the extra permutations in the time ordering of the fields become more important. Since they contribute mainly for negative delays they tend to symmetrize the observed experimental trace.

As mentioned before, next to the chirp rate the system dynamics determines the importance of the additional terms as well. In these extra terms there is no rephasing possible, since no inversion of a coherent superposition state of the system occurs. When there is considerable inhomogeneous broadening in the system (on the time scale in which the field sweeps through the resonances), dephasing/rephasing processes dominate the signal and the extra terms contribute

Springer Series in Chemical Physics, Vol. 55 **Ultrafast Phenomena VIII**
Editors: J.-L. Martin · A. Migus · G.A. Mourou · A.H. Zewail

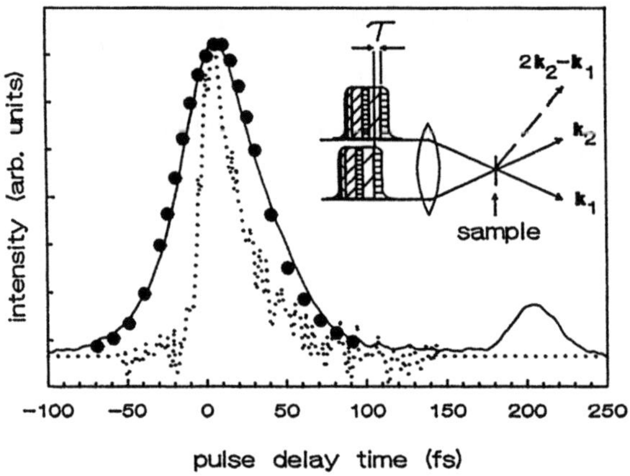

Fig 1: Two-pulse delayed scattering signals with 9 fs compressed pulses (dotted trace) and 1.5 ps chirped pulses (solid trace) of resorufin in liquid DMSO. The beam geometry is shown in the insert. Black dots are the calculated signal.

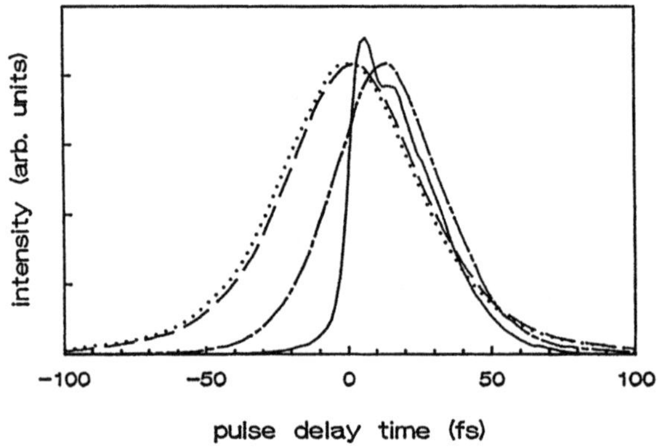

Fig 2: Calculated signal traces for chirped nonlinear scattering with different chirp rates. The system dynamics is of resorufin in DMSO. The traces are:

solid:	100 cm^{-1}/fs,
dot–dashed:	10 cm^{-1}/fs,
dashed:	1 cm^{-1}/fs,
dotted:	0.1 cm^{-1}/fs.

negligibly compared to the original ones. The result is again an asymmetric signal trace. The symmetrization of the signal thus is a sensitive probe to the inhomogeneity that is present.

In Fig. 2 calculated delayed signals are shown as a function of chirp rate. The non–Markovian dephasing dynamics of this system is known from the impulsive scattering experiment [4]. Due to the lack of inhomogeneity in this system, the traces for slow chirp rates are rather symmetric around zero delay: all field permutations contribute significantly. When the chirp rate is made faster the signal shifts to positive delays, a clear signature of the fact that rephasing processes begin to dominate the signal. For the fastest chirp rate shown here the contribution from the field permuted diagrams begins to peak around zero delay. They still contribute for small delays and give rise there to what can be called a coherent artifact.

The calculated signal for the experimental chirp rate of 2.6 cm^{-1}/fs is shown in Fig. 1 as the black dots. Here, a slightly modified stochastic model was used, to take account of the solvent shift that occurs in a liquid state system during the prolonged chirped interaction. Details on this so–called Brownian oscillator model can be found in another contribution to this conference [5]. It suffices here to note that no fitting parameter whatsoever was involved in the calculation. This proves that chirped nonlinear scattering is a well–defined process, that can be used to eludicate important dynamical aspects in complicated systems. Especially the comparison with impulsive scattering results can be very fruitful in that respect.

Springer Series in Chemical Physics, Vol. 55 **Ultrafast Phenomena VIII**
Editors: J.-L. Martin · A. Migus · G.A. Mourou · A.H. Zewail

References

1. H. Nakatsuka, D. Grischkowsky and A.C. Balant, Phys. Rev. Lett. **47**, 910 (1981).
2. T. Tokizaki, Y. Ishida and T. Yajima, Opt. Commun. **71**, 355 (1989).
3. E.T.J. Nibbering, D.A. Wiersma and K. Duppen, Phys. Rev. Lett. **68**, 514 (1992).
4. E.T.J. Nibbering, D.A. Wiersma and K. Duppen, Phys. Rev. Lett. **66**, 2464 (1991).
5. D.A. Wiersma, E.T.J. Nibbering and K. Duppen, This Conference.

Multiple Excitation Pulse, Multiple Probe Pulse Femtosecond Spectroscopy

G.P. Wiederrecht[1], *W. Wang*[1], *K.A. Nelson*[1], *A.M. Weiner*[2], *and D.E. Leaird*[2]

[1]Department of Chemistry, Massachusetts Institute of Technology, 77 Massachusetts Avenue, Cambridge, MA 02139, USA
[2]Bell Communications Research, 331 Newman Springs Road, Red Bank, NJ 07701-7040, USA

Multiple excitation pulse impulsive stimulated Raman scattering is used to resonantly drive the lowest frequency A_1 polariton in $LiTaO_3$. Multiple probe pulse experiments are described which permit single-shot, real-time femtosecond spectroscopy.

The use of multiple femtosecond excitation pulses for increased control over material behavior has been demonstrated experimentally [1-3] and also discussed at some length theoretically [4]. Applications demonstrated to date include vibrational amplification in crystal lattices and manipulation of molecular electronic and vibrational coherences. One of the primary goals is increased control over chemical or structural rearrangements. Some of the experimental challenges involved include electronic and vibrational dephasing, which limit the extent to which the effects of the first pulse in a sequence are "remembered" by a sample as additional pulses arrive, and unwanted optical effects which lead to sample damage or other undesirable responses and which limit the light intensity that can be used.

Here we report new multiple-pulse impulsive stimulated Raman spectroscopy (ISRS) experiments. We also report preliminary results of experiments in which multiple probe pulses, arriving at different times after excitation, were used so that time-resolved data could be collected in one "shot" of the laser. The purpose of these experiments is to record data from light-sensitive samples with high excitation intensities without adverse effects of sample damage.

In our original multiple pulse ISRS experiments, selected lattice modes in an α-perylene organic molecular crystal were driven repetitively, and thereby amplified, by successive pulses whose timing was controlled through a pulse-shaping method [5]. Part of the purpose of that experiment was to achieve larger coherent phonon amplitudes than could be reached with a single pulse by circumventing optical damage through two photon absorption. Unfortunately this goal was not fully realized because of cumulative damage which is initiated through single-photon absorption. In recent experiments on ionic crystals, we have been able to more extensively amplify lattice vibrational responses. Figure 1 shows multiple-pulse ISRS data from a $LiTaO_3$ crystal at 120K. At this temperature, photorefractive damage limits the excitation intensity to levels too low to yield single-pulse ISRS data with acceptable signal to noise. Using multiple-pulse excitation, the total optical energy deposited into the crystal could be increased by almost an order of magnitude. With multiple excitation pulses, sufficient phonon amplitudes were generated to yield the data shown in Fig. 1, from which the vibrational frequency and dephasing rate could be determined. Classical and quantum mechanical analysis of mode selective multiple-pulse ISRS excitation yield essentially identical results and will be published in the near future.

To maximize the extent of phonon amplification in crystalline samples, additional cumulative damage mechanisms must be circumvented. This is ordinarily difficult because time-resolved data are collected not in real time but through many repetitions of the excitation-probe sequence with various probe delays. Each excitation-probe sequence is used to determine the sample's response at a single temporal delay, and data such as that in Fig. 1 are collected "point by point" on the time axis with many (usually thousands of) excitation-probe repetitions. To circumvent cumulative damage during data acquisition, we have developed a "single-shot" femtosecond spectrometer in which the excitation pulse (or pulse sequence) is followed by many probe pulses, each variably delayed by a different amount and each detected separately to provide many points on the time axis. The beamsplitting apparatus for producing many (16) probe pulses is illustrated schematically in Fig. 2a, and preliminary optical Kerr effect data from

Springer Series in Chemical Physics, Vol. 55 **Ultrafast Phenomena VIII**
Editors: J.-L. Martin · A. Migus · G.A. Mourou · A.H. Zewail

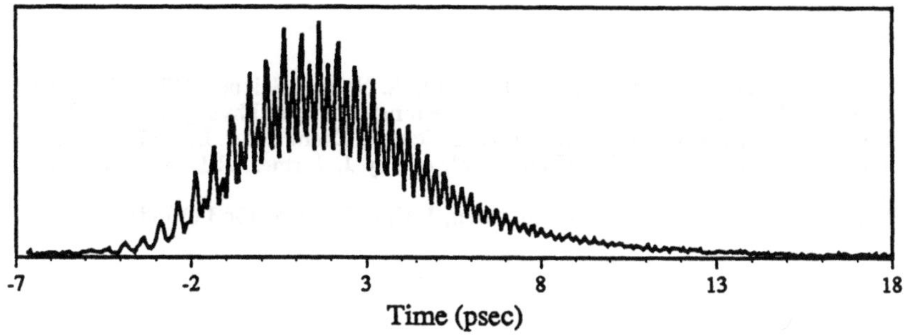

Figure 1. Multiple pulse ISRS scan of the lowest frequency A_1 polariton in $LiTaO_3$ at 120K. The center of the excitation pulse sequence is at zero of the time axis.

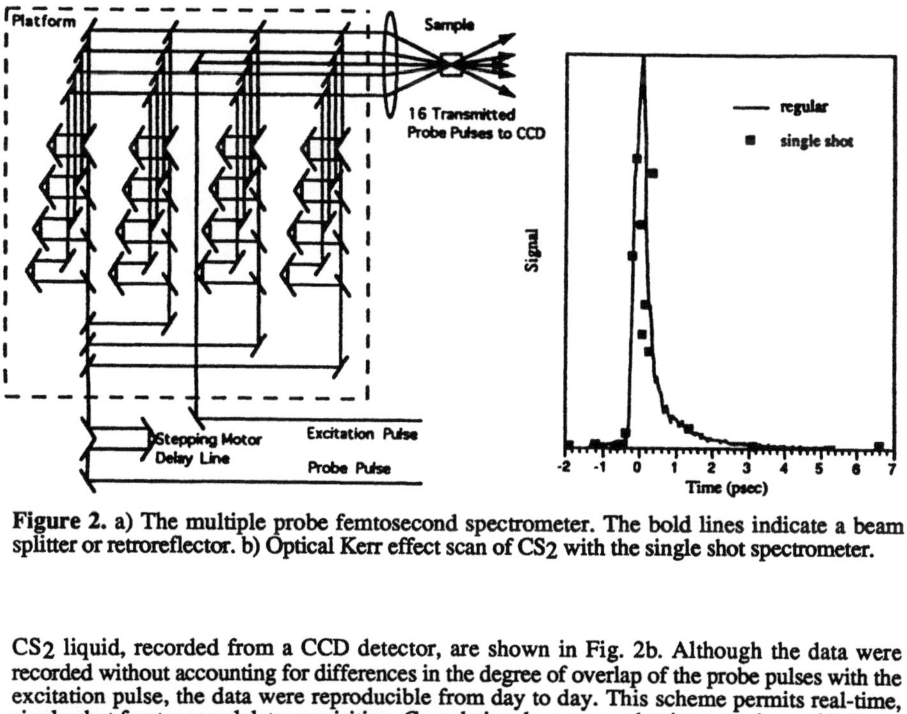

Figure 2. a) The multiple probe femtosecond spectrometer. The bold lines indicate a beam splitter or retroreflector. b) Optical Kerr effect scan of CS_2 with the single shot spectrometer.

CS_2 liquid, recorded from a CCD detector, are shown in Fig. 2b. Although the data were recorded without accounting for differences in the degree of overlap of the probe pulses with the excitation pulse, the data were reproducible from day to day. This scheme permits real-time, single-shot femtosecond data acquisition. Cumulative damage mechanisms can be avoided, and data could be collected even if mechanical damage (which is slow) is caused by a single excitation pulse. Thus the use of multiple probe as well as excitation pulses can play a key role in enhancing pulsed optical manipulation of matter.

Acknowledgements

This work was supported by NSF Grant No. DMR-9002279 and ONR Grant No. N00014-90-J-4070.

References

1. A.M. Weiner, D.E. Leaird, G.P. Wiederrecht, and K.A. Nelson, Science 247, 1317 (1990).
2. N.F. Scherer, A.J. Ruggiero, M. Du, and G.R. Fleming, J. Chem. Phys. 93, 856 (1990).
3. J.J. Gerdy, M. Dantus, R.M. Bowman, and A.H. Zewail, Chem. Phys. Lett.171, 1 (1990).
4. See for example "Mode Selective Chemistry", ed. by J. Jortner, R.D. Levine, and B. Pullman (Kluwer, 1991).
5. A.M. Weiner, J.P. Heritage, and E.M. Kirschner, J. Opt. Soc. B5, 1563 (1988).

Stimulated Emission Pumping and Selective Excitation by Adiabatic Passage with Frequency-Modulated Picosecond Laser Pulses

J.S. Melinger[1,2], *A. Hariharan*[1], *S.R. Gandhi*[1], *and W.S. Warren*[1]

[1]Chemistry Department, Princeton University, Princeton, NJ 08544, USA
[2]Naval Research Laboratory, Code 4613, Washington, DC 20375, USA

Here we demonstrate that picosecond frequency modulated laser pulses produce more efficient population transfer, and more selective excitation than equivalent (same spectral width) nearly transform limited laser pulses [1,2]. We also note the "robustness", or insensitivity of the properties of the material excitation to the parameters of the laser field, that is inherent to the adiabatic process. The robustness obtainable by adiabatic passage may ultimately prove useful to laboratory applications of theoretical schemes for laser-control of chemical reaction dynamics.

Frequency modulated laser pulses with a moderately linear sweep were produced by chirping transform-limited laser pulses (≤ 3 ps) in an optical fiber (yielding 15 ps pulses), or by double passing the transform limited laser pulses through a diffraction grating pair (yielding 20 ps pulses). The laser pulses were amplified to energies of several tens of microjoules.

Enhanced population transfer in Na vapor by stimulated emission pumping was obtained by using frequency modulated laser pulses. The 5s $\leftarrow$ 3s transition was excited by two-photon absorption with unmodulated laser pump pulses. The traces in Fig. 1 show the resulting 5s $\rightarrow$ 3p fluorescence decays from a 100 nsec after the application of either a frequency modulated dump pulse or an unmodulated dump pulse. From these fluorescence traces it is concluded that greater than 85% population transfer was observed when the modulated pulses were applied, compared to roughly 40% transfer (saturation) for unmodulated pulses of equivalent bandwidth. The enhanced population transfer persisted as the laser power of the frequency modulated dump pulse was varied by two orders of magnitude. Less than 100% population transfer may be attributed to incomplete excitation in the wings of the laser spatial profile where the Rabi frequency is not strong enough to achieve adiabatic following.

Frequency modulated laser pulses also permit an excitation profile that is narrower than the spectrum of the laser pulse. Here, the center frequency of the modulated pulse is tuned to the 3p $\leftarrow$ 3s transition in Na vapor so that each of the D-line transitions is nearly equally overlapped by the laser spectrum ($\Delta\tau$ = 15 ps; $\Delta\nu \approx 50$ cm^{-1}). A probe pulse (5 ps; $\Delta\nu$ = 2 cm^{-1}) was

Springer Series in Chemical Physics, Vol. 55 Ultrafast Phenomena VIII
Editors: J.-L. Martin · A. Migus · G.A. Mourou · A.H. Zewail

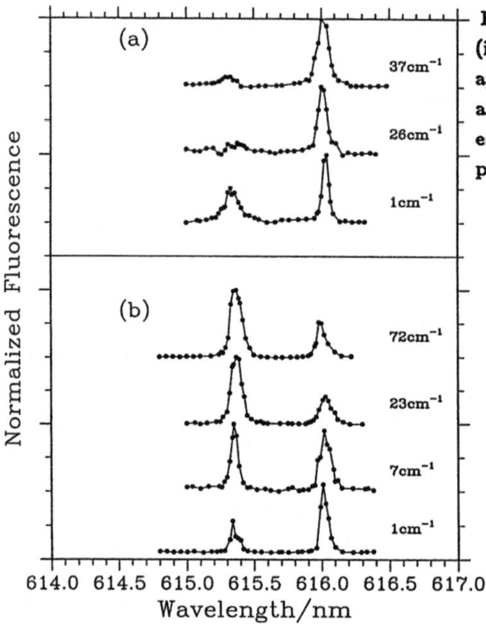

Figure 1. $5s \rightarrow 3p$ **fluorescence (inverted signal) in Na vapor after application of unmodulated dump pulse (a), and frequency modulated dump pulse (b). In each panel lower curve is fluorescence from pump pulse only.**

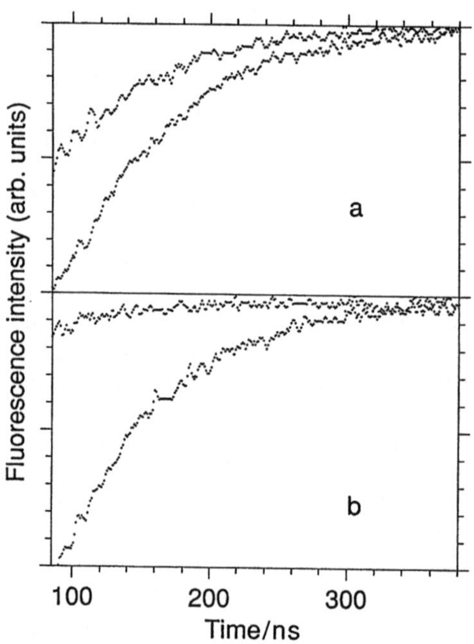

Figure 2. $5s \rightarrow 3p$ **fluorescence in Na vapor showing selective excitation of D-lines by frequency modulated pulses as a function of Rabi-frequency. Red to blue sweep (a), blue to red sweep (b).**

tuned through the 5s ← 3p transitions to detect population in either the $^2P_{1/2}$ or $^2P_{3/2}$ state. Fig. 2 shows fluorescence excitation profiles for both "red to blue" and "blue to red" frequency sweeps on the laser pulse as a function of Rabi frequency.

At low Rabi frequency, the excitation profile reflects the relative oscillator strength of the D-lines. As the Rabi frequency increases to allow adiabatic passage, the degree of excitation of individual states shifts dramatically. For the red to blue sweep the $^2P_{1/2}$ state ($\approx$ 615. 3 nm) is preferentially excited, while for the blue to red sweep the $^2P_{3/2}$ state ($\approx$ 616. 1 nm) is almost exclusively excited. The selectivity persists even in the limit where the Rabi frequency substantially exceeds the D-line splitting.

References:

1. W.S. Warren, Science **242**, 878 (1988).
2. J.S. Melinger, A. Hariharan, S. R. Gandhi, and W.S. Warren, J. Chem. Phys. **95**, 2210 (1991).

A Subpicosecond Optical Sampling System

J.D. Kafka, J.W. Pieterse, and M.L. Watts

Advanced R&D Group, Spectra-Physics Lasers, Inc.,
1330 Terra Bella Ave., Mountain View, CA 94039, USA

Abstract. We have constructed a pump-probe system with independently tunable pump and probe pulses and demonstrated a system resolution of 150 fsec.

1. Introduction

Many ultrafast spectroscopic applications require the use of two independently tunable lasers to generate the pump and probe pulses. Recently, a Ti:sapphire laser was used to pump an optical parametric oscillator [1] or a dye laser [2] and synchronized tunable subpicosecond pulses were generated. In this paper, we describe a subpicosecond optical sampling (SOS) technique for performing pump-probe experiments and demonstrate a system response of 150 fsec. [3]

In a pump-probe experiment, the relative delay between the pump and the probe is varied and the intensity of the transmitted or scattered probe is recorded as a function of this delay time. One way to generate a linear increase in delay time [4] is to set the pump and probe lasers at slightly different repetition rates, F and F + ΔF. In our experiments F is about 80 MHz and ΔF is 80 Hz. The delay between one probe pulse and the next will be 12 nsec. The change in the relative delay time between the pump and probe, however, will be 12 fsec. Thus the ratio between the actual time recorded on the oscilloscope and the delay time is F/ΔF or 1×10^6. If we wish to average many traces, we need a "start" signal that is extremely accurate. We obtain this signal by cross correlating the two lasers in a doubling crystal. Thus, while the jitter at the end of the sampling trace may be many picoseconds, we are able to erase this error and start the next trace with an accuracy of less than a pulse width.

2. Experimental Results

We have demonstrated the SOS technique using two regeneratively mode-locked Ti:sapphire lasers pumped by the all lines output from two cw Argon ion lasers. One is pumped by 8 W and produces 1 W of power while the other is pumped by 5 W and produces 0.5 W. In principle, since the pump source is operated cw, a single Argon ion laser could be used as shown in Fig. 1. Each of the

Springer Series in Chemical Physics, Vol. 55 **Ultrafast Phenomena VIII**
Editors: J.-L. Martin · A. Migus · G.A. Mourou · A.H. Zewail

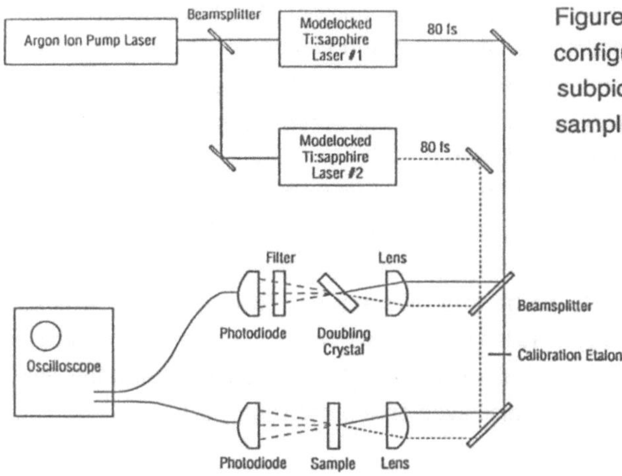

Figure 1. The experimental configuration for the subpicosecond optical sampling (SOS) system.

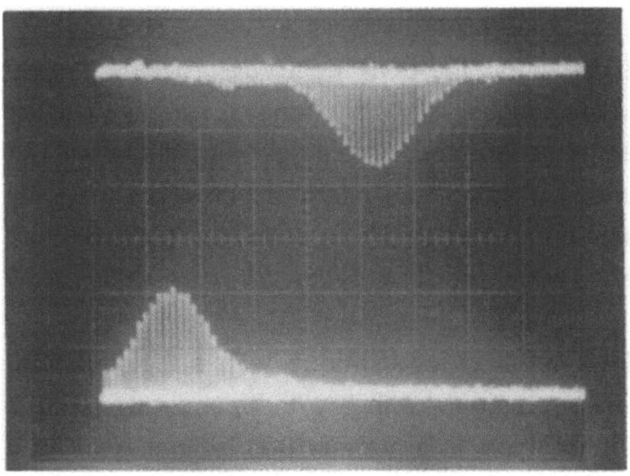

Figure 2. Oscilloscope traces with the start signal shown in the lower trace and the experimental signal shown (inverted) in the upper trace.

lasers produces 80 fsec pulses which broaden to about 100 fsec by the time they reach the experiment due to group velocity dispersion. The repetition rates of the lasers are both about 80 MHz. We set the frequency offset to be 80 Hz by adjusting the end mirror of one of the lasers. Over a period of a few minutes, the repetition rate of each laser varies only a few Hz. The two lasers are combined using a 50% beam splitter as shown in Fig. 1. One beam from each laser is focused into a doubling crystal and the cross correlation signal is observed with a photo-diode. This signal is the start signal for the sampling trace and is used to trigger the oscilloscope. The other two beams are focused into a sample and the probe transmission is observed on the oscilloscope with the use of a second photodiode.

To demonstrate the ultimate time resolution of the SOS system, we used a second frequency doubling crystal as our sample. The outputs from the two photodiodes are shown in Fig. 2. The lower trace is the start signal and the upper trace (which is inverted) is the signal from the experiment. We have added a 320 fsec delay to the experimental signal with a 200 μm calibration etalon. The time scale is 100 nsec per division and the individual pulses are separated by 12 nsec. Each pulse, however, corresponds to a 12 fsec increment in the delay time. The cross correlation width is less than 150 fsec.

If our sample was a piece of GaAs, we might want to measure the time dependent change in reflectivity of the probe. The pump and probe would be focused onto the sample and the reflected probe intensity would be measured by the diode and oscilloscope. To probe at a different wavelength, the probe laser would be tuned and the cavity length would be adjusted to return to the 80 Hz frequency offset. The doubling crystal angle would then be adjusted and the new time dependent reflectivity would be displayed on the oscilloscope.

3. Discussion

In addition to superior time resolution, the SOS technique provides several advantages over other pump-probe techniques. Rapidly changing samples can be investigated. [4] A zero timing mark is generated by the trigger signal and will not change as either of the lasers is tuned. Finally, the system response is constantly available in the form of the start signal and can easily be used for deconvolution. By making use of this deconvolution, we expect that decay times as short as 20 fsec should be retrievable.

In conclusion, we have demonstrated a technique with subpicosecond resolution for doing pump-probe measurements with two independently tunable lasers. The SOS technique eliminates noise in a manner that is analogous to a lock-in amplifier. We discriminate against both the low frequency amplitude noise and the low frequency phase noise from the laser. The SOS technique is also similar to a vernier in time where delay times of femtosecond duration are expanded to the nanosecond time scale before measurement.

References

[1] G. Mak, Q. Fu and H. M. van Driel, Appl. Phys. Lett. **60,** 542 (1992) and W. S. Pelouch, P. E. Powers and C. L. Tang, in *Digest of Conference on Lasers and Electro-Optics*, (Optical Society of America, Washington, D. C., 1992), Post-deadline Paper CPD14.

[2] W. H. Knox and F. A. Beisser, in *Digest of Conference on Lasers and Electro-Optics*, (Optical Society of America, Washington, D. C., 1992), Postdeadline Paper CPD13.

[3] J. D. Kafka, J. W. Pieterse and M. L. Watts, to be published in Opt. Lett.

[4] P. A. Elzinga, R. J. Kneisler, F. E. Lytle, Y. Jiang, G. B. King and N. M. Laurendeau, Appl. Opt. **26**, 4303 (1987) and R. J. Kneisler, F. E. Lytle, G. J. Fiechtner, Y. Jiang, G. B. King and N. M. Laurendeau, Opt. Lett. **14**, 260 (1989).

Femtosecond Sagnac Interferometry

J.-C. Diels, P. Dorn, M. Lai, W. Rudolph, and X.M. Zhao

Department of Physics and Astronomy, The University of New Mexico, Albuquerque, NM 87131, USA

Abstract. A fs ring dye laser is shown to be a powerful instrument for sensing small rotation, observing vibrations of small amplitude, and inducing atomic coherences.

The two counterpropagating beams of a fs ring dye laser are not coupled in phase. This unexpected property makes this laser an exceptional sensitive device to detect any differential change in cavity perimeter for the two directions of propagation. A first manifestation of this absence of coupling in this laser is its gyrcscopic response *which has no dead band.* The two outputs corresponding to the counterpropagating pulse trains are made to interfere on a photodiode (which requires accurate timing in order to have the two train of pulses coincident on the detector). The envelope of the interfering pulse trains show a beat note of frequency $\Delta\nu = a + b\Omega$, where Ω is the rotation rate of the table in $^0/s$ and $b = 31,170$ is a geometry parameter of the ring cavity. The small bias note a has been reduced to zero by accurate cavity adjustment.

The absence of dead-band comes as a surprise, in view of the large scattering at the crossing point of the counterpropagating pulses in the absorber jet. Indeed, we measure a field scattering coefficient of $r = 10^{-3}$, which should be sufficient to "lock" this gyro (absence of beat note) up to rotation rates as high as $7^0/s$ (or no beat note smaller than 0.2 MHz should be observed). The minimum beat note measurable is limited to 20 Hz by mechanical noise (vibrations of the post mounted mirror mounts of the laser and detection interferometer). The waveform of the beat note is approximately sinusoidal as shown in Fig. 1. A 20 Hz beat frequency corresponds to a difference in cavity perimeters of $\Delta P = P \times \Delta\nu/\nu = 0.2\ 10^{-12}$ m, or a differential phase shift of $\Delta\phi = 3\ 10^{-7}$. Experimental tests have shown the absence of measurable dead band to be due to a combination of transit time broadening and degenerate four wave mixing. The latter effect was demonstrated previously [1] and linked to a reduction in dead band [2] The former effect is demonstrated by creating a dead band through introduction of a scatterer at the other pulse crossing point in the cavity. The dead band is eliminated by motion of the scatterer.

The fs laser is not only merily a rotation sensor, but a sensitive detector of any small asymmetry in cavity length for the two senses of rotation. The beat note at zero rotation rate – referred to as "bias" – is due to an intensity imbalance between both directions in the gain jet, which, because of the nonlinear index of the saturable gain, results in a different optical length of the cavity for both directions.

Springer Series in Chemical Physics, Vol. 55 Ultrafast Phenomena VIII
Editors: J.-L. Martin · A. Migus · G.A. Mourou · A.H. Zewail

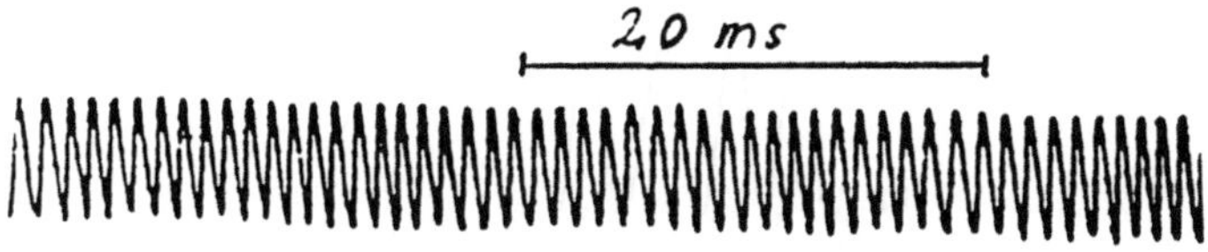

Figure 1: Waveform of the beat note at a small rotation rate

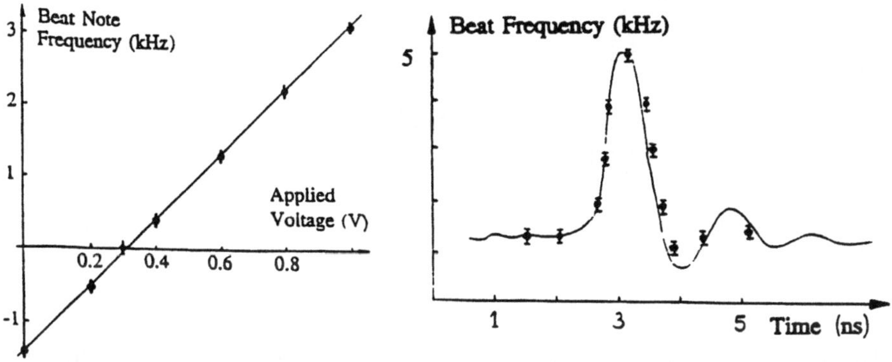

Figure 2: Beat note versus amplitude of the electrical pulse (a), and versus the optical delay of the pulse impinging on the photodiode (b)

A cavity asymmetry can be introduced by various means such as rotation, Fresnel drag, Faraday effect, or mechanically. For example, the following effects will contribute to a signal – i.e. a change in beat note between the counter-propagating pulse trains –:

1. change in *phase shift upon reflection*
2. induced absorption
3. electrostriction (change in volume under electric field)
4. thermal expansion

The exceptional phase and optical length sensitivity can be exploited to investigate the response of ultrafast photodetectors. To demonstrate the concept, we have inserted a 1 cm aperture Pockel cell (3 kV for $\lambda/2$ rotation at 620 nm) in the cavity. An electrical pulse from an avalanche photodiode detecting one of the laser outputs is applied to the cell. An appropriate optical delay ensures temporal coincidence of the electrical pulses and one of the cavity pulses at the electro-optic (EO) crystal. Because the other cavity pulse always reaches the cell between electrical pulses, the two counterpropagating intracavity beams experience different indices in the cell. Therefore, the bias of the beat frequency of the ring laser is modulated by the amplitude and polarity of the electrical pulse. Fig. 2a is a plot of the bias versus applied pulse voltage. By varying the optical delay, one can record the temporal response of the detector-EO combination (Fig. 2b).

The application possibilities of this unique device can be further increased by implementing a coupled cavity to house the samples to be measured. To

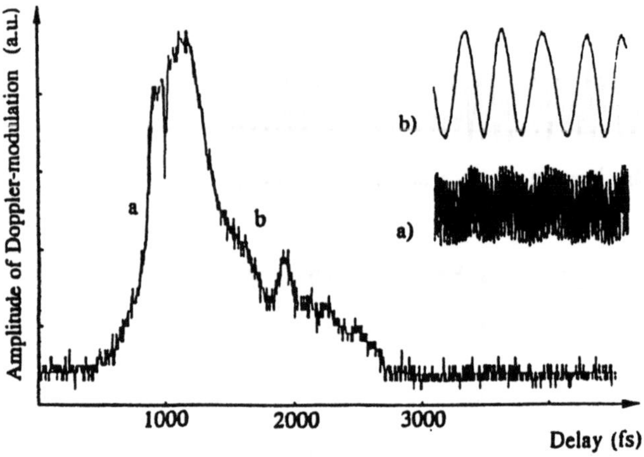

Figure 3: Amplitude of the Doppler oscillation vs delay of the backscattered pulse

demonstrate the sensitivity of the beat note with respect to changes in the coupled resonator we performed the following experiment. The output of one propagation direction was reflected back into the main resonator in the direction of the counterpropagating pulse where the delay was adjusted so that both pulses superpose. Then the backreflecting mirror of the coupled cavity was moved with constant velocity ($v \approx 100\mu m/sec$). The result was a modulation of the original beat note with the Doppler frequency, see Fig.3. When the backreflected pulse precedes the pulse circulating in the main cavity the original beat note vanishes and only the Doppler oscillation is observed. The amplitude of the latter decreases with increasing delay. We tentatively attribute the existence of the Doppler beat over more than 2ps delay to a coherent coupling in the gain medium (Rh6G) for which vibrational relaxation times in the electronic ground state of few ps have been measured [3].

From the large number of other possible applications let us briefly mention two in somewhat more detail. (i) The rotating laser and its linear Sagnac response is an ideal instrument to induce and investigate (intracavity) atomic coherences between sublevels in "Λ level" or "V level" structures" which have been proposed for lasing without inversion [4] and giant index of refraction[5]. The laser is tuned to the main atomic resonance. Plots of beat note versus rotation rate will differ from linear, because, as the resonance is approached, there will be a different index for both pulses arriving in sequence in the resonant atomic mediun. (ii) A change in bias frequency of 100 Hz (which is easily observable) corresponds to a difference in optical path of only 1 pm! This sensitivity suggests the analysis of microscopic shock waves induced by a fs laser pulse impinging on certain materials which are mechanically attached to intracavity reflectors for the circulating pulses (example: Cr layer on a glass substrate which is overcoated by dielectric reflecting layers and which can be excited through the back of this mirror by a suitably timed fs pulse).

References

[1] J.-C. Diels and I. C. McMichael. Degenerate four-wave mixing of femtosecond pulses in an absorbing dye jet. *Journal of the Opt. Soc. of Am.*, B3:535–543, 1986.

[2] J.-C. Diels and I. C. McMichael. Influence of wavefront conjugated coupling in the operation of a laser gyro. *Optics Letters*, 6:219–221, 1981.

[3] D. Ricard and J. Ducuing. Vibrational relaxation in the ground electronic state of large organic molecules in solution. *J. Chem. Phys.*, 62:3616–3619, 1975.

[4] M. O. Scully, S. Zhu, and A. Gavrieliedes. *Physical Review Letters*, 62:2813–2816, 1989.

[5] M. O. Scully. Enhancement of the index of refraction via quantum coherence. *Physical Review Letters*, 67:1855–1858, 1991.

Femtosecond Time-Gated Imaging of Translucent Objects Hidden in Highly Scattering Media

K.M. Yoo, B.B. Das, F. Liu, Q. Xing, and R.R. Alfano

Departments of Electrical Engineering and Physics, Institute for Ultrafast Spectroscopy and Lasers, Photonics Application Laboratory, The City College of the City University of New York, New York, NY 10031, USA

Opaque objects hidden in random media have been shown to be imaged using ultrafast laser pulses and ultrafast detection techniques[1-8]. Recently, research efforts have been made to develop this time-resolved technique for imaging breast tumor. In this paper, we show how a *translucent* object hidden in a highly scattering random medium can be located using either the ballistic or the snake component of the transmitted laser pulse.

An ultrafast laser pulse propagating through a random medium of discrete scatterers splits into a ballistic and a diffuse component[9-11]. The ballistic component arises from the coherent interference of scattered light in the forward direction and propagates straight through the medium. The attenuation and the speed[11] of the ballistic pulse are strongly dependent on the scattering characteristics of the random medium. Thus, the presence of a translucent object with scattering characteristics different from its surrounding medium will change the intensity as well as the arrival time of the ballistic pulse. These changes in the ballistic pulse are monitored by time-resolved detection systems and can be used to locate a hidden translucent object.

To illustrate the imaging technique, a sample was made of a small compartment placed inside a glass cell as shown in Fig. 1(a). The thickness of the cell and the compartment are 16 mm and 6 mm, respectively. The center compartment acts as a translucent object while its surrounding acts as a scattering medium. Various random scattering media are prepared by suspension of latex beads of diameter 0.091 μm in water. The volume concentration of the latex beads of the surrounding medium is 1.67% and that of the object is varied.

Ultrafast laser pulses of 100 fs at 630 nm were generated from a colliding pulse mode-locked dye laser and were amplified by a copper vapor laser to 1 μJ/pulse at a pulse repetition rate of 6.5 kHz. The laser beam was narrowed down to a diameter of 200 μm and was incident on either region A or region B. The ballistic component of the transmitted pulse was measured using cross-correlation second harmonic generation by potassium dihydrogen phosphate and lock-in amplifier. The temporal profile (intensity of the second harmonic cross-correlation) of the ballistic pulse transmitted through the medium alone when the beam was incident at region A is plotted by the solid curve in Fig. 1(b). When the object was made more strongly scattering with bead concentration of 2.8%, the ballistic pulse was found to arrive at a significantly later time as shown by the dotted curve.

Fig. 1(c) shows the cross-correlation intensities of the ballistic pulse passing through the object with different bead concentrations. The reference pulse for cross-correlation was adjusted so that its arrival overlapped in time with the arrival of the ballistic pulse passing through the surrounding medium, i.e., at the peak of the solid curve of Fig. 1(b). The sample was placed on a motorized translation stage and was scanned perpendicular across the laser beam at regions A and B. The cross-correlation intensities were recorded by a lock-in amplifier. This cross-correlation intensity depends on the intensity of the transmitted ballistic pulse and the

Springer Series in Chemical Physics, Vol. 55 Ultrafast Phenomena VIII
Editors: J.-L. Martin · A. Migus · G.A. Mourou · A.H. Zewail

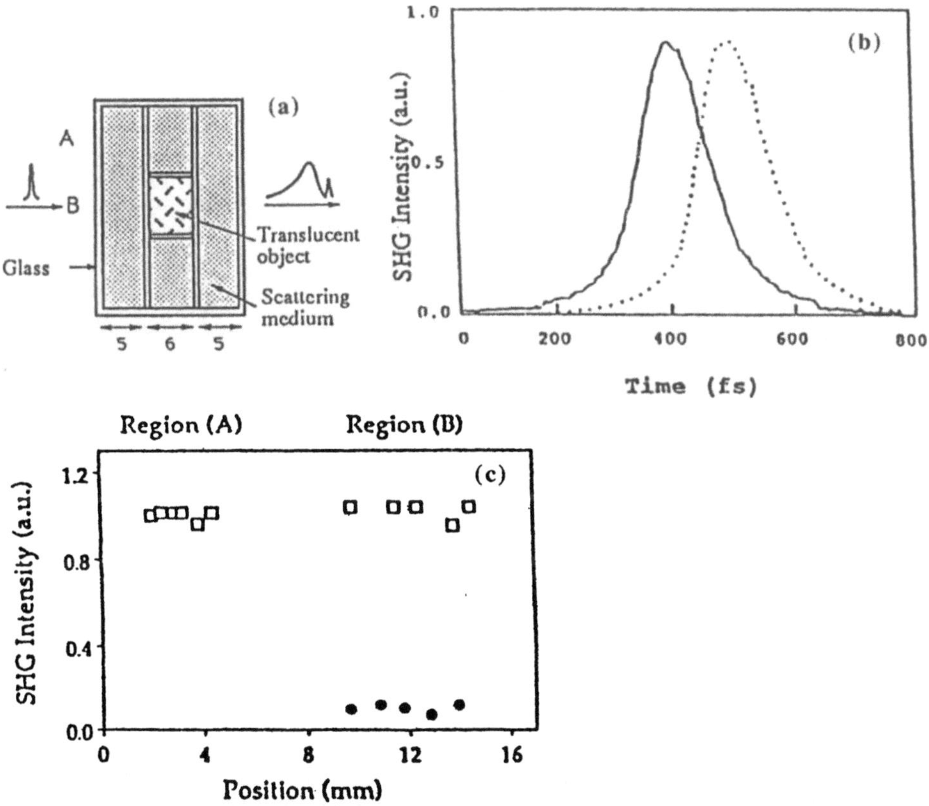

Fig. 1 (a) Structure of the scattering sample. (b) Temporal profiles of the ballistic pulse transmitted through region A (1.67% latex bead) (solid curve), and region B with a translucent object (2.8% latex bead) (dotted curve). (c) Second harmonic cross-correlation intensity when the sample is scanned perpendicular to the laser pulse at regions A and B. The transmitted pulse is time gated at the peak of ballistic pulse through the scattering medium, i.e. at peak of the solid curve of (b). Squares and filled circles are for object with 1.67% and 2.8% latex beads, respectively.

delay time with respect to the reference pulse. The measured cross-correlation intensity around region A is plotted by the squares, while the intensities around region B with the object of bead concentration 1.67 % (same as the medium) and 2.8 % are plotted by squares and filled circles, respectively. This intensity reduction shows the presence of 2.8% object (filled circles). The cross-correlation intensity is reduced by both the delay time of arrival and the intensity reduction of the ballistic pulse.

The ballistic pulse has been shown to be very sensitive in detecting a translucent object of slightly different scattering characteristics from its surrounding medium. However, the ballistic pulse will become too weak to be detected in a thick random discrete medium and in biological tissues. In these cases, the early portion of the diffuse pulse (snake photon) has to be used for imaging, though with a lower resolution. The snake photon traverses a small zigzag path slightly away from the straight line path across the medium. Snake photons are found to be dominant in biological tissues and decrease exponentially with the thickness of the tissue.

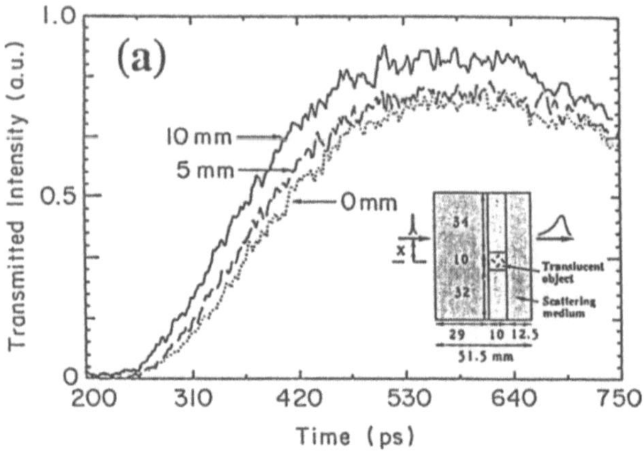

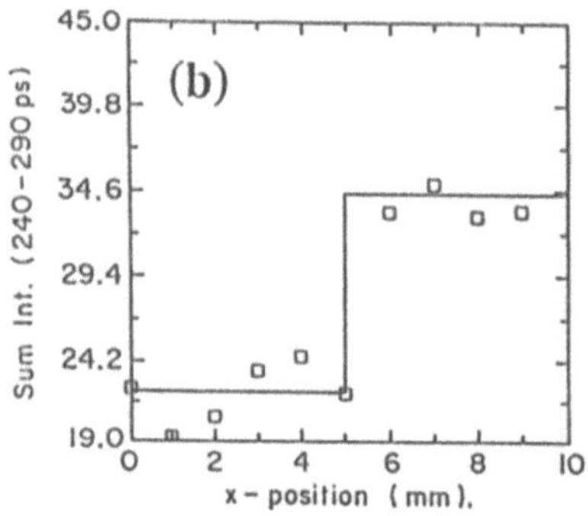

Fig. 2 (a) Transmitted pulses through the sample at positions x = 0, 5, and 10 mm as shown on the curve. Inset: the structure and the dimension of the sample. (b) The integrated intensity from 240 to 290 ps of the transmitted pulse through the sample at various positions of incidence x from the center of the object.

The basis of snake photon imaging of a translucent object hidden inside a scattering medium is that the temporal profile of the diffuse component of the transmitted pulse depends strongly on the spatial distribution of photon transport mean free paths in the scattering medium. The transport mean free path, l_t, is the mean distance of the random walk step the scattered photon has undergone in the medium. The shorter the l_t the stronger is the scattering medium. The presence of an object with a l_t different from that of the medium will affect the temporal profile of the transmitted pulse. The snake photons travel in nearly straight line paths and carry information (scattering characteristics) along these paths across the medium, and can easily be measured selectively using current ultrafast technologies. The change in the snake photon intensity can be used to determine the change in scattering characteristics along a path line across the medium. By scanning the medium across a narrow laser beam in steps and selectively detecting only the snake photons in each step, the position of a translucent object hidden in a scattering medium can be located.

The followings illustrate an example of the snake photon imaging. Fig. 2(a) presents the transmitted pulse profiles measured by a synchroscan streak camera when 100 fs laser pulses were incident at various points on a sample. The sample (Inset) resembles a tumor hidden in a breast, where the transport mean free path of the translucent object and its surrounding medium are 2.2 mm and 4.4 mm, respectively. The intensities of the transmitted pulses at early time are lower for the pulse passing through the object (dotted curve) compared to the pulse passing through the medium alone (solid curve).

By scanning the sample across the laser beam at 1 mm step size and simultaneously detecting the snake photon, the presence of the translucent object can be located. The integrated intensities of transmitted pulses from 240 to 290 ps (snake photons) as a function of the positions (x) of incident beam on the sample are plotted by square points shown in Fig. 2(b). The solid line is traced to guide the eye on the location of the object. The depression shown by squares from points x = 0 to 5 mm matches the location and the size of the translucent object.

In conclusion, the ability of locating translucent objects hidden in a highly scattering medium shows that ultrafast laser pulse and ultrafast detection technologies may be developed for imaging breast tumor.

References

[1] S. Andersson-Engels, R. Berg, S. Svanberg, and O. Jarlman, Opt. Lett. 15, 1179 (1990)
[2] H. Chen, Y. Chen, D. Dilworth, E. Leith, J. Lopez, and J. Valdmanis, Opt. Lett. 16, 487 (1991)
[3] J. C. Hebden, R. A. Kruger, and K. S. Wong, Appl. Opt. 30, 788 (1991)
[4] E. N. Leith, C. Chen, H. Chen, Y. Chen, J. Lopez, P. C. Sun, and D. Dilworth, Opt. Lett. 16, 1820 (1991)
[5] M. D. Duncan, R. Mahon, L. L. Tankersley, and J. Reintjes, Opt. Lett. 16, 1868 (1991)
[6] L. M. Wang, Y. Liu, P. P. Ho, and R. R. Alfano, Science 253, 769 (1991)
[7] K. M. Yoo, B. B. Das, and R. R. Alfano, Opt. Lett. 17, 958 (1992)
[8] K. M. Yoo, Q. Xing, and R. R. Alfano, Opt. Lett. 16, 1019 (1990)
[9] K. M. Yoo and R. R. Alfano, Opt. Lett. 15, 320 (1990)
[10] A. Ishimaru and S. H. Hong, Radio Sci. 10, 637 (1975)
[11] F. Liu, K. M. Yoo, and R. R. Alfano, Opt. Lett. 16, 351 (1991)

Femtosecond Waveform Processing via Spectral Holography

A.M. Weiner, D.E. Leaird, D.H. Reitze, and E.G. Paek

Bellcore, 331 Newman Springs Road, Red Bank, NJ 07701-7040, USA

Abstract. Storage, recall, and processing of shaped femtosecond waveforms is achieved via spectral holography within a grating-and-lens pulse shaping apparatus.

1. Introduction

Compared to the advanced technology now available for generating femtosecond optical pulses, the ability to manipulate and process ultrashort pulses is still rather primitive. Recently, a number of techniques for shaping and synthesizing femtosecond pulse waveforms have been developed.[1],[2] These techniques rely on *linear filtering* of the individual optical frequency components contained within an ultrashort pulse. Although these techniques can be very effective, certain basic signal processing operations, such as time reversal, correlation, and convolution, cannot be achieved by *linear filtering*. In this paper we discuss the use of spectral holography to perform storage, recall, and *nonlinear filtering* of shaped femtosecond pulses. In particular, we demonstrate time-reversal, correlation, convolution, and signal processing of femtosecond optical waveforms[3].

Our experiments constitute a time- and optical-frequency-domain analog of off-axis spatial holography. In spatial holography information on a spatially patterned signal beam is recorded as a set of fringes arising due to interference with a spatially uniform reference beam. In our work the reference is a short femtosecond pulse with a broad and regular spectrum. The signal is a shaped pulse with information patterned onto the spectrum. The experiments are performed by directing reference and signal beams into a modified femtosecond pulse shaping apparatus [1], in which the patterned mask usually used for pulse shaping is replaced by a holographic recording material. As proposed by Mazurenko[4], spectral holograms are written by recording the interference between matching frequency components from signal and reference beams (Fig. 1a). Different frequency components are recorded at different spatial positions of the hologram, which in our experiments is a thermoplastic plate. Readout of the spectral hologram is accomplished by illuminating the thermoplastic plate with a short test (or read-out) pulse (Fig. 1b). Because the thermoplastic plate is a *thin* holographic medium, Bragg matching is not required for readout. The output from the pulse shaping apparatus corresponds either to a real or a time-reversed reconstruction of the initial signal waveform, depending on the experimental geometry, just as reconstruction of off-axis spatial holograms can yield either real or conjugate images. Correlation and convolution operations can be performed by using a shaped test beam for read-out.

Conceptually, our work is related to experiments demonstrating storage and recall of optical waveforms via spectral hole burning and photon echoes in inhomogeneously broadened resonant absorbers (usually held at cryogenic temperatures)[5]. Recent experiments using time-domain holography have also demonstrated storage and recall of femtosecond temporal waveforms[6].

In our experiments the output field is given (in the frequency domain) by:

$$E_{out}(\omega) \sim E_t(\omega)E_r^*(\omega)E_s(\omega)e^{i\alpha\tau} + E_t(\omega)E_r(\omega)E_s^*(\omega)e^{i\beta\tau}. \quad (1)$$

Here $E_t(\omega)$, $E_r(\omega)$, and $E_s(\omega)$ are the complex spectral amplitudes of the test, reference, and signal

Springer Series in Chemical Physics, Vol. 55 **Ultrafast Phenomena VIII**
Editors: J.-L. Martin · A. Migus · G.A. Mourou · A.H. Zewail

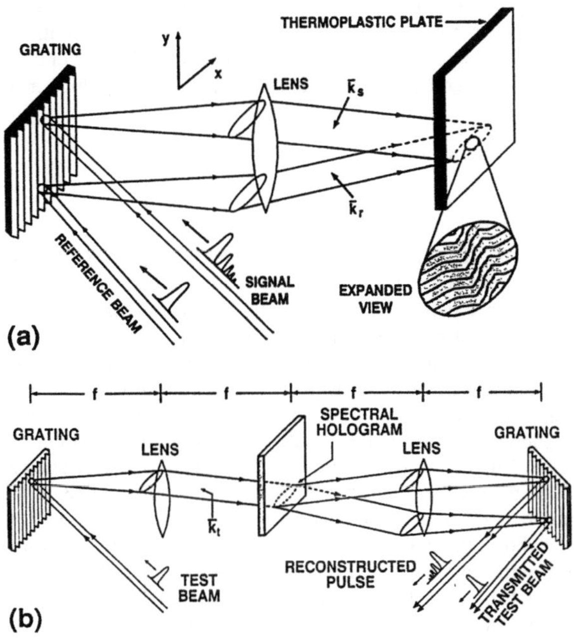

Fig. 1 Schematic illustration of femtosecond spectral holography. (a) Setup for recording. (b) Setup for readout.

fields, respectively, $\alpha=\mathbf{k}_t-\mathbf{k}_r+\mathbf{k}_s$, $\beta=\mathbf{k}_t+\mathbf{k}_r-\mathbf{k}_s$, and $\mathbf{k}_t$, $\mathbf{k}_r$, and $\mathbf{k}_s$ are the propagation vectors just prior to the hologram. The envelopes of the reconstructed output pulses are given by the Fourier transform of eq. (1), as follows:

$$e_{out}(t) \sim e_t(t)*e_r(-t)*e_s(t)e^{i\alpha r} + e_t(t)*e_r(t)*e_s(-t)e^{i\beta r}, \quad (2)$$

where $e_{out}(t)$, $e_t(t)$, $e_r(t)$, and $e_s(t)$ are the complex electric field amplitudes of the output, test, reference, and signal beams, respectively, and the $*$ sign denotes convolution. When both test and reference beams consist of unshaped pulses with durations short compared to the duration of the shaped signal pulse, the first term in eq. (2) represents a real reconstructed signal pulse emerging from the hologram in the α direction, whereas the second term signifies a time-reversed reconstruction propagating along β. In both cases the sharpest temporal features of the original signal pulse may be blurred due to the convolution with test and reference pulses. Furthermore, if the test beam itself is a shaped pulse, then signal processing operations can be achieved. Some of the possible output waveforms are listed below:

- recall of shaped waveforms: $E_{out}(\omega) \sim E_s(\omega)$.
- time-reversal: $E_{out}(\omega) \sim E_s(\omega)^*$.
- convolution: $E_{out}(\omega) \sim E_t(\omega)E_s(\omega)$. The convolution involves the electric field amplitudes of test and signal waveforms (as opposed to the intensity profiles).
- correlation: $E_{out}(\omega) \sim E_t(\omega)E_s^*(\omega)$. Again, this is the correlation of the corresponding electric field amplitudes. The correlation is sensitive to phase.
- matched filtering: $E_{out}(\omega) \sim E_s(\omega)E_s^*(\omega)$. This can be obtained when the test waveform is identical to the signal waveform. The output waveform becomes the electric field autocorrelation of the signal. For a signal of the form $E_s(\omega)$, a linear filter of the form $E_s^*(\omega)$

gives the maximum possible output peak intensity -- hence the term matched filter. Matched filtering can be used to compress a phase-modulated signal to the minimum pulse duration. In addition to its use for pulse compression and dispersion compensation in ultrafast optics, matched filtering is used, e.g., in optical signal processing for spatial pattern recognition and in receivers for chirped radar.

2. Experiments

An example of our data is plotted in Fig. 2. The signal pulse was generated by passing 75-fsec pulses from a CPM dye laser through a second pulse shaping apparatus containing a multielement liquid crystal phase modulator[1]; the reference pulse was taken directly from the CPM laser. The modulator was programmed for a cubic spectral phase variation, leading in the time domain to a nonsymmetric pulse distortion and a long, oscillatory tail (Fig. 2a). Figures 2b and 2c show measurements of reconstructed pulses read-out by using unshaped, 75-fsec test pulses in the real and time-reversed experimental geometries, respectively. These data demonstrate storage, recall, and time-reversal of shaped femtosecond waveforms. Figure 2d shows the waveform read-out by a test pulse identical to the signal pulse. The output waveform, which is the *field* autocorrelation of the signal waveform, is a short and featureless, compressed femtosecond pulse. This is an example of matched filtering -- the phase modulation of the input pulse is completely compensated by the matched filter. Finally, Fig. 2e shows the output for a test pulse which is a time-reversed version of the signal; in this case, the autoconvolution of a time-reversed version of the original signal waveform is generated. The effects of the autoconvolution are clearly evident in the data: the main pulse is broadened, and the oscillatory tail becomes substantially longer.

Many applications of femtosecond spectral holography are possible. For example, one can store femtosecond waveforms in an encrypted format and subsequently recover the original, unscrambled data. Consider a signal with a spectrum given by $E_s(\omega)=E_{data}(\omega)E_{noise}(\omega)$, where

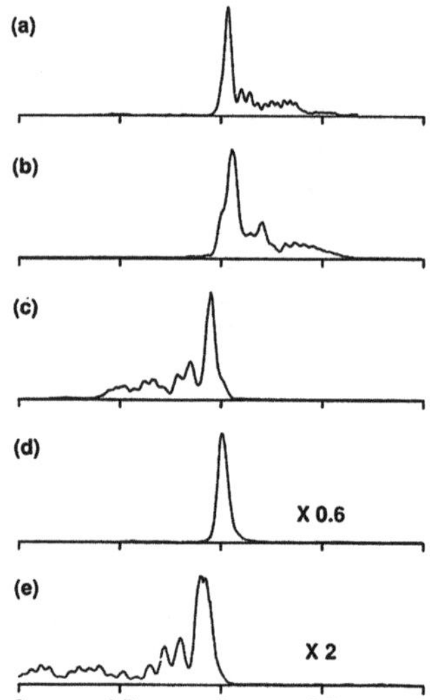

Fig. 2 Intensity cross-correlation measurements, showing holographic storage and recall of a signal pulse distorted via cubic spectral phase modulation.

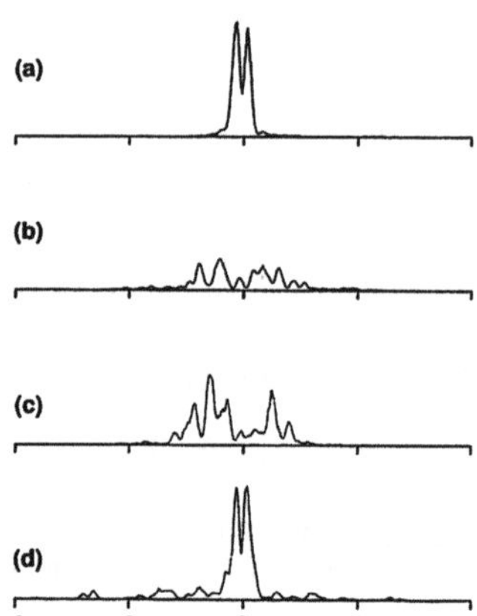

Fig. 3 Intensity cross-correlation measurements, showing storage and read-out of encrypted femtosecond waveforms.

$E_{data}(\omega)$ and $E_{noise}(\omega)$ are the desired data and a pseudorandom phase code, respectively. The data can be recovered from a stored spectral hologram by using a test field with $E_t(\omega)=E_{noise}(\omega)$. In that case one can obtain an output waveform with $E_{out}(\omega)\sim E_{data}(\omega)|E_{noise}(\omega)|^2$, which reduces to $E_{data}(\omega)$ if the encrypting noise field is purely phase modulated. On the other hand, if the encrypting code is not known, the data waveform cannot be uncovered.

The results of experiments demonstrating this concept are shown in Fig. 3. The input data waveform, shown in Fig. 3(a), is an "odd pulse"[1], corresponding to a spectrum with a π phase shift at its center. The coding waveform (Fig. 3(b)) is created by using a pseudorandom phase mask to randomize the spectral phases[7]. A spectral hologram corresponding to the encrypted data is recorded and then read-out with various test waveforms. When the test waveform is a short, unshaped pulse, the output waveform (Fig. 3(c)) bears no resemblance to the input odd pulse. On the other hand, when the test waveform is identical to the coding/noise waveform, the original data waveform is reconstructed.

These experiments are similar conceptually to a number of experiments which have been performed in the spatial domain. For example, holographic techniques have been used to perform imaging through distorting media. If the test beam experiences the same distortions as the signal beam, the distortions cancel and the reconstructed beam is a faithful replica of the original signal beam [8].

3. Further applications: dispersion compensation

One very interesting potential application of spectral holography is for dispersion compensation. We have seen that spectral holography can be used to generate matched filters which cancel the phase variation of incoming pulses, thus compressing chirped pulses to the bandwidth limit. A possible setup is sketched in Fig. 4. A modulated stream of ultrashort pulses is transmitted through an optical fiber. The power is kept sufficiently low that nonlinear effects are not important. At the output of the fiber, the pulses are broadened due to group velocity dispersion. These pulses are then input as the signal beam into a spectral holography setup. A mode-locked local oscillator provides a synchronized train of bandwidth-limited femtosecond pulses (synchronized to the incoming data stream) for the reference beam. The interference between signal and reference forms a spectral hologram of the phase variations imposed by the dispersive transmission channel. Subsequent passage of the dispersed pulses through the holography setup restores the original femtosecond data stream. One advantage of this approach is that the phase variation need not be known beforehand, since the holographic process automatically produces a matched filter corresponding to the incoming phase variation.

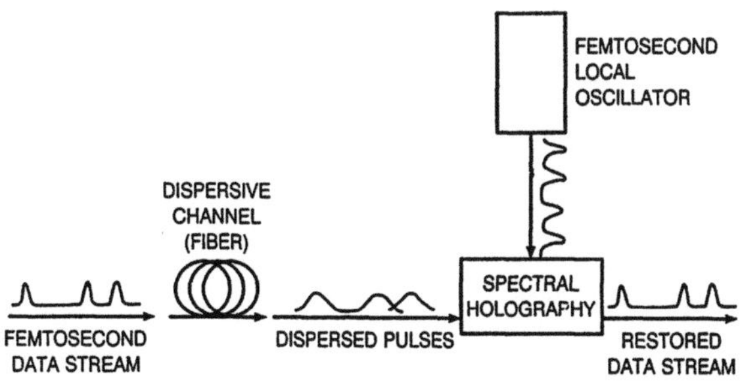

Fig. 4 Schematic diagram illustrating the use of spectral holography to compensate for dispersion in fibers.

4. Summary

In summary, we have demonstrated storage and recall of shaped optical pulses via spectral holography in a femtosecond pulse shaping apparatus. Furthermore, we have demonstrated time-reversal and convolution of ultrashort pulse waveforms, operations which cannot be achieved by traditional ultrashort pulse techniques. Holographic matched filtering of femtosecond waveforms has also been discussed. Our work extends powerful holographic signal processing techniques, traditionally used in the spatial domain, into the ultrafast time domain.

References

1. A.M. Weiner, J.P. Heritage, and E.M. Kirschner, J. Opt. Soc. Am. **B5**, 1563 (1988); A.M. Weiner et al, Opt. Lett. **15**, 326 (1990).
2. M. Haner and W.S. Warren, Appl. Phys. Lett. **52**, 1458 (1988); K. Ema and F. Shimizu, Jpn. J. Appl. Phys. **29**, L631 (1990).
3. A.M. Weiner, D.E. Leaird, D.H. Reitze, and E.G. Paek, Opt. Lett. **17**, 224 (1992).
4. Yu. T. Mazurenko, Appl. Phys. **B50**, 101 (1990).
5. W.R. Babbitt and T.W. Mossberg, Opt. Commun. **65**, 185 (1988); M.K. Kim and R. Kachru, Opt. Lett. **14**, 423 (1989); A. Rebane, J. Aavikso, and J. Kuhl, Appl. Phys. Lett. **54**, 93 (1989); V.L. da Silva et al, Opt. Lett. **16**, 1340 (1991).
6. C. Joubert, M.L. Roblin, and R. Grousson, Appl. Opt. **28**, 4604 (1989); L.H. Acioli, M. Ulman, E.P. Ippen, J.G. Fujimoto, H. Kong, B.S. Chen, and M. Cronin-Golomb, Opt. Lett. **16**, 1984 (1991).
7. A.M. Weiner, J.P. Heritage, and J.A. Salehi, Opt. Lett. **13**, 300 (1988).
8. R.J. Collier, C.B. Burckhardt, and L.H. Lin, Optical Holography (Academic Press, San Diego, 1971), chapter 13.

The Chronocyclic Representation of Ultrashort Light Pulses

J. Paye

Laboratoire d'Optique Appliquée, ENSTA, Ecole Polytechnique,
Centre de l'Yvette, F-91120 Palaiseau, France

Abstract. A new representation of light pulses in the time-frequency phase-space based on the Wigner distribution function gives a broader view of their behavior and greatly improves the understanding of ultrashort-pulse measurements.

1. Introduction

The usual description of an ultrashort light pulse is based on the complex electric field as a function of either time or frequency. These two descriptions are linked by Fourier transformation. As a consequence, the time-frequency uncertainty relation sets a lower limit to the product of the root-mean square (rms) time duration and the rms spectral width. Any alteration of the temporal shape of a pulse also changes its spectral profile. The complex field representation using amplitude and phase cannot provide a simple picture of the consequences of the uncertainty relation. Therefore, an alternative description is highly desirable to understand more intuitively the behavior of ultrashort pulses.

2. The Chronocyclic Intensity

In quantum mechanics, the wavefunction description of a particle is far from intuitive. As an alternative, the Wigner distribution function [1] describes a quantum particle by a pseudo-density of probability in the position-momentum phase-space. The uncertainty relation is naturally apparent in this representation through a lower limit for the area covered by the Wigner function in phase-space. This mathematical tool can be applied to the complex electric field $\mathcal{E}(t)$ of a light pulse:

$$W(t,\omega)=\int_{-\infty}^{+\infty}\mathcal{E}\left(t+\frac{t'}{2}\right)\mathcal{E}^{*}\left(t-\frac{t'}{2}\right)\exp[i\omega t']dt' .$$

This is a real function defined in the time-frequency space which can take negative values. The integration of this function over ω yields the temporal intensity of the pulse, whereas its integration over t gives the spectrum. Double integration over phase-space gives the energy of the pulse. In analogy with temporal intensity which yields the energy of a pulse when integrated over time, we propose to call $W(t,\omega)$ the chronocyclic intensity where the word *chronocyclic* emphasizes that W is a function both of time and angular frequency. The chronocyclic intensity of a pulse completely determines the complex electric field (within an unimportant constant phase factor).

This way of representing a pulse provides more information about the pulse than can be displayed by conventional plots of temporal and spectral intensity [2]. In the left-hand side of Fig. 1, we show the chronocyclic intensity of a Gaussian pulse: it is a bell-shaped surface in phase-space. At the two edges of the plot we have put the corresponding temporal and spectral intensities. We numerically computed the propagation of this pulse in a Kerr nonlinear medium (with no dispersion). The effect of self-phase modulation appears in the chronocyclic representation as plotted on the right-hand side of Fig. 1. Added to the well-known spectral broadening,

Springer Series in Chemical Physics, Vol. 55 Ultrafast Phenomena VIII
Editors: J.-L. Martin · A. Migus · G.A. Mourou · A.H. Zewail

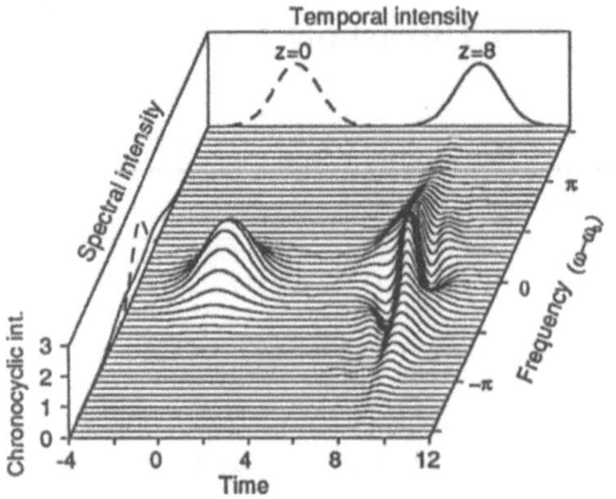

Figure 1 :
Chronocyclic intensity of a Gaussian pulse (left) and of the same pulse after self-phase modulation (right).

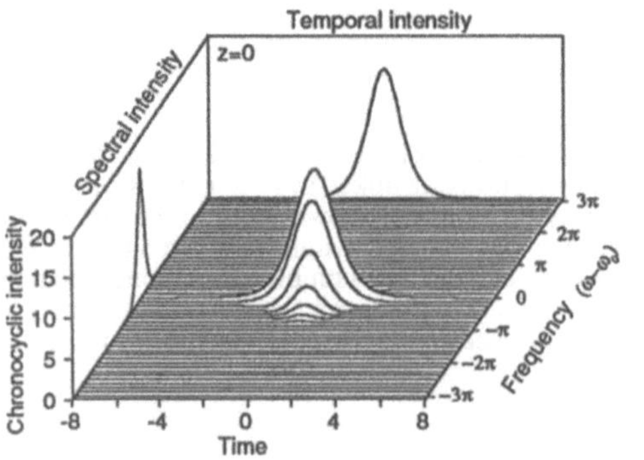

Figure 2 :
Chronocyclic representation of the second-order soliton at the beginning of a soliton period.

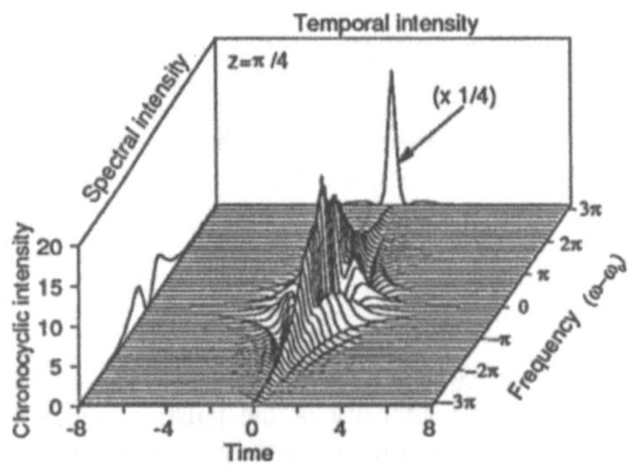

Figure 3 :
Chronocyclic representation of the second-order soliton at half a soliton period.

a shear (parallel to the time axis) of the central part of this pulse is apparent. This shear is the signature of a linear frequency chirp in the chronocyclic representation. The possibility of pulse shortening by chirp correction of a self-modulated pulse is thus clearly displayed.

When dispersion and nonlinearity of opposite signs act together as in an optical fiber in the anomalous-dispersion regime, it is well known that optical solitons can propagate [3]. The second-order soliton propagates with a periodic narrowing and broadening of its temporal shape. To illustrate this behavior in the chronocyclic representation, in Fig. 2 we show the second-order soliton at the beginning of the soliton period, where its electric field is a hyperbolic secant. In Fig. 3, we present the chronocyclic intensity of this pulse after a propagation over half a soliton period.

3. Chronocyclic Pulse Measurements

A light pulse is usually characterized by either temporal (auto- and cross-correlations) or spectral measurements. These measurements, however, provide only partial information about the actual shape of the pulse. We propose a method for a rigorously exact determination of the pulse amplitude and phase based on the chronocyclic representation of ultrashort pulses.

The most interesting property of the chronocyclic representation comes in its application to nonlinear optics. In a pulse-mixing experiment, the squared magnitude of the nonlinear polarization of any order can be written as a convolution of the chronocyclic intensities of the input pulses. Therefore, most nonlinear experiments with ultrashort pulses find a natural description in this conceptual framework. For instance, a cross-correlation experiment mixes two pulses in a nonlinear crystal with a delay τ between them. It is relatively simple to measure the spectrum, $|\tilde{\mathcal{E}}_s(\tau,\omega_s)|^2$, of the sum-frequency pulse as a function of τ and the sum-frequency ω_s [4]. We have simulated in Fig. 4 the result of such an experiment using the pulses in Fig. 2 and Fig. 3.

In the limit of a thin nonlinear crystal, we proved [2] that:

$$|\tilde{\mathcal{E}}_s(\tau,\omega_s)|^2 \propto \int_{-\infty}^{+\infty}\int_{-\infty}^{+\infty} W_1(t,\omega)W_2(t-\tau,\omega_s-\omega)\,dt\,d\omega\ .$$

If $W_2(t,\omega)$ is known, then $W_1(t,\omega)$ can be exactly determined. With this formula, it is possible to retrieve, from the experimentally measurable surface of Fig. 4, the complete knowledge of the pulse shown in Fig. 3. Thus, we have described, for the first time, a method for measurement of the electric field of an unknown pulse which is rigorously correct. It also does not depend on the duration or the spectral width of the reference pulse; the only requirement is that the chronocyclic intensity of the reference pulse be well known (e. g. a chirp-corrected pulse at the output of a femtosecond oscillator).

In a similar way, we have also characterized the information contained in a spectrally-resolved second-order autocorrelation. Higher order autocorrelation schemes could also be analyzed in this framework.

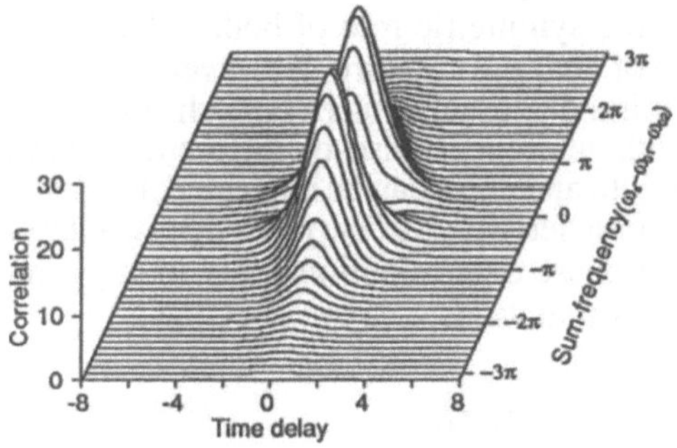

Figure 4 : Simulated spectrally-resolved cross-correlation of the second-order soliton at half-period (Fig. 3) with the same soliton at the beginning of the soliton period (Fig. 2).

References

[1] M. Hillery, R.F. O'Connell, M.O. Scully and E.P. Wigner, Phys. Rep. **106**, 121 (1984).
[2] J. Paye, Submitted to IEEE J. Quant. Elect. (1992).
[3] L.F. Mollenauer, R.H. Stolen and J.P. Gordon, Phys. Rev. Lett. **45**, 1095 (1980).
[4] J.-P. Foing, J.-P. Likforman and M. Joffre , this conference.

Femtosecond Pulse Phase Measurement by Spectrally Resolved Up-Conversion

J.-P. Foing[1,2], *J.-P. Likforman*[1], *and M. Joffre*[1]

[1]Laboratoire d'Optique Appliquée, CNRS URA 1406, ENSTA-Ecole Polytechnique, Centre de l'Yvette, F-91120 Palaiseau, France
[2]Laboratoire Lasers Ultra-Stables, ETCA/CREA/Lasers, 16 bis av. Prieur de la Côte d'Or, F-94114 Arcueil, France

We report the direct determination of the phase of femtosecond light pulses by spectrally resolved up-conversion. The correlation signal spectrum is recorded for different delays between the probed pulse to characterize and a reference one. The phase derivative of the probed pulse is then retrieved considering for each UV wavelength the delay for which its intensity is maximum (fig.1). Basically this technique is a frequency sampling of the probed pulse and therefore needs a spectrally narrow reference beam. This is clearly seen in the case of linearly chirped gaussian pulses which can be explicitely carried out. The delay curve is a line of equation:

$$\tau = \omega \frac{\Phi''\Delta\omega^2 - \phi''\delta\omega^2}{\Delta\omega^2 + \delta\omega^2}, \tag{1}$$

where $\Delta\omega$ is the spectral full-width half-maximum of the studied pulse, Φ'' its linear chirp, and τ is the delay between this pulse and the reference pulse of width $\delta\omega$ and linear chirp ϕ''. Due to the symmetric role of both fields in the correlation, the line slope is a function of both pulses chirp. However we notice that (i) in the simple case of linearly chirped gaussian pulses both chirps can be determined from two correlations with reference pulse of different spectral width, (ii) if the reference pulse bandwidth and chirp can be neglected, the delay curve is within an additive constant the first derivative of the probed pulse phase: $\tau \approx \omega\Phi''$. These limitations for a direct phase determination apply to any kind of pulses as we show through different simulations.

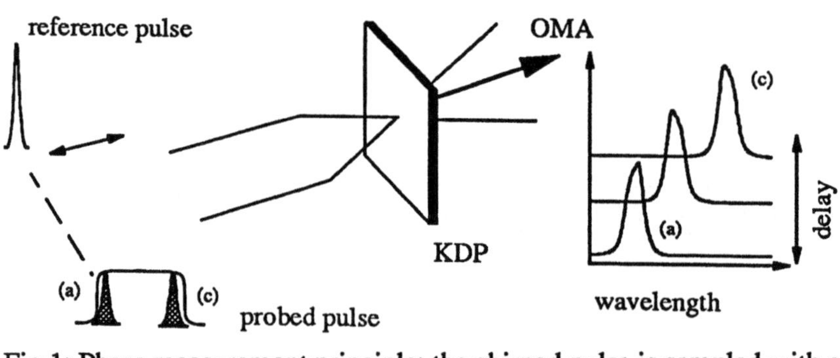

Fig.1: Phase measurement principle: the chirped pulse is sampled with a narrow bandwidth reference pulse. By changing the delay between pulses, the different wavelengths composing the pulse are successively probed.

Springer Series in Chemical Physics, Vol. 55 Ultrafast Phenomena VIII
Editors: J.-L. Martin · A. Migus · G.A. Mourou · A.H. Zewail

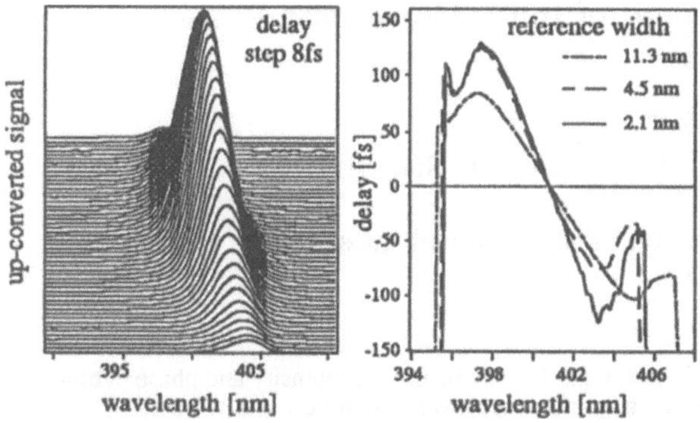

Fig.2: Experimental correlation of a chirped Ti:Sapphire oscillator and delay curves deduced for different reference pulses. The slope evolution with the reference pulse spectral width follows eq.1.

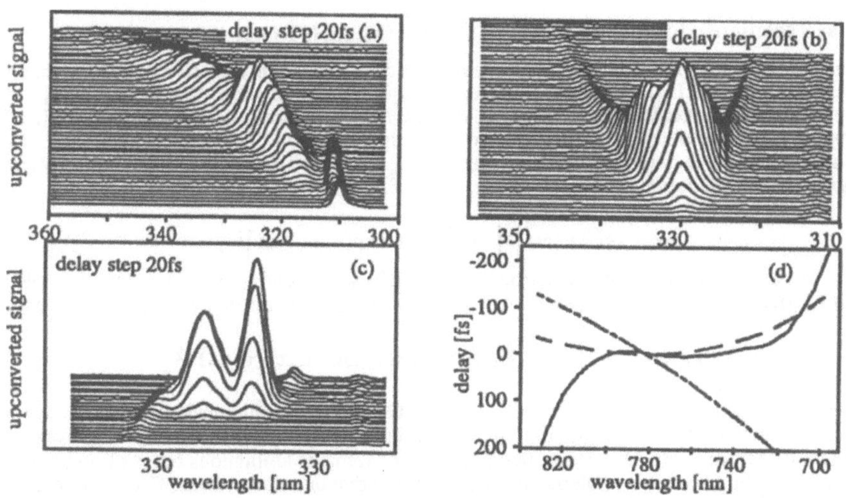

Fig.3: Control of continuum compression by up-conversion. The original linear chirp (a) is corrected with prisms (b) and to higher order with a combination of prisms and gratings (c). Delay curves deduced (d).

These results are experimentally illustrated (i) by the phase determination of a linearly chirped Ti:Sapphire oscillator. The reference pulse is then part of the oscillator spectrum selected in the Fourier plane of a compression line, set to minimize the pulse duration (fig.2); (ii) by the control of broadband continuum compression. The reference pulse is then part of the fundamental pulse used to generate the continuum (fig.3).

Compared with techniques recently proposed [1,2], this correlation method has respectively the advantages to be background free, to measure the pulse effectively used in further experiment and to allow the characterization of very low energy pulses since the reference beam can be any powerfull pulse spectrally narrow.

REFERENCES

1 T.F. Albrecht, K. Seibert and H. Kurz; Opt. Comm. **84**, 223 (1991)
2 J.L.A. Chilla et O.E. Martinez; Opt. Lett. **16**, 39 (1991)

Single-Shot Measurement of the Intensity and Phase of a Femtosecond Pulse

D.J. Kane[1] *and R. Trebino*[2]

[1]Southwest Sciences, Inc., 1570 Pacheco St., Suite E11, Santa Fe, NM 87501, USA

[2]Combustion Research Facility, Sandia National Labs., Livermore, CA 94551, USA

Abstract. We introduce a simple technique for measuring the intensity and phase evolution of an arbitrary ultrashort pulse using phase-retrieval techniques.

1. The Technique: Frequency-Resolved Optical Gating (FROG)

We have developed a method for measuring the full, time-dependent intensity and phase, that is, the full electric field, E(t), of an individual, arbitrary femtosecond pulse. Our technique[1] uses an optical-gate arrangement (See Figs. 1, 2). A pulse is split, and one variably delayed replica of the pulse gates the other in an instantaneously responding nonlinear-optical medium. The resulting signal-pulse electric field is given here by $E_{sig}(t,\tau) \propto E(t)\,|E(t-\tau)|^2$. Frequency-resolving the gated pulse, we obtain the signal, I_{FROG}, as a function of frequency, ω, and delay, τ:

$$I_{FROG}(\omega,\tau) \propto \left| \int E_{sig}(t,\tau)\, \exp(-i\omega t)\, dt \right|^2 , \tag{1}$$

which can be considered to be a spectrogram[2] of E(t) with variable-delay window $|E(t-\tau)|^2$.

If $E_{sig}(t,\Omega)$ is the Fourier transform of $E_{sig}(t,\tau)$ with regard to the delay variable, τ, we have:

$$I_{FROG}(\omega,\tau) \propto \left| \iint E_{sig}(t,\Omega)\, \exp(-i\omega t - i\Omega\tau)\, dt\, d\Omega \right|^2 . \tag{2}$$

The problem of inverting Eq. (2) to find $E_{sig}(t,\Omega)$ is the two-dimensional phase-retrieval problem—known to have a unique solution in essentially all cases[3,4]. It is then straightforward to obtain E(t) from $E_{sig}(t,\Omega)$. A constant phase factor remains ambiguous but is of no interest.

To obtain E(t), we have implemented a simple iterative one-dimensional Fourier-transform algorithm, which involves Fourier transforming back and forth between $E_{sig}(t,\tau)$ and $E_{sig}(\omega,\tau)$. See Fig. 3. In the ω-domain, we replace the magnitude with the square root of $I_{FROG}(\omega,\tau)$. In the t-domain, we use a novel constraint, generating the (k+1)st iteration for $E_{sig}(t,\tau)$ by first setting:

$$E^{(k+1)}(t) = \int E_{sig}^{(k)}(t,\tau)\, d\tau = \left[E^{(k)}(t) \int |E^{(k)}(t-\tau)|^2\, d\tau \propto E^{(k)}(t) \right] . \tag{3}$$

Using $E_{sig}^{(k+1)}(t,\tau) \propto E^{(k+1)}(t)\,|E^{(k+1)}(t-\tau)|^2$, we then construct the (k+1)st iteration for $E_{sig}(t,\tau)$.

We have obtained E-fields from theoretical FROG traces using this algorithm for most commonly encountered pulses. In all cases, convergence to the correct field occurs, yielding rms error values per point ≤ 0.1%. We are at present investigating pathological pulse shapes.

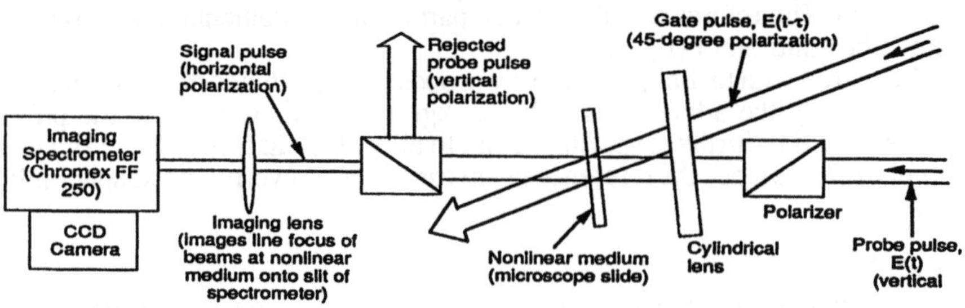

Figure 1. FROG apparatus. (The spatially filtered beams are crossed at an angle, and the cylindrical lens yields a line focus, so that the delay between the pulses varies across the sample.)

Springer Series in Chemical Physics, Vol. 55 Ultrafast Phenomena VIII
Editors: J.-L. Martin · A. Migus · G.A. Mourou · A.H. Zewail

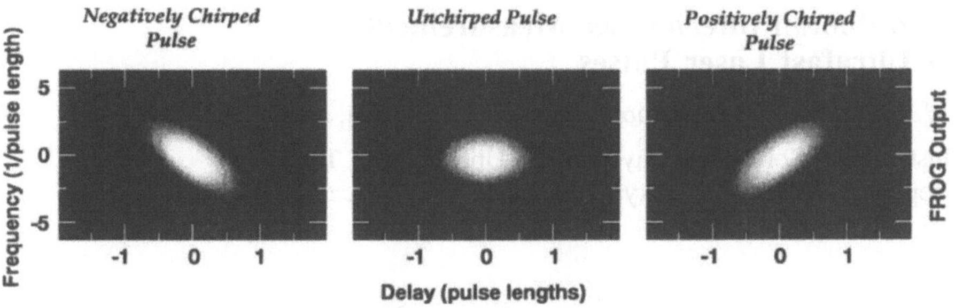

Figure 2. Theoretical FROG traces for negatively chirped, transform-limited, and positively chirped Gaussian-intensity pulses. Here we can see that both the magnitude and the sign of chirp are easily distinguished. For smoothly varying pulse intensities, the shape of the trace indicates the instantaneous frequency vs. time for the pulse.

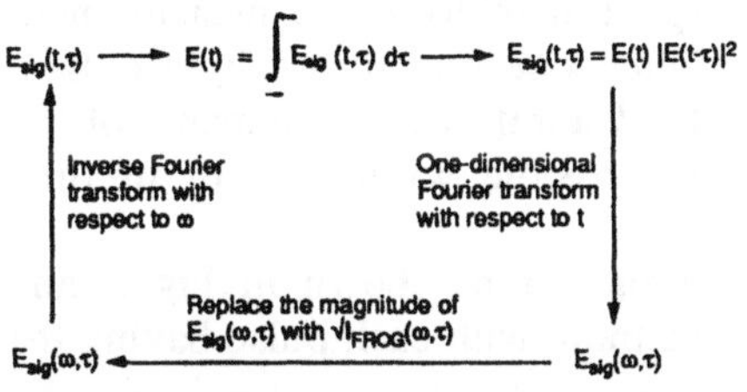

Figure 3. Iterative Fourier-transform algorithm to invert FROG traces to obtain the electric field vs. time of an ultrashort pulse. The Fourier transforms are one dimensional. The algorithm uses the experimental trace to replace the magnitude in the frequency domain, and it uses the known mathematical form of the signal field as the constraint in the time domain. A good initial guess is obtainable directly from the FROG trace, but noise appears to work as well, especially for more complex pulse shapes.

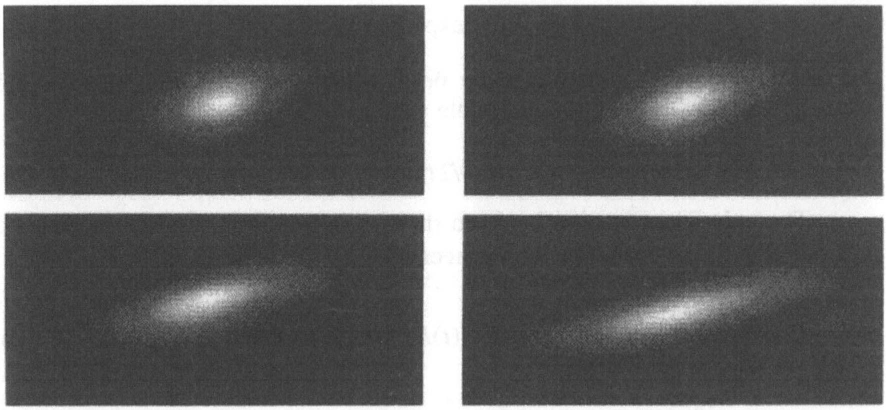

Figure 4. Experimental density plots of single-shot FROG signal intensity vs. delay (horizontal axis, increasing upward, full scale ≈ 700 fsec) and frequency (vertical axis, increasing upward, full scale ≈ 40 nm) for pulses of various chirps. The shape indicates frequency vs. time; here linear positive chirp (longer wavelengths occurring earlier) is indicated. Upper left: nearly transform-limited pulse; each additional trace involves an additional 2.5 mm of SF-10 glass path length. This progression of pulses differs from that in Fig. 2 in that, here, the bandwidth remains constant while the pulse length varies, whereas in Fig. 2 the pulselength is identical for all three plots while the bandwidth varies with the magnitude of the chirp.

Figure 4 shows experimental FROG traces of ~10-μJ, ~200-fsec pulses of varying chirp by varying the amounts of glass in the beam. Measured chirp clearly increases with increasing glass.

2. References

1. D. J. Kane and R. Trebino, submitted to J. Quant. Electron.
2. R. A. Altes, J. Acoust. Soc. Amer., **67**, 1232 (1980).
3. R. Barakat and G. Newsam, J. Math. Phys., **25**, 3190 (1984).
4. J. R. Fienup, J. Opt. Soc. Amer., **4**, 118 (1987).

Two-Photon Interference Measurement of Ultrafast Laser Pulses

M. Matsuoka, Y. Miyamoto, T. Kuga, M. Baba, and Y. Li

Institute for Solid State Physics, The University of Tokyo,
Roppongi, Minato-ku, Tokyo 106, Japan

Recently novel two-photon interference experiments have been carried out, in which a Mach-Zehnder interferometer is combined with two detectors to obtain higher time resolution than that of the photon detectors used[1~3].

We propose a new scheme of two-photon interference measurement which provides the intensity correlation of envelope function of light pulses, for example, laser pulses.[4] With this method ultrafast measurements of the coherence time and the pulse duration is possible without limit in time resolution.

Let us consider a Michelson interferometer as shown in Fig.1, and assume that the incident laser field is a pulse train, with each pulse having the profile

$$E(t) = C\exp(-t^2/\tau_p^2)A(t)\exp(-i\omega_0 t), \tag{1}$$

where $A(t)$ denotes a steady random field, which gives rise to a steady-field coherence time τ_c through the ensemble average:

$$\langle A^*(t)A(t+\tau)\rangle_A = \exp(-t^2/2\tau_c^2). \tag{2}$$

The coincidence count between the two detectors D_1 and D_2, measured as a function of the corner cube displacement $\delta\tau$, is expressed as

$$Count_{12}(\delta\tau) = K\int_{-\infty}^{\infty}dt\int_{-T_R}^{T_R}d\tau\,\langle E_1{}^*(t)E_2{}^*(t+\tau)E_2(t+\tau)E_1(t)\rangle_{A,\phi}\,. \tag{3}$$

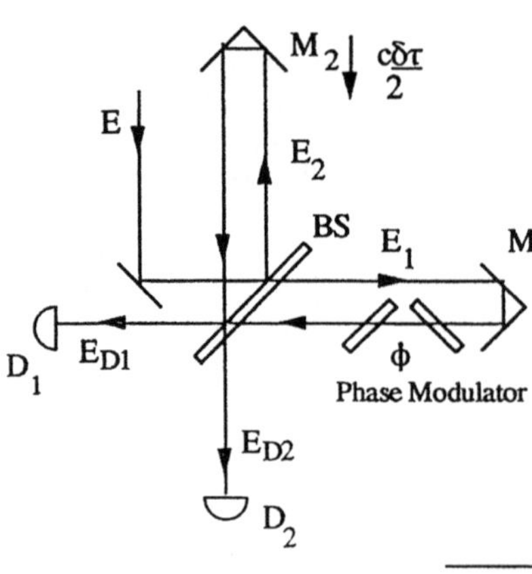

Fig. 1

Springer Series in Chemical Physics, Vol. 55 Ultrafast Phenomena VIII
Editors: J.-L. Martin · A. Migus · G.A. Mourou · A.H. Zewail

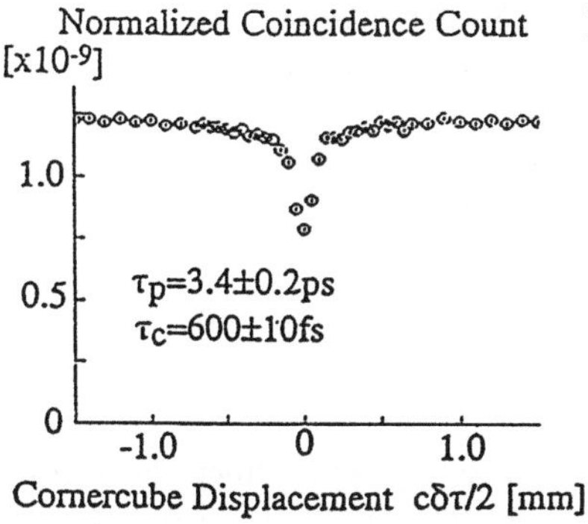

Fig. 2

T_R is the detector response time, and $\langle\ \rangle_{A,\phi}$ denotes ensemble averages with respect to $A(t)$ and ϕ, the latter being the slowly varying random phase introduced by the phase modulator.

Choosing T_R so that it is long compared with τ_p and τ_c but shorter than the pulse separation time, and assuming Gaussian random statistics for $A(t)$, we obtain

$$Count_{12}(\delta\tau) = K'\left(1 - \frac{1}{2}\exp[-(\delta\tau)^2/\tau^{*2}] + \frac{\tau^*}{\tau_p}\{1 - \frac{1}{2}\exp[-(\delta\tau)^2/\tau_p^2]\}\right), \quad (4)$$

$$1/\tau^{*2} = 1/\tau_p^2 + 1/\tau_c^2, \quad (5)$$

where τ^* denotes the coherence time of the pulse field. Thus it is possible to evaluate both τ_p and τ^* (or τ_c) from the resulting 'dip' in the coincidence count rate curve. Experiments using the output pulses of a cw mode-locked dye laser have been performed, with results (Fig. 2) in good agreement with both theory and the conventional second harmonic autocorrelation measurement.

REFERENCES

1. C. K. Hong, Z. Y. Ou, and L. Mandel: Phys. Rev. Lett. 59 (1987) 2044.
2. Z. Y. Ou, E. C. Gage, B. E. Magill, and L. Mandel: Opt. Commun. 69 (1988) 1.
3. Z. Y. Ou, E. C. Gage, B. E. Magill, and L. Mandel: J. Opt. Soc.Am. B 6 (1989) 100.
4. M, Matsuoka, Y.Miyamoto, T. Kuga, and M. Baba: Proceedings of Shanghai International Symposium on Quantum Optics, March, 1992 (SPIE, 1992).

Picosecond Single-Shot Pulse-Shape Measurement by Stochastic Sampling of Detected Photon Times

N. Adams, C. Bovet, E. Rossa, and A. Simonin

European Organization for Nuclear Research (CERN), CH-1211 Geneva 23, Switzerland

Abstract. Light pulses to be analysed are stored in a fibre ring system from which single photons, extracted at random, are detected and histogrammed on a periodical time basis in order to provide a measurement of the pulse shape.

1. Introduction

Sampling technique is frequently used in order to measure the shape of repetitive pulses. It needs a synchronous trigger plus a variable delay and a peak sensing device which often sets the limit on the time resolution . The best sampling oscilloscope has a rise time in the range of 10 ps but cannot be used for single shot measurement.

A streak camera with a display of the pulse on a memory tube makes it possible to reach a resolution of 1 ps. However, the use of a streak camera is cumbersome and cannot be done on line because of the time needed for digital image evaluation. In some cases, transient recorders may be used, but the passband of such apparatuses is at present only of 6 GHz, which is insufficient for studying very short incoherent light pulses.

An alternative procedure [1],[2] is to measure the times of occurrence of events whose probability is proportional to the pulse height. The histogram of these occurrences reproduces the pulse shape, provided the event probability is small enough for multiple event rate to be negligible. This method, called here *stochastic sampling,* is attractive because it measures the time of flight between a detected photon and with a base time clock; and that this time measurements can easily be performed with TDCs or TACs with a resolution better than 10 ps. In order to speed up the sampling by orders of magnitude, and still allow for single shot measurements,it is proposed to store the light pulses in a system of optical fibres which will then deliver single photons periodically.

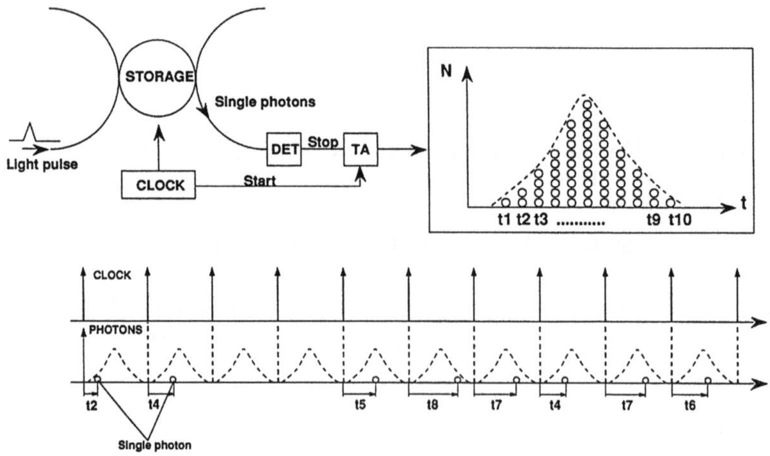

Fig. 1 Principle of single shot measurement

Springer Series in Chemical Physics, Vol. 55 Ultrafast Phenomena VIII
Editors: J.-L. Martin · A. Migus · G.A. Mourou · A.H. Zewail

2.1 Sampling procedure

The principle of stochastic sampling can be summarized as following. A time of flight counter is started by a clock and the stop is set by the detection of the first photon extracted at random from the light pulses. Let $g(t)$ be the normalized photon distribution. A histogram constructed with the observation of the first detected photons has the shape: $h(t) = g(t) \exp\{-\lambda \int_0^t g(x)dx\}$, where λ is the expectation value for the number of photons detected from each light pulse. When $\lambda \ll 1$ the histogram $h(t)$ represents very closely the pulse shape $g(t)$. But the shape can easily be corrected if a higher detection probability $\lambda \approx 1$ is used in order to reduce the sampling time. Another large improvement can be obtained by the use of photon storage rings to keep the photon pulse over a long time interval and create a pseudo-periodic function from which many single photons can be observed over a much longer time interval.

2.2 Photon storage rings

Monomode fibres are now available with an attenuation of only 0.2 dB/km and with a dispersion at 1318 nm of about 1 ps/km/nm which is adequate to keep the pulse shape during $\approx 10\ \mu s$. Couplers have been found with typical performances of 1 – 10% coupling, 1% loss and 90 – 98% transmission. With the help of such couplers, various schemes of storage rings can be designed where the loops are closed by using low loss connectors. Two couplers with k=1% can be connected to form a storage ring, delivering pulses which are 10^{-4} times smaller than the input pulses, and the decay rate would be of 4% per turn (2% coupling plus 2% loss).

A time base must be provided during the whole sampling period, with a stability of ≈ 2 ps which is feasible with a quartz clock. The ring periods must also be adapted to within 10^{-7} of the clock frequency. This has been achieved in three steps: i) the fibres have been cut to an accuracy of 10^{-4}, ii) further mechanical adjustment was made by polishing the mating surfaces at the connectors and iii) the final tuning and the compensation of drift is reached with individual thermal control of the fibre rings (ΔT=0.1 K corresponds to 10^{-7}).

2.3 Detection of single photons

To detect the single photons a germanium Avalanche Photo Diode (APD) is used. The APD was originally developed for long-haul high speed fibre-optic communications, at gains of 10-30, many orders of magnitude too low for our purposes. However, by biasing the APD well above its breakdown voltage several nanoseconds before the photon arrives, the gain may be made extremely high, i.e. the resulting avalanche continues indefinitely unless *quenched* externally. Germanium APD's are inherently noisy and cannot be used for single photon detection at room temperature.

For this reason the APD is cooled to -200° C by immersion in liquid nitrogen. When pulse biased correctly the APD gives a very large signal (200-500 mV). We have performed a large number of experiments to test the various aspects of the APD for our unique application : using the diode in single photon mode for timing. Running the APD at low repetition rates, it has been found to have the following characteristics :i) avalanche probability of 0.5 for each photo-electron, ii) noise background count of several percent (this is spontaneous thermal breakdown of the diode, even at -200° C!), iii) inherent time jitter of 20-30 ps FWHM caused by the noisiness of the avalanche. This will increase the apparent size of a 100 ps FWHM light pulse by only 5 ps.

3. Results

Synchrotron light photons from CERN large electron collider have been fed into a monomode fibre and used for stochastic sampling of bunch length, thus demonstrating the quality of the detector and the proper functioning of the process. Devices for photon storage have been assembled with two and three fibre rings, with total losses ranging between 2% and 4% per ring. Their synchronization to within 10^{-7} has been ensured and the effect of dispersion after 10 μs has been measured and found negligible, when using an adequate spectral filter.

References

[1] C. Bovet et al. Single shot bunch length measurement at LEP by stochastic sampling of synchrotron light photons, Proc. EPAC, 1990, p. 762.

[2] A. Simonin, Etude ... de la forme temporelle d'une impulsion de lumière très brève issue du collisionneur LEP par effet synchrotron, Thèse , Université Paris XI, (2 déc.1991).

Integrated Devices for Single Picosecond Pulse Measurements

V. Gerbe, M. Cuzin, M.C. Gentet, and J. Lajzerowicz

LETI (CEA-Technologies Avancées), B.P. 85X,
F-38041 Grenoble Cedex, France

An autocorrelator has been developped as an X-ray monitor to measure the bunch length of the CERN synchrotron in Geneva [1]. An extented application is the temporal measurement of single laser pulses. Using the same technology, we are now developping an optoelectronic sampler.

The autocorrelator is a device able to get several points of the autocorrelation function of a single pulse (light, X-rays, ionizing particles). The device is made up with metallic striplines combined with a fast photoconductor thin film (see figure 1). When interacting with a photoconductor element connected to the bias, an electrical pulse, image of the light pulse, is created and propagates in a propagation line. Other photoconductor elements, connected to charge amplifiers, are lightenned at the same time. For a photoconductor element distant of τ_i (propagation time) from the biased element, the collected charge is proportionnal to $\int f(t).f(t+\tau_i)dt$ the value for τ_i of the autocorrelation function of the signal [2]. This is true only if the recombination time of the photoconductor is short enough compared to the pulse length. On one hand, the resistance reached by the photoconductor element must be small enough to get a high electrical signal in the propagation line and large enough for the propagating signal not to be affected by the reading (few kΩs for a 50 Ω line).

Figure 1 (b) represents the autocorrelator realized to measure the longitudinal profile of the electron bunch going around in CERN synchrotron. In this case, the synchrotron radiation interacts with the photoconductor elements. CdTe is used for a better X-ray absorption. It has been MOCVD deposited and the carrier lifetime is in the order of 10 ps [3]. Associated with an adapted electronics, the device allows to measure the profile of the bunch every 22 µs (one round trip, 45 kHz) with a time resolution better than 10 ps. On figure 2 is presented a 80 ps laser pulse autocorrelation function obtained with this device. With 1 ps lifetime semiconductor films, a similar device could be realized able to measure single pulses down to 10 ps every 10 µs (up to 100 kHz).

The optoelectronic sampler is a device able to get several points of a single electrical pulse. As illustrated on figure 3, the sampler consists in a synchronized switching of n photoconductor elements with a short laser pulse and the reading of the collected electrical charges.

The device is studied through three interacting directions: -1- design and tests of the structure, -2- choice and characterization of the material, -3- modelisation of the whole device.

-1- A test-device has been designed with different patterns devoted to the study of different photoconductor elements geometries, of the propagation in the propagation line and of the propagation and coupling in the sampling lines. This test device includes an optoelectronic sampler with a pitch of 5 ps.

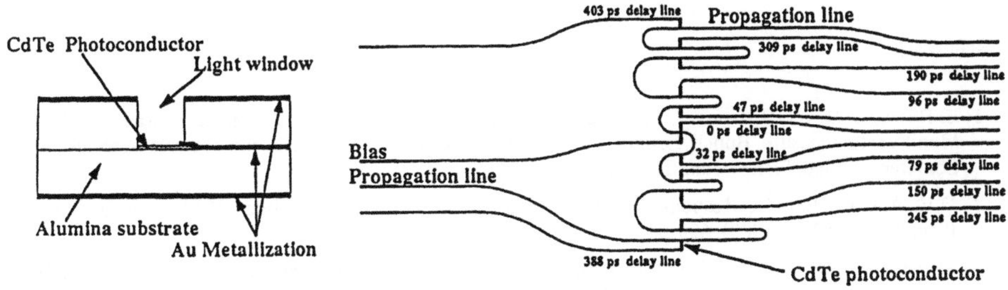

Figure 1: (a) structure and (b) design of an autocorrelator

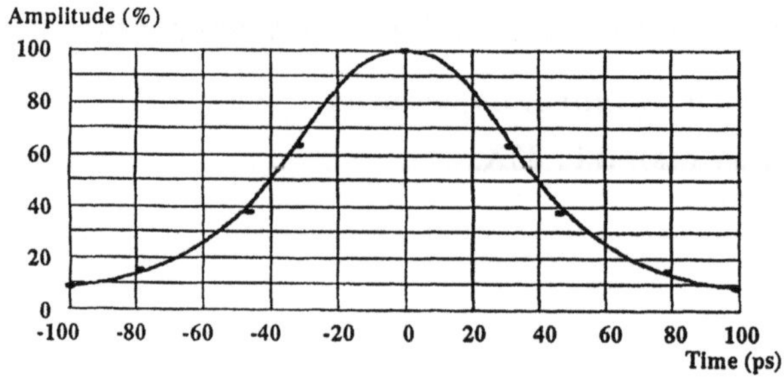

Figure 2: 80 ps laser autocorrelation function from measurement with the autocorrelator

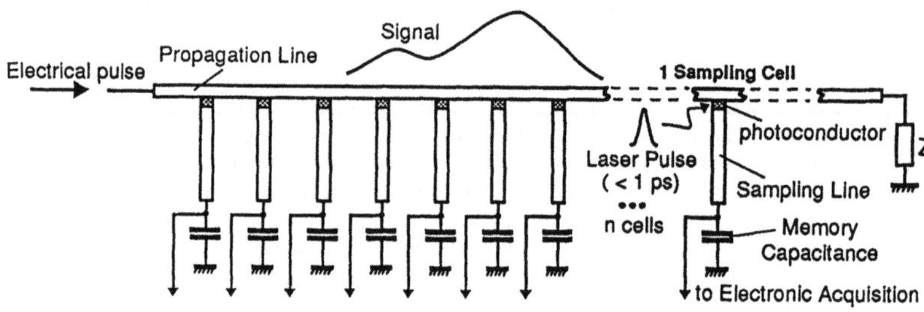

Figure 3: principle of the optoelectronic sampler

-2- The ideal photoconductor material is expected to be deposited on alumina or sapphire, to have carriers lifetime as fast as 1 ps, to show a high dark resistivity and an acceptable mobility. Under laser illumination, the resistance of the photoconductive elements must be as low as possible for the charge collection to be optimized. Studies are carried on with oxygen implanted MOCVD CdTe films and low temperature MBE deposited CdTe and GaAs. We use different characterizations including Hall measurements, time-resolved photoluminescence and reflectivity.

-3- In the considered frequency-domain (DC-100GHz), the modelisation of the device must include dispersion effect, conductor and dielectric losses, radiation phenomena and parasitic couplings. Moreover, we have to use time-domain full-wave analysis for such broadband devices. Unfortunately, there is a lack in existing tools and we consider to develop a time-domain model of the structure.

The optimization of the three above points should lead to the realization of an optoelectronic sampler with a time resolution of 5 ps (corresponding to bandwidth up to 40 GHz). We think that this sampling method could be extended down to 1 ps resolution (~200 GHz). The most critical point seems to be the synchronization between the laser pulse and the single electrical signal.

Acknowledgements — This work has been performed in collaboration with CERN, LEP laboratory for the autocorrelator and with CEA/DAM Limeil, instrumentation at PHEBUS for the optoelectronic sampler.

References :

[1] E. Rossa, C. Bovet, B. Jenny, J. Spanggaard, E. Jeanclaude, M. Cuzin, M.C. Gentet, C. Ravel "X-Ray Monitors to measure Bunch Length and Vertical Profile at LEP" European Particle Accelerator Conference, Nice, France, June 12-16, 1990

[2] D.H. Auston "Impulse Response of Photoconductors in Transmission Lines" IEEE J. of Quantum Electron. vol.19(4) (1983), 639-648

[3] M. Cuzin, C. Pierrat, E. Rossa "Mesures picosecondes avec des couches polycristallines de CdTe photoconducteur" NIM vol.A283 (1989), 310-317

The C850X Ultrafast Streak Camera: An Instrument to Study Spatially and Temporally Subpicosecond Laser–Matter Interaction

A. Mens[1], *R. Sauneuf*[1], *D. Schirmann*[1], *R. Verrecchia*[1], *P. Audebert*[2], *J.C. Gauthier*[2], *J.P. Geindre*[2], *A. Antonetti*[3], *J.P. Chambaret*[3], *G. Hamoniaux*[3], and *A. Mysyrowicz*[3]

[1]C.E.A. Centre d'Etudes de Limeil-Valenton, F-94195 Villeneuve St.-Georges Cedex, France

[2]Laboratoire de Physique des Milieux Ionisés, Ecole Polytechnique, F-91128 Palaiseau, France

[3]Laboratoire d'Optique Appliquée, ENSTA, Ecole Polytechnique, F-91128 Palaiseau, France

Introduction:

Streak cameras used in laser-matter interaction experiments by CEA at CEL-V have been the subject of years of tube development with the help of Philips Components (now Philips Photonics). In particular the concept of bilamellar electron optics was introduced to simultaneously improve the spatial and temporal performances of streak tubes (1).

The last type of bilamellar tube is P850X (fig.1). The prototype of camera which uses it , C850X, has been designed at CEL-V and transformed into an industrial product by ARP. We recall the calculations of its dynamic spatial and temporal resolutions.

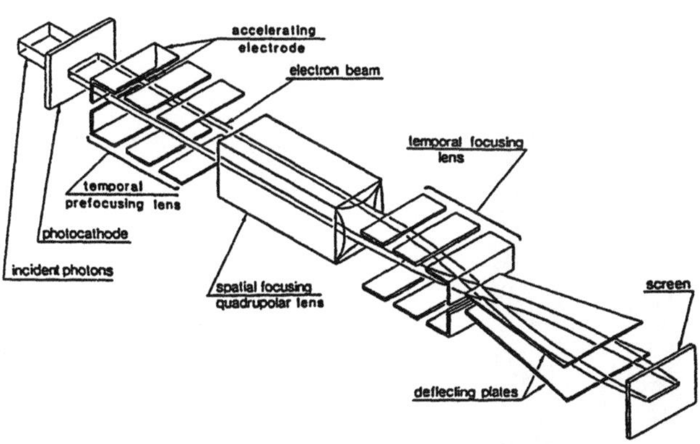

Figure 1
The P850X tube

Resolution in the UV range:

The LOA has a dye laser which produces 60 femtoseconds (FWHM) pulses at a 620 nm wavelength. One of them has been frequency doubled and focused on a Pd photocathode equipped with a 16 lp/mm resolution chart. The response of the camera is about 2 ps under these conditions, and the contrast of the chart shows that the spatial resolution is far better than 16 lp/mm. We show the

Springer Series in Chemical Physics, Vol. 55 Ultrafast Phenomena VIII
Editors: J.-L. Martin · A. Migus · G.A. Mourou · A.H. Zewail

experimental setup and discuss the fact that these values are not the limits of the camera.

Resolution in the X-ray range:

This laser is also able to deliver 1 mJ pulses with a 20 Hz repetition rate. A single pulse is focused on a target with a focal spot diameter better than 10 μm, producing a flux of 10^{17}W/cm^2 on it.

Its X-ray emission is observed through absorbers which select energies above 1.5 keV. Theory predicts that their duration, depending on the nature of the target, is some tens or hundreds of femtoseconds. The experimental results given by the camera (such as that of fig.2, obtained with a Cu target and a CsI photocathode) are pulses of rise time and FWHM below 2 picoseconds, which can therefore be taken for an upper value of the temporal resolution of the camera.

In this paper we show results obtained with various targets and photocathodes, and the conclusions that we can make on temporal and spatial resolutions, which are at the best level achieved so far (3), and on the dynamic range of this instrument.

We also demonstrate (for the first time at energies over 1 keV) its interest to study subpicosecond laser-matter interaction, by proving the effect of the laser pulse contrast on the level of the X-ray emission, but also on its duration, which confirms theoretical predictions (4).

Figure 2
Experimental image (a)
Cross section along
temporal axis (b)

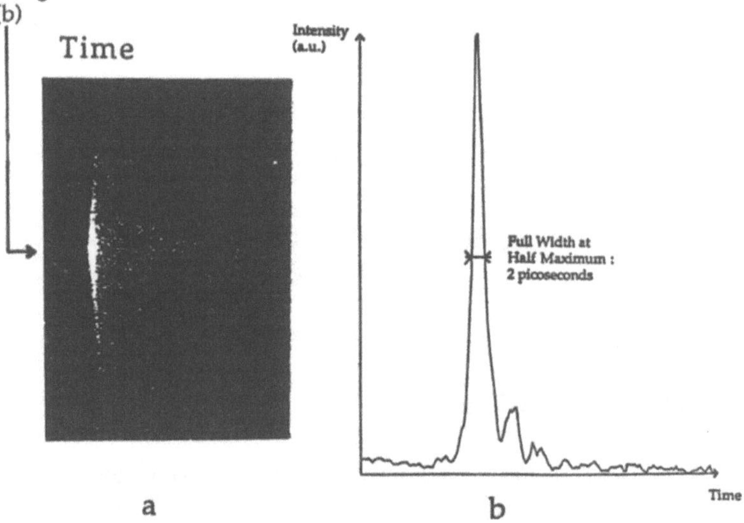

References:

1."P600/650 X-ray streak camera with optimized spatio-temporal resolution" B.Boutry et al., High speed photography, San Diego, SPIE Vol.348 (1982)

2."C850X picosecond high resolution streak camera" A.Mens et al., High speed photography and photonics, Cambridge, SPIE Vol.1358 (1988)

3."X-ray streak camera with 2 ps response" M.M.Murnane et al., Appl.Phys.Lett.56 (20) 14 May 1990

4."X-ray and optical diagnostics of femtosecond laser-produced plasmas" P. Audebert et al., this conference

Distortion of a 6 fs Pulse in the Focus of a BK7 Lens

Zs. Bor and Z.L. Horváth

Department of Optics and Quantum Electronics, JATE University,
Dóm tér 9, H-6720 Szeged, Hungary

It was shown in [1-3] that the chromatic aberration of a lens causes a distortion of a femtosecond pulse front. In this paper the behaviour of the pulse front in the vicinity of the focal plane of a lens is described by using wave optics. For the calculation the input pulse (6 fs, 620 nm) is decomposed into monochromatic plane waves, which are converted by the lens into Airy-type of diffraction patterns. Due to chromatic aberration the position of the diffraction patterns are slightly shifted with respect to each other along the optical axis. The pulse front is obtained as a result of the superposition of diffraction patterns of the components [4,5].

The calculations were made for a BK7 lenses having the same focal length (f=30 mm) but different diameters (d=1,2,4,7 and 10 mm). The total input energy of the pulse was regarded to be constant. Spatially homogeneous illumination was assumed.

In agreement with ref [1] the broadening of the pulse in the focus is proportional to the square of the diameter of the lens. For d=10 mm in the focus (z=0) the input pulse broadened to 60.4 fs (Fig. 1e). For smaller diameters, as expected from usual diffraction, the angular size of the spot is larger (Fig. 1a). Therefore the highest intensity was obtained for an intermediate (d=4 mm) diameter (Fig. 1c).

In Fig. 1n and 1o in front of the main pulse an intensive spike, called boundary wave pulse can be seen. This pulse is caused by the interference of the boundary waves generated by the edge of the aperture of the lens.

This work has been supported by the OTKA Foundation (No 3055 and 3056) and the U.S.-Hungarian Science and Technology Joint Fund (No 061/90).

References

1. Z. Bor, J. Mod. Opt. **35** (1988) 1907
2. S. Szatmari, G. Kuhnle, Opt. Commun. **69** (1988) 60
3. Z. Bor, Z. Gogolak amd G. Szabo, Opt. Lett. **14** (1989) 862
4. Z. Bor and Z. L. Horváth, "Dye Lasers 25 Years" Ed.: M. Stuke, Topics in Appl. Phys. **vol. 70** Springer-Verlag 1992, p. 87
5. Z. Bor and Z. L. Horváth, "Distortion of Femtosecond Pulses in Lenses. Wave Optical Description", accepted for publication in Opt. Commun.

Springer Series in Chemical Physics, Vol. 55 **Ultrafast Phenomena VIII**
Editors: J.-L. Martin · A. Migus · G.A. Mourou · A.H. Zewail

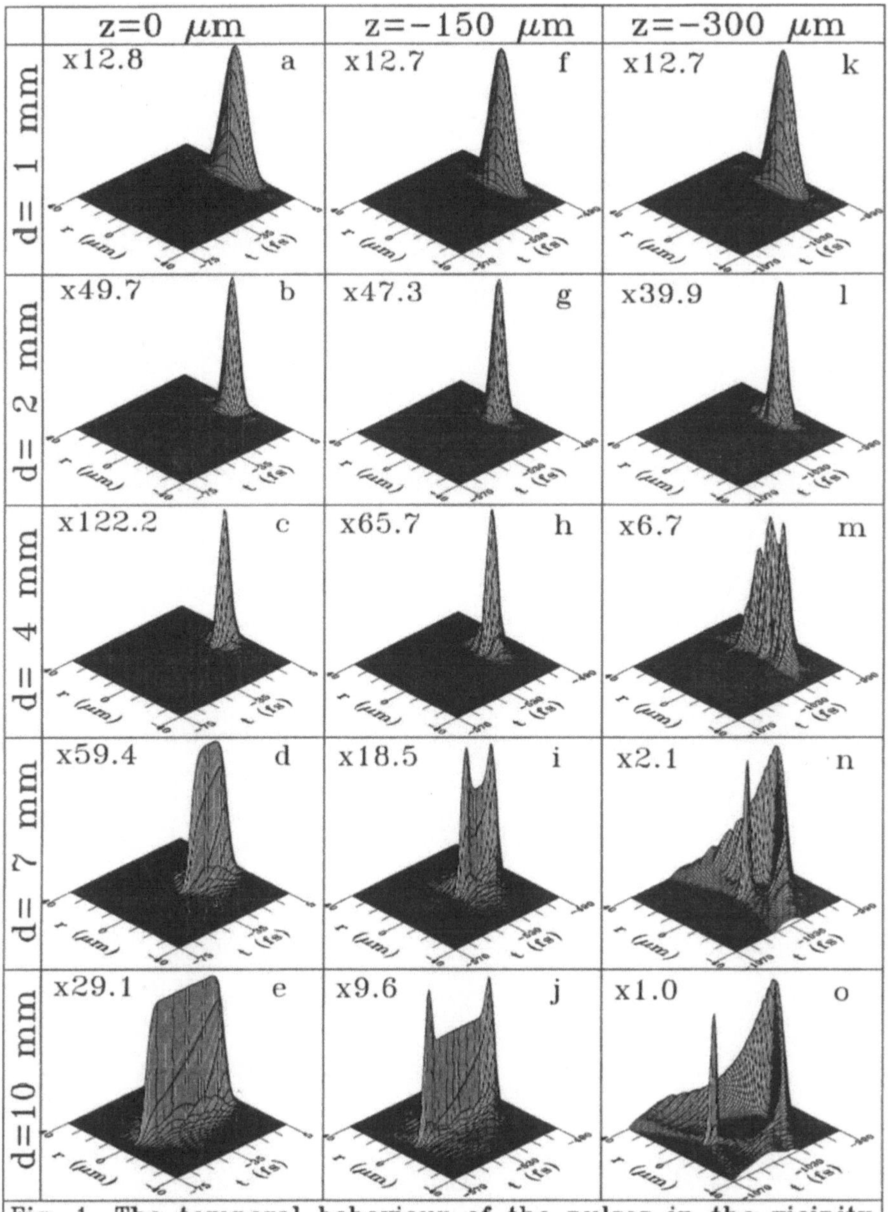

Fig. 1. The temporal behaviour of the pulses in the vicinity of the focus. r and z are the distances from the optical axis and from the focal plane respectively. The numbers on the insets indicate the value of the peak intensity in relative units.

Part IV

Tools: Sources and Amplifiers

Modelocking, Stabilizing, and Starting Ultrashort Pulse Lasers

E.P. Ippen

Department of Electrical Engineering and Computer Science,
Research Laboratory of Electronics, Massachusetts Institute of Technology,
Cambridge, MA 02139, USA

Introduction: Ultrashort pulse lasers have for many years relied upon the use of real saturable absorber elements to achieve passive modelocking. But, since real saturable absorbers have real recovery times, femtosecond pulse generation has required the use of gain media that also saturated dynamically. Pulse formation in this way is called slow saturable absorber modelocking[1]. The CPM dye laser, in which pulses as short as 28 fs have been produced[2], is testimony to how well it works. Recently new methods for modelocking have been discovered that work with a wide range of solid-state lasers. These methods utilize ultrafast reactive nonlinearities to create an effective, fast saturable absorber action. Dynamic gain saturation is not required. Pulses as short as 17 fs from Ti:sapphire[3,4] and Nd:glass[5] systems are being reported. These recent advances have prompted a re-visiting of the theory of fast saturable absorber modelocking[6] as well as new analyses of how it initiates and stabilizes. In this paper we discuss some elements of the understanding that has developed.

Modelocking: The basic requirement for modelocking is that higher gain be available to a pulse in the laser than to the cw background. Some amplitude modulation, either active or passive must operate to create a window of net gain for the peak of the pulse and net loss for the wings. In the start-up phase some random fluctuation grows in energy and is shortened on every roundtrip by this windowing process. Consider first the shortening process. It is different for different types of modelocking. An active modulator multiplies the pulse by the same modulation function every time. So, as the pulse gets shorter the shortening effect of the modulation decreases. In slow absorber modelocking both the absorber and the gain saturate with the integral of pulse intensity. The window that is created depends only on the pulse energy; the shortening effect stays the same, independent of pulsewidth. In fast absorber modelocking the nonlinear shaping is proportional to the instantaneous intensity; so that the effect becomes stronger as the pulse gets shorter. It is easy to derive analytical expressions (within the perturbational, parabolic shaping approximation[6]) that show pulse shortening proportional to τ^2, a constant, and $1/\tau$, respectively, for the three cases. Figure 1 illustrates this comparison of types in terms of the fractional change in pulsewidth per roundtrip (i.e. the pulse shortening velocity (PSV)[7]) as a function of inverse pulsewidth. The hatched region of the figure indicates where the shortening begins to be limited by bandwidth and group velocity dispersion (GVD). This simple picture becomes less accurate for ultrashort pulse durations where the interaction of self-phase modulation (SPM) and GVD becomes significant; but it reveals quite a bit about the early transient behavior. The fast saturable absorber produces very little shortening per pass at first; small competing effects can prevent self-starting. Once started, however, the fast absorber becomes a powerful shortening mechanism. Numerical calculations show that a PSV of only several percent produces strong modelocking.

Figure 2 illustrates several methods for creating a fast saturable absorber effect with a reactive (index of refraction) nonlinearity. The coupled-cavity configuration was

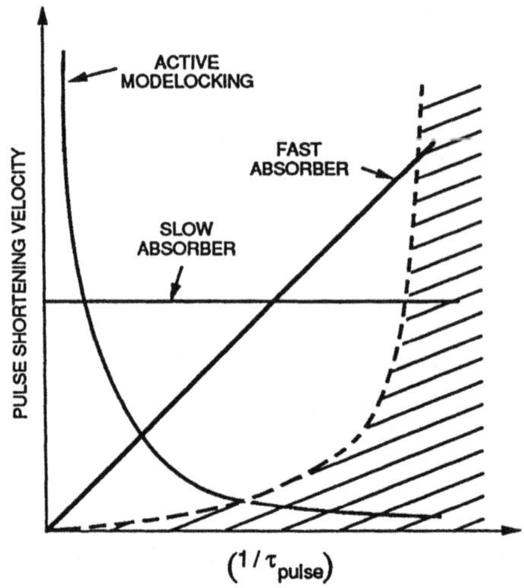

Fig. 1. The fractional pulse shortening per roundtrip vs. inverse pulsewidth, for three types of modelocking.

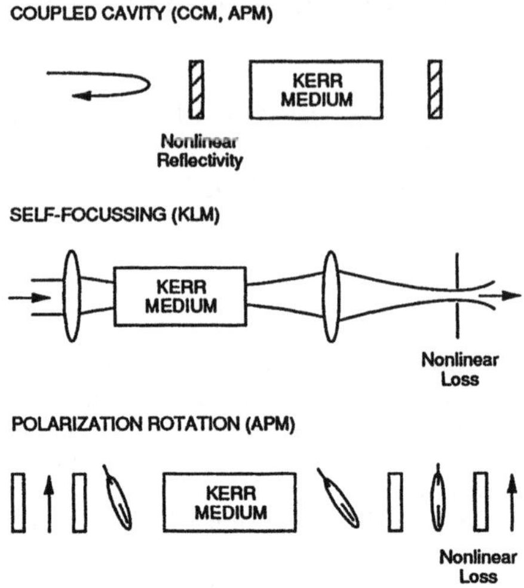

Fig. 2. Device configurations for producing a fast saturable absorber effect with nonlinear refractive index.

investigated first in conjunction with the soliton laser[8] and then discovered to be effective in non-soliton lasers as well[9,10]. Its pulse shaping properties have been described in terms of additive pulse modelocking (APM)[11,12] whose applications include F-center[11,13], Ti:sapphire[14,15], Cr:forsterite[16], Nd:YAG[17,18], Nd:YLF[19,20], Nd:glass[21], and fiber lasers[22]. Self-modelocking in a single cavity was first demonstrated in a dispersion-compensated Ti:sapphire laser[23]. Its realization via dynamic self-focussing[24] has been developed commercially and dubbed Kerr lens modelocking (KLM)[25]. Other recent applications include Cr:LiSAF[26,27] and Cr:forsterite[28]. Finally, nonlinear polarization rotation, which may also be thought of as APM of two polarization components in the same cavity, has been realized most effectively in fiber lasers[29]. The advantages that all of these methods have over real saturable absorbers are essentially instantaneous response, low insertion loss, and designable nonlinearity. In addition to introducing self-amplitude modulation, however, they also introduce a related[30] quantity of self-phase modulation. This can shorten pulses further; it also introduces stability problems.

Steady-state Pulsewidth: By the time the pulse reaches the steady state it must reproduce itself exactly after each transit; i.e. the multiplicative effect of all the elements must equal one. To describe this process in detail, one may have to resort to numerical computation; but to obtain an approximate (but analytical!) solution, we turn to the master equation approach developed by Haus[31,32]. We assume that the changes produced by the elements are small and therefore additive, and we obtain a second order nonlinear differential equation describing the steady state pulse amplitude a:

$$\left[-j\psi-(\ell+jx)+g\left(1+\frac{1}{\Omega_g^2}\frac{d^2}{dt^2}\right)+jD\frac{d^2}{dt^2}+(\gamma-j\delta)|a|^2\right]a=0\,.$$

Here ψ and x are phase shifts per pass necessary for closure, ℓ and g are the small signal loss and gain, the real and imaginary second derivative terms (Fourier transforms of

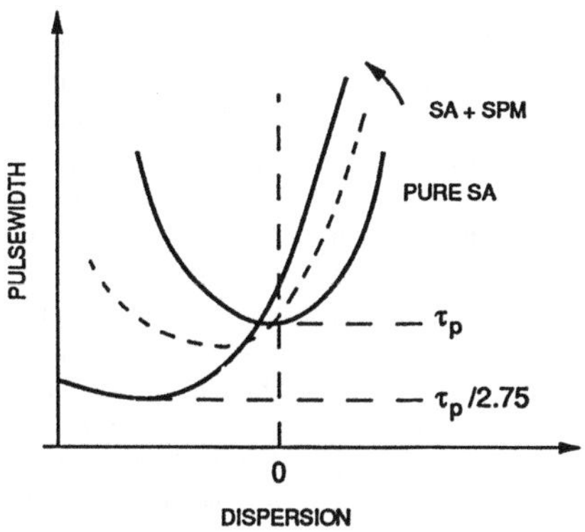

Fig. 3. Steady state pulsewidth vs. GVD for cases with and without SPM.

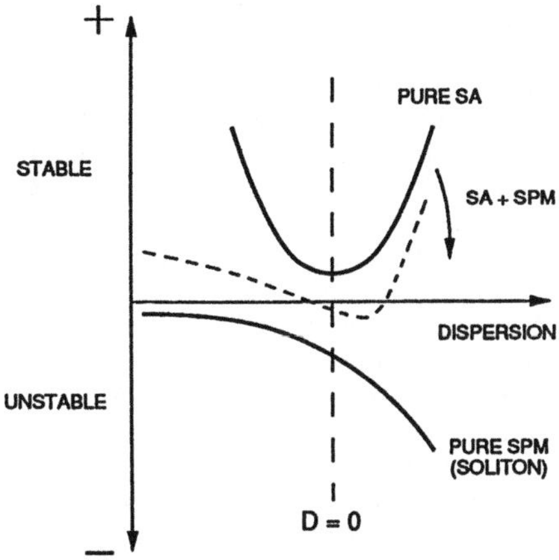

Fig. 4. The stability of the steady-state pulse solution with respect to cw.

quadratic phase dependences in the frequency domain) produce temporal spreading due to bandwidth-limiting and GVD, and γ and δ are coefficients for saturable absorber action and SPM (both proportional to pulse intensity). Remarkably, this equation has an analytic solution[33]:

$$a = A\,\mathrm{sech}(t/\tau)\exp[j\beta\, ln\, \mathrm{sech}(t/\tau)] \;,$$

where β is a chirp parameter and τ is the pulsewidth. One obtains equations for each by simply plugging the solution form into the master equation. Figure 3 illustrates how the pulsewidth τ varies with GVD for a fixed cavity bandwidth Ω_g. For a zero SPM and GVD it has a minimum value of $\tau = 4g/\gamma W \Omega_g^2$, where W is the pulse energy. As SPM is added to the equation, the point of minimum pulsewidth moves to negative GVD where the chirp is compensated. It is interesting to note, however, that by introducing soliton effects, one can reduce the pulse duration by at most a factor of 2.75 over what it would be with the self-amplitude modulation alone[32].

Stability: It is important to point out that the solutions thus predicted by the master equation are not always stable with respect to cw oscillation. The condition for stability is not just that a steady-state pulseshape be produced but that there must be net loss both before and after the pulse. To elucidate this condition we plot g_{pulse} - g_{cw} in Figure 4. From it we deduce another interesting fact: a pure soliton laser (i.e. one without any saturable absorber) is always unstable; that is, cw has higher gain and will always take over. The saturable absorber not only determines the ultimate pulsewidth, it is required for stability. SPM can drive the system unstable unless sufficient GVD is introduced at the same time.

Higher order effects: The master equation incorporates several approximations that are certainly violated by some practical laser systems. To get things right, one has to resort to numerical simulation[34,35]. Nevertheless, the effects of higher order terms can be discussed in the context of the analytical model described above. Consider some of them: a) At high powers the saturable absorber saturates fully, so that its modulating effect no longer follows the pulseshape. It still provides a window of gain for stability, but this window becomes flatter and extended in time. Soliton shaping can keep the pulse somewhat shorter than the width of the amplitude shaping window, but the steady-state pulse duration will be longer than that predicted by the analytical theory alone. b) In high power solid-state lasers and in fiber lasers, SPM produces phase shifts that can no longer be considered small. The laser is pushed far into the soliton regime. There it is more susceptible to pulse instability and soliton quantization[36]. c) At very short pulsewidths, third-order dispersion has been shown to be a limiting factor[37–39]. It should be minimized. In the master equation formalism, it introduces a third-order time-derivative term for which no analytical solution has yet been found. Numerical simulations[40] demonstrate that pulses become longer and asymmetrical in shape. Narrow-band features observed in the output spectrum may also be a result of third-order dispersion[40]. d) When the changes produced by individual elements is large, the pulse may have a different duration in different parts of the cavity, and the ordering of the elements can be important. This is particularly true with strong SPM. A laser operating under these conditions has been dubbed a solitary laser[34]. The pulse duration may be approximated by[34] $\tau = (3.53D/\delta W) + \alpha\delta W$ where α is an empirical constant. The first term is the prediction of the master equation and the second term gives the increase due to large changes. Also, the strong periodic modulation of the pulse scatters energy out of the pulse into the cw background. This too can generate narrow-band spectral components in the output[41].

Self-starting: Self-starting is problem inherent to fast saturable absorber systems. As Figure 1 illustrates, they have a very slow pulse-shortening-velocity when the pulse is long. In theory, if you wait long enough a short pulse should develop out of any initial fluctuation. In practice, this does not always happen. One reason can be dynamic gain saturation. When it is significant, only fluctuations shorter than a given duration will develop[42]. Another reason may be competing pulse scattering processes. If the pulse does not shorten significantly within a cavity mode coherence time[43], it will be dispersed. This sets a power threshold for self-starting. Mode pulling due to spurious reflections is a prime culprit for producing a short coherence time[44]. One can try to maximize longitudinal mode beating and to keep reflections to a minimum. Still, there is not yet any completely satisfying way to ensure self-starting. Current methods (other than banging on the table) include synch-pumping, active modulation, tilting plates, moving mirrors, and adding a real saturable absorber. Each has its drawbacks. The development of practical, reliable ultrashort pulse sources may depend upon solving this self-starting problem.

The author would like to acknowledge collaboration on this work with H.A. Haus and J.G. Fujimoto, valuable discussions with J. Moores, J. Jacobson and J.Chen, and support from A.F.O.S.R., N.S.F. and the Joint Services Electronics Program.

References

[1] H.A. Haus, IEEE J. Quant. Electron. **12**, 169 (1976).
[2] J.A. Valdmanis, R. L. Fork, and J. P. Gordon, Opt. Lett. **9**, 522 (1985).
[3] C.P. Huang, M.T. Asaki, S. Backus, N. Nathel, M.M. Murnane, and H.C. Kapteyn, ICUP, paper ThE2, these proceedings.
[4] F. Krausz, Ch. Spielmann, T. Brabec, and E. Wintner, ICUP, paper MA3, these proceedings.
[5] A.J. Schmidt, ICUP, paper MA2, these proceedings.
[6] H.A. Haus, J. Appl. Phys. **46**, 3049 (1975).
[7] M.S. Stix and E.P. Ippen, IEEE J. Quant. Electron. **19**, 520 (1983).

[8] L.F. Mollenauer and R.H. Stolen, Opt. Lett. **9**, 13 (1984).
[9] K.J. Blow and B.P. Nelson, Opt. Lett. **13**, 1026 (1988).
[10] P.N. Kean, X. Zhu, D.W. Crust, R.S. Grant, N. Langford, and W. Sibbett, Opt. Lett. **14**, 39 (1989).
[11] M. Mark, L.Y. Liu, K. Hall, H. Haus, and E. Ippen, Opt. Lett. **14**, 48 (1989).
[12] E.P. Ippen, H.A. Haus, and L.Y. Liu, J. Opt. Soc. Am. B **6**, 1736 (1989).
[13] C.P. Yakymyshyn, J.F. Pinto, and C.R. Pollock, Opt. Lett. **14**, 621 (1989).
[14] J. Goodberlet, J. Wang, J. Fujimoto, and P. Schulz, Opt. Lett. **14**, 1125 (1989).
[15] P.M.W. French, J.A.R. Williams, and J.R. Taylor, Opt. Lett. **14**, 686 (1989).
[16] A. Sennaroglu, T.J. Carrig, and C.R. Pollack, CLEO'92, paper CTuI2.
[17] J. Goodberlet, J. Jacobson, J.G. Fujimoto, P.A. Schulz, and T.Y. Fan, Opt. Lett. **15**, 553 (1990).
[18] L.Y. Liu, J.M. Huxley, E.P. Ippen, and H.A. Haus, Opt. Lett. **15**, 553 (1990).
[19] J.M. Liu and J.K. Chee, Opt. Lett. **15**, 685 (1990).
[20] G.P.A. Malcolm, P.F. Curley, and A.I. Ferguson, Opt. Lett. **15**, 1303 (1990).
[21] F. Krausz, Ch. Spielmann, T. Brabec, E. Wintner, and A.J. Schmidt, Opt. Lett. **15**, 737 (1990).
[22] M.E. Fermann, M. Hofer, F. Haberl, A.J. Schmidt, and L. Turi, Opt. Lett. **16**, 244 (1991).
[23] D.E. Spence and W. Sibbett, J. Opt. Soc. Am. B **8**, 2053 (1991).
[24] F. Salin, J. Squier, and M. Piche, Opt. Lett. **16**, 1674 (1991).
[25] D.K. Negus, L. Spinelli, N. Goldblatt, and G. Feuget, OSA Proc. on Advanced Solid State Lasers **10**, 120 (1991).
[26] N.H. Rizvi, P.M.W. French, and J.R. Taylor, Opt. Lett. **17**, 877 (1992).
[27] P. Li Kam Wa, B.H.T. Choi, H.S. Wang, and A. Miller, CLEO'92.
[28] A. Seas, V. Petricevic and R.R. Alfano, Opt. Lett. **17**, 937 (1992).
[29] M. Hofer, M.E. Fermann, F. Haberl, M.H. Ober, and A. Schmidt, Opt. Lett. **16**, 502 (1992).
[30] H.A. Haus, J.G. Fujimoto, and E.P. Ippen, IEEE J. Quant. Electron., (1992).
[31] H.A. Haus, **Waves and Fields in Optoelectronics**, Prentice-Hall, 1984.
[32] H.A. Haus, J.G. Fujimoto, and E.P. Ippen, J. Opt. Soc. Am. B **8**, 2068 (1991).
[33] O.E. Martinez, R.L. Fork, and J.P. Gordon, J. Opt. Soc. Am. B **2**, 723 (1985).
[34] T. Brabec, Ch. Spielmann, and F. Krausz, Opt. Lett. **16**, 1961 (1991).
[35] J. L.A. Chilla and O.E. Martinez, CLEO'92, paper CTuP4.
[36] D.J. Richardson, A.B. Grudinin, and D.N. Payne, CLEO'92, paper JWA5.
[37] C.P. Huang, H.C. Kaptsyn, J.W. McIntosh, and M.M. Murnane, Opt. Lett. **17**, 139 (1992).
[38] F. Krausz, Ch. Spielman, T. Brabec, E. Wintner, and A.J. Schmidt, Opt. Lett. **17**, 204 (1992).
[39] J.M. Jacobson, A.G. Jacobson, K. Naganuma, H.A. Haus, and J.G. Fujimoto, CLEO'92, paper CTuU2.
[40] J. Moores, private communication.
[41] N. Pandit, D.U. Noske, S.M.J. Kelly, and J.R. Taylor, CLEO'92, paper JWA7.
[42] E.P. Ippen, L.Y. Liu, and H.A. Haus, Opt. Lett. **16**, 183 (1990).
[43] K. Krausz, T. Brabec, and Ch. Spielmann, Opt. Lett. **16**, 135 (1991).
[44] H.A. Haus and E.P. Ippen, Opt. Lett. **16**, 1331 (1991).

17 fs Pulses from a Mode-Locked Ti:Sapphire Laser

C.P. Huang[1], *M.T. Asaki*[1], *S. Backus*[1], *H. Nathel*[2], *H.C. Kapteyn*[1], *and M.M. Murnane*[1]

[1]Department of Physics, Washington State University, Pullman, WA 99164-2814, USA
[2]Lawrence Livermore National Laboratory, Livermore, CA 94550, USA

Abstract

Optimization of higher order intracavity dispersion compensation in a self-modelocked Ti:sapphire laser resulted in the generation of 17 femtosecond pulses at 810 nm with 43 nm FWHM bandwidth.

1. Experiment

We have generated 17 femtosecond duration pulses directly from a self mode-locked ti:sapphire laser.[1] The laser output has wavelength centered at 817 nm, a pulse repetition rate of 80 MHz, and an average power of 500 mW. By minimizing the amount of material inside the laser cavity and choosing the correct glass for the intracavity prism pair, third-order dispersion in the laser can be significantly reduced compared with previous designs, resulting in a dramatic reduction in pulse width. In previous work, we demonstrated the generation of 32 femtosecond pulses from a mode-locked ti:sapphire laser.[2] In this paper, we report further progress in minimizing intracavity dispersion in a self mode-locked ti:sapphire laser. The generation of sub-20 fs pulses directly from a laser is unprecedented. This laser is reliable, and there is no degradation in performance, stability, or ease of use compared with longer pulse-duration ti:sapphire lasers. Our result demonstrates that the exceptionally broad bandwidth of ti:sapphire, which is large enough to support <<10 fs pulses,[3] can be utilized to generate pulses shorter than has been possible with any other type of laser material to date.

The laser cavity consists of two end mirrors, two 10 cm radius of curvature mirrors focusing into a ti:sapphire rod, and a pair of intracavity prisms for dispersion compensation. The ti:sapphire crystal is 9 mm long, 0.11% doped material with FOM>150 (Union Carbide Inc.). The 10 cm mirrors are Spectra-Physics, Inc. 800-900 nm range mirrors for a 3900 CW ti:sapphire laser. The ti:sapphire crystal was pumped by 5 Watts of all-lines green from an argon-ion laser, through one of the 10 cm mirrors, and focused by a 12.5 cm f.l. lens. The two end-mirrors of the cavity are flat, with transmission of 6.5% and 2.5% and a center wavelength of ~790 nm. The total cavity length is 185 cm, with 85 cm in one arm of the laser, and 90 cm on the side of the dispersion compensating prisms. No apertures are used in the laser cavity to induce mode-locking, although there is some clipping of the beam at the apex of the prisms.

The prisms used in this laser are made of Schott LaKL21 glass, cut to an apex angle (63°) such that the angle of incidence corresponds to Brewster at minimum deviation. The prism separation was 51 cm. This prism glass was chosen because

Springer Series in Chemical Physics, Vol. 55 **Ultrafast Phenomena VIII**
Editors: J.-L. Martin · A. Migus · G.A. Mourou · A.H. Zewail

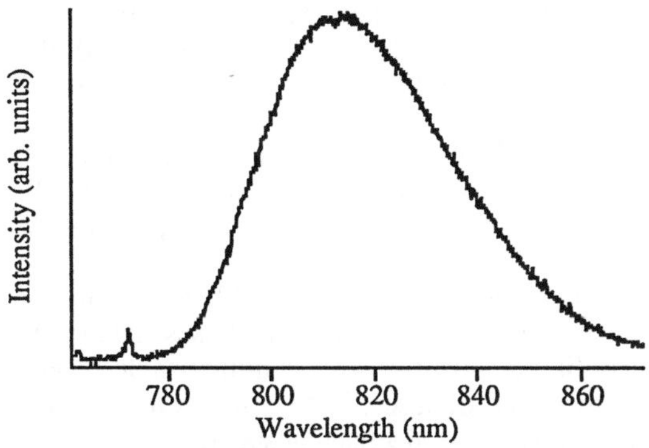

Figure 1: Wavelength spectrum of laser output from the undispersed end of the laser. The FWHM is 42.5 nm, with a center wavelength of 817 nm.

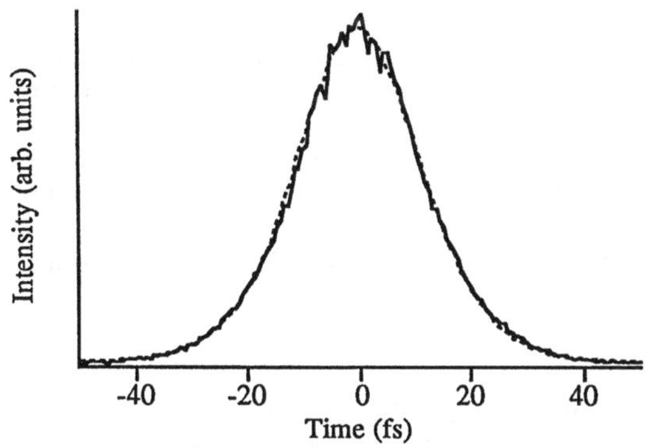

Figure 2: Autocorrelation trace of 16.9 femtosecond pulse (solid line). Also shown is a $Sech^2(t)$ fit to the data (dashed line).

(1) it is a relatively stable, hard glass which can be polished optically flat; (2) the prism separation required for compensation of the linear (second-order) dispersion introduced by the ti:sapphire rod can be accommodated in a 1.5-2.0 meter cavity length; and (3), of the glasses which fit the first two criteria, LaKL21 introduces the minimum amount of third-order dispersion in the laser cavity, according to our calculations.

Figure 1 shows the spectrum of the output of this laser through the spectrally-undispersed end. The spectrum full-width at half-maximum is 42.5 nm. This bandwidth corresponds to a transform-limited $sech^2$ pulse duration of 16.5 fs. Since the pulses from the laser disperse as they pass through the substrate of the output coupler (3/8" fused silica), we used an extracavity double-pass prism compressor arrangement (BK7 prisms 121 cm apart).

Figure 2 shows a representative autocorrelation trace, along with a fit to the autocorrelation of a $sech^2$ pulse. This fit yields an autocorrelation FWHM of 26.1 fs, corresponding to a pulse width of 16.9 fs. The time-bandwidth product of the laser output is thus $\delta\nu\delta\tau = 0.323$. We estimate that the extracavity prism separation is just sufficient to compensate for the output coupler and the autocorrelator optics; thus it appears that there is little chirp on the pulse at the point where it exits the cavity.

The output power of the laser was 370 mW from this end of the laser, and 130 mW from the dispersed end, resulting in a total output power of 500 mW for 5 W pump power. The output mode was primarily a round TEM00 mode, with <5% light in higher-order modes.

Once the prism configuration is optimized for wide-bandwidth operation, the laser consistently generates pulses of bandwidth > 40 nm, and the pulses can always be recompressed to near the transform limit. Thus, the laser is quite usable at sub-20 fs pulsewidths. As is usual with this type of laser, it does not self-start in a pulsed mode when lasing is interrupted by a sizable mechanical shock.

In conclusion, We have demonstrated a simple and reliable laser capable of generating pulses of unprecedented short duration. Our results confirm the fact that linear third-order dispersion is the primary process limiting pulse duration in these lasers. Further improvement in pulse duration is likely given that this third-order dispersion is readily computable. For many applications, this laser should be a desirable (i.e. much lower cost, and much high stability and pulse repetition rate) substitute for ultrashort laser systems based on amplification, continuum generation, and pulse compression.[4]

This project was supported by National Science Foundation grant ECS-9204033 and PHY-9158373, by funds from Washington State University, by the U.S. Department of the Air Force, by generous donations from Tektronix Inc. and Union Carbide, and through a collaboration with Lawrence Livermore National Laboratory (W-7405-ENG-48).

1. D. E. Spence, P. N. Kean, W. Sibbett, Opt. Lett. **16**, 42 (1991).
2. C.-P. Huang, H. C. Kapteyn, J. W. McIntosh, M. M. Murnane, Opt. Lett. **17**, 139 (1992).
3. P. F. Moulton, J. Opt. Soc. Am. B **3**, 125 (1986).
4. P. C. Becker, H. L. Fragnito, R. L. Fork, F. A. Beisser, C. V. Shank, Appl. Phys. Lett. **54**, 411 (1989).

Design Considerations for Femtosecond Ti:Sapphire Oscillators

Ch. Spielmann, P.F. Curley, T. Brabec, E. Wintner, A.J. Schmidt, and F. Krausz

Quantenelektronik und Lasertechnik, Technische Universität Wien, Gusshausstr. 27/359/3, A-1040 Wien, Austria

Introduction

Since the discovery of self-mode locked operation in Ti:sapphire oscillators,[1] there has been rapid development in the generation of sub-100fs optical pulses. The gain bandwidth of Ti:sapphire extends over $> 200nm$ and has the potential for supporting pulses of less than 10fs. In this paper we discuss the experimental limitations to ultrashort pulse generation and present some guidelines for the optimization of femtosecond oscillators.

Solitary Lasers

In these short pulse lasers the attainable pulse generation is determined by the critical interplay between intracavity self-phase modulation (SPM) in Kerr media, and negative group delay dispersion (GDD). In contrast to soliton lasers,[2,3] where these effects occur simultaneously in an optical fibre, solitary lasers such as Kerr Lens Modelocked (KLM) Ti:sapphire have separate, discrete elements of SPM and GDD, (Fig. 1). As a consequence of this separate pulse shaping, instabilities arise requiring stabilisation by amplitude modulation, and the steady-state pulse is longer than the corresponding soliton pulse length. A semi-empiricial solution[4] to this solitary model yields, for the solitary pulse duration:-

$$\tau(z) = \frac{3.53|D|}{\phi W} + \alpha(z)\phi W \tag{1}$$

Where D (< 0) is the net round trip GDD, W is the intracavity pulse energy, and ϕ is the round trip nonlinear phase shift (W^{-1}) due to the Kerr medium. The constant $\alpha(z)$ is dependent on the cavity position z, and has the value $\alpha_1 = 0.10$ at the output coupler of the dispersive cavity end (OC1), and $\alpha_2 = 0.25$ at the non-dispersive end (OC2). It

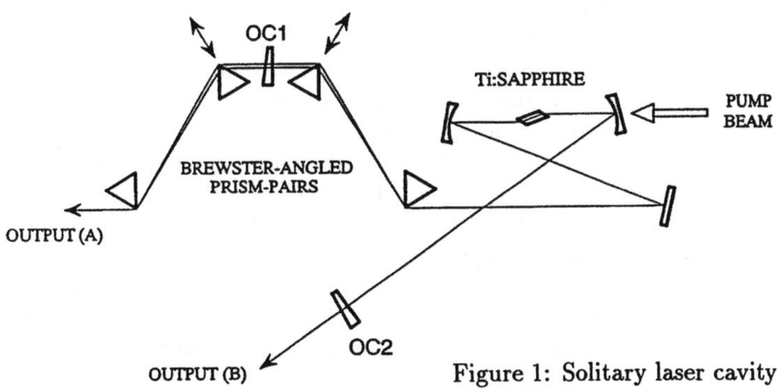

Figure 1: Solitary laser cavity schematic

Springer Series in Chemical Physics, Vol. 55 Ultrafast Phenomena VIII
Editors: J.-L. Martin · A. Migus · G.A. Mourou · A.H. Zewail

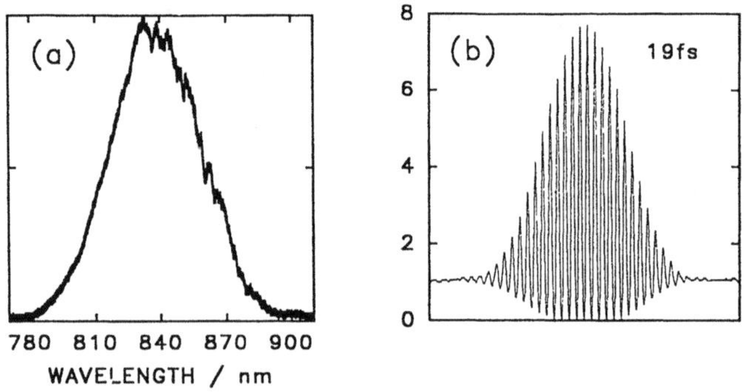

Figure 2

can be shown generally[5] that the pulses at the cavity ends are bandwidth limited, and therefore the pulse spectrum must also change as a function of cavity position. Hence the conventional description of mode-locked lasers in terms of phase-locked resonator modes is not directly applicable to these systems. The primary implication of the above results is that the shortest pulses are present at the output of the dispersive cavity end. These theoretical predictions have been investigated experimentally using a synchronously pumped Ti:sapphire laser (Fig 1). Employing an 8mm long Ti:sapphire crystal and a F2 prism pair with a separation of 40cm, where $|D| \approx 500 fs^2$, W=46nJ and $\phi \approx 1.9 \times 10^{-6} W^{-1}$, the predicted pulse durations are 30fs and 43fs for the dispersive and non dispersive cavity ends, respectively. Experimentally, the pulse durations are measured as 35fs and 48fs, respectively. The spectra are recorded simultaneously using a CCD multichannel spectrum analyser, with FWHM widths of 21.0nm and 14.7nm for the dispersive and non dispersive cavity ends, respectively. The corresponding time-bandwidth products are $\tau\Delta\nu$=0.35 and 0.36, respectively.[5]

Cavity Design Considerations.

For sub-50fs pulse generation the adverse effects of third-order dispersion (TOD) D_3 become dominant and must also be considered. Theoretical investigations[6] have indicated for stable bandwidth-limited pulse generation the condition,

$$|D_3| < 0.2|D|\tau \tag{2}$$

must be satisfied. The main source of third order dispersion is the intracavity prism pair, and the conclusion that can be drawn from Eq.(2) is that in order to achieve a low τ, the prism pair must have a low third-order dispersion relative to second order dispersion. Table 1 shows dispersion ratios for prism pairs of common materials.[7] Note the reduction of the magnitude of net TOD with increasing ϕ''/ϕ'''.

From table 1, it is evident that the lower the second order dispersion of a prism pair the higher the ratio ϕ''/ϕ'''. Therefore the overall intracavity material dispersion has to be minimised to allow the use of low dispersion prism pairs.

Using fused silica prisms (65cm separation), and a 4mm Ti:sapphire rod, the intracavity TOD could be reduced to $D_3 = -(3 \pm 1) \times 10^2 fs^3$, which allowed a further reduction of $|D|$ to $(1.5 \pm 0.5) \times 10^2 fs^2$. For $W = 60nJ$, and $\phi = 1.5 \times 10^{-6} W^{-1}$, the predicted solitary pulse duration is $(15 \pm 3) fs$. At 3W absorbed pump, the laser produced $\approx 200mW$ output, near symmetric spectra, Fig.2(a), pulse durations below 20fs (sech^2 assumed), Fig.2(b), and $\tau\Delta\nu \approx 0.4$ over the wavelength range 795-850nm. Further reduction of $|D|$ allowed the generation of shorter pulses, Figs.3(a)(b). We assume the structure in the pulse wings

TABLE 1. ($\lambda = 780nm$)

Prism Material	$\phi''(\omega)/\phi'''(\omega)$	For D=0 (L_{sapph}=1cm, L_{prism}=0.4cm)	
		Prism separation	Residual TOD
SF10 (n=1.72)	0.41 fs^{-1}	20 cm	-6.8 × 10^3 fs^3
F2 (n=1.61)	0.46 fs^{-1}	31 cm	-4.5 × 10^3 fs^3
BK7 (n=1.51)	0.73 fs^{-1}	67 cm	-1.4 × 10^3 fs^3
SiO_2 (n=1.46)	0.84 fs^{-1}	82 cm	-0.9 × 10^3 fs^3

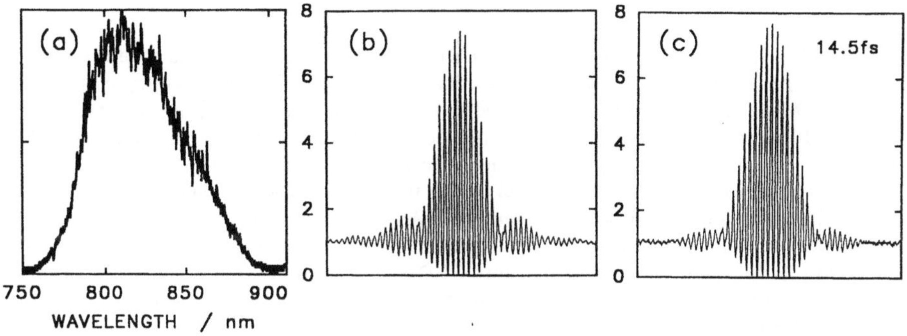

Figure 3

relates to a nonlinear chirp arising from the discrete nature of the pulse shaping in the cavity. This is confirmed by a reduction in the wing structure coinciding with a reduction of the pump power Fig.3(c), (output at OC1=120mW). Assuming a $sech^2$ pulse-shape the pulse duration from Fig.3(c) is ≈ 14.5fs. If we could exploit the complete lasing bandwidth of ≈ $70nm$, pulses of 10fs should be possible. We believe this could be achieved by further optimisation of the intracavity passive amplitude modulation.
In summary, we have presented a brief description of solitary laser operation and have presented a comparison between experiment and theory. The primary parameters controlling attainable pulse duration have been discussed with general guidelines given for achieving short pulses in solitary lasers.

References

1. D. E. Spence, P. N. Kean, W. Sibbett, Opt. Lett. **16**, 42 (1991)
2. L. F. Mollenauer, R. H Stolen, Opt. Lett, **9**, 13 (1984)
3. H. A. Haus, M. N. Islam., IEEE. J. Quantum Electron. **QE-21**, 1172 (1985)
4. T. Brabec, Ch. Spielmann, F. Krausz, Opt. Lett. **16**, 1962 (1991)
5. F. Krausz, M. E. Fermann, T. Brabec, P. F. Curley, M. Hofer, M. H. Ober, Ch. Spielmann, E. Wintner, and A. J. Schmidt, to be publ. in IEEE J. Quantum Electron., **QE-28**, Oct. 1992
6. T. Brabec, Ch. Spielmann, F. Krausz, Opt. Lett. **17**, 748 (1992)
7. Schott Glass Catalogue, Schott, Germany.

Self-Mode-Locked Cr^{3+}:$LiCaAlF_6$ and Cr^{3+}:$LiSrAlF_6$ Lasers

A. Miller[1], *P. Li Kam Wa*[1], *B.H.T. Chai*[1], *J.M. Evans*[2], *and W. Sibbett*[2]

[1]Center for Research in Electro-Optics and Lasers, University of Central Florida, 12424 Research Parkway, Orlando, FL 32826, USA

[2]Department of Physics, University of St. Andrews, St. Andrews, Fife KY16 9SS, Scotland

Cr^{3+}:$LiCaAlF_6$ (LiCAF) and Cr^{3+}:$LiSrAlF_6$ (LiSAF) are two new solid state materials[1] offering large fluorescence bandwidths for ultrashort pulse generation and tunability in the near infrared. Mode-locked operation is expected to be similar to Ti:sapphire[2], but red absorption bands and longer upper state lifetimes give these crystals the advantage of being readily diode pumpable. We have previously reported mode-locking of LiSAF and Cr^{3+}:$LiSr_{0.8}Ca_{0.2}AlF_6$ (LiSCAF) lasers with the production of 150fs pulses tunable from 800 to 900nm[3,4] when pumped using 1 Watt of red light from a cw krypton laser. An acousto-optic modulator and two prisms to compensate for group velocity dispersion (GVD) were incorporated within the cavity. Subsequently, 50fs pulses were reported from a GVD compensated LiSAF laser pumped at 488nm from an argon laser, with a dye jet within the cavity to initiate and sustain mode-locking[5]. In this paper we will describe the first mode-locked operation of LiCAF and the first self-mode-locking in both LiCAF and LiSAF resulting from self-phase modulation and self-focussing.

LiCAF and LiSAF crystals were grown by the Czochralski pulling technique with 2 at. % of Cr in the melt. Crystals were cut at Brewster angle for light polarized parallel to the c-axis to give a 15mm physical path length of the beam in the crystals which were contained within X-fold cavities containing two GVD compensating prisms. The LiCAF was pumped using a krypton laser operating on both the 647 and 676nm red lines while the LiSAF was pumped using the 476nm blue line from an argon laser. The nonlinear refractive indices of these fluoride crystals are approximately a factor of 4 smaller than Ti:sapphire.

The fluorescence of LiCAF is centered at 763nm with a peak emission cross-section of $1.3 \times 10^{-20} cm^2$ and a spontaneous lifetime of 175μs. The laser mirrors of the cavity comprised a standard mid-band set (nominally 800 to 900nm) from a Schwartz E-O cw Ti:sapphire laser. The pump wavelength was well matched to the red absorption band of LiCAF and produced cw lasing with up to 400mW output and a slope efficiency of 22% utilizing a 4% output coupler. Self-mode-locked operation was achieved without the need for any intracavity active or passive modulator except for the laser crystal itself, when a 1% output coupler was employed. Gentle rocking of the output coupler initiated the ultrashort pulse formation. With the output coupler stationary, stable mode-locking continued until a perturbation caused the laser to revert back to the cw state. Intensity and essentially chirp-free interferometric autocorrelations implied pulse widths of 170fs, fig. 1. A measured bandwidth of 4.5nm gave a time-bandwidth product of 0.36 assuming a $sech^2$ temporal profile. The average output power when self-mode-locked at 800nm was around 100mW at 90MHz with 2.5W of pump power. This wavelength of operation was red shifted from the peak of the emission band due to the mirror set employed. Further improvements in self-mode-locked operation of this laser is expected if optimized mirrors are employed which better cover the emission band of this crystal between 720 and 840nm.

In the LiSAF case, the resonator configuration incorporated both intra- and extra-cavity GVD compensating prisms and a mid-band mirror set from a Spectra-Physics cw Ti:sapphire laser. The emission band is broader and red shifted compared to LiCAF, being centered at 800nm with an emission cross-section of $4.8 \times 10^{-20} cm^2$. The upper state spontaneous lifetime is 67μs, but this falls rapidly just above room temperature. This can limit the maximum pumping level of LiSAF for given heat sinking conditions. The absorption spectra of LiSAF and LiCAF are similar with prominent red and blue absorption bands. In this work, the LiSAF crystal was

Springer Series in Chemical Physics, Vol. 55 Ultrafast Phenomena VIII
Editors: J.-L. Martin · A. Migus · G.A. Mourou · A.H. Zewail

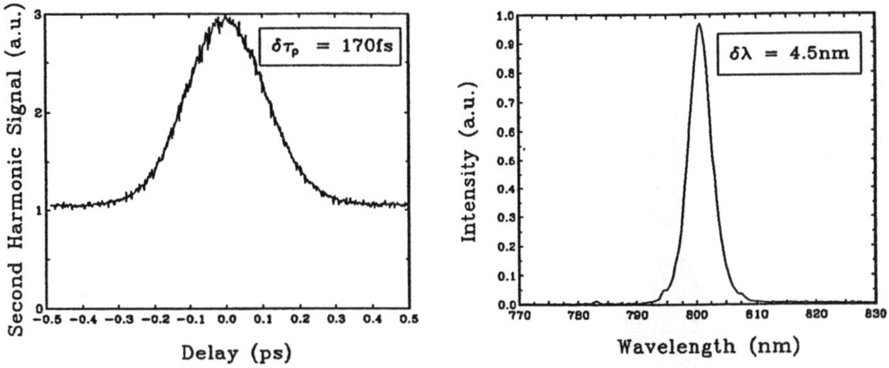

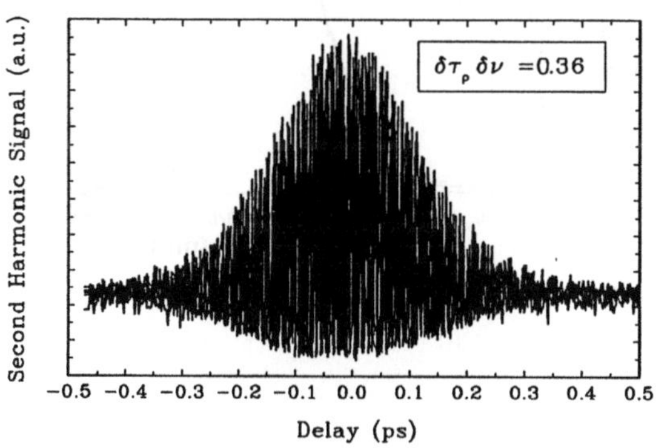

Figure 1 Characteristics of self-mode-locked LiCAF laser

pumped in its blue absorption band using the 476nm line from an argon-ion laser. For an output coupling of 3.5%, the cw laser threshold was around 300mW. At a pump power of 2W, the typical output power was 200mW. In order to initiate mode-locked operation, an acousto-optic modulator (NEOS) was inserted into the cavity. The modulator was weakly driven at 42.24MHz, which was midway between two resonances, by a signal derived from the initial mode-beating of the laser resonator. Once stable mode-locking was established, its self-sustaining characteristics could be demonstrated by the removal of the regenerative RF signal applied to the modulator. The stable self-mode-locked operation threshold was observed at around 50mW output power. The tuning range was 800 to 920nm which was limited by the optics available for this work. Chirp-free interferometric autocorrelation traces were obtained. The spectral bandwidth of the output pulses was 16nm. Theoretical fitting to intensity and interferometric autocorrelation profiles gave 50fs pulses with a time bandwidth product of 0.32 at an optimized prism separation of 23cm intracavity and 20cm extracavity. Pumping with 2W produced a mode-locked output power of 150mW.

In summary, we report the first mode-locked operation of Cr:LiCAF and self-mode-locking of Cr:LiCAF and Cr:LiSAF. This confirms that the generic nature of the self-mode-locking mechanism in Ti:sapphire can be used in other solid state laser materials as demonstrated in these colquiriite structured fluorides. The potential for efficient diode pumped ultrashort pulse systems makes the fluorides very attractive.

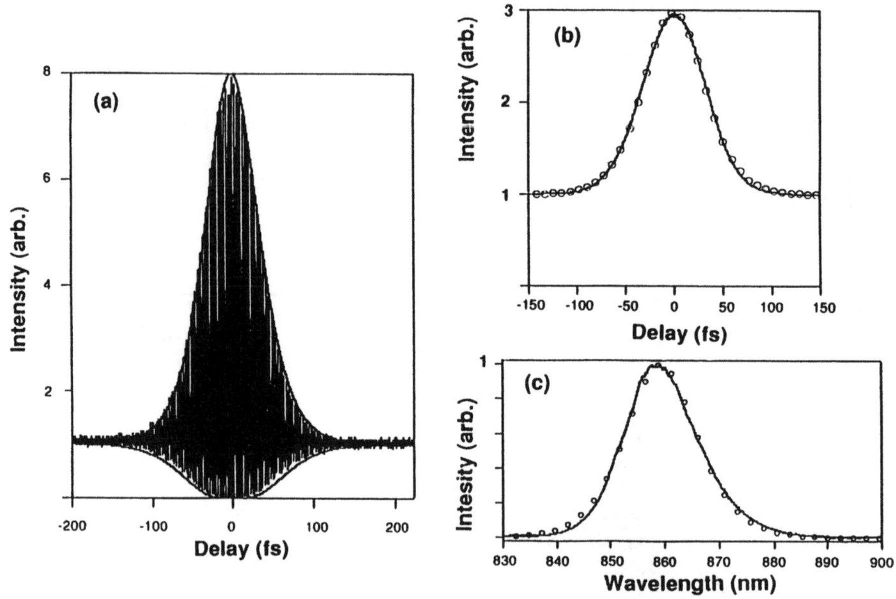

Figure 2 Characteristics of self-mode-locked LiSAF laser

Acknowledgements

This work was funded by the U.S. Defense Sciences Office of the Defense Advanced Research Projects Agency (DARPA), the U.S. Army Research Office, the Florida High Technology and Industrial Council and the U.K. Science and Engineering Research Council.

References

1. S. A. Payne, L. L. Chase, L. K. Smith, W. L Kway, H. W. Newkirk, J. Appl. Phys. 66 1051 (1989)
2. D. E. Spence, P. N. Kean, W. Sibbett, Opt. Lett., 16, 324 (1991)
3. A. Miller, P. Li Kam Wa, B. H. T. Chai, E. W. Van Stryland, Opt. Lett., 17, 195 (1992)
4. P. Li Kam Wa, B. H. T. Chai, A. Miller, Electron. Lett., 27, 2351 (1991)
5. N. H. Rizvi, P. M. W. French, J. R. Taylor, Eighth International Conference on Ultrafast Phenomena, Antibes, June 1992.

Sub-50 fs Pulse Generation from a Self-Starting CW Passively Mode-Locked Cr:LiSrAlF$_6$ Laser

N.H. Rizvi, P.M.W. French, and J.R. Taylor

Femtosecond Optics Group, Physics Department, Imperial College, London SW7 2BZ, UK

Abstract. A continuous wave, argon-ion pumped, Cr^{3+}:LiSrAlF$_6$ laser is reported for the first time. This laser has been passively mode-locked in a self-starting configuration exploiting the nonlinear refractive index of the laser medium together with a saturable absorber dye. Pulses as short as 33 fs have been generated with absorbed pump powers as low as 500 mW. The first cw mode-locked Cr^{4+}:YAG laser is also reported, pumped at room temperature by a Nd:YAG laser.

Introduction

Femtosecond laser physics has been transformed by the advent of the Titanium-doped sapphire laser which has been been mode-locked by conventional and novel means [e.g. 1-5], yielding pulses as short as 25 fs [5]. Consequently the dye laser will be largely superceded in ultrafast laser applications. We report here on two broadly tunable, room-temperature Chromium-doped lasers that continue this trend and which, once they are fully developed, may also eventually supplant the (ultrafast) colour-center laser.

Cr^{3+}:LiSrAlF$_6$ is an attractive alternative to Ti:sapphire, lasing across almost the same spectral range (760-1000 nm) but exhibiting a longer upper-state lifetime and red-shifted absorption profile, permitting it to be efficiently pumped by laser diodes at 670 nm and by flash-lamps or arc-lamps. It has a greater energy storage capacity and, as a tunable oscillator, its lower lasing threshold and ability to be diode-pumped make it potentially more useful than Ti:sapphire. Mode-locked operation and intracavity frequency doubling would permit the construction of a compact, diode-pumped efficient blue laser source.

The observation of self-mode-locking [2] in Ti:Al_2O_3 has led to the realisation that the nonlinear refractive index of the laser medium can be used to provide both amplitude modulation, through self-focussing or "Kerr Lens Mode-locking" [3], and phase modulation which, together with group velocity dispersion, can lead to further pulse compression. These mode-locking techniques are generally applicable to vibronic solid-state laser media eg. krypton ion-pumped Cr:LiSrCaAlF$_6$ [6].

Here we report on a c.w. Cr^{3+}:LiSrAlF$_6$ laser pumped, for the first time, by the 488 nm output from a low power argon ion laser. This laser was passively mode-locked using a dye jet stream of neocyanine. We also present preliminary results from our c.w. mode-locked Cr^{4+}:YAG laser operating at 1.44 μm at room temperature.

Experimental Results on Cr^{3+}:LiSrAlF$_6$ Laser

The c.w. Cr:LiSAF laser was pumped by the 488 nm output from a low power argon-ion laser. The laser crystal, supplied by Lightning Optical Corporation, was 23 mm in length and had a nominal concentration of Cr^{3+} ions of ~ 1.5% by weight. It was Brewster-angled and orientated such that the E-vector of the laser radiation was parallel to the its c-axis. This configuration led to 68% of the pump radiation being absorbed by the crystal. An AR-coated 10 cm focal length lens was used to pump a standard four-mirror, astigmatically-compensated cavity for initial slope efficiency measurements. All mirrors had single-stack dielectric

Editors: J.-L. Martin · A. Migus · G.A. Mourou · A.H. Zewail

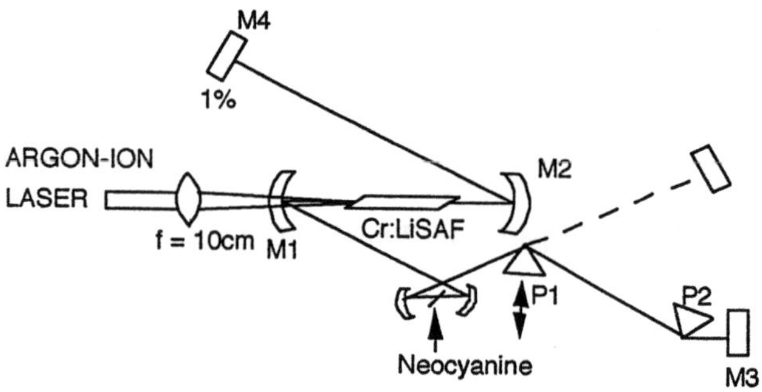

Figure 1. Cavity configuration for passively mode-locked Cr^{3+}:LiSrAlF$_6$

coatings. The slope efficiency was measured to be 9.5% at 830 nm using a 1% output coupler, with output powers of 135 mW being obtained at 1.3 W of absorbed pump power. No deleterious thermal effects were observed up to these power levels. With the introduction of a 0.8 mm thick single-plate birefringent filter, this laser was tuned from 799 nm to 873 nm. The tuning range was limited on the long wavelength side by the dielectric mirror coatings used.

The four-mirror cavity was then modified to include an additional folded section, comprising 50 mm radius of curvature mirrors, which contained a 50 μm thick jet of neocyanine in ethylene glycol. Figure 1 shows the cavity configuration. Stable self-starting mode-locking was achieved at an absorber concentration of 2 x 10^{-5}M. The cw lasing threshold of this cavity containing the saturable absorber was 305 mW of absorbed pump power. This configuration produced pulses of 9 ps duration - substantially shorter than the recovery time of neocyanine in ethylene glycol. The insertion of a SF10 prism pair for intracavity dispersion compensation, however, resulted in the routine generation of femtosecond pulses for absorbed pump powers of 900 mW and above. This femtosecond regime was self-starting and exhibited excellent stability. Tuning of the mode-locked laser was achieved by translating a vertical slit between prism P2 and mirror M3. Femtosecond pulses were produced from 819 nm to 844 nm. To minimise the contribution from intracavity third order dispersion, the glass path through the prisms and the prism separation were kept to a minimum. The prism separation used was 26cm. After optimisation of the second order and minimisation of the third order dispersion of the cavity, sub-100 fs pulses were routinely generated. The shortest pulses (~50 fs) were obtained at the minimum intra-cavity glass path for which stable pulse trains were observed. Increasing the glass path produced stable trains of broader pulses while decreasing the glass path led to the break-up of the femtosecond pulse trains. These observations are qualitatively similar to those of reference [4].

In order to generate yet shorter pulses, it was necessary to further reduce the intracavity third order dispersion [4,5]. This was achieved by replacing the SF10 prisms with F2 prisms [4]. For a prism separation of 40 cm, pulses as short as 33 fs duration were generated (assuming a sech^2 pulse profile), for absorbed pump powers of only 500 mW. This power level is well within the range of red (670 nm) diode lasers. The time-bandwidth product of these pulses was 0.39. Figure 2 shows a typical interferometric autocorrelation trace.

The role of the various pulse shaping mechanisms in this Cr:LiSrAlF$_6$ laser was investigated. The fast saturable absorber in the laser cavity appears to provide a weak modulation, sufficient to initiate the pulse formation, which also tends to stabilize the laser by suppressing the evolution of unwanted satellite structures. Using a fast saturable absorber, rather than relying on Kerr Lens mode-locking alone, means that the laser is not as critically sensitive to alignment and can be routinely switched on and run with femtosecond pulse generation. Evidence of self-focussing acting to compress the pulses, however, has been noted and this mechanism certainly plays a role in the generation of sub-100 fs pulses. It is necessary to incorporate a slit in the cavity to discriminate against the low intensity c.w. laser operation.

As well as neocyanine, we have also used DDCI-4 (1,1 diethyl 4,4 dicarbocyanine iodide) to provide self-starting and assist the mode-locking of the Cr^{3+}:LiSrAlF$_6$ laser. Sub-100 fs pulses were again readily generated.

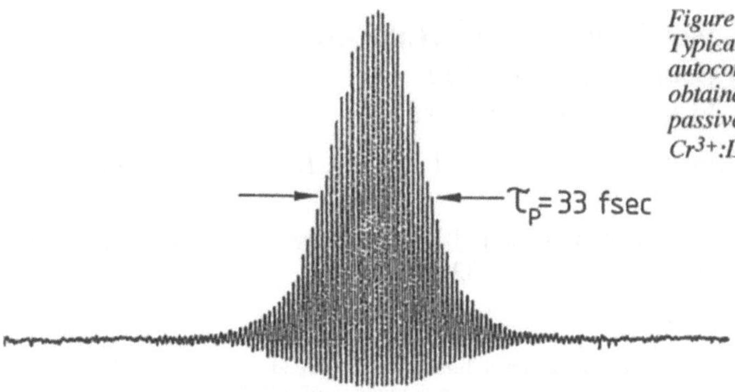

Figure 2
Typical interferometric autocorrelation trace obtained from the passively mode-locked Cr^{3+}:LiSrAlF$_6$ laser

Experimental Results on Cr^{4+}:YAG Laser

We anticipate that the pulse-forming mechanisms considered above will be applied to new solid-state lasers tuning from 1.1 to 1.6 µm -the spectral region usually covered by cryogenic colour-center lasers. Two promising media are Cr^{4+}:Forsterite and Cr^{4+}:YAG. We report here the first cw mode-locked Cr^{4+}:YAG laser. Cr^{4+}:YAG can lase from 1.35-1.56 µm and is a stable room-temperature laser which may be pumped at 1.06 µm by a (diode-pumped) Nd:YAG laser [7]. In our initial experiments using an acousto-optic modulator, detector-limited pulses < 50 ps duration have been generated at 1.44 µm at repetition rates up to 500 MHz. The Cr^{4+}:YAG laser was a 20 mm long Brewster-angled rod which was water-cooled and pumped at room-temperature by 10 W of 1.06 µm radiation from a c.w. Nd:YAG laser. Synchronous pumping with 100 ps 1.06 µm pulses resulted in pulses of ~200 ps duration from the Cr^{4+}:YAG laser. We are currently working on exploiting Kerr Lens mode-locking in this laser.

References

1) N.Sarukura, Y.Ishida, H.Nakano, Opt.Lett. **16**, 153 (1991)
2) D.E.Spence, P.N.Kean, W.Sibbett, Opt.Lett. **16**, 42 (1991)
3) D.K.Negus, L.Spinelli, N.Goldblatt, G.Feugnet, Proc. OSA Meeting on Advanced Solid State Lasers, 120 (1991)
4) F.Krausz, Ch. Spielmann, T.Brabec, E.Wintner, A.J.Schmidt, Opt.Lett. **17**, 204 (1992)
5) C-P Huang, H. C. Kapteyn, J. W. McIntosh and M. M. Murnane, Opt.Lett. **17**, 139 (1992)
6) A. Miller, P. Li Kam Wa, B. H. T. Chai and E. W. Van Stryland Opt.Lett. **17**, 195 (1992).
7) A. V. Shestakov, N. I. Borodin, V. A. Zhitnyuk, A. G. Ohrimtchyuk and V. P. Gapontsev, Conference on Lasers and Electro-optics, Paper CPDP11 (1991)

CW Krypton-Laser Pumped Cr^{3+}:$LiSrAlF_6$ and Cr^{3+}:$LiSr_{0.8}Ca_{0.2}AlF_6$ Crystals Produce 150 fs Mode-Locked Pulses

A. Miller, P. Li Kam Wa, H.S. Wang, S.L. Ayres, E.W. Van Stryland, and B.H.T. Chai

Center for Research in Electro-Optics and Lasers, University of Central Florida, 12424 Research Parkway, Orlando, FL 32826, USA

Mode-locked operation of wavelength tunable Cr^{3+} doped $LiSrAlF_6$ (Cr:LiSAF) and $LiSr_{0.8}Ca_{0.2}AlF_6$ (Cr:LiSCAF) lasers at a repetition frequency of 82MHz has been achieved by optical pumping with a cw krypton laser. An acousto-optic modelocker was used to initiate the pulse formation and pulses of 150fs duration were directly produced from both lasers when an intra-cavity two prism sequence was incorporated for cavity dispersion compensation. The low threshold characteristics of these lasers make Cr doped LiSAF and LiSCAF crystals promising solid state laser materials in the realization of compact, diode pumped, mode-locked, wavelength tunable laser systems.

The chromium concentration in both the Cr:LiSAF and Cr:LiSCAF crystals grown by the Czochralski pulling technique, was nominally 2%(atomic) in the melt in both cases. Each crystal was cut into a rectangular block 6x6x20mm^3 with a and c axes in the plane of the end faces. The ends were further cut and polished at Brewster's angle for light polarized parallel to c, giving a physical path length of 15mm in the crystal. CW operation of the LiSAF laser has previously been achieved by chopping the krypton pumping beam(1). Pulsed lasing action in these crystals with higher Cr doping has been demonstrated previously by flashlamp pumping(2,3).

Either one of the Brewster cut crystals was mounted in a Schwartz CW Titan laser and was pumped by a Laser Ionics cw krypton laser operating on both the 647 and 676nm red lines (power ratio 4:1). With either crystal, the threshold input power for cw lasing was found to be 210mW and the output power peaked at around 300mW with 1W of pump. As the pump power was increased further, the output power remained unchanged and decreased when the input power exceeded 1.5W. This is attributed to the reduction in upper state lifetime of the crystal that occurs above room temperature(3). Wavelength tuning from 800 to 900nm was achieved using a birefringent filter. This tuning range was restricted by the bandwidth of the mirror coatings used in our laser and should extend to 1μm by choosing the appropriate optics.

For modelocked operation, a NEOS high-Q, acousto-optic (A-O) modulator, broadband anti-reflection coated for an 800nm center wavelength gave a cavity loss modulation frequency of 82MHz. The output coupler had a transmission of 1% in the 800 to 900nm range and 60° prisms made from SF10 glass were used in the dispersion compensating arrangement. Wavelength tuning was accomplished without a birefringent filter by a variable-aperture slit placed between the back mirror and the prism closest to it.

The mode-locked pulses were monitored by a real time autocorrelator that was readily modified from a zero-background autocorrelation signal to a co-linear 1:3 intensity autocorrelation which could also give interferometric data. The pulse train was monitored using a fast photodiode and a quarter meter monochromator was used to observe the spectral content of the pulses. The threshold pump power level was again around 210mW from the krypton laser and stable operation of the laser was achieved at about 1W pump power with an average output power of 50mW. By carefully adjusting the prisms to compensate for self-phase modulation in the crystal and group velocity dispersion in the laser cavity, the output pulse width could be minimized. The likeness of the two crystals was again highlighted by the similarities in the mode-locked characteristics. The autocorrelation trace recorded at the minimum pulse width for either crystal, is shown in fig.1 for an optimized prism separation of 48cm. A fitted sech^2 pulse shape with 225fs full width at half maximum power(FWHM) implies a laser pulse duration of 150fs. The spectral content of the pulses shown in fig.2, has a bandwidth of 6nm (FWHM). The time-bandwidth product, $\Delta\tau_p\Delta\upsilon$ of 0.36 (compared to the Fourier transform limit of 0.32) indicates that the pulses are almost chirp free as is evidenced in the interferometric autocorrelation traces of fig.3.

Springer Series in Chemical Physics, Vol. 55 Ultrafast Phenomena VIII
Editors: J.-L. Martin · A. Migus · G.A. Mourou · A.H. Zewail

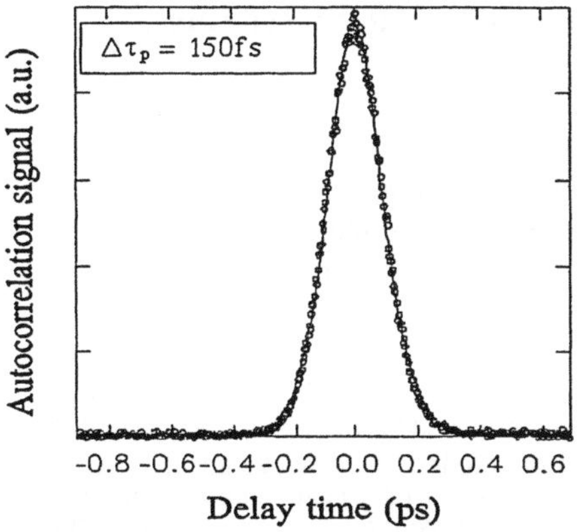

Figure 1 Intensity autocorrelation of modelocked pulses from Cr:LiSCAF laser

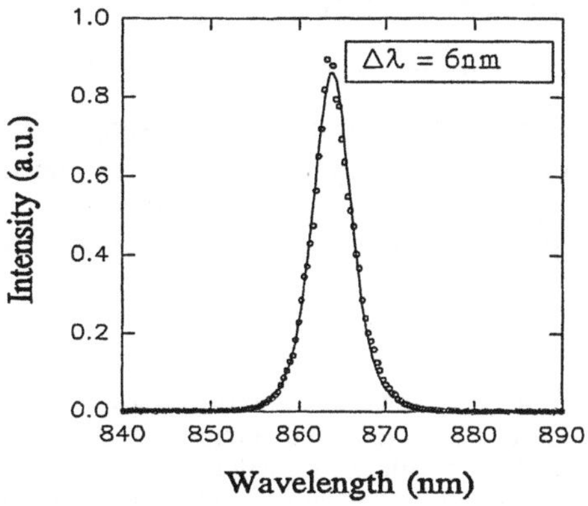

Figure 2 Spectral bandwidth of modelocked pulses from Cr:LiSCAF laser

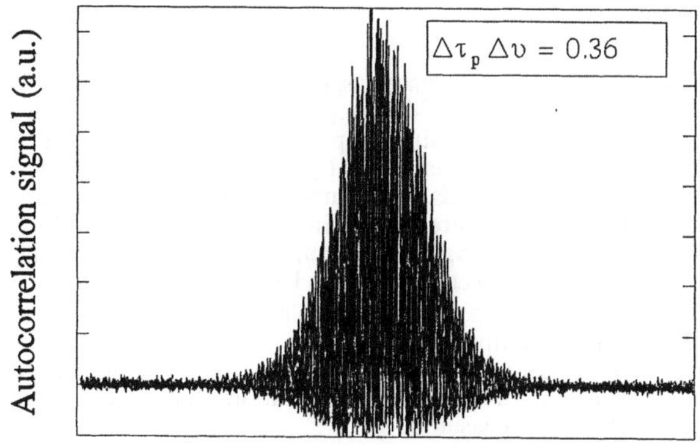

Figure 3 Interferometric autocorrelation of modelocked pulses from Cr:LiSCAF laser

The output spatial mode profile was viewed by directing the laser beam (heavily attenuated) onto the sensitive area of a computer camera with the lens removed. In both cases, a slight ellipticity was apparent in the beam shape but otherwise the laser beam retained an essentially TEM_{00} mode profile.

Preliminary experiments with a Cr:LiCAF laser, have produced stable chirp free self-mode-locked pulses of 170fs duration. The laser cavity design was similar to the one described earlier except for the absence of an acousto-optic modelocker within the cavity. The spectral bandwidth of the pulses was around 4.5nm and is believed to be limited by the bandwidth of the mirrors employed in our experiments. The self-mode-locking action was obtained in LiCAF because of the higher optical intensity in the crystal. The crystal could be pumped harder partly because the spontaneous emission lifetime is only affected when the crystal is allowed to warm up to elevated temperatures, and partly as a result of improved crystal growth.

References

1. S.A. Payne, L.L. Chase, L.K. Smith, L.K. Way, H.W. Newkirk, J. Appl. Phys. 66, 1051 (1989).
2. Stalder M., Chai B.H.T., Bass M., Appl. Phys. Lett., 58, 3, 1991.
3. Chai B.H.T., Stalder M., Bass M., submitted to Opt. Lett.

60-fs Chromium-Doped Forsterite (Cr^{4+}:Mg_2SiO_4) Laser

A. Seas, V. Petričević, and R.R. Alfano

Institute for Ultrafast Spectroscopy and Lasers, and
Center for Analysis of Structures and Interfaces, Department of Physics
and Electrical Engineering, The City College and Graduate School
of the City University of New York, New York, NY 10031, USA

Abstract. Generation of 60-fs pulses from a cw mode-locked Cr:forsterite (Cr^{4+}:Mg_2SiO_4) laser is reported. The forsterite laser was actively mode-locked using an acousto-optic modulator with two Brewster high-dispersion glass prisms for intra-cavity chirp compensation. Transform-limited sub-100-fs pulses were routinely generated, tunable over 1230-1280 nm. The shortest pulses of 60-fs pulsewidth were measured.

1. Introduction

Chromium-doped forsterite is a tunable solid-state laser based on Cr^{4+} as laser-active ion with wide tuning range in the near infrared from 1167-1345 nm.[1] Recently, actively mode-locked[2] and self mode-locked[3] operation of chromium-doped forsterite laser has been demonstrated, producing picosecond pulses. Over the past several years research was directed towards generating ultrashort pulses by various methods from tunable solid-state lasers. Sub-100-fs pulses were obtained directly from a Ti:sapphire laser using self-mode-locking[4-6], colliding-pulse mode-locking[7], synchronous pumping[8], and coupled-cavity mode-locking using multiple quantum wells.[9] Actively mode-locked Ti:sapphire laser produced transform-limited 150-fs pulses[10] and 150-fs pulses were generated from a Cr^{3+}:$LiSrAlF_6$ laser using active mode-locking with intracavity dispersion compensation.[11]

We report on the generation of femtosecond pulses from an actively mode-locked chromium-doped forsterite laser with intracavity dispersion compensation. Pulses as short as 60 fs full-width-at-half-maximum (FWHM) were measured.

2. Experiment

The laser design is shown in Fig. 1.[12] The Brewster-angle-cut forsterite crystal was placed in a four-mirror, Z-fold astigmatically compensated cavity. The combination of mirrors used was: a flat back mirror, two 10-cm-radius folding mirrors, and a flat output coupler. The transmission of the output coupler was varied between 1% and 5% at the lasing wavelength, while the folding mirrors and the back mirror had reflectivity R=99.9% for the 1200-1300 nm range. The shortest pulses were obtained using a 1% output coupler. The length of the sample was 0.9 cm and the absorption coefficient at the pump wavelength of 1064 nm was α=0.65 cm^{-1}. The laser crystal was wrapped in indium foil and mounted in a copper block which was cooled by a single-stage thermoelectric cooler. The Cr:forsterite crystal was pumped by a continuous-wave Nd:YAG laser. The pump beam was focused by a 7.5-cm lens through the 10-cm-radius folding mirror into the crystal. The 1064-nm pump power incident on the forsterite crystal was 4.7 W, of which 1.9 W was absorbed by the crystal. The output of the forsterite laser was monitored with a fast germanium detector and an oscilloscope. The pulsewidth was measured with a real-time background free autocorrelator. The bandwidth of the mode-locked forsterite laser was measured using a monochromator equipped with PbS detector.

Actively mode-locked operation of the forsterite laser was achieved when an acousto-optic modulator ML was inserted in the basic Z-fold cavity, without prisms. A stable train of 6-ps

Springer Series in Chemical Physics, Vol. 55 Ultrafast Phenomena VIII
Editors: J.-L. Martin · A. Migus · G.A. Mourou · A.H. Zewail

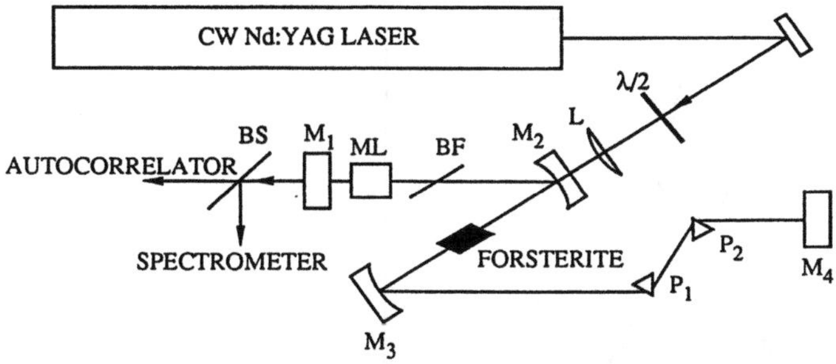

Figure 1: Schematic diagram of the experimental arrangement for actively mode-locked operation of Cr:forsterite laser: $\lambda/2$, half-wave plate for 1064 nm; L, focusing lens; M_1, output mirror, M_2, M_3, 10-cm-radius folding mirrors; M_4, back mirror; ML, acousto-optic modulator; BF, birefringent tuning plate; BS, beam splitter; P_1 and P_2, Schott SF-14 glass prisms.

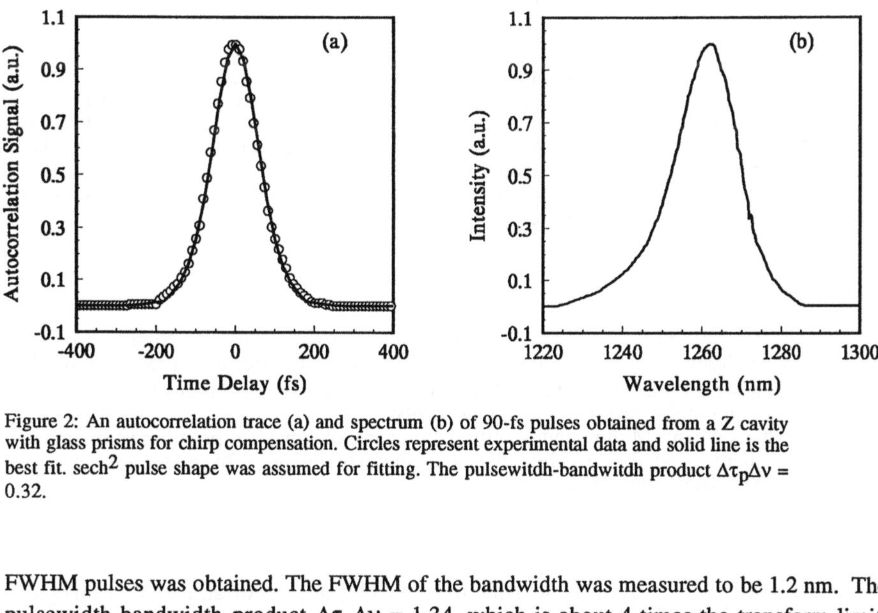

Figure 2: An autocorrelation trace (a) and spectrum (b) of 90-fs pulses obtained from a Z cavity with glass prisms for chirp compensation. Circles represent experimental data and solid line is the best fit. $sech^2$ pulse shape was assumed for fitting. The pulsewitdh-bandwitdh product $\Delta\tau_p\Delta\nu$ = 0.32.

FWHM pulses was obtained. The FWHM of the bandwidth was measured to be 1.2 nm. The pulsewidth-bandwidth product $\Delta\tau_p\Delta\nu = 1.34$, which is about 4 times the transform limit, indicating that the pulses from the mode-locked forsterite laser were strongly chirped, due to self-phase modulation and group velocity dispersion in the forsterite crystal.

To compensate for dispersion, a pair of high-dispersion Brewster prisms were inserted in the cavity, as shown in Fig. 1. The distance between the prisms was varied until shortest pulsewidths were measured. This resulted in significant reduction of pulsewidth, down to under 100 fs, with a spectral width of the order of 20 nm. An autocorrelation trace and the corresponding spectrum of a typical pulse are shown in Fig. 2 (a) and (b). The pulsewidth shown is 90 fs and the bandwidth is 19 nm. The pulsewidth-bandwidth product $\Delta\tau_p\Delta\nu = 0.32$, indicating transform-limited pulses for $sech^2$ pulse. Very often, pulses shorter than 90 fs were observed. Autocorrelation trace presented in Fig. 3 shows a pulse of less than 60 fs FWHM.

The actively mode-locked forsterite laser was tuned using an intracavity single-crystal quartz birefringent plate as shown in Fig. 1. With only one combination of laser mirrors the laser output was continuously tuned between 1230 - 1280 nm. The power output of 85 mW was measured, for 1.9 W of absorbed pump power. The pulsewidth and output power did not change significantly over the tuning range.

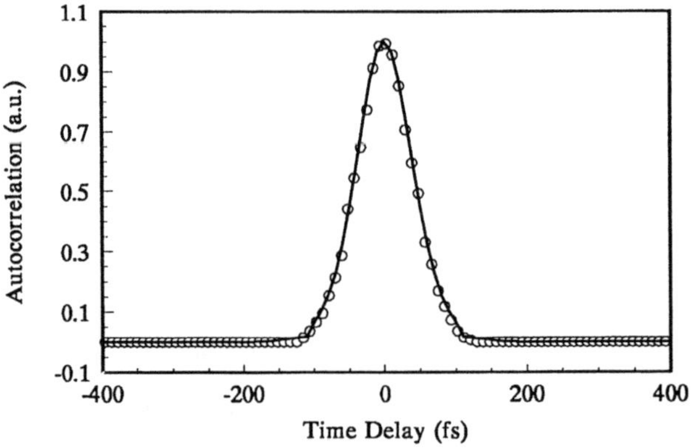

Figure 3: An autocorrelation trace of 60-fs pulses (assuming sech^2 pulse shape). Circles represent experimental data and solid line is the best fit.

Stable sub-100-fs pulses were monitored when the RF power was disconnected from the acoustooptic modulator. Within the first 30 seconds no change in the output was observed, i. e. stable sub-100-fs pulses were generated without any external modulation. The mode-locked operation usually ceased after this initial period due to mechanical disturbances. This is an indication that the Cr:forsterite laser actually operated in a self-mode locked regime, where active mode-locking only sets the conditions necessary for self-mode-locked operation by producing intense optical fields in the cavity.

Conclusion

Stable sub-100-femtosecond transform-limited pulses were generated from a chromium-doped forsterite laser using active mode-locking. The shortest pulses measured were of 60 fs FWHM, assuming a sech^2 pulse shape. The tuning range was of the order of 50 nm, limited only by the coating of the mirrors used and the cavity losses.

We thank Mitsui Mining and Smelting Company of Japan for providing the forsterite crystal and Coherent, Inc. for providing the acoustooptic modulator. This research is supported in part by NSF, NASA and ARO.

References

1. V. Petričević, S. K. Gayen, R. R. Alfano, K. Yamagishi, H. Anzai, and Y. Yamaguchi, Appl. Phys. Lett. **52**, 1040 (1988).
2. A. Seas, V. Petričević, and R. R. Alfano, Opt. Lett. **16**, 1668 (1991).
3. T. J. Carrig, A. Sennaroglu, and C. R. Pollock, in *Advanced Sold-State Lasers Technical Digest*, 1992 (Optical Society of America, Washington, D.C., 1992), pp. 29-31.
4. D. E. Spence, P. N. Kean, and W. Sibbet, Opt. Lett. **16**, 42 (1991).
5. C. P. Huang, H. C. Kapteyn, J. W. McIntosh, and M. M. Murnane, Opt. Lett. **17**, 139 (1992).
6. N. H. Rizvi, P. M. W. French, and J. R. Taylor, Opt. Lett. **17**, 279 (1992).
7. K. Naganuma and K. Mogi, Opt. Lett. **16**, 738 (1991).
8. F. Krausz, Ch. Spielmann, T. Brabec, E. Wintner, and A. J. Schmidt, Opt. Lett. **17**, 204 (1992).
9. U. Keller, G. W. 't Hooft, W. H. Knox, and J. E. Cunningham, Opt. Lett. **16**, 1022 (1991).
10. P. F. Curley and A. I. Ferguson, Opt. Lett. **16**, 1016 (1991).
11. A. Miller, P. LiKamWa, B. H. T. Chai, and E. W. Van Stryland, Opt. Lett. **17**, 195 (1992).
12. A. Seas, V. Petričević, and R. R. Alfano, to be published in Opt. Lett..

Femtosecond Pulses from Nd:Glass Lasers

A.J. Schmidt, M.H. Ober, M. Hofer, M.E. Fermann, F. Krausz, T. Brabec, Ch. Spielmann, and E. Wintner

Abteilung Quantenelektronik and Lasertechnik, Technische Universität Wien, Gusshausstr. 27/359/9, A-1040 Wien, Austria

Introduction

During the last decade the passively mode locked dye laser [1] has been the only widely used source of femtosecond optical pulses. Recent advances in ultrashort pulse generation with solid state lasers [2,3] opened up a new era in femtosecond pulse technology. Single mode rare–earth–doped fibers [4] have recently been shown to be an attractive medium for the generation of sub 100 fs pulses using nonlinear polarization evolution [5]. Additionally efficient passive mode–locking of bulk Nd:glass lasers [6,7] was demonstrated by means of Additive Pulse Mode–locking (APM) [8,9]. In our experiments these lasers are pumped by a Krypton laser or a 150 mW laser diode. The best performance in a low-gain APM system is achieved with a dispersively balanced Michelson interferometer [7]. In this APM laser the Kerr–nonlinearity of optical fibers is utilized for mode locking. The usefulness of these techniques is demonstrated by the generation of sub–100–fsec pulses in bulk Nd:glass and Nd:fiber lasers.

Femtosecond Passive Mode Locking of a Nd:bulk laser by a Nonlinear Interferometer

In this section we report on dispersively compensated additive–pulse mode locking using a Michelson–cavity configuration [7]. Dispersion compensation is shown to improve steady–state laser performance, facilitate self–starting, and reduce relaxation oscillation instabilities. Using Nd:phosphate glass as the active material we demonstrate the generation of stable femtosecond pulses at considerably reduced self–starting threshold in comparison to the conventional APM technique. Consider a laser cavity terminated into a nonlinear Michelson interferometer as illustrated in Fig. 1. This Michelson cavity configuration is superior to the conventional coupled–cavity arrangement owing to its greater compactness due to a smaller number of cavity components and to a reduction of the physical extension of the system. Furthermore, the single–beam output leads to the production of more intensive output pulses. The beam splitter BS transmits a small part of the intracavity power into the weakly–coupled nonlinear arm M5–BS containing a short length of a single mode optical fiber. The greater part of the power is fed into the dispersive linear arm M4–BS having a net negative GVD. At sufficiently high intracavity powers the laser immediately becomes mode locked when the relative phase is properly adjusted. Without dispersion compensation ($D \approx 0$) the laser produces strongly chirped pulses (≈ 1 ps) as predicted by numerical investigations [10]. When dispersion in the linear interferometer arm is optimized, nearly bandwidth–limited femtosecond pulses are generated. Compensation can be achieved by a single pair of prisms separated by a distance approximately equal to the fiber length allowing a compact optical setup.

Dispersion compensation also reduces the self–starting threshold significantly. The threshold intra–cavity average power is reduced by more than a factor of three with the

Springer Series in Chemical Physics, Vol. 55 Ultrafast Phenomena VIII
Editors: J.-L. Martin · A. Migus · G.A. Mourou · A.H. Zewail

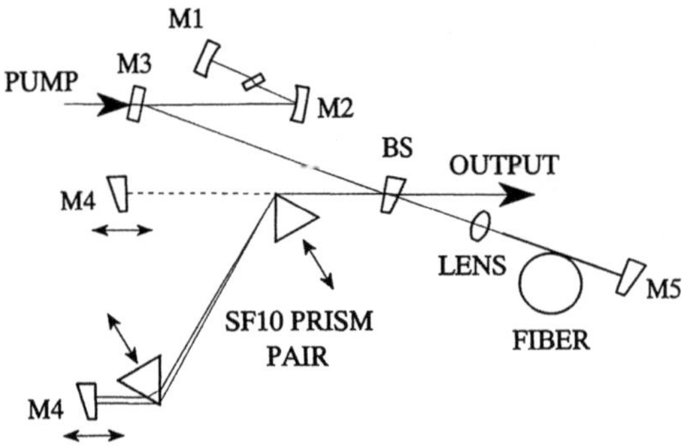

Figure 1: Schematic diagram of the passively mode–locked laser (M 1-5), high–reflectivity mirrors at the laser–wavelength; (BS) 16% transmitting beam splitter. The dashed beam indicates the laser without dispersion compensation.

introduction of negative GVD pointing to its significance even in the initial pulse evolution. The fiber length is an important scaling parameter for both threshold power and steady–state pulse duration. Long fiber lengths allow low power operation at the expense of an increase in pulse width, whereas short fiber lengths support shorter pulses at the cost of higher thresholds for passive mode locking. In our experiments the shortest pulses have been generated at $l_f = 18\ cm$ with $D = -3.8 \times 10^{-3}\ ps^2$. Careful optimization of system parameters might bring the phase settings for minimum pulse width and pedestal–free pulse generation closer together. Nevertheless, the 88 fs pulses represent the shortest optical pulses produced in a passively mode-locked bulk Nd:glass laser to date. Note that the short fiber length improves the stability of the laser as well. Owing to the reduced extension of the interferometer, stable long–term operation can be achieved with a servo–loop providing active stabilization.

Passively Mode–Locked Fiber Laser

Although in principle a large number of passive amplitude modulation mechanisms could be employed for passive mode–locking of fiber lasers, only two systems have been studied so far due to their unique features and simplicity. The first system employs a nonlinear amplifying loop mirror(NALM) [11–15] where passive amplitude modulation is obtained by placing the gain medium asymmetrically inside a nonlinear Sagnac fiber ring interferometer and thus breaking its reciprocity. The reflection or transmittion coefficient then becomes a function of pulse intensity. The NALM incorporated in a laser is shown in Fig. 2. Note that operation at the optimum dispersion point is achieved by adjusting the overall cavity dispersion with the grating pair. For a phase bias of $= \pi/2$ the passive amplitude modulation for large gain values can approach the magnitude of passive frequency modulation. This is an exceptional property for an all optically passively mode–locked solid–state laser as in all these systems demonstrated to date self–phase modulation is typically much stronger than the self–amplitude modulation. Since passive amplitude modulation is maximized in this system, it is particularly advantageous for the generation of the shortest self–starting pulses possible. The operation principle of the second system and its practical implementation are shown in Fig. 3. Here nonlinear polarization evolution in a weakly birefringent fiber is employed for passive amplitude modulation [5].

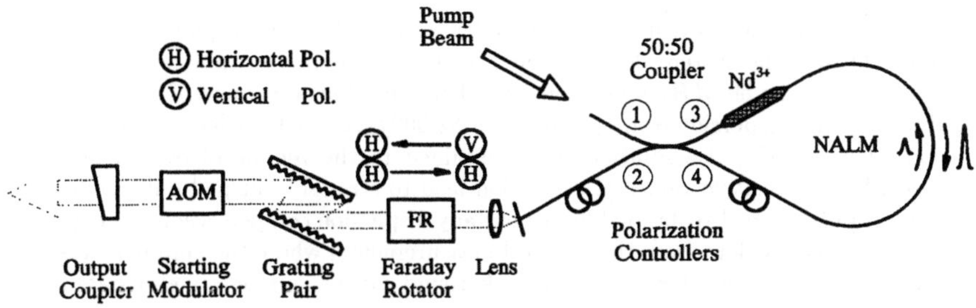

Figure 2: Passively mode–locked Nd:fiber laser using a NALM in reflection for amplitude modulation and employing dispersion compensation to obtain soliton–like pulse shaping. (System 1)

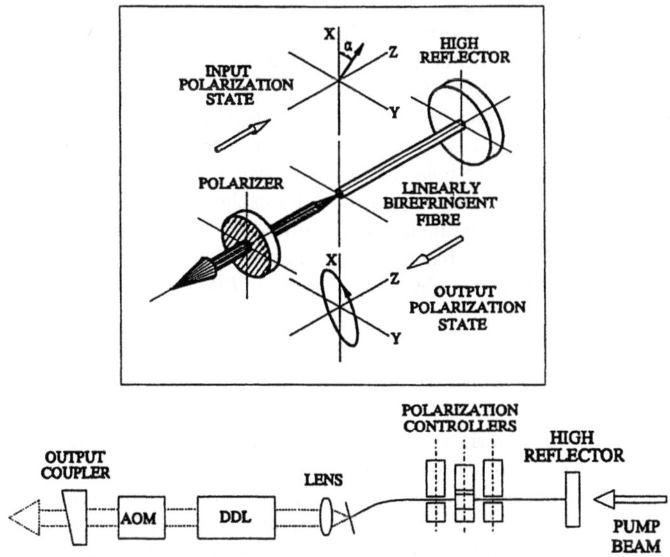

Figure 3: Nonlinear polarization evolution as applied to passive mode–locking of a Nd:fiber laser (System 2). The principle of operation is shown in the insert.

Passive amplitude modulation is reduced in comparison to system 1 and thus system 2 is closer to conventional solid–state femto–second oscillators such as the Ti:sapphire and the color center laser. In particular we can expect stable operation of system 2 over a larger range of dispersion parameters than with system 1 due to the reduced amplitude modulation. In addition to the features of the fiber laser discussed above, the high intra-cavity gain and nonlinearities allow a very large difference between R_0 and $R_0 + \kappa P$ in the cavity. Whereas nonlinear pulse propagation modulates R_0 by no more than 10% in bulk lasers, in fiber lasers the modulation can be close to 100% [5]. Equally, this allows to modelock the full bandwidth of glass lasers, i.e. pulse widths as short as 38 $fsec$ are obtainable.

So far detailed studies of pulse formation are only available for fiber system 2 [5]. In these a laser with a 20 cm long Nd:fiber operating at 1.06 μm with positive fiber group–velocity dispersion was investigated. With an absorbed pump power of 350 mW at around 800 nm, the system delivered 50 $fsec$ transform–limited pulses with a pulse energy of 400 pJ and a 50 MHz repetition rate. The special arrangement of fiber system 2 allowed to vary the dispersion from +11.000 to $-32.000 fsec^2$ without loosing modelocked operation. Unchirped pulse solutions were obtained in the regime of overall negative dispersion whereas positive values of dispersion lead to strongly–chirped pulse solutions. In particular, pulses are also obtainable without any dispersion compensation. The quality of the pulses is also evident from the spectral measurements, where the complete absence of spectral spikes demonstrates that a cw background can be completely eliminated in a modelocked fiber laser.

A complete noise characterization of system 2 pumped with a Krypton laser [16] and with a 150mW laser diode [17] were recently performed. Additional frequency stabilization was employed by locking the round–trip frequency to a temperature stabilized quartz oscillator [18]. Operation below the amplitude noise of the Krypton pump laser was obtained by maximizing the passive amplitude modulation. For low frequencies the phase noise was determined by the noise of the reference oscillator, but the phase noise may be even lower for high frequencies. The RMS amplitude fluctuations σ_A in system 2 were measured as 0.4% in a frequency band from 30 Hz- 100 kHz for the krypton pumped system and the corresponding RMS timing jitter σ_J was obtained as 110 $fsec$. For the diode pumped system the RMS noise was as low as 0.018% in a frequency band from 15 Hz- 100 kHz. To our knowledge, these noise figures are lower than obtained for any other femtosecond oscillator demonstrated to date. Due to its exellent noise performance and long term stability (no additional adjustments were required for hours) the fiber laser was implemented as a master oscillator of a regenerative amplifier yielding 300 $fsec$ pulse duration and 13 μJ pulse energy [19] and might even be used in TW systems [20].

Conclusions

In conclusion, we have presented a variety of different mode–locking techniques for Nd:glass systems. The generation of optical pulses as short as 38 $fsec$ is demonstrated. The mode–locked pulses of the fiber system 2 exhibit a high stability to environmental perturbations and have significantly less amplitude noise and timing jitter then other femtosecond sources. Owing to the simplicity and solid state nature of the system, the described performance is easily reproduced from day to day, a feature making this novel class of femtosecond sources especially attractive for practical applications.

The authors want to thank F. Haberl for participation in the experiments. This work was supported by the Fonds zur Förderung der wissenschaflichen Forschung, in Österreich, projects no. P8024–TEC and P8391–TEC and by the Jubiläumsfonds der österreichischen Nationalbank, project no. 4049.

References

1. R. L.Fork, B. I. Green, C. V. Shank, Appl. Phys. Lett. **38**, 671, (1981)

2. I.N. Duling, Opt. & Phot. News, **3**, 5, 29, (1992)

3. W.H. Knox, Opt. & Phot. News, **3**, 5, 10, (1992)

4. S.B. Poole, D.N. Payne and M.E. Fermann, Electron. Lett., **21**, 737, (1985)

5. M. Hofer, M. H. Ober, F. Haberl, and M. E. Fermann, IEEE J. Quantum Electron. **QE-28**, 720, (1992)

6. F. Krausz, Ch. Spielmann, T. Brabec, E. Wintner, and A. J. Schmidt, Opt. Lett. **15**, 1082, (1990)

7. Ch. Spielmann, F. Krausz, T. Brabec, E. Wintner, and A. J. Schmidt, Appl. Phys. Lett. **58**, 2470, (1991)

8. J. Mark, L.Y. Liu, K.L. Hall, H.A. Haus and E.P. Ippen, Opt. Lett., **14**, 48, (1989)

9. P.N. Kean, X. Zhu, D.W. Crust, R.S. Grant, N. Langford and W. Sibbet, Opt. Lett., **14**, 39, (1989)

10. T. Brabec, F. Krausz, M. Budil, Ch. Spielmann, and E. Wintner, J. Opt. Soc. Am. B **9**, 1818, (1991)

11. M.E. Fermann, F. Haberl, M. Hofer and H. Hochreiter, Opt. Lett., **15**, 752, (1990)

12. M.E. Fermann, M. Hofer, F. Haberl A.J. Schmidt and L. Turi, Opt. Lett., **16**, 244, (1991)

13. I.N. Duling, Opt. Lett., **16**, 539, (1991)

14. I.N. Duling, Elektr. Lett., **27**, 544, (1991)

15. D.J. Richardson, R.I. Lamming, D.N. Payne, V.I. Matsas, M.W. Phillips, Electron. Lett., **27**, 544, (1991)

16. F.Haberl, M.H.Ober, M.Hofer, M.E.Fermann, E.Wintner and A.J.Schmidt, IEEE Photon.Technol.Lett., **3**, 1071, (1991)

17. M.H. Ober, F. Haberl and M.E. Fermann, Appl. Phys. Lett., **60**, 2177, (1992)

18. M.J.W. Rodwell, D.M. Bloom and K.J. Weingarten, IEEE J. Quantum Electron., **QE-25**, 817, (1989)

19. M.Hofer, M.H.Ober, F.Haberl, M.E.Fermann, E.R.Taylor and J. Wang, Opt.Lett., **17**, June, (1992)

20. C. Sauteret, D. Husson, G. Thiell, S. Seynec, S. Gary, A. Migus, and G. Mourou, Opt. Lett., **16**, 238, (1991)

A Diode-Pumped Picosecond Oscillator at 1053 nm

I.P. Mercer[1], *Z. Chang*[2], *M.R.G. Miller*[3], *C.N. Danson*[3], *C.B. Edwards*[3], *and M.H.R. Hutchinson*[1]

[1]The Blackett Laboratory, Imperical College, London SW7 2BZ, UK
[2]Royal Society Visitor, Xian Institute of Fine Mechanics, Xian, Peoples Rep. of China
[3]Central Laser Facility, Rutherford Appleton Laboratory, Chilton, Didcot, Oxon OX11 OQX, UK

We report on the development of a novel diode-pumped modelocked laser oscillator operating on the 1053 nm transition in Nd:YLF. The device has been used to generate 1.6 ps pulses for injection into a high power Nd:glass laser system producing multi-terawatt output using Chirped Pulse Amplification (CPA) techniques [1].

Diode pumping offers important advantages over competing techniques for the generation of picosecond laser pulses at 1μ [2]. The narrow band diode output is temperature tuned to the peak of the absorption in the Nd:YLF medium, giving extremely high pumping efficiency, low power consumption, and highly stable C.W. operation over extended periods (many hours). These characteristics, coupled with low cost and compact mechanical design, make this system an ideal front end driver for CPA systems based on phosphate glass laser amplifier technology.

A schematic of the coupled cavity oscillator is shown in figure 1. A 3W CW laser diode array (SDL 2482) was coupled to a 10 mm long Nd:YLF rod through a dichroic multilayer dielectric coating via optics designed to correct for the

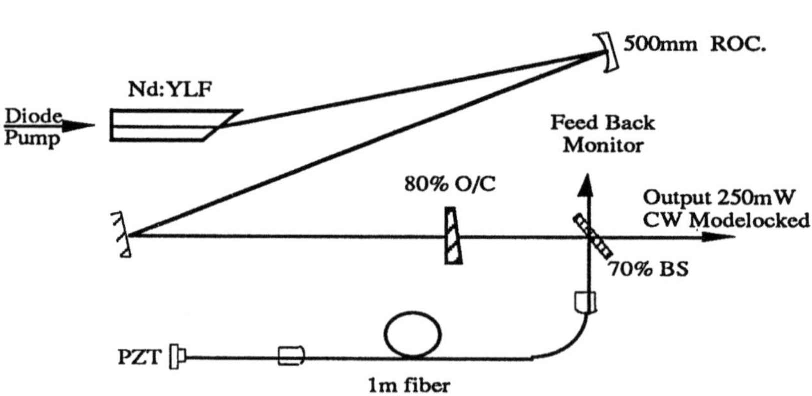

Figure 1: Schematic of arrangement for the A.P.M. oscillator.

Springer Series in Chemical Physics, Vol. 55 Ultrafast Phenomena VIII
Editors: J.-L. Martin · A. Migus · G.A. Mourou · A.H. Zewail

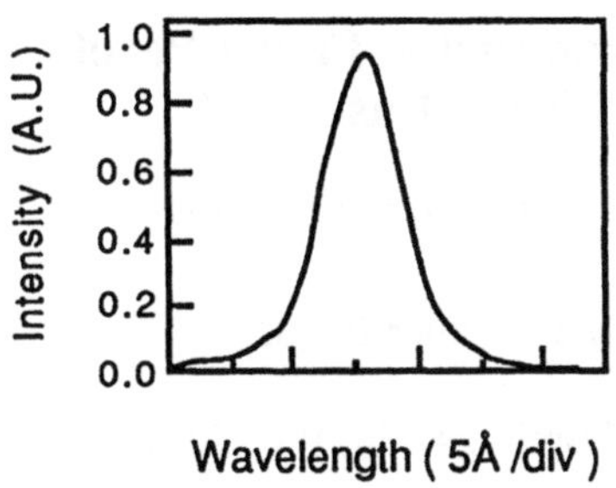

Fig 2: Bandwidth of 7 Å.

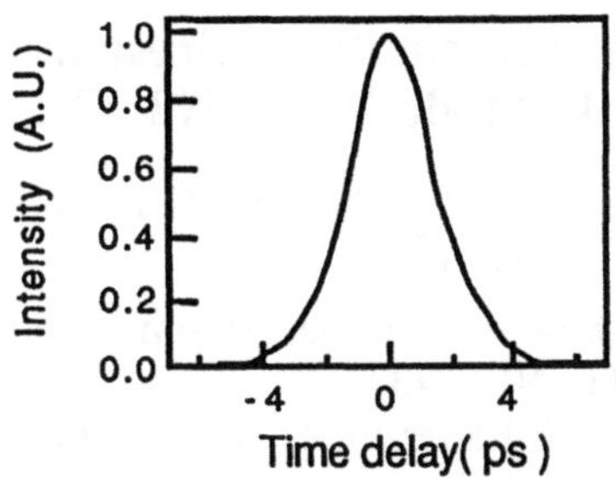

Fig 3: Corresponding Autocorrelation representing a 2.2 ps pulse width.

astigmatic output mode pattern of the diode array. The Nd:YLF rod was cut at Brewsters angle at the output to select the correct polarisation of the output for maximum discrimination between the 1053 nm and 1047 nm gain lines in Nd:YLF.

The shortest pulse length obtained form the oscillator was 1.6 ps, comparable with the best performance of Nd:YLF operating at 1047 nm [3]. The CW output power was 250 mW corresponding to a single pulse energy of 1.8 μJ. The most stable operating regime delivered a 2.2 ps pulse length accompanied by a 210 GHz bandwidth (fig.2,3), corresponding to the transform limit for a Gaussian pulse shape.

A Michelson interferometer arrangement, including dispersion compensation has also been constructed. It delivers a higher CW output power of 350 mW, but otherwise has similar characteristics.

The authors acknowledge the valuable assistance of the Optoelectronics Research Centre at Southampton University during the course of this work.

References

1. Chang et al., 1991 Annual Report to the Laser facility Committee, Rutherford Appleton Laboratory Publications. In preparation.
2. J.M.Liu and J.K.Chee; Opt. Lett. **1 5**, 685 (1990).
3. G.P.A. Malcolm, P.F. Curley, and A.I.Ferguson; Opt. Lett. **1 5**, 1303 (1990).

A New Intracavity Antiresonant Semiconductor Fabry-Perot Passively Mode-Locks Nd:YLF and Nd:YAG Lasers

U. Keller[1], *D.A.B. Miller*[1], *G.D. Boyd*[1], *T.H. Chiu*[1], *J.F. Ferguson*[1], *and M.T. Asom*[2]

[1]AT&T Bell Laboratories, Room 4B-523, Crawfords Corner Road, Holmdel, NJ 07733

[2]AT&T Bell Laboratories, 9999 Hamilton Blvd., Breinigsville, PA 18031, USA

Abstract. We introduce a new low-loss intracavity semiconductor Fabry-Perot saturable absorber operated at antiresonance (A-FPSA) to both start and sustain stable mode locking of cw pumped Nd:YLF and Nd:YAG lasers. With this new A-FPSA inside a Nd:YLF / Nd:YAG laser cavity, we produce self-starting 3.3 ps / 9.2 ps cw mode locked pulses at a 1.047 µm / 1.064 µm wavelength with an average output power of 700 mW / 790 mW when cw pumped with 2 W at 792 nm / 808 nm. An A-FPSA can be used as an continuous starting mechanism for femtosecond KLM/self-focusing mode locking and can provide broadband tunability.

Solid-state lasers with small gain cross sections require a fast and low-loss saturable absorber for passive mode locking. Intracavity semiconductors introduce too much loss, have too small a saturation intensity and have problems withstanding the high intracavity peak intensities. One solution is to put the semiconductor inside a low-Q coupled cavity (RPM) [1, 2]. An intracavity reactive nonlinearity, such as self-focusing, has been used to produce fast saturable-absorber-like mode locking that is, however, not self-starting [3]. We have demonstrated that weak RPM can continuously start KLM/self-focusing mode locking [4, 5].

We introduce a new low-loss intracavity saturable absorber element, an Antiresonant Fabry Perot Saturable Absorber (A-FPSA), which has a low-temperature MBE grown InGaAs/GaAs semiconductor saturable absorber, with a bandgap ≈ 1.05 µm, monolithically integrated between two reflecting mirrors [6]. The top reflector is a TiO_2/SiO_2 dielectric mirror with a 98% reflectivity (Fig. 1). The use of the A-FPSA effectively transforms the semiconductor to a high saturation intensity, low-loss absorber as required. This scheme is simpler and more compact than RPM, and avoids issues associated with coupled cavity length control. However, the intracavity A-FPSA can be explained in terms of coupled cavity mode locking. In fact, A-FPSA corresponds to "monolithic" RPM (Fig. 2)

At antiresonance the intensity inside the Fabry Perot is always smaller than the incident intensity. Therefore, we achieve a high damage threshold, a high effective saturation intensity, and a low insertion loss (top reflector is 98%). Because of the large free spectral range of ≈ 138 nm, we obtain no bandwidth limitation, insensitivity to thermal loading, loose design tolerances, and the possibility of designing a broadband saturable absorber. In contrast, a FPSA operated at resonance has a higher intensity inside the Fabry Perot, has very critical design tolerances and can exhibit bistability effects, which are detrimental for our application.

With this new A-FPSA inside a standing-wave Nd:YLF or a Nd:YAG laser cavity end-pumped by a Ti:sapphire laser, we produce stable self-starting 3.3 ps [6] (Fig. 3a) or 9.2 ps (Fig. 3b) cw mode

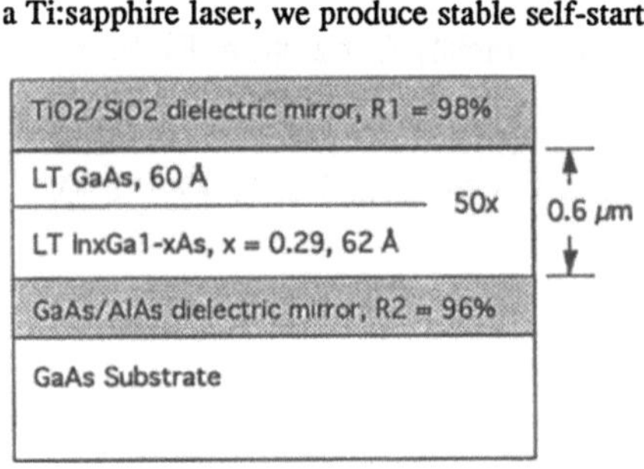

Fig. 1: A-FPSA design

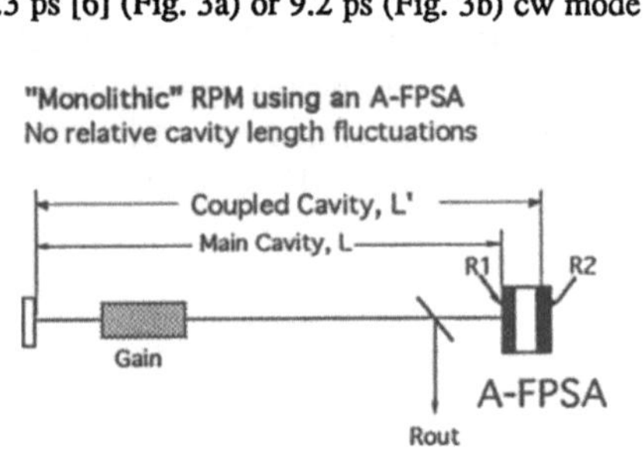

Fig. 2: "monolithic" RPM

Springer Series in Chemical Physics, Vol. 55 Ultrafast Phenomena VIII
Editors: J.-L. Martin · A. Migus · G.A. Mourou · A.H. Zewail

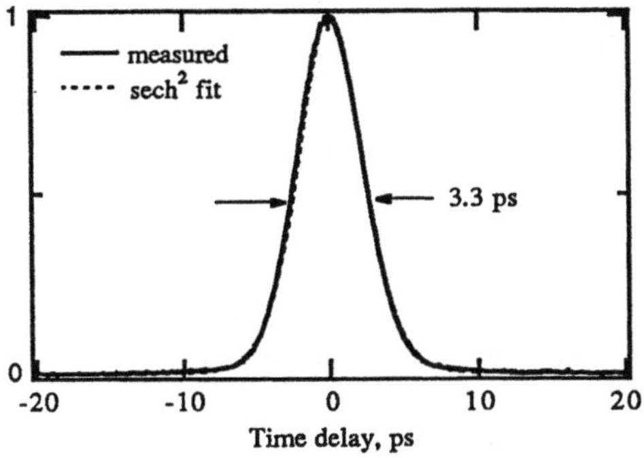

Fig. 3a: Nd:YLF laser

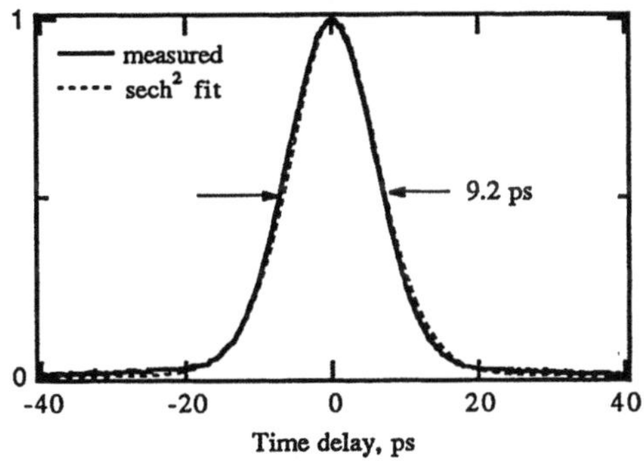

Fig. 3b: Nd:YAG laser

locked pulses at a 220 MHz repetition rate. With a 2% output coupler, the slope efficiencies are 35% and 40%, and the pump thresholds are 43 mW and 32 mW respectively. We show below that no self-focusing effects assist the mode locking process and that this A-FPSA provides the dominant mode locking mechanism.

The saturable absorber inside the Fabry Perot has a subpicosecond time component due to carrier thermalization and a slower component (≈ 50 ps) due to carrier recombination that is longer than the steady-state pulse duration but much shorter than the pulse repetition period (≈ 4.4 ns). Therefore, when the laser is cw mode locked, the absorption bleaching, and hence the reflectivity, is increased due to the increased numbers of carriers generated within one laser pulse duration. Although the change in the semiconductor's absorption is only a few percent [2], and the change in reflectivity of the A-FPSA is further reduced by the 98% top mirror, this change is still large enough to strongly mode lock the laser.

The pulse energy density on the A-FPSA sets the pump power required to start the mode locking process. The minimum pulse energy density on the A-FPSA is ≈ 2 mJ/cm^2 and ≈ 3 mJ/cm^2 for the Nd:YAG and Nd:YLF laser respectively. This has been confirmed by changing the pulse repetition rate, the mode-size on the A-FPSA, and the output coupler. Also at an intermediate pump power the Nd:YLF laser is stably self-Q-switched, producing 4 ps cw mode locked pulses within a 1.4 µs Q-switched pulse at the relaxation oscillation rate of the Nd:YLF laser. At higher pump power the self-Q-switching disappears and the Nd:YLF laser is stably cw mode locked. In contrast, no self-Q-switching was seen with the Nd:YAG laser due to the shorter upper state lifetime.

Self-focusing inside the laser cavity does not significantly contribute to the mode locking process because no intracavity aperture has been used [4] and the laser is not misaligned [3]. In addition, while decreasing the cavity mode size in the Nd:YAG crystal from 135µm x 300µm to 50µm x 100µm, thereby increasing any possible self-focusing contribution to the mode locking process, we did not observe a reduction in the pulse duration. Indeed, ABCD matrix calculations [7] determine that at an intracavity peak power of ≈ 100 kW (i.e. 4 W average output power), the beam radius only changes ≈ 0.1 µm. Additionally, the nonlinear phase shift due to self-phase modulation in the Nd:YAG crystal is typically ≈ 10^{-3} radians.

For femtosecond pulse generation it is more efficient to use self-focusing mode locking, which can be started by weak RPM [4]. Additionally, we have demonstrated a new broadband semiconductor saturable absorber, with which we extended the tunability of an RPM Ti:sapphire laser to over 100 nm

[8]. Because of the direct analogy between A-FPSA mode locking and RPM, we expect that an A-FPSA self-starts broadband femtosecond mode-locking. Broadband starting can be obtained because the Fabry-Perot saturable absorber is operated at antiresonance. Hence, one end mirror in the laser cavity could be replaced by this nonlinear, low-loss and compact mirror which continuously starts self-focusing modelocking.

[1] U. Keller, W. H. Knox, H. Roskos, *Opt. Lett.* **15**, 1377 (1990).
[2] U. Keller, T. H. Chiu, *IEEE J. Quantum Electron.* **28**, July (1992).
[3] D. E. Spence, P. N. Kean, W. Sibbett, *Opt. Lett.* **16**, 42 (1991).
[4] U. Keller, G. W. 'tHooft, W. H. Knox, J. E. Cunningham, *Opt. Lett.* **16**, 1022 (1991).
[5] U. Keller, W. H. Knox, G. W. 'tHooft, *IEEE J. Quantum Electron.* **28**, September (1992).
[6] U. Keller, D. A. B. Miller, G. D. Boyd, T. H. Chiu, J. F. Ferguson, M. T. Asom, *Opt. Lett.* **17**, 505 (1992).
[7] F. Salin, J. Squier, M. Piché, *Opt. Lett.* **16**, 1674 (1991).
[8] G. R. Jacobovitz-Veselka, U. Keller, and M. T. Asom, CLEO paper CTuP6

CW Mode-Locked Singly-Resonant Optical Parametric Oscillator Pumped by a Ti:Sapphire Laser

A. Nebel, U. Socha, and R. Beigang

Fachbereich Physik, Universität Kaiserslautern,
Erwin-Schrödinger-Str. 46, W-6750 Kaiserslautern, Fed. Rep. of Germany

Abstract. Operation of a singly resonant optical parametric oscillator pumped by a cw mode locked Ti:Sapphire laser is reported using $KTiOPO_4$ as the nonlinear crystal. With an average pump power of 1.65 W at a repetition rate of 82 MHz an output power of more than 500 mW was obtained for the signal wave only. The internal power conversion efficiency reached 45 % and the OPO was tunable from 1.05 μm to 1.22 μm by tuning the pump laser wavelength. The pulse length of the OPO pulses was measured to be 1.2 ps with a pulse–bandwidth product of 0.35.

Optical parametric oscillators (OPO) are very attractive light sources for the generation of tunable ultrashort light pulses in the near infrared spectral region. In particular, cw mode locked systems synchronously pumped by a mode locked pump source can produce tunable picosecond and sub picosecond pulses with a high repetition rate [1–3].

Here we report on the demonstration of a ps–OPO with a KTP nonlinear crystal synchronously pumped by a cw mode locked Ti:Sapphire laser. Tuning of the pump laser wavelength results in tunable OPO output without changing the OPO–resonator configuration.

The KTP crystal is placed at the waist of a 4 mirror ring cavity. The curvature of the two spherical mirrors was varied from 50 mm to 150 mm. The output coupler transmission was varied from 0.5 % to 15 %. For precision adjustment of the cavity length, the plane high reflector was mounted on a piezo electric transducer and a precision translation stage. The pump beam was focused collinearly through one of the spherical mirrors with lenses of different focal length (80 mm < f < 200 mm).

The pump laser was a commercial cw mode locked Ti: Sapphire laser with an average power of 1.65 W at a repetition rate of 82 MHz (Spectra Physics: Tsunami). The pulse length was around 1.4 ps with transform limited pulses in the wavelength range from 720 nm < λ_p < 850 nm.

The KTP crystal was cut for type II phase matching in the xy–plane (e → e + o). In this plane and for the pump wavelengths

Springer Series in Chemical Physics, Vol. 55 Ultrafast Phenomena VIII
Editors: J.-L. Martin · A. Migus · G.A. Mourou · A.H. Zewail

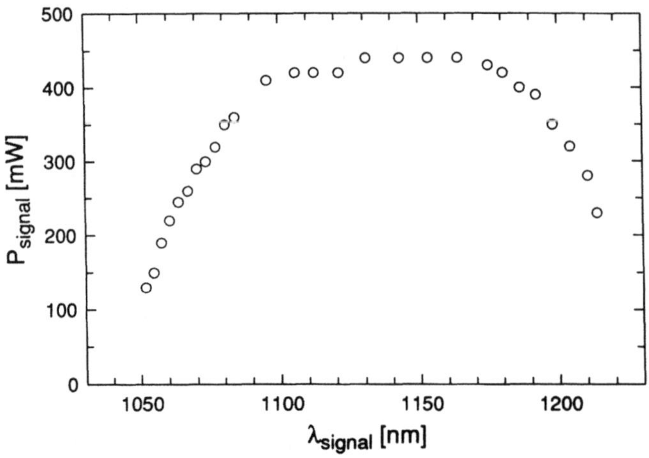

Fig. 1: Signal wave tuning curve of the KTP–OPO

used noncritical phase matching becomes possible ($\varphi = 0^0$, $\vartheta = 90^0$) with a maximum figure of merit compared to all other phase matching conditions. The crystal was antireflection coated on both sides for a center wavelength of 1150 mm. The residual losses from the crystal surfaces were less than approximately 0.5 % for both sides. The extraordinary (e) polarized signal wave was resonated.

A typical tuning curve of the KTP–OPO for the signal wave is shown in Fig. 1. The OPO can be tuned from 1050 nm up to 1220 nm by changing the pump wavelength from 720 nm to 850 nm.

Average powers of up to 440 mW were obtained over a wide tuning range. This corresponds to an overall efficiency ($(P_{idler} + P_{signal}) : P_{pump}$) of almost 40 %.

The idler wavelength extended from 2.3 μm to 2.8 μm with a maximum average output power of 190 mW. Oscillation threshold of the OPO was at 390 mW average pump power. The OPO yielded a slope efficiency of 31.5 %.

The pulse length of the OPO pulses was measured with a background free autocorrelation. The width of 1.2 ps together with a spectral bandwidth of 303 GHz indicates ($\Delta\nu \cdot \Delta t = 0.35$) almost Fourier limited pulses.

Utilizing the total tuning range of the pump laser from 720 nm to 1050 nm, tunable picosecond OPO pulses in the wavelength range from 1040 nm to 1500 nm can be generated with a high efficiency.

References

1 G. Mak, Q. Fu, and H.M. van Driel
Appl. Phys. Lett. 60, 542 (1992)
2 D.C. Edelstein, E.S. Wachmann, and C.L. Tang
Appl. Phys. Lett. 54, 1728 (1989)
3 M.J. McCarthy and D.C. Hanna
Opt. Lett. 17, 402 (1992)

70 fs, High-Average Power, CW Infrared Optical Parametric Oscillator

G. Mak, Q. Fu, and H.M. van Driel

Department of Physics, University of Toronto, Toronto, Canada M5S-1A7

Abstract. We demonstrate a 76 MHz optical parametric oscillator, synchronously pumped by a Kerr-lens mode-locked Ti:sapphire laser, and which is tunable from 0.9 to 4.5 μm with average power >175 mW and pulse width as short as 62 fs. The output beam can be frequency doubled so that together with the Ti:sapphire laser one has a femtosecond source tunable over the range 0.36-4.5 μm.

1. Introduction

Most investigations of subpicosecond phenomena have relied upon visible, and near-visible, pulsed femtosecond dye lasers with limited tuning range. Optical parametric oscillators (OPOs) have long been considered as devices for extending the tuning range while maintaining the high repetition rate but the lack of high power, low noise pump sources has prevented significant progress in this area. Recently Tang and co-workers [1] demonstrated an OPO pumped intracavity of a colliding-pulsed mode-locked laser which is tunable over certain intervals in the 0.75-4.5 μm range, has pulse width as short as 105 fsec, and average power up to 3 mW. This year we were the first to establish the proof-of concept of using a cw femtosecond laser [2] to construct an externally and synchronously pumped, high power and high repetition rate OPO potentially tunable from 0.9 to 4.5 μm. The pump source was a commercial dye laser and the OPO was capable of producing 140 fs pulses with average powers up to 30 mW in the signal beam with $1.2 < \lambda < 1.34$ mm, a range limited by the particular mirror set. Here we show that the replacement of the dye pump laser with a Kerr lens mode-locked Ti:sapphire laser increases the average power of the signal beam in the OPO to 150 mW and decreases the pulse width to below 70 fs [3].

2. The Ti:sapphire pumped OPO

The singly resonant OPO which is based on a 1.5 mm thick KTP crystal mounted at an intracavity focus of a ring resonator is shown in Fig. 1. The crystal is cut for type II non-collinear phase matching ($o \Rightarrow o + e$) with the signal wave being resonant. The wavelength can be tuned by rotating the KTP crystal in the cavity plane. The crystal is pumped by 110 fs pulses at $\lambda = 765$ nm from a 76 MHz Ti:sapphire laser operating with an average power of 800 mW. With two different mirror sets for the OPO we have demonstrated tuning from 1.2 to 1.34 μm and from 1.45 to 1.6 μm for the signal beam with the idler tuning from 2.1 to 1.78 μm and 1.62 to 1.46 μm; with additional mirror sets, the signal and idler wavelengths should be tunable from 0.9 to 4.5 μm. The average output power in the signal beam is as high as 200 mW with pulses as short as 149 fs (sech^2 pulse). The maximum idler beam pulse power is approximately 100 mW while the pulse width is < 400 fs, a value limited by group velocity dispersion effects in the KTP crystal.

The peak intensity in our system is as high as 40 GWcm^{-2} at the crystal and results in strong self phase modulation and chirping of the signal beam pulses. These pulses were compressed with little loss of power by an external prism pair to less than 62 fs as indicated by the interferometric autocorrelation trace of Fig. 2. It is likely that the pulses are even shorter than

Springer Series in Chemical Physics, Vol. 55 Ultrafast Phenomena VIII
Editors: J.-L. Martin · A. Migus · G.A. Mourou · A.H. Zewail

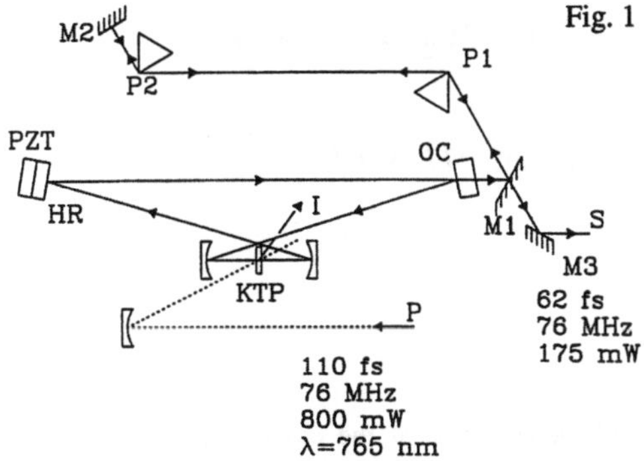

Fig. 1 Schematic diagram of the OPO.

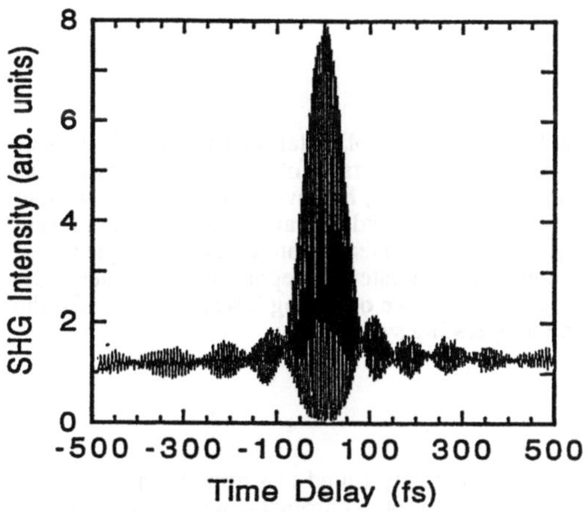

Fig. 2 Interferometric autocorrelation trace of a compressed 1.3 μm pulse.

this because the autocorrelation measurement is influenced by group velocity dispersion effects in a 300 μm thick KDP doubling crystal. The low frequency rms noise of the pulse trains from the OPO is less than 1%. The high peak power of the intracavity beams produces non-phase-matched but visible second harmonic generation and sum frequency mixing in the KTP crystal. External phase-matched second harmonic generation of the signal beam has resulted in tunable femtosecond pulses with $0.6 < \lambda < 0.67$ μm, below that covered by the Ti:sapphire laser, with average powers of 10's of milliwatts. When the potential full tunability of the OPO is utilized it should be possible to have tunable, femtosecond pulses in the region 0.36 - 4.5 μm.

3. References

1. E. S. Wachmann, D. C. Edelstein, and C. L. Tang, Opt. Lett., **15**, 136 (1990).

2. G. Mak., Q. Fu, and H. M. van Driel, App. Phys. Lett. **60**, 542 (1992).

3. Q. Fu, G. Mak, and H.M. van Driel, Opt. Lett. (in press), (1992).

Femtosecond Intracavity Dispersion Measurements

W.H. Knox

AT&T Bell Laboratories, 4B-415, Holmdel, NJ 07733, USA

Abstract. By measuring the wavelength dependence of the cavity repetition rate in CW-pumped passively modelocked lasers, the complete intracavity dispersion can be directly obtained "in-situ". We introduce this technique and show results for Ti:Sapphire lasers.

1. Introduction

In the new class of CW-pumped widely tunable femtosecond solid-state self-modelocked lasers [1], dispersive phenomena represent an important limit. Dispersion of individual elements has been measured by interferometric techniques previously [2], however it is also necessary to measure the total cavity dispersion in a "live" or modelocked operating state. In the present article, we discuss a recently developed Frequency-Domain Dispersion (FDD) technique which provides such information [3]. These measurements illustrate the importance of higher order dispersion and its effects on the modelocking physics in the operating laser, and show for the first time the dispersive behavior of a complete operating system.

2. Experiment

The basis of the FDD method [3] is as follows: the cavity round trip time at any particular wavelength is exactly the total group delay of a quasi-monochromatic wave packet at that wavelength, then clearly *by accurately measuring the cavity repetition rate as a function of wavelength while tuning the laser, we obtain the group delay as a function of wavelength directly.* A CW-pumped Ti:Sapphire laser (modified Spectra-Physics Model 3900) is modelocked by self-focusing and resonantly started with a dye jet [1]. A pair of SF-10 brewster angle prisms placed 52 cm apart provide dispersion compensation. A photodiode, amplifier and digital frequency counter is used to monitor the cavity repetition rate, and this data is converted to time delay : $\delta T = -\delta\nu/\nu_o^2$. Fig. 1a shows data for four different prism positions. Translation of the prisms changes the total dispersion dramatically, as expected. The dispersion parameter is defined: $\partial T/\partial\lambda = 2 \cdot l \cdot D$, and for a whole system, we would measure a sum : $\partial T/\partial\lambda = 2 \cdot \sum_i l_i D_i$. Figure 1b shows the system dispersion for two prism settings. We note that stable modelocked operation is obtained with group delay increasing with wavelength, which corresponds to "negative" dispersion in the range of 1.5 to 10.5 fs/nm, which is qualitatively in agreement with theory [4]. Outside this range, the spectrum exhibits CW breakthrough. The large variation of the dispersion is a result of primarily third order dispersion from the SF-10 prisms. This explains the limited single-parameter tuning range of the Ti:Sapphire laser, and also the pulsewidth variation with the tuning. Figure 2a shows the group delay for another Ti:Sapphire laser (Coherent MIRA), and Figure 2b shows a polynomial fit to the data, again indicating large third order phase contribution.

A self-modelocked femtosecond laser is a complicated nonlinear self-organizing system, and in principle a number of different effects could contribute to these signals. We find that the group delay varies by less that 1 fs when the pulsewidth is varied from 100 fs to 300 fs by adding an intracavity loss, indicating that FDD signals are dominated by the "cold cavity" dispersion due to linear (intensity-independent) accumulated phase effects. This is reasonable, since for 100 fs pulses the nonlinear phase from the Ti:Sapphire rod is $B = \frac{2\cdot\pi}{\lambda}\int n_2 I\, dl \sim \pi$, therefore we

Springer Series in Chemical Physics, Vol. 55 Ultrafast Phenomena VIII
Editors: J.-L. Martin · A. Migus · G.A. Mourou · A.H. Zewail

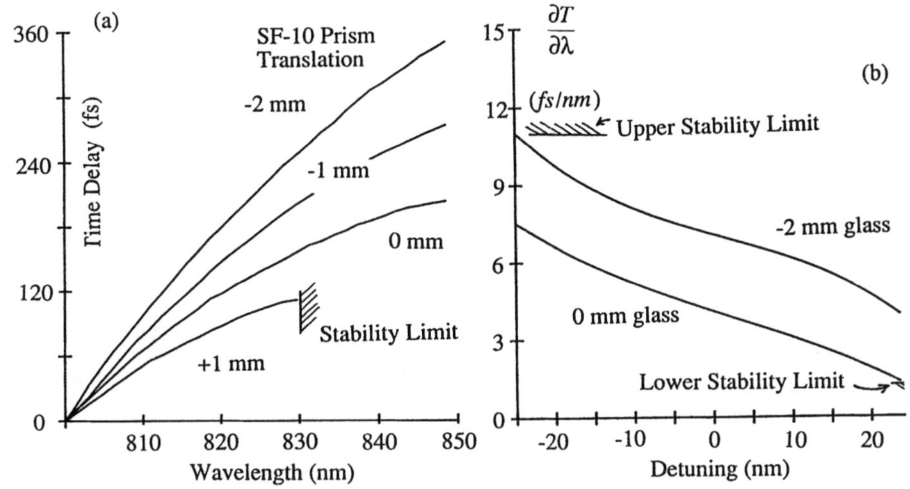

Figure 1. (a) group delay for four prism positions, (b) total system dispersion

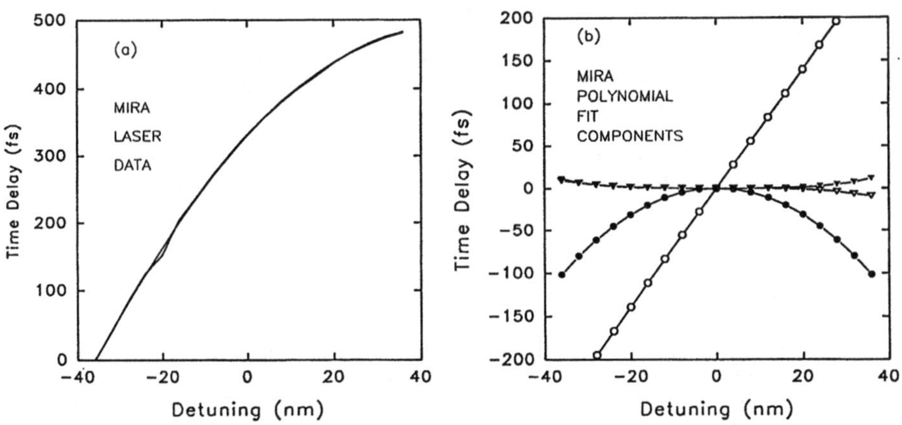

Figure 2. (a) Group delay for MIRA laser, and (b) polynomial decomposition of data showing individual components.

expect a change of only 1.5 fs in the round trip time between low and high intensity operation modes. Since this is essentially a linear technique FDD can also be used for intracavity spectroscopy of additional elements

[1] D.E. Spence, P.N. Kean and W. Sibbett, Opt. Lett. **16**, 42 (1991); N. Sarukura, Y. Ishida and N. Nakano, Opt. Lett. **16** 153 (1991).
[2] W.H. Knox, N.M. Pearson, K.D. Li and C.A. Hirlimann, Opt. Lett. **13**, 574 (1988); M. Beck, I.A. Walmsley and J.D. Kafka, IEEE JQE **27**, 2074 (1991); K. Naganuma, K. Mogi and H. Yamada, Opt. Lett. **15**, 393 (1990).
[3] W.H. Knox, Opt. Lett. **17**, 514 (1992).
[4] H.A. Haus, J.G. Fujimoto and E.P. Ippen, JOSA **B8**, 2068 (1991).

Time Synchronization Measurements Between Two Self-Modelocked Ti:Sapphire Lasers

D.E. Spence[1], *W.E. Sleat*[1], *J.M. Evans*[1], *W. Sibbett*[1], and *J.D. Kafka*[2]

[1]J.F. Allen Physics Research Laboratories, Department of Physics and Astronomy, University of St. Andrews, North Haugh, St. Andrews KY16 9SS, Scotland

[2]Spectra-Physics Lasers, Inc., Mountain View, CA 94039, USA

Abstract. The relative timing jitter of a self-modelocked Ti:sapphire laser has been measured. This jitter has been reduced by locking the laser repetition frequency to an electronic reference oscillator. The technique has been extended to two similar lasers and the relative timing jitter between them has been estimated.

1. Introduction

Using the approach of self-modelocking, picosecond or femtosecond pulses can be produced from a Ti:sapphire (Ti:Al_2O_3) oscillator over a wide tuning range[1]. Techniques such as electro-optic sampling require that the phase noise on the laser output be kept to a minimum so that an adequate signal-to-noise ratio can be achieved using a minimum sample time. In this paper we describe a method which can be used to reduce the timing jitter on the output of passively modelocked lasers. When applied to the self-modelocked Ti:Al_2O_3 laser, this timing jitter was reduced by a factor of 40 at frequencies below 500 Hz and by by a factor of 2 - 3 at higher frequencies. The scheme was then extended to two similar lasers and the relative timing jitter between them was measured using a synchronously operating electron optical streak camera.

2. Experiment and Results

The pulse repetition rate of a passively modelocked laser is determined by the cavity length so that any random movement of the cavity components will lead to a corresponding variation in the laser repetition frequency, which manifests itself as phase noise, or pulse timing jitter on the modelocked output. This noise can be evaluated quantitatively from the rf power spectrum of the modelocked pulse sequence[2].

The laser configuration was similar to the regeneratively initiated, self-modelocked laser described elsewhere[3]. The single sideband phase noise spectrum for the unlocked laser was determined from rf power spectrum measurements. The noise spectrum obtained was characteristic of passively modelocked lasers in which the noise power showed a progressive decrease as the frequency offset from the carrier increased. Above a few kilohertz, the recorded noise spectrum had fallen to the level of the noise floor of the measurement system. The rms pulse timing jitter (calculated using the 1st and 10th harmonics) in this configuration, was: 8.9 ps (100 - 500 Hz), 0.85 ps (0.5 - 5 kHz) and 0.47 ps (5 - 50 kHz).

By phase locking the laser repetition frequency to a low phase noise reference oscillator, the timing jitter was reduced[4,5]. The noise measurements were repeated for the phase locked laser and the single sideband noise spectrum obtained from the 1st and 20th harmonics is illustrated in figure 1. The calculated rms timing jitter was: 0.35 ps (100 - 500 Hz), 0.25 ps (0.5 - 5 kHz) and 0.24 ps (5 - 50 kHz). The remaining predominant noise contributions were attributed primarily to environmental mechanical and acoustic noise, as well as electronic power supply noise and phase errors in the servo loop.

Two similar lasers were phase locked to a common reference oscillator which was also used to provide the deflection signal for a synchronously operating streak camera (Photochron IIC). By simultaneously monitoring the output from both lasers on the streak camera, the relative

Springer Series in Chemical Physics, Vol. 55 Ultrafast Phenomena VIII
Editors: J.-L. Martin · A. Migus · G.A. Mourou · A.H. Zewail

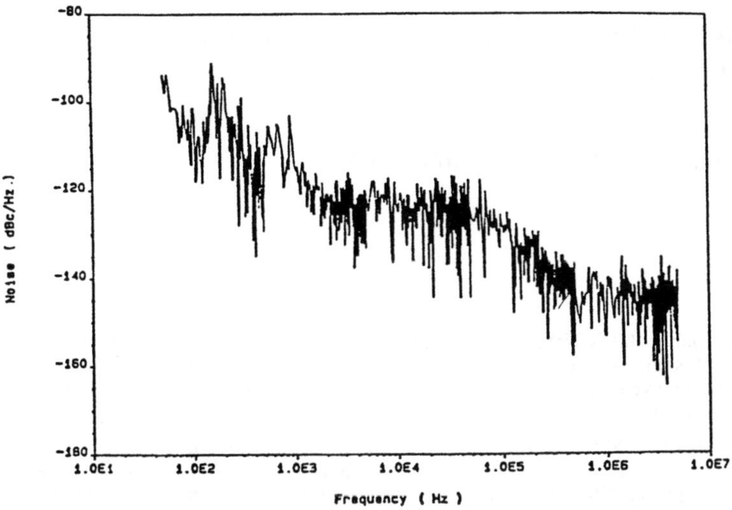

Figure 1. Single sideband phase noise spectrum for the phase locked laser.

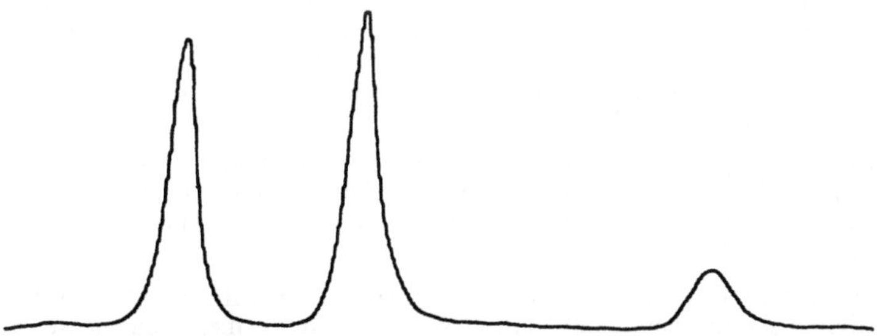

Figure 2. Streak camera traces showing one pulse from each laser. The delay between the 2nd and 3rd pulses is 20 ps, and the duration of the streak profiles was 1.8 ps.

timing jitter between the lasers and the reference oscillator could be determined. The streak profiles are illustrated in figure 2 and have a measured duration of 1.8 ps, which represented the resolution limit of the camera.

Conclusions

The phase noise or pulse timing jitter of the self-modelocked $Ti:Al_2O_3$ laser has been measured. By phase locking the laser cavity to an electronic reference oscillator, the timing jitter was reduced by a factor of 40 over the 100 - 500 Hz frequency range and by a factor of 2 - 3 over higher frequencies. Finally, the technique was extended to two self-modelocked lasers with the relative timing jitter being reduced below 1.8 ps.

References

1. D.E. Spence, P.N. Kean and W. Sibbett, Opt. Lett. **16**, 42, (1991).
2. D. von der Linde, Appl. Phys. **B 39**, 201, (1986).
3. D.E. Spence, J.M. Evans, W.E. Sleat and W. Sibbett, Opt. Lett. **16**, 1762, (1991).
4. S.B. Darack, D.R. Dykaar and G.T. Harvey, Opt. Lett. 16, 1677, (1991).
5. D.R. Walker, D.W. Crust, W.E. Sleat and W. Sibbett, IEEE. J. Quant. Electron. **28**, 289, (1992).

Femtosecond Synchronous Pumping of Dye Lasers with <100 fs Jitter

W.H. Knox and F.A. Beisser

AT&T Bell Laboratories, Holmdel, NJ 07733, USA

Abstract. A tunable 100 fs Ti:Sapphire laser is used to synchronously pump a HITCI dye laser, resulting in 140 fs pulses independently tunable from 840-900 nm with <100 fs jitter. This new source makes two-wavelength experiments possible, and also produces new wavelengths in the visible and mid-infrared.

Recently, pulses of ~100 fs duration of high average power which are widely tunable in the near-infrared have become available using various schemes with CW-pumped Ti:Sapphire lasers. Extension of high repetition rate (100 MHz) techniques to two or more wavelengths is interesting, and optical parametric oscillators (OPO) pumped by femtosecond pulses have produced femtosecond pulses at a number of wavelengths [1]. In contrast to OPOs, however, dyes are incoherent energy converters requiring no critical phase-matching, small wavelength shifts are easily obtained, the resonant interaction requires only a thin gain medium thus minimizing dispersion and walkoff effects, and alignment is made easy by the bright luminescence. In this paper, we discuss experiments in which a dye laser produces a tunable pulse train which exhibits <100 fs jitter with respect to the pump pulse train [2].

A self-modelocked Ti:Sapphire laser synchronously pumps a 300 micron thick dye jet with pulses of width of ~100 fs and average power of ~150 mW, tunable between 750-800 nm (Figure 1). The dye jet is placed in a three mirror cavity, the length of which is controlled with a low-voltage piezoelectric stack, and the pump threshold is ~60 mW.

The sum cross-correlation between the pump pulses and the dye pulses in $LiIO_3$ is 200 fs FWHM (Fig. 2a), which is slightly shorter than the dye laser FWHM. From this measurement we can conclude that the jitter is <100 fs. With the present configuration, femtosecond synchronous pumping is extended to a new and qualitatively different regime, as the pump pulse width is less than the dye laser pulse width. Therefore, the rise of the pulse is determined by the rapid gain rise, and the decay of the pulse is determined by the rapid gain depletion, since the pulse energy is a significant fraction of the dye stored energy. This is effectively a "fast saturable gain"

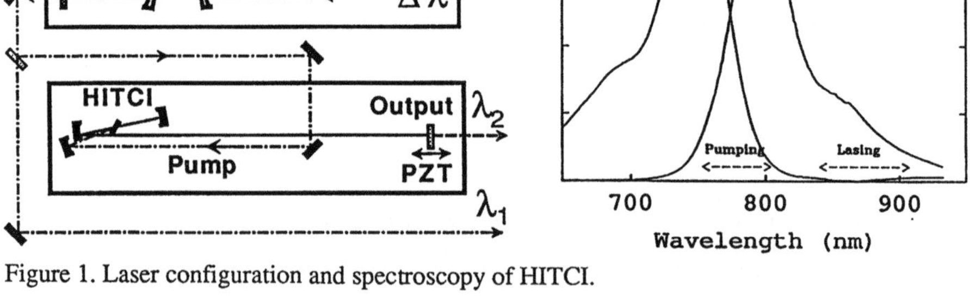

Figure 1. Laser configuration and spectroscopy of HITCI.

Springer Series in Chemical Physics, Vol. 55 Ultrafast Phenomena VIII
Editors: J.-L. Martin · A. Migus · G.A. Mourou · A.H. Zewail

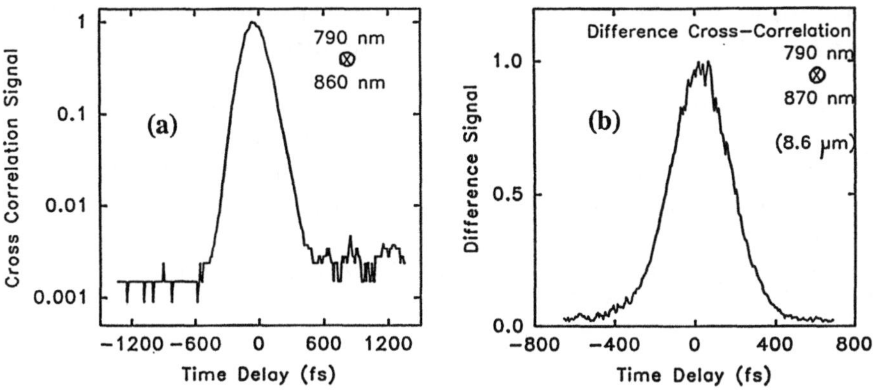

Figure 2 (a) sum cross-correlation in $LiIO_3$, and (b) difference cross-correlation in $AgGaS_2$ at 8.6 microns wavelength.

As indicated, the dye laser lases when the pump wavelength is in the range 750-800 nm, limited at the short range by the Ti:Sapphire mirror set and the long end by the dye absorption wing. The HITCI laser emits pulses of width ~140 fs which can be tuned from 840-900 nm, with the short range limited by re-absorption and the long end limited by the gain spectrum of the HITCI. Over a certain length range, the dye laser spectrum self-adjusts to compensate for cavity length fluctuations, and a simple active-feedback loop [3] is used to lock the cavity length [2]. modelocking. Once the dye pulse has been emitted, the pump pulse is already turned off, therefore satellite pulses are impossible with dyes which have fast vibrational relaxation times. In the case of dyes with slower vibrational relaxation times, spectral holeburning [4] could cause re-pumping and produce satellite pulses.

Difference generation in a $AgGaS_2$ crystal directly yields femtosecond pulses tunable in the mid-infrared (MIR) between 4 and 18 microns (Figure 2b). Dispersion and walkoff effects will make the MIR pulsewidth slightly longer than the sum cross-correlation.

In other embodiments, visible dyes can be pumped with the second harmonic of the Ti:Sapphire, or two NIR dye lasers could pumped. Shorter pulsewidths are certainly possible, although the limits of this technique are not clear.

---

[1] D. C. Edelstein, E. S. Wachmann and C. L. Tang, Appl. Phys. Lett. **54**, 1728 (1989); G. Mak, Q. Fu and H. van Driel, Appl. Phys. Lett. **60**, 542 (1992).
[2] W. H. Knox and F.A. Beisser, Opt. Lett., **17**, 1012 (1992).
[3] J. Chesnoy and L. Fini, Opt. Lett. **11**, 635 (1986).
[4] C.V. Shank, R.L. Fork, C.H. Brito-Cruz and W.H. Knox, Proc. Ultrafast Phenomena V, Springer Verlag, NY (1986), p. 179.

Development of High Average Power Femtosecond Amplifiers Based on Ti:, Cr: and Nd:Doped Materials

J. Squier[1], *S. Coe*[1], *G. Mourou*[1], *D. Harter*[2], *and F. Salin*[3]

[1]Center for Ultrafast Optical Science, University of Michigan, 2200 Bonisteel, IST Bldg., Rm. 1006, Ann Arbor, MI 48109-2099, USA

[2]Allied Signal Corporation, 101 Columbia Turnpike and Park Avenue, Morristown, NJ 07960, USA

[3]Institut d'Optique Théorique et Appliquée, Université Paris-Sud, B.P. 147, F-91403 Orsay Cedex, France

Abstract. We have developed a novel series of femtosecond amplifiers based on a combination of materials including Nd:glass, alexandrite, Ti:Al_2O_3, and Cr:$LiSrAlF_6$ with the intent of increasing the average power capability of high inensity laser sources.

High field physics studies which require intensities of 10^{16} W/cm^2 and greater can now be preformed with compact solid-state laser sources which utilize techniques such as chirped pulse amplification to safely extract the stored energy from the gain media. To date, we have built multiple femtosecond chirped pulse amplification sources which are based on a variety of solid-state materials, including Nd:glass, alexandrite, Ti:Al_2O_3, and Cr:$LiSrAlF_6$. The most interesting (and promising) systems, are not based on any single material, but rather a hybrid scheme where these materials are used as a combination, taking advantages of the individual strengths of each material, and compensating for individual weaknesses. Such hybrid sources are not only capable of amplifying the shortest pulses, but of increasing the average power capability over present designs.

Ti:Al_2O_3 has a demonstrated tunability range of 690-1100 nm which is the broadest of the materials listed above. The lifetime of 3.2 μs is relatively short however, and makes this material difficult to flashlamp pump. For this reason, frequency doubled Nd:YAG or Nd:YLF are the most convenient pump sources for Ti:Al_2O_3. Ti:Al_2O_3 regenerative amplifiers using these pump sources have met with excellent success [1]. With an additional amplifier, pumped by the same source, 100 mJ output has been obtained [2]. To go beyond the 100 mJ level flashlamp pumped dye lasers have been used and joule energies have been demonstrated [3]. To obtain higher energies from Ti:Al_2O_3, better pump sources are needed. The alternative, is to take advantage of the high gain and broad bandwidth of Ti:Al_2O_3, and use it as the preamplifier for materials which are better suited for high energy amplifiers.

Alexandrite has the largest storage capability of the aforementioned solid-state materials, thus it is a good high energy amplifier. We have amplified 100 fs pulses from our Ti:Al_2O_3/alexandrite system to the 50 mJ level, and in an all alexandrite system, 1 ns pulses have been amplified to 1 J. When used as an amplifier in this manner, the energy storage capability of alexandrite, which is one of the greatest strengths of this material, cannot be fully exploited. Therefore we have

Springer Series in Chemical Physics, Vol. 55 Ultrafast Phenomena VIII
Editors: J.-L. Martin · A. Migus · G.A. Mourou · A.H. Zewail

also begun investigating using alexandrite as a pump source itself. Free running alexandrite lasers can produce 5-15 J in a 50-100 µs pulse. This pulse width is readily integrated by materials such as Cr:LiSrAlF$_6$ and Nd:glass.

Cr:LiSrAlF$_6$ has a cross-section and fluorescent lifetime which are between those of Ti:Al$_2$O$_3$ and alexandrite. The main difficulty with Cr:LiSrAlF$_6$ is its inability to handle high thermal loading. The heat load from flash lamp pumping can be reduced by longitudinal laser pumping. This is due to a better pump-to-lasing mode match and a smaller quantum defect. We have developed a gain-guided, unstable Cr:LiSrAlF$_6$ amplifier which is pumped by an alexandrite laser. 100 fs pulses from a self-modelocked Ti:Al$_2$O$_3$ laser have been amplified from 2 pJ to 10 mJ and recompressed to produce streak camera limited pulses. A four-pass, laser pumped Cr:LiSrAlF$_6$ power amplifier with a small signal gain of 20 has also been demonstrated [4].

The primary advantage of Nd:glass is that it can be fabricated in large quantities, with excellent optical quality. The two major limitations with Nd:glass amplifiers are narrow bandwidth, and poor thermal characteristics. In order to overcome the bandwidth limitations, we use a Ti:Al$_2$O$_3$ regenerative amplifier (centered at 1.053 µm) to amplify the chirped pulse from the picojoule to the millijoule level [5]. The millijoule pulse is then amplified through a series of flashlamp pumped Nd:glass amplifiers. The present laser configuration produces 400 fs, 640 mJ pulses at a .03 Hz repetition rate.

Although the Ti:Al$_2$O$_3$ regenerative amplifier has successfully reduced the gain narrowing over the first 10^9 orders of amplification, we still experience approximately a 30% reduction in bandwidth over the last 10^3 orders of amplification. To address this problem, and simultaneously increase the repetition rate of the system we have begun investigating alexandrite pumped, Nd:glass amplifiers [4].

The viability of this technique has first been demonstrated by building an alexandrite pumped Nd:glass regenerative amplifier. We have amplified picojoule pulses to the 5 mJ level in this configuration. Moreover, we have been able to conveniently extend the overall gain bandwidth of this system by using a combination of silicate and phosphate glasses. Present work includes extending this technique so that it is compatible with our present Ti:Al$_2$O$_3$/Nd:glass system.

From our results with Cr:LiSrAlF$_6$ and Nd:glass pumped with alexandrite, it appears that compact, one laser head systems which can have energies in excess of the one joule limitation of most of today's present designs are possible. In addition, the careful attention given to reducing the thermal load of the amplifiers ensures that the average power capability of these materials is optimized.

References

1. J. Squier, F.Salin, G. Mourou, D.Harter, Opt. Lett. **16**, 324 (1991).
2. J. Kemetec, J. Macklin, J. Young, Opt. Lett., **16**, 1001 (1991).
3. A. Sullivan, H. Hamster, H. Kapteyn, S. Gordon, W. White, H.Nathel, R.Blair, R. Falcone, Opt. Lett. **16**, 1406 (1991).
4. J. Squier, G. Mourou, D. Harter, in *Digest of Meeting on Advanced Solid-State Lasers* (Optical Society of America, Washington, D.C., 1992), paper MF3.
5. F. Salin, C. Rouyer, J. Squier, S. Coe, G. Mourou, Opt. Comm. 84, **67** (1991).

Femtosecond Pulse Amplification and Continuum Generation at >250 kHz with a Ti:Sapphire Regenerative Amplifier

T.B. Norris

Center for Ultrafast Optical Science, University of Michigan, 2200 Bonisteel, Ann Arbor, Michigan 48109-2099, USA

Amplification of femtosecond pulses to the microjoule level at high (>1 kHz) repetition rates has generally been accomplished using dye amplifiers pulse-pumped by copper-vapor lasers or by frequency-doubled Nd:YAG regenerative amplifiers [1,2]. The repetition rate of these systems is generally limited by the repetition rate of available pulsed pump sources. In this work, I demonstrate that the superior energy storage of Ti:sapphire allows one to amplify femtosecond pulses to the microjoule level in a *cw*-pumped regenerative amplifier. The gain storage time of Ti:sapphire is about 3 μs, which makes it possible to efficiently extract the energy stored by the pump at repetition rates greater than 250 kHz. Q-switching, injection, and cavity-dumping of the amplifier are done acousto-optically in order to achieve the very high repetition rate [3].

The operation of the laser system is as follows. The output of a cw argon-ion laser is split so that 6W is used to pump a self-mode-locked Ti:sapphire oscillator, and 12W is used to pump the regenerative amplifier. The oscillator produces 75-fs pulses around 800 nm at 76 MHz, with an average power of 200 mW. A fraction of the oscillator output (typically about 10-20%) is aligned through a Faraday isolator and injected into the amplifier cavity .

The amplifier cavity (as shown in fig. 1) is essentially similar to the oscillator cavity, i.e. a folded astigmatically-compensated cavity, except that one end mirror is replaced by a cavity-dumper (a pair of focussing mirrors with a TeO_2 Bragg cell at focus); a second Bragg cell (quartz) is included in the cavity as a Q-switch.

A schematic of the operation of the amplifier is given in fig. 2, showing the timing of the RF to the acousto-optic cells, and an idealized schematic of the pulse train buildup in the cavity. The Ti:sapphire crystal is pumped by the argon laser, establishing a population inversion. The cavity is held below threshold by RF power applied to the Q-switch. The RF is rapidly (in about 100 ns) turned off, Q-switching the cavity. When the Q-switch is open, a short (5 ns) RF pulse is applied

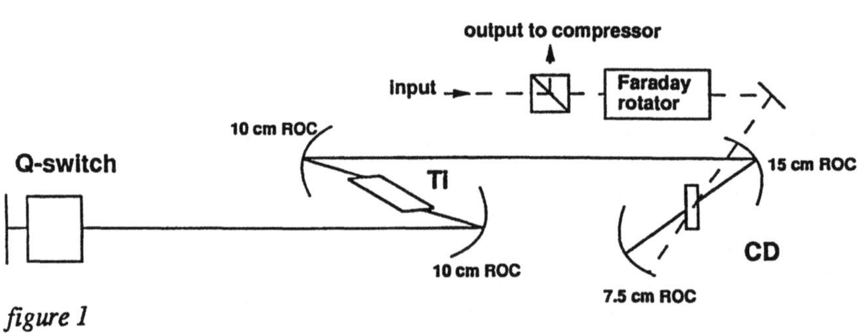

figure 1

Springer Series in Chemical Physics, Vol. 55 Ultrafast Phenomena VIII
Editors: J.-L. Martin · A. Migus · G.A. Mourou · A.H. Zewail

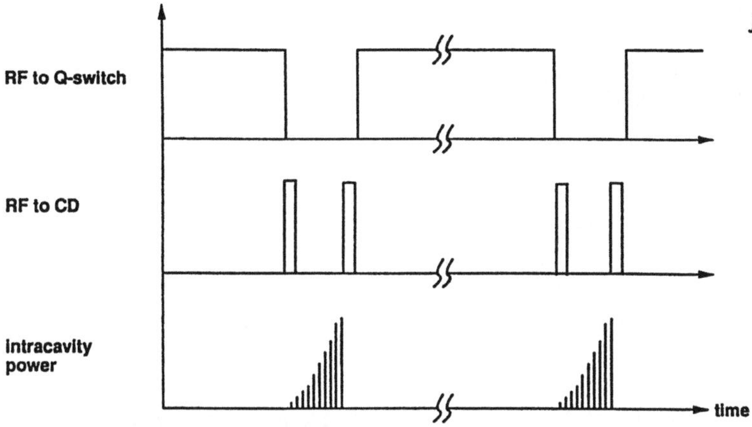

figure 2

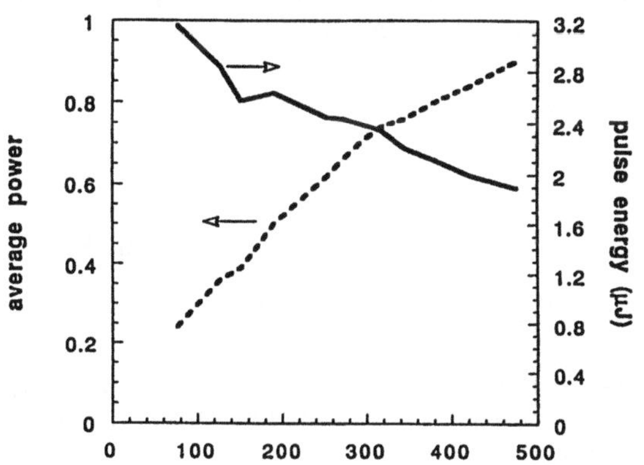

figure 3

to the cavity-dumper Bragg cell to switch a single oscillator pulse into the cavity. After about 50 round trips in the cavity, the pulse has depleted the gain, with a pulse energy of 3.4 μJ at 250 kHz. A second RF pulse is then applied to the cavity-dumper to eject the pulse from the cavity. The cavity length of the amplifier is matched to that of the oscillator so that the RF to the cavity-dumper cell will have the correct phase relative to the optical pulse for both the injection and dumping pulses. A dumping efficiency of about 70% is attained, yielding a pulse energy of 2.4 μJ. A plot of the amplifier output energy and average power versus repetition rate is shown in fig. 3. Pulse energies greater than 2 μJ are obtained for repetition rates up to 400 kHz.

As the pulse circulates inside the amplifier cavity, group-velocity dispersion causes the pulse to stretch to about 10 ps by the time the gain is saturated. This pulse stretching is required, since the peak power of the stretched pulse is low enough to prevent nonlinear effects (or even damage) inside the cavity. After the pulse is cavity-dumped, it must be recompressed. This is accomplished with a set of six equilateral SF-10 flint prisms, as shown in fig. 4 [4]. A double pass through the compressor with a separation between the prism sets of 215 cm results in a recompressed pulse width of 130 fs. With a compressor throughput of 60%, a

figure 4

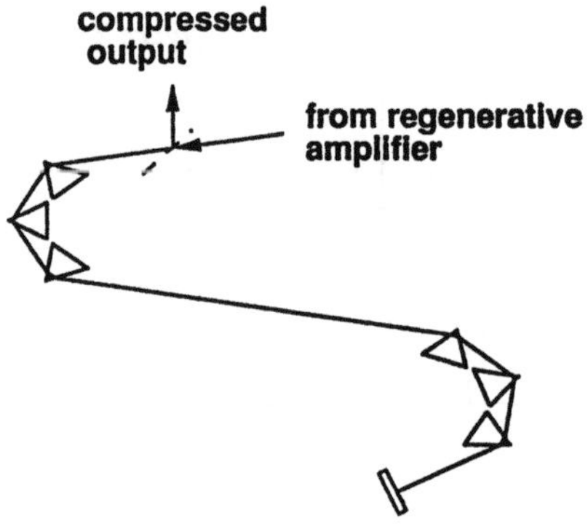

compressed pulse energy of 1.4 μJ at 250 kHz is obtained, which is sufficient to generate a strong white light continuum in an 0.5-mm jet of ethylene glycol. The continuum extends from wavelengths below 400 nm to above 1 μm (detector-limited).

This work was supported by NSF PHY-8920108.

1. W.H. Knox, M.C. Downer, R.L. Fork, and C.V. Shank, Opt. Lett. **9**, 552 (1984).
2. I.N. Duling III, T. Norris, T. Sizer II, P. Bado, and G.A. Mourou, J. Opt. Soc. Am. B **2**, 616 (1985).
3. The scheme used here is essentially similar to that developed independently by A.F. Ruggiero, N.F. Schere, G.M. Mitchell, G.R. Fleming, and J.N. Hogan, J. Opt. Soc. Am. B **8**, 2061 (1991).
4. M. Hofer, M.E. Ferman, F. Haberl, M.H. Ober, and A.J. Schmidt, Opt. Lett. **16**, 502 (1991).

Millijoule Femtosecond Pulse Amplification in Ti:Al_2O_3 at Multi-kHz Repetition Rates

F. Salin[1], *J. Squier*[2], *G. Mourou*[2], *and G. Vaillancourt*[3]

[1]Institut d'Optique Théorique et Appliquée, Université Paris-Sud, B.P. 147, F-91403 Orsay Cedex, France
[2]Center for Ultrafast Optical Science University of Michigan, 2200 Bonisteel, Ann Arbor, MI 48209-2099, USA
[3]Laser Photonics, Inc., 436 A Central Avenue, Bohemia, NY 11716, USA

Abstract: We amplified femtosecond pulses at multi-kHz repetition rates in a Ti:Al_2O_3 regenerative amplifier. Energies up to 1 mJ at 1 kHz and spectral continuum with 1W of average power have been obtained.

We present here a Ti:Al_2O_3 based regenerative amplifier pumped by a Q-switched Nd:YLF laser which is used to amplify femtosecond pulses from a self-modelocked Ti:Al_2O_3 oscillator at repetition rates up to 7.2 kHz. In order to fully exploit the storage capability of Ti:Al_2O_3 we used chirped pulse amplification[1]. The experimental set-up is shown in Fig.1. The Ti:Al_2O_3 amplifier cavity is a folded, astigmatically compensated resonator. The pump laser is a newly designed, cw-pumped, acoustooptically Q-switched, Nd:YLF laser from Laser Photonics (model KYD-101). This laser runs on the 1.047 µm Nd:YLF line and is intracavity frequency doubled. The frequency doubled light contains up to 10 mJ per pulse at 1 kHz, and 1 mJ at 10 kHz in a uniformly distributed multimode output beam. The pulse duration is 250-400 ns, ideally suited for pumping Ti:Al_2O_3. With a 20% output coupler the output from the Ti:Al_2O_3 laser was 1.7 W in a TEM_{00} mode for 4.6 W at 523 nm incident on the crystal. This corresponds to 35% conversion efficiency demonstrating the excellent coupling of the pump into the cavity.

The output coupler was then replaced by a 4 m radius of curvature high reflector. In addition, a broadband thin film polarizer and a KD*P Pockels cell (Medox Electro-Optics) were introduced in the cavity. The Pockels cell is oriented to give a static quater-wave birefringence, so that the cavity Q is low until the gain has reached its maximum value, which is at the end of the pump pulse. The oscillator used in this experiment was a self-modelocked Ti:Al_2O_3 laser[1], pumped by an Argon-ion laser. It produces 100 mW, 90 fs pulses at 82 MHz. It is tunable from 730 to 800 nm and was used at 764 nm in the present experiment. The pulse from the oscillator is stretched 5000 times using a diffractive stretcher [1] consisting of two 2380 lines/mm gratings separated by a 0.5m focal length mirror. The stretched pulse is injected in the regenerative amplifier by applying a quarter wave voltage step to the intracavity Pockels cell. After the pulse has built up to its maximum energy, an additional quarter-wave voltage is applied, and the pulse is cavity dumped from the laser. The output pulse energy as a function of the energy stored in the Ti:Al_2O_3 crystal is given in Fig.2. Note that these data lie on a straight line with a 25% slope, although they have been recorded at different repetition rates and pump powers. The maximum average power produced by our amplifier was 1.6 W for 9 W pump power (6.75 W absorbed in the crystal) at 1 kHz. This corresponds to 1.6 millijoule per pulse. The pump power was limited only by thermal effects in the uncooled Ti:Al_2O_3 crystal. We also tried the amplifier at different repetition rates. At 7.2 kHz the amplified pulses average power was 0.9 W corresponding

Springer Series in Chemical Physics, Vol. 55 Ultrafast Phenomena VIII
Editors: J.-L. Martin · A. Migus · G.A. Mourou · A.H. Zewail

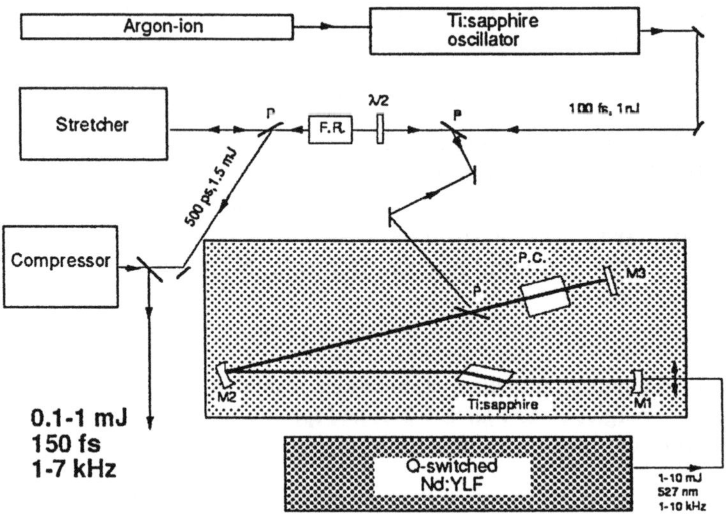

Fig.1: Experimental set-up. F.R, Faraday rotator; P, polarizer; λ/2, half-wave plate; M1,M2, 0.5 m radius of curvature mirrors; M3, 4 m radius mirror.

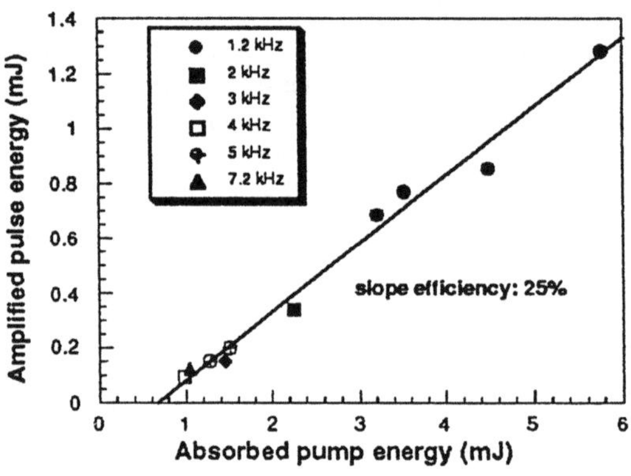

Fig.2: Amplified pulse energy measured at the compressor input versus 523 nm pump pulse energy absorbed in the Ti:Al_2O_3 crystal.

to 125 μJ per pulse for 10 W pump power (7.5 W absorbed). The amplified pulse stability was excellent with about ±2% fluctuations peak to peak.

The amplified pulse was then compressed by double passing a 2380 lines/mm grating pair. A single shot autocorrelator was used to measure the pulse. A typical autocorrelation function is given in Fig.3. The overall transmission of the compressor was 66%. The maximum energy after compression that we have obtained was 1 mJ at 1.2 kHz. The amplified pulse was then focused on a 1 cm piece of glass using a 400 mm focal length lens. The piece of glass was kept 5 cm from the focal point to avoid damage. The high intensity achieved in the glass led to small scale self-focusing and spectral continuum generation. All the incident energy was converted and we obtained 1 W in a perfectly white beam[2]. The continuum extended from the near UV, limited by the transparency range of the glass, to the near infrared. As expected from continuum generation far above threshold

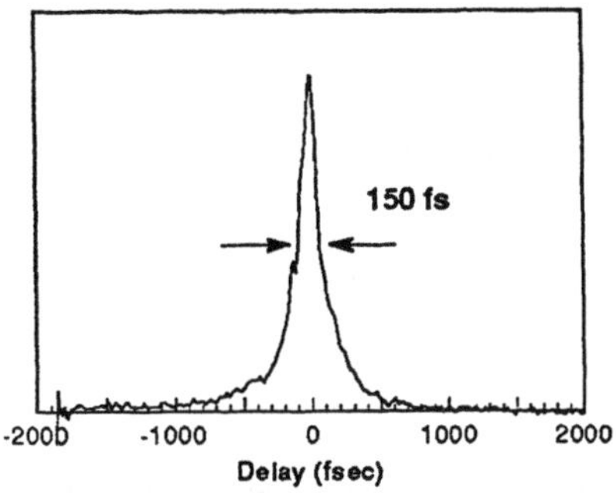

Fig.3: Autocorrelation function of the amplified pulses. About 16 single-shot traces are averaged.

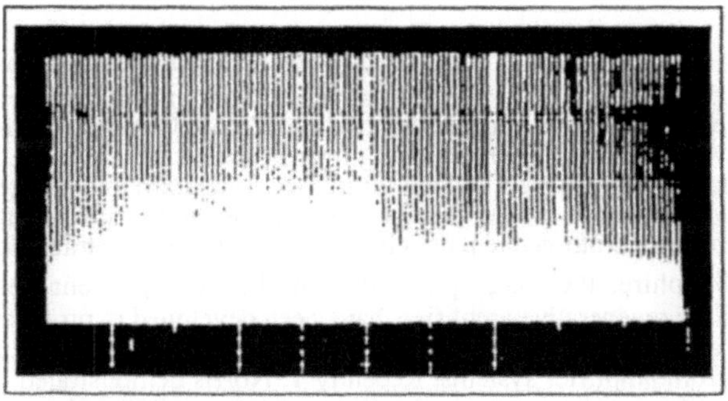

Fig.4: Pulse train corresponding to a blue portion of the spectral continuum (λ~400 nm). The horizontal scale is 10 ms per division and the laser repetition rate is 1.2 kHz.

the stability is excellent (±3% fluctuations peak to peak at ~400nm). When the piece of glass in which the continuum was generated was kept far enough from the focal point, a total conversion to a white beam was observed without damage to the glass (Fig.4).

In conclusion, we have demonstrated the operation of a femtosecond kHz amplifier producing 1 W of amplified 150 fs pulses. This average power is almost hundred times higher than previous femtosecond amplifiers. The excellent stability (±1%) obtained from the cw-pumped Nd:YLF pump laser leads to stable continuum generation. Furthermore as the amplified pulse can be focused on spots only a few microns wide, intensities in excess of 10^{16} W/cm^2 should be attainable making it possible to do high experiments at kilohertz repetition rates.

1. J. Squier, F. Salin, G. Mourou and D. Harter, Opt. Lett. **16**,324 (1991).

2. F. Salin, J. Squier, G. Mourou and G. Vaillancourt, Opt. Lett. **16** ,1964 (1991).

High Repetition Rate CW Pumped Cr:LiSAF Regenerative Amplifier

F. Balembois, P. Georges, F. Salin, G. Roger, and A. Brun

Institut d'Optique Théorique et Appliquée,
Unité de Recherche Associée au CNRS No. 14, Université Paris-Sud,
B.P. 147, F-91403 Orsay Cedex, France

Abstract: We have developed a cw Krypton ion pumped Cr:LiSAF regenerative amplifier producing 10 μJ energy pulses at 835 nm with a 5 kHz repetition rate.

Since its first demonstration by Payne et al at the Lawrence Livermore National Laboratory [1], the Cr^{3+}:$LiSrAlF_6$ (LiSAF) crystal has proved to be a promising laser medium. This crystal combines a large fluorescence bandwidth (750-950 nm) with an absorption band in the red between 600 and 700 nm making it particularly suitable for the development of an all solid state diode pumped tunable laser. Due to its extremely large emission band, Cr:LiSAF is well suited to the generation and amplification of ultrashort pulses. Self mode-locking to produce femtosecond pulses has already been demonstrated in Cr:LiSAF [2,3]. Unlike Ti:Sapphire, the long upper state lifetime (67 μs) enables flashlamp pumping and Cr:LiSAF regenerative amplifiers have been developed to produce pulses of several millijoules at low repetition rates [4]. There is now great interest in increasing the repetition rate from amplifier systems. Recently T. Norris demonstrated a regenerative Ti:Sapphire amplifier working at a repetition rate up to 250 kHz and producing 2 μJ amplified pulses [5]. Considering that the long upper state lifetime of the Cr:LiSAF allows a higher energy storage compared to Ti:Sapphire, we have developed a Cr:LiSAF regenerative amplifier pumped by a cw Krypton ion laser. This amplifier operates at 5 kHz and produces 10 μJ energy pulses.

The schematic of the amplifier is as follows [fig.1]. A 15 mm long, Brewster angle cut Cr:LiSAF crystal (from Lightning Optical Corporation) was used in a three mirrors linear cavity. These mirrors had a high reflective dielectric coating centered at 820 nm. The length of the cavity was around 1 meter corresponding to a 7 ns cavity round trip time. The crystal was pumped by the red lines of a cw Krypton ion laser and its absorption was around 97%. In order to reduce the thermal load, the pump laser was chopped at 50 Hz with a 1/6 duty cycle. A Pockels cell and a broadband polarizer were used to trap and dump the pulses. Without injection (i.e. in a free running Q:switched configuration), the amplifier produced 10 μJ, 7 ns pulses at 5 kHz around 830 nm and for a pump power of 3.6 W on the Cr:LiSAF crystal. The spectrum of these pulses was 20 nm wide showing that the amplifier can support femtosecond pulses. Note that the repetition rate is limited to 5 kHz by the Pockels cell and it should be possible to operate at 14 kHz (corresponding to the inverse of the upper state lifetime) without any decrease of the pulse energy.

We have tested the regenerative amplification by using a picosecond Ti:Sapphire oscillator producing pulses around 10 ps at 835 nm. A fraction of the oscillator output was aligned through a Faraday isolator and injected into the amplifier cavity by applying a quarter-wave voltage on the Pockels cell [Fig.1]. Initially the Pockels cell was oriented to give a quarter wave of static birefringence in order to prevent any lasing effect in the cavity. Figure 2 shows the evolution of the intracavity pulse energy with the number of round trips in the cavity showing that the pulses made 70 round trips (500 ns) before reaching saturation. The pulse was then extracted from the cavity by applying a second quarter wave voltage on the Pockels cell [Fig.3]. The energy of the dumped pulses was 10 μJ after the Faraday rotator. Figure 4 shows the spectrum of the output with and without the injection showing an efficient seeding of the amplifier by the picosecond pulses.

Springer Series in Chemical Physics, Vol. 55 **Ultrafast Phenomena VIII**
Editors: J.-L. Martin · A. Migus · G.A. Mourou · A.H. Zewail

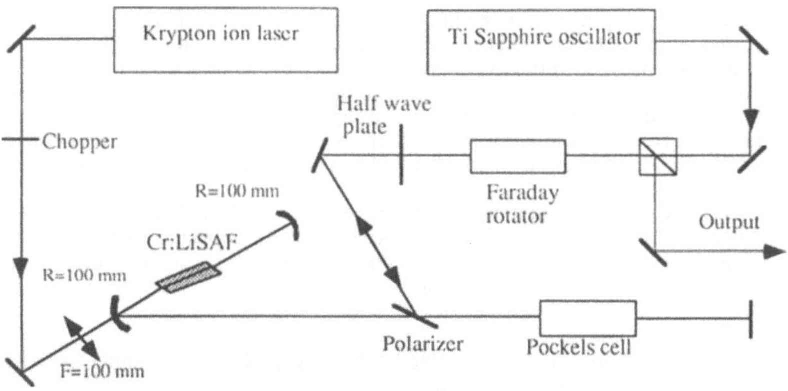

Fig 1: Schematic of the regenerative amplifier.

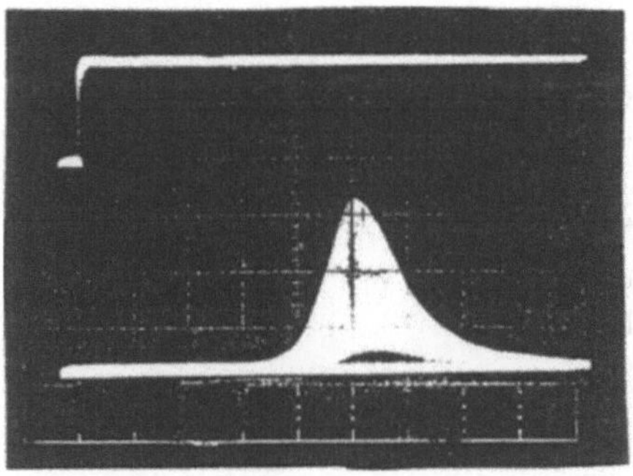

Fig.2: Evolution of the intracavity pulse energy (the upper trace is the high voltage step applied on the Pockels cell, 100 ns per scale).

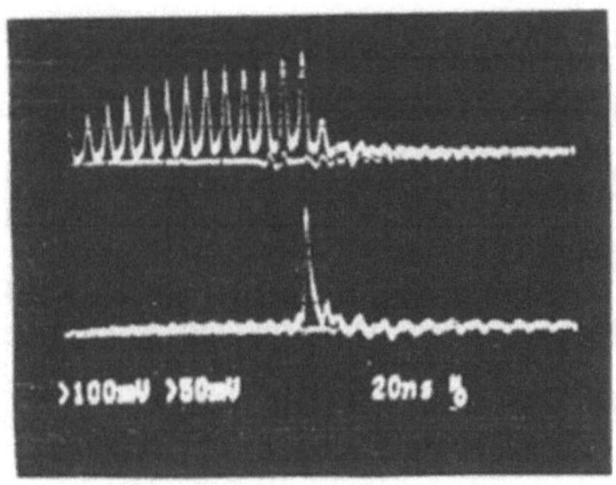

Fig.3: Extraction of the pulses.

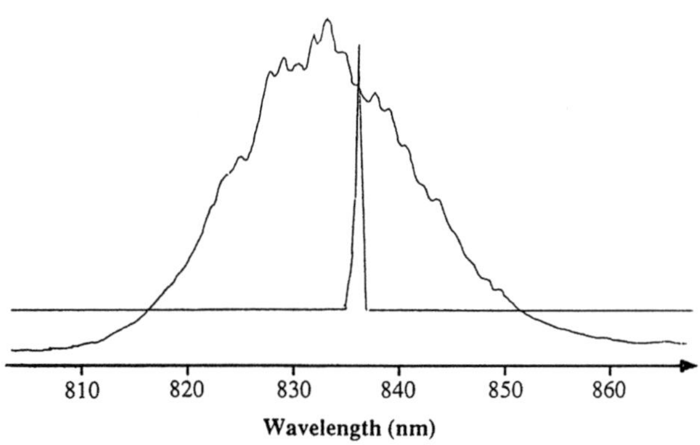

Fig.4: Output spectrum with and without injection.

This regenerative amplifier is expected to produce the same energy level when injected by femtosecond pulses because of the broad spectrum in the free runing configuration. In that case, the intracavity group velocity dispersion would stretch the pulses to around 10 ps. This stretching will decrease the peak power of the pulse and so prevent nonlinear effects in the intracavity components. Furthermore, the nonlinear refractive index n_2 is five times lower for Cr:LiSAF than for Ti:Sapphire, so we believe that the Chirped Pulses Amplification technic [6] will be unnecessary for our system.

In conclusion, we believe that this high repetition rate Cr:LiSAF regenerative amplifier used in combination with a self mode-locked Cr:LiSAF oscillator will lead in the future to an **all solid state** (including the pump lasers) femtosecond system producing up to 5 μJ pulses at around a 10 kHz repetition rate.

Acknoledgments : The authors want to thank the LOA-ENSTA (Palaiseau) and especially J.P. Chambaret for the loan of the Pockels cell.

References :

[1] S. Payne, L. Chase, L. Smith, W. Kway and H. Newkirk
J. App. Phys, 66 1051 (1989).
[2] A. Miller, P. LiKamWa, B. Chai and E. Van Stryland
Opt. Lett. 17, 195 (1992).
[3] N. Rizi, P. French and J. Taylor
Ultrafast Phenomena 8, paper MB3 (8-12 june 1992).
[4] T. Norris
Postdeadline paper N° CPD12, CLEO (1992).
[5] M. Perry, D. Strickland, T. Ditmire and F. Patterson
Opt. Lett. 17,604 (1992)
[6] D. Stickland and G. Mourou
Opt. Comm. 56, 219 (1985)

18 fs Pulse Generation by a Single Excimer-Laser-Pumped Pulsed Dye Laser

P. Simon[1], *C. Jordan*[1], *and S. Szatmari*[2]

[1]Laser-Laboratorium Göttingen e.V., P.O. Box 2619, W-3400 Göttingen, Fed. Rep. of Germany
[2]Max-Planck-Institut für biophys. Chemie, Abt. Laserphysik, P.O. Box 2841, W-3400 Göttingen, Fed. Rep. of Germany

The shortest pulses directly obtained by CPM dye lasers are in the range of 30 fs. To obtain even shorter pulses, propagation through nonlinear media in conjunction with pulse compression is normally applied, including chirping in single mode fibers [1,2,3] and continuum generation [2,3]. For the applications of these techniques previous amplification of the dye laser pulses is needed, necessitating the use of various pump lasers with exact synchronism. When additional amplification of the fs pulses is needed up to the μJ range, which evidently involves increased complexity, the minimum pulse duration is limited by the gain width of dyes to typically 10 - 20 fs.

Here we report on the generation of pulses less than 20 fs duration and μJ energy, by a single excimer laser pumped dye laser (Fig. 1).

These output parameters are comparable to those of cw laser based systems and have been achieved by a proper combination of continuum generation and/or fiber-chirping compression techniques with pulsed dye lasers. This makes it possible to use the same pump laser for pulse generation and for amplification.

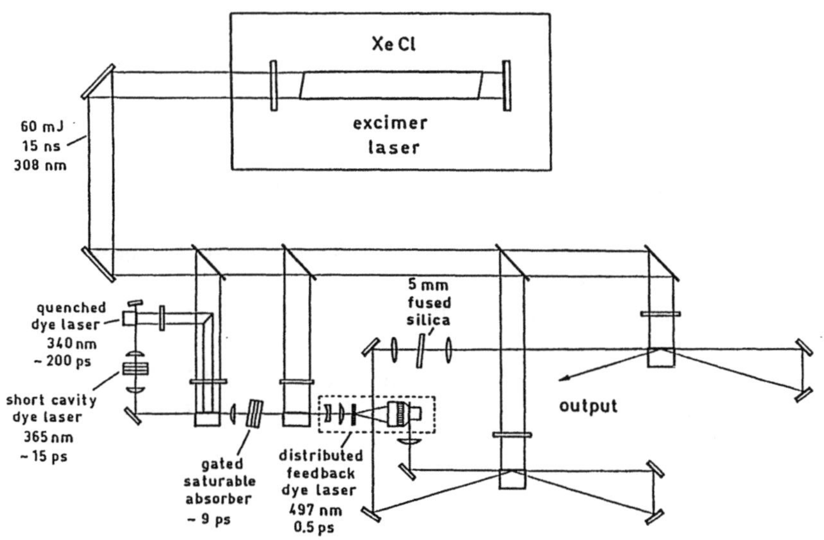

Fig. 1. Experimental arrangement

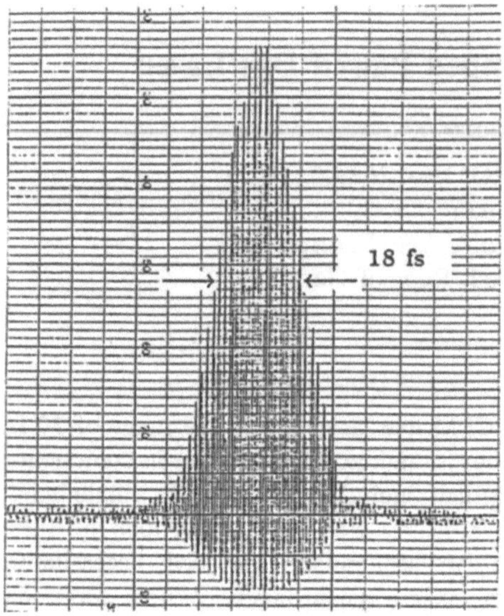

Fig. 2. Autocorrelation trace of pulses at 530 nm

Despite of the relatively long duration (500 fs) of the pulses directly generated by our excimer-pumped dye laser [4], less than 30 fs and 40 fs pulses were obtained by continuum generation [5] and fiber-grating compression, respectively. Using a combination of grating and prism compressors for the arrangement shown in Fig.1, pulses as short as 18 fs were generated (Fig.2).

Continuum generation at three different pulse durations, 500 fs, 200 fs, and 30 fs has been tested. It was found that at longer pulse durations there is a smaller spectral spread, less modulation in the spectrum around the generation wavelength, and a more linear chirp associated with the continuum generation process.

References:

1. R.L.Fork, C.H.Brito Cruz, P.C. Becker, and C.V. Shank: Opt. Lett. 12, 483 (1987)
2. P.C. Becker, H.L. Fragnito, R.L. Fork, F.A. Beisser, and C.V. Shank: Appl. Phys. Lett. 54, 411 (1989)
3. R.W. Schoenlein, J.-Y. Bigot, M.T. Portella, and C.V. Shank: Appl. Phys. Lett. 58, 801 (1991)
4. S. Szatmari and F.P. Schäfer: Optics Commun. 68, 196 (1988)
5. P. Simon, S Szatmari, and F.P. Schäfer: Opt. Lett. 16, 1569 (1991)

Monolithic CPM Diode Lasers

M.C. Wu, Y.K. Chen, T. Tanbun-Ek, and R.A. Logan

AT&T Bell Laboratories, Room 1C-301,
600 Mountain Avenue, Murray Hill, NJ 07974, USA

Abstract. Monolithic mode-locked diode lasers offers many advantages for generating ultrashort optical pulses. We have successfully integrated a colliding pulse mode-locked (CPM) laser onto a single piece of semiconductor. Sub-picosecond optical pulses with transform-limited time-bandwidth products and ultra-high repetition frequencies (30 GHz to 350 GHz) are generated.

1. Introduction

Colliding pulse mode-locking (CPM) [1] has been the most common technique to generate ultrashort optical pulses in passive mode-locked dye lasers. In many applications such as ultrafast all-optical switching [2,3], time-division-multiplexed optical communications systems [4], and soliton transmission in optical fibers [5], a compact and stable source generating short optical pulses is needed. By building the whole mode-locked laser on a single piece of semiconductor, the short-pulse source becomes extremely compact. Since there are no mechanical moving parts, the tedious optical alignment process is completely eliminated. Thermal and mechanical stabilities are also greatly improved. A further advantage of the monolithic mode-locked diode lasers is their capability to generate optical pulses at ultra-high repetition frequencies. Because the cavity lengths of monolithic diode lasers can be scaled down to less than a few hundred micrometers, continuous pulse trains with repetition rates near 1 THz can be obtained. This is very useful for applications in optical generation of millimeter waves and sub-millimeter waves.

Several schemes of mode-locking have been implemented in monolithic mode-locked lasers [6-9]. By using a novel monolithic CPM diode laser, we have successfully generated the first sub-picosecond optical pulses with transform-limited time-bandwidth products [10,11]. The pulse width of 640 fs is the shortest reported for monolithic mode-locked lasers. In this paper, we will review the recent progress in monolithic CPM quantum well lasers. Other schemes of monolithic mode-locked diode lasers will also be discussed.

2. Hybrid Monolithic CPM Diode Lasers

The schematic diagram of the hybrid monolithic CPM quantum well laser is shown in Fig. 1. The optical cavity consists of a buried heterostructure InGaAs/

Springer Series in Chemical Physics, Vol. 55 **Ultrafast Phenomena VIII**
Editors: J.-L. Martin · A. Migus · G.A. Mourou · A.H. Zewail

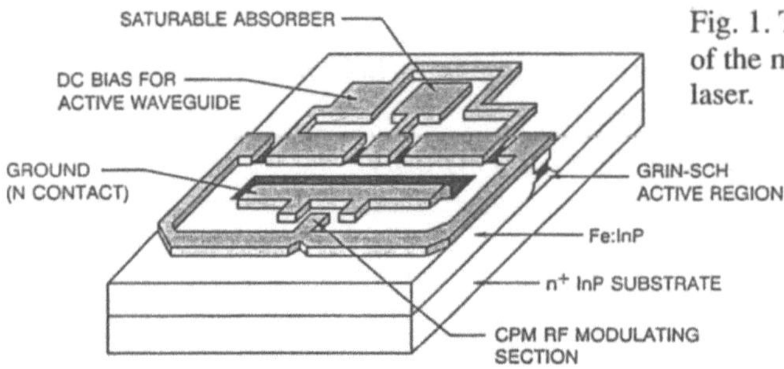

Fig. 1. The schematic diagram of the monolithic CPM diode laser.

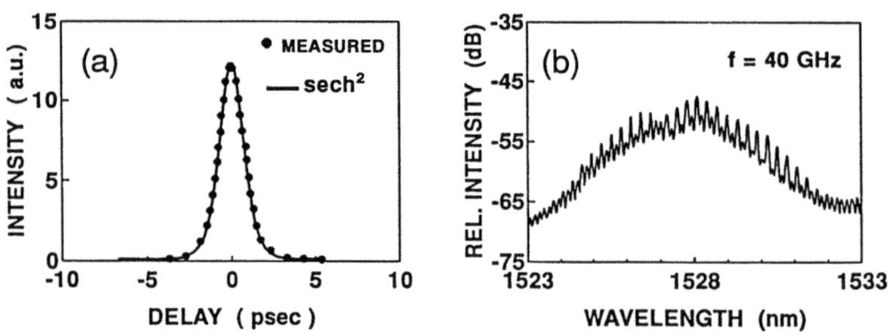

Fig. 2. The autocorrelation trace and the optical spectrum of the hybrid monolithic CPM laser with 40 GHz repetition frequency.

InGaAsP quantum-well laser with cleaved mirrors. The linear cavity is divided into five sections: The saturable absorber is located in the center of the cavity for dual-pulse CPM operation. The optical pulses are synchronized with electrical clocks by the two optical modulators near the facets. To ensure the precise timing between the two modulators, they are connected to the center rf feed by integrated microstrip transmission lines. The absorber and the modulators are linked by active waveguides. The various sections are defined by separate top electrodes. The active quantum wells remain continuous throughout the linear cavity to minimize undesired intra-cavity reflections.

In hybrid CPM operation, a pair of optical pulses are initiated from the end modulators. The two pulses then propagate and collide in the center saturable absorber. The coherent interaction effectively shortens and stabilize the pulses [12]. The modulators thus force the laser to operate in CPM mode. The autocorrelation trace and the optical spectrum of a 2.1-mm-long hybrid monolithic CPM laser is shown in Fig. 2(a) and (b), respectively. The autocorrelation trace agrees very well with that of a $sech^2$ pulse shape. The full-width-at-half-maximum (FWHM) is 0.95 ps. The optical spectral width is 2.4 nm, which yields a time-bandwidth product of 0.32, which is very close to the transform-limited value of 0.31.

3. Passive Monolithic CPM Lasers

The monolithic CPM laser can also be passively mode-locked in the absence of external rf sources. A necessary condition for passive mode-locking is that the saturable absorber cross section (σ_A) is larger than the gain cross section (σ_G)[13]. This is satisfied by reverse-biasing the absorber quantum wells and forward-biasing the gain quantum wells [6]. However, the ratio of σ_A/σ_G can not be unlimitedly increased in monolithic diode lasers which use the same quantum well materials for both absorber and gain media. The monolithic CPM scheme plays a particularly important role here because it can further increase the ratio of effective σ_A/σ_G by a factor of three [12]. Shorter pulses and wider stability range are therefore expected for monolithic CPM diode lasers.

The repetition frequency of the passive CPM laser is not limited by the available rf sources. Very high repetition rate can be obtained by scaling down the cavity length. Figure 3 shows the autocorrelation traces and the optical spectra for the passive monolithic CPM lasers with repetition frequencies of 40 GHz, 80 GHz, and 160 GHz, respectively. The corresponding cavity lengths are 2.1 mm, 1.0 mm, and 0.5 mm. Very good agreement with the sech^2 pulse shapes are obtained at all frequencies. The FWHM pulse widths are 1.1 ps, 0.83 ps and 1.0 ps, respectively. Nearly transform-limited time-bandwidth products are obtained in all cases. The suppression of the every other modes in the optical spectrum results from the dual pulse nature of the CPM. By further shortening the cavity length to 250 μm, a record high repetition frequency of 350 GHz is achieved (Fig. 4). The time-bandwidth product remains nearly transform-limited. The FWHM pulse width of 640 fs is the shortest reported for passive mode-locked semiconductor lasers without external compression.

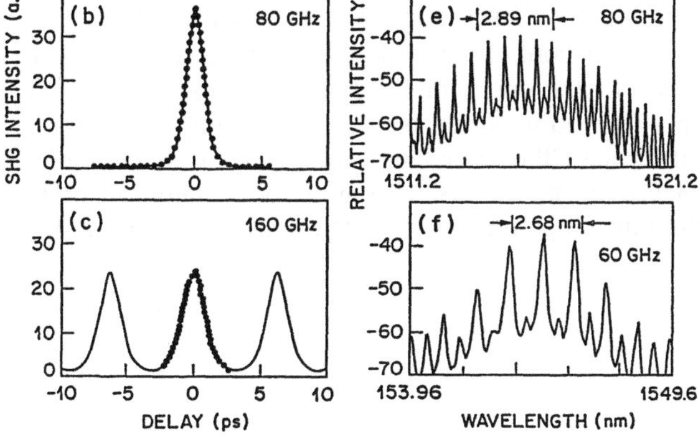

Fig. 3. The autocorrelation trace and the optical spectra of monolithic CPM lasers with repetition rates of 40 GHz, 80 GHz, and 160 GHz.

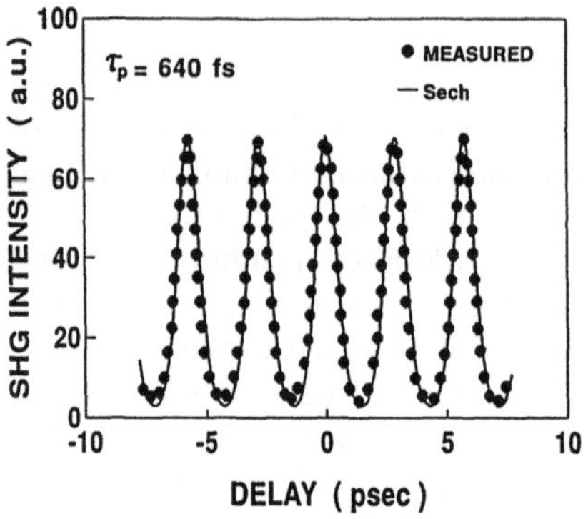

Fig. 4. The autocorrelation trace of the passive monolithic CPM laser with 350 GHz repetition rate. The FWHM pulse width is 640 fs.

4. Stability Ranges and Self-Pulsation

It is known that a laser with intra-cavity saturable absorber tend to oscillate at low frequencies (~ 0.1 GHz to a few GHz). This oscillation is called self-pulsation. A necessary condition for self-pulsation to occur is that the saturable absorber cross section is larger than the gain cross section. As mentioned earlier, this is also the necessary condition for passive mode-locking. As a result, the laser bias ranges for mode-locking and self-pulsation often overlap with each other. Because the occurrence of self-pulsation disturbs or even quenches the mode-locking, it is obviously desirable to have the mode-locking and the self-pulsation ranges separated as wide as possible. Previous study showed that in monolithic tandem-contact diode lasers it was very difficult to generate very short pulses without simultaneous self-pulsation [14].

In addition to producing shorter pulses and wider stability ranges, the CPM configuration also helps differentiate the mode-locking range against self-pulsation. As discussed earlier, the CPM configuration increases the effective absorber cross section, which translates the stable mode-locked region to lower pumping levels. On the other hand, the absorber cross section of self-pulsation is not sensitive to the relative position of the absorber in the cavity because the pulses generated by the self-pulsation are much longer than the round-trip time of the cavity. As a result, maximum delineation of the mode-locking and the self-pulsation ranges is achieved by the CPM configuration. Experimentally, we have compared the bias ranges of two lasers: one with CPM configuration and the other with the saturable absorber near an uncoated facet (Fig. 5). Much wider mode-locking range is observed for the CPM laser. The various ranges are determined by simultaneously characterizing the pulses with an autocorrelator, an optical spectrum analyzer, a sampling scope and an rf spectrum analyzer with a high-speed photodetector.

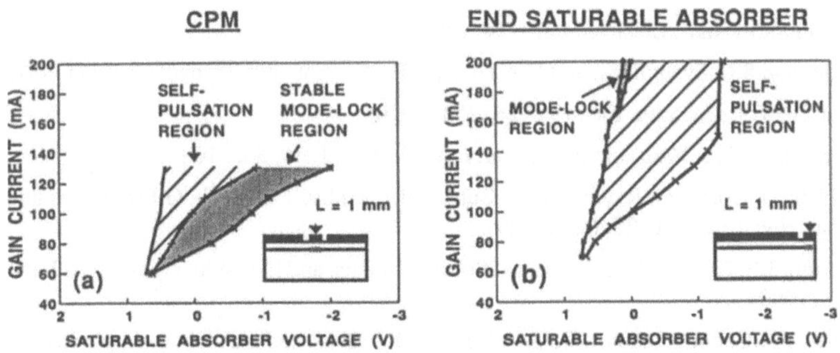

Fig. 5. The mode-locking and the self-pulsation ranges of (a) a monolithic CPM laser and (b) a similar laser with end saturable absorber.

5. Other Monolithic Mode-locked Diode Lasers

Many other schemes of monolithic mode-locked diode lasers are demonstrated recently. By placing the saturable absorber near the high-reflection coated facet, the "self-CPM" laser is shown to have similar performance to the CPM laser with center saturable absorber and twice cavity length [15]. Using the photonic integrated circuit technologies, active mode-lock lasers with low repetition rate (~ 8 GHz) were demonstrated by integrating a long passive waveguide and a distributed Bragg reflector [16] or using a ring cavity [17]. Passive mode-locking at 5.9 GHz (pulse width = 4.8 ps) has also been achieved with a 6.6-mm-long cavity [18].

6. Conclusion

We have demonstrated the cw hybrid and passive mode-locking of a novel monolithic colliding pulse mode-locked (CPM) semiconductor quantum well laser. The CPM scheme produces shorter pulses as well as increases its immunity to the low-frequency self-pulsation noise. Sub-picosecond optical pulses with transform-limited time-bandwidth products have been generated with repetition frequencies from 30 GHz to 350 GHz. A record short pulse width of 640 fs is achieved without any external compression.

REFERENCES

[1] R.L.Fork, B.I.Greene, and C.V.Shank, Appl. Phys. Lett., **38**, 671 (1981).

[2] N.R.Whitakar, H.Avramopoulos, P.M.French, M.C.Gabriel, R.E.LaMarche, D.J.Digiovanni, and H.M.Presby, Optics Lett., **16**, 1838 (1991).

[3] P.A.Andrekson, N.A.Olsson, J.R.Simpson, D.J.Digiovanni, P.A.Morton, T.Tanbun-Ek, R.A.Logan, and K.W.Wecht, OFC 1992, postdeadline paper.

[4] Takada and Saruwatari, Electron. Lett., **24**, 1406 (1988).

[5] L.F.Mollenaur, S.G.Evanglides, and H.A.Haus, IEEE J. Lightwave Technol., **9**, 194 (1991).

[6] K.Y. Lau, IEEE J. Quantum Electron., **26**, 250 (1990).

[7] S. Sanders, L. Eng, J. Paslaski, and A. Yariv, Appl. Phys. Lett., **56**, 310 (1990).

[8] R.S. Tucker, U. Koren, G. Raybon, C.A. Burrus, B.I. Miller, T.L. Koch, G.Eisenstein, and A. Shahar, Electron. Lett., **25**, 622 (1989).

[9] P.A. Morton, J.E. Bowers, L.A. Koszi, M. Soler, J. Lopata, and D.P. Witt, Appl. Phys. Lett., **56**, 111 (1990).

[10] M.C.Wu, Y.K.Chen, T.Tanbun-Ek, R.A.Logan, M.A.Chin, and G.Raybon, Appl. Phys. Lett., **57**, 759 (1990).

[11] Y.K.Chen, M.C.Wu, T.Tanbun-Ek, R.A.Logan, and M.A.Chin, Appl. Phys. Lett., **58**, 1253 (1991).

[12] M.S. Stix and E.P. Ippen, IEEE J. Quan. Elec., **QE-19**, 520 (1983).

[13] H.A. Haus, IEEE J. Quantum Electron., **QE-12**, 169 (1976).

[14] K.Y.Lau and J.Paslaski, IEEE Photon. Technol. Lett., **3**, 974 (1992). see also J.Paslaski and K.Y. Lau, Appl. Phys. Lett., **59**, 7 (1991),

[15] D.J.Derickson, J.G.Wasserbauer, R.J.Jelkey, A.Mar, J.Bowers, D.Coblentz, R.A.Logan, and T.Tanbun-Ek, *Optical Fiber Communication Conference*, (Optical Society of America, Washington DC) 1992.

[16] P.B.Hansen, G.Raybon, U.Koren, B.I.Miller, M.G.Young, M.Chien, C.A.Burrus, and R.C.Alferness, IEEE Photon. Technol. Lett., **3**, 215 (1992).

[17] P.B.Hansen, G.Raybon, M.D.Chien, U.Koren, B.I.Miller, M.G.Young, J.-M.Verdiell, and C.A.Burrus, IEEE Photon. Technol. Lett., **3**, 215 (1992).

[18] L.R.Brovelli, H.Jackel, and L.H.Melchior, *Conference on Lasers and Electro-Optics*, (Optical Society of America, Washington DC) 1992, p. 408.

Ultrashort Pulse Generation from High-Power Arrays Using Intracavity Nonlinearities

L.Y. Pang[1], *J.G. Fujimoto*[1], *and E.S. Kintzer*[2]

[1]Department of Electrical Engineering and Computer Science, Research Laboratory of Electronics, Massachusetts Institute of Technology, Cambridge, MA 02139, USA
[2]Lincoln Laboratory, Massachusetts Institute of Technology, Lexington, MA 02173, USA

The development of high power semiconductor diode laser arrays makes possible the use of new mode-locking techniques which utilize high intracavity intensities to produce nonlinearities. Ultrashort pulse generation can be investigated and optimized using external cavity designs incorporating a controllable amount of self phase modulation (SPM), group velocity dispersion (GVD), and fast saturable absorption.

Previous researchers have demonstrated mode locking of diode laser arrays using active modulation[1] as well as passive saturable absorbers[2]. However, the shortest pulse durations that have been achieved are 26 to 5 picoseconds. Pulses as short as 0.58 picosecond with 30 mW peak power have been generated by active mode locking of a single stripe diode in an external cavity[3]. However, the peak power available from single stripe diodes is limited. Thus, to obtain both high peak power and subpicosecond pulse durations, external pulse compressors and amplifiers have been required. Using a hybrid mode locking technique with an external traveling wave amplifier and a grating pair compressor, 207 fs pulses with over 165 watts peak power have recently been reported[4].

Here we present a new approach for mode locking of high power diode arrays using intracavity nonlinearities. The laser operates with an external cavity containing a single mode fiber and a dispersion compensating prism sequence. The single mode fiber provides SPM, positive GVD and artifitially creates fast saturable absorber-like action througth nonlinear elliptical polarization rotation[5]. The prisms provide variable negative GVD. By optimizing these combined intracavity nonlinearities, enhanced pulse shortening can be achieved[6].

The schematic of the experimental setup is shown in Fig. 1. The gain medium was a 40-stripe AlGaAs MQW diode laser array with an active area 400 μm wide and 500 μm long operating at 0.8 μm center wavelength. The front facet was AR coated and the back facet was high reflection coated. The array aperture was mode matched into a single mode fiber using a cylindrical lens and objective lenses. The spot sizes were

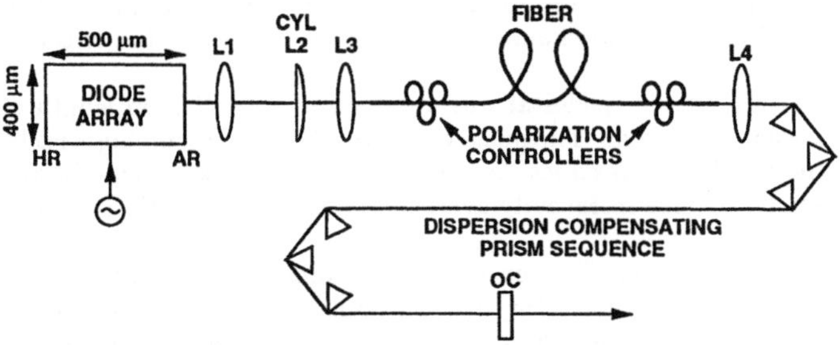

Fig. 1. Schematic of the hybrid mode-locked diode array external cavity.

Springer Series in Chemical Physics, Vol. 55 Ultrafast Phenomena VIII
Editors: J.-L. Martin · A. Migus · G.A. Mourou · A.H. Zewail

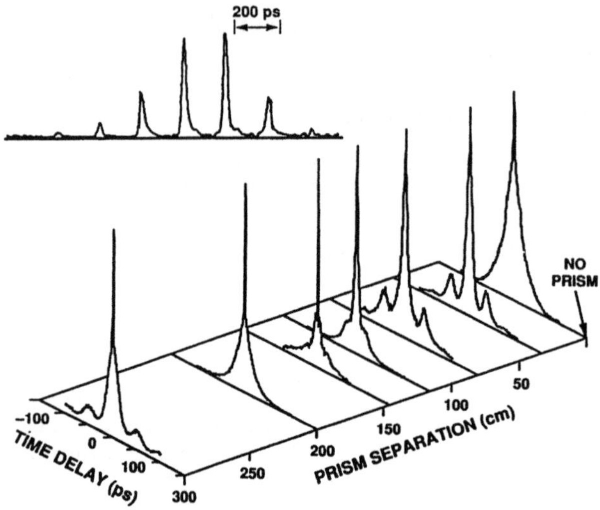

Fig. 2. Variation of the pulse shape as a function of intracavity prism separation. The autocorrelation traces have been scaled to the same height for display purposes. The insert is a typical sampling oscilloscope output of the mode-locked pulse train.

calculated using ABCD matrix analysis for the mode dimensions in the horizontal and vertical directions. For this analysis, the array can be treated as a broad area amplifier. The output from the fiber was collimated and directed through a six SF10 Brewster prism sequence. The beam was then retroreflected from a output coupler (OC).

The laser was actively mode locked by DC biasing and RF modulating near 446 MHz, corresponding to a harmonic of the external cavity. To enhance pulse shaping, we detuned the mode-locked frequency to be slightly higher than the cavity round-trip frequency and used the reflection from one of the fiber facets for intracavity pulse seeding[7]. This resulted in a multiple pulse train within the modulation envelope as shown in insert in Fig. 2.

To study the effect of total cavity GVD on the mode-locked laser performance, the length of fiber was chosen to be 300 cm and the prism spacing was varied. Figure 2 displays the autocorrelation traces of the pulses as a function of dispersion. Without prisms, the mode locked pulses were relatively long. With small prism separations, the total cavity GVD was positive and the laser generated a shorter pulse with a satellite. As the overall GVD was reduced and made slightly negative, a pulse with minimum duration was observed. We believe that dynamic gain saturation in the diode contributes to the pedestal observed on the pulse. The saturation energy for a subpicosecond pulse is less than for a picosecond pulse because of gain compression produced by nonequilibrium carrier temperatures[8]. This tends to favor the formation of satellite pulses or pedestals. Reduction of the pedestal should be possible by increasing the effective saturable absorption.

The shortest pulse with a smaller pedestal was obtained using a 450-cm length of fiber with a prism separation of 300 cm. The pulse width was 0.68 ps (assuming a $sech^2$ pulse shape) with a time-bandwidth product $\Delta\tau\Delta\nu$ of 0.56. The average mode-locked output power was 13.5 mW and the intracavity average power was 180 mW. The peak power was estimated to be 4.5 watts.

The effect of intracavity SPM and GVD on pulse shortening was also investigated by varying the prism spacing for two fiber lengths. The pulse width versus the prism separation for a 300-cm and a 450-cm length of fiber are plotted in Fig. 3. Appreciable pulse shortening was obtained with larger amounts of SPM in the negative GVD region.

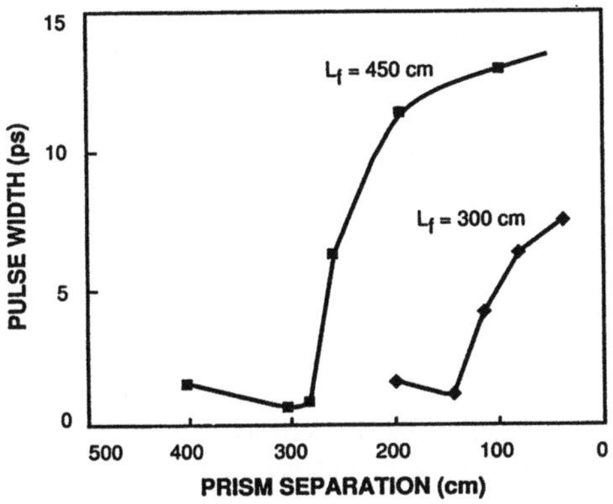

Fig. 3. The dependence of pulse duration on prism separation for two fiber lengths.

Using the 450 cm fiber, the minimum attainable pulse duration was 0.68 ps while the shortest pulse obtained with the 300-cm fiber was 1.1 ps.

The overall cavity GVD had a strong effect on short pulse generation for both cases shown in Fig. 3. For positive GVD, the cavity generated pulses that were relatively long and highly chirped. With increasing positive GVD, the pulse width and chirp increased. In the case of the 450-cm fiber, the value of $\Delta\tau\Delta\nu$ varied from 4.5 to 3.4. Pulse shortening in this regime was produced by the bandwidth limiting element which filtered the high and low frequency components of the chirped pulses. Minimum pulse durations were obtained for both cases when the total cavity GVD value was negative. Varying the total cavity GVD had a very small effect on the pulse durations in this operating regime. The measured $\Delta\tau\Delta\nu$ was in the range of 0.54 to 0.83. Pulses generated in this region were very short and nearly chirp-free. A soliton-like pulse shaping mechanism was responsible for the pulse shortening.

The observed pulse shortening characteristics agree qualitatively with theoretical models[6]. The sign of net cavity GVD of a mode-locked laser is very critical in determining ultrashort pulse performance. Minimum duration and relatively chirp-free pulses require total cavity GVD to be negative. Increasing SPM can introduce additional pulse shortening in a strong negative GVD region.

In summary, we have demonstrated a new approach for developing compact efficient high power ultrashort pulsed laser sources. Extension of these techniques may hold the promise for achieving high power femtosecond generation in a wide range of diode laser systems. This work was sponsored by the Department of the Air Force.

References

[1] H.Masuda and A. Takada, Electron. Lett. **25**, 1418 (1989).
[2] J.H. Zarrabi et al. Appl. Phys. Lett. **59**, 1526 (1991).
[3] S. W. Corzine et al., Appl. Phys. Lett. **52**, 348 (1988).
[4] P. J. Delfyett et al., Opt. Lett. **17**, 670 (1992).
[5] M. Hofer et al., Opt. Lett. **16**, 502, (1991).
[6] H.A. Haus et al. J. Opt. Soc. Am. **8**, 2068 (1991).
[7] J.E. Bowers et al., IEEE J. Quan. Electron. **25**, 2621 (1989).
[8] M. P. Kesler and E. P. Ippen, Appl. Phys. Lett. **51**, 1765 (1987).

100-Gbps Response of Microcavity Lasers

H. Yokoyama, Y. Nambu, and T. Shimizu

Opto-Electronics Research Laboratories, NEC Corporation,
34 Miyukigaoka, Tsukuba 305, Japan

Abstract. Extremely short photon lifetimes of wavelength-sized microcavity lasers provide ultrafast response capability. 100 gigabit-per-second response of a planar microcavity laser is demonstrated with optical-pulse-train excitation.

1. Ultrafast Response Capability of Microcavity Lasers

A lot of research interest is presently focused on microcavity lasers. These structures are resonators having at least one dimension of the order of a wavelength. So far, attention has been paid mainly to controlling spontaneous emission and the threshold-less feature of the microcavity laser [1], i.e. the advantage of low power consumption. However, the microcavity laser offers another attractive feature: the ultrafast response of possibly over 100 gigabit-per-second (Gbps). One reason is due to the extremely short photon lifetime in an appropriately designed microcavity accompanied by the extremely short cavity length. The other reason for the response speed increase is the cavity-enhanced spontaneous emission rate. Assuming a linear gain, the resonant relaxation oscillation frequency is inversely proportional to the square-root of both lifetimes. A three-dimensionally or two-dimensionally confined microcavity could induce a marked reduction in the spontaneous emission lifetime.

In a planar microcavity, although the enhancement in the spontaneous emission rate is at most a factor of ~2 [2], ultrafast response is still expected. For example, a symmetric Fabry-Perot cavity having a 0.99 reflectivity and $\lambda/2$ length gives a 0.2 ps photon lifetime. This value is ~1/10 of a conventional diode laser. Assuming a 1 ns spontaneous emission lifetime and an excitation level of 20 times higher than the threshold, the above photon lifetime gives a relaxation oscillation frequency of ~100 GHz. Under these conditions, large signal analysis has shown that the light output follows a series of 100 Gbps excitation pulses without serious distortions [3].

2. Experimental

Experimental verification of this ultrafast response capability was carried out using planar microcavities containing a rhodamine 6G dye solution and a GaAs multipule-quantum-well (MQW) structure as laser medium. In this experiment, the mirror reflectivity and the mirror separation were respectively chosen to be 0.985 and $\lambda/2$ for the dye microcavity (λ for the MQW microcavity). Figure 1 shows the single-shot streak-camera traces for the excitation pulse trains and the output pulse trains from the dye microcavity simulating 50 Gbps response. Pseudo-trains of four excitation pulses were generated by dividing and combining single picosecond laser pulses (λ = 480 nm, pulse duration = 7 ps, pulse energy = 10 μJ) using two stages of variable delay Michelson interferometers. The figure shows that the microcavity laser output followed the

Springer Series in Chemical Physics, Vol. 55 Ultrafast Phenomena VIII
Editors: J.-L. Martin · A. Migus · G.A. Mourou · A.H. Zewail

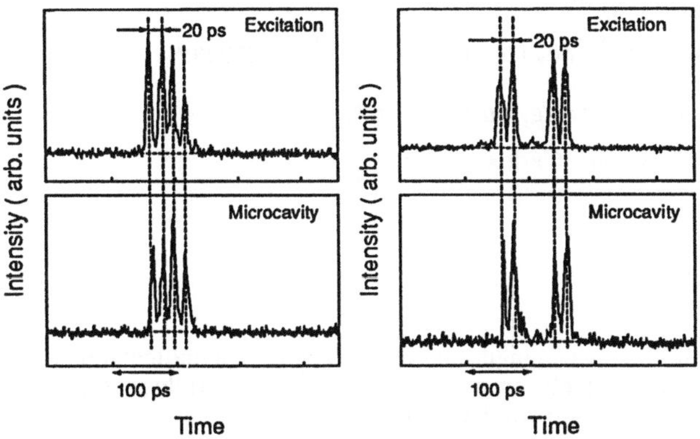

Fig. 1 Single-shot streak-camera traces of outputs from a dye microcavity laser showing 50 Gbps response with optical pulse pseudo-train excitation. The wavelength of the excitation optical pulses is ~480 nm, pulse energy is ~10 μJ.

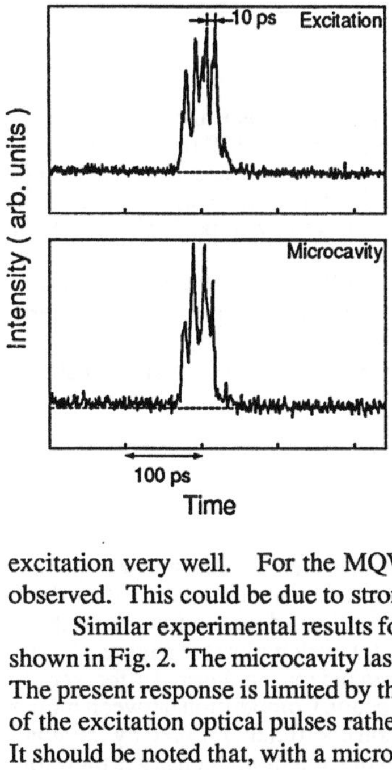

Fig. 2 Single-shot streak-camera traces corresponding to 100 Gbps response of a dye microcavity. The conditions for excitation optical pulses are the same as those for Fig. 1.

excitation very well. For the MQW microcavity, a similar but slightly poorer response was observed. This could be due to strong gain nonlinearities in semiconductors.

Similar experimental results for a dye microcavity simulating 100 Gbps response are also shown in Fig. 2. The microcavity laser still follows the pseudo-trains of excitation optical pulses. The present response is limited by the temporal resolution of the streak camera and the duration of the excitation optical pulses rather than the ultimate capability of the microcavity.

It should be noted that, with a microcavity laser, the spectral width is well restricted by a single resonance peak of the cavity (~1 nm width) even under ultrafast modulation. The present result indicates the potential of microcavity lasers potential for ultrafast operation of over 100 Gbps which cannot be achieved with a conventional size laser.

References

1. H. Yokoyama, M. Suzuki, and Y. Nambu, Appl. Phys. Lett. **58**, 2598 (1991).
2. S. D. Brorson, H. Yokoyama, and E. P. Ippen, IEEE J. Quantum Electron. **26**, 1492 (1990).
3. Y. Nambu and H. Yokoyama, paper presented at the Quantum Electronics and Laser Science Conference, Baltimore MD, May 1991, paper JThB1.

Sequential Laser Emission in Multiple Quantum Well Vertical-Cavity Structures

C. Tanguy, J.-L. Oudar, B. Sermage, and R. Azoulay

CNET, B.P. 107, F-92225 Bagneux Cedex, France

Abstract: Laser emission at different wavelengths, corresponding to longitudinal modes of the Fabry-Perot cavity, occur successively on a picosecond timescale, following pulsed optical pumping. Higher-energy modes are emitted first.

Due to their reduced dimensions, microcavity lasers exhibit cavity lifetimes in the picosecond range. Since, in solids, many phenomena occur on this timescale, possible new effects could affect the dynamics of laser emission, which has been investigated in this work.

1. Experimental set-up

A vertical structure originally designed for bistable operation has been used to study the competition between longitudinal modes. The active zone consists of 130 periods of 100 Å wide GaAs wells and 100 Å wide $Al_{0.3}Ga_{0.7}As$ barriers. The top mirror has 7 periods of $\lambda/4$ $AlAs/Al_{0.1}Ga_{0.9}As$ layers, whereas the bottom mirror (on the substrate side) is made of 14 periods. An Ar^+ mode-locked laser is used to synchronously pump a pyridine dye laser, which can deliver 5 ps (FWHM) wide pulses, tunable between 710 and 760 nm, at a repetition rate of 80 MHz, and with a maximum intensity slightly inferior to 100 mW. The excitation beam is tightly focused (spot diameter ~ 5 μm) on the sample held at room temperature. No lateral confinement of the carriers is provided, and the samples are not coupled to any heat sink: the lasing modes are thus gain guided. The emitted light is directed onto a Hamamatsu streak camera operating in synchroscan mode.

2. Results

Microcavity lasers usually show monomode operation because of their large free spectral range. In the case of intense ultrashort pumping, several longitudinal modes of the Fabry-Perot cavity may lie in the gain curve, which leads to possible multimode emission. We experimentally observed such an emission at two or more distinct wavelengths (see Figure 1), with a threshold of about 25 mW of average power. We have checked that no thermal effects could be incriminated by varying the duty cycle of the excitation beam. Competition between modes is greatly reduced because of the large spontaneous emission coefficient implied by the small cavity volume. A smooth transition threshold as a function of excitation intensity is therefore expected, and is indeed observed. It is crucial to have access to both time and energy domains near threshold, since the integrated spectrum cannot distinguish a short burst of light from the luminescence signal typical of under threshold behavior.

An unexpected feature appears in the time-resolved study of laser emission dynamics: the lines are emitted successively, the higher-energy line being emitted first (see Figure 2). Emission at λ_1=804 nm and λ_2=828 nm corresponds to the recombination of carriers belonging to the *same* n=1 subband of the quantum wells. The large delays (up to 20 ps) between emission at λ_1 and λ_2 occur when the line intensities are roughly equal. A longer cavity lifetime at λ_1 than at λ_2 can explain a larger signal at λ_1. A higher gain at λ_1 could also be obtained for a carrier density such that the band gap renormalization will bring the quantum

Springer Series in Chemical Physics, Vol. 55 Ultrafast Phenomena VIII
Editors: J.-L. Martin · A. Migus · G.A. Mourou · A.H. Zewail

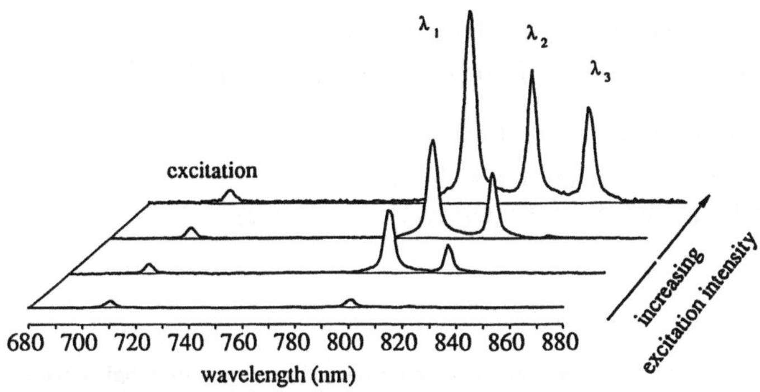

Figure 1: variation of emitted spectrum with excitation intensity.

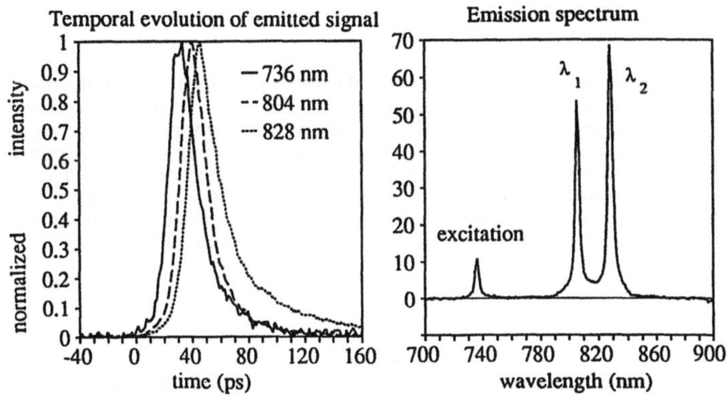

Figure 2: spectral and temporal dependence of laser emission.

well n=2 level (and the corresponding gain enhancement) closer to the transition energy. Laser emission lasts only for less than the experimental time resolution (15-20 ps), since the photon population buildup is not instantaneous. This time delay of a few picoseconds strongly decreases with excitation intensity.

A standard rate equation model for two photon modes and constant losses has been developed. However, its numerical integration could never reproduce the sequential emission for emission intensities of the same order of magnitude. This suggests that the net gain at the shorter wavelength λ_1 can be larger than the one at λ_2. On the timescale involved, it seems unlikely that intraband relaxation effects explain such large delays. Chirp effects can also be ruled out, since only a slight chirp is observed at each line for the highest available excitation intensities. Along with a gain saturation with carrier density, some carrier density-dependent losses could arise due to the self-defocusing properties of the transverse carrier profile; this effect should be greater at λ_2 than at λ_1.

Experimental Analysis of Gain Modulation in Sub-Picosecond (~0.45 ps) Mode-Locked Laser Diodes

N. Stelmakh, J.-M. Lourtioz, and D. Pascal

Institut d'Electronique Fondamentale, URA 22 du CNRS,
Bât. 220, Université Paris-Sud, F-91405 Orsay Cedex, France

Abstract. We propose a non-perturbative method of probing the gain medium of subpicosecond mode-locked laser diodes. An estimation is given for the saturation level of the laser gain.

1. Introduction

Mode-locked laser diodes present important applications in the field of optical communications and are an interesting alternative to more conventional (sub)picosecond lasers for time-resolved spectroscopy and optical testing of various materials. During last years, many papers have been addressed to the experimental demonstration of laser diode mode-locking using different optical schemes and different types of diodes [1-3]. However, the mechanisms at the origin of *coherent* subpicosecond mode-locked pulses in laser diodes are not clearly understood. In this paper, we propose a non-invasive method for probing the gain medium in mode-locking regime. Results are presented for a passively mode-locked AlGaAs laser emitting pulses *as short as ~450 fs*.

2. Experimental set-up

Fig. 1 shows the experimental set-up. Passive mode-locking is achieved with a saturable absorber created by ion implantation through one of the diode facets. A ~12 cm long external cavity is used which includes an end-reflector (R=98%), a microscope objective and a polarizing beamsplitter. The latter serves to couple a weak CW probe Ti-sapphire laser beam into the gain medium. The CW beam is polarized perpendicularly to the laser diode beam in order to minimize its influence on the mode-locking process. The gain and refractive index variations of the semiconductor medium lead to probe beam modulations which are measured by time-resolved spectroscopy. A streak camera and a diffraction grating are used for this purpose. A polarizing prism is used to balance the intensities of the two laser beams (CW and pulsed). A 1280 MHz generator followed by a frequency divider (1/16) serves to synchronize the streak sweep unit with the mode-locked laser pulses. The latter are simultaneously measured by autocorrelation.

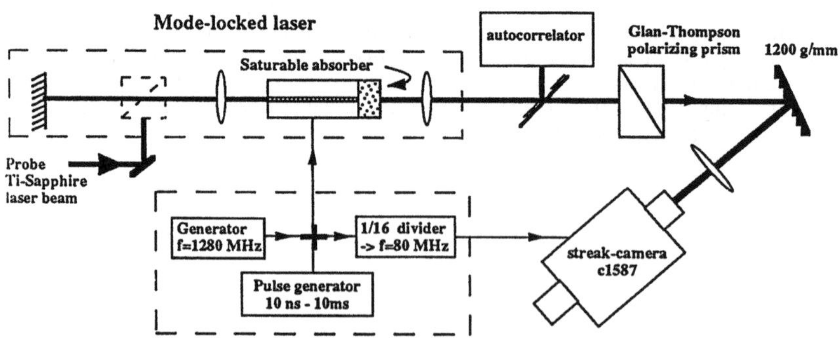

Fig. 1: Experimental set-up

Springer Series in Chemical Physics, Vol. 55 Ultrafast Phenomena VIII
Editors: J.-L. Martin · A. Migus · G.A. Mourou · A.H. Zewail

3. Experimental results and discussion

Fig. 2 shows the autocorrelation trace of a train of coherent mode-locked pulses emitted for certain tunings of the external cavity. The time separation of the pulses (≈ 8 ps) is simply the roundtrip time in the diode cavity, T_c. The width of the autocorrelation trace is ~700 fs (full-width half-maximum FWHM), which corresponds to mode-locked pulses of about 450 fs. The spectral width is measured to be 18 Å (FWHM) and the time-bandwidth product is 0.33.

Fig. 3 (left) shows the time-resolved spectroscopic images of the laser pulses. Two trains of pulses are represented, the time separation (780 ps) corresponding to the roundtrip time in the external cavity. The straight line at the top is related to the CW laser whose emission wavelength, λ_{Ti}, is 30 Å downshifted from the laser diode wavelength. Fig.3 (right) shows spectral profiles obtained around λ_{Ti} by accumulating in temporal windows T_0 = 80 ps. Profile (a) corresponds to a mode-locked situation. Profile (b) is obtained when the external cavity is slightly misaligned (no mode-locking). The bottom profile is obtained by substracting (b) from (a). Modulation sidebands ~3 Å apart from the main peak are clearly visible in this last spectrum with an average sideband amplitude $(\delta\mu_1+\delta\mu_2)/2$ ≈ 0.2%. Theoretically, this modulation amplitude can be approximated by $[(T_c/T_0).(g_0/2).(I/I_s).n]^2$, where T_c and T_0 have been previously defined, g_0 is the unsaturated gain per path, I/I_s, the level of saturation and n, the effective number of pulses in a mode-locked pulse train. The application of previous formula leads to I/I_s ≈ 0.7. A pulse shortening of ≈ 20 fs can then be estimated per roundtrip in the diode cavity [4], not far from the value of pulse lengthening due to dispersion effects.

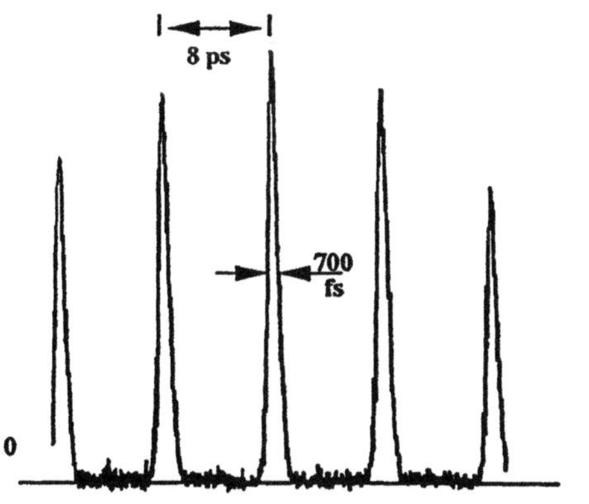

Fig. 2: Autocorrelation trace of mode-locked pulses

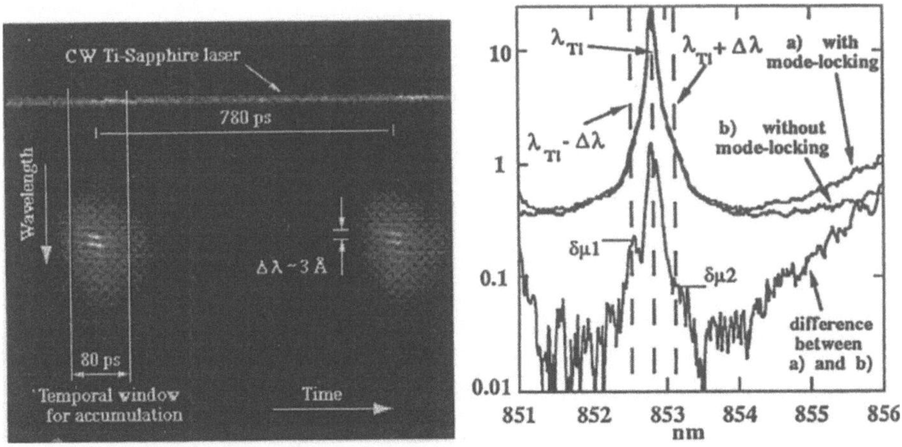

Fig. 3 : Time-resolved spectroscopic images of laser pulses (left) and spectral profiles associated with the probe Ti: sapphire laser signal

References

[1] J.P. Van der Ziel et al., Appl. Phys. Lett. 39, 525 (1981)
[2] H. Haus and Y. Silberberg, J. Opt. Soc. Am. B. 2, 1237 (1985)
[3] J.E. Bowers et al., IEEE J. Quantum Electron. QE-25, 1426 (1989)
[4] A.E. Siegman, "Lasers", University Science Books (1990)

Generation of Stable Pulse Trains with a Passively Modelocked Er-Fiber Laser

M.E. Fermann, M.J. Andrejco, Y. Silberberg, and A.M. Weiner

Bellcore, 331 Newman Springs Road, Red Bank, NJ 07701-7040, USA

Abstract. The generation of stable pulse trains with a modelocked Er-fiber laser is demonstrated. Near bandwidth-limited pulses of 180 fsec FWHM width and a pulse energy of 200 pJ are obtained.

Passively modelocked Er- fiber lasers are currently of great interest as compact solid-state sources of ultrashort pulses at the important telecommunications wavelength near 1.55 μm [1-4]. In these lasers the pulses are generated by passive amplitude modulation in a soliton-supporting fiber, i.e. a fiber with negative group-velocity dispersion (-GVD). However, since the soliton period of the generated pulses is typically comparable to the cavity length the pulse experiences large phase changes per round trip which leads to a characteristic instability and the formation of side-bands in the pulse-spectrum [3]. It is generally believed that this instability also prevents the formation of stable pulse trains in Er-fibers, which typically exhibit a chaotic behavior.

Here we show that by resorting to a comparatively high-Q dispersion-compensated cavity, stable pulse train formation is possible in Er-fibers. In this we use an Er-fiber with a slightly positive dispersion (we estimate β_2=+4000 +- 3000 fsec^2/m) and compensate the pulse-chirp intracavity with a single ZnS prism pair (β_2=-16500 fsec^2/m). The use of ZnS prisms at 1.55 μm has the additional advantage that the 3rd-order dispersion coefficient from the silica fiber may be partially compensated. Passive amplitude modulation is obtained by nonlinear polarization evolution [5]. The cavity set-up is shown in Fig. 1. The fiber length was 0.68 m with a core diameter of 4 μm. The steady-state pulse formation was passive and initiated with a weak modulation from an acousto-optic modulator. With a 50 % output coupler cw modelocked powers up to 10 mW were obtained when pumping with 0.3 W from an Argon laser at 528 nm.

Unlike most all-fiber Er lasers, the modelocked pulse trains were stable with a stability range from 5 to 10 mW of cw modelocked power. A photograph of a stable pulse train is shown in Fig. 2. Higher modelocked powers (up to 20 mW) generated chaotic pulse trains or pulse bunches. The shortest pulses were obtained with a prism separation of 1.22 m and had a typical FWHM width of 180 fsec and a time bandwidth product of 0.36 assuming a sech^2-shape. An autocorrelation of the pulse train is shown in Fig. 3, where the background was smaller than 10^{-5}. The pulse spectrum is given in Fig. 4. We believe that the structure on the pulse spectrum results from residual intra-cavity reflections, which also generated small satellite pulses with pulse powers up to 10^{-4} of the main pulse.

We believe that the good stability of the cavity arises from the relatively high Q and the linearization of the chirp through the interaction of positive dispersion and nonlinearity in the fiber. In contrast in a highly nonlinear passively modelocked all-fiber soliton laser there is no mechanism for chirp linearization and excess energy is shed into a dispersive-wave background which interacts with the main pulse.

In conclusion we have produced the shortest pulses from a modelocked Er fiber laser to date and demonstrated the formation of stable fsec pulses in Er-fibers for the first time without resorting to coupled cavities.

Springer Series in Chemical Physics, Vol. 55 **Ultrafast Phenomena VIII**
Editors: J.-L. Martin · A. Migus · G.A. Mourou · A.H. Zewail

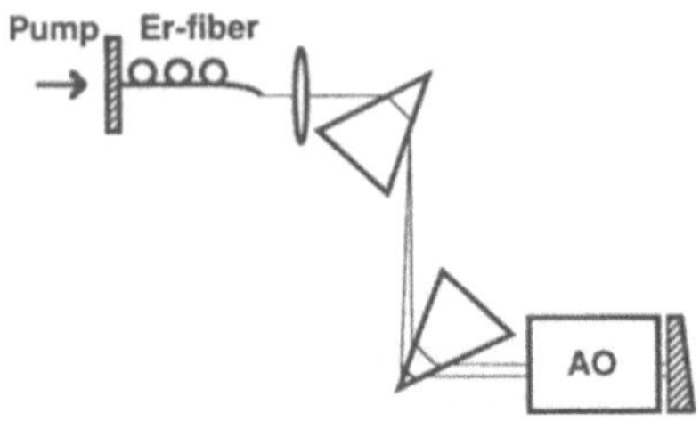

Fig. 1 Experimental set-up of passively modelocked Er fiber laser.

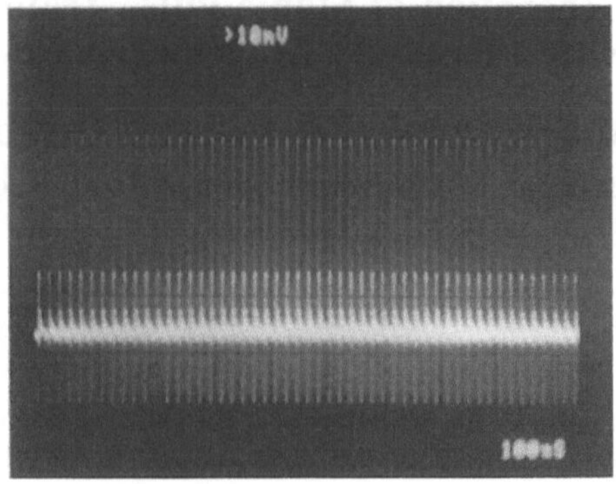

Fig. 2 Photograph of stable pulse train in the Er fiber laser.

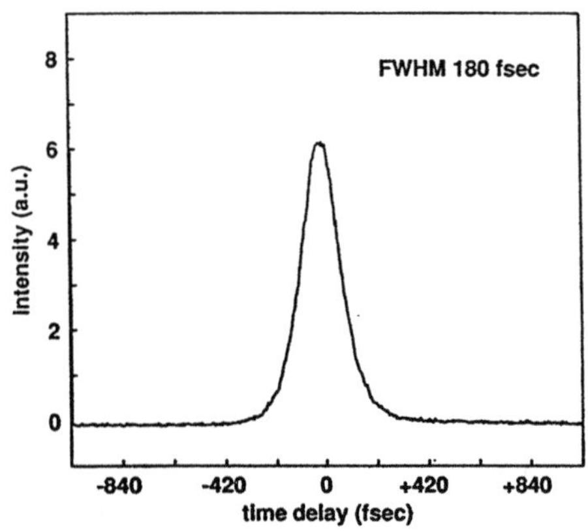

Fig. 3 Autocorrelation trace of generated pulses. The FWHM width is 180 fsec.

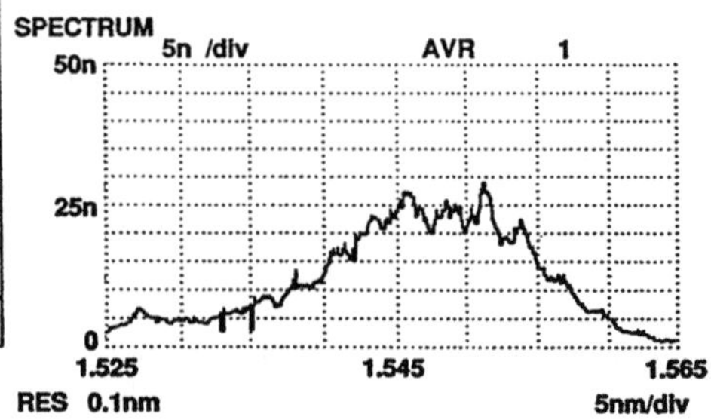

Fig. 4 Spectrum of generated pulses. The FWHM width is 16 nm.

Acknowledgements

M.E. Fermann gratefully acknowledges financial support from the Alexander von Humboldt Stiftung.

References

1. I.N. Duling, Electron. Lett., **27**, 545 (1991).
2. D. Richardson and A.B. Grudinin, Conf. on Lasers and Electro-Optics, CLEO 92, Anaheim, paper JWA5.
3. E. Yoshida Y, Kimura and M. Nakazawa, Appl. Phys. Lett., **60**, 932, (1992).
4. N. Pandit, D.U. Noske, S.M.J. Kelly and J.R. Taylor, Conf. on Lasers and Electro-Optics, CLEO 92, Anaheim, paper JWA7.
5. M. Hofer, M.E. Fermann, F. Haberl, M.H. Ober and A.J. Schmidt, Opt. Lett., **16**, 502 (1991).

Generation of Pairs of Solitons in an All-Fibre, Femtosecond Soliton Source

D.J. Richardson[1], *V.V. Afanasjev*[2], *A.B. Grudinin*[2], *and D.N. Payne*[1]

[1]The Optoelectronics Research Centre, Southampton University, Southampton, UK

[2]Russian Academy of Sciences, General Physics Institute, Moscow, Russia

Abstract. We demonstrate both experimentally and theoretically the generation of single and pairs of sub-100 fs soliton pulses from a passively mode-locked, all-fibre module.

1. Introduction

In recent years much work has been devoted to the development of convenient solid-state laser sources of ultrashort pulses (1,2). Passive generation of femtosecond pulses in optical fibres is of particular interest. In this paper we report for the first time the generation of pairs of solitons in an all-fibre femtosecond fibre laser/amplifier combination.

2. Experiment

The experimental configuration (fig.1) consists of two main components: a master oscillator based on the Figure-8 fibre laser and an external erbium-doped fibre amplifier (EDFA) [3]. Although ultimately potentially diode pumpable, for convenience in our experiments both units were pumped from a common 3W Ti:Sapphire laser operating at 980nm. With only 20 mW of pump power, approximately 10 bandwidth-limited soliton pulses of 450 fs duration, at 1558 nm, circulated within the laser cavity. The cavity round trip frequency was 5 MHz. The average laser output power coupled into the amplifier was 80 μW (laser output coupling α_c=30%). The amplifier pump power was varied between 320-410 mW.

With an initial amplifier length of 5.4 m (NA = 0.15, λ_{co} = 1230 nm and Er^{3+} doping level of 800 ppm) and pump power of 320 mW, a total power gain of 29dB was obtained and pulse compression from 450 fs down to 90 fs was observed at the amplifier output (Fig.2), together with simultaneous Soliton Self Frequency Shift (SSFS) from 1558 nm down to 1590 nm. (inset). The time-bandwidth product for these pulses is 0.33, in reasonable agreement with that expected for a sech^2x pulse shape.

In order to study the nonlinear dynamics of pulse propagation along an amplifier, the erbium-doped fibre was gradually cut-back from the initial starting length of 5.4 m. A graph of the experimentally-

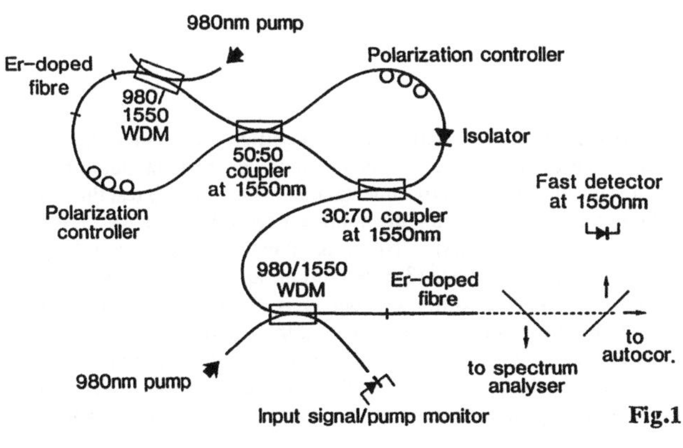

Fig.1 Experimental configuration.

Springer Series in Chemical Physics, Vol. 55 **Ultrafast Phenomena VIII**
Editors: J.-L. Martin · A. Migus · G.A. Mourou · A.H. Zewail

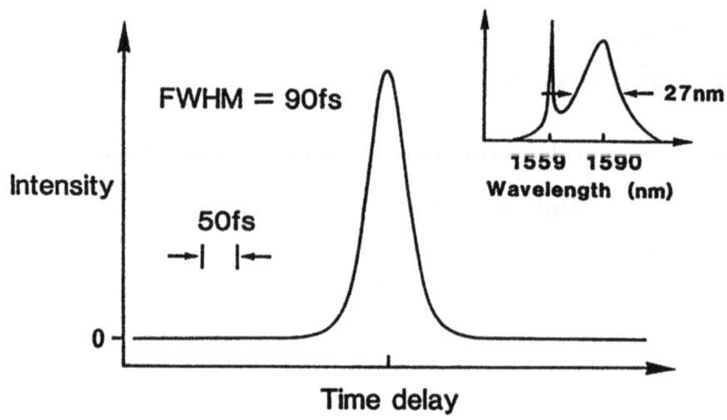

Fig.2 Autocorrelation trace and spectra of 90 fs solitons obtained at the end of the 5.5m long EDFA section

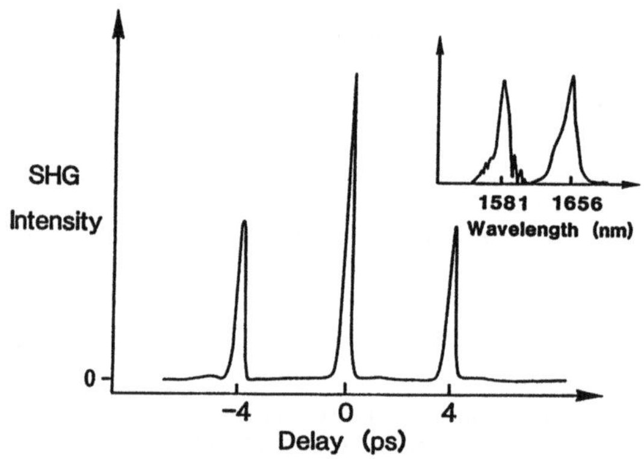

Fig.3 Auto-correlation trace and spectra of a pair of spectrally and temporally separated 100 fs solitons.

determined pulse duration against amplifier length (asterisks) is shown in Fig.4, where one can see, that the pulse undergoes a dramatic change in duration (by a factor of 10 within a distance of 20 cm) after propagating 2.8m within the amplifier. The shortest single pulse at the amplifier output (50 fs at 1590nm) was obtained at a pump power of 300 mW with a fibre length of 3m. On increasing the amplifier gain multiple coloured soliton generation was observed.

The three-level nature of erbium-doped fibre amplifiers and the effect of SSFS provide us with an elegant way to generate individual, or even sets of spectrally 'cleaned' coloured solitons. By using an over-long amplifier the excess (unpumped) length acts as an absorber for wavelengths within the original signal band, whilst passing with minimal attenuation the SSFS, red-shifted components which fall outside the erbium absorption band. Fig.3 shows the experimental spectrum and autocorrelation trace obtained at the output of a 6m long fibre amplifier, pumped with 300mW at 980nm. We have obtained two 'cleaned' solitons with approximately the same intensity, but with a 75nm wavelength separation.

3. Theoretical Model

Although the effects which determine the general behaviour of the amplified pulse compression are readily understandable, the complete set of cut-back data (output signal power, residual pump power, signal correlation function and spectra) provide us with experimental date with which to test theoretical models. We have modelled the data by numerically solving the nonlinear schrodinger equation for the system, introducing a z-dependence for the amplifier gain. Note, that due to the output coupling and the discrete change in dispersion between the laser fibre (3.5 ps/nm.km) and the amplifier fibre (13 ps/nm.km) the input pulses to the amplifier have a soliton number $N \approx 0.2$-0.3 with respect to the amplifier fibre and as such can be considered as almost linear pulses at the amplifier input. The solid curves in Fig.4 represent theoretical estimates of the pulse duration against amplifier length which are in good agreement with the experimental results (curve 1 in Fig.4). The curves (1-4) in Fig.4 show results for different values of the input pulse power, defined by the laser output coupler (a_c=30% for

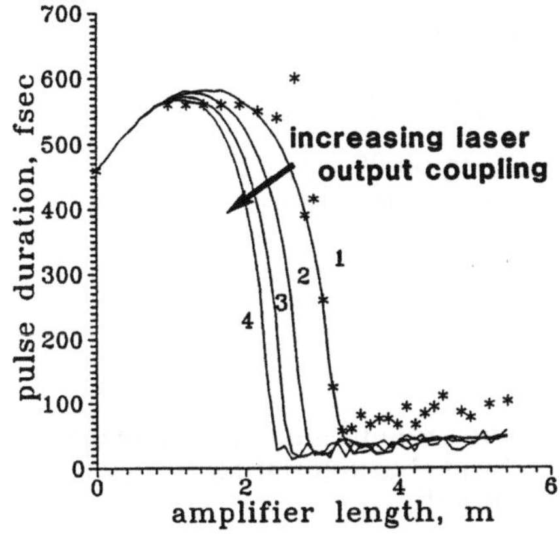

Fig.4 Effect of increasing oscillator coupling on pulse evolution along EDFA

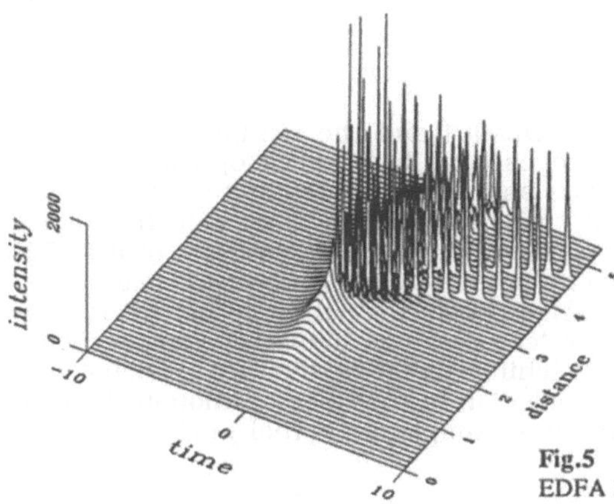

Fig.5 Dynamics of pulse evolution along EDFA for generation of pairs of solitons.

curve 1, 50% for curve 2, 70% for 3 and 90% for 4). One can see that an increase in the input soliton number results in a shorter compression length. Fig.5 shows the dynamics of the pulse evolution for an overlong amplifier. Note the almost linear pulse amplification process before the rapid compression and dual coloured soliton generation.

4. Conclusion

Our results show that it is possible to generate reasonably clean individual (or sets) of fundamental solitons starting from an almost linear pulse. In our opinion this is important as it allows us to generate sub-picosecond, fundamental solitons from for example, a combination of a laser diode and an EDFA.

References

1. M.E. Fermann, M. Hofer, F. Haberl, A.J. Schmidt: Optics Letters, 16, 244, 1991.
2. D.E. Spence, P.N. Kean, W. Sibbett, Optics Letters: 16, 42, 1991.
3. D.J. Richardson, A.B. Grudinin, D.N. Payne: Paper TUG6, Proceedings OFC (San Jose), 1992.
4. V.V. Afanasjev, V.N. Serkin, V.A. Vysloukh: "Amplification and compression of femtosecond optical solitons in active fibres", Sov. Lightwave Commun., to be published.

Nonlinear Loop Mirrors in Fiber Lasers

I.N. Duling III[1], *C.J. Chen*[2], *P.K. Wai*[2], *and C.R. Menyuk*[2]

[1]Naval Research Laboratory. Code 6570, Washington 20375, USA
[2]University of Maryland, Baltimore County Campus, Baltimore, MD, USA

Abstract. The nonlinear loop mirror plays a key role in short pulse formation in the figure eight fiber laser. The switching characteristics of these mirrors and their effect on the laser are discussed.

1. Introduction

The nonlinear optical loop mirror (NOLM)[1] and the nonlinear amplifying loop mirror (NALM)[2] have been proposed as high speed optical switches. Another application of these devices is as intensity dependent elements in all fiber laser cavities. One particularly successful application is in the Figure Eight Laser (F8L).[3] This laser, pictured in Figure 1, consists of a NALM whose output has been coupled to its input through an optical isolator and an output coupler. The laser is capable of producing pulses shorter than 300 fs and may possibly produce repetition rates as high as 150 GHz. Passive mode-locking is accomplished by the switching action of the NALM along with pulse shaping due to nonlinear propagation in the fiber. Although the concept of the operation of the F8L is rather simple, its actual working exhibits a rich phenomenology.

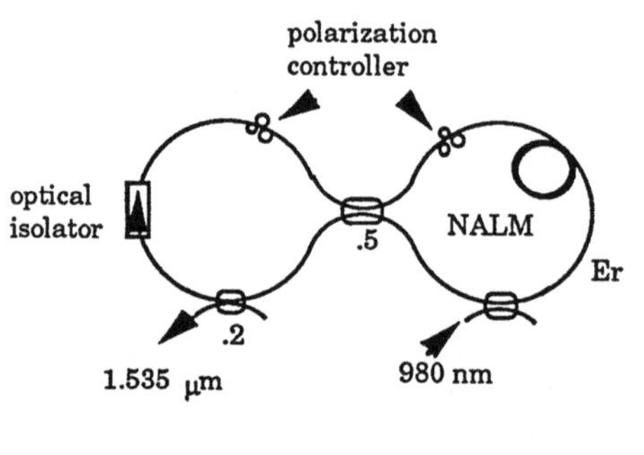

Figure 1. Layout of the figure eight fiber laser. The couplers are labeled with their power splitting ratio.

2. Nonlinear Optical Loop Mirror

Since one of the primary elements in the dynamics of the laser is the loop mirror, this element has been numerically modeled taking into account the full nonlinear propagation and gain saturation. In contrast to their behavior in the normal dispersion regime, pulses of lower intensity broaden as they go through the loop mirror, and higher intensity pulses shorten. Figure 2 shows the dependence of the transmission and the relative output pulsewidth for a soliton incident on a NOLM.

Springer Series in Chemical Physics, Vol. 55 Ultrafast Phenomena VIII
Editors: J.-L. Martin · A. Migus · G.A. Mourou · A.H. Zewail

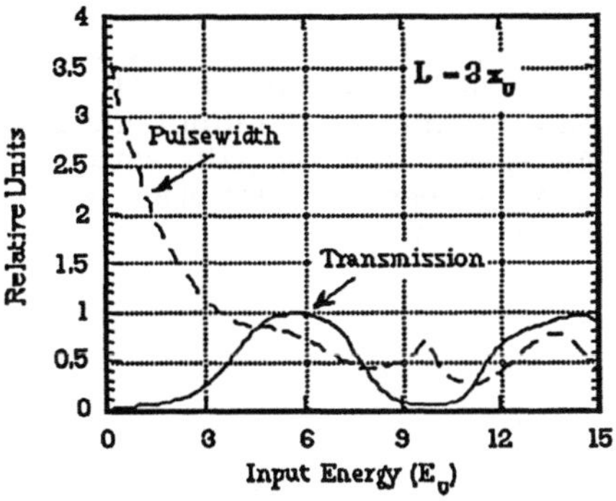

Figure 2. Relative pulse width and transmission for a soliton pulse incident on a NOLM as a function of energy measured in units of the fundamental soliton energy for a loop of length 3 z_0, where z_0 is the characteristic soliton period.

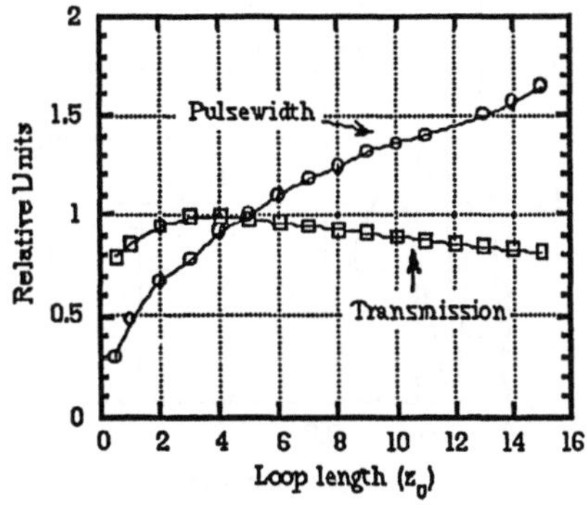

Figure 3. Transmission and relative pulsewidth as a function of loop length at the energy of the first transmission maximum. Both the maximum transmission and the point where the pulse maintains its pulsewidth occur at about 4-5 z_0.

For this length of loop in normalized units, the maximum transmission occurs for an input energy between 5 and 6 times the fundamental soliton energy.

If the laser is operating with pulses which match the criterion of minimum loss, then the the pulse will emerge with a duration which is .75 times its input duration. This decrease in width, however, will increase the length of the switching loop in normalized units, since

$$z_0 = .322\pi^2 c\tau^2/\lambda^2 |D|, \tag{1}$$

where τ is the pulse width and D is the fiber dispersion.

If the value of the peak transmission and the relative output pulsewidth at the first transmission maximum is plotted, Figure 3, a likely operating point of the laser can be easily identified.

At a loop length of 4-5 z_0 the transmission is near the maximum and the pulsewidth emerges from the mirror with the same width as it entered. This analysis ignores the effects of the external loop and the pulse shaping which may occur there.

3. Nonlinear Amplifying Loop Mirror

For the case of the NALM, the intensity mismatch in the switching loop is governed by the saturated gain. The gain in a CW laser is generally dictated by the loss in the cavity, but since the light reflected from the loop mirror is ejected from the cavity, the gain becomes a dynamic quantity.

Assuming that the pulse energy only changes slowly, the gain saturation can be calculated from the input pulse energy, an assumed repetition rate,

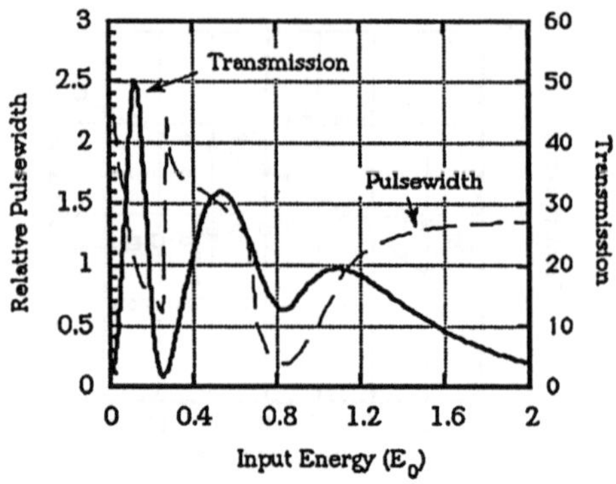

Figure 4. Switching characteristic of a NALM in the soliton regime showing the varying rate of phase accumulation as the input energy is increased and the gain saturates, and the early transmission values greater than 1.

and the calculated behavior of the NALM. The fundamental differences between the NALM and the NOLM are seen when Figure 4 is compared to Figure 2. Since the NALM has gain, the transmission can be much greater than one. Additionally, the differential phase accumulation with increasing pulse energy is no longer constant, but depends on the amount of gain. As a consequence the oscillation period of the transmission curve increases with increasing saturation.

It is easy to understand, then, why the self-starting occurs with many pulses in the cavity. When the first noise peak switches the mirror, the switching threshold is very low (allowing self-starting), but the gain immediately increases the pulse energy to a large factor more than the switching energy. On the next pass, the pulse is far above the first switching peak, and breaks into multiple pulses from the action of the switching mirror and the soliton pulse shaping. As the gain saturates, the pulse breakup process slows until only enough pulses are present of sufficient energy to reach steady state. In some sense the NALM in saturation operates much like a NOLM.

The F8L has a rich phenomenology. Much of the interdependence can only be addressed by numerical analysis. A complete model of the F8L has been developed and is being currently investigated.

The work of CJC, PKW, and CRM was supported by DOE, NSF and computation performed at SDSC. The work of IND is supported by the Office of the Chief of Naval Research.

References

[1] N. J. Doran and D. Wood, Opt. Lett. **13**, 56 (1988).

[2] M. E. Fermann, F. Haberl, M. Hofer and H. Hochreiter, Opt. Lett. **15**, 752 (1990).

[3] I. N. Duling, OSA Annual Meeting 1990, PD4., D. J. Richardson, R. I. Laming, D. N. Payne, V. Matsas and M. W. Phillips, Electron. Lett. **27**, 542 (1991)., and I. N. Duling, Electron. Lett. **27**, 544 (1991).

Temporal Characteristics of the Ytterbium–Erbium Figure-8 Laser

I.Yu. Khrushchev, A.B. Grudinin, and E.M. Dianov

General Physics Institute, Russian Academy of Sciences, 38 Vavilov Str., 117942 Moscow, Russia

Modes of operation of the Figure-8 laser have been studied experimentally. It was pulses or 850 fs solitons.

1. Introduction

The Figure-8 laser configuration has recently been the object of considerable study [1,2]. It has been shown that the laser can operate in at least two distinctly different modes: in a 'soliton' regime producing relatively chirp-free, transform-limited, sub-picosecond pulses and in a 'square' pulse regime characterised by long ($\approx$150ps) pulses, having a broad optical bandwidth (>35 nm)[3]. We report here for the first time an investigation into the temporal structure of the output radiation occuring during soliton generation from a 1064 nm pumped, Yb^{3+}/Er^{3+} co-doped fibre based, Figure-8 laser, using an autocorrelator and a streak-camera with a time resolution 1.5ps.

2.Experiment

The experimental configuration was similar to that described in [3]. The amplifier section of the NALM loop consisted of 4m Yb^{3+}-Er^{3+} codoped fibre [4], enabling us to exploit a conventional YAG laser operating at 1.064 μm as a pump source. The output power dependence upon pump power (Fig.1) reveals a hysteresis having a stair-case structure due to energy quantization [5].

Regions A, B and C corresponds to cw, noise pulse and stable soliton regimes of the laser operation respectively as noted in [5]. Note the different slope efficiencies within each regime. Oscilloscope traces are shown at Figs.2A, 2B.

The spectra and autocorrelation traces of the soliton pulses are presented in Figs.3A, 3B and those of the noise pulses - in Figs.3C, 3D respectively. Single shot, streak camera traces of the soliton bunches (Fig.2C) show a noisy structure with an envelope of about 200ps. Traces of stable soliton trains (not presented here) demonstrate either randomly

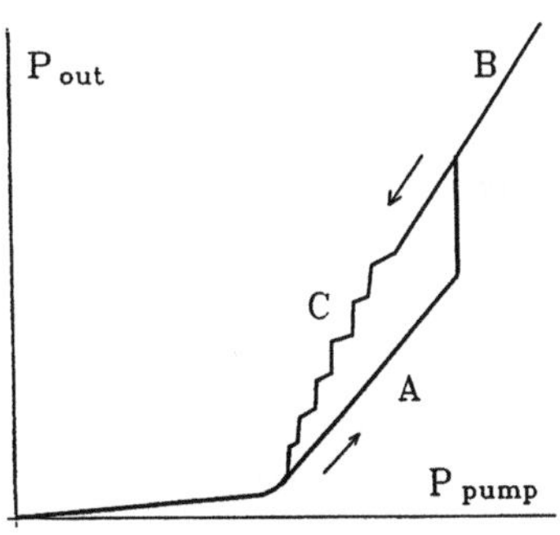

Fig.1. Output power of the fiber laser as a function of pump power.

Springer Series in Chemical Physics, Vol. 55 Ultrafast Phenomena VIII
Editors: J.-L. Martin · A. Migus · G.A. Mourou · A.H. Zewail

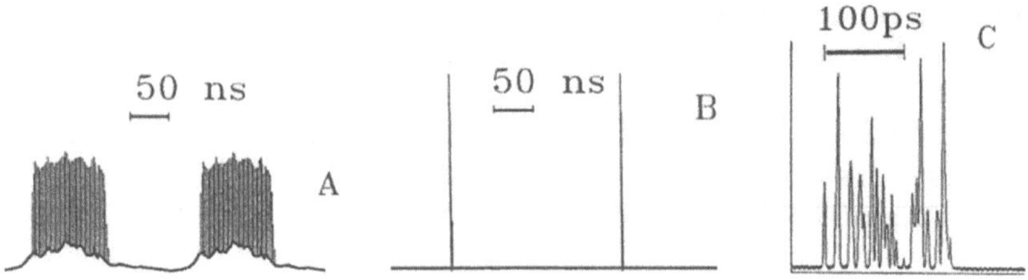

Fig.2. Oscilloscope traces of soliton bunches (A) and noise pulses (B) and camera trace of noise pulse (C).

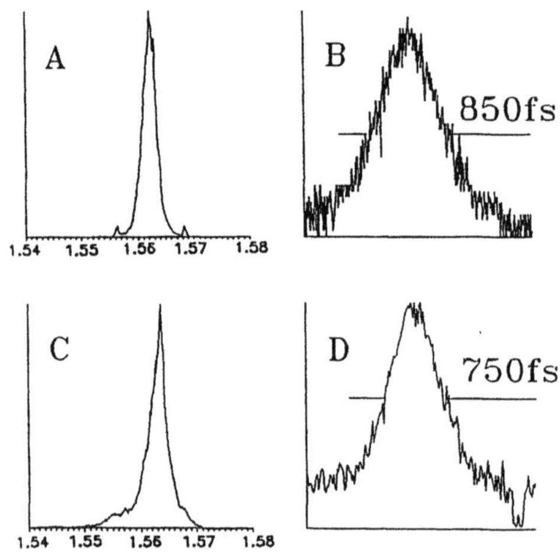

Fig.3. Spectra and autocorrelation traces of solitons (A,B) and noise pulses (C,D).

spaced, discrete pulses, or quasiperiodic trains of pulses with repetition rates of up to 0.5 GHz. As of yet broadband "square" pulse generation has not been observed.

3.Conclusion

In summary, we observed two modes of solitons generation in the Figure-8 laser: solitons bunches and soliton noise, similar to Raman solitons generation.

References

1.I.N.Duling III, Proc. of the OSA Annual Meeting, PDP4, 1990
2.A.G.Bulushev et al., Sov.Lightwave Comm., **1**, 133, 1991
3.D.J.Richardson et al., Electron.Lett., **27**, 542, 1991
4.J.E.Townsend et al., Electron.Lett., **27**, 1558, 1991
5.A.B.Grudinin et al., Electron.Lett., **28**, 67, 1992

Generation of 1.7 ps Solitons by Amplification of Pulses from a Laser Diode with Saturable Absorber in an Erbium-Doped Fibre

I.Yu. Khrushchev[1], *A.B. Grudinin*[1], *E.M. Dianov*[1], *D.V. Kuksenkov*[2], *and E.L. Portnoy*[2]

[1]General Physics Institute, Russian Academy of Sciences, 38 Vavilov Str., 117942 Moscow, Russia
[2]Physico-Technical Institute, Russian Academy of Sciences, 26 Politechnicheskaya Str., 194021 St. Petersburg, Russia

1.7 ps solitons were obtained by 30dB amplification of a signal from a laser diode with a fast saturable absorber and a two-cascade erbium-doped fibre amplifier.

1. Introduction

The combination of a picosecond laser diode, operating at 1550nm wavelength region and an erbium-doped fiber amplifier is an attractive proposition since it allows one to realize a miniature source of ultrashort pulses. In our paper we present the results of an experimental study of such a source which is capable of generating picosecond solitons. The unit is potentially diode pumpuble and may be adapted for the generation of sub - 100fs pulses.

2. Experiment

In our experiment picosecond pulses from an injection mesastripe InGaAs/InP gain-switched laser with a fast saturable absorber [1] were launched into an amplifier, consisting of erbium-doped fibres EF1 and EF2 (Fig.1). Isolators ISO1 and ISO2 prevented the injection of EFs spontaneous emission into the laser diode.
The spectra of the input and amplified signals (Fig.2) reveal linewidth narrowing due to the linear amplification process. The peak power of the pulses at the output of EF1 and EF2 was 20W, giving 30 dB net power gain.

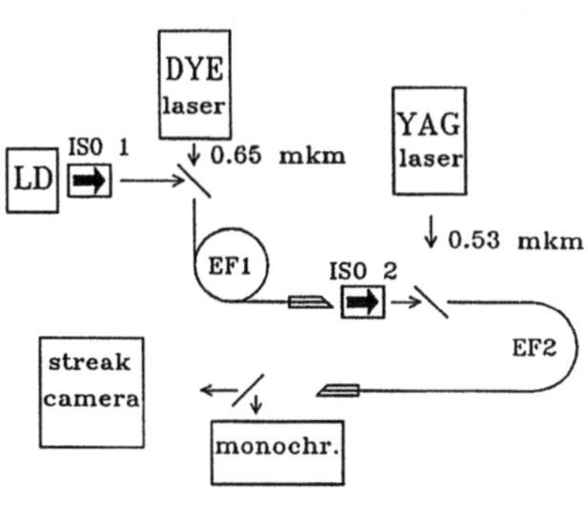

Fig.1. Scheme of experiment.

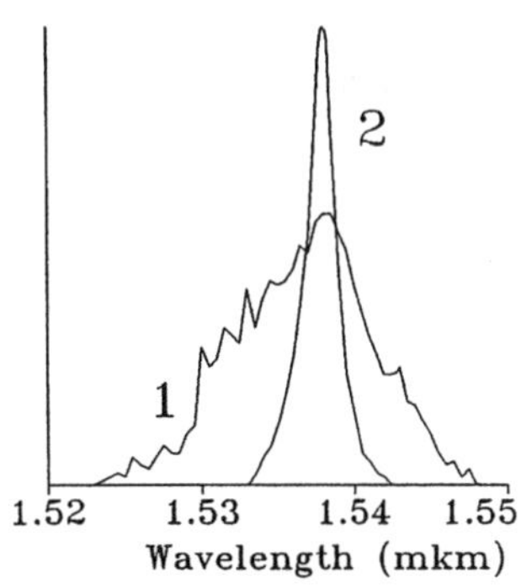

Fig.2. Spectra of initial (1) and amplified (2) signals.

Springer Series in Chemical Physics, Vol. 55 **Ultrafast Phenomena VIII**
Editors: J.-L. Martin · A. Migus · G.A. Mourou · A.H. Zewail

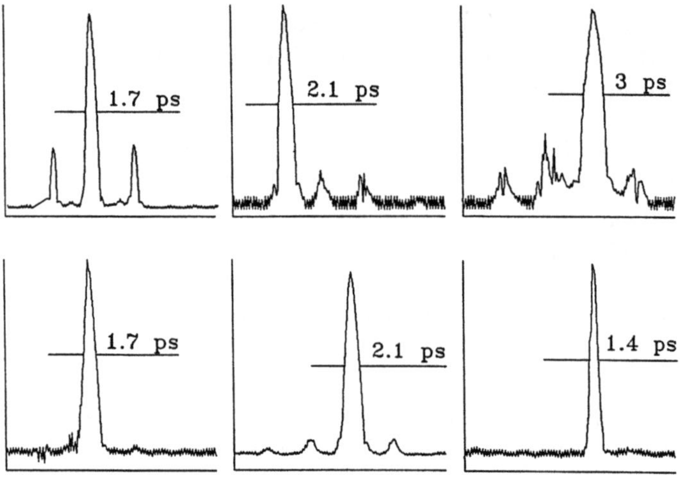

Fig.3. Camera traces of amplified signal without isolator 1 (upper row) and with isolator 1 (lower row).

Fig.3 shows tipical pulse shapes af the amplified signal with (Fig.3B) and without (Fig.3A) isolator ISO1. The time-bandwidth product of the amplified pulses was 0.4. Note the existence of satellites having 6ps temporal period (which is equal to the round trip transit time of LD cavity) and also that when the amplifier and the LD are not isolated (Fig.3A), the injection of spontaneous emission to the LD leads to enhanced satellite growth.

3. Conclusion

We have successfully generated 1.7ps pulses at 1535nm with a peak power corresponding to fundamental solitons, using a laser diode and an erbium-doped fibre amplifier.

References

1. E.L.Portnoy, A.V.Chelnokov, Proc. of the XII International Conference on Semiconductor Lasers, p.140, Davos, 1990.

Part V

High Intensity and Nonlinear Effects

Generation of Ultra-Intense Pulses and Applications

G. Mourou

Center for Ultrafast Optical Science, University of Michigan,
2200 Bonisteel, Ann Arbor, MI 48109-2099, USA

We review the development of ultra intense lasers, capable of producing 10^{18} W/cm^2 in the near future. We describe the potential applications, including the observation of relativistic effects, *i.e.*, harmonic generation, self-focusing *etc.*, the acceleration of electrons, the generation of collimated X-ray and γ-ray beams by nonlinear Compton scattering, observation of nonlinear quantum electrodynamic dynamic effects, and ignition of thermonuclear fusion.

1 Introduction

It is now possible with the technique of Chirped Pulse Amplification (CPA) [1-2] to build compact solid-state lasers that produce ultrashort pulses with intensities three to four orders of magnitude higher than was previously possible. These pulses have multi-terawatt peak power, and when focused can produce intensities in the range of 10^{18} W/cm^2. The electric field at the laser focus for 1-μm light at this intensity is approximately 3 x 10^{10} V/cm. Since this greatly exceeds the Coulomb electric field seen by the valence electrons, it results in both collisionless ionization without tunneling and highly nonlinear interactions with bound electrons. The field is high enough, in fact, to cause plasma electrons to oscillate at relativistic velocities, and thus exhibit highly nonlinear motion due to the magnetic component of the Lorentz force. For the first time, this allows the study of nonlinear optics involving free electrons. The interaction of these high peak power pulses from relativistic and superrelativistic electron beams will make possible the generation of X-ray and γ -ray beams, as well as critical fields necessary to observe nonlinear quantum electrodynamic effects. These pulses also have extremely short time durations, in the range of 100 fs to 1 ps. Since this is shorter than the timescales of either significant hydro-dynamic motion or thermal equilibration, short-scale length, solid-density, and non-equilibrium plasmas may be produced. Furthermore, the high-average powers of these lasers make possible the use of sampling techniques to study inherently statistical phenomena. Applications of these novel laser interactions include the generation of coherent XUV radiation, the ignition of laser-fusion targets, and the acceleration of electrons.

Springer Series in Chemical Physics, Vol. 55 Ultrafast Phenomena VIII
Editors: J.-L. Martin · A. Migus · G.A. Mourou · A.H. Zewail

2 CPA Terawatt Lasers

The CPA technique requires some impressive pulse manipulations. First, the generation of an extremely temporally clean short pulse. Second, a stretching of this pulse by 1000 to 12000 times. Third, an amplification by a factor of 10^{10}, without pulse spectrum degradation. And finally, a pulse recompression by another factor of 1000 to 12000 times. After all these elaborate manipulations, the laser pulse has to be close to diffraction limit, as short as the initial pulse, and still be temporally clean with a peak-to-background of at least 10^6:1. We will now discuss in detail each of these problems and their solutions.

2.1 The Generation of Ultra-Clean Pulses

These solutions are presently embodied in the terawatt Ti:sapphire/Nd:glass systems at the University of Michigan and at the CEA-Limeil in France. The system consists of a 100-MHz mode-locked Nd:YLF oscillator producing 40-ps pulses. The output is sent into an 800-m, single-mode fiber. Through self-phase modulation and group velocity dispersion the bandwidth and pulse width are increased to 42 Å and 110 ps respectively. The pulse is then recompressed and sent to a pulse cleaner to improve the contrast ration [3]. A high contrast is required to prevent the formation of a pre-plasma during the experiment. The pulse cleaner consists of a birefringent fiber between high-quality polarizers. Owing to the nonlinear index of refraction in the fiber, the state of polarization of the light becomes intensity dependent. This can be used to discriminate between the low and the high-intensity part of the pulse with a discrimination ratio of 10^6:1. Since the threshold for pre-plasma formation is $\sim 10^{12}$ W/cm^2, this allows intensities up to 10^{18} W/cm^2 to be used in experiments. The transmission of the pulse cleaner for high intensity is of nearly unity. The pulse after the grating-pair compressor has been measured over eight orders of magnitude with a high-dynamic range autocorrelator. A spectacular improvement can be obtained by using this pulse cleaning technique. This technique departs from previously demonstrated ones, which use either Pockels cells [5] or saturable absorbers [6]. It is purely passive with a femtosecond-time response, unlike Pockels cells; and it does not degrade the beam, unlike saturable absorbers.

2.2 Pulse Stretching and Compression

Before amplification, the pulse must be stretched. For a large stretcher/compressor ratio, R, the group-velocity dispersion of the fiber cannot be used for stretching because it cannot be exactly compensated (over all orders) by that of the grating pair compression stage. The mis-matching between the two group velocity dispersions can lead to some unacceptable background. In

1987, Martinez proposed a grating pair compressor composed of gratings in an anti-parallel configuration to compress pulses between 1.3 and 1.6 μm, stretched by negative group velocity dispersion in the fiber [7]. We demonstrated [8] that this grating configuration could also be used as a matched stretcher for our applications, with wavelengths shorter than 1.3 μm. We recently demonstrated [9] that this grating-based stretcher/compressor permits stretching/compression ratio greater than 10^4. This system is now one of the basic components of the CPA technique.

2.3 Large-Bandwidth Amplification

The stretched pulse is amplified before compression from the subnanojoule to the joule level, which is a factor of more than ten orders of magnitude. Our initial systems were composed entirely of Nd:glass amplifiers, which have a relatively narrow gain bandwidth ($(\Delta \lambda)g = 250$ Å), limiting the final pulse duration to approximately 1 ps. Most of the gain narrowing occurs at the front end of our amplifier system, where the pulse is amplified by a generative amplifier from the subnanojoule to the millijoule level (gain of 10^7). This problem was recognized early, and we proposed to use a combination of Nd:glass to broaden the gain bandwidth, a suggestion that was eventually used by others [10]. We later replaced the Nd:glass by Ti:sapphire in the regenerative amplifier [11]. Ti:sapphire has an extraordinary, large gain bandwidth ($(\Delta \lambda)_g \sim 2{,}000$ Å) peaking at 8,000 Å and extending up to 1 μm. By this replacement, the gain-bandwidth problem encountered with Nd:glass is completely eliminated, and the final pulse width is limited only by the subsequent amplification stages, which only amplify the pulse by a factor of 10^3, from the millijoule to the joule level. After the last amplifier the pulse has a duration of 1 ns and an energy level of 1 J at Michigan and 50 J at Limeil. Finally, the pulse is compressed by a double-pass grating system that is matched with the grating pulse stretcher. In order to achieve the transform-limited pulse duration, the angles of incidence on the stretcher and on the compressor have to be identical within a fraction of a degree.

After compression the pulse is 400 fs in duration and has an energy ranging from 1 J (Michigan) to 25 J (Limeil) corresponding to peak power of 2 TW (Michigan) and 55 TW (Limeil). The pulse contrast has been measured at the University of Michigan by a third-order autocorrelator [12] and is better than 10^5:1. The beam spatial quality has been measured to be better than twice the diffraction limit on the Michigan system. On the Limeil system, at this power level, beam spatial quality is limited not so much by the laser but by the nonlinear propagation in air to the experimental chamber. We are currently working on installing vacuum beam lines to mitigate this problem. On this system intensities as high as 10^{19}-10^{20} W/cm are expected.

2.4 Towards 10-TW Table-Top (T^4) Systems at a Hertz

For applications requiring both high-average powers and energetic pulses, two laser-pumped systems, alexandrite-pumped Nd:glass and alexandrite-pumped

Cr:LiSAF, are under development [14]. Flashlamp-pumped glass systems provided the necessary energies, on the order of 1 J, but are limited in repetition rate because of thermal effects. Laser-pumping can deposit the pump energy more efficiently. For instance, the average power of the current flashlamp-pumped T^3 system is on the order of 30 mW. On the other hand, we are developing an alexandrite-pumped Nd:glass system, which will potentially have an average power of 1-10 W. Several factors make this combinations of solid-state materials well matched. First, the Nd:glass has a long upper-level lifetime, ranging from 300 to 500 μs, which is well matched to the pulse width of free-running alexandrite, 100 μs. Second, good optical quality Nd:glass covering the spectrum from 1.04-1.09 μm can be obtained, making possible the amplification of pulses as short as 30 fs with the proper choice of glasses. Third, the absorbtion band of Nd:glass is around 750 to 800 nm, well matched to Alexandrite's tunable range. Finally, alexandrite can deliver an average power of up to 100 W with tens-of-joules energy per pulse. We hope in the near future to have terawatt pulses at a few hertz, corresponding to greater than 1-W average power.

Similarly, we are developing an alexandrite-pumped Cr:LiSAF system [14], which takes advantage of the superior upper-level lifetime (70 μs) and thermal properties of the Cr:LiSAF. It should be capable of even greater repetition rates and still maintain joule-level energies. To date, millijoule CPA systems at 2-5 Hz have been demonstrated in both materials [14], by our group and researchers at Allied Signal Corp.

3. Ultra High Intensity Applications

Figure 1 describes qualitatively, over 30 orders of magnitude, the threshold of different physical phenomena encountered at increasing laser intensites.

3.1 10^6 W/cm^2 Nonlinear Optics of Bound Electrons (15).

At these intensities the laser field corresponds to a small but significant fraction of the Coulomb field. The nonlinear response of electrons is giving rise to a variety of phenomena, harmonic generation, optical rectification, self-focusing, *etc.* After the initial discovery at the University of Michigan by P. Franken (15) and his coworkers, these effects have been thoroughly studied over the past 30 years and constitute the bulk of the field of nonlinear optics [16].

3.2 10^{13}-10^{15} W/cm^2 Multiphoton Ionization [17-18] and Above Threshold Ionization [19]

At these intensities the laser field becomes comparable to the Coulomb field. For hydrogen this field is 3×10^9 V/cm corresponding to $I=10^{16}$ W/cm^2.

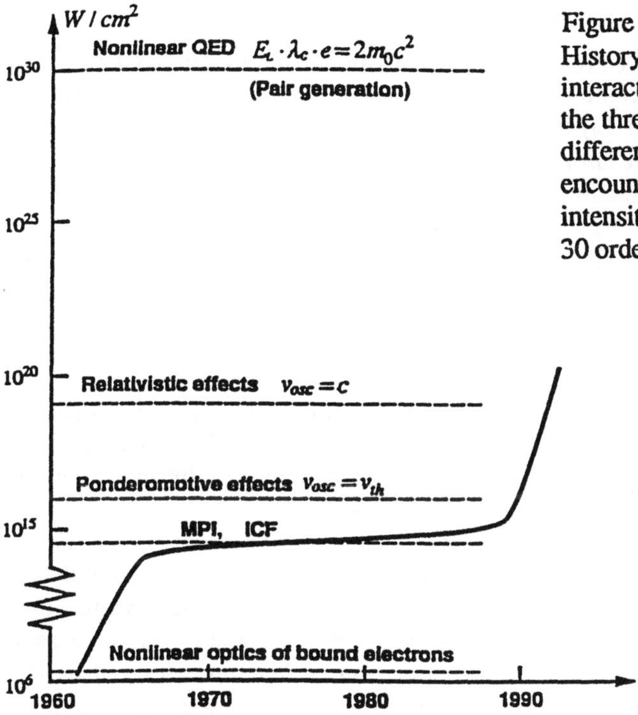

Figure 1
History of laser-matter interaction emphasizing the thresholds of the different physical phenomena encountered as the laser intensity is increased over 30 orders of magnitude.

Perturbation theory can not be applied. At these intensities, highly nonlinear multiphoton processes, tunneling ionization, above threshold ionization, and high-harmonic generation are observable. Harmonics greater than a hundred have been generated [20-21] exhibiting roughly the same intensity. It is interesting to note that in the time-domain these harmonics for 1-μm light correspond to electromagnetic pulses at the generation point of 30×10^{-18} s or 30 as, separated by 3 fs. The next challenge will be the propagation of these pulses, maintaining the phase matching between the fundamental and its harmonics.

3.3 10^{16} W/cm^2 Large Light-Pressure Effects [22]

Light pressure should play an important role in plasma dynamics when the gradient of the ponderomotive pressure, $n_e m (v^2_{os})$, becomes comparable to the gradient of the plasma thermal pressure, (3/2) $n_e k T_e$. In the laser interactions with solid targets, the gradients are roughly comparable, and only the ratio of the quiver velocity (v_o=eE/mω) to the thermal velocity ($v_e=(kT_e/2m)^{1/2}$) matters. For the laser-plasma parameters discussed in those interactions (T_e~200 eV and λ=1μm), this ratio is approximately unity if E_o again equals a value close to the Coulomb field. At somewhat higher fields, the enormous ponderomotive pressures of a short-pulse laser may provide a novel means for igniting thermonuclear reactions [23].

3.4 10^{18} W/cm^2: Relativistic Effects at 1 μm.

The mass of an electron should become modulated by relativistic effects, and thus its motion becomes highly nonlinear, when v_o approaches c, the speed of light. The value of E_o for $\lambda = 1$ μm is found to be equal to the highest attainable with current laser technology (3 x 10^{10} V/cm or I=10^{18} W/cm^2). This results in the generation of relativistic harmonic radiation [24], relativistic self-focusing [25] and large amplitude wakefield plasma waves with extremely high-longitudinal [26] [27] electric fields that may be used to accelerate electrons.

3.5 10^{30} W/cm^2: Nonlinear Quantum Electrodynamics

Finally, quantum electrodynamics effects, such as pair production, will become important when the work done by the laser electric field over a distance of a Compton wavelength, λ=h/mc, equals the rest mass of an electron, mc^2. This requires that E_o equals a value that will be difficult to achieve even in the forseeable future (3x10^{16} V/cm or I=10^{30} W/cm^2).

Although the field strength required for the catastrophic production of electron-positron pairs from the vacuum is six orders of magnitude higher than is currently available with even the most intense lasers, the observation of a statistically significant number of pairs is predicted to occur at fields as low as $E_o \sim 10^{14}$ V/cm. However, the interaction of the terawatt pulse with superrelativistic electron may provide a way to reach this formidable E-field. We know that the electric field in a frame moving with a relativistic electron beam is increased by γ where γ is the relativistic factor associated with the electron beam. For instance for a 20-GeV electron beam, the field is increased by a factor of 4x10^4, making the threshold for observation of pair-production attainable. Due to the compactness of today's high-peak-power lasers this type of experiment has been proposed [28] and will be carried out on the Stanford linear accelerator.

1. D. Strickland and G. Mourou, Opt. Commun., 56, 219.221 (1985).
2. P. Maine, D. Strickland, *et al.*, IEEE J. Quantum Electron. QE-24, 398 (1988).
3. J. L. Tapie and G. Mourou, Opt. Lett. 17,136 (1992).
4. R. Fedosejevs, R. Ottoman, *et al.*, Appl. Phys. B, 50, 79 (1990).
5. K. Yamakawa, H. Shiraga, *et al.*, Opt. Lett. 16, 1593 (1991).
6. Y.H. Chuang, D.D. Meyerhofer, *et al.*, J. Opt. Soc. Am. B. 8, 1226 (1991).
7. O. E. Martinez, IEEE J. Quantum Electron. QE-23, p. 59 (1987).
8. M. Pessot, P. Maine, *et al.*, Opt. Commun. 62, 419 (1987).
9. F. Salin, J. Squier, *et al.*, "Very Large Stretching of Ultrashort Pulses," Appl. Opt.
10. M.D. Perry, F. G. Patterson, *et al.*, Opt. Lett. 7, 381 (1990).

11. F. Salin, C. Rouyer, *et al.*, Opt. Commun. 84, 67 (1991).
12. G. Albrecht, A. Antonetti, *et al.*, Opt. Commun. 40, 59 (1981).
13 J. Kmetec, J. J. Macklin, *et al.*, Opt. Lett. 16, 1001 (1991).
14. J. Squier, G. Mourou, *et al.*, "Alexandrite-Pumped Chronium and Neodynium Materiaals for Chirped Pulse Amplification," Digest of Meeting on Advanced Solid State Lasers (Optical Society of America, Washington, DC, 1991) p. 114.
15. P.A. Franken, A.E. Hill, *et al.*, Phys. Rev. Lett. 7, 118 (1961)
16. N. Bloembergen, Nonlinear Optics, Benjamin, New York (1965)
17. G. Voronov and N. Delone, J.E.T.P. Lett 1, 66 (1965)
18. P. Agostini, G. Barjot, *et al.*, IEEE J. Quant. Electron QE-4 667 (1968).
19. P. Agostini, F. Fabre, *et al.*, Phys. Rev. Lett. 42, 1127 (1979).
20. X.F. Li, A. H'Huillier, *et al.*, Phys. Rev. A 39, 5751 (1989).
21. J. Kmetec, J.F. Macklin, *et al.*, private communication.
22. D. Umstadter, X. Liu, *et al.*, "Density Profile Steepening by the Ponderomotive Force of an Intense Picosecond Laser," in *OSA Proceedings on Short Wavelength Coherent Radiation: Generation and Applications, 1991*, P. Bucksbaum and N. Ceglio, eds. (Optical Society of America, Washington DC, 1991), 11, 55-7.
23. M. Campbell, private communication.
24. E.S. Sarachik and G.T. Schappert, Phys. Rev. D 1, 2738 (1970).
25. C. Max, J. Arons, *et al.*, Phys. Rev. Lett., 33, 209 (1974); P. Sprangle, C.M. Tang, *et al.*, IEEE Trans Plasma Science, PS-15, 145 (1987).
26. P. Sprangle, E. Esarey, *et al.*, Appl. Phys. Lett. 53, 2146 (1988); E. Esarey, A. Ting, P. Sprangle, and G. Joyce, Controlled Fusion 12, 191 (1989).
27. T. Tajima and J.M. Dawson, Phys. Rev Lett. 43, 267 (1979); L. M. Gorbunov and V.I. Kirsanov, Zh. Eksp. Teor. Fiz. 93,509 (1987) [Sov. Phys. JETP 66, 290 (1987)].
28. K.T. McDonald and A.C. Melissinos, Proposal for the Study of QED at Critical Field Strength Experiment E-144 at SLAC.

Generation of 50 TW Femtosecond Pulses in a Nd-Glass Chain

C. Rouyer, E. Mazataud, I. Allais, A. Pierre, and S. Seznec

Commissariat à l'Energie Atomique, Centre d'Etudes de Limeil-Valenton, F-94195 Villeneuve-Saint-Georges Cedex, France

Abstract. Femtosecond pulses in the 50 - 60 TW power range have been generated at 1064 nm using the Chirped Pulse Amplification technique applied to a 90 mm output aperture Nd:Silicate glass amplification line.

1. Introduction

High power chain such devoted to inertial confinement fusion (I.C.F.) such as Nova or Phebus are able to deliver tens of kilojoules pulses in the nanosecond regime. It is therefore tempting to associate them to the chirped pulse amplification (CPA) technique. Intensities higher than 10^{20} W/cm^2 might be obtained. We have developed a first step toward that goal using a medium size Nd-silicate-glass power chain (P102) combined with Titanium Sapphire (Ti:S) technology. In the following we will present the generation of 24 J in a 440 fs pulse duration and discuss the fundamental limitations of the technique in terms of pulse duration and best focusing.

2. From picosecond to femtosecond oscillators

In the recent years there has been much progress in solid state sources generating ultrashort pulses able to feed a high power Nd chain. For instance two years ago our laboratory used a fiber compression technique based on a CW mode-locked Nd-YAG laser coupled to a medium size Nd-silicate glass amplifier chain (P102), generating 20 TW pulses of 1.2 ps duration [1]. However this method does not allow to get pulses shorter than 1 ps and presents instability and poor intensity contrast ratio. To improve this contrast ratio we applied the double compression technique using the non linear birefringence of the fiber pulse cleaner [2]. Clean 500 fs pulses have then been obtained at the final output corresponding to 37 TW peak power. However the breakthrough came from the developpement of Ti:S self modelocked oscillators generating 100 fs pulses, and for our project to their extension to the spectral domain of Nd-glass amplifiers, allowing therefore the application of the true C.P.A. concept.

Our system starts from a Ti:S femtosecond laser (Coherent Mira) extending the emission up to 1.1 µm. The pulse duration is 100 fs and the average laser power 300 mW. The beam then goes into a stretcher consisting of a grating pair (1740 grooves/mm) working at Littrow and an optical afocal system. Pulses stretched from 100 fs to 1.1 ns are sent into a Ti:Sapphire regenerative amplifier before being amplified up to 50 J in the P102 power chain. Finally 24 J, 450 fs pulses are

Springer Series in Chemical Physics, Vol. 55 Ultrafast Phenomena VIII
Editors: J.-L. Martin · A. Migus · G.A. Mourou · A.H. Zewail

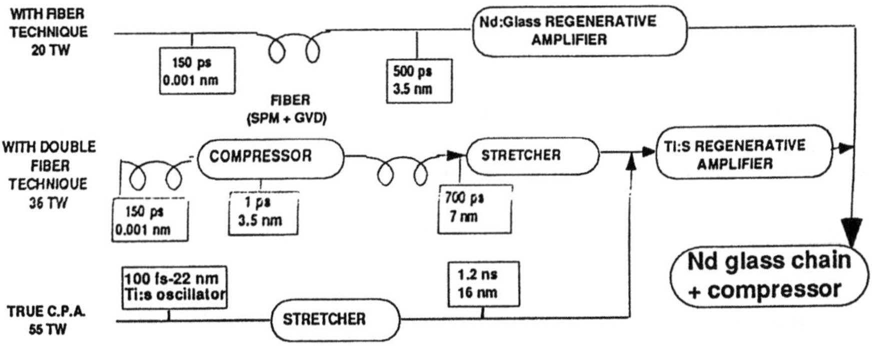

Fig. 1 : Evolution of picosecond and femtosecond oscillators at C.E.L-V.

obtained after compression in a double pass grating setup, corresponding to the emission of 55 TW optical peak power.

3. Limitation in the pulse amplification

Bandwidth of the amplifying medium and gain narrowing gives the ultimate limit of the pulse duration. The Ti:Sapphire oscillator produces 90 fs - 20 nm spectral bandwidth pulses. The stretcher clips the spectrum to about 15 nm, and our Ti:S regenerative amplifier boosts the pulse energy from a few nanojoule to 100 microjoule without any gain narrowing. The chirped pulse is then sent in the Nd silicate glass power chain. As expected the gain narrowing effect is very apparent as soon as the pulse propagates in the Nd-glass amplifiers. First the gain bandwidth (FWHM) of silicate rods is 18 nm as measured on the fluorescence spectrum. In theory this medium has the potential of amplifying pulses shorter than 300 fs. However gain narrowing effect reduce the bandwidth of the amplified pulse. We checked that during propagation and amplification the pulse spectrum decreased following closely the theoritical law:

$$\Delta\lambda = \Delta\lambda_0 \,.\, \left[1 + \beta \,.\, \mathrm{Ln}(G) \,.\, \left(\Delta\lambda_0 / \Delta\lambda_{Fluo}\right)^2\right]^{-\frac{1}{2}}$$

where $\Delta\lambda$, $\Delta\lambda_0$, $\Delta\lambda_{Fluo}$ are respectively the output, input and fluorescence spectra, G is the gain, and $\beta \cong 1$ depends on the spectral shift of the gain.

Eventhough the pulse bandwidth decreases significantly during its amplification from the milijoule to the 60 J level it could be compressed under 500 fs duration. The availability of different kind of glasses with shifted emission curves opens the possiblity of enlarging the emission bandwidth so that pulses of less than 300 fs should be available at the 100 J level. We should notice that a more drastic limit comes from the limited size (< 35 cm) of the commercially available gratings.

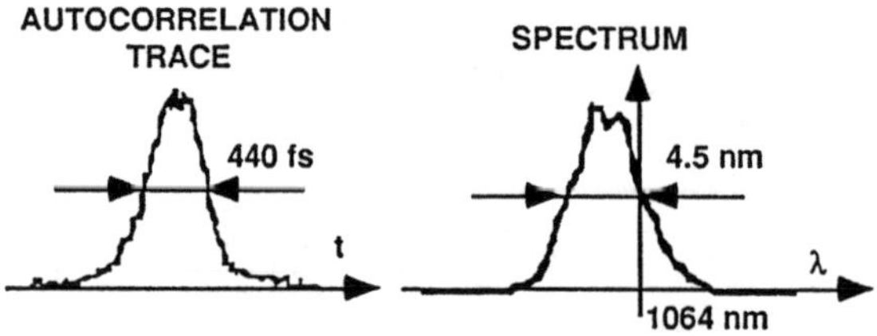

Fig. 2 : Temporal and spectral shape of 55 TW pulses (24 J, 440 fs).

4. Conclusion

We have demonstrated the capability of our Nd-glass power chain to produce pulses above 50 TW peak power. The 100 TW level should be reached by getting shorter pulses using different rod glasses in the amplifying chain and the kilojoule level is possible using the Phebus laser. For many applications one key parameter is the intensity on target. We have studied the focusability of such intense beams and especially the effects due to linear and non linear wavefront distorsions [3]. In particular we have shown that including the path in the grating pair, the beam should propagate into vacuum or low n_2 gas, and be focused with reflective optics. This should allow to generate intensities as high as 10^{19}-10^{20} W/cm^2 in a near future.

5. Acknowledgements

We thank C. Sauteret, G. Mourou, and A. Migus for constant support and J.L. Tapié (Coherent) for his input in some parts of this work.

References.

[1] C. Sauteret, D. Husson, G. Thiell, S. Seznec, S. Gary, A. Migus, and G. Mourou, Opt. Lett., **16**, 238, 1991.
[2] J.L. Tapié and G. Mourou, Opt. Lett., **17**, 1, 1992.
[3] S. Seznec, C. Sauteret, S. Gary, E. Béchir, J.L. Bocher and A. Migus, Opt. Commun., **87**, 331, 1992.

All-Solid Femtosecond Oscillator–Amplifier Laser Chain with 100 mJ per Pulse

C. Le Blanc, G. Grillon, J.P. Chambaret, G. Boyer, M. Franco, A. Mysyrowicz, and A. Antonetti*

Laboratoire d'Optique Appliquée, ENSTA-Ecole Polytechnique, Batterie de l'Yvette, F-91120 Palaiseau, France
*Also at QUANTEL, Avenue de l'Atlantique, Z.A. de Courtaboeuf, B.P. 23, F-91941 Les Ulis Cedex, France

Abstract : A compact CPA laser system relying entirely on Ti:Sapphire technology is described. The energy content of each pulse is amplified up to 100 mJ at 10 Hz while keeping a high peak to background contrast.

We describe a femtosecond laser system relying entirely on Solid State technology, capable of delivering femtosecond optical pulses with an energy of 100 mJ per pulse at a repetition rate of 10 Hz .

As an oscillator, we use a self mode-locking Ti:Sapphire linear cavity [1] pumped with 7 Watts from a CW argon ion laser. In contrast to previous report [2,3], our system does not use optical fibers or dyes. As a consequence, the emitted pulses have very clean Fourier transform limited temporal shapes of 80 fs duration. Each pulse in the output train (repetition rate 100 MHz) has an energie of 4 nJ. The transverse beam profile is a gaussian TEM_{00} mode.

Single pulses are extracted from the train with a Pockels cell (contrast ratio C=500). The pulses are then temporally stretched to 400 ps with a pair of gratings (2000 groves/mm) with positive group velocity dispersion and fed to a first amplifier system. In this first amplifier stage, the pulses cross eight times a 2 cm long Ti:Sapphire rod cut at Brewster angle and pumped longitudinally from both sides with the second harmonic of the output of a Nd:YAG laser. Green pump energy is 70 mJ per pulse. The overall optical gain is more than 10^6, bringing the energy content of each pulse to the millijoule level. The good spatial quality of the beam (TEM_{00}) is preserved. Background ASE energy is low representing less than 5% of the total.

In the second four pass amplifier stage, a 3 cm long Ti:Sapphire rod is pumped from both sides with 400 mJ of green light from the second harmonic of the same Nd:YAG laser. In order to have a good spatial overlap between pump and amplified beams, a small angle $\Theta=1°$ is chosen between successive paths through the rod. We obtain an energy per pulse of 100 mJ at the output, at a repetition rate of 10 Hz and the beam quality is still preserved.

We have compressed these optical pulses at the millijoule level and pulses of 150fs have been obtained. Large (15x15cm) grating are now operating to compress the full power. We will present data on contrast with the ASE background as well as focal spot measurements and we will discuss the limitations in the intensity delivered by this system.

We have also investigated the performances of the new promising materials such as LiSAF and LiCAF [4,5]. In particular, the amplification obtain with flash-pumped LISAF will be

Springer Series in Chemical Physics, Vol. 55 Ultrafast Phenomena VIII
Editors: J.-L. Martin · A. Migus · G.A. Mourou · A.H. Zewail

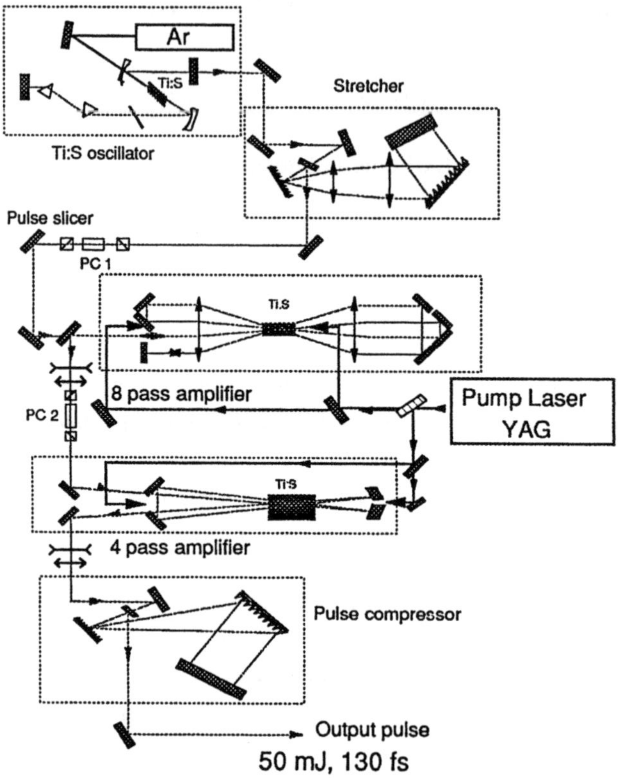

Schematic of the Ti:S amplifier producing 10^{18} w/cm^2 pulses

presented. The final objectives of the system as well as the projected experiments will be described.

References.

[1] D.E. Spence, P.N. Kean, and W. Sibbett, Opt. Lett. **16**, 42 (1991)

[2] J.D. Kmetec, J.J. Macklin, and J.F. Young, Opt.Lett. **16**, 1001 (1991)

[3] A. Sullivan, H. Hamster, H.C. Kapteyn, S. Gordon, W. White, H. Nathel, R.J. Blair, and R.W. Falcone, Opt.Lett. **16**, 1406 (1991)

[4] M. Stalder, B. Chai, and M. Bass, Appl.Phys.Lett. **58**, 216 (1991)

[5] S. Payne, L. Chase, H. Newkirk, L. Smith, and W. Krupke, IEEE J. Quantum Electron. vol.QE.24, p 2243, 1988

Development of a High Intensity Femtosecond LiSAF Laser

M.C. Richardson, P. Beaud, B.H.T. Chai, E. Miesak, Y.-F. Chen, and V. Yanovsky

Center for Research in Electro-Optics and Lasers, University of Central Florida, 12424 Research Parkway, Orlando, FL 32826, USA

Abstract. *We report Fourier-transform limited 110fs Gaussian pulses with a time-bandwidth product of 0.44 from a flashlamp-pumped Cr:LiSAF regenerative amplifier using chirped pulse amplification.*

Cr-doped LiSAF ($LiSrAlF_6:Cr^{3+}$) has many attractive characteristics for the amplification of femtosecond optical pulses. It now can be grown by Czochralski methods in large boules from which rod amplifiers can be fabricated. It has a high gain cross-section ($\sim 5 \cdot 10^{-20} cm^2$) and a very large gain bandwidth (~220nm) [1]. Small-signal gain values as high as $0.17 cm^{-1}$ have recently been measured [2]. The peak emission wavelength of Cr:LiSAF is 850nm, overlapping well that of a mode-locked femtosecond Ti:Sapphire laser which can be used as an initial pulse source. Cr:LiSAF has a long fluorescence lifetime (~67µsec) [1], permitting efficient pumping with conventional xenon flashlamps [3]. The latter is an important feature to the design a compact, low cost high-intensity femtosecond laser system using chirped pulse amplification (CPA) [4].

The regenerative amplifier, shown in Fig. 1, comprises an 65% reflecting output mirror, a Q-switch unit and a 65mm long 3%-doped Cr:LiSAF rod of 4mm diameter. The rod is pumped in a dual flashlamp laser head with a maximum pump energy of 50J and a flashlamp duration of 130 µs. A single chirped pulse of 1 nJ energy is injected into the regenerative

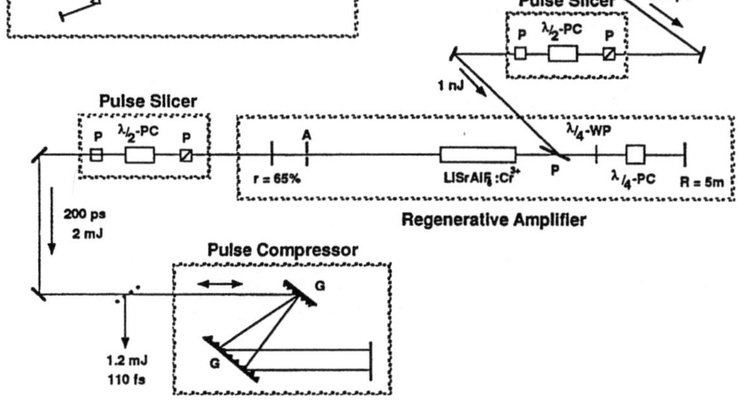

Fig. 1: Experimental set-up: BF: birefringent filter, S: slit, G: grating, f: lens, P: polarizer, PC: Pockels-cell, WP: waveplate.

Springer Series in Chemical Physics, Vol. 55 **Ultrafast Phenomena VIII**
Editors: J.-L. Martin · A. Migus · G.A. Mourou · A.H. Zewail

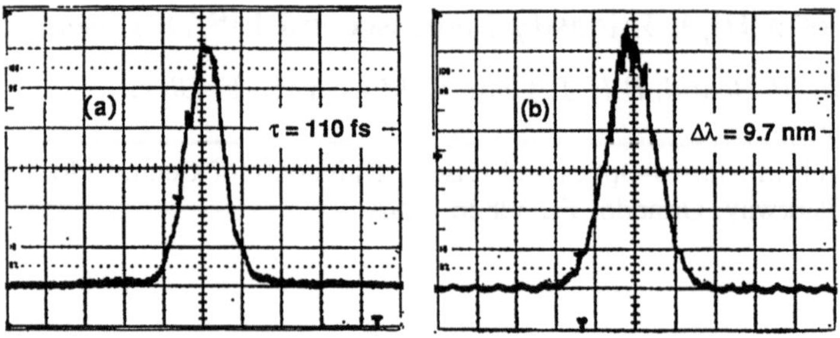

Fig. 2: Single shot (a) autocorrelation and (b) spectrum of the recompressed 1.2 mJ pulse.

amplifier through a thin film polarizer. The 85fs input pulse is produced by a self-mode-locked Ti:Sapphire laser consisting of a 7.5mm Brewster angled $Al_2O_3:Ti^{3+}$ crystal pumped by an Argon ion laser in an X-type cavity containing a prism pair for group velocity dispersion compensation and a slit as a fast saturable absorber. The laser reliably produces 85fs $sech^2$-shaped pulses of 50kW peak power at 855nm with a time-bandwidth product of 0.4. These pulses are expanded ~2400 times to 200ps by a double passage through an anti-parallel grating pair containing a 1:1 telescope. The gratings are gold coated 1800-grooves/mm holographic gratings and separated by 115 cm. Two achromatic, anti-reflection coated lenses of 40cm focal length form the telescope between the gratings. A single pulse is selected out of the pulse train by a half wave Pockels-cell optical gate and coupled into the regenerative amplifier. A digital logic circuit is used to control synchronism of the optical gate and the Q-switch with the Ti:Sapphire-oscillator.

The regenerative amplifier produces an output pulse train with an envelope of 80 ns duration and an overall energy of 25mJ. It takes 35 roundtrips to reach gain saturation resulting in a total amplification of $\sim 10^7$. A second pulse slicer is used to select a single pulse of 2mJ energy. The stability of the pulse energy is better than +/-5%. This pulse is then recompressed in a double pass through a parallel grating-pair identical to those used in the pulse stretcher. The single shot autocorrelation and spectrum of the recompressed pulse is shown in Fig.2. Both traces correspond close to a Gaussian. Due to spectral wing clipping in the pulse expander and to the limited bandwidth of the regenerative amplifier cavity the initial spectral bandwidth of 11.6nm narrows to 9.7nm. The pulse duration is 110fs (assuming a Gaussian pulse) resulting in a time-bandwidth product of 0.44 which corresponds to that expected from a Fourier-transform limited Gaussian pulse.

In order to further amplify these pulses we have started to fabricate Cr:LiSAF rods of 25 mm diameter and >100 mm length. Preliminary measurements show that two of these rods - used in a 4pass amplification scheme - will permit to amplify femtosecond pulses up to the Joule level.

This work is supported by the State of Florida and the National Science Foundation under Grant # ECS-9113726. One of the authors P. Beaud is supported by the Swiss National Science Foundation.

[1] S. A. Payne, L. L. Chase, L. K. Smith, W. L. Kway, and H. W. Newkirk, J. Appl. Phys. **66**, 1051, 1989
[2] P. Beaud, Y-F. Chen, B. Chai, and M. Richardson, Opt. Lett., to be published.
[3] M. Stalder, B. H. T. Chai, and M. Bass, Appl. Phys. Lett. **58**, 216, 1991
[4] P. Maine, D. Strickland, P. Bado, M. Pessot, and G. Mourou, IEEE J. Quantum Electron. **QE-24**, 398, 1988

Contrasted Behaviors of Stark-Induced Resonances in Multiphoton Ionization of Krypton

E. Mevel[1], *R. Trainham*[1], *J. Breger*[1], *G. Petite*[1], *P. Agostini*[1], *J.P. Chambaret*[2], *A. Migus*[2], *and A. Antonetti*[2]

[1]DRECAM/SRSIM CE Saclay, F-91191 Gif-sur-Yvette, France
[2]LOA-ENSTA, Ecole Polytechnique, F-91120 Palaiseau, France

Abstract. Multiphoton Ionization of krypton with 617 nm, 100 fs pulses is investigated over a large intensity range. Near appearance intensity, electron energy spectra display Stark-induced resonances.When intensity is increased,the spectra broaden and the structures are progressively washed out. At still higher intensities, a strong resonance emerges, corresponding to the 4f state whose static detuning is 2.94 eV, a value larger than the photon energy.

1. Introduction

Many experiments [1] have shown that electron energy spectra from Multiphoton Ionization (MPI) and Above-Threshold Ionization (ATI) of atoms by intense subpicosecond pulses reveal Stark-induced resonances and provide an elegant and unique method to investigate AC-Stark shifts of strongly perturbed excited states.

Under short pulse condition, each structure of a typical ATI spectrum showing sequences of peaks separated by the photon energy, splits into a series of narrow substructures. They arise from ionization enhancement via excited states AC-Stark shifted into resonance with an harmonic of the laser field. If the resonant state AC-Stark shift is nearly equal to the ionization threshold shift (as normally the case for Rydberg states) the corresponding subpeak occurs at an energy which is the photon energy reduced by the binding energy of the bare state.

The aim of this paper is to report an experimental investigation of photoelectron energy spectra of krypton whose MPI by 2.01 eV photons provides a particularly interesting configuration.

2. Experimental results and discussion

The laser is based on a CPM oscillator amplified in five dye Bethune cells pumped by an injected single-mode YAG laser. It delivers 1 mJ, 100 fs pulses and can be focused to a few micrometers, thus giving rise to intensities up to 10^{17} W.cm^{-2}, only a fraction of which is necessary for the present work.

The photoelectrons are energy-analyzed in a 1m time-of-flight spectrometer maintained at a background pressure of 10^{-9} torr. The electrons are collected by a parabolic mirror with an efficiency larger than 50%. The resolution is better than 50 meV at 1 eV.

For an intensity of 4.1 10^{13} W.cm^{-2}, the spectrum shows structures (Fig.1a) which can tentatively be assigned to 5d, 7s, 6d, 7d, 5p and 8s. When the intensity is increased the spectrum progressively broaden to a continuum. Above 9 10^{13} W.cm^{-2}, surprisingly, strong peaks emerge from the background (fig.1b). They are clearly identified as due to the 4f and 5f states for which the static detuning is 2.94 and 2.63 eV respectively.

This sequence, discrete structure-broad continuum-discrete structure, is at variance with what is observed in MPI of H, Xe or Ar and is probably due to two circumstances peculiar to krypton and the photon energy used here. The first one is that the 6d, 7d and 8s are about one photon above the 5p thus coupled through dipole interaction which results in shifts different from the threshold shift. Not only this makes the assignment of the structures (*) in Fig.1a uncertain but also might be responsible for the fact that the corresponding part of the spectrum

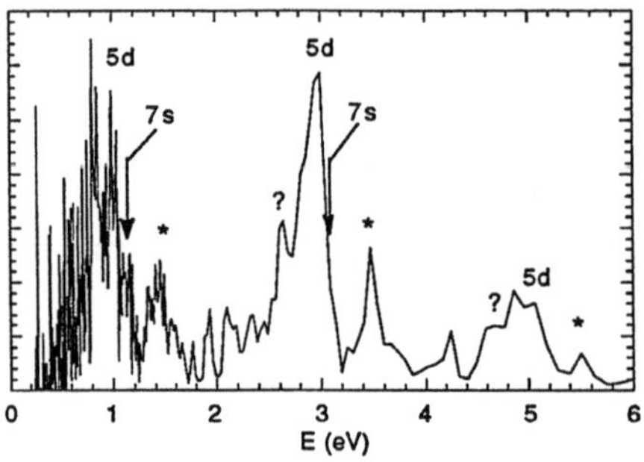

Fig. 1a: Electron energy spectrum from krypton at 4.1 10^{13} W.cm^{-2}. * points out for 6d, 7d, 8s and 5p. ? indicates an unidentified substructure.

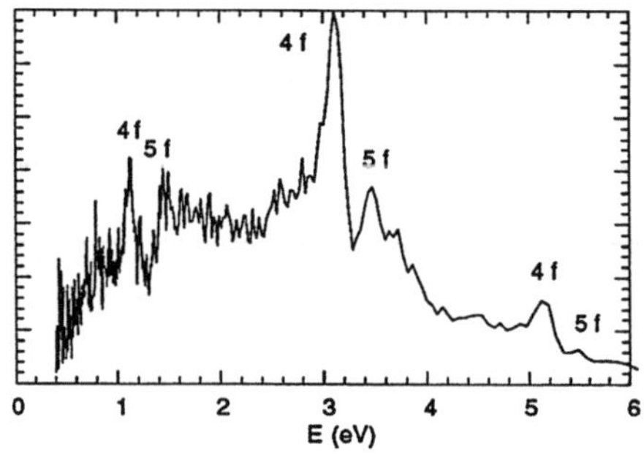

Fig. 1b: Electron energy spectrum from krypton at 1.3 10^{14} W.cm^{-2}

drastically changes when the peak intensity is varied: if one assumes that the states are actually populated during the rise of the pulse and ionized at different intensities during the rest of the pulse, as recently suggested by *M.P. de Boer and H.G. Muller* [2], electrons would be released with different energies thus washing out the structures. Let us stress that if ionization takes place only at I_r, as previously proposed by *Freeman et al.* [3], virtually no change in the resonance width is expected when the peak intensity is varied. The second one is the exceptionally large static detuning for the f states which implies that the corresponding resonances occur at high intensity, thus increasing the amplitude of the resonant structures and reenforcing the effect of the propensity rule $\Delta l = +1$ in dipole transitions [4] which favors high angular momentum states. Of course, the appearance of such highly shifted resonances is possible because, fortunately, 7-photon resonances with s and d states are not strong enough to saturate the ionization process (that is deplete the ground state) before 8-photon resonances with p and f can appear.

Because of this quite unusual behavior, Krypton offers an interesting testing ground for the theory of MPI of complex atoms in strong fields currently in progress [5].

This work is supported by the European Community under the grant No.SC1000103.

References

1. R. R. Freeman et al. *Advances in Atomic, Molecular and Optical Physics* M. Gavrila Ed. (Academic, Orlando, in press). See also H. G. Muller, P. Agostini and G. Petite same volume.
2. M. de Boer, H. G. Muller, Phys. Rev. Lett. **4**, 2747 (1992)
3. T. J. McIlrath, R. R. Freeman, W. E. Cooke, L. D. van Woerkom, Phys. Rev. A **40**, 2770 (1989)
4. U. Fano, Phys. Rev. A 32, 617 (1985)
5. M. Dörr , Invited paper, ECAMP 4 , Riga (1992)

Phase-Dependent Ionization Using an Intense Two-Color Light Field

D. Schumacher[1], *M.P. de Boer*[2], *H.G. Muller*[2], *R.R. Jones*[1], *and P.H. Bucksbaum*[1]

[1]Physics Department, University of Michigan, Ann Arbor, MI 48109-1120, USA

[2]FOM-Institute for Atomic and Molecular Physics, Kruislaan 407, NL-1098 SJ Amsterdam, The Netherlands

Abstract. We have studied Multi-Photon Ionization (MPI) in Xenon using a field made up of two sub-picosecond laser pulses, one the second harmonic of the other, as a function of phase shift between the two pulses. We observe a phase dependent variation in the amplitude of the various peaks occurring in the photoelectron spectrum. We compare these results to data taken with 150 ps pulses.

1. Introduction

Recent similar experiments have used either two pulses of incommensurate frequencies, or pulses that were in the "long-pulse" regime[1] where ponderomotive effects dominate the final kinetic energies of ejected photoelectrons. The sub-picosecond pulses of this experiment allow us to identify laser induced resonances during the ionization process[2] and hence specific ionization pathways. In an intense field, an electron can acquire a significant probability to ionize within a single optical cycle, with ionization most likely during the peak field. In a two-color field, the amplitude of the peak electric field is phase dependent as is the variation of the total electric field in time. The ponderomotive potential is phase independent however, being an averaged quantity. Thus studying phase variations allows us to determine the utility of using the ponderomotive potential to predict light induced shifts in an atom. Moreover, the final state wavefunction of an electron in a field will be phase dependent in this kind of experiment, and could also modify the ionization process. Finally, phase experiments can serve as sensitive tests of various models of ionization (ie. tunneling vs multi-photon pictures)[3].

2. The Experiment

We generated 100 fs, 10 mJ laser pulses at 770 nm using a Ti:Sapphire based laser system employing chirped pulse amplification[4]. This light was then divided into two components of orthogonal polarization displaced in time using two KD*P crystals but following the same path. The leading component was frequency doubled using a 1 mm thick KD*P crystal yielding a pulse at the second harmonic with the same polarization as the trailing fundamental pulse. The system was designed so that after suffering passage through a (necessary) thin film polarizer and calcium fluoride window, which both have positive group velocity dispersion, the two pulses were overlapped in time inside our vacuum chamber where they were focussed with a normal incidence spherical mirror. Any photoelectrons produced were measured using a time-of-flight spectrometer. The relative phase was varied by tilting one of the first two KD*P crystals.

Springer Series in Chemical Physics, Vol. 55 Ultrafast Phenomena VIII
Editors: J.-L. Martin · A. Migus · G.A. Mourou · A.H. Zewail

3. Results

The spectrum averaged over two cycles of phase variation of the fundamental is shown in Figure 1. Figure 2 plots the variation of the first two peaks as a function of phase. The variation of the next two peaks may show a periodic variation, but the remaining ones do not. The total integrated signal does, however. The fine structure atop the third peak has not been clearly identified, but may be an f Rydberg series starting with 5f. There is no evidence for variation of peak position or width with phase, supporting the view that the fine structure resonances represent time-averaged shifts in the intermediate states. These data were collected with the peak intensity of the fundamental and second harmonic each at roughly $5 * 10^{13}\ W/cm^2$. The fundamental alone was unable to ionize the xenon. Note, as well, the second harmonic cannot produce signal at all the peak locations present in Figure 1 alone. The fundamental pulse length was 300 fs before passing the calcium fluoride window. The pulse width of the second harmonic was not measured.

Using 150 ps pulses by bypassing the compressor in our laser system, we saw no phase variation in the first threshold peak, or in the total signal. The second through fourth peaks did show variation, however. The variation shifted 180 degrees in phase from peak to peak.

This work was supported by the National Science Foundation.

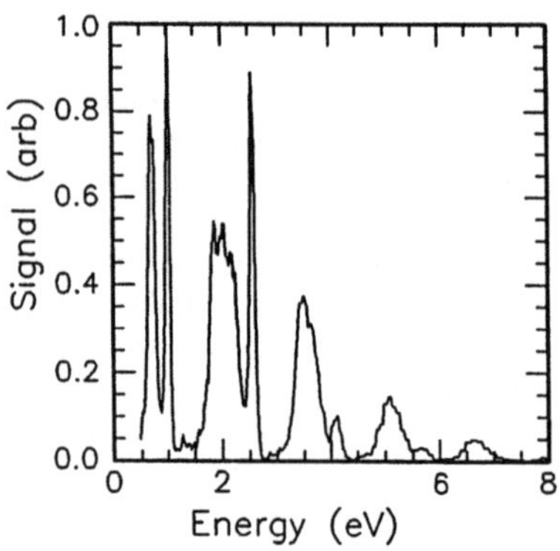

Fig. 1: Spectra summed over all phases.

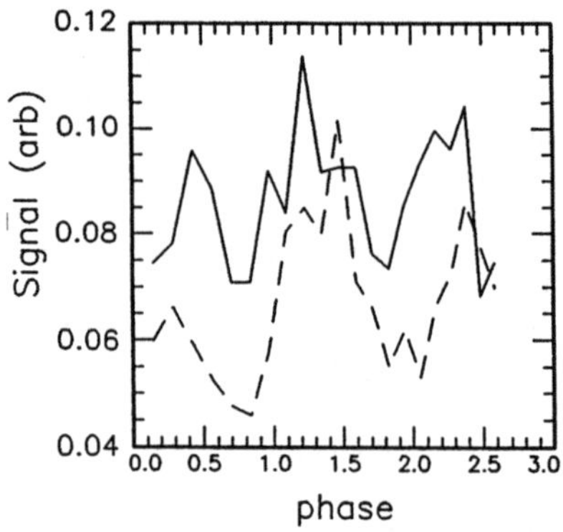

Fig. 2: Phase dependent signal of the first (solid) and second (dashed) peaks

REFERENCES

1. H.G. Muller, P.H. Bucksbaum, D.W. Schumacher, A. Zavriyev, J. Phys. B **23**, 2761 (1990)

2. R.R. Freeman, P.H. Bucksbaum, H. Milchberg, S. Darack, D. Schumacher, M.E. Geusic, Physical Review Letters **59**, 1092 (1987).

3. K.J. Schafer, K.C. Kulander, (to be published in Phys. Rev. A).

4. J. Squier, F. Salin, G. Mourou, D.J. Harter, Optics Letters **16**, 324 (1991).

Stabilization of Atoms in Ultra-Intense Laser Pulses: A Classical Model

A. Maquet, T. Ménis, R. Taïeb, and V. Véniard

Laboratoire de Chimie Physique (URA 176),
Université Pierre et Marie Curie,
11, Rue Pierre et Marie Curie, F-75231 Paris Cedex 05, France

Recent theoretical studies,[1,2] have suggested that atomic stabilization could be observed in very strong high-frequency laser pulses, with peak intensities beyond the atomic unit of intensity: $I_0 = 3.5\ 10^{16} W/cm^2$. These predictions contradict the commonly accepted view [3], that bound systems would not survive to the rise of the pulse as they would be ionized well before the pulse has reached its maximum of intensity. Several of these recent studies are based on numerical simulations performed for a simplified 1-dimension model, proposed a few years ago [2]. Within this framework, a one-electron atomic system is modelled through the use of a "soft Coulomb potential" of the general form (atomic units will be used):

$$V(a,x) = -(a^2 + x^2)^{-1/2}, \tag{1}$$

which correctly reproduces the long range of the Coulomb interaction and has the advantage to suppress the Coulomb singularity at the origin. It is understood that the x-axis is oriented along the linear polarization of the laser field, the Hamilton function for the atom-laser system being :

$$H = p^2/2 + V(a,x) + F(t)\, x \sin \omega t, \tag{2}$$

where F(t) is the amplitude of the laser electric field with frequency ω. Note that, in the calculations reported so far, the parameter a has been chosen to be equal to unity, [2].

Several quantum [2],[4], as well as classical [5], numerical simulations performed on this model indicate that atoms:

i) could be stabilized even at much lower frequencies than originally predicted by the Gavrila-Kaminski model [1]. A typical frequency, often discussed in the literature is $\omega = 0.8$ au, so that the ratio between the photon energy and the binding energy is $\omega/E_0 \approx 1.2$.

ii) could even survive to the rise of the laser pulse intensity. As already mentioned, this is a crucial point to assess in order to discuss the possibility to expose an "atom", i.e. a bound electron, to ultra-intense laser fields [3].

It has also been shown, through a classical calculation [5], that stabilization is not so easily observed in the case of a real 3-dimension Coulomb potential. This feature of the 3-D model was then ascribed to the fact that much larger regions of the phase space were accessible to the electron, as compared to the 1-D case. We want to point out here that, the main reason why the electron explores larger regions of the phase space, thus being able to get large enough momenta to be ionized, lies in the presence of the Coulomb singularity at the origin. As a result of its existence, the electron can experience much stronger accelerations when it comes close to the nucleus, thus acquiring the necessary energy to be ionized. This suggests that models, either 1-D or 3-D, in which the singularity at the origin is smoothed can lead to overestimate the extent of the stabilization. We support our assertion with several numerical simulations based on a classical Monte Carlo trajectory approach much similar to the ones described in [5,6].

Springer Series in Chemical Physics, Vol. 55 Ultrafast Phenomena VIII
Editors: J.-L. Martin · A. Migus · G.A. Mourou · A.H. Zewail

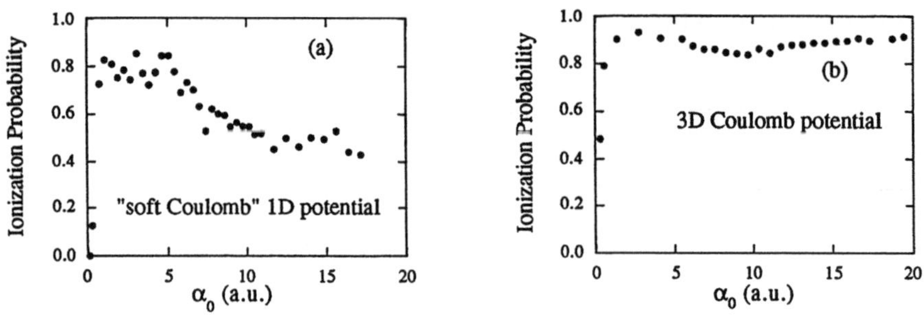

In the figures are shown the variations of the ionization probability as a function of the parameter $\alpha_0 = F_0/\omega^2$, (which is proportional to the field strength intensity and is characteristic of the "dichotomy" of the atom in the field, [1]), as they are obtained from: (a) the 1D potential V(1,x) in Eq. (1); (b) a 3D Coulomb potential. In both cases the ratio $\omega/E_0 \approx 1.2$ and the laser pulse has a durarion of 50 laser periods. It is clearly observed that the "soft Coulomb potential" exhibits more stabilization than the more realistic Coulomb potential. A detailed analysis shows that this significant difference can be ascribed to the existence of the Coulomb singularity at the origin. These results cast some doubts on the conclusions drawn from simplified 1D atomic models, about stabilization in the low frequency domain.

References.

[1] M. Pont and M. Gavrila, Phys. Rev. Lett. **65**, 2362 (1990) and references therein.
[2] Q. Su, J. H. Eberly and J. Javanainen, Phys. Rev. Lett. **64**, 862 (1990) and references therein.
[3] P. Lambropoulos, Phys. Rev. Lett. **20**, 2141 (1985).
[4] V. C. Reed, P. L. Knight and K. Burnett, Phys. Rev. Lett. **67**, 1415 (1991)
[5] J. Grochmalicki, M. Lewenstein and K. Rzazewski, Phys. Rev. Lett. **66**, 1038 (1991).
[6] G. Bandarage, A. Maquet and J. Cooper, Phys. Rev. A **41**, 1744 (1990) and G. Bandarage, A. Maquet, T. Ménis, R. Taïeb, V. Véniard and J. Cooper, to be published.

Inertially Confined Molecular Ions

M. Laberge, P. Dietrich, and P.B. Corkum

National Research Council of Canada, Ottawa, Ontario K1A 0R6, Canada

Over the last decade, the behavior of atoms in intense laser fields has been under keen investigation. With the development of high power, ultrashort pulse lasers, it is now possible to perform the analogous molecular ionization experiments [1]. This is because ionization can occur on a time that is short compared to the dissociation time of even highly charged molecular ions. The subsequent Coulomb explosion gives the ion fragments a characteristic kinetic energy which can be studied in a time of flight (TOF) mass spectrometer. In this paper, we analyze the TOF spectrum in order to obtain spectroscopic information about the highly charged molecular ions. Such data were previously unavailable because of the very short lifetime of these species. We also performed a correlation experiment to confirm our interpretation of the TOF spectrum.

We use 90 fs, 300 µJ pulses at $\lambda = 615$ nm. These pulses are focused into 5 x 10^{-7} Torr of jet cooled iodine vapor with an F/3 parabolic mirror. The intensity at the focus is estimated to be ~ 10^{16} W/cm^2. Two small apertures in the TOF spectrometer select only the ions that are emitted with a small transverse velocity.

The relevant portions of the TOF spectrum (averaged over 1000 shots) are shown in Figure 1. The horizontal scale under each ions species (I^{3+}, I^{2+} I^+ and I_2^+) represents the initial energy of the fragments. The I_2^+ peak is centered on zero initial energy and has a width of 0.001 eV (T ≈ 10° K if the width is limited by thermal motion). This value is the maximum energy resolution of this

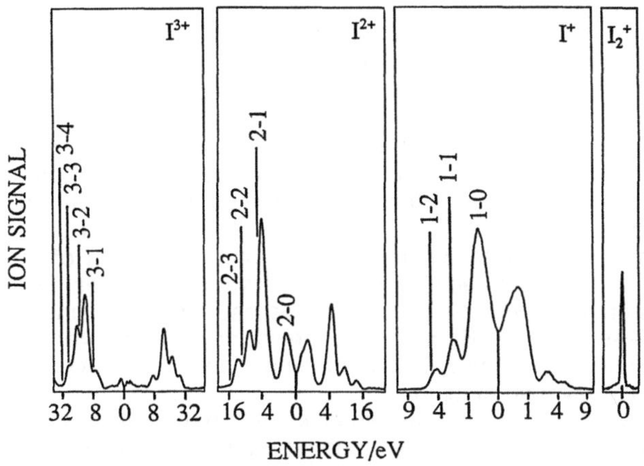

Fig. 1: Time of flight spectrum for the I^{3+}, I^{2+}, I^+ and I_2^+ ions. The vertical bars labeled a-b shows the Coulomb energy for $I_2^{(a+b)+} \rightarrow I^{a+} + I^{b+}$. For dissociation producing a neutral, the corresponding peak is indicated with no vertical bar. The I_2^+ peak has a width of 0.001 eV.

spectrometer. The other peaks are produced when the I_2^{n+} (n = 1 - 7) molecular ions dissociate, ejecting the resulting atomic ions or neutrals with some energy. Only the ion fragments are detected. To understand the structure in the TOF spectrum, note that the two apertures in the spectrometer select the ions that are ejected directly towards the detector (and arrive earlier than if they had zero initial kinetic energy) or backwards (and arrive later). The vertical lines in Figure 1 labelled (a-b) show the Coulomb energy that would be release for $I_2^{(a+b)+} \rightarrow I^{a+} + I^{b+}$, when starting from an interatomic distance of 0.266 nm (the equilibrium internuclear separation of the I_2 molecule). The lowest energy peaks for I^+ and I^{2+} on the TOF spectrum have much less energy than a Coulomb explosion can produce. These peaks correspond to $I_2^+ \rightarrow I^+ + I^o$ and $I_2^{2+} \rightarrow I^{2+} + I^o$. There is a one-to-one correspondence between the higher energy peaks and the corresponding Coulomb explosion energy. However, the observed peaks have somewhat less energy than the Coulomb energy, particularly for the explosion of higher charged states. Also, the linewidth of the peaks (~ 2 eV) are much broader than would be expected for the Coulomb explosion model where the linewidth should be instrument limited (0.001 eV). Therefore, it is essential to confirm the peak assignment.

We performed an experiment to observe the correlation that exists between two points of the TOF spectrum corresponding to two fragments from the same exploding molecule. For example, in the explosion $I_2^{3+} \rightarrow I^{2+} + I^+$, we should have a large probability (limited by the detection efficiency) of detecting a forward going I^{2+} if we detect a backward I^+.

In order to increase the detection efficiency, the apertures in the TOF spectrometer were removed. The pressure, final spot size, and intensity, were adjusted to detect one (or more) iodine ion only for one out of twenty laser shots. If two ions are detected, a point is plotted on the graph with the x and y coordinate being the TOF of the first and second ions respectively. Events with one, or more than two ions are discarded. Figure 2 shows the results of the correlation experiment. The averaged TOF spectrum is shown along the graph axis. With the apertures removed all ions, regardless of their initial explosion direction, are detected. One might expect to see a structureless TOF spectrum. However, independent of the apparatus, ions are detected only around the direction of the laser electric field [1] and the structures in the TOF spectrum are still visible in Figure 2, although with much less resolution.

Two ions coming from a single Coulomb explosion always have the same energy and opposite direction. They should produce a correlation point along the lines indicated on Figure 2. These groups of points are clearly visible above the background noise which is due to uncorrelated two ions events. Also, the correlation signal due to the explosion of a certain charged state is centered on the corresponding peak in the average TOF spectrum, confirming our original assignment of the peaks. Of particular interest is the presence of both the symmetric and asymmetric decay channels of I_2^{4+} ($\rightarrow I^{2+} + I^{2+}$ and $\rightarrow I^{3+} + I^{+}$) on both the TOF spectrum and the correlation map. Significant population of the excited (asymmetric) level is important because the lower level experiences more Coulomb repulsion and a population inversion can be produced for a few hundred femtoseconds after the laser pulse. Because of the strong charge transfer coupling between these levels, such a new way of creating a population inversion could produce an interesting new gain medium.

Having established the peak assignment we can now explore the spectroscopy of highly charged molecular ions. However, it is important to determine the origin of the linewidth and of the deviations of the kinetic energy from the Coulomb energy. We identify six phenomena that can affect the time of flight spectrum. They are: (1) the initial thermal energy of the molecule, (2) the amount of transverse velocity accepted by the apparatus, (3) nuclear dynamics that can occur on the time-scale of the pulse, (4) excitation of the molecule by a laser prepulse,

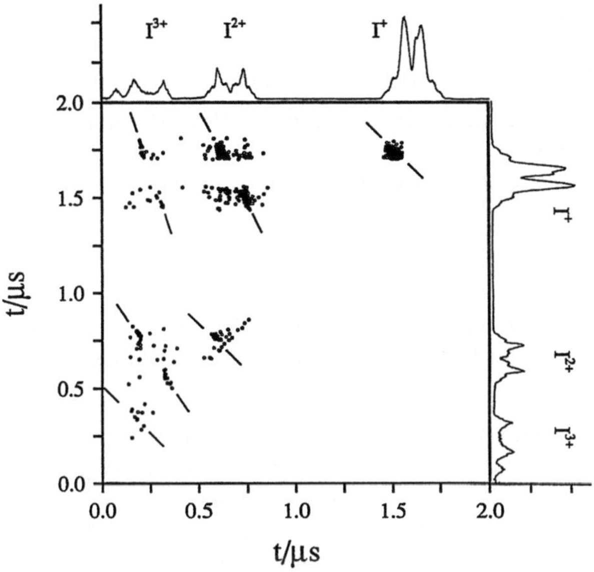

Fig. 2: Correlation map for I_2 explosion. The time of flight spectrum is shown along the axis. The different groups of points represent the following Coulomb explosion (from left to right, top to bottom) 3-1, 2-1, 1-1; 3-2, 2-2 and 3-3.

(5) the molecular polarizability and (6) the intrinsic energy structure of the molecular ion. Points 1 and 2 can be easily eliminated by using a supersonic nozzle to cool the I_2 gas, and by using sufficiently small apertures in the extraction plates of the TOF spectrometer.

Point 3 and 4 were investigated experimentally by varying the pulse duration and the energy in the pre-pulse. A computer model of these processes using tunnel ionization, classical trajectories in a Coulomb potential and pre-excitation to the B state of the iodine molecules was also developed. Although points 3 and 4 have an effect on the TOF spectrum, it was found to be insufficient to explain the observed discrepancy between the experimental and Coulomb kinetic energy.

The polarizability of molecular ion (point 5) allows some coupling between the laser field and the kinetic energy of the ionic fragments. This could lead to some of the observed broadening of the peaks. Finally, at short internuclear distances, the molecular ions can have an intrinsic energy structure (point 6) that can reduce the potential below the Coulomb energy and produce peak broadening if the individual energy levels are not well resolved.

In conclusion, we observed the Coulomb explosion of highly charged iodine molecular ions. Two pulse experiments, where the first pulse excite the molecule to a known repulsive curve and the second ionizes it, should allow us to determine the full potential energy curve for any of the accessible states. Also, the existence of a symmetric and asymmetric decay channel for the I_2^{4+} ions may give rise to a new inversion mechanism.

Reference:

[1] D.T. Strickland, Y. Beaudoin, P. Dietrich and P.B. Corkum, Phys. Rev. Lett. **68**, 2755 (1992)

A Femtosecond Lightning Rod

X.M. Zhao[1], *C.Y. Yeh*[1], *J.-C. Diels*[1], *and C.Y. Wang*[2]

[1]Department of Physics and Astronomy, The University of New Mexico, Albuquerque, NM 87131, USA

[2]Department of Applied Physics, Columbia University, New York, NY 10027, USA

Abstract. Experimental data and a theoretical model are presented to demonstrate the use of fs ultraviolet laser pulse for triggering lightning.

Can laser light be used to trigger lightning? Previous attempts [1] with high power CO_2 laser have been unsuccessfull, because the high density plasma created by avalanche ionization with the intense infrared laser is *opaque to the beam*. The method under investigation here uses a fs UV pulse at 248 nm to create a moderate ($10^{13}/cm^3$) electron density in a narrow (a few mm in diameter) vertical column of air. The charges are injected in the high field region between ground and cloud (at least 100 m altitude). The motion of the charges in the "external" atmospheric field result in a field enhancement (up to tenfold) at the "tip" of the cylinder of charges. A long (up to 1 μs) visible laser pulse has to be sent simultaneously through the same path to prevent the creation of negative ions O^- and O_2^-.

There are two main reasons for using fs pulses. First, the high intensity of fs pulses at 248 nm is needed to create free electrons by three photon ionization of oxygen and four photon ionization of nitrogen. Second, a mechanism of pulse compression in air is needed to compensate peak power depletion associated with the multiphoton process. The dispersion of air at 248 nm is sufficient for effective compression of downchirped fs pulses over a 50 m optical path. Kerr effect in diamond (n_2 is negative at that wavelength) is used for downchirping (instantaneous frequency decreasing with time during the pulse) the 200 fs pulse.

We have measured the multiphoton ionization cross sections for oxygen and nitrogen in low pressure ($\ll 1$ mtorr) cell to be $\sigma^{(3)} = 1.28 \times 10^{-23} J^{-3} s^2 cm^6$ and $\sigma^{(4)} = 4 \times 10^{-34} J^{-4} s^3 cm^8$, respectively. These numbers imply that a 200 fs pulse with an energy density of $1 mJ/cm^3$ will extract $1.8 \times 10^{12} e^-/cm^3$ from atmospheric oxygen, and $10^{12} e^-/cm^3$ from atmospheric nitrogen.

A scaled down laboratory model of the pre-lightning conditions is simulated with a pair of electrodes spaced 25 cm apart and charged to 100 kV. The radiation is introduced through a profiled hole in the electrodes. The fs UV pulses are produced by amplification in a KrF laser of frequency trippled 744 nm pulses. The latter are obtained by three stage amplification of the radiation of an hybridly pumped antiresonant ring dye laser (LDS721 in EPH as gain dye and DDI in ethylene glycol as absorber). Laser induced discharge as been observed as shown in Fig. 1.

Springer Series in Chemical Physics, Vol. 55 Ultrafast Phenomena VIII
Editors: J.-L. Martin · A. Migus · G.A. Mourou · A.H. Zewail

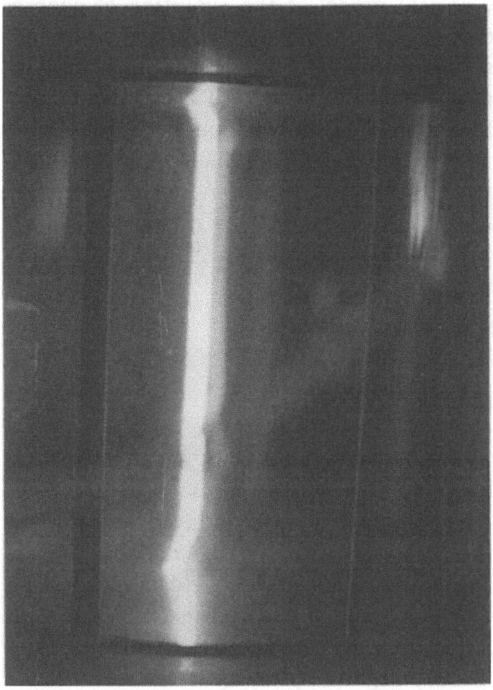

Figure 1: UV ultrashort pulses induced lightning discharge in a HV testing cell.

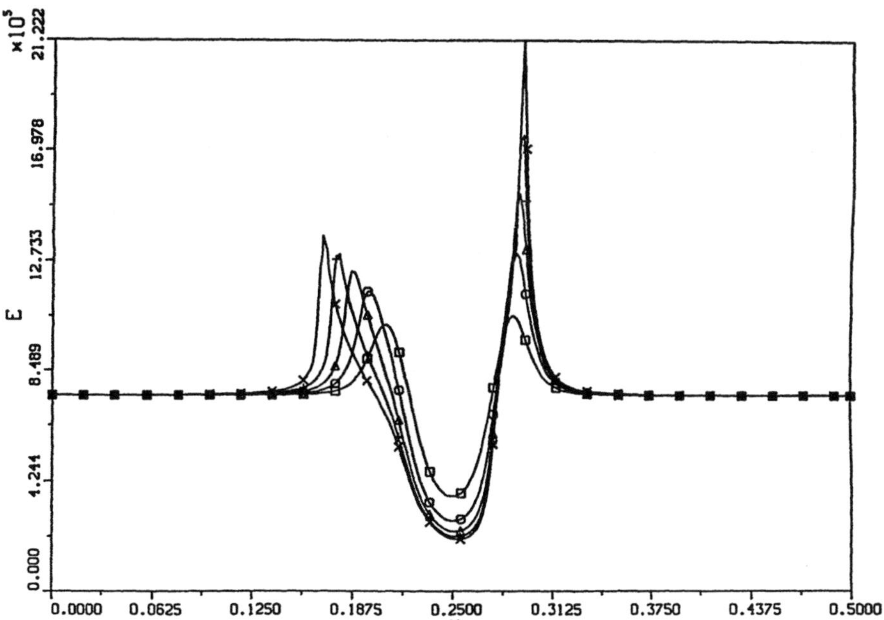

Figure 2: Electric field enhancement in the 25 cm testing cell. The curves of □, ○, △, +, and × correspond to successive time sequences of 14.6 μs.

A "quasi-three dimensional" numerical model of the evolution of charges in the applied field has been constructed. The charges are assumed to be uniformly seeded in a narrow cylinder of finite height and radius located with its axis parallel to the externally applied electric field. All attachment, photodetachment, photoionization, diffusion and recombination rates have been included in the model as described in a Poission's equation and thte continuity equation:

$$\frac{\partial N_e}{\partial t} + \frac{\partial (W_e N_e)}{\partial z} = \alpha N_e - \eta N_e - \beta_{ei} N_e N_p + \gamma_l N_n + \frac{\partial}{\partial z}\left(D_e \frac{\partial N_e}{\partial z}\right), \quad (1)$$

$$\frac{\partial N_p}{\partial t} + \frac{\partial (W_p N_p)}{\partial z} = \alpha N_e - \beta_{ei} N_e N_p - \beta_{ii} N_n N_p, \quad (2)$$

$$\frac{\partial N_n}{\partial t} + \frac{\partial (W_n N_n)}{\partial z} = \eta N_e - \beta_{ii} N_n N_p - \gamma_l N_n, \quad (3)$$

where N_s are electron, positive and negative-ion densities; W_s are their drifting velocities, respectively; α, η, β_{ei}, β_{ii}, γ_l and D_e are the ionization, attachment, electron-ion recombination, ion-ion recombination, laser-induced detachment and electron-diffusion coefficients, respectively and z is the distance in the direction of propagation. The calculated evolution of the field in the testing cell is shown in Fig. 2. As we can see the local field has been enhanced several times. Under this enhanced field, electrons are accelerated, multiplied and reach the avalanche threshold for the discharge of lightning.

References

[1] Jr. C. W. Schubert. *The Laser Lightning Rod System: a Feasibility Study.* Technical Report AFFDL-TR-78-60, Air Force Flight Dynamics Laboratory, 1977.

Plasma Physics with Ultra-Short and Ultra-Intense Laser Pulses

T.W. Johnston[1], Y. Beaudoin[1], M. Chaker[1], C.Y. Côté[1], J.C. Kieffer[1], J.P. Matte[1], H. Pépin[1], C.Y. Chien[2], S. Coe[2], G. Mourou[2], and D. Umstadter[2]

[1]INRS-Energie et Matériaux, Université du Québec, 1650 Montée Sainte-Julie, Varennes, Québec, Canada J3X 1S2
[2]Center for Ultrafast Optical Sciences, University of Michigan, Ann Arbor, MI 48109, USA

Abstract. The study of ultra-intense and ultra-short laser light pulses produced from "Table Top" compressed-pulse laser systems and their interaction with the plasmas produced from solid targets or from gas is an explosively growing field. This effort should result in the production of bright ultra-short pulse X-ray sources for rapid diagnostics with many applications in chemistry and related fields. Success here will, however, require considerable improvement in the understanding of target interactions for optimum results, which in turn implies major improvements in diagnostics and in simulation/theory. Although this effort is still in the preliminary phases, the progress so far is encouraging. As far as pure research on high-field laser plasma interaction is concerned, preliminary experiments are being made and one can hope that the theory for plasma physics of waves with relativistic intensity will soon be receiving many of the experimental tests which have been so long awaited.

1. Introduction.

With the laser powers and intensities now available in systems of modest "Table-Top" size, obtained via pulse compression to subpicosecond durations[1], many groups have been able to enter the field of high laser intensities, previously limited to the very large lasers giving pulse lengths of the nanosecond range. In the plasma research using such systems, two themes have emerged.

The first theme, using solid targets, is that of the production of very short pulses of X-ray [2] from a small area, offering the likelihood of very fast X-ray diagnostics with picosecond resolution for many experiments[3], particularly in chemistry and biology. The challenge here is to produce a useful amount of short-pulse X-ray energy in affordable systems.

The other major theme in ultrafast and ultraintense laser-plasma interaction includes the fascinating basic laser-plasma phenomena that arise when laser fields are so strong that relativistic electron effects become important (equivalent to $I\lambda^2$ values exceeding 10^{18} Watts/cm^2-micron^2). In underdense plasmas formed from gaseous targets, there is ongoing work on wave harmonic production by relativistic effects, self-focusing and strongly coupled stimulated Raman scattering.

2. Short pulse X-rays.

Powerful short pulse lasers have been available for some time[1], but the resulting X-ray pulses had not been shown to be in the picosecond range. Then Murnane et al. [2]

Springer Series in Chemical Physics, Vol. 55 Ultrafast Phenomena VIII
Editors: J.-L. Martin · A. Migus · G.A. Mourou · A.H. Zewail

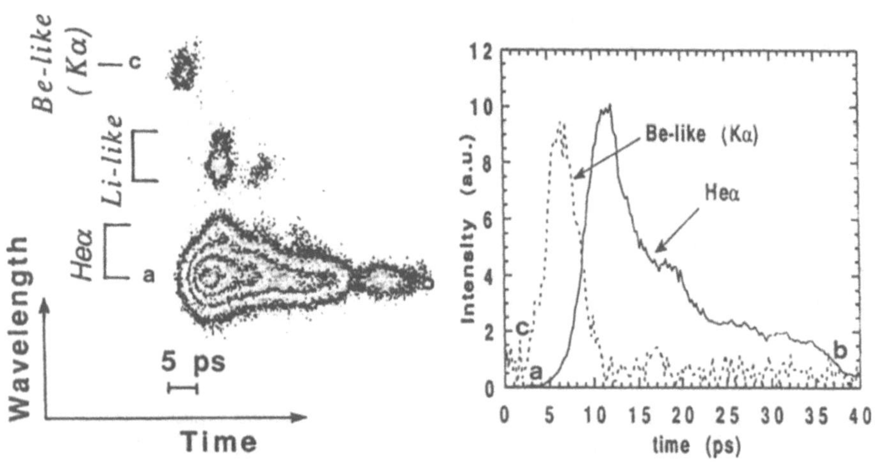

Fig. 1 An example[8] of improved time resolution for an experiment shot, with irradiating laser wavelength of 1.06μ.

demonstrated (using their improvements in X-ray streak camera technology) that X-ray pulses likely as short as 1.5 psec could indeed be obtained with subpicosecond laser pulses. However, from proof-of-principle to a useful instrument much remains to be done[4].

Given a suitable laser pulse source, the chief requirement for production of a short X-ray pulse is to keep the density and its gradient very high, so that the X-ray emission will drop rapidly due to thermal conduction inward and rapid expansion outward. Either prepulse (premature leakage of energy at very low relative intensities over a nanosecond or so) or excess fluence[5] (energy/area over the laser pulse time) produce extended plasmas which cool too slowly for rapid X-ray pulse termination. Contrast values (prepulse intensity / pulse intensity) of better than 10^{-6} or so are necessary and attainable[6]. Apart from this density control, the target must not become over-ionized, since the excessively ionized ions (i.e. more than He-like) can recombine into He-like ions and radiate during the cooling phase. At excessive intensity, anomalously hot electrons will aggravate the unfavorable effects of ablation on X-ray pulse duration. (The same issues arise - hot electrons and strong ponderomotive force effects - that were closely investigated when long-wavelength CO_2 laser light was still in contention as a laser fusion driver[7].)

All this entails modelling improvements that require experimental testing of the basic models, hence the need for the very best and fastest experimental diagnostics. Recently Kieffer et al.[8] (See Fig.1) have applied the techniques previously used for the *overall* X-ray emission[9] to obtain picosecond X-ray *spectra*, which will allow much improved diagnostics of the target dynamics. Other diagnostic techniques discussed at this Conference[10] will, one hopes, also play important roles in this effort.

The need for these diagnostics is obvious when one notes that, while the modelling[2] of the target physics at the "modest" intensities of 10^{14} watts/cm^2 (or less) is "well understood", this is only so *provided* the laser absorption is *legislated* to agree with experiment. This really means the early absorption is poorly understood, being badly underestimated if calculated with the available physics. (Though such *ad hoc* recipes may suffice for soft X-ray production, they will certainly not address recent results[11] on very energetic (MeV) X-ray emissions from solid targets.) Finally, in order to predict the X-ray yield in detail with steep gradients, it may be necessary to do electron kinetic modelling (perhaps even in two dimensions for small spot sizes[12]), of the kind required in the recent X-ray line polarization measurements which demonstrated electron distribution function anisotropy in short-pulse interactions[13].

3. Laser-plasma interactions at relativistic intensities.

The various effects to be expected as laser intensities become such as to induce relativistic effects in electrons have long been confined to the ideal world of theory and simulation, and only recenly have experiments become possible on such diverse plasma phenomena as (to date) coherent odd-harmonic production, self-focusing and the plasma version of stimulated Raman scattering (SRS). Since the phenomena of interest are for sub-critical plasma densities (densities usually rather less than the critical value of about $10^{21}\lambda^{-2}$ electrons/cm^3/micron^{-2}), the experiments are usually done using gas targets.

The harmonic emission from individual atom effects have already been well demonstrated[14], but the real interest is in the free electron incoherent emission and collective plasma effects on coherent harmonic emission. High harmonic generation from plasma effects at very high intensity has already been seen, using solid targets and very powerful CO_2 laser beams[15]. (Harmonics (and half-harmonics) were seen up to the 46th for intensities greater than 10^{15} W/cm^2, 10µ wavelength, near-nanosecond pulse durations. The densities involved, however, were believed to be in the neighbourhood of the critical density for the harmonics measured.) While calculations have been done on both incoherent[16] and coherent[17] odd-harmonic generation, experimental plasma or free electron results from gas targets have yet to appear in the refereed literature[18].

The usual self-focusing due to molecular χ_3 that has been analyzed for many years is well known to behave differently for very short pulse lengths[19], but the phenomenon that is of interest here is a self-focusing effect involving free electrons in relativistically strong laser fields with large charge density differences between electrons and ions[20]. This analysis in effect assumes that ions do not move (i.e. infinite mass) while electrons are in equilibrium (zero mass), however realistic mass ratio effects should be significant. Results have recently appeared[21] that purport to be due to such relativistic electron focusing. One point to note in this phenomenon is that the electrons are essentially excluded from the regions where the light is focused most intensely, so the strong field effects must therefore be manifested by interaction with the ions remaining in the concentrated beam or by other interactions as the self-focused beam impinges on a solid target as a novel type of intense beam delivery system.

Very recently a short-pulse version of the now classic laser-plasma SRS instability has been seen experimentally[22], but with a novel frequency signature whose form differs dramatically (being actually spread to the blue side of the incident laser wavelength) at the highest intensities from the classic SRS signature (downshift by about the plasma frequency). This result has been explained as being due to the fact that the pulse is so short that the entire pulse is well inside the approriate density region, and the SRS light exits through an interaction boundary which is moving with the laser pulse at nearly light velocity. For strongly driven instabilities this produces a previously unexpected frequency change along the lines of that both observed and explained[23] in connection with the microwave frequency upshift produced via rapidly moving ionization fronts.

Whether by direct beam focusing or with assistance from self-focusing, relativistic-intensity beams on solid targets raise the possibility of moving a solid density boundary essentially by radiation pressure rather than by the normal ablation process. (However, in a rapidly-rising pulse over a large area, an electromagnetic version of the Rayleigh-Taylor instability would probably obtain.) Such a process could apparently be of use in some laser fusion concepts (as mentioned by Mourou and Umstadter[1]). While fast electron production would be a sure sign of high intensity, experimental evidence for any real radiation pressure effects would probably have to be quite indirect, failing a new diagnostic concept.

4. References.

[1] As summarized in G. Mourou and D. Umstadter, Physics of Fluids B, (to be published in 1992), (text of their invited paper at 1991 Meeting of the Division of Plasma Physics of the American Plasma Physics Society), and G. Mourou, paper TuA1, this conference. P. Maine, D. Strickland, P. Bado M. Pessot and G. Mourou, IEEE J. Quant. El. **24**, 398 (1988).

[2] As summarized in M.M. Murnane, H.C. Kapteyn, M.D. Rosen and R.W. Falcone, "Ultrafast X-ray Pulses from Laser-produced Plasmas", Science, **251**,531-536, (1991). See also: F. Brunel, Phys. Rev. Lett.**59**, 52 (1987). H. Milchberg, R.R. Freeman, S.C. Davey and R. More, Phys. Rev. Lett. **61**, 2364 (1988). J.C. Kieffer, P. Audebert, M. Chaker, J.P. Matte, H. Pépin, T.W. Johnston, P. Maine, D. Meyerhofer, J. Delettrez, D. Strickland, P. Bado and G. Mourou, Phys. Rev. Lett. **62**(7), 760 (1989), R.Fedosejevs, R.Ottmann, R.Sigel, G. Kühnle, S. Szatmári and F.P. Schäfer, Phys. Rev. Lett. **64**(11) 1250 (1990). P. Gibbon and A.R. Bell, Phys. Rev. Lett. **68**, 1535 (1992). A. Zigler, P.G. Burkhalter, D.J. Nagel, M.D. Rosen, K. Boyer, G. Gibson, T.S. Luk, A. McPherson and C.K. Rhodes, Appl. Phys. Lett. **59**(5), 534-536 (1991).

[3] For instance, see J.P. Bergsma, M.H. Coladonato, P.M. Eldston, J.D. Khan and K.R. Wilson, J. Chem. Phys. **84**, 6151 (1986).

[4] Paper TuE3 at this Conference: by J.P. Geindre, P. Audebert, J.C. Gauthier,A. Mysyrowicz, G. Grillon, J.P. Chambaret and A. Antonetti.

[5] H. Milchberg, I. Lyubomirsky and C.G. Durfee III, Phys. Rev. Lett. **67** 2654 (1991).

[6] Y. Beaudoin, C.Y. Chien, S. Coe, J.L. Tapié and G. Mourou, Optics Lett. (to be published). J.L. Tapié and G. Mourou, Optics Lett. **17**, 136 (1992). M.D. Perry, F.G. Patterson and J. Weston, Optics Lett. **15**,381 (1990). K. Yamanaka, C.P Barty, H. Shiraga and Y. Kato, IEEE J. Quantum Electron. **QE-27**, 288 (1991).

[7] N.H. Burnett and G.D. Enright, Can. J. Phys. **64**(8) 920-931 (1986),S.J. Gitomer, R.D. Jones, F. Begay, A.W. Ehler, J.F. Kephart and R. Kristal, Phys. Fluids **29**(8) 2679-2688 (1986). But see also D. Umstadter, X.Liu, J.S. Coe and C.Y. Chien, in *Proceedings on Short Wavelength Coherent Radiation*, Eds. H. Bucksbaum and N. Ceglio, Optical Society of America, Vol. 11, 276-280 (1991)

[8] J.C. Kieffer, Y. Beaudoin, C.Y. Côté, M. Chaker, H. Pépin, C.Y. Chien, S. Coe and G. Mourou in *Proceedings of the Third International Colloquium on X-ray Lasers*, Schliersee, 1992, (to be published) IOP.

[9] M.M. Murnane, H.C. Kapteyn, and R.W. Falcone, Phys.Rev. Lett. **62**,155 (1989).

[10] M.H. Sher, U. Mohideen, H.W.K. Tom, O.R. Wood II,G.D. Aumiller, T.J. McIlrath, J. Bokor and R.R. Freeman, Paper TuE4 at this conference.

[11] J.D.Kmetec, C.L. Gordon III, J.J. Macklin, B.E. Lemoff, G.S. Brown and S.E. Harris, Phys. Rev. Lett. **68**(10) 1527 (1992).

[12] G.J.Rickard, A.R. Bell and E.M. Epperlein, Phys. Rev. Lett. **62**(3), 2687 (1989).

[13] J.C. Kieffer, J.P. Matte, H. Pépin, M.Chaker, Y. Beaudoin, T.W. Johnston, C.Y. Chien, S. Coe, G. Mourou and J. Dubau, Phys. Rev. Lett. **68**(4) 480 (1992).

[14] S. Augst, C.L. Moore, J. Peatross, D.D. Meyerhofer, *Proceedings on Short Wavelength Coherent Radiation*, Eds. H. Bucksbaum and N. Ceglio, Optical Society of America, Vol. 11, 23-27 (1991), X.F. Li, A. L'Huillier, M. Ferray, L.A. Lampré and G. Mainfray, Phys. Rev. A **39**, 5751-5761 (1989).

[15] R.L. Carman, C.K. Rhodes, R.F. Benjamin Phys. Rev. A **24**, 2649 (1981).
[16] E.S.Sarachik and G.T. Schappert, Phys. Rev. D **1**,2738 2753 (1970), C.I. Castillo-Herrera and T.W. Johnston (submitted to IEEE Trans. Plasma Sci., for special issue February 1993).
[17] E. Esarey, P. Sprangle, J. Krall and G. Joyce, paper TuA4 at this Conference. See also: P. Sprangle, E. Esarey, and A. Ting, Phys. Rev. A, **41**(8) 4463-4469 (1990), P. Sprangle, E. Esarey, and A. Ting, Phys. Rev. Lett.**64**(17) 2011-2014 (1990). For back-scattered odd-harmonic light, see E.Esarey and P. Sprangle, Phys. Rev. A, **45**(8), 5872-5882 (1992). As yet unpublished but available as Princeton Plasma Physics Laboratory (PPPL) Reports: J.M. Rax and N.J. Fisch, PPPL-2817 (1992), J.M. Rax and N.J. Fisch, PPPL-2836 (1992) and J.M. Rax, PPPL-2838 (1992).
[18] X. Liu and D. Umstadter, (submitted IEEE Trans. Plasma Sci. special issue Feb. 1993).
[19] P.B Corkum, C. Rolland and T. Srinivasan-Rao, Phys. Rev. Lett. **57**(18) 2268-71 (1986),P.B Corkum and C. Rolland IEEE J. Quant. Electron. **25**(12) 2634-2639 (1989).
[20] G.Z.Sun, E. Ott, Y.C. Lee, and P. Guzdar, Phys. Fluids, **30**(2) 526-532 (1987), A.B. Borisov, A.V. Boriskey, V.V. Khorobkhin, A.M. Prokhorov, C.K. Rhodes and O.B. Shiryaev, Phys. Rev. Lett. **65**(14) 1753-6 (1990), A.B. Borisov, A.V. Boriskey, O.B. Shiryaev, V.V. Khorobkhin, A.M. Prokhorov, J.C. Solem, T.S. Luk, K. Boyer and C.K. Rhodes, Phys. Rev. A **45**(8) 5830-5845 (1992).
[21] A.B. Borisov, A.V. Boriskey, V.V. Khorobkhin, A.M. Prokhorov, O.B. Shiryaev, X. M. Shi, T.S. Luk, A. McPherson, J.C. Solem, K. Boyer and C.K. Rhodes, **68**(15) 2309-2312 (1992), also X. Liu and D. Umstadter (submitted for publication).
[22] C.B. Darrow, C. Coverdale, M.D. Perry, W.B. Mori, C. Clayton, K. Marsh and C. Joshi, (submitted for publication, 1992).
[23] R.L. Savage Jr., W.B. Mori, R.P Brogle and C. Joshi, paper TuA5 at this Conference. See also S.C. Wilks, J.M. Dawson and W.B. Mori Phys. Rev. Lett. **61**(3) 337-340 (1988), C.J. Joshi, C.E. Clayton, K. Marsh, D.B. Hopkins, A. Sessler and D. Whittum IEEE Trans. Plasma Sci. **PS-18**(5) 814-818 (1990), R.L. Savage Jr., C. Joshi and W.B.Mori, Phys. Rev. Lett. **68**(7) 946-9 (1992), W.B. Mori Phys. Rev. A **44**(8) 5118-5121 (1991).

X-Rays Generated by Femtosecond Laser-Produced Plasmas

J.P. Geindre[1], *P. P. Audebert*[1], *A. Rousse*[1], *F. Falliès*[1], *J.C. Gauthier*[1], *A. Mysyrowicz*[2], *G. Grillon*[2], *J.P. Chambaret*[2], *A. Antonetti*[2], *A. Mens*[3], *R. Verrecchia*[3], *R. Sauneuf*[3], *and P. Schirman*[3]

[1]Laboratoire PMI, Ecole Polytechnique, F-91128 Palaiseau, France
[2]Laboratoire d'Optique Appliquée, ENSTA-X,
Batterie de l'Yvette, F-91120 Palaiseau, France
[3]CEN Limeil-Valenton, F-94195 Villeneuve-St.-Georges, France

Abstract. We review some recent experiments that we have performed at the LOA on femtosecond laser-matter interaction with aluminum and silicon targets. Reflection and scattering of the incident laser and time-resolved Schlieren images have been correlated with x-ray diode measurements as a function of the ASE pedestal fluence. X-ray spectroscopy has been used to measure the plasma parameters. Hot electron target heating has been studied. Supra-kilovolt x-rays probed by a 2 ps resolution streak camera exhibit a sub-picosecond resolution-limited pulse duration.

1. Introduction and experimental system

With high intensity femtosecond lasers it is possible to create plasmas with near solid densities. Such plasmas can bring valuable informations on the dynamics of dense hot solids. They appear also very promising as a source of sub-picosecond x-ray pulses. The availability of such short duration x-rays can have a significant impact in the field of plasma and solid state physics.

Our dye laser system (CPM oscillator with 5 amplifier stages) delivers pulses of 80 fs duration at 620 nm with a maximum irradiance of 10^{17} W/cm^2 on target. This short pulse is accompanied by a 3 ns pedestal of amplified spontaneous emission (ASE) containing less than 5% of the total pulse energy. This provides an intensity contrast ratio of 10^{-7} on target. We monitor the ASE energy and the temporal position of the main pulse relative to the pedestal with the help of a fast diode and a 7 Ghz oscilloscope. In the present experiments, the short pulse laser intensity was limited to 10^{16} W/cm^2 and the laser beam incident angle was 7° to target normal. Specularly reflected light and scattering from the target were monitored by visible diodes to infer the absorption. Scattering was measured over the angular range 10 to 70° from target normal by a wide angle lens coupled to an imaging CCD camera. An XRD diode filtered for energies above 1 keV monitored the plasma x-ray emission. An independent laser beam of 30 fs duration at 590 nm was used as a probe beam for plasma shadow diagnostics. A Schlieren technique using an imaging lens and a knife edge [1] allowed to study the expansion dynamics of the plasma. Time-integrated x-ray spectra in the 1.5-2 keV energy range were obtained from a high-resolution spectrograph fitted with two curved PET crystals in the Von Hamos configuration. A 2ps resolution fast streak camera was used to measure the pulse duration of the x-rays at energies above 1 keV.

2. Experimental results

Figure 1 shows the results of optical diode measurements of the specularly reflected laser light (Fig.1a), the x-ray signal for energies above 1 keV (Fig.1b) and the total scattering signal in the 10 to 70 degrees angular range from target normal (Fig.1c) for a 1 μm aluminum layer deposited on a glass substrate as a function of the ASE fluence.

Below 0.2 J/cm^2, the reflection is close to 90% and the scattering and x-ray signals vanish. By increasing the ASE fluence to 0.7 J/cm^2, the reflection signal drops sharply, the

Springer Series in Chemical Physics, Vol. 55 Ultrafast Phenomena VIII
Editors: J.-L. Martin · A. Migus · G.A. Mourou · A.H. Zewail

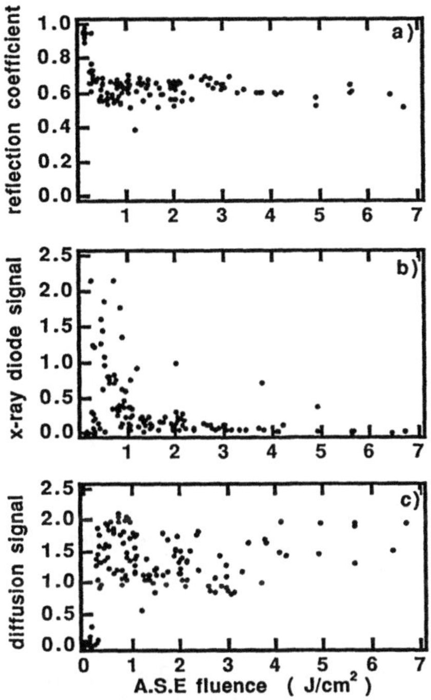

Figure 1: Reflection (a), x-ray signal (b) and scattering signal (c) for a 1μm aluminum layer as a function of ASE fluence. The 80fs laser intensity is 10^{16} W/cm^2.

scattering signal appears suddenly and the x-ray signal presents a pronounced maximum. At higher ASE fluences, the specular reflectivity and the angularly-integrated scattering are almost constant and the x-ray signal decreases by an order of magnitude. Angularly-resolved measurements show a strong decrease of the scattering signal at large angles similar in shape to the x-ray signal.

The measured spatial distributions of reflectivity and scattering accross the focal spot show clearly an increase of the roughness of the target surface, revealed by a sharp maximum of large angle scattering and a smearing-out of the residual reflectivity over twice the original focal diameter for fluences below 1 J/cm^2. This is followed at larger fluences by the formation of a pre-plasma at the center of the focal spot which reflects efficiently the incoming laser over a diameter much smaller than the original focal spot size. At the same time, large angle scattering is greatly reduced. Time-resolved Schlieren results clearly show the onset of ASE-induced overdense pre-plasma formation at fluence levels above 3 J/cm^2. Previously reported laser-induced deformation of reflective surfaces and studies of the dynamics of femtosecond laser surface melting [2,3] are consistent with the present observations.

The laser energy penetration depth have been studied by varying the aluminum thickness and was found to be of the order of 1000 Å, significantly larger than the skin depth associated with the laser. A detailed analysis of Kα line emission from different ion stages shows that hot electrons created during the interaction heat the bulk of the target to temperatures of 15-20 eV. Relating the intensity of the Kα silicon line of the target substrate to the hot electron deposited energy, we found hot electron energies of about 5-10 keV. The width of the aluminum Heα line varied by more than a factor of 2 when the laser intensity was changed from 10^{16} to 10^{17} W/cm^2. Taking into account instrumental and opacity broadening, an electron density of 10^{23} cm^{-3} was inferred from the data at 10^{17} W/cm^2. The temperature deduced from the satellite to resonance line ratio was 400±100 eV. The pulse width of x-ray emission (hν>1 keV) from aluminum and copper targets was measured at the optimum ASE fluence and at 10^{16} W/cm^2

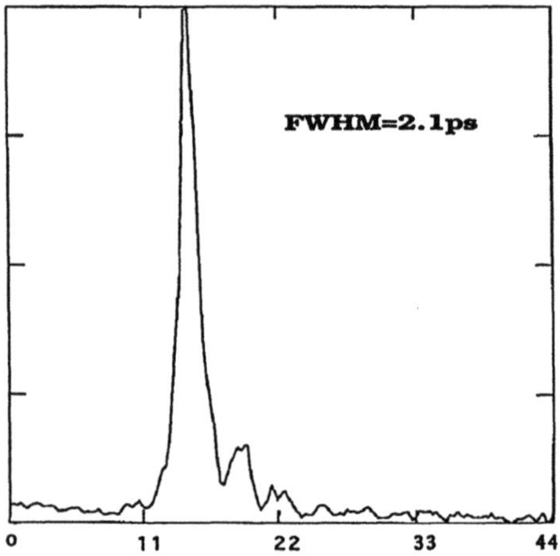

Figure 2: Time-resolved x-rays above 1 keV from a Cu target. Streak camera resolution 2ps. Laser intensity $10^{16} W/cm^2$.

laser intensity using an ultrafast streak camera [5]. Results are shown in Fig.2 for copper. The FWHM of the pulse is about 2.1 ps, a value which is comparable to the UV measured streak camera resolution. XRAD radiative hydrodynamic calculations [4] predict a pulse duration of 300-400 fs for photons energies around 1.2 keV (L-band), in qualitative agreement with these observations.

In conclusion, our results show that the x-ray signal is maximized for an ASE fluence close to the damage threshold of the target material, well below pre-plasma formation. Target conditioning by the ASE prepulse does not prevent target heating above critical density. An ultrashort x-ray pulse is produced at energies above 1 keV. These results suggest that ASE-produced target roughness may be efficient in ultrafast laser absorption.

[1] R. Benattar, J.P. Geindre, P. Audebert, J.C. Gauthier, A. Mysyrowicz, J.P. Chambaret and A. Antonetti, Optics Communications **88**, 376 (1982).
[2] C. Koo and R.E Slusher, Appl. Phys. Lett. **28**,614 (1976).
[3] M.C. Downer, R.L. Fork and C.V. Shank, J. Opt. Soc. Am. **B2**, 595 (1985).
[4] J.C. Gauthier and J.P. Geindre, PMI Reports n° 1971 and 1974, (1988).
[5] A. Mens, R. Verrecchia, R. Sauneuf and P.Schirman, 19th International Congress on High Speed Photography and Photonics (Cambridge 1990).

K-Shell Emission from 100 fs Laser-Produced Plasmas Created from Porous Aluminum Targets*

R. Shepherd[1], *D. Price*[1], *B. White*[1], *S. Gordan*[2], *A. Osterheld*[1], *R. Walling*[1], *D. Slaughter*[1], *and R. Stewart*[1]

[1]Lawrence Livermore National Laboratory, P.O. Box 808, Livermore, CA 94550, USA

[2]Department of Physics, University of California, Berkeley, CA 94720, USA

1. Introduction

The interaction of ultra-short pulse (≤ 1 ps) lasers with solid matter are currently under investigation at numerous laboratories around the world [1-3]. The motivation for many of these studies is the production of near solid density, high temperature plasmas. The plasmas are produced by rapid absorption of the laser before hydrodynamic expansion. The enthusiasm for high energy-density matter using this technique has been tempered by the difficulty of coupling high intensity laser light into solid surfaces. Numerous theoretical and experimental papers have addressed ultra-short pulse absorption, heating, and x-ray emission from flat solid targets at high intensities ($I \geq 10^{16}$ W/cm^2)[4-7]. Additionally, work has been reported on the increased hard X-ray (> 1keV) yield in porous gold [8]. We report the first study of the K-shell emission of ultra-short pulse heated porous aluminum . The intensity dependent K-shell emission is used to determine the density and temperature of the plasma while the specular and non-specular reflected light are monitored to infer the absorption. A qualitative comparison of the K-shell emission indicates a much high temperature in the porous aluminum than that of the flat targets.

2. 0 The Experiment

The porous aluminum consists of clusters of aluminum particles (≈ 1 μm in diameter) , with typically 200 Å between particles. The density is ≈ 1% solid density, with the voids assumed to remain unfilled during laser irradiation.

The laser used in this study is a 150 femtosecond (FWHM) Ti:Sapphire laser capable of producing 50 mJ focused to a 5 μm spot at normal incidence. To minimize the effects of pre-pulse, an IR 140 saturable absorber was added after the last power amplifier. The addition of the saturable absorber limited our peak intensity to ≈ 10^{18} W/cm^2.

The aluminum K-shell data was collected with Von Hamos crystal spectrograph. The spectrograph uses a 7.62 x 1.9 cm potassium acid phthalate (KAP, 2d=26.623 Å) 5 cm cylindrically curved crystal . The detector resolving power is $E/\Delta E \approx 2000$. The specular and non-specular reflected light was monitored using filtered silicon PIN diodes. The angular position of the non-specular diodes could be varied between 20° and 80° in the plane of incidence of the laser beam. The non-specular 1 ω, 3/2 ω, and 2ω light was monitored with a Czerny-Turner visible spectrometer viewing the plasma at 60° above the target normal.

* Work performed under the auspices of the U. S. Department of Energy by Lawrence Livermore National Laboratory under contract number W-7405-ENG-48.

2.1 Data and Analysis

The temperature was determined by minimizing the difference between the experimentally measured spectrum and a synthetically generated spectrum using the quasi-steady state equilibrium model [9]. A plot of the temperature as a function of laser intensity is displayed in figure 1. The temperature is noted as having a weak dependence on laser intensity. The electron density was inferred using Stark broadening of the $1s4p\text{-}1s^2$ line . By comparing the measured line width to the calculated line width [10], the density was estimated and plotted as a function of laser intensity. A qualitative comparison of the relative thermal temperature achieved in the porous aluminum versus that of the flat aluminum indicated that under similar irradiance (4×10^{17} W/cm^2), the $1s2p\text{-}1s^2$ emission was a factor of four greater in the porous aluminum with a factor of ten fewer shots. Furthermore, no emission from Rydberg states higher than the $1s2p\text{-}1s^2$ was seen in the flat aluminum. Estimates of continuum lowering due to higher densities could not account for the lack of emission from the higher Rydberg states in the flat aluminum.

The data collected with the visible diodes indicates the relative amount of specular and non-specular reflected light was noticeably higher when shooting the flat targets. The visible spectrometer indicates 4-10 times as much 2ω light was generated when the laser interacts with the flat targets as opposed to the porous targets. A small amount of 1ω was seen with flat target experiments as compared to no detectable amounts with the porous targets. The $3/2\ \omega$ from both porous and flat targets was negligible when the saturable absorber was used .

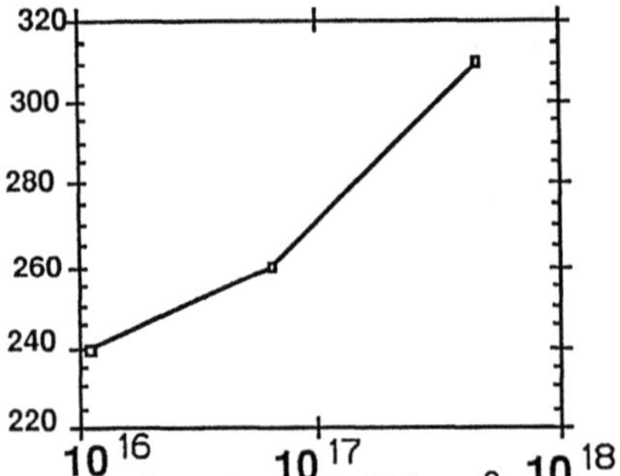

Figure 1
Porous Aluminum Electron Temperature

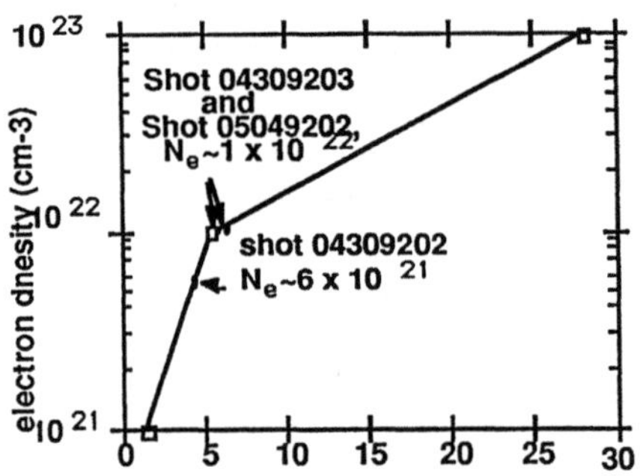

Figure 2
Porous Aluminum Electron Density

3. Conclusions

We have done the first measurement of the temperature and density of porous targets using spectroscopic techniques. The measurements indicate temperatures between 240 eV and 310 eV (depending on irradiance) and a density of $\approx 10^{22}$ e/cm^3. Work is currently being done to increase the density of the targets.

4. References

1. J.A Cobble, et al., Phys. Rev. A., **39**, 454 (1989).
2. M. M. Murname, H. Kapteyn, and R. Falcone, Phys. Rev. Lett. **62**, **155** (1989).
3. A. Zigler, P. G. Burkhalter, D. A. Newman, D. J. Nagel, T. S. Luk, A. McPherson , C. K. Rhodes, and M, D. Rosen, Bull. Am. Phys. Soc. **35**, 1943 (1990).
4. P. Gibson, and A.R. Bell, Phy. Rev. Lett., **68**, 1535 (1992).
5. S. C. Rae, and K. Burnett, Phy. Rev. A, **44**, 3835, (1991).
6. J. C. Kieffer, et. Al., Phys. Rev. Lett., **62**, 760 (1989).
7. H.M. Milchberg, R.R. Freeman, Opt. Soc. Am. B., **6**, 1351,(1989).
8. M. Murname, et. al., in *Short-Wavelength Coherent Radiation: Generation and Applications*, P. Bucksbaum and N. Ceglio editors, O.S.A,(1991).
9. V.A. Boiko, et. Al., Soviet J. Plasma Phys., **10**, 82 (1984).
10. Calculations were done with RATION, a spectral line shape code provided by Dick Lee at Lawrence LIvermore National Laboratory.

Kilovolt X-Ray Emission from Femtosecond Laser-Produced Plasmas

G. Jenke[1], *H. Schüler*[1], *T. Engers*[1], *D. von der Linde*[1], *I. Uschmann*[2], *E. Förster*[2], *and K. Gäbel*[2]

[1]Institut für Laser- und Plasmaphysik, Universität (GHS) Essen, W-4300 Essen, Fed. Rep. of Germany

[2]Institut für Quantenelektronik der Friedrich-Schiller-Universität Jena, Max-Wien-Platz 1, O-6900 Jena, Fed. Rep. of Germany

Abstract. Strong K_α line emission from plasmas generated with amplified dye laser pulses of 100 fs and up to 10^{17} W/cm^2 focussed on solid aluminium targets is observed. Photon numbers of approximately 5×10^7 from a smooth aluminium surface are measured with calibrated X-ray photographic films and sets of Al-coated mylar foils. The X-ray emission spectra in the regime of 1 keV is observed with thin bent mica crystall in a "von Hamos" configuration.

It is now well established that microplasmas with very interesting properties can be generated by tightly focussing femtosecond laser pulses on the surface of solid targets. One of the most exiting aspects of these plasmas is that they represent a point source of X-ray emission, possibly with subpicosecond pulse duration. Most of the previous work on X-ray emission from plasmas produced by femtosecond laser pulses was concerned with the soft X-ray regime, roughly around 100 eV. Here we report observations of X-ray emission in the regime above 1 keV.

In our work we used p-polarized amplified dye laser pulses incident at 45 degrees with respect to the surface normal of the targets ($\lambda = 620$ nm, pulse duration approximately 100 fs or less, a few mJ total energy). The pulses were focussed onto the surface of solid targets to a spot diameter of 2 to 4 μm. Peak intensities up to 10^{17} W/cm^2 were attained. Typically 0.1 – 0.2 percent of the total energy is amplified spontaneous emission (ASE) which limits the peak-to-background intensity ratio to approximately 10^7.

Using carefully calibrated X-ray photographic films and sets of Al-coated mylar foils the spectrally integrated X-ray yield (2 nm > λ > 0.4 nm) was studied for a variety of targets and laser conditions. A typical example is the measured number of X-ray photons of approximately 5×10^7 from an aluminium target with a smooth surface for a 3 mJ laser pulse. Targets with rough surfaces gave much higher X-ray yield; a greater amount of ASE also increased the X-ray emission, but apparently in both situations the spectral distribution is quit different.

A cylindrically bent thin mica crystal in a von Hamos configuration with either the (002) or the (004) Bragg reflection was used to measure the X-ray spectra from aluminium samples with smooth surfaces. A typical example is shown in Fig. 1. It can be seen that the spectrum is dominated by line emission. The two strongest lines can be atributed to Al K_α emission ($\lambda_1 = 0.834$ nm) and to He-like Al^{11+} ($\lambda = 0.776$ nm). Our observations are in agreement with the results of Mysyrowicz et al. [1].

We carefully examined the variation of the reflectivity of the mica crystal over the spectral range of interest. A factor of 1.3 was determined for the ratio $R(\lambda_1)/R(\lambda_2)$. Thus Al K_α is indeed the strongest line of the spectrum in Fig. 1.

In a preliminary experiment a quartz crystal with spherical curvature (radius 15.2 cm) was used to collect X-ray photons ($\lambda = 0.668$ nm) from the plasma to focus the emission

Springer Series in Chemical Physics, Vol. 55 Ultrafast Phenomena VIII
Editors: J.-L. Martin · A. Migus · G.A. Mourou · A.H. Zewail

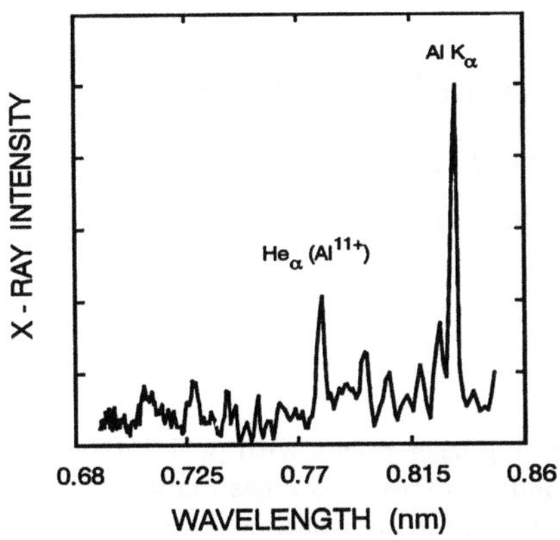

Figure 1: X-ray spectrum from aluminium

on a target (X-ray film). An X-ray spot less than 100 μm in diameter could be produced. Actually a much smaller spot is expected, but in this preliminary experiment the location of the focal plane of the mirror was not precisely known.

Finally, it is worth noting that the measured yield would suggest X-ray peak intensities at the target roughly of the order of 10^{11} to 10^{12} W/cm^2, if the X-ray pulse duration was indeed comparable with the laser pulse duration.

[1] A. Mysyrowicz, J.-P. Chambaret, A. Antonetti, P. Audebert, J.-P. Geindre, J.-C. Gauthier, "Femtosecond X-Ray Emission from Laser Irradiated Al-Targets", Proc. Ultrafast Phenomena in Spectroscopy UPS '91, ed. by A. Lauberau (IOP Adam Hilger Publ., Bristol, 1992)

Ultrafast Spectroscopy of Plasmas Generated by Superintense Femtosecond Laser Pulses

D. von der Linde, H. Schüler, H. Schulz, and T. Engers

Institut für Laser- und Plasmaphysik, Universität (GHS) Essen, W-4300 Essen, Fed. Rep. of Germany

Abstract. Microplasmas are generated by focussing laser pulses with peak intensities up to $10^{17}\,W/cm^2$ on solid materials. Femtosecond time–resolved measurements of the optical reflectivity and the second harmonic provide new information on the ultrafast plasma dynamics.

The interaction of femtosecond laser pulses with solid matter is studied at peak intensities up to $10^{17}\,W/cm^2$. At these high intensities the material is ionized and electrons are heated to high temperatures [1] in a time which is given by the duration of the laser pulse. Because the time is too short for significant transport of mass and energy it is expected that the interaction results in the formation of a thin plasma layer of approximately solid state density which expands rapidly away from the surface.

For a detailed study of the formation and the subsequent expansion of the plasma we have used femtosecond time–resolved pump–probe spectroscopy to measure the optical reflectivity and the generation of the second harmonic (SH). We have characterized the time dependence, the polarization properties, and the frequency spectra of both the fundamental and the SH of the reflected probe pulse.

By means of time–resolved reflectivity measurements we are able to directly monitor the transformation of solid material into a plasma. Figure 1 shows the time–dependence of the optical reflectivity of a plasma generated by a 60 fs pulse ($\lambda = 620$ nm) focussed on a fused silica target ($\approx 10^{16}\,W/cm^2$), measured with p–polarized probe pulses (same wavelength and duration). The pump/probe intensity ratio is greater than 100. Within less than 100 fs the reflectivity rises from $\approx 1\,\%$, which is the optical reflectivity of the unperturbed solid target, to $\approx 20\,\%$, which is consistent with the expected optical reflectivity of a plasma. This result is clear evidence of a peak–to–background intensity ratio of the pump pulse large enough that no preplasma is formed; the pump pulse interacts with cold solid material turning it into a plasma in less than 100 fs.

Second harmonic experiments provide complementary information [2] not available, e.g. from simple reflectivity measurements. The top curve in Fig. 2 shows the p–polarized, specularly reflected SH as a function of the delay time between the pump pulse and the *p–polarized* probe pulse. A distinct temporal variation is observed. At $\Delta t = 0$ SH generation is very weak. A sharp maximum appears at 170 fs, followed by a much broader secondary maximum around 3 ps and a decay in ≈ 10 ps. The inset in Fig. 2 shows the fully resolved sharp peak. This peak consists of a temporal substructure with components comparable in duration with the pulse width.

The bottom curve in Fig. 2 depicts the time–dependence of the reflected SH for probe pulses with *s–polarization*. The polarization of the reflected SH is also p–type. The absense of the sharp SH peak that was seen with p–polarized probe pulses is quite remarkable. The ratio of the SH conversion efficiency for p– and s–polarization was 16.

We believe that the different time–characteristics of the SH for p– and s–polarized probe pulses represent direct evidence of resonance effects (resonant absorption [3]). It

Springer Series in Chemical Physics, Vol. 55 Ultrafast Phenomena VIII
Editors: J.-L. Martin · A. Migus · G.A. Mourou · A.H. Zewail

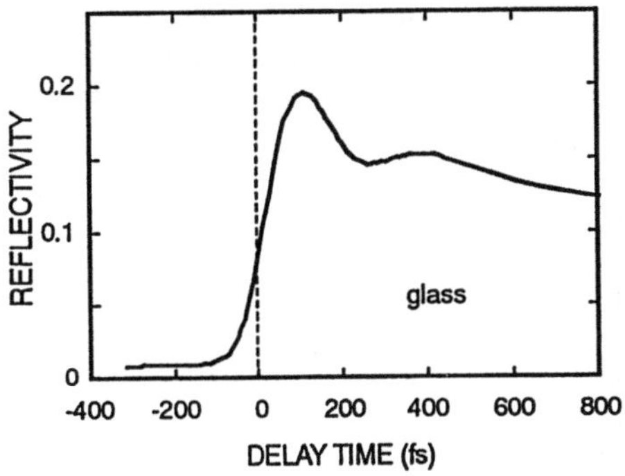

Figure 1: Optical reflectivity of a plasma formed on a glass sample (p–polarized probe pulse incident at 45 degrees).

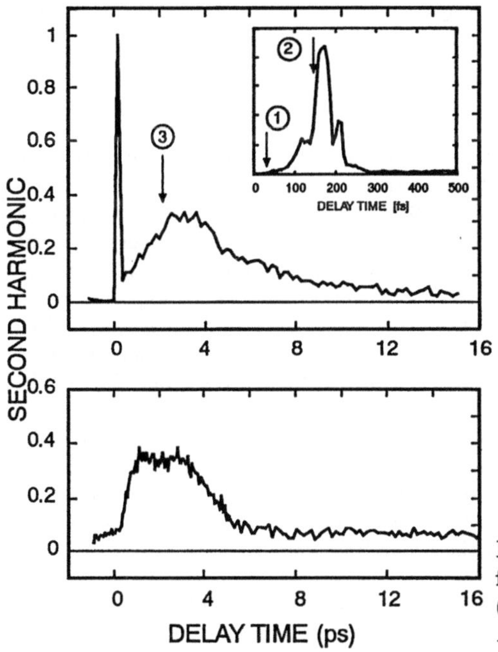

Figure 2: SH from glass vs. delay time for 65 fs p-polarized (top) and s-polarized (bottom) probe pulse.
Arrows: times of spectra in Fig. 3

is well known that in an inhomogeneous plasma longitudinal electron plasma oscillations are excited when the incident light possesses an electric field component parallel to the density gradient, which is the case for p–polarized (but not for s–polarized) light. Under certain conditions a strong resonant enhancement of the electronic current and of the electric field component in the direction of the density gradient can occur at the critical density, where the local plasma frequency is equal to the frequency of the light. It can be shown that during the expansion the plasma passes through some condition of optimum resonance enhancement, which occurs for a plasma thickness (scale length) of about 5 – 10 % of the wavelength of the light. We attribute the observed sharp maximum of SH generation near $\Delta t = 170$ fs to the occurence of such an optimum resonance enhancement condition.

This interpretation is further supported by time–resolved measurements of the frequency spectra of the SH. Figure 3 depicts the SH spectra of the reflected pulse for three different delay times. The delay times of the spectra are marked by arrows in

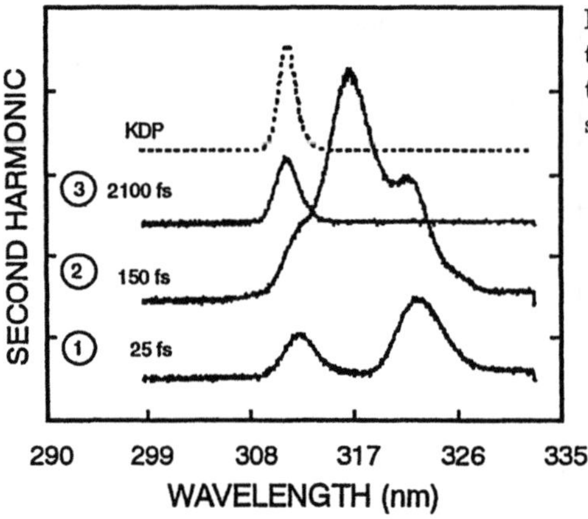

Figure 3: Spectra of the reflected SH of the probe pulse for three different delay times. The dotted curve is a reference spectrum from KDP.

Fig. 2. Also shown is a second harmonic reference spectrum generated in a 100 μm thick KDP crystal. It can be seen that the spectrum measured at a time corresponding to the sharp temporal maximum ($\Delta t = 150\,fs$) possesses distinct structure, being strongly broadened and shifted to longer wavelength. This SH spectrum suggests that pronounced fluctuations of the electronic density occur in this stage of the plasma evolution. Strong perturbations of the electronic distribution are indeed expected when resonant absorption plays a role [4]. On the other hand the spectrum corresponding to the broad temporal feature ($\Delta t = 2.1\,ps$) is very similar to the KDP reference spectrum with the exception of a small blue shift similar to the observed blue shift of the fundamental.

Evidence of ponderomotive interaction is seen when the intensity of the probe pulse is increased ($4.4 \times 10^{14}\,W/cm^2$). We observe a distinct temporal modulation of the optical reflectivity and the SH on a time scale of a few picoseconds. This effect is tentatively attributed to modifications of the plasma density profile caused by ponderomotive interaction of the intense probe pulse with the expanding plasma.

This research was financially supported by the Deutsche Forschungsgemeinschaft.

References

[1] H. M. Milchberg, R. R. Freeman, S. C. Davey and R. M. More, Phys. Rev. Lett. **61**, 2364 (1988)

[2] N. G. Basov, Y. A. Zadharenkov, N. N. Zorev, A. A. Rupasov, G. V. Sklizkov and A. S. Shikanov, *Heating and compression of thermonuclear targets by laser beam*, (Cambridge University Press, 1986), p. 189

[3] N. G. Denisov, Sov. Phys. JETP **4**, 544 (1957)

[4] J. P. Freidberg, R. W. Mitchel, R. L. Morse and L. I. Rudsinski, Phys. Rev. Lett. **28**, 795 (1972)

Picosecond Soft-X-Ray Pulse Length Measurement by Pump–Probe Absorption Spectroscopy

M.H. Sher[1], *U. Mohideen*[1], *H.W.K. Tom*[1], *O.R. Wood II*[1], *G.D. Aumiller*[1], *D.L. Windt*[2], *W.K. Waskiewicz*[2], *J. Sugar*[3], *T.J. McIlrath*[4], and *R.R. Freeman*[1]

[1]AT&T Bell Laboratories, Crawford's Corner Rd., Holmdel, NJ 07733, USA
[2]AT&T Bell Laboratories, 600 Mountain Ave., Murray Hill, NJ 07974, USA
[3]National Institute of Standards and Technology, Gaithersburg, MD, USA
[4]Institute of Phys. Sciences and Technology, University of Maryland, College Park, MD, USA

Abstract. We have demonstrated picosecond time-scale, soft-x-ray continuum, pump-probe absorption spectroscopy. We have measured the temporal profile of a laser-generated x-ray pulse and the absorption spectrum of multiphoton-ionized Kr near 90 eV.

Current developments in high power subpicosecond lasers are making possible a new class of experiments in which matter may be pumped or probed with picosecond or even subpicosecond pulses of laser-plasma-generated x-rays. The application of these x-rays to time resolved absorption spectroscopy involves solving a variety of problems and requires accurate knowledge of the x-ray pulse duration. We have demonstrated a picosecond time scale, soft-x-ray continuum absorption spectroscopy technique; we are using it to measure the temporal profile of laser-generated soft-x-rays and to do absorption spectroscopy of previously unobserved inner shell excitations. The pulse length measurement is based on the rapid but permanent switching of the transmission of an x-ray absorbing medium by multiphoton ionization with a 200 fsec laser pulse. We use Kr gas as the absorbing medium and monitor the transmission on two ~90 eV transitions from the neutral ground state. When the Kr is ionized these absorption features vanish. The change in transmitted energy as a function of the time delay between the ionizing and x-ray pulses is then the integral of the x-ray pulse shape. In principle, the resolution of this technique is limited only by the ionizing pulse duration which can be much better than the current state-of-the-art for x-ray streak cameras [1]. An advantage of this method over the transient Stark shift method reported previously [2] is that it is useful for soft-x-ray pulse lengths ranging from the resolution limit, to the recombination time of the absorbing medium.

The experiment is shown schematically in Fig. 1. Two 815 nm laser beams of 200 fsec pulse duration are produced by a Ti:Sapphire, chirped pulse amplification laser system [3]. The plasma-producing beam, containing approximately 5 mJ, is focused by an f# = 8 lens onto a rotating, electroplated gold target. The plasma formed on the target radiates soft-x-rays. A fraction of these x-rays are collected (f# = 15) by a concave Mo/Si multilayer mirror, focused

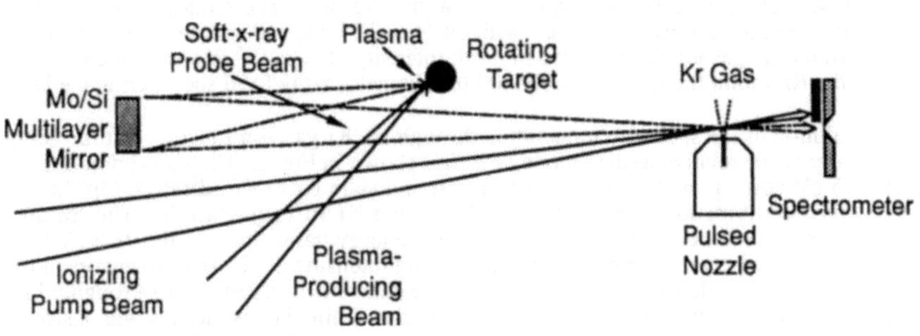

Figure 1. Schematic of experiment.

Springer Series in Chemical Physics, Vol. 55 **Ultrafast Phenomena VIII**
Editors: J.-L. Martin · A. Migus · G.A. Mourou · A.H. Zewail

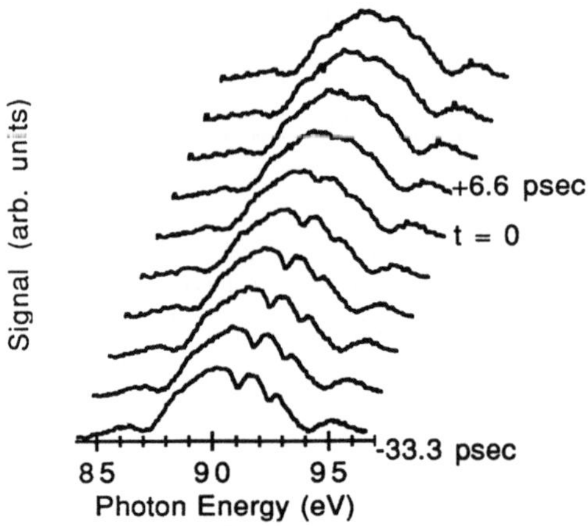

Figure 2. Evolution of the spectrum transmitted through the Kr gas as a function of time delay of the soft-x-ray probe relative to the ionizing beam. Plasma producing intensity is $\sim 1.1 \times 10^{16}$ W cm^{-2}.

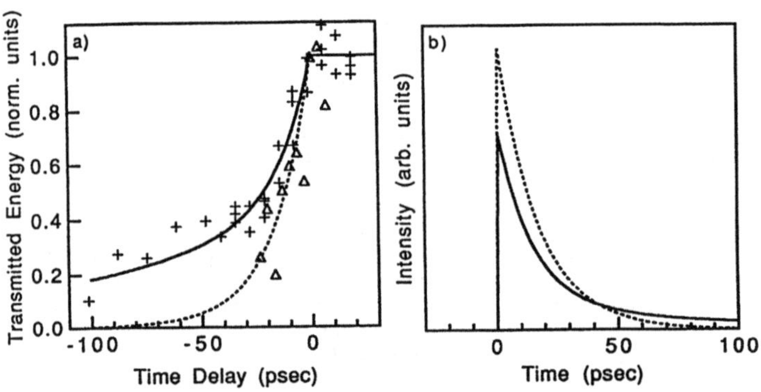

Figure 3. a) Change in transmitted energy on the 91.2 eV line (normalized to the total change for $t > 0$) vs. time delay between ionizing and probe pulses for plasma-producing intensity of $\sim 1.1 \times 10^{16}$ W cm^{-2} (crosses) and $\sim 0.4 \times 10^{16}$ W cm^{-2} (triangles). b) 91.2 eV pulse shapes (normalized to the same total energy content) calculated from the curves in a).

through a pulsed jet of Kr gas, and dispersed in a grazing incidence monochromater. The ionizing beam contains 20 mJ, is focused with an f# = 33 lens, and overlaps the soft-x-ray probe beam in the Kr gas jet. This predicts a peak intensity of $\sim 2 \times 10^{15}$ W cm^{-2} which should ionize all of the Kr within the approximately 100 μm diameter x-ray focal spot during the laser pulse.

The evolution of the spectrum transmitted through the Kr gas as a function of time delay of the soft-x-ray probe relative to the ionizing beam is shown in Fig. 2. The absorption lines at 91.2 and 92.5 eV correspond to transitions from ground state Kr to Kr $3d^9(5/2)4s^24p^65p$ and to an unresolved combination of Kr $3d^9(5/2)4s^24p^66p$ and Kr $3d^9(3/2)4s^24p^65p$. The features are suppressed near $t = 0$. Fig. 3a shows the change in energy transmitted on the 91.2 eV line, as a function of delay at two different plasma-producing intensities. The smooth curves in the figure are exponential or double exponential fits to the data. At 1.1×10^{16} W cm^{-2} roughly half of the 91.2 eV energy decays with a 15 psec time constant, while the rest decays with a slower (~115 psec) tail. At the lower intensity all of the energy seems to decay with an approximately 15 psec time constant, but signal to noise problems are currently preventing an accurate

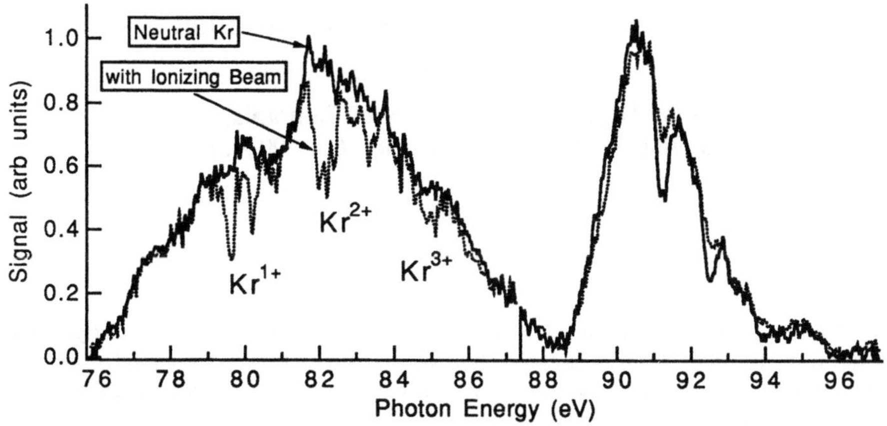

Figure 4. Spectrum of neutral and ionized Kr taken with double stack Mo/Si mirror.

measurement . Fig. 3b shows the soft-x-ray pulse shapes calculated from Fig. 3a. Again, the early time behavior is similar at the two intensities, but the higher intensity pulse includes a slow tail which contains about half the energy.

The reason behind the measured pulse lengths is not currently clear. The fact that this data is spectrally resolved at ~90 eV, while previous measurements [1] have been spectrally integrated up to > 1 keV, is certainly significant, but may not completely account for the observed long decay times. Two potential causes are the uncharacterized target surface and the possibility that the intensity contrast of the laser pulse is not sufficient to allow the 800 nm energy to be deposited at solid density. The effect of these important parameters can now be investigated and optimized.

In order to measure the absorption of 3d excitations in the ion as well as neutral Kr, a double stack Mo/Si multilayer was fabricated to provide reflectivity in two separate wavelength regions. Figure 4 shows the spectra obtained with this mirror for both neutral Kr and with the probe ~30 psec after the ionizing pulse. Upon ionization, the two absorption lines of neutral Kr are almost entirely removed while many new absorption lines appear between 79 and 87 eV. Computer calculations indicate that these lines are transition arrays of the form Kr^{n+} $3d^{10}4s^24p^{6-n}$ -> Kr^{n+} $3d^94s^24p^{7-n}$ for n = 1, 2, and 3.

The present experiment provides spectroscopic data on previously unobserved core excitations in ionized Kr as well as a way to measure soft-x-ray pulse duration to femtosecond resolution. Both results are made possible by the synchronization of short soft-x-ray pulses to high intensity laser pulses. The availability of x-ray mirrors to the carbon K-edge combined with further shortening of the x-ray pulse will allow a variety new atomic and molecular spectra as well as experiments in plasma and perhaps chemical dynamics.

References

[1] M.M. Murnane, H.C. Kapteyn, and R.W. Falcone, Appl. Phys. Lett. **56**, 1948 (1990).

[2] L.D. van Woerkom, R.R. Freeman, S. Davey, W.E. Cooke, and T.J. McIlrath, OSA Topical Meeting: High Energy Density Physics with Subpicosecond Laser Pulses, Snowbird UT Sept. 11-13, 1989, PDP-3.

[3] M.H. Sher, U. Mohideen, H.W.K. Tom, W.H. Knox, O.R. Wood II, J. Bokor, and R.R. Freeman, in preparation.

Photon Acceleration via Laser-Produced Ionization Fronts*

R.L. Savage Jr., R.P. Brogle, W.B. Mori, and C. Joshi

Electrical Engineering Department, University of California,
Los Angeles, 405 Hilgard Avenue, Los Angeles, CA 90024, USA

Abstract. Microwave radiation has been upshifted in frequency and compressed in duration by more than a factor of five via its interaction with a relativistically propagating, underdense ionization front. The experimental observations are in good agreement with theoretical predictions.

1. Introduction

When a pulse of electromagnetic radiation reflects from a moving mirror, its frequency and duration are altered by the relativistic Doppler effect. For an approaching mirror the frequency increases and the pulsewidth decreases by a factor of $(1+\beta)/(1-\beta)$, where β is the velocity of the mirror divided by the speed of light in vacuum, c. It has been proposed that a similar frequency shift and temporal compression would occur upon reflection from a rapidly-propagating ionization front. An ionization front is simply a boundary between a region of neutral gas and a region of plasma. This type of boundary is produced when an ionizing laser pulse propagates through a volume of neutral gas. Unlike a physical mirror, the velocity of a laser-produced ionization front can be very close to c. Thus if incident radiation were reflected from such a front it would be upshifted by many orders of magnitude [1].

In general, electromagnetic radiation of frequency ω_0 will be reflected from a plasma boundary if $\omega_p > \omega_0$ (an overdense plasma) and will be transmitted through the boundary if $\omega_p < \omega_0$ (an underdense plasma). Here ω_p is the characteristic plasma frequency which is given by $\omega_p \approx 2\pi \times 9000\sqrt{n}$, where n is the plasma density in cm^{-3}. In order to determine whether an ionization front will reflect or transmit incident radiation, one must compare the plasma and radiation frequencies in the rest frame of the ionization front. In this frame the radiation frequency is greater by a factor of $[(1+\beta)/(1-\beta)]^{1/2}$ but the plasma frequency, which is Lorentz invariant, is unchanged. Thus extremely large plasma densities are required for a highly relativistic front to be overdense.

However, it has recently been proposed [2] and experimentally confirmed [3] that even when the front is underdense, large frequency upshifts and temporal compressions may be observed. In this case the radiation is transmitted into the plasma, and because the boundary is moving, its frequency is upshifted. For highly relativistic fronts the upshift is given by $\omega_{up} \approx \omega_0(1+\omega_p^2/4\omega_0^2)$.

---

* This work was supported by the U.S. Department of Energy under grant number DE-FG03-91-ER12114.

Springer Series in Chemical Physics, Vol. 55 Ultrafast Phenomena VIII
Editors: J.-L. Martin · A. Migus · G.A. Mourou · A.H. Zewail

2. Experiment

Our experimental arrangement is shown in Fig. 1. Microwave radiation is fed into a straight section of WR-19 waveguide through a 3dB directional coupler. The waveguide is sealed at each end with 0.1 mm-thick quartz windows which, because they are much less than a wavelength in thickness, reflect very little of the microwave radiation. The guide is evacuated and filled with azulene (chosen because it is easily ionized by 266 nm radiation) vapor at pressures of up to a few hundred millitorr. The microwave source is a 35 GHz pulsed magnetron with a peak power of 10 kW and a pulsewidth of 300 ns. A highly relativistic ionization front is created by photo-ionization as a laser pulse (40 mJ, 50 ps, 266 nm) propagates down the azulene-filled waveguide. This front overtakes the slower-traveling microwaves which propagate at $0.4c$ in the WR-19 waveguide. Theory predicts that the frequency shift will be proportional to the plasma density of the front. Also, because the upshifted radiation propagates at a higher velocity (hence the term "photon acceleration"), the overtaken wave is compressed in duration.

Fig. 2 shows the temporal evolution of a typical upshifted pulse recorded using a broadband diode detector and a 1 GHz bandwidth oscilloscope. This pulse was upshifted from 35 GHz to over 174 GHz (measured using a cutoff waveguide) and compressed to less than 0.4 ns in duration . The spectra of the upshifted pulses were measured using a microwave diffraction grating. Fig. 3 shows an upshifted spectrum for an ionization front density of 5×10^{12} cm^{-3} (measured using a microwave interferometer).

Fig. 4 shows the center frequency of the upshifted spectra plotted vs. front density. The error bars represent the widths of the measured spectra (FWHM) and the solid line is the theoretical prediction. Fig. 5 shows the upshifted signal pulsewidths as a function of front density plotted with the theory curve. The points at higher densities are detector-limited to ~ 0.4 ns.

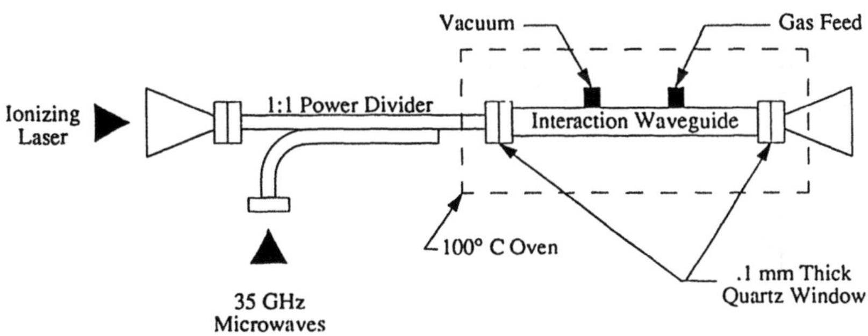

Figure 1: Schematic of experimental set-up.

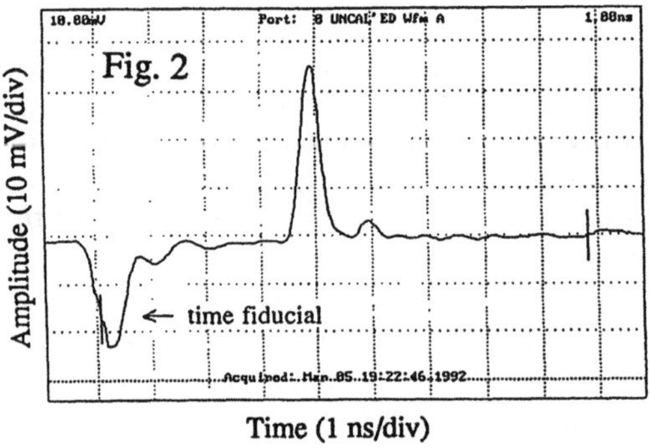

Figure 2: Temporal width of upshifted radiation for a front density of 8×10^{13} cm^{-3}.

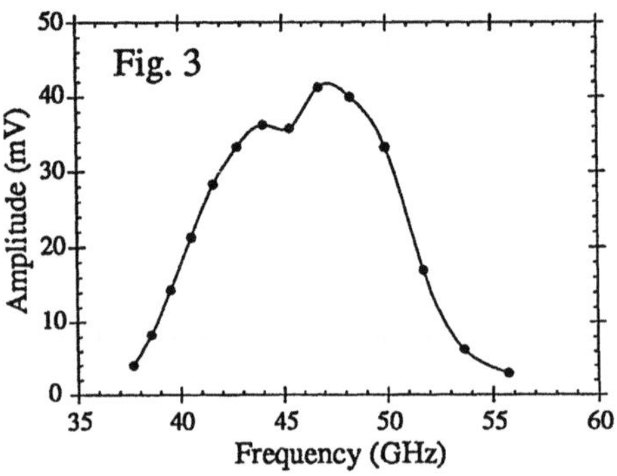

Figure 3: Frequency spectrum of upshifted radiation for a front density of 5×10^{12} cm^{-3}.

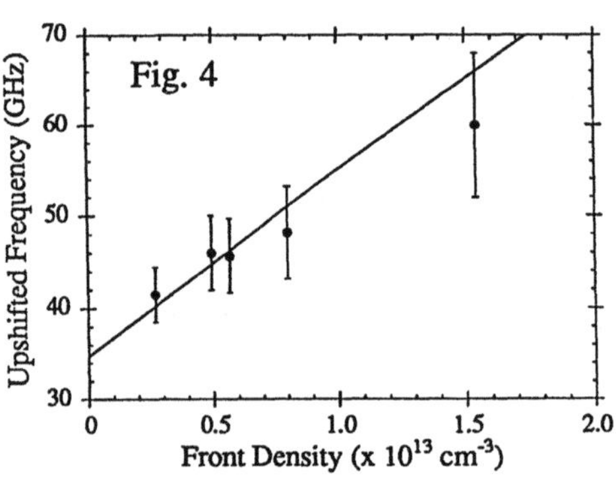

Figure 4: Measured upshifted frequencies vs. front density.

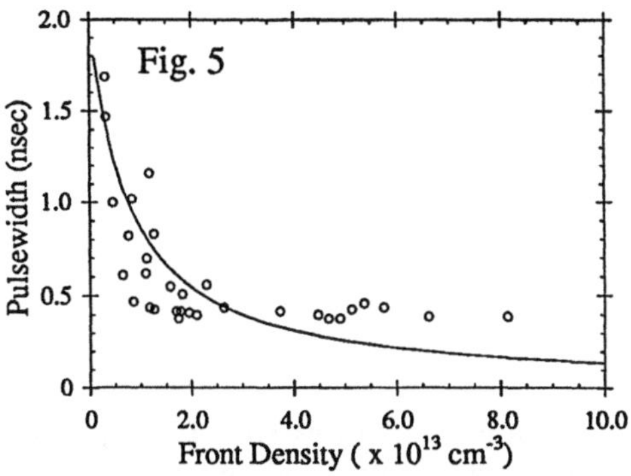

Figure 5: Measured upshifted pulsewidths vs. front density. The solid lines are the theoretical predictions.

3. Conclusions

By sending an intense ionizing laser pulse through a cavity of neutral azulene gas, we created a relativistically propagating, underdense ionization front. When this front impinged upon a train of microwaves, we observed large (up to a factor of five) frequency upshifts and pulse compressions. By varying the neutral gas pressure (and thus the front density), we continuously tuned the frequency and pulsewidth of the upshifted radiation. We expect that broadly tunable sources based on this technique will have many applications, including time-resolved spectroscopy. By utilizing higher frequency source radiation and higher density ionization fronts this technique has the potential to provide a new class of tunable, short pulse, coherent radiation sources operating from the microwave regime to the vacuum ultra-violet.

References

[1] V. I. Semenova, Sov. Radiophys. **10**, 599 (1967) and M. Lampe, E. Ott, and J. H. Walker, Phys. Fluids **21**, 42 (1978).

[2] W.B. Mori, Phys. Rev. A **44**, 5118 (1991).

[3] R. L. Savage Jr., C. Joshi, and W. B. Mori, Phys. Rev. Lett. **68**, 946 (1992).

Propagation of Intense Laser Pulses in Plasmas

E. Esarey, P. Sprangle, J. Krall, and G. Joyce

Beam Physics Branch, Plasma Physics Division,
Naval Research Laboratory, Washington, DC 20375-5000, USA

Abstract. An axisymmetric, relativistic fluid model describing the propagation of intense laser pulses in plasmas has been formulated and numerically evaluated. Relativistic guiding is ineffective in preventing the diffractive spreading of short laser pulses. Laser pulses can be propagated over many Rayleigh lengths by use of a plasma density channel or by tailoring the pulse profile.

The propagation of intense laser pulses in underdense plasmas may have widespread importance in a number of areas, including laser wakefield acceleration [1]-[3]. The recent development of compact terawatt lasers capable of providing ultrashort pulses ($\lesssim$ 1 ps) of ultrahigh intensities ($\gtrsim 10^{18}$ W/cm^2) and moderate energies ($\gtrsim$ 10 J) gives additional impetus to these applications. In vacuum, the focused laser propagation distance is limited to a few Rayleigh lengths, $Z_R = \pi r_{L0}^2/\lambda$, where r_{L0} and λ are the minimum spot size and wavelength of the laser, respectively. In plasmas, nonlinear and relativistic effects associated with intense lasers can significantly modify the propagation characteristics of the laser [1]-[3]. The large differences between the laser wavelength and other characteristic axial lengths in the system, i.e., propagation distance, pulse length and plasma wavelength, make the direct numerical integration of the dynamical equations over extended distances impractical.

A fully nonlinear, relativistic, 2D–axisymmetric laser–plasma propagation model has been formulated and numerically evaluated for ultra–intense laser pulses in plasmas. The formulation has a number of unique features which allow for numerical simulations to be carried out over extended laser propagation distances. The appropriate Maxwell–fluid equations are recast into a convenient form by (i) performing a change of variables from the laboratory frame (r = radial position, z = position along axis of propagation, t = time) to the speed of light frame ($r, \zeta = z - ct, \tau = t$), (ii) applying the quasi–static approximation (neglecting $\partial/\partial\tau$ when determining the plasma fluid response) [2], and (iii) averaging over the fast spatial scale length, i.e., the laser wavelength. The resulting equations are used to study the (i) failure of relativistic focusing for short laser pulses, (ii) optical guiding of tailored laser pulses, and (iii) use of plasma density channels to guide intense laser pulses.

Springer Series in Chemical Physics, Vol. 55 **Ultrafast Phenomena VIII**
Editors: J.-L. Martin · A. Migus · G.A. Mourou · A.H. Zewail

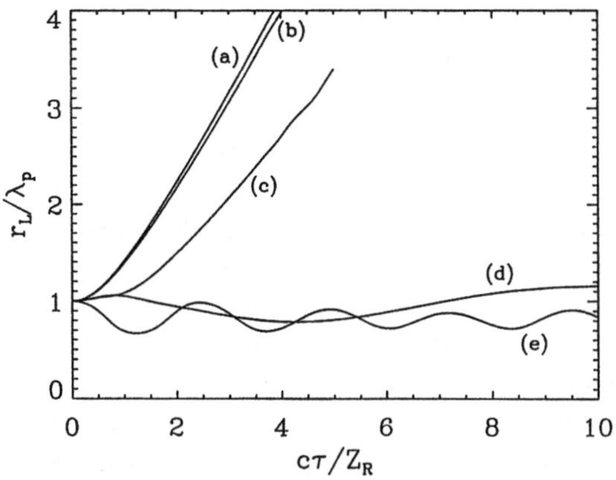

Fig. 1. Normalized laser spot size r_L/λ_p at the pulse center versus normalized propagation distance $c\tau/Z_R$, for (a) vacuum diffraction, (b) an ultrashort pulse with $L=\lambda_p/4$, (c) a short pulse with $L=\lambda_p$, (d) a shaped pulse and (e) a channel-guided pulse.

Reference [2] showed theoretically that relativistic guiding, which requires laser powers $P \geq P_{crit}[\mathrm{GW}] = 17(\lambda_p/\lambda)^2$, does not occur for short pulse lengths $L < \lambda_p$, where $\lambda_p[\mu\mathrm{m}] = 3.3 \times 10^{10}(n_0[\mathrm{cm}^{-3}])^{-1/2}$ is the plasma wavelength and n_0 is the initial plasma density. For short pulses the plasma can not collectively respond to modify the refractive index. Simulation results of short pulse propagation are shown in Fig. 1 for the parameters $\lambda_p = 0.03$ cm, $r_L = \lambda_p$ (Gaussian radial profile), $\lambda = 1$ μm ($Z_R = 28$ cm) and $P = P_{crit}$. The initial axial laser profile is given by $|a(\zeta)| = a_0 \sin(-\pi\zeta/L)$ for $0 < -\zeta < L$, where $a_0 = 0.9$ for the above parameters. The parameter $|a|$ is related to the laser intensity I_0 by $|a| = 8.5 \times 10^{10}\lambda[\mu\mathrm{m}](I_0[\mathrm{W/cm}^2])^{1/2}$. Simulations are performed for two laser pulse lengths, $L = \lambda_p$ (1 ps) and $L = \lambda_p/4$ (0.25 ps). The spot size r_L at the pulse center versus propagation distance $c\tau$ is shown in Fig. 1 for (a) the vacuum diffraction case, (b) the $L = \lambda_p/4$ pulse and (c) the $L = \lambda_p$ pulse. The $L = \lambda_p/4$ pulse diffracts almost as if in vacuum. The $L = \lambda_p$ pulse experiences some initial guiding before diffracting.

A tailored laser pulse can propagate many Rayleigh lengths without significantly altering its original profile. Consider a long pulse, $L > \lambda_p$, in which the spot size is tapered from a large value at the front to a small value at the back, so that the laser power, $P \sim r_L^2|a|^2$, is constant throughout the pulse and equal to P_{crit}. The leading portion ($\ll \lambda_p$) of the pulse will diffract as if in vacuum [2]. However, since r_L is large at the front of the pulse, the Rayleigh length is also large. Hence, the locally large spot size allows the pulse front to propagate a long distance, whereas the body of the pulse will be relativistically guided. Simulations of a tailored pulse have been performed in which $|a|$ increases from 0.09 to 0.9 over a length $L_{rise} = 2\lambda_p$, $P = P_{crit}$ throughout the pulse ($\lambda_p = 0.03$ cm, $\lambda = 1$ μm) and r_L decreases from $10\lambda_p$ to λ_p. At peak intensity the vacuum diffraction length is $Z_R = 28$ cm. The effectiveness of pulse tailoring can be seen by the $r_L(c\tau)$ plot in Fig. 1(d). Further simulations indicate that at $c\tau = 24Z_R$ the pulse is distorted but largely intact.

A preformed plasma density channel can guide short, intense laser pulses. Analysis of the wave equation in the weak pulse limit indicates that a parabolic density channel will guide a Gaussian laser beam provided that the depth of the density channel [3] is $\Delta n = 1/\pi r_e r_L^2$, where $\Delta n = n_0(r_L) - n_0(0)$ and r_e is the classical electron radius. A simulation of channel guiding has been performed for the $L = \lambda_p$ pulse (discussed above) propagating in a parabolic density channel with $\Delta n = 1.3 \times 10^{15}$ cm^{-3} and $n_0(0) = 1.2 \times 10^{16}$ cm^{-3}. Guiding is confirmed by the $r_L(c\tau)$ plot in Fig. 1(e). The $r_L(c\tau)$ oscillations indicate a slight mismatch between the laser and channel parameters. This is caused by the laser pulse further reducing the density in the region of peak intensity. The axial wakefield generated by this pulse at $c\tau = 10Z_R$ has a peak accelerating gradient of 4.6 GeV/m. Additional simulations have shown that this wakefield traps and accelerates a trailing electron bunch to high energies.

This work supported by the U.S. Department of Energy and the U.S. Office of Naval Research.

[1] P. Sprangle, E. Esarey, A. Ting, and G. Joyce, Appl. Phys. Lett. **53**, 2146 (1988); E. Esarey, A. Ting, P. Sprangle, and G. Joyce, Comments Plasma Phys. Controlled Fusion **12**, 191 (1989).

[2] P. Sprangle, E. Esarey, and A. Ting, Phys. Rev. Lett. **64**, 2011 (1990); Phys. Rev. A **41**, 4463 (1990); A. Ting, E. Esarey, and P. Sprangle, Phys. Fluids B **2**, 1390 (1990).

[3] P. Sprangle and E. Esarey, Phys. Fluids B, July (1992).

Ponderomotive Steepening in Short-Scale-Length Laser-Plasmas

D. Umstadter and X. Liu

Center for Ultrafast Optical Science, University of Michigan,
Ann Arbor, MI 48109-2099, USA

When an intense laser interacts with a solid target, a thermal pressure gradient, inversely proportional to the density scale length (d), causes a rapid expansion of the plasma into the vacuum. This force is opposed by the pressure gradient of the laser light, known as the ponderomotive force. When the laser light is reflected from the plasma, twice its momentum is imparted to the plasma at the reflection point, near the critical surface, z_c. Here we report experimental results indicating that the ponderomotive force may significantly impede the thermal expansion of even an extremely short density scale length plasma, one in which $d << \lambda$, the laser wavelength. This limit is studied experimentally by using short-pulse (400 fs and 1 ps) lasers.

The Doppler shift of light reflected from the critical surface is proportional to the surface's expansion velocity, $\Delta\lambda/\lambda = -2(v_{\rm exp}/c)\cos\theta$,where θ is the angle of incidence. In the case where the laser's ponderomotive force is neglected, as has been done in Ref. [1], the plasma temperature T can be inferred from the Doppler shift measurement. However, under our experimental conditions the ponderomotive force is comparable to the plasma thermal force, and a simple relation between $v_{\rm exp}$ and T no longer exists. For short-pulse laser-plasma interaction, where the scalelengths of the plasma and the laser ponderomotive potential are approximately equal, the condition that the two forces are approximately equal is given by[2]:

$$m\langle v_{\rm os}^2\rangle/2kT_e \equiv 3.2\times10^{-13}I_{\rm inc}\lambda_0^2/kT_e \sim 1 \qquad (1)$$

($I_{\rm inc}$ is in W/cm^2, λ_0 is in μm, and kT_e is in eV).

In the experiment, laser pulses of wavelength 1.05 μm were generated in a Nd:glass laser system based on the chirped pulse amplification technique[3]. Two sets of experiments were performed: (1) a self-reflected Doppler shift measurement with 1-ps pulses; and (2) a pump-probe Doppler shift measurement using 400-fs pulses obtained after the laser system was upgraded to include a titanium:sapphire regenerative amplifier. The intensity peak-to-background contrast ratio of the laser pulses was measured to be 10^3:1 for the 1-ps case, and 5×10^5 : 1 for the 400-fs case. For either case, the contrast ratio was further approximately squared by frequency-doubling ($\lambda = 0.53$ μm) the laser pulses using a KDP crystal. The maximum energy of the 0.53 μm pulse was about 80 mJ ($I_{\rm max} = 1.6\times10^{16}$ W/cm^2 for 400-fs pulses and $I_{\rm max} = 6\times10^{15}$ W/cm^2 for 1-ps pulses). The laser was then focused at normal incidence ($\theta = 0°$) in a vacuum chamber onto various solid targets, which were rastered between shots.

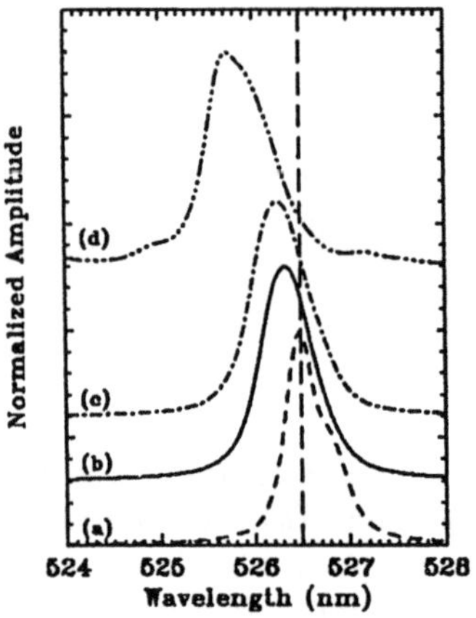

Figure 1: Spectrum of pump light: (a) incident on the target; (b) reflected from the moving critical surface, $I \simeq 2 \times 10^{15}$ W/cm^2; (c) calculated numerically with the ponderomotive force included; and (d) calculated numerically without the ponderomotive force.

We first show the results obtained from the spectra of the self-reflected 1-ps pump incident on aluminum, as shown in Fig. 1. The measured Doppler shift agrees quite well with a Doppler shift predicted by the simulation[2], shown in Fig. 1(c). Fig. 1(d) shows the Doppler shift calculated in the simulation without including the ponderomotive force; it is three times greater than the measured shift (Fig. 1(b)). This translates into a difference in ZkT_e of a factor of approximately ten. This disagreement comes as no surprise, since from Eq. 1, using our laser parameters and the temperature obtained from the simulation, $kT_e \sim 200$ eV, we would expect the free expansion model to be inaccurate.

We now present a pump-probe experiment using 400-fs pulses incident on a molybdenum target. The 0.5-μm probe pulse, beam-split from the pump, with final intensity less by a factor of 10^{-4}, was incident at 25° to the target normal. The centers of their focal spots were aligned to within 20% of the pump's focal spot with a 5×-magnification imaging lens and displayed on a digitizing CCD camera. Spectra were recorded as the time delay was varied. The time when they overlapped, $t = 0$, was determined to within approximately half a pulse width ($\sim$ 200-fs) accuracy by observing interference fringes formed by beating the probe and scattered pump light.

Figure 2 shows $\Delta\lambda$ as a function of delay at pump laser intensity $I \sim 10^{16}$ W/cm^2. For clarity, the Gaussian temporal profile of the pump laser is also shown on a normalized linear scale. Five shots were fired at each time delay, and Fig. 2 shows the average shift of these shots. Beginning with negative delay times, corresponding to the probe preceding the pump, as the delay time in Fig. 2 is increased, or as the delay between the pump and probe is decreased, the Doppler shift is observed to increase. This is attributed to an expected increase in the plasma temperature as the pump laser intensity increases. Evidence of a Doppler shift several pulse widths (2 ps) before the peak of the pump may be explained by the fact that a plasma is produced at an intensity that is several orders of magnitude ($\sim$ 4) below the pump's peak intensity, and the foot of the pulse at this intensity is 2 ps before the peak. A local minimum is observed in the Doppler shift centered near $t = 0$, when the pump and probe beam overlap

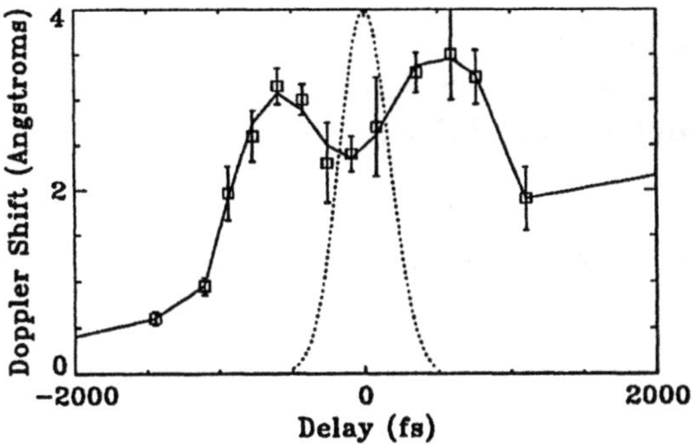

Figure 2: The Doppler shift of the reflected probe beam, due to motion of the critical surface [n_c(probe) = $0.8n_c$(pump)], as a function of delay between the pump and probe (solid line). The normalized Gaussian temporal profile of the laser is superimposed for clarity (dotted line).

in time. This is consistent with the expected effect of the ponderomotive force, which also maximizes at $t = 0$, when the pump intensity maximizes. Finally, when the probe follows the pump by a time equal to approximately a pulse width, the shift begins to decrease. This is expected since after the peak of the pump intensity, plasma heating decreases and both expansion and diffusion lead to rapid cooling.

In conclusion, the ponderomotive force will affect those processes that depend strongly on the evolution of the density profile if the conditions given by Eq. 1 are met. Specifically, the simple relation between the Doppler shift of the reflected pump laser light and the electron temperature, as given by the free expansion model, does not apply in this case. Resonance absorption of the laser light — because it depends strongly on the density scale length near the critical surface — and expansion cooling of the plasma are important processes that are affected.

This work was partially funded by the National Science Foundation Center for Ultrafast Optical Science, contract #PHY8920108.

References

[1] H. M. Milchberg, R. R. Freeman, and S. C. Davey, Phys. Rev. Lett. **61**, 2364 (1988); H. M. Milchberg, and R. R. Freeman, Physical Review A **41**, 2211 (1990).

[2] D. Umstadter, X. Liu, J. S. Coe, and C. Y. Chien, in *OSA Proceedings on Short Wavelength Coherent Radiation: Generation and Applications*, P. Bucksbaum and N. Ceglio, eds. (Optical Society of America, Washington, D.C., 1991), **11**, pp. 55-57.

[3] P. Maine, D. Strickland, P. Bado, M. Pessot and G. Mourou, IEEE J. Quantum Electron. **QE-24**, 398 (1988).

Possibility of Experimental Studies of Nonlinear Quantum Electrodynamics Effects Using High Power Ultrashort Laser Pulses

P.G. Kryukov

P.N. Lebedev Physics Institute, 117924 Moscow, Russia

Abstract. Investigation of interaction of electrons and γ - quanta with very strong electromagnetic fields having the strength about $F_0 = m^2 c^3/eh = 1.32 \times 10^{16}$ V/cm (characteristic QED field strength) is of principal interest for fundamental physics. Such a field performs the work equal to the electron rest energy (mc^2) at its Compton wavelenght (h/mc). Therefore one may think that intrinsic proprties of an electron would be manifested due to action the field $\approx F_0$ on parts of it. If such a field is produced with a laser then frequency and wavelenght must have values $\nu \approx mc^2/h$ and $\lambda \approx h/mc$ in the rest frame of the electron i.e. defined by fundamental constants and electron mass. That is way so important to examine experimentally relatively simple situations compatible with detailed quantum electrodynamics interpretaion. Proposed experiments are based on bringing a high-energy electron or photon beam into collision with a high-intensity laser beam.

1. Introduction

The recent development of terawatt-class laser systems [1,2,3] permits to achieve fields strengths of about 10^{11}V/cm at the beam focussing. On the other hand, there are a few electron accelerators having a electron energy of 30-50-GeV (e.g. HERA(DESY) and LEP(CERN) storage rings for electrons). At collision a high-energy electron with a laser beam the field strength would be transformed with Lorentz factor γ and to exceed of F_0. But there are many difficalties encountered in realization this idea and we would like to demonstrate how to overcome them.

The high laser field is achived over a focal spot of a few optical wavelengths and during the time less than an ultrashort pulse duration. On the other hand, an electron beam consists of bunches having duration of about 100 ps and trancverse dimensions of tens of micrometers. A very complex problem arises to make coincidence with space and time the both beams of two devices are operated independently of one another, namely, a high intense laser and an unique electron accelerator. It is desirable to organize experiments on nonlinear QED effects so that they would not disrupt planned operation of the accelerator. At last, those experimental data should be measured in which high strong field effects are exhibited.

2. A proposal sceme of experiments on interaction of electrons and γ-quanta with a high strong laser field

To begin with a problem of the laser beam focussing into the bunch tranjectory. It is possible to use detection of γ-quanta which are produced fron Compton scattering for monitoring of the coincidence of the beams at a remote control of the focussing optics. Perhaps, the most complex problem is ensuring of the temporal coindence a high-powerful laser pulse (200-500 fs) and an electron bunch ($\approx$100 ps). For solution of this problem one may invoke that a storage ring for electrons and a high-powerful laser can be operated in a regime of the high stable repetition frequency. In this case one must use a CW ultrashort pulse laser with minimal timing jitter [4,5,6] as a master ocsillator. Its repetition frequency must be made a multiple of the electron bunch orbit one. At an intermediate amplification the repetition frequency must bé equal the electron bunch orbit frequency (48 kHz for HERA and 11 kHz for LEP) and an energy of the amplified pulse must be of 10-50 μJ. Such parameters are quite realizable [7]. Using an optical variable time delay and registration of the γ-quanta number one may define the coincidence moment of both beams and to switch an final amplifier on during the next revolution of the bunch. By the way, the procedure of the measerment of the γ-quanta number via the

Springer Series in Chemical Physics, Vol. 55 Ultrafast Phenomena VIII
Editors: J.-L. Martin · A. Migus · G.A. Mourou · A.H. Zewail

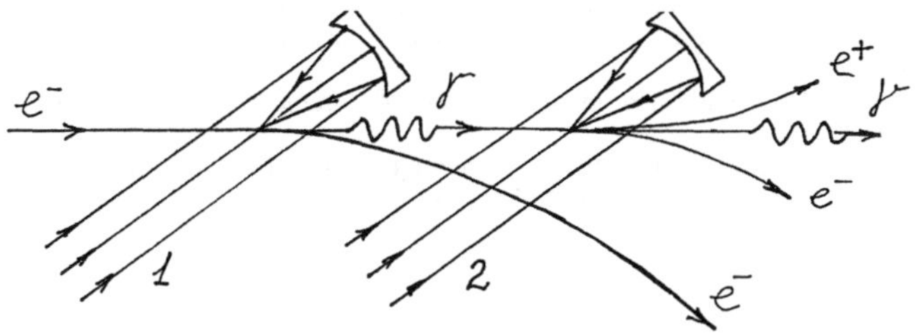

Fig.1 Block diagram of the proposal experiment on $e^+ e^-$ pair creation by light in the strong laser field. The laser beam (1) is used for γ-quanta producing. The $e^+ e^-$ pairs are created in the laser beam (2) focus.

time delay may be cosidered as a useful method of the bunch dimension characterization by a subpicosecond sampling.

Because a strong laser field area is far less than the bunch dimension, a very small part of total number of electrons (partically about 10^{-7}) will be involved into the interaction. It means that the proposal experiments may be performed with a negligible influence on a planned operation of a storage ring.

The most accessible for experimental investigation is the process of nonliner Compton scattering of laser photons on electrons:

$$n\,h\omega_l + e \rightarrow h\omega_\gamma + e\ .$$

By registration of the energy and the coordinate of γ-quantum it would be defined higher harmonic radiation n.

Another effect is $e^+ e^-$ pair creation by light in the strong laser field:

$$h\omega_\gamma \rightarrow e^+ e^-\ .$$

In this experiment (Fig.1) a beam of γ-quanta which produced due to Compton scattering is used. Generation of such a beam may be considered as a useful byproduct of the proposal experiment. Estimation shows that the γ - quanta beam can wave the following parameters:

- high flux about 10^4 photons (energy about $50\,\mu$J);
- nearly monochromatic energy spectrum;
- angular divergence about 10^{-5};
- almost 100% polarization, defined with a laser beam polarization.

This γ-quanta beam source may be cosidired as a "γ-ray laser". At the interaction γ-quanta with the second laser focus the $e^+ e^-$ pairs will be created. It is expected that this interaction with the strong laser field depends on the beam polarization [8]. By variation and controlling both laser beam polarization it is possible to study a nonlinear QED effect, namely the dependence the $e^+ e^-$ pairs creation via the polarizations of the γ-quantum and the strong electromagnetic wave.

I thank V.I. Ritus, A.I. Nikishov and V.I. Sergeenko for fruitful discussion.

References

1. A.J.Taylor, C.R.Tallman, J.P.Roberts, C.S.Lester, T.R.Gosnell, P.H.Y.Lee and, G.A.Kyrala, Opt.Lett.15,39 (1990).
2. M.Watanabe, K.Hata, T.Adachi, R.Nodomi, and S.Watanabe, Opt.Lett.15,845 (1990).
3. C.Sauteret, G.Mainfrau, G.Mouruo, Laser Focus World 10, 85 (1990).
4. D.von der Linde, Appl.Phys. B39, 201 (1986).

5. M.J.W.Rodwell, K.J.Weingarten, D.M. Bloom, T.Baer, and B.H.Kolner, Opt.Lett.11, 638 (1986).
6. G.T.Harvey, M.S.Heutmaker, P.R.Smith, M.C.Nuss, U.Keller, and J.A.Valdmanis, IEEE J.Quantum Electron.27, 295 (1991).
7. F.Salin, J.Squier, G.Mourou, and G.Vallancourt, in *Digest on the 8th International Conference on Ultrafast Phenimena* (1992), paper MB5.
8. V.I.Ritus, "Quantum Effects of the Interaction of Elementary Particles with an Intensc Electromagnetic Field", J. of Sov. Laser Research, vol. 6, №5 (1985).

Soliton-Like Self-Trapping of Three-Dimensional Patterns

A. Barthelemy[1], *C. Froehly*[1], *M. Shalaby*[1], *P. Donnat*[2], *J. Paye*[3], *and A. Migus*[2;3]

[1]IRCOM/Optique, URA CNRS, Faculté des Sciences, F-87060 Limoges, France

[2]Commissariat à l'Energie Atomique, CEL-V, F-94195 Villeneuve-St.-Georges, France

[3]Laboratoire d'Optique Appliquée, CNRS URA 1406, ENSTA-Ecole Polytechnique, F-91120 Palaiseau, France

Abstract: Description of experiments on self-focusing, through nonlinear liquids, of pulsed laser beams exhibiting various geometrical shapes, some of them giving rise to stable self-trapping similar to soliton beam propagation.

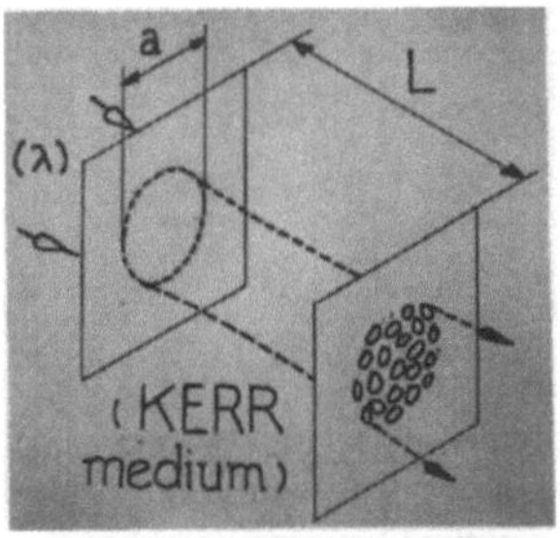

Figure-1 : Multimode break-up of high power, smoothly bell-shaped, laser beams through intensity-dependent refractive index gradients (optical Kerr non-linearity)(ref.1,2,3). Each spot of the ouput pattern carries a few "critical power" $P_c=\lambda^2/\gamma$; λ= λ laser wavelength;$\gamma=\Delta n/I$ Δn : refractive index variation induced by laser illumination I.

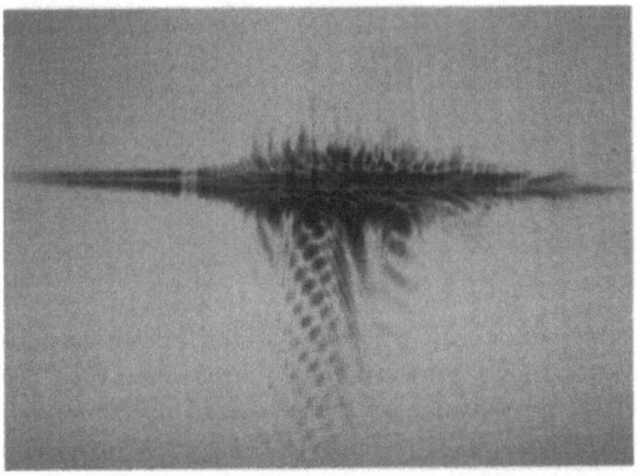

Figure-2: Chaotic break-up and self-frequency modulation of a planar "laser sheet"; output beam spectral analysis (horizontally: transverse position; vertically: wavelength display where longer -Stokes- wavelengths are below the central laser line).

Springer Series in Chemical Physics, Vol. 55 **Ultrafast Phenomena VIII**
Editors: J.-L. Martin · A. Migus · G.A. Mourou · A.H. Zewail

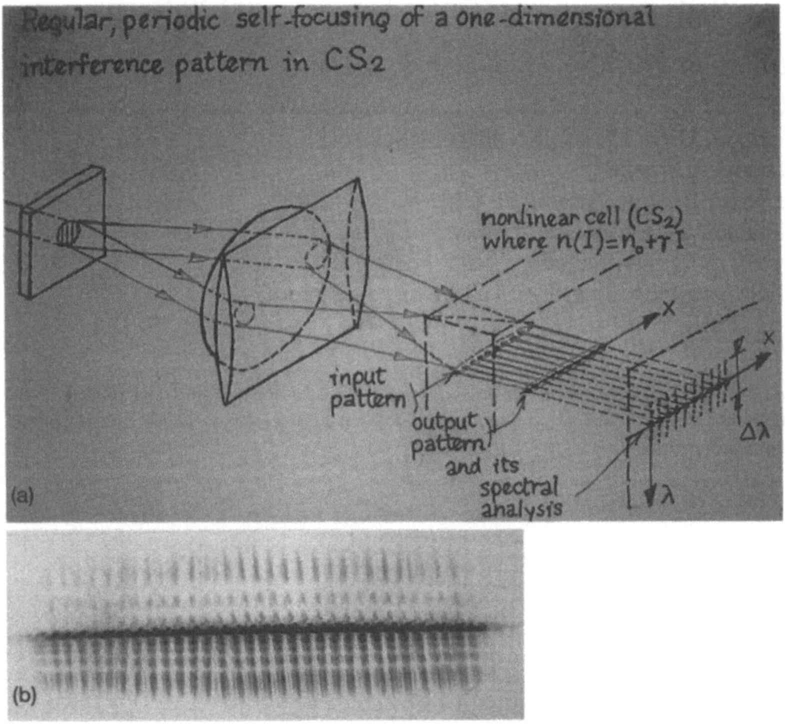

Figure 3:Regular periodic self-focusing of a planar "laser sheet" sinusoidally modulated by two beams interference fringes; (a) experimental arrangement, (b) spectral analysis of the output (Stokes side below the central laser line). It demonstrates the strong nonlinear coupling between neighbouring fringes (ref.4-5).

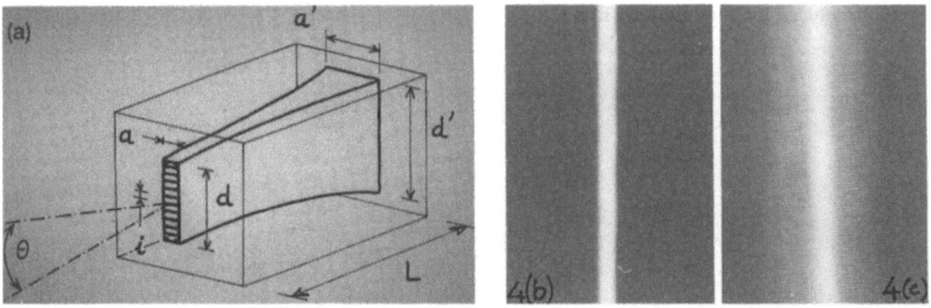

Figures 4 (a,b,c) : Below self-focusing intensity, stable self-trapping of the above pattern (sinusoidally modulated planar light sheet) may take place when intensity-induced convergence balances diffraction-induced divergence in the interference fringes plane - here horizontally - (ref.6,7,8): (a) low intensity pattern, horizontal divergence a'/L ; (b) input; (c) temporally integrated output photograph showing superposition of low intensity (laser pulse wings) wide divergence pattern with soliton intensity (laser pulse top) zero-divergence pattern. Spectral analysis (not shown here) demonstrates maximum self-phase shift of soliton beam to be less than 3π or 4π, before the sudden rise of stimulated Raman scattering (SRS).

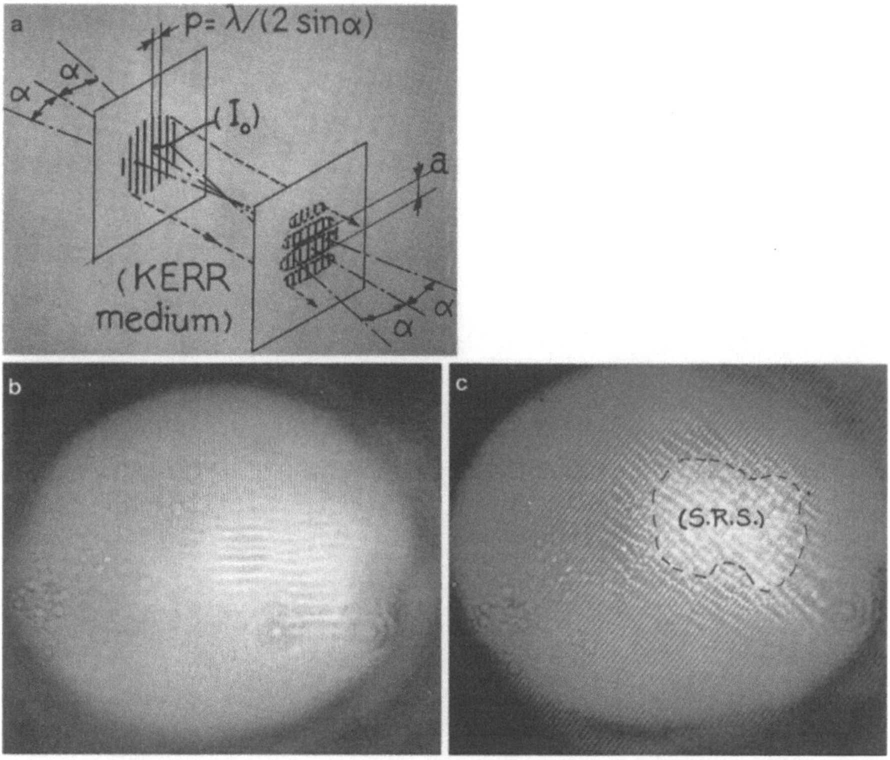

Figures 5 (a,b,c) : Sinusoidal interference fringe field of circular cross-section much larger than soliton width "a" of figure 4 (a) ; one-dimensional modulational instability grows up as regular stripes perpendicularly to interference fringes (ref.9-10). Each stripe carries about one soliton intensity (b and c) . At very high intensities self-focusing spots exhibit regular alignments, still orthogonal to interference fringes. Regions of strong illumination generate Stimulated Raman Scattering. (Photographs (b) and (c) as well as fig.6.c kindly supplied by H. Maillotte and J. Monneret , Lab. Duffieux, Besancon Univ., France).

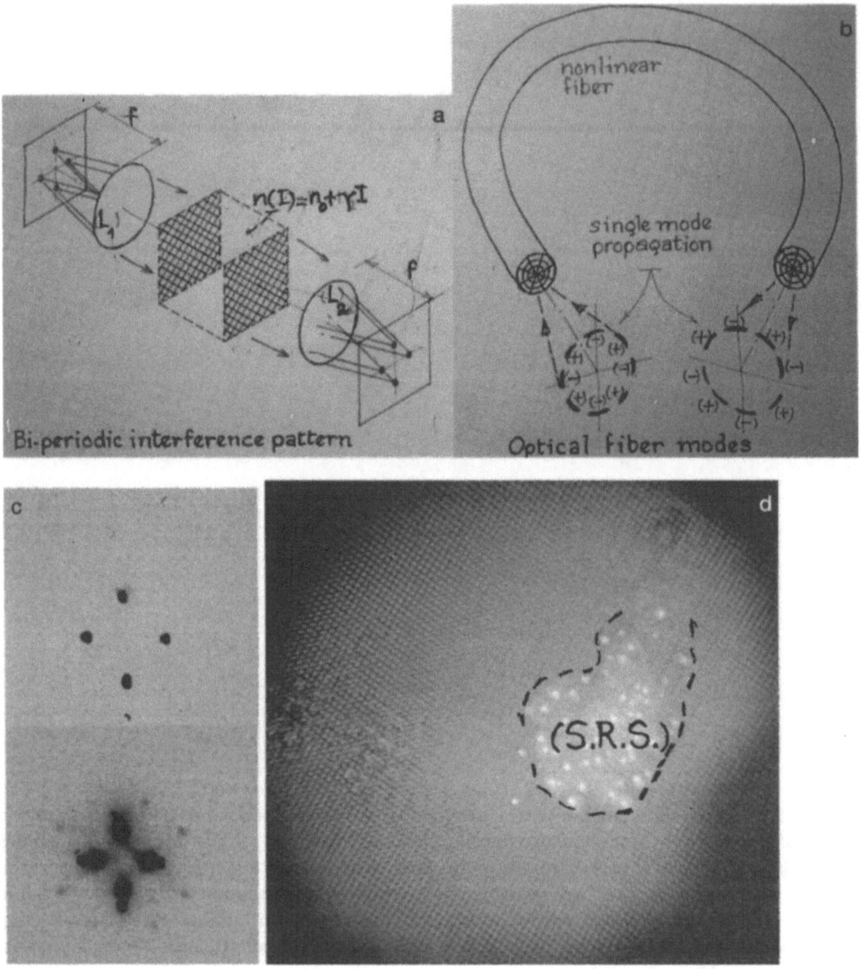

Figures 6 (a,b,c,d): Sharp modulations along both transverse orthogonal directions (by interference or coupling to a mode of a waveguide, cf.fig.6.a and b, or even speckle (ref.11-12)) introduce strong bi-dimensional diffraction, thus nonlinear modulational instability vanishes as shown by regularity of both nearfield and farfield outputs. At very high intensities, far field spots attract each other (see the two photographs fig.6.c): this is self-deflection through average index gradients generated in the four beams overlapping region ; irregularly distributed Stimulated Raman Scattering spots are generated(fig.6.d). At still higher intensities, regular "coherent" coupling of these spots may again be observed, giving rise to spatially bi-periodic "distributed feedback Raman laser" coherent emission (cf ref.13).

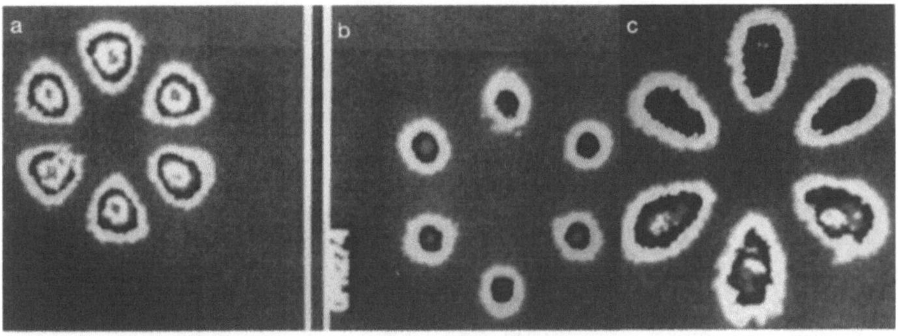

7-a: input 7-b: self-trapped output 7-c: linear output

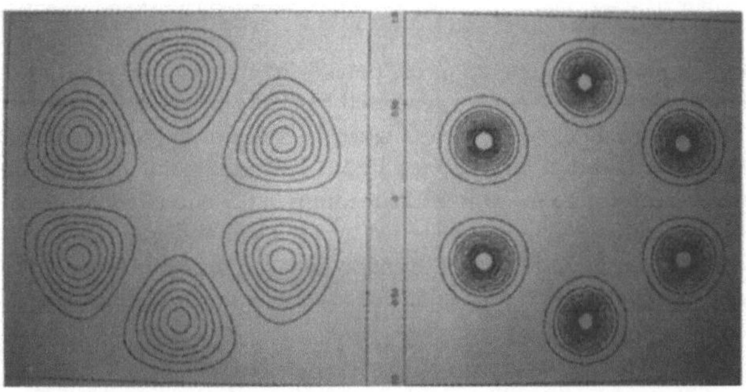

Figures 7 (a,b,c,d): "Rolling up" patterns of fig. 4 (a and b) in a circular configuration generates "tubular beams" azimuthally modulated , like Laguerre-Gaussian beams of lasers or parabolic graded index multimode fibers. Such tubular beams really exhibit stable self-trapping at well definite intensity (fig.7.a) (ref.14). Generation of such patterns was performed by filtering techniques due to P. Facq and coworkers (ref.15), that could be advantageously replaced by intracavity modal filtering of the laser emission itself (ref.16). Dependence of self-trapping on the number of azimuthal nodes and on petal ellipticity is under study. Numerical simulation (fig.7.d) yields excellent qualitative agreement with experiments (fig.7.c) bottom. Suppression of azimuthal modulation (ring-shaped input) again leads to quasi-periodic break-up as pointed out twenty years ago by Campillo and coworkers (ref.9), and considered theoretically by Feit and Fleck (ref.17). Stable petal-like patterns could be seen as "spiralling solitons" of ref.18.

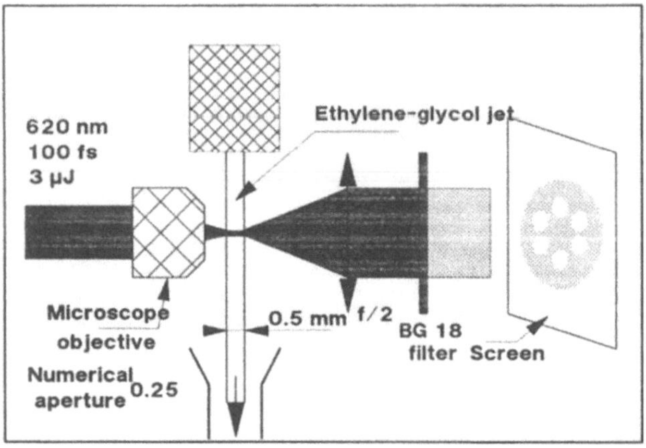

Figure 8 : Experimental set-up for spontaneous generation of "petal " modes of the non-linear Schrödinger equation. Pulses, 100 fs duration at 620 nm are generated in a CPM dye laser and further amplified in dye amplifiers to the 100 mJ range. Great care is taken in order to have a very high energy stability (<3% rms) and uniform spatial profile (see fig.9). This last point is achieved through spatial filtering. The beam is focused into the non linear medium (a 0.5 mm ethylene glycol flowing jet) using a microscope objective. Emerging light is collected by a f/2 lens and goes to a screen where the pattern is seen and recorded. A colored filter is inserted to enhance the contrast since the most intense part of the beam undergoes some spectral broadening due to continuum generation.

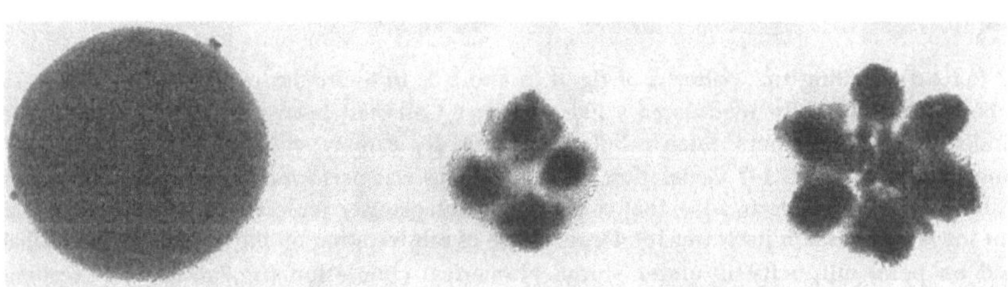

Figure 9: Far field pattern obtained respectively (left to right) at the low intensity regime, for 20 MW and 30 MW peak powers. Such spontaneously appearing modes have been obtained with as many as N=16 petals. N is always an even number, suggesting as expected that two adjacent spots are in phase opposition. The corresponding repulsive interaction due to Kerr nonlinearity explains the stability of this structure. Such a behavior has been obtained numerically in ref.17 but starting from a ring mode and an azimuthal periodical phase perturbation. In our experiment, if the number of petals is initially linear with power (roughly one critical power per spot) this linear law saturates for larger values of N. Although the initial beam looks uniform, the asymetric "petal" pattern reproduces itself identically from one laser shot to the next one, showing a breaking of symmetry in the jet or due to defects in the initial beam. The numerical simulation whose results are shown in fig.7.d does not take into account temporal effects and therefore cannot predict the spectral broadening occuring during propagation.

References:

1/ P.L. Kelley; Phys. Rev. Lett., **15**, p.1085, 1965.

2/ J.H. Marburger; Prog. Quant. Electron., **4**, p.35, 1975.

3/ V.I. Bespalov and V.I. Talanov; " Filamentary structure of light beams in nonlinear liquids", Soviet JETP Lett., **3**, p.307, 1966.

4/ M. Vampouille, B. Colombeau and C. Froehly; "Application du contrôle de l'autofocalisation dans le CS2 au raccourcissement d'impulsions laser picosecondes", Opt. and Quant. Electron., **14**, pp. 253-261, 1982,

5/ C. Froehly, B. Colombeau and M. Vampouille; "Shaping and Analysis of Picosecond Light Pulses" in Progress in Optics , vol. **XX**, E. Wolf editor, pp. 65-153, 1983.

6/ A. Barthelemy, B. Colombeau, C. Froehly, F. Reynaud and M. Vampouille; "Stable soliton and non-soliton 3D propagation of intense light by multiple beams interference in Kerr materials", OSA Annual Meeting, Boston, MA, 1990, Technical Digest Series, 1990, vol.**15**, p.167.

7/ V. E. Zakharov and A. B. Shabat; "Exact theory of two-dimensional self-focusing and one-dimensional self-modulation of waves in nonlinear media", Sov. Phys. JETP, **34**,1, pp.62-69, 1972.

8/ A. Barthelemy, S. Maneuf and C. Froehly; "Propagation soliton et autoconfinement de faisceaux lasers par non linearité optique de Kerr", Opt. Comm., **55**, 3, 1985.

9/ A.J. Campillo, S.L. Shapiro and B.R. Suydam; " Relationship of self-focusing to spatial instability modes", Appl. Phys. Lett., **24**, 4, pp.178-180, 1974.

10/ A.J. Campillo, S.L. Shapiro and B.R. Suydam; " Periodic breakup of optical beams due to self-focusing", Appl. Phys. Lett., **23**,11, pp.628-630, 1973.

11/ A.A. Mak, V.A. Serebryakov and V.E. Yashin; "Suppression of self-focusing in spatially incoherent light beams", Sov. Tech. Phys. Lett., **6**,2 pp.57-58, 1980.

12/ H. Maillotte, J. Monneret and C. Froehly; "Noise free phase conjugation of high power single mode laser beams by stimulated Brillouin scattering", Opt. Comm., **77**, 2-3,pp.241-246, 1990.

13/ H. Maillotte, J. Monneret and C. Froehly; "Distributed feedback picosecond Raman laser emission from CS2"; Opt. Comm., **79**, 3-4, pp.259-266, 1990.

14/ A. Barthelemy, R. De La Fuente, C. Froehly, O. Guy, M. Shalaby, F. Reynaud; "Soliton self-trapping of light beams"; Nonlinear Guided-wave Phenomena, Cambridge, UK, 1991, Technical Digest Series, **15**, pp.380-383, 1991

15/ P. Facq, P. Fournet and J. Arnaud; "Observation of tubular modes in multimode graded-index optical fibers", Electron. Lett., **16**, pp. 648-650, 1980.

16/ W.W. Rigrod; "Isolation of an axi-symmetrical optical resonator modes"; Appl. Phys. Rev., **2**, 1963.

17/ M. D. Feit and J.A. Fleck, Jr; "Beam non paraxiality, filament formation and beam breakup in the self-focusing of optical beams", J. Opt. Soc. of Am. B, **5**, 3, pp. 633-640, 1988.

18/ L. Polodian, A.W. Snyder and D.J. Mitchell; "Spiralling spatial solitons", Opt. Comm., **85**, pp. 59-62, 1991.

Physical Origins of the Spectral Continuum: Self-Focusing, Self-Trapping and Cerenkov Radiation

F. Salin[1], *J. Watson*[1], *J.-F. Cormier*[2], *P. Georges*[1], *and A. Brun*[1]

[1]Institut d'Optique Théorique et Appliquée, CNRS URA-14, Université Paris-Sud, B.P. 147, F-91403 Orsay Cedex, France
[2]Centre d'Optique, Photonique et Laser, Université Laval, Québec, G1K 7P4 Canada

Abstract: We show that femtosecond spectral continuum is initiated by self-focusing leading to Cerenkov radiation and self-trapping of broadband radiation.

Although many authors have studied continuum generation since its discovery 22 years ago[1], no universal explanation has been found and most of the models fail to explain the spatial properties of the emission. In this paper we show that at least two mechanisms including a Cerenkov-based effect play a role in continuum generation and that self-focusing is the fundamental trigger for this spectacular phenomenon.

Our experiment consisted of focusing intense femtosecond pulses generated by a Ti:sapphire laser chain[2] onto various glasses and into water and observing the characteristics of the spectral continuum (SC) thus generated. Typically, the pulses were sub-100 fs with an energy of 2-3 mJ and a wavelength of 780 nm. The observations and data reported here are for the specific case of SC generation in a 6 mm piece of BK7 glass which had been placed at 15 cm in front of the beam focus. However, no qualitative differences were observed in the properties of the emission when other media such as water and different glasses were tried.

First, we found that there is a threshold for continuum generation and that a progressive increase of the input intensity does not lead to a progressive increase of the spectral width. We imaged the output surface of the glass sample and observed that the onset of spectral continuum generation was always linked to the appearance of white filaments in the sample. The position of the filaments was perfectly stable and only the number of filaments changed from one shot to the next. From the image of the sample exit surface we found that the size of the filaments was constant at about 8 μm. Moreover, these filaments were spatially coherent and interference fringes could be observed between spatially overlapping images of the filaments. We also found that up to laser intensities a few times threshold, the SC generation was accompanied by a ring emission like a circular rainbow. A typical SC beam shape recorded through a 580 nm interferential filter is given in Fig.1.

Using different filters, we found that the ring size as well as the central spot size increased when the wavelength decreased and that no ring was present for wavelengths longer than the pump wavelength. We would like to emphasize the fact that the SC spatial properties, and in particular its divergence, are totally independent of those of the pump beam. We observed a diverging SC when the sample was placed in a converging beam and the divergence of this SC was independent of whether it was placed in a converging or a diverging beam. Furthermore, we focused the beam with a cylindrical lens and observed the same circular pattern for the SC. When the intensity was further increased, the gap between the central spot and the first ring disappeared and a white beam surrounded by a bluer ring was observed. At the same time, on the sample surface we observed a progressive merging of the filaments into a broad white spot.

Springer Series in Chemical Physics, Vol. 55 Ultrafast Phenomena VIII
Editors: J.-L. Martin · A. Migus · G.A. Mourou · A.H. Zewail

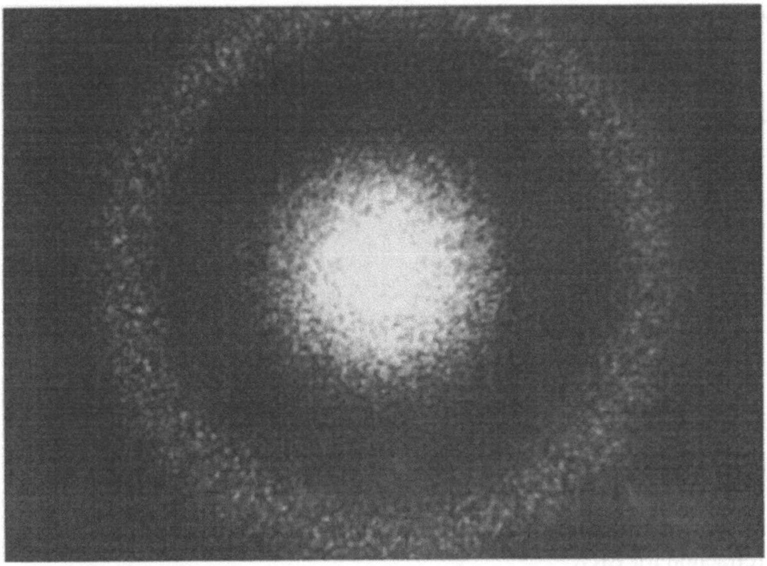

Figure 1: Spectral continuum beam shape as recorded through a 580 nm interferential filter.

Starting from the observation that SC generation was always accompanied by filamentation, we propose self-focusing as the fundamental process for SC generation. This idea has already been proposed following studies of SC generation in gases[3]. In the case of self-focusing of short pulses, it has been shown that the filaments correspond to a moving focus[4]. The leading edge of the pulse will focus further away from the input surface than the peak of the pulse. Since the peak of the pulse enters the sample later than the leading edge, the focus will move through the sample at a speed which largely exceeds the speed of light in the medium. Since the intensity at focus is very large, these moving foci induce a "nonlinear polarization wave" propagating at a speed greater than the speed of light. Such a moving polarization induces Cerenkov radiation propagating at an angle θ_c from the filament axis. The angle is given by the fact that the radiation phase velocity equals the incident beam group velocity in the medium: $\cos\theta_c = n_L/n(\omega)$, where n_L is the index of refraction at the incident wavelength and $n(\omega)$ is the index at the radiated frequency. I. Golub was the first to propose Cerenkov radiation at the filament surface to explain the conical emission in SC generation[5]. We checked this explanation by plotting the angle of emission of different wavelengths filtered through interferometric filters as a function of wavelength. The result is given in Fig.2 along with the theoretical prediction for BK7. Note that Cerenkov radiation predicts the right evolution (blue farther from the axis) and leads to a good agreement with experimental data. It also explains why we do not see rings for wavelengths longer than the laser wavelength. Finally, since it adds radiation only to the high frequency part of the SC, it contributes to the the spectral asymmetry of the SC. The central spot in Fig.1, however, cannot be explained by Cerenkov radiation.

In summary, we show that spectral continuum generation is initiated by self-focusing. We identify at least two types of emission participating in the white light generation. One is the Cerenkov effect induced by the moving focus and the other one could

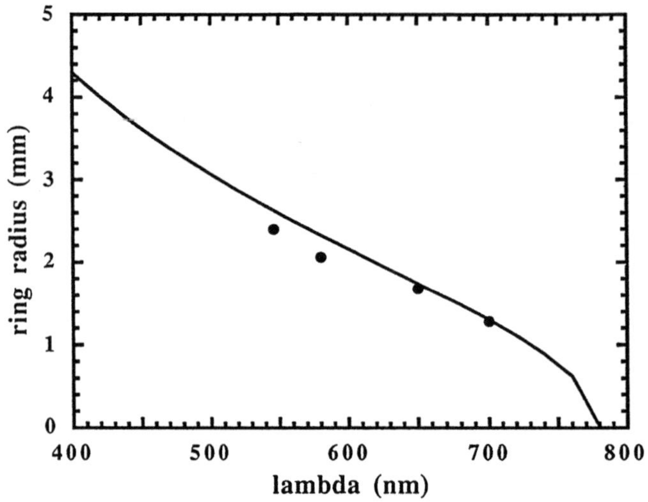

Figure 2: Cerenkov ring size in BK7. The solid line is the theoretical prediction using the index of refraction for BK7.

be self-trapping of self-phase modulated light. Better understanding of the SC physics could be used to produce even shorter pulses (less than 6 fs) by compressing the spatially coherent part of the continuum.

1. R.R. Alfano and S.L. Shapiro, Phys. Rev. Lett. **24**, 584 (1970).
2. J.Squier, F. Salin, G. Mourou and D. Harter, Opt. Lett. **16**, 324 (1991).
3. P.B. Corkum and C. Rolland J. Opt. Soc. Am. **B4**, 256 (1987).
4. Y.R. Shen, *The Principles of Nonlinear Optics*, Wiley, New York, 1984 (Ch. 17).
5. I. Golub, Opt.Lett. **15**, 305 (1990).

Diffraction and Focussing of Spectral Energy in a Two-Photon Process

B. Broers[1], *L.D. Noordam*[1], *and H.B. van Linden van den Heuvell*[1,2]

[1]FOM-Institute AMOLF, Kruislaan 407,
NL-1098 SJ Amsterdam, The Netherlands
[2]Van der Waals-Zeeman Lab., Valckenierstraat 65,
NL-1018 XE Amsterdam, The Netherlands

Abstract. An analogy between a two-photon process with a chirped pulse and Fresnel diffraction from a slit is pointed out. With the help of this analogy it is demonstrated experimentally how the idea of a Fresnel zone plate can be used in a two-photon process to focus spectral energy of sub-picosecond pulses onto a small fraction of the large bandwidth of the original pulse.

1. Analogy between a two-photon process and Fresnel diffraction

Consider a two-photon process driven by a pulsed field, $E(t)$. The field that drives the two-photon process is then proportional to $E^2(t)$. This multiplication in the time domain implies a convolution in the frequency domain: $E^{(2)}(\omega) = \int d\omega' E(\omega')E(\omega - \omega')$, where $E(\omega)$ denotes the original pulse in the frequency domain. If one now calculates the two-photon yield at exactly twice the central frequency ω_0 of a pulse with a square-shaped power spectrum, then $|E^{(2)}(\omega)|^2$ takes the form of a simple integration over all frequencies within the bandwidth of the pulse, *taking into account the phase of each frequency component.* This implies that the phase profile $\phi(\omega)$ of the pulse, i.e. its chirp, is relevant. The amount of chirp determines whether the various 'paths' $(\omega_0 - \delta, \omega_0 + \delta)$ adding up to $2\omega_0$, interfere constructively or destructively, so it determines the total transition probability at $2\omega_0$. For a quadratic phase profile, $\phi(\omega) = \alpha(\omega - \omega_0)^2$, this is schematically depicted in Fig 1a.

Exactly the same type of 'path integral' is encountered when Fresnel diffraction from a slit is studied (Fig 1b). To calculate the diffracted intensity at the central point O on the screen, one has to integrate the contributions from all parts of the slit, again taking into account the phase of each contribution. So we can make the following identifications: bandwidth of the pulse $(\Delta\omega)$ $\longleftrightarrow$ slit width, power spectrum of the pulse $(|E(\omega)|^2)$ $\longleftrightarrow$ intensity distribution over the slit, and phase profile of the pulse (chirp) $\longleftrightarrow$ phase profile over the slit. This last identification can be made more explicit: $\alpha \longleftrightarrow \frac{1}{\lambda}(\frac{1}{L_1} + \frac{1}{L_2})$, where λ is the wavelength of the light, L_1 the distance from light source to slit, and L_2 the distance from slit to screen (Fig 1b). This formulation is not surprising, since α measures the curvature of the phase front of the pulse, and $(\frac{1}{L_1} + \frac{1}{L_2})$ is the summed geometric curvature of the wave fronts (=phase fronts) at the slit.

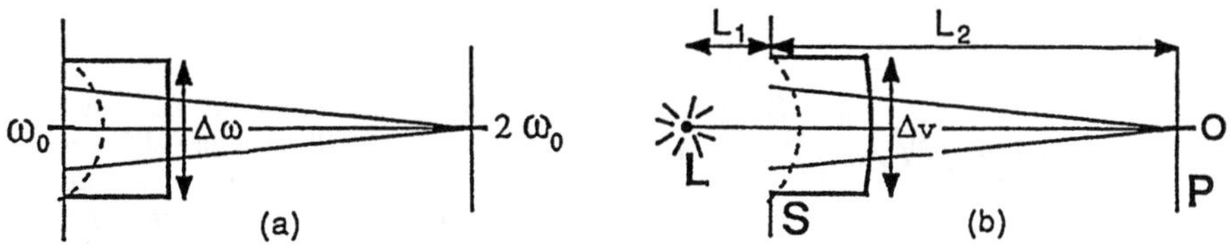

Fig 1. (a) *The two-photon process: full and dashed curves denote the power spectrum and phase profile of the fundamental pulse, respectively. Each frequency component contributes with its own phase to the spectral energy at* $2\omega_0$. (b) *Fresnel diffraction: full and dashed curves denote the intensity distribution and phase profile over the slit respectively. Each part of the slit contributes with its own phase to the intensity at the central part of the screen, O.*

Springer Series in Chemical Physics, Vol. 55 Ultrafast Phenomena VIII
Editors: J.-L. Martin · A. Migus · G.A. Mourou · A.H. Zewail

2. Spectral Diffraction and Focussing

If one considers all possible paths from L to O through the slit, it is clear that all are longer than the direct straight path from L to O. Some paths, however, have a phase difference of $\approx 2\pi n$ (n integer) with respect to the direct one, so they will interfere constructively with the direct path, while others have a phase difference of $\approx 2\pi(n+1)$, resulting in destructive interference. By placing a so-called Fresnel zone plate in the slit, which blocks those zones through which paths go that would contribute negatively, one greatly enhances the intensity in O. Translating this idea to the situation of a two-photon process with a chirped pulse, gives the following: block those frequencies that would contribute negatively to the spectral energy at $2\omega_0$. This will give rise to 'focussing' of spectral energy onto $2\omega_0$. This idea is demonstrated in Fig 2 for second harmonic generation in KDP.

Another example is given in Fig 3, where the results of two-photon wavelength scans over the Rydberg series in Rubidium are shown. It clearly demonstrates the unique combination of high yield and good spectral resolution in a two-photon process that can be obtained with the spectrally focussed pulse.

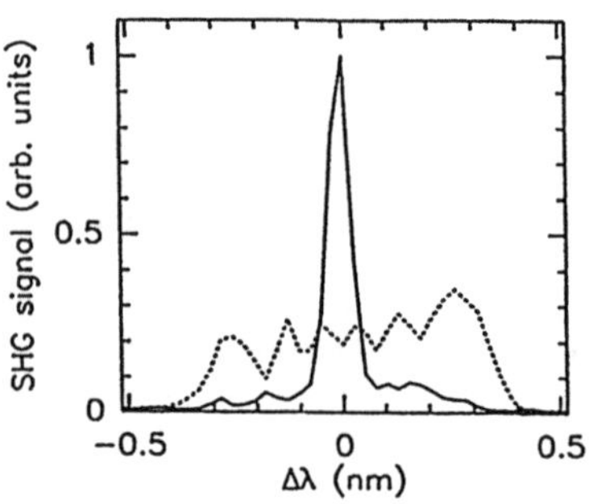

Fig 2. *Power spectra of frequency-doubled light resulting from two different pulses:* dotted*: chirped pulse with square-shaped power spectrum ($\Delta\lambda$ = 1.57nm around λ = 597nm, chirp-free pulse duration 666fs). The wiggles can be seen as 'diffraction fringes'.* full*: same chirp, but even 'Fresnel zones' of the fundamental spectrum blocked with a 'zone plate', resulting in 'focussing' of the energy onto $2\omega_0$.*

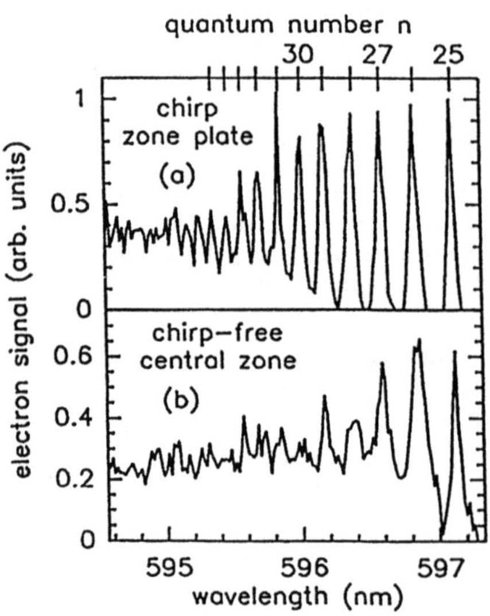

Fig 3. *Two-photon wavelength scans over Rydberg series in Rb with two different pulses:* (a) *chirped pulse with zone plate: 'spectrally focussed' result.* (b) *chirp-free pulse, bandwidth equal to central zone of zone plate. Note the better yield and resolution with the spectrally focussed pulse.*

Efficient Raman Conversion of Femtosecond UV Light Pulses

K.A. Stankov and Y.-W. Lee

Max-Planck-Institut für biophysikalische Chemie, Abt. Laserphysik, Am Fassberg, Postfach 2841, W-3400 Göttingen, Fed. Rep. of Germany

500-fs pulses at 248 nm were shifted up to the fifth Stokes in methane. The efficiency was 18% for the first, and 15% for the second Stokes.

Stimulated Raman shifting in gases is widely used for generation of coherent radiation at new frequencies. The method is of great importance in the UV range where the frequency doubling, tripling or mixing in the nonlinear crystals fails because of the absorption and because the complexity and low efficiency of the heat pipes with metal vapours is prohibitive. Many authors have reported efficient Raman conversion of nanosecond pulses from excimer lasers [1] as well as of picosecond pulses at various wavelengths [2]. However, the long relaxation time in the gaseous Raman medium (several tens to hundreds picoseconds) has been always considered as an obstacle for efficient conversion of femtosecond light pulses, due to the highly transient nature of the stimulated Raman process [3]. The appearance of other competitive nonlinear processes such as self-phase modulation, multi-photon absorption and ionization which may gain over the stimulated Raman scattering has also been thought as a limitation of the method.

Here we report, for the first time to our knowledge, on efficient stimulated Raman conversion of ultraviolet femtosecond pulses. In the course of investigation of a recently proposed method for spatially-dispersed stimulated Raman scattering [4], we obtained efficient conversion of 500-fs long pulses at 248 nm into several Stokes lines. The light source was a femtosecond laser system [5], delivering several millijoules energy in the short pulse. These pulses, however, are chirped and can be compressed down to 100 fs.

The laser radiation was attenuated to below 0.5 mJ and focused by a 30-cm focal length lens into the 50-cm long Raman cell filled with methane at 20 bar. At this pressure the transverse relaxation time of the methane gas was approximately 59 ps. Behind the cell, the beam was recollimated by an identical lens at the exit of the cell and dispersed by means of a quartz prism. For pumping energy of 400 μJ, we obtained 72 μJ (18% conversion) into the first Stokes at 267 nm, and 60 μJ (15% conversion) into the second Stokes (290 nm). Streak camera measurements gave resolution-limited identical 4.3 ps durations at all involved wavelengths. The spectral measurements were performed by a laboratory-made grating spectrograph, equipped with a 256-elements diode array. A smooth spectrum of 1.6 nm and 3.7 nm for the first and second Stokes, correspondingly, was recorded. The spectra shown in Figure 1 imply pulse durations considerably shorter than the pump pulse and possibly compressible to several tens of femtoseconds .

It is noteworthy that anti-Stokes components were not observed in this experiment. With the highest pump available, up to the fifth Stokes could be visually detected. At this pump level, the spectrum of the first and second Stokes was very broad (6 - 7 nm) and irregular.

We attribute this high conversion efficiency to optimization of the pump intensity and geometry. The attempts to achieve higher efficiency were unsuccessful.

Springer Series in Chemical Physics, Vol. 55 Ultrafast Phenomena VIII
Editors: J.-L. Martin · A. Migus · G.A. Mourou · A.H. Zewail

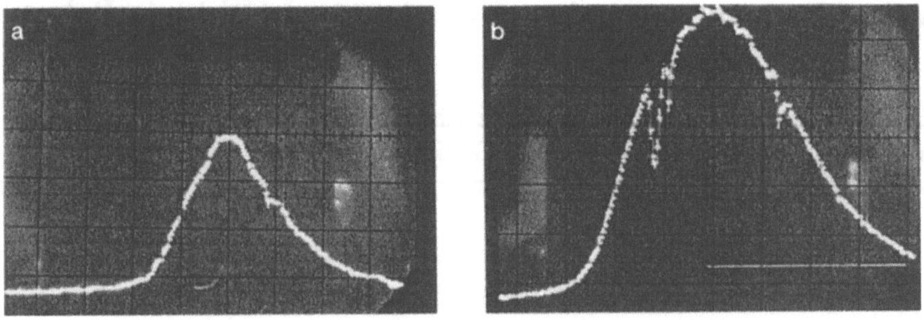

Fig.1. Spectrum of the first (a) and the second Stokes (b). Vertical scale: intensity, arb. units; horizontal scale: 0.91 nm/div. The dips in the spectra are artifacts, due to imperfections of the diode array.

Further experiments are aimed at autocorrelation measurements of the Stokes pulses and scattering in other gases.

In conclusion, we have demonstrated that efficient Raman conversion is possible even with femtosecond light pulses which are almost three orders of magnitude shorter than the transverse relaxation time of the Raman medium. However, in order to achieve conversion efficiencies which are comparable with those for pulses in the nanosecond range, new techniques are needed. A promising approach is the so-called spatially-dispersed Raman scattering [4], which is analogous to the technique of the spatially-dispersed amplification of ultrashort laser pulses [6].

References:

1. J.R.Murray, J.Goldhar, D.Eimerl, and A.Szöke, IEEE J. Quant.Electr. **15** 342 (1979) , and references therein.
2. Ch.Wilkerson, Jr., E.Sekreta, and J.P.Reilly, Appl.Opt. **30** 3855 (1991)
3. R.L.Carman, F.Shimizu, C.S.Wang, and N.Bloembergen, Phys.Rev. A **2**, 60 (1970) .
4. K.A.Stankov, unpublished result.
5. S.Szatmari and F.P.Schäfer, Opt.Commun. **68** 196 (1988)
6. I.P.Christov and M.B.Danailov, Opt. Commun. **84** 61 (1991)

Organic Crystalline Fiber for Efficient Compression of Femtosecond Laser Pulses

M. Yamashita

Department of Engineering Science, Faculty of Engineering, Hokkaido University, Kita 13-jyo, Nishi 8-chyome, Kita-ku, Sapporo 060, Japan

Abstract. We show that organic crystals with large second-order nonlinearity in the nonresonant wavelength region have femtosecond response, large nonlinear refractive indexes, and their crystal cored fibers are attractive femtosecond pulse compressors.

Compression of femtosecond optical pulses using a single-mode fused silica fiber (glass fiber) always needs high amplification of laser output pulses because of a low nonlinear refractive index n_{2G} of its fiber [1]. This leads to some problems concerning the pulse stability in time and space, the pulse repetition rate, the optical damages to the glass fiber, and the precise adjustment of a big amplifier system. In order to solve these problems we inverstigate femtosecond-response, highly third-order nonlinear organic crystalline fibers.

Recently, many second-order nonlinear organic crystals have been developed for effective second-harmonic generation of semiconductor laser pulses. However, their third-order nonlinearities and femtosecond responses are not yet made clear. Then, we have examined the relation between the largest second-order nonlinear coefficients $|d_{ijk}|$ and third-order nonlinear susceptibilities $\chi^{(3)}$ of already measured nonlinear crystals and glasses [2], as shown in Fig. 1. We have selected only the materials whose second- and third-order nonlinearities were both measured in the nonresonant wavelength region. This means that their nonlinearities arise from the electronic polarization and respond to femtosecond pulses. From Fig. 1 we find that, the third-order nonlinear susceptibility increases in proportional to the 1.7th power of the second-order nonlinear coefficient, which is qualitatively explained by a theory. This fact predicts that,

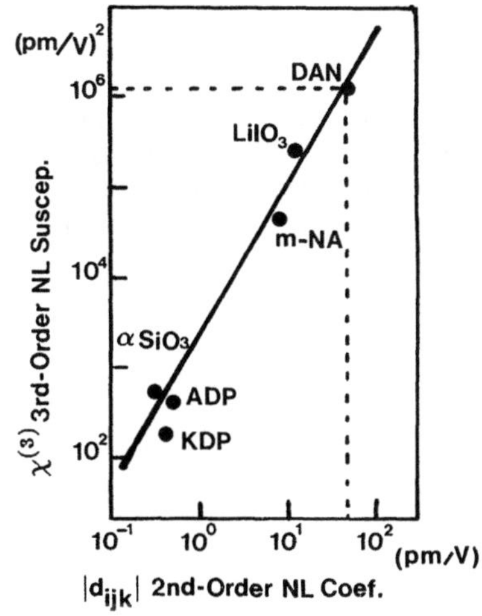

Fig. 1. $\chi^{(3)} - |d_{ijk}|$ relation in the nonresonant wavelength region.

Springer Series in Chemical Physics, Vol. 55 Ultrafast Phenomena VIII
Editors: J.-L. Martin · A. Migus · G.A. Mourou · A.H. Zewail

the nonresonant third-order nonlinear susceptibilities of the large second-order nonlinear organic crystals, whose nonlinear polarization arise from the π-electron cloud distribution, are larger than about the fourth power of that of the fused silica glass.

Basesd on these considerations, we have measured a femtosecond-response, nonlinear refractive index $n_{2\,\mathrm{DAN}}$ of a 4-(N,N-dimethylamino)-3-acetamidonitrobenzene (DAN) single-crystal cored fiber with a 3.9 mm length [3]. The spectral broadening of a 250-fs pulse due to dispersive self phase modulation (SPM) in the fiber, arising from the nonlinear refractive index effect, has been measured in comparison with that of the glass fiber. The result has shown that the fiber output spectrum for the input pulse of the 500 W low power has been effectively broadened from 3.3 to 12.1 nm by a frequency upchirp due to dispersive SPM in the DAN fiber. On the other hand, for a glass fiber with a 4 mm length a much higher power of 500 kW has been needed for a spectral broadening of 19.8 nm (at the same input pulse duration). Using a ratio equation $\Delta\nu_{\mathrm{DAN}}/\Delta\nu_{\mathrm{G}}$ of the spectral width broadened in the DAN fiber to that in the glass fiber during nonlinear pulse propagation [4], we estimate an ultrafast $n_{2\,\mathrm{DAN}}$ to be $1.7 \times 10^4 \times n_{2\,\mathrm{G}}$ from those results. This is 1.3×10^2 times as large as that of a femtosecond component of an $n_{2\,\mathrm{N}}$ of liquid nitrobenzene which is the same molecular family as DAN. In addition, this result corresponds to the above mentioned 1.7th power relation of $\chi^{(3)}$ and $|d_{ijk}|$, as shown by a dotted line in Fig. 1.

Femtosecond output pulses from a CPM dye laser have been directly compressed using a 5-mm-long DAN fiber followed by a grating pair. The output pulse duration and average power of the CPM laser have been 39 fs and 28 mW at the central wavelength of 623 nm with the spectral width of 11.0 nm, respectively. The total transmission efficiency including the coupling and propagetion losses of the DAN fiber has been typically 18%. While monitoring the laser output and compressed pulses by two autocorrelators and the fiber output spectrum by an optical spectrum analyzer, the grating separation has been carefully adjusted to get the optimum amount of negative group velocity dispersion. Consequently, the compressed pulse duration of 22 fs has been generated with the spectral broadening of 17.9 nm [5]. This is the shortest pulse, to our knowledge, among the pulse compressed without any amplification of laser output pulses, and the first application of an organic fiber to pulse compression.

This result suggests that, the application of organic fibers to direct compression of femtosecond output pulses from a mode-locked Ti:sapphire laser with a higher peak power, will generate shorter pulses. Moreover, the application to semiconductor laser pulse compression will bring the more interesting results. We have calculated in details the dependence of the compressed pulse duration on the output peak power from a picosecond semiconductor laser, using a nonlinear Schrödinger equation. The computor calculation has shown that, a 200 mW peak power pulse with a 20 ps duration, which is available, can be compressed directly to 500 fs.

[1] R. L. Fork et al., *Opt. Lett.* **12**, 483(1987). [2] S. Singh et al., in *"Handbook of Laser Science andTechnology III"* ed. M. J. Weber (CRC, 1986). [3] M. Yamashita et al., *Appl. Phys. Lett.* **57**, 1301(1990). [4] J. Herrman et al, *"Lasers for Ultrashort Light Pulses"* (North-Holland, 1987), p.253. [5] M. Yamashita et al., *Appl. Phys. Lett.* **58**, 2727(1991).

Nonlinear Temporal Diffraction in Optical Fibers

G.R. Boyer[1], *M.K. Jackson*[1,2], *J. Paye*[1], *M.A. Franco*[1], *and A. Mysyrowicz*[1]

[1]Laboratoire d'Optique Appliquée, ENSTA, Ecole Polytechnique, Centre de l'Yvette, F-91120 Palaiseau, France

[2]Department of Electrical Engineering, University of British Columbia, 2356 Main Mall, Vancouver, BC V6T-1Z4, Canada

ABSTRACT: We describe and discuss an experiment in which two temporally stretched femtosecond pulses interact in a Kerr-like nonlinear medium to generate new pulses of comparable duration. In contrast to photon-echo experiments, both advanced and retarded pulses are observed. The results can be explained by drawing an analogy between four-wave mixing in real space and diffraction in the time domain.

A colliding-pulse mode-locked dye laser amplified by a copper-vapor laser at 6.5kHz is used to provide the pulses for the experiment. The pulse width is 65fs and the spectrum is centered at 620nm. The pulse train is divided in a Michelson interferometer into two equal-amplitude parts which are recombined with a variable time delay ΔT of the order of 1 to 10ps between them. The pulses are then temporally stretched using two antiparallel gratings separated by a telescope of -1 magnification before being launched into a single-mode fiber. As the stretched pulse length of 90ps is much longer than the delay ΔT used in the experiment, the pulses temporally overlap in the fiber. At the output of the fiber another grating pair is used to compress the pulses. The recompressor is adjusted to compensate both the initial chirp introduced by the stretcher and the linear group velocity dispersion due to the fiber. After recompression, the pulses are analyzed with an autocorrelator.

In Fig.1 we show the autocorrelation of the recompressed output measured after propagation in a 4.4m non-polarization-preserving fiber for a delay between the pulses of 2ps. Three different pulse energies were used, with intensity I_0 corresponding to an energy of 96pJ per unstretched pulse. The autocorrelation for an intensity of I_0 shows the three peaks expected for the autocorrelation of a pulse pair. At an intensity of $2I_0$ two small satellite peaks can be seen in the autocorrelation, and at $4I_0$ the amplitude and number of satellite peaks has increased; the satellite peaks are separated by multiples of ΔT from the main peak. The peaks at ±2ps become greater with respect to the main pulse as the intensity increases. The autocorrelation for $4I_0$ shows seven peaks, suggesting the presence of additional pulses in the recompressed output. We have seen similar behavior with a 4m polarization-preserving fiber, and have observed up to 13 peaks in the autocorrelation. The spacing between the peaks in the autocorrelation increases as the delay ΔT at the input increases.

We have numerically calculated the effects of linear group-velocity dispersion and self-phase modulation on the evolution of the total electric field envelope function $A(z,t)$ in the fiber. In Fig.2 we show the calculated intensity$|A(z,t)|^2$ as a function of time, at various locations in the experimental apparatus, for an intensity $4I_0$. Fig.2(a) shows the pulse pair before stretching; the pulse separation is 2ps and the pulse width is 65fs. Fig.2(b) shows the intensity of the stretched input to the fiber, near the maximum intensity. The full-width at half maximum of the stretched pulses is 90ps, and their temporal overlap gives rise to interference. Because the chirp is linear, the beat frequency is constant across the pulse, and leads to the regular fringes seen in Fig.2(b).

Springer Series in Chemical Physics, Vol. 55 **Ultrafast Phenomena VIII**
Editors: J.-L. Martin · A. Migus · G.A. Mourou · A.H. Zewail

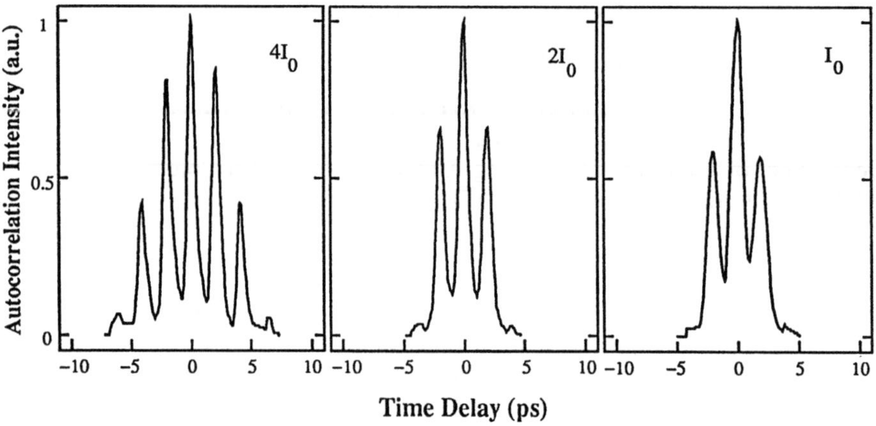

Fig.1 Experimental autocorrelation for three pulse intensities.

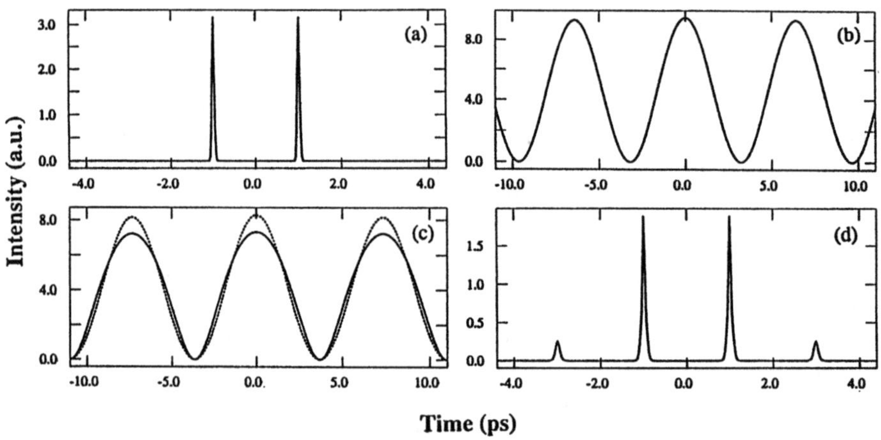

Fig.2 Calculated intensities at various points in the experimental apparatus: (a) unstretched input, (b) stretched input to fiber, (c) uncompressed output from fiber including Kerr-like nonlinearity (solid line) and without fiber nonlinearity (dashed line), (d) recompressed output.

In Fig.2(c) we show the intensity at the output of the fiber, taking account of the Kerr-like nonlinearity, before recompression. The fiber nonlinearity used is $n_2 = 3.1 \times 10^{-20} m^2/W$, the fiber dispersion is $\beta_2 = 6.5 \times 10^{-26} s^2/m$, and the core area is $9 \times 10^{-12} m^2$. The peak stretched input power is 16W. For comparison the dashed line in Fig.2(c) shows the intensity for the case of linear propagation. The nonlinearity leads to a sharpening of the fringe minima, and a distortion of the sinusoidal interference fringes. Fig.2(d) shows the recompressed output, which would be identical to Fig.2(a) in the absence of nonlinearity. The nonlinearity creates side peaks preceding and following the two input pulses by ΔT.

Fig.3 shows the calculated intensity autocorrelation traces corresponding to Fig.1. The calculated autocorrelation traces have been convolved with a 1ps Gaussian response function

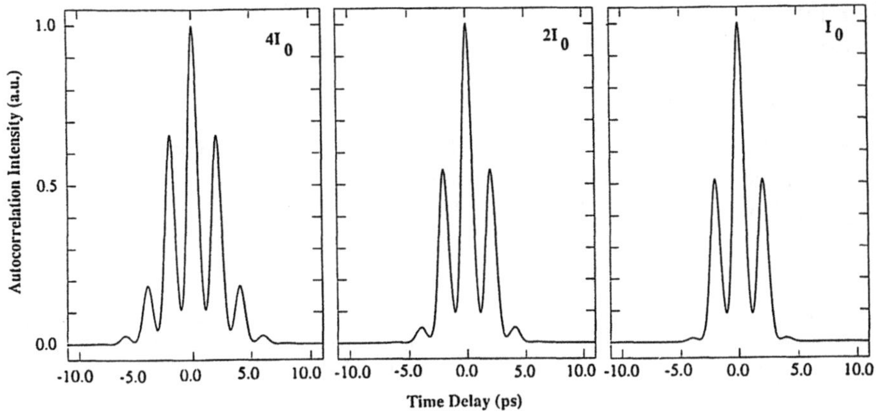

Fig.3 Calculated autocorrelation for the same three intensities shown in Fig.1.

to account for the experimental autocorrelator resolution. The comparison of Figs.1 and 3 is striking. Both curves show the presence of small peaks separated from the central peak by multiples of the delay between the input pulses. The calculated autocorrelations show clearly the increase of the height of the satellite peaks with respect to the main peak for increasing intensity, and agree well with the experimental observations. The presence of many peaks in the experimentally observed autocorrelation trace is consistent with the formation of pre- and post-pulses in the recompressed intensity as seen in Fig.2(d).

It is interesting to explore the analogy between the present experiment, which involves the temporal overlap of two pulses in a nonlinear medium, and experiments involving spatial overlap in a nonlinear medium. As is well-known, the nonlinear interaction of two intense noncollinear spatially overlapping beams leads to the generation of new beams in different directions. These new beams correspond to photons diffracted by the index (or absorption) grating created by the interference of the two beams. By analogy, we interpret the peaks that are formed before and after the two input pulses as being due to "temporal diffraction" of photons from the two input pulses. As we see in Fig.2(b), the interference of the two stretched pulses leads to temporally-modulated intensity which changes the index of refraction, leading to dark-soliton-like pulse shaping [1]. Whereas the experimental conditions of Ref.1 rely on the fiber dispersion to chirp the input pulses and create interferences between them, in our experiment this chirp is controlled and introduced by the stretcher before the pulses are launched into the fiber. The nonlinear propagation in the fiber results in the formation of pre- and post-pulses in the recompressed intensity, as seen in Fig.2(d).

The authors would like to acknowledge stimulating discussions with A. Migus, and technical assistance from J.-C. Pourny. This work was supported by contract 908B039LAB from the CNET. One of us (M.K.J.) would like to acknowledge support from the Natural Sciences and Engineering Research Council of Canada.

References

[1] J.E. Rothenburg and H.K. Heinrich, Opt. Lett. **17**, 261(1992).

Generation of a Soliton Pulse Train in an Optical Fibre Using Two CW Single-Frequency Diode Lasers

S.V. Chernikov[1], *J.R. Taylor*[2], *P.V. Mamyshev*[1], *and E.M. Dianov*[1]

[1]General Physics Institute, 38 Vavilov Str., 117942 Moscow, Russia
[2]Femtosecond Optics Group, Physics Department, Imperial College, London SW7 2BZ, UK

Abstract. Tunable repetition rate (80 - 130 GHz) soliton pulse trains, with pulse durations in the range 1.5 - 3.0 picoseconds at 1.53 μm have been derived from cw DFB diode laser sources, amplified in an erbium doped fibre and propagated in a special optical fibre which exhibited a decreasing dispersion over its length.

1. Introduction

The generation of simple sources of high repetition soliton pulse trains has particular relevance to future all-optical communication systems. Of the several techniques, see for example the references in [1], it has been proposed that the cw beat signal derived from two distinct frequencies can be used for the generation of a cw train of solitons, at repetition rates in the Gigahertz regime [1,2]. The transformation of the beat signal into a soliton train results from the nonlinear propagation in a fibre exhibiting a slow amplification or decreasing dispersion over its length. The combined effect of self phase modulation and group velocity dispersion, influenced by a weak amplification, or a decreasing fibre dispersion, leads to an adiabatic reshaping and pulse compression of the sinusoidal beat frequency into a train of solitons, where the repetition frequency is determined by the frequency difference of the two cw input modes. It has been theoretically predicted that using this technique, a cw train of high quality solitons can be generated with a negligible pedestal component. Here we describe the experimental realization of this technique.

2. Experimental

A schematic of the experimental arrangement is shown in figure 1.

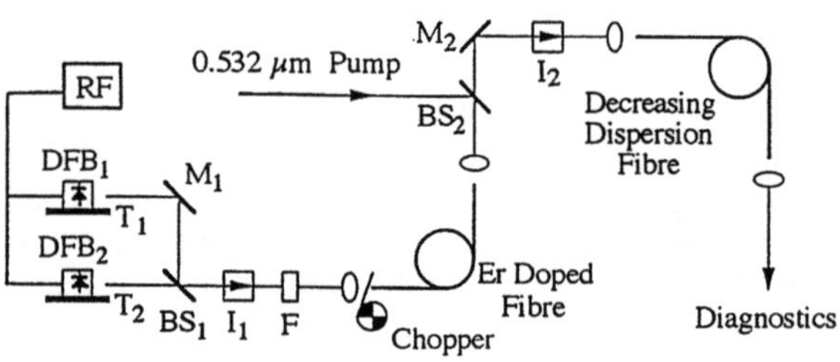

Figure 1. Experimental schematic

Two cw DFB diode lasers operating around 1.53 μm were used to generate the beat frequency, the period of which could be varied by changing the stabilized relative temperature of the devices. Amplification of the beat signal to nonlinear power levels was achieved in a single through pass through a conventional erbium doped fibre. The nonlinear transformation was undertaken in a specially fabricated [3], fibre which had a dispersion decreasing from 6.5 ps/nm.km to 1.1 ps/nm.km over its 2.2 km length. In order to completely suppress the effects

Springer Series in Chemical Physics, Vol. 55 Ultrafast Phenomena VIII
Editors: J.-L. Martin · A. Migus · G.A. Mourou · A.H. Zewail

of stimulated Brillouin scattering, the output from the DFB lasers was phase modulated around 100 MHz via the device drive current.

3. Results

Input and output spectra to the tapered fibre were recorded, see figure 2(a) and (d) respectively. In the time domain, both linear autocorrelation functions at input and output (see figure 2 (b) and (e) respectively) were made using the Michelson Interferometric technique. Conventional background-free SHG intensity autocorrelations were correspondingly taken (figure 2(c) and (f)). The representative results in figure 2 were recorded at a soliton pulse train repetition rate of 95 GHz (ie wavelength difference of cw DFB laser signals $\Delta\lambda = 0.74$ nm). Analysis of the results of figure 2 indicated a good fit to a train of fundamental solitons with durations of 1.8 psec with negligible pedestal component.

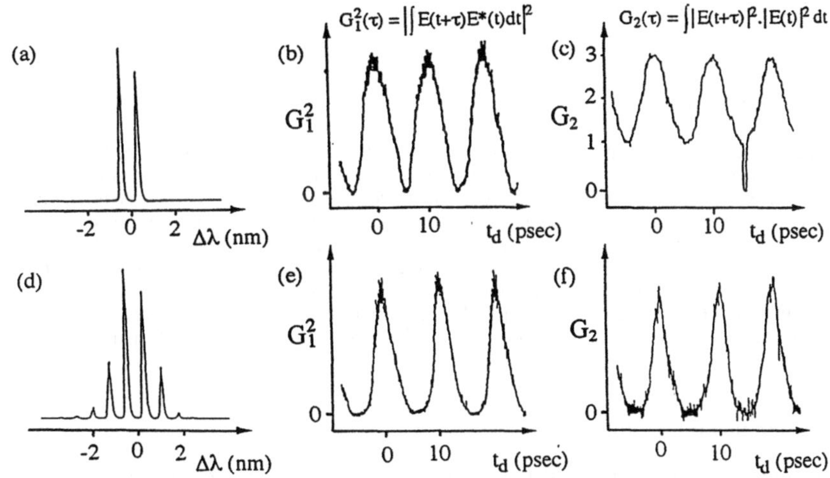

Figure 2. Input signal to tapered fibre (a) spectrum, (b) linear autocorrelation and (c) background free SHG intensity autocorrelation. The corresponding measurements at the output are shown (d) - (f) respectively

4. Conclusions

The experimental generation of a train of high repetition rate solitons from a sinusoidal beat signal through propagation in a fibre with decreasing dispersion has shown good agreement with theoretical prediction. Through operating at various input powers and frequency separations from those described above, trains of solitons with durations in the range 1.5-3.0 psec at repetition rates of 80 - 130 GHz were readily generated [4].

5. References

[1] P.V.Mamyshev, S.V.Chernikov & E.M.Dianov; IEEE J.Quant. Elect. 27, 2347 (1991)
[2] E.M.Dianov, P.V.Mamyshev, A.M.Prokhorov, S.V.Chernikov; Opt.Lett 14, 1008 (1989)
[3] V.A. Bogatyrev et al.; J. Lightwave Technology LT9, 561 (1991)
[4] S.V. Chernikov, J.R.Taylor, P.V.Mamyshev, E.M.Dianov; Electron. Lett. 28, 931 (1992)

Experimental Investigation of Dark Solitons Interaction

Ph. Emplit, J.-P. Hamaide, and M. Haelterman*

Université Libre de Bruxelles, Service d'Optique et d'Acoustique,
50 av. F.D. Roosevelt, CP 194/5, B-1050 Bruxelles, Belgium
*Permanent address: Alcatel Alsthom Recherche, DPE,
Route de Nozay, F-91460 Marcoussis, France

Abstract. We investigate the nonlinear propagation of pairs of picosecond dark pulses over one kilometer single-mode fiber in the near infrared. A repulsive force between black soliton-like pulses is observed, resulting from their interaction.

1. Introduction

Temporal dark optical solitons are stable odd solutions of the Nonlinear Schroedinger Equation describing the propagation, of a spatially single-mode plane wave in presence of normal (or positive) group velocity dispersion and Self-induced Phase Modulation (SPM). After some pioneering experimental demonstrations in the visible region of the spectrum, such propagation of temporal dark signal in single-mode optical fibers has received much attention [1,2] because of its potential telecommunication applications. Recently, single dark pulse propagation experiments, with telecommunication compatible parameters (picosecond duration, infrared wavelength and 1km fiber), were reported [3]. In this paper, as a first step to a study of soliton sequences, we present experimental observations of the propagation of dark pulse pairs.

2. Experimental results

The experimental configuration is depicted in Fig. 1. The picosecond pulses of a mode-locked Ti : Sapphire laser are temporally shaped by a spatial amplitude and phase filtering method, to provide a large square background pulse presenting, centered, a pair of dark pulses. The temporal delay between both dark pulses can be adjusted. By choosing "black" pulses [1], we know that each dark pulse, if isolated, would propagate with a group velocity identical to that of the square background pulse [1]. So, a modification of such a behavior when both pulses propagate will be called a linear (or nonlinear) interaction effect. The effective core area of the fiber is $S_{eff} = 28.3\ \mu m^2$ and its dispersion coefficient D is -91 ps/nm km.

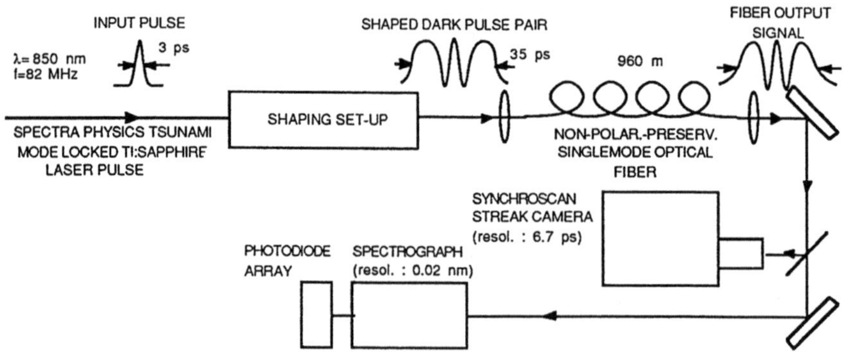

Fig.1: Experimental set-up.

Springer Series in Chemical Physics, Vol. 55 **Ultrafast Phenomena VIII**
Editors: J.-L. Martin · A. Migus · G.A. Mourou · A.H. Zewail

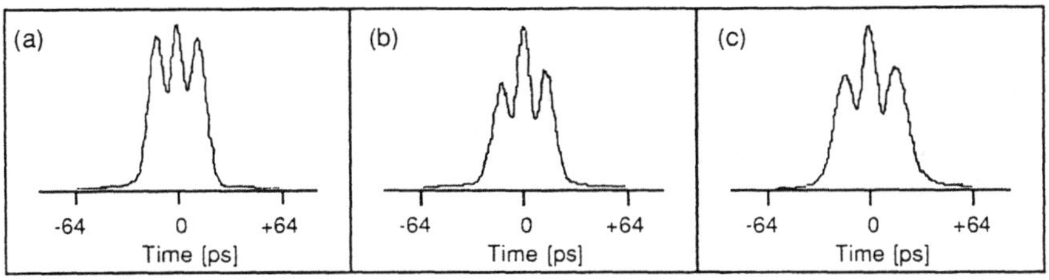

Fig.2 Temporal intensity profiles; (a) Fiber input (dark pulse separation 2.3 T_0); (b) Fiber output (linear regime); (c) Fiber output (nonlinear regime)

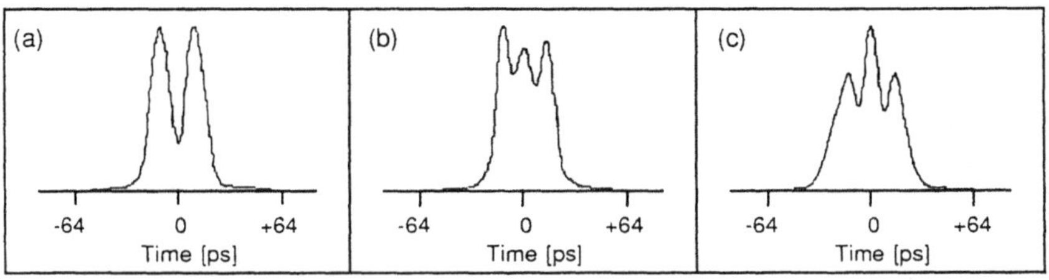

Fig.3 Temporal intensity profiles; (a) Fiber input (dark pulse separation 0.9 T_0); (b) Fiber output (linear regime); (c) Fiber output (nonlinear regime)

The first set of measurements is dedicated to the propagation of a pair of black odd pulses, with individual temporal FWHM $T_0 = 5.5$ ps. Their centers are originally separated by 12 ps, it is 2.3 T_0 (v. Fig.2a). In these experimental conditions, the usual soliton normalizing fiber length Z_0 is 440 m and the soliton peak power $P_0 = I_0.S_{eff}$ is 510 mW. At the output of the fiber (length = 2.2 Z_0), the pulse separation has already reached 15 ps (or 2.7 T_0) at low peak power ($P_{peak} = 8$ mW), as a result of the linear propagation (v. Fig.2b). When P_{peak} is set to 930 mW, at a soliton order of 1.3, the pulse separation remains quite similar and no extra nonlinear repulsion between the pulses is observed (v. Fig.2c).

For the second experiment ($T_0 = 5.1$ ps, $Z_0 = 370$ m, $P_0 = 605$ mW), the black odd pulses are set closer (temporally delayed by 4.6 ps or 0.9 T_0). The proximity of both pulse centers results in an effective absence of "separation" between them (P_{max} between pulses = 0.15 P_{peak}) and makes them temporally undistinguishable on Fig.3.a. Their existence is however confirmed by the spectral analysis. At the fiber output (2.6 Z_0), the linear propagation (P_{peak} = 1 mW) still reveals a large dispersion of the central peak and the delay between the dark pulses is 13 ps (or 2.6 T_0)(v Fig.3.b). Increasing the peak power P_{peak} to 915 mw (soliton order 1.2), pushes the black pulses separation to 16 ps (or 3.2 T_0)(v. Fig.3.c), which appears significantly larger than in the linear case.

3. Conclusion

In order to conclude this preliminary study, we effectively observe an interaction force between black odd pulses when propagating through a 2 Z_0 fiber length. This force is repulsive and decreases when separating the initial pulses. However, we point out that this behavior finds its principal contribution in the linear propagation properties and is not only a direct consequence of the nonlinear SPM effect. The essential role of the SPM is to prevent the dispersion of the black pulses themselves, when the soliton conditions on peak power and time duration are applied.

This research was partially supported by the PAI P3-047 convention of the Belgian Government and partially by the 2.4506.90 grant of the FRFC.

4. References

[1] A. M. Weiner, in *Optical solitons : theory and experiment* , edited by J.R.Taylor (Cambridge University Press, Cambridge,in press).

[2] Y. Kivshar, IEEE J. Quantum Electron. (to be published).

[3] Ph. Emplit, J. P. Hamaide and M. Haelterman, in the Digest of the NLGWP'91, Proc. of the Opt. Soc. of Am. vol 15, 248 (1991)

Femtosecond Pulse Propagation in Erbium-Doped Single-Mode Fibers

J.M. Hickmann[1], *A.S.L. Gomes*[1], *C.B. de Araújo*[1], and *A.S. Gouveia-Neto*[2]

[1]Departamento de Física, Universidade Federal de Pernambuco, Recife, PE, 50739, Brazil

[2]Departamento de Física, Universidade Federal de Alagoas, Maceió, AL, 57000, Brazil

We describe experimental results and numerical simulations of femtosecond pulse propagation in Erbium doped fibers including the effect of the intrapulse Raman scattering term.

Nonlinear pulse propagation and amplification in Erbium doped fibers (EDF) has arisen considerable attention [1,2]. Pulse propagation without amplification can be affected by effects such as self-phase modulation (SPM), intrapulse stimulated Raman scattering (ISRS) and by the dopant susceptibility. Recently Zhu and Sibbett [2] described results of pulse propagation in EDF in the femtosecond regime, without amplification, but the observed spectral and temporal features have not been fully explained. Here we report on a similar experiment which is further supported by numerical simulations. Pulses with $130fs$ centered at $1515nm$, were generated with a time-bandwidth product of 0.45 through a soliton-Raman process in an optical fiber (minimum dispersion at $1360nm$) pumped at $1320nm$ with a $Nd:YAG$ laser [3]. The pulses were launched through $3m$ of EDF (core diameter: $6.5\mu m$, Er concentration of $\sim 0.01M$). The output spectrum and autocorrelation of the pulses are shown in fig. 1(a-c) for two different incident peak powers.

Since the central wavelength of the pump pulse does not coincide with the center transition of the Erbium ions, one has besides gain dispersion, a detuning effect [1]. In the spectra of figs. 1(a,b), the valley are related to the dopant absorption, which splits the spectrum in two parts. The autocorrelation of the output pulse shown in fig.1(c) corroborates the two-peaked spectrum by its three-peaked shape. The field envelope

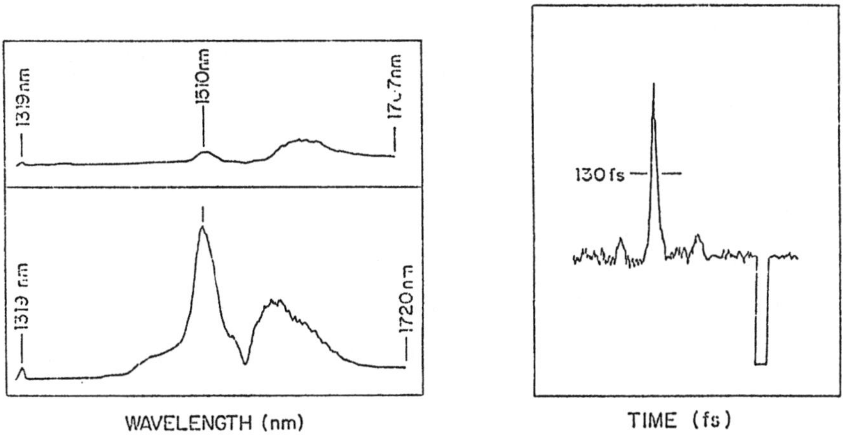

Fig.1 (a) Spectrum after propagation through the Erbium doped fiber for $P_p = 30W$; (b) $P_p = 0.5kW$; (c) Autocorrelation corresponding to $P_p = 0.5kW$.

Springer Series in Chemical Physics, Vol. 55 Ultrafast Phenomena VIII
Editors: J.-L. Martin · A. Migus · G.A. Mourou · A.H. Zewail

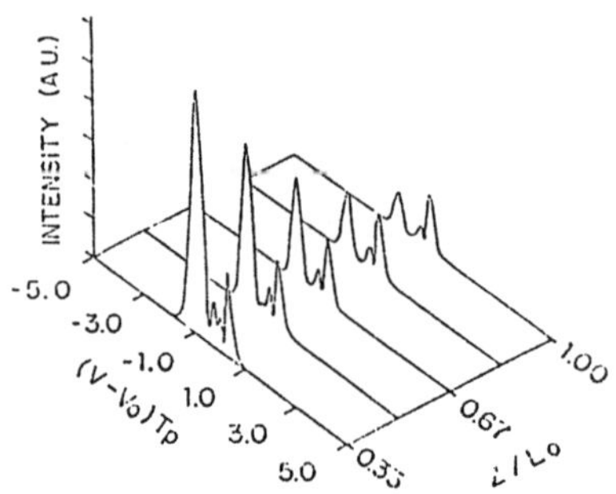

Fig.2 Spectral evolution calculated from equation (1) for $N \sim 2$.

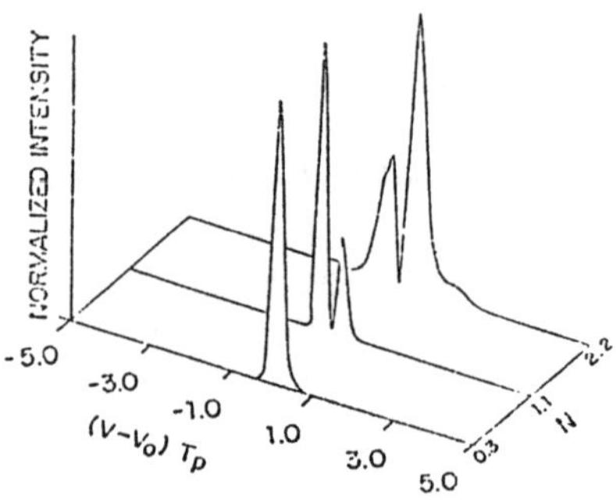

Fig.3 Spectral variation as a function of the soliton order for a fixed fiber length.

amplitude is described by the following equation

$$i\frac{\partial A}{\partial Z} = \beta_1^{im}\frac{\partial A}{\partial T} + \frac{1}{2}\beta_2\frac{\partial^2 A}{\partial T^2} - \frac{g_0}{2}\frac{\delta - i}{1+\delta^2}A - \gamma|A|^2 A - T_R\frac{\partial |A|^2}{\partial T}A \quad , \tag{1}$$

where the first term on the right-hand side represents the imaginary part of the group velocity, associated with the gain dispersion. The second term represents the group velocity dispersion and, the third term includes the detuning and the gain (or loss). The fourth term accounts for the SPM and the last term takes into account the ISRS, which is important in this pulsewidth regime. Using the fiber and dopant parameters, we simulated the pulse propagation for the various soliton orders. Fig.2 shows the spectral evolution as a function of normalized fiber length for a soliton of order $N \sim 2$, where only the final stages of the pulse evolution is shown. A shift to low frequencies can be seen, due to the ISRS term, in agreement with ref. [2]. The change in the intensity ratio of the two peaks in Fig.2 is illustrated in Fig.3 which compares favorably with the results of Fig.1. This behavior has not been seen in ref.[2], and it is due to the ISRS term, which becomes more important for higher values of N, and manifests itself in a shorter propagation distance. As the ISRS shifts the frequency to lower values, where the Er^{3+} absorption is higher, the high frequency side of the spectrum can keep more energy than the low frequency one. Notice that if the ISRS term is neglected, the calculated features do not agree with the observed experimental behavior.

References

[1] G.P. Agrawal, Phys.Rev. A 44, 7493 (1991); Opt. Lett. 15, 224 (1990).

[2] X.Zhu and W. Sibbett, IEEE J.Quant. Electron. 27, 101 (1991).

[3] A.S.Gouveia-Neto, A.S.L.Gomes and J.R.Taylor, IEEE J Quant Electron.QE-24, 332 (1988).

Compression of Pulses from Soliton Fibre Lasers in a Dispersion-Decreasing Fibre

S.V. Chernikov[2], *D.J. Richardson*[1], *E.M. Dianov*[2], *and D.N. Payne*[1]

[1]Optoelectronics Research Centre, Southampton University, UK
[2] General Physics Institute, Russian Academy of Sciences, 38 Vavilov Str., 117942 Moscow, Russia

We report high-quality adiabatic compression of 2-4 psec pulses from a passive all-fibre soliton ring laser operating at 1550 nm down to 250 fsec in a 1.6 km dispersion decreasing fibre (DDF). Compression of 600 fsec solitons from a Figure-8 fibre laser down to 100 fs in a 100m DDF was also obtained.

1. Introduction

Dispersion decreasing fibres (DDF) are promising tools for the compression of soliton pulses. It has been shown theoretically [1-3] that picosecond fundamental solitons can be adiabatically compressed during propagation along a fibre with slowly decreasing chromatic dispersion, the effect is similar to that of soliton compression during adiabatic soliton amplification. In the DDF technique the pulse remains a fundamental soliton as it propagates and conserves the input pulse energy. Its duration therefore decreases in proportion to the absolute value of dispersion. The behavior of femtosecond solitons in DDF is more complicated due to the Raman Self-Frequency Shift (RSFS) and third-order dispersion effects [4].

In this paper we report for the first time a demonstration of the adiabatic compression of both picosecond and subpicosecond solitons in DDFs. A passive mode-locked erbium fibre ring laser [5] and a Figure-8 [6] laser were used as sources of picosecond and subpicosecond soliton pulses respectively. 2-4 ps solitons from the ring laser were compresed upto 230 fs in a 1.6 km DDF (DDF1) having dispersion of 10 ps/nm/km and 0.5 ps/nm/km at 1550 nm at the input and output, respectively. Compression of 600 fs solitons from a Figure-8 laser down to 100 fs was achieved in a 100 m DDF (DDF2) which had dispersion tapered from 10 ps/nm/km to 1.4 ps/nm/km over its length. 600 fs solitons were also adiabatically compressed down to 200 fs using the 1.6 km DDF1.

2. Picosecond soliton compression

The fibre ring laser whose operation is based on nonlinear polarisation evolution [6] consists of an erbium-doped fibre, a 180m section of lo-Bi fibre a polarization sensitive isolator and two polarization controllers (see Fig.1).In order to ensure sufficient laser output power the 70% output coupler was placed directly after the laser amplifier. By appropriate setting of the polarisation controllers and pump power, transform-limited soliton pulses with durations in the range 2-5 ps could be obtained at 1557 nm. A tunable attenuator was spliced between the laser output and the input to DDF1 enabling the pulse energy at the DDF to be matched to that of a fundamental soliton at the DDF1 input. Fig.3a and b shows typical autocorrelation functions and spectra at the DDF1 input and output respectively. 3.5 ps soliton pulses were compressed down to a duration of 230 fs, corresponding to a compression factor of 17. The compressed pulses are seen to have no pedestal and to remain bandwidth-limited. The compression process is accompanied by a spectral shift of ~10 nm due to RSFS. Note, that the spectral features observed in Fig. 3d at the original signal wavelength are associated with the generation of a low level dispersive wave along with the solitons within the passive laser cavity. The additional anti-

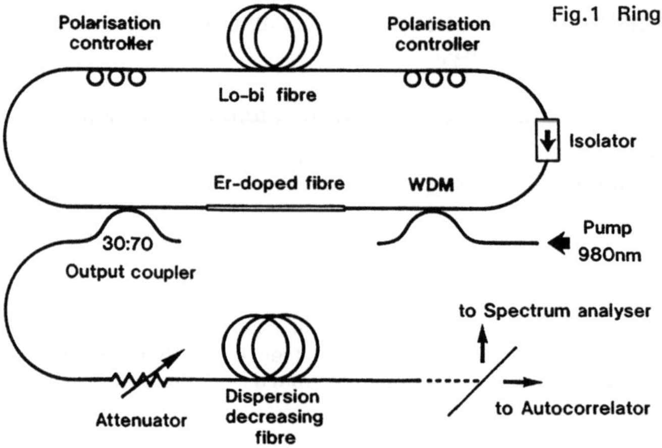

Fig.1 Ring laser configuration.

stokes component at ≈ 1520 nm is associated with soliton propagation in the vicinity of the zero-dispersion point and the effect of third-order dispersion [4]. The output soliton duration did not depend significantly on the input soliton pulse width in the tuning range 2 to 4 ps as expected due to the pulse width stabilisation of femtosecond soliton compression in DDFs [4].

3. Subpicosecond soliton compression

The Figure-8 laser [6] was constructed of fibre with dispersion D = 3.5 ps/nm/km. The 90% output coupler was placed immediately after the gain section. The isolator loop was

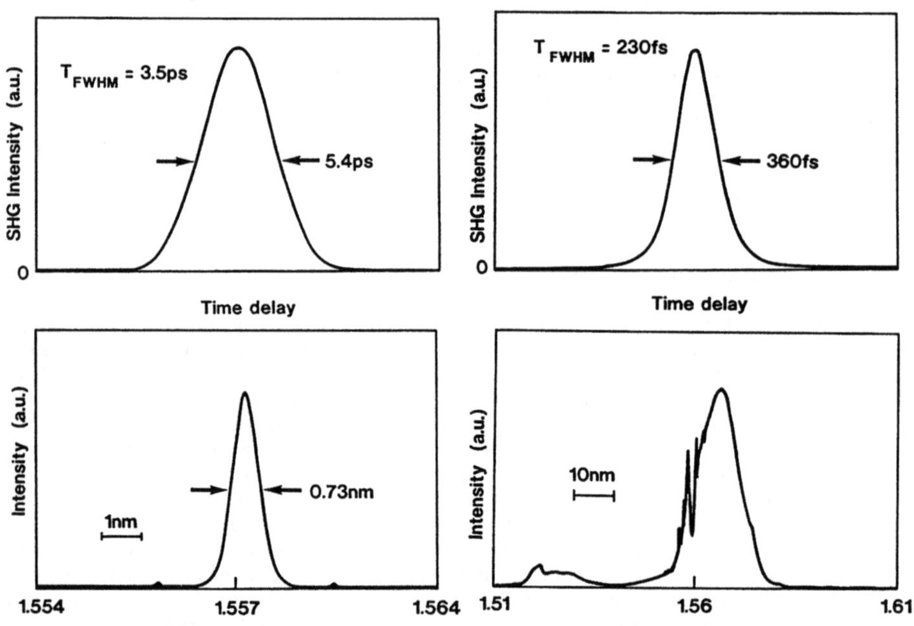

Fig.2 (a) 3.5ps pulses at DDF input. (b) 230 fs pulses at 1.6 km DDF output.

5m long and the total cavity length 40m (see Fig.2). The laser generated 600 fs transform-limited pulses at 1558 nm (Fig.4a). The output passed through an optical isolator and attenuator into the DDF. Adiabatic soliton compression down to 100 fs and corresponding to a compression factor of 6 was obtained due to propagation in the 100 m DDF2 (Fig.4b). The additional spectral line at 1.536 μm in Fig.4b is simply an additional cw component from the laser. The compression of the 600 fs pulses in the 1.6 km DDF1 resulted in compression to 200 fs and a corresponding spectral shift of 25 nm. However, the basic features of the soliton compression remain the same as in the picosecond pulse case discussed earlier.

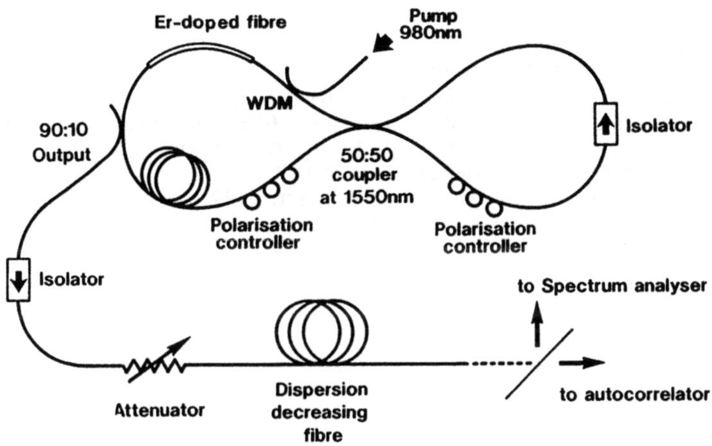

Fig.3 Figure-8 laser configuration.

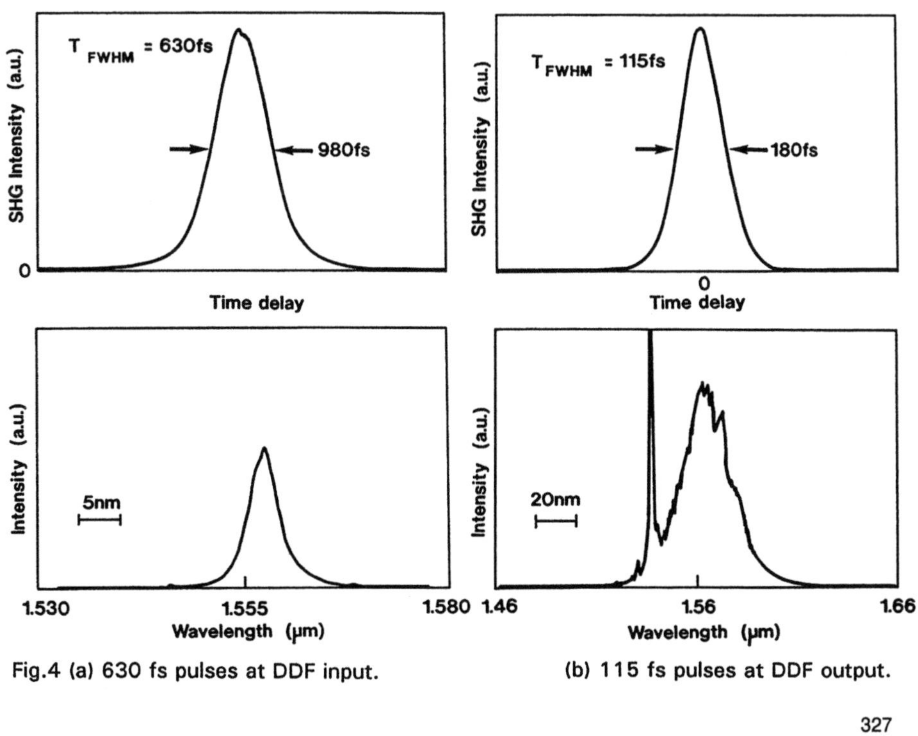

Fig.4 (a) 630 fs pulses at DDF input. (b) 115 fs pulses at DDF output.

4. Conclusion

We have demonstrated high-quality adiabatic soliton compression for both picosecond and subpicosecond solitons in a DDF. Compression factors of upto 17 have been obtained. In our opinion DDFs have great potential as simple passive pulse compressors for use in conjunction with various mode-locked fibre soliton lasers.

References.

1. K.Tajima, Opt.Lett. **12**, 54 (1986).
2. H.H.Kuehl, J.Opt.Soc.Am. B **5**, 709 (1988).
3. V.A.Bogatyrev et al. IEEE J Lightwave Technol. **9**, 561 (1991).
4. S.V.Chernikov and P.V.Mamyshev, J.Opt.Soc.Am. B **8**, 1633 (1991); E.M.Dianov et al.,in *Topical Meeting on Nonlinear Guided Wave Phenomena: Physics and Applications,*vol.2 of 1989 OSA Techn. Digest Series, p.157.
5. V.J. Matsas et al. Accepted for publication, Electron. Lett. 1992.
6. D.J. Richardson et al. Electron. Lett. **27**, 542 1991.

Part VI

Metals, Surfaces and Materials

Observation of the Thermalization of Electrons in a Metal Excited by Femtosecond Optical Pulses

W.S. Fann[1], *R. Storz*[1], *H.W.K. Tom*[1], *and J. Bokor*[2]

[1]AT&T Bell Laboratories, Holmdel, NJ 07733, USA
[2]AT&T Bell Laboratories, Murray Hill, NJ 07974, USA

Abstract. Using 270 fs time-resolved photoemission, we observe the evolution of the electron energy distribution from the nascent distribution to the thermalized Fermi-Dirac distribution in a thin Au film excited with 180 fs laser pulses. Depending on the excitation level, thermalization times as long as ~1 ps were observed. As a result, the relaxation of nonequilibrium electrons is found to be inadequately described by the standard electron-phonon coupling model.

1. Introduction

Because the electron-phonon energy relaxation time in metals is of the order of picoseconds, it is possible to excite metals with femtosecond optical pulses and generate extremely hot electron energy distributions while leaving the lattice relatively cold. Several groups [1-3] have used time-resolved visible reflection spectroscopy to follow the equilibration of the electronic system after femtosecond pulsed excitation. Assuming that the thermalization time is much shorter than the electron-phonon coupling time, the optical data has been used to deduce the electron-phonon coupling constant, G, with a model [4] that assumes that the electrons are always well-characterized by an electron temperature, T_e (e.g., the electrons are instantaneously thermalized to a Fermi-Dirac distribution). However, no direct experimental measurement of the form of electron energy distribution has been reported to justify this assumption. In fact, no quantitative measurement of the electron temperature has been reported either. The optical transient reflectivity experiments [1-3] were only interpreted in terms of *relative* temperature changes.

By using time-resolved photoemission spectroscopy, we have directly measured the time evolution of the electron energy distribution in femtosecond laser excited thin Au films. The distribution was measured with a time-resolution of 270 fs and an electron energy resolution of 0.04 eV. We have been able to directly observe evidence of the nascent (as photoexcited) electron distribution, and the time-evolution from the nascent to a thermalized Fermi-Dirac distribution. Thermalization times as long as ~1 ps were observed. Hence the relaxation of non-equilibrium electrons excited by ultrashort pulses can not be described by the standard electron-phonon coupling model which assumes instantaneous thermalization, e.g.,~10 fs [4]. The breakdown of the model is illustrated by attempting to use it to fit the measured time dependence of the temperature extracted from the best fit of a Fermi-Dirac function to the measured distribution. It is of interest to note that Anisimov, et al. [4] originally proposed the use of photoemission spectroscopy to study the dynamics of picosecond laser-heated metals.

2. Experiment

The laser system consists of a 2-jet prism-compensated synchronously-pumped dye oscillator, and an excimer laser pumped amplifier chain operating at 100 Hz. The laser was operated at a photon energy of 1.84 eV (674 nm wavelength) for these experiments. The measured amplified FWHM pulsewidth was 180 fs. A UV probe pulse (5.52 eV) was generated by frequency-tripling the 1.84 eV pulse in two sequential BBO crystals. The UV pulse FWHM was 270 fs, deduced by observing a 350 fs FWHM cross-correlation with the fundamental. The heating and UV pulse were overlapped on a 300 Å thick Au film held in a vacuum chamber at 5×10^{-11} Torr. Photoemitted electrons were detected with a time-of-flight electron spectrometer. We used a

Springer Series in Chemical Physics, Vol. 55 Ultrafast Phenomena VIII
Editors: J.-L. Martin · A. Migus · G.A. Mourou · A.H. Zewail

relatively large probe beam spot size, ~1mm^2, to eliminate space-charge broadening of the spectrum while maintaining an adequate photoelectron count rate. Since the work function of gold is 5.1 eV, the 5.52 eV probe photons produce photoelectrons by single-photon photoemission.

3. Results

Figure 1 shows the time-evolution of the distribution function for an absorbed fluence of 120 ± 40 μJ/cm^2. The energy is referenced to the Fermi-level, E_F. To directly compare the data with the Fermi-Dirac function, the spectrum plotted is the electron counts per unit energy, dN(E)/dE, divided by the normalized joint density of states (JDOS). The JDOS takes into account the energy-dependent differences in initial and final density of states and in oscillator strengths. The procedure for experimentally deriving the JDOS, and fitting the distributions extracted from the data to a Fermi-Dirac function has been published previously [5].

The data clearly exhibit a two-component distribution. The main part of the distribution is Fermi-Dirac, but there is a non-thermal component extending out to 1.8 eV. In Fig. 1, the best-fit Fermi-Dirac function is shown as the dashed line. At t= 130 fs, the non-thermal component reaches its maximum value as a plateau from 0.7 to 1.8 eV. We can easily see the one photon cutoff at 1.8 eV, which is a signature of the nascent distribution. The possibility must be considered that the plateau region is due to direct, coherent, two-photon photoemission from levels below E_F. This "artifact" should vary in time as the cross-correlation of the pump and probe pulses and be essentially flat from 0 to 1.84 eV. Using a density matrix calculation, we find that at t=0 fs, the ratio of the two-photon coherent contribution to the contribution arising from single-photon photoemission from the real excited population of non-thermal electrons is roughly equal to the ratio between the dephasing time, T_2, and the energy relaxation time, T_1. Since, in general, $T_1 \geq 2T_2$, the maximum artifact is 1/2 of the observed signal. Thus the maximum artifact is ~0.0005 at t=0 fs.

At t=400 fs the best-fit Fermi-Dirac temperature reaches its maximum, 710K. The maximum temperature calculated assuming no energy transfer to the lattice would be 860K. At this time 30% of excitation energy still remains in the non-thermal tail.

According to Fermi liquid theory [6], the energy relaxation time should scale as $(E - E_F)^{-2}$, which is consistent with the data. Even at t=670 fs, a significant population resides in the non-thermal component of the distribution. Yet the thermalized Fermi-Dirac component of the

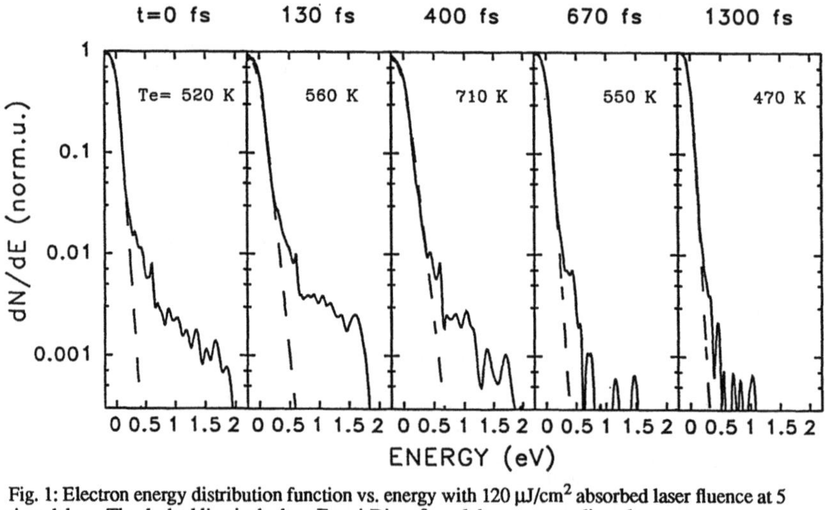

Fig. 1: Electron energy distribution function vs. energy with 120 μJ/cm^2 absorbed laser fluence at 5 time delays. The dashed line is the best Fermi-Dirac fit and the corresponding electron temperature, T_e, is shown. The vertical scale is in units of density of states.

distribution has already begun to cool. It is quite clear that the assumption of instantaneous electron thermalization followed by electron-phonon mediated cooling of the electron distribution is not valid in this experiment.

4. Discussion

We now examine the relaxation dynamics of the electron temperature for the Fermi-Dirac component of the distribution and compare it with the usual electron-phonon coupling model [4]. The time-evolution of the electron and lattice temperatures are given by the coupled differential equations [4]:

$$C_e(T_e)\frac{\partial T_e}{\partial t} = \frac{\partial}{\partial x}\left(\kappa\frac{\partial T_e}{\partial x}\right) - G(T_e - T_l) \quad (1a)$$

$$C_l\frac{\partial T_l}{\partial t} = G(T_e - T_l) \quad (1b)$$

where T_e and T_l are the electron and lattice temperatures, $C_e(T_e)$ is the electronic heat capacity, C_l is the lattice heat capacity, κ is the thermal conductivity, and G is the electron-phonon coupling constant.

Figure 2 shows the nominal electron temperatures extracted from fits to the measured spectra as solid circles. In addition, we show the calculated electron temperatures obtained by numerically solving Eqs. 1. For gold $C_e = \gamma T_e$ with $\gamma = 71.5\ Jm^{-3}K^{-2}$, $C_l = 3.5 \times 10^6\ Jm^{-3}K^{-1}$, and

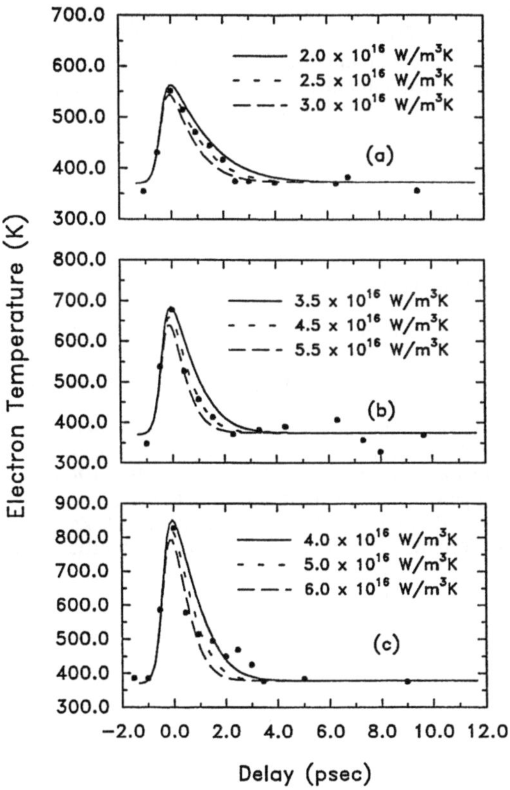

Fig.2: Time dependence of best-fit electron temperature for three heating laser pulse fluences. Fits used the values of G shown and heating laser fluences of (a) 0.09±0.03 and fitted 0.01 mJ/cm^2, (b) 0.16±0.03 and fitted 0.21 mJ/cm^2, and (c) 0.2±0.06 and fitted 0.37 mJ/cm^2.

$\kappa = 310\ Wm^{-1}\ K^{-1}$. Due to experimental uncertainties, for the data of Fig. 2, the t=0 point was arbitrarily chosen to correspond to the time of maximum electron temperature for both the data and calculation. .

No single value of G can fit all three sets of data shown in Fig. 2. However, once the Fermi-Dirac distribution is obtained (at this excitation level, $t > 800$ fs), then a single value of $G = (3.0 \pm 0.5) \times 10^{16}\ W\ m^{-3}\ K^{-1}$ can be used to fit all of the data. This is not a surprising result since Eq. 1a does not properly take into account 1) the finite time for the nascent electrons to relax to the Fermi-Dirac distribution, and 2) the possibility of an increased rate of energy transfer to the lattice when the electron energy distribution is non-Fermi-Dirac. In order for Eqs. 1 to be valid, the time required to relax to the Fermi-Dirac distribution must be much faster than the time required for electron-phonon scattering. This assumption is clearly not valid in view of Fig. 1. On the other hand, once the system is thermalized, the relaxation dynamics is well described by the model with a single value of G.

The values of the absorbed fluence used in the calculations for Fig. 2b and Fig. 2c are somewhat higher than the measured fluence. This is attributable to the fact that the calculation ignores the energy which resides in the non-thermalized component of the distribution

These results help to clarify a previous experiment on femtosecond electronic heat transport dynamics in thin gold films [7]. In that experiment, it appeared that there was a component of non-thermalized electrons which exhibited ballistic transport across a 1000-2000 Å thick Au film. We can now understand that these are the electrons comprising the hot tail which we observe in the present experiment.

5. Conclusions

This study shows that the thermalization of electrons in a metal occurs over a time scale much longer than previously assumed. We observe a thermalization time as long as 1 ps. Our results strongly suggest that ultrashort pulsed excitation of metals produces large numbers of electrons with the nascent energy distribution for up to a few hundred fs. Hot electrons are believed to play a crucial role in enhancing photochemical reaction rates for adsorbed molecules on metal surfaces irradiated with ultrashort pulses [8]. This may open up new possibilities for laser-surface photochemistry.

References

1. H.E. Elsayed-Ali, T.B. Norris, M.A. Pessot, and G.A. Mourou, Phys. Rev. Lett. **58**, 1212 (1987).
2. R.W. Schoenlein, W. Z. Lin, J.G. Fujimoto, and G.L. Eesley, Phys. Rev. Lett. **58,** 1680 (1987).
3. S. D. Brorson, A. Kazeroonian, J. S. Moodera, d. W. Face, T. K. Cheng, E. P. Ippen, M. S. Dresselhaus, and G. Dresselhaus, Phys. Rev. Lett. **64,** 2172 (1990).
4. S. I. Anisimov, B. L. Kapeliovich and T.L. Perelman, Sov. Phys. JETP **39,** 375 (1975).
5. W.S. Fann, R.H. Storz, H.W.K. Tom, and J. Bokor, Phys. Rev. Lett. **68,** 2834 (1992).
6. D. Pine and P. Nozieres, *The Theory of Quantum Liquids,* (W. A. Benjamin, Inc. 1966).
7. S.D. Brorson, J. G. Fujimoto, and E. P. Ippen, Phys. Rev. Lett. **59,** 1962 (1987).
8. F. Budde, T. F. Heinz, M. M. T. Loy, J. A. Misewich, F. de Roughemont, and H. Zacharias, Phys. Rev. Lett. **66,** 3024 (1991); J. A. Prybyla, H. W. K. Tom, and G. D. Aumiller, Phys. Rev. Lett. **68,** 503 (1992).

Femtosecond Thermionic Emission: Experiment, Analytical Theory, and Particle Simulations

M.C. Downer[1], *D.M. Riffe*[1], *X.Y. Wang*[1], *J.L. Erskine*[1], *D.L. Fisher*[1], *T. Tajima*[1], *and R.M. More*[2]

[1]Physics Department, The University of Texas, Austin, TX 78712, USA
[2]Physics Department, Lawrence Livermore National Laboratory, Livermore, CA 94550, USA

Abstract. Simultaneous measurements of reflectivity, total electron yield, and electron-energy distributions, modeled analytically and with particle-simulation codes, characterize space-charge limited ultrafast thermionic emission from metals.

High-intensity femtosecond (fs) light pulses provide the means to excite, on a subpicosecond time scale, the electrons in a solid to temperatures T_e greatly in excess of the lattice vibrational temperature T_l [1]. For excitation of the electron gas to temperatures k_BT_e approaching ϕ (ϕ is the workfunction of the metal), subpicosecond thermionic emission [2] (and possibly thermally assisted photoemission [3]) supersede multiphoton photoemission [4] as the major electron-emission process and is thus a candidate for evaluating the temperature T_e under intense fs radiation. However, the space-charge potential associated with the emitted electron packet, in addition to limiting the total emitted yield, distorts the measured distribution from its initial thermal character. Here we achieve semiquantitative assessment of the yield and distributions, including the strong space charge fields, by simultaneously measuring the incident-light reflectivity, the total electron yield, and electron energy distributions curves (EDC's) of the emitted electrons from three very different single-crystal surfaces —Ag, Al and W — under UHV conditions. A simple quasi-one-dimensional (1D) analytical model and a more rigorous quasi-two-dimensional (2D) particle in a cell (PIC) simulation [5] have been developed to understand the measured yield and EDC's.

10-Hz YAG-amplified 2.0-eV CPM pulses provide the nonequilibrium excitation of the electrons. The average intensity/pulse is varied up to 2.6×10^{11} W/cm^2 (43 mJ/cm^2 fluence), below the cumulative-damage threshold of the metal surfaces. The total yield/pulse is determined from the average current from the sample to ground measured with a solid-state electrometer, while the electron-energy distributions are obtained from time-of-flight spectra measured with a multichannel plate coupled to a transient digitizer [6]. The EDC's considered here are for normal emission.

We find that the temperature-dependence of the total yield is in good agreement with the total yield calculated from the 1D analytical model, which is derived from a space-charge-limited extension of the Richardson-Dushman rate equation for thermionic emission. For a square-wave temperature pulse of duration τ, the total number N of emitted electrons is given by [7]

$$N \approx 460R(k_BT_e)\log\{1+7.76\times10^6R\tau(k_BT_e)\exp(-\phi/(k_BT_e)]\}. \qquad (1)$$

Here R is the radius in microns of the spot being heated; k_BT_e and ϕ are in eV. For reasonably high temperatures (>0.2 eV for a typical metal) the yield is very nearly linear with temperature. Figure 1 has plotted this theoretical yield from Eq. (1) using a measured value of R=158 μm and assumed values of τ=1 picosecond (ps) and ϕ=4.5 eV. Note that the yield is not very sensitive to the last two values since they appear inside the log term, and is thus a simple and direct indicator of the nonequilibrium rise T_e. Also shown are measured yields for Ag, Al and W. Temperatures for the three metals were determined calorimetrically using the measured reflectivities and solving standard coupled equations for the electron and lattice temperatures [8]. Given the simplicity of the model, the agreement in both the turn-on in emission near 0.25 eV and the linearity for higher temperatures is striking. The higher yield for Ag compared to the other two metals is likely due to an enhanced thermal emission associated with surface-plasmon heating of

Editors: J.-L. Martin · A. Migus · G.A. Mourou · A.H. Zewail

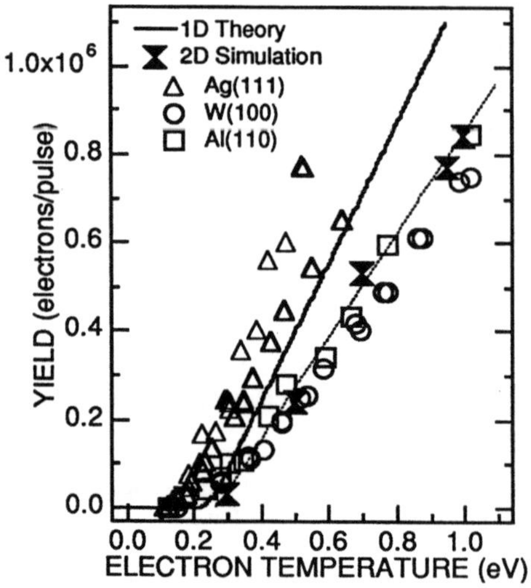

FIG. 1. Experimental, 1D analytic, and quasi-2D computer-simulated total yields vs electron temperature.

imperfections in the visibly scratched Ag surface. Also plotted are the 2D PIC simulation results. Both the high-temperature linearity and an extrapolated threshold near 0.3 eV are apparent in the PIC results which confirm the space-charge-limited thermionic nature of the electron emission.

The general result for the electron EDC's from the experimental and both theoretical investigations is that the space-charge fields produce significant numbers of electrons with much higher energies than a thermal distribution at T_e. The left-hand panel of Fig. 2 shows an experimental EDC (solid curve) for excitation of Al(110) to an electron temperature of 0.7 eV. This is compared with an EDC which would impinge upon the detector for a purely thermal distribution at the same temperature (dotted curve). While the position of the low-energy peaks at ~0.5 eV are reasonably close, the experimental distribution is characterized by a much slower decrease in intensity towards higher kinetic energy (KE). Although not shown here, significant intensity is measured for kinetic energies as high as 40 eV. The experimental distribution also shows a surface-state related feature at ~1.5 eV KE. This sharp feature is highly sensitive to surface preparation (in contrast to the rest of the spectrum) and is overall much weaker than the thermionic emission. It is probably due either to multiphoton photoemission resonantly enhanced by an image-potential state [9] or photoemission from an image-potential state which has become thermally populated by the thermally excited electrons.

The center panel shows a distribution calculated with the analytical 1D model (solid curve) and a comparison with the 0.7 eV thermal EDC (dotted curve). It clearly shows an increased emission towards higher KE caused by coulombic expansion of the emitted electron packet. However, the 1D result suppresses the lowest KE electrons compared to the experimental data. This suppression of the low KE piece of the distribution is undoubtedly related to the fact that in the 1D calculation all of the Coulombic potential energy is transformed into kinetic energyalong the direction normal to the surface. That is, coulombic expansion parallel to the surface is not considered.

The last panel of Fig. 2 compares the result of a 2D-PIC simulation with a thermal EDC at kT_e = 0.7 eV (dotted curve). The PIC EDC was generated by a T_e pulse which follows the calculated temporal evolution of T_e. PIC simulations with other temperature-pulse shapes show that only the peak temperature T_e (and not the pulse shape) is important in determining the yield and distribution. Thus, the use of a square-wave temperature pulse in the 1D model is justified. Again, the simulation shows that space-charge effects lead to significant numbers of electrons with perpendicular KE's greater than thermally possible. The quasi-2D nature of the simulation makes substantial improvement over the 1D analytic theory in re-

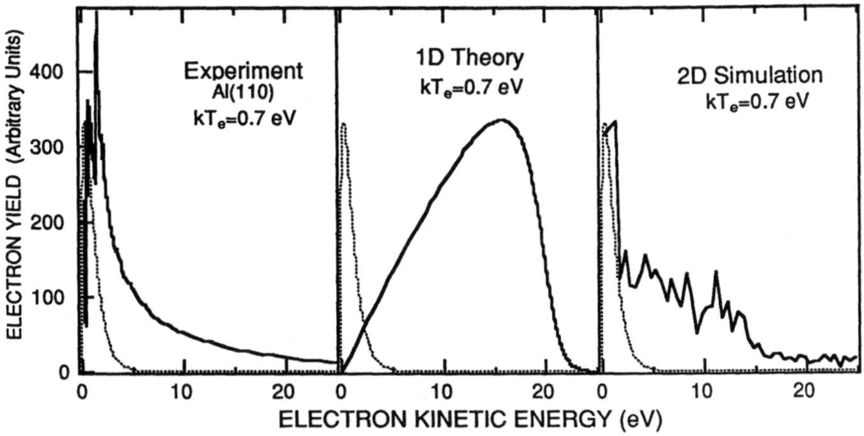

Fig. 2. Experimental, 1D analytic and quasi-2D PIC simulation EDC's (solid lines), each compared to a thermal distribution (dotted curve) at the same temperature.

producing the shape of the experimental EDC by correcting the low-energy suppression seen in the 1D theory. Work at present is aimed towards understanding the remaining discrepancy.

In summary, the quanitative features of the measured total yield is well accounted for by the 1D extension of the Richardson-Dushman equation and the quasi-2D PIC somputer simulations. The shape of the experimental EDC is fairly well reproduced. The 1D and 2D calculations both clearly exhibit the space-charge effect of producing higher KE electrons than thermally possible given the electron temperature T_e.

This work was supported by the Air Force Office of Scientific Research (Contract F49620-89-C-0044), the Office of Naval Research (Contract N00014-88-K-0663), the National Science Foundation (DMR-8858388 and Science & Technology Center Program Grant CHE-8920120), and the Robert A. Welch Foundation (Grant F-1038).

References

[1] H. M. Milchberg, R. R. Freeman, S. C. Davey, and R. M. More, Phys. Rev. Lett. **61**, 2364 (1988).
[2] W. C. Banyai, D. C. Anacker, X. Y. Wang, D. H. Reitze, G. B. Focht, M. C. Downer and J. L. Erskine, in *Ultrafast Phenomena VII* (Springer Verlag, 1990), p.116.
[3] R. Yen, J. Liu, and N. Bloembergen, Optics Communication **35**, 227 (1980).
[4] J. G. Fujimoto, J. M. Liu, E. P. Ippen, and N. Bloembergen, Phys. Rev. Letter. **53**, 1837 (1984).
[5] T. Tajima, *Computational Plasma Physics*, (Addison-Wesley, NY, 1989).
[6] D. C. Anacker and J. L. Erskine, Rev. Sci. Instrum. **62**, 1246 (1991).
[7] R. M. More, X. Y. Wang, D. M. Riffe, M. C. Downer, and J. L. Erskine (unpublished).
[8] S. I. Anisimov, B. L. Kapeliovich, and T. L. Perel'man, Sov. Phys.-JETP **39**, 375 (1974).
[9] W. S. Fann, R. Storz, and J. Bokor, Phys. Rev. B **44**, 10980 (1991).

Electron–Electron Dynamics Observed in Femtosecond Thermoreflection Measurements on Noble Metals

R.H.M. Groeneveld[1], *R. Sprik*[1], *and Ad. Lagendijk*[1;2]

[1]Van der Waals-Zeeman Laboratorium der Universiteit van Amsterdam, Valckenierstraat 65–67, NL-1018 XE Amsterdam, The Netherlands
[2]FOM Instituut voor Atoom- en Molecuulfysica, Kruislaan 407, NL-1098 SJ Amsterdam, The Netherlands

The relaxation dynamics of laser-heated electrons has been studied extensively in metals and (hiT$_c$) superconductors[1] by various femtosecond thermomodulation techniques. These measurements give new insights in the electronic properties and provide direct information about the electron-phonon (e-p) coupling strength. A nearly unexplored type of dynamics is the (e-e) dynamics *within* the electron gas in a laser-heated metal. The extremely small e-e collision time ($\sim$ 10 fs), associated with the highly nonthermal electron distribution during excitation, makes direct observation a challenging task. We report the first experimental evidence of e-e dynamics in thermoreflection measurements by its direct effect on the e-p energy relaxation process for Ag and Au in the regime of low temperature and low laser-energy density[2]. The well-established two-temperature model cannot account for our results and a new model has been succesfully applied.

Our pump-probe setup is based on a standard colliding pulse modelocked laser and incorporates the surface-plasmon polariton (SPP) resonance of a thin metal film[3]. This experimental technique offers an extremely high sensitivity to the change of reflectivity ($\Delta R/R \sim 10^{-4}$) at weak pump-energy densities $\sim$ 1 Jcm^{-3}. The SPP's excited by the pump pulse decay into hot electron-hole pair which redistribute the excess energy by e-e collisions. The electron gas transfers energy to the lattice by e-p collisions in a typical time τ_E: the electron-phonon energy relaxation time ($\sim$ 700 fs at 300 K). The hot-electron induced change of the dielectric constant $\Delta\epsilon$ of the metal is detected as a change of the reflectivity $\Delta R/R(t)$ for various delay times t. We can show that, when the laser frequency lies below the interband transition frequency, $\Delta\epsilon$ is proportional to the total excess energy E contained in the electron gas. As this is the case for Ag and Au, we interpret the measured subpicosecond decay time of $\Delta R/R$ as being the time τ_E. We have investigated 45 nm thick polycrystalline Ag films and 30 nm single-crystalline Au films. The measured relaxation time τ_E decreases monotonously when the temperature is lowered from 300 K to 10 K (dots of Fig 1). In addition, the measured relaxation time did not alter when we reduced the deposited energy density from 1.3 to 0.3 Jcm^{-3}.

We compare our data with the generally accepted *two-temperature model* (TTM), in which the hot electron gas is characterized by a temperature T_e, which is higher than the lattice temperature T_i. The underlying assumption is that the electron and phonon subsystems are each maintained in a thermalized state by e-e and p-p scattering, respectively. In the TTM, the time evolution of the temperatures is described by two coupled heat equations[1,4]. From the TTM, we derived an effective e-p energy relaxation time $\tau_E(T_i)$ as a function of lattice temperature T_i at various deposited laser energy-densities (see Fig 1). The relaxation time of the TTM can be made to match the experimental data point at 300 K. However, at lower temperatures a large discrepancy arises, which could not be due to the small but finite amount of deposited laser energy in our sample. Lowering the laser energy had no effect on τ_E whereas the TTM predicted a large effect (curve C $\rightarrow$ B). We conclude that the e-p energy relaxation process occurs significantly slower at low temperatures compared to the TTM.

Our observations can be readily understood as a consequence of the nonthermal distribution of electrons. We present simple intuitive arguments and the results of a full numerical calculation of the Boltzmann equation including e-e and e-p dynamics simultaneously, which explain that incomplete thermalization of the electron gas greatly reduces the e-p energy relaxation rate at

Springer Series in Chemical Physics, Vol. 55 Ultrafast Phenomena VIII
Editors: J.-L. Martin · A. Migus · G.A. Mourou · A.H. Zewail

Comparison between the experimental data (dots) and the calculated electron-phonon (e-p) energy relaxation time τ_E versus lattice temperature.

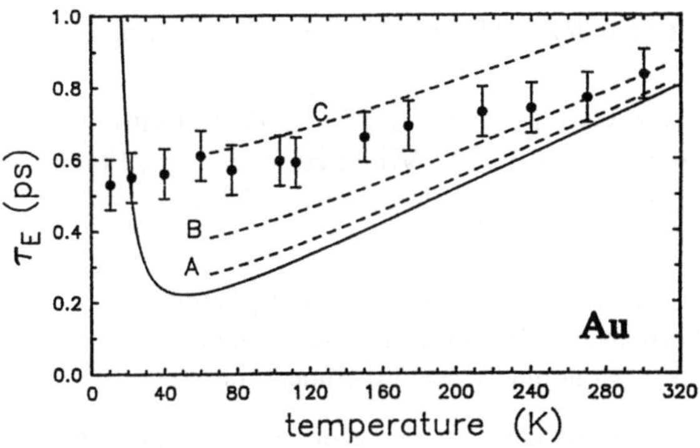

Figure 1: *Solution of the two-temperature model for various amounts of laser-energy densities: 0, 0.3, 1.3, and 5.2 Jcm^{-3} for the solid curve and the dashed curves A,B, and C, respectively.*

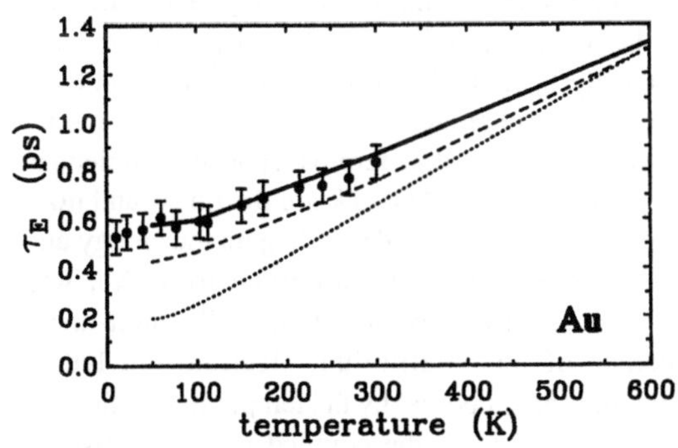

Figure 2: *Solution of the Boltzmann equation which accounts for both e-e and e-p scattering. The e-p coupling strength is a constant. Nevertheless, τ_E varies when the e-e scattering rate K varies. The dotted, dashed and solid lines represent $K = \infty, 0.2,$ and 0.1 $fs^{-1}eV^{-2}$, respectively.*

low temperatures. In Fig. 2, the calculated relaxation time $\tau_E(T_i)$ is depicted as a function of the lattice temperature T_i at a constant e-p coupling strength and for various e-e scattering rates K. The finite e-e scattering rate K of 0.1–0.2 $fs^{-1}eV^{-2}$ reproduces our measurements and is in good agreement with extracted scattering rates from resistivity measurements[5]. Moreover, the effect of the energy density on the calculated relaxation time was nearly absent, which is in agreement with our experimental results.

References

[1] S.D. Brorson, A. Kazeroonian, J.S. Moodera, D.W. Face, T.K. Cheng, E.P. Ippen, M.S. Dresselhaus, and G. Dresselhaus, Phys. Rev. Lett. **64**, 2172 (1990).
[2] R.H.M. Groeneveld, R. Sprik, and A. Lagendijk, Phys. Rev. B. Rapid Communications **45**, 5079 (1992).
[3] R.H.M. Groeneveld, R. Sprik, and A. Lagendijk, Phys. Rev. Lett. **64**, 784 (1990).
[4] M.I. Kaganov, I.M. Lifshitz, and L.V. Tanatarov, Sov. Phys. JETP **4**, 173 (1957).
[5] M. Kaveh, and N. Wiser, Adv. Phys. **33**, 257 (1984).

Inversion of Single- and Two-Photon Photoelectric Sensitivities of Metals in the Femtosecond Range

J.P. Girardeau-Montaut[1], *C. Girardeau-Montaut*[1], *S.D. Moustaïzis*[2], *and C. Fotakis*[2]

[1]Laboratoire des Interactions Laser-Matériau, Université Claude Bernard, Lyon I, 43, bd. du 11 November 1918, F-69622 Villeurbanne Cedex, France
[2]FORTH/IESL, P.O. Box 1527, GR-71110 Heraklion, Crete, Greece

We report for the first time, the inversion of single -and 2-photon photoelectric sensitivities of Au. This property of metals can be observed only in the femtosecond range, as confirmed by our most recent experiments on W.

We report single- and 2- photon photoelectric emission sensitivities of pure-polycristaline Au and W, measured at 248 nm and 496 nm respectively. The radiation source was a femtosecond KrF excimer laser [1] supplying 3-5 Hz repetition rate laser pulses with 450 fs duration and peak intensity I_{ω} up to few 100 GW/cm^2, i.e., the laser damage threshold of most of metals in femtosecond range. In agreement with our model [2] and previous experiments [3] our work confirms that femtosecond photo-electric sensitivity of metals varies as $S_{\omega} \alpha I_{\omega}^{k-1}$, where k > N. N and k are the linear and non-linear photoelectric emission process order, respectively. k is depending on the energy and duration of the laser pulse. This non-linear behavior, first observed from Au, was confirmed later from W, in both cases N = 1, 2 as show in Figure 1. The non-linear growing of S_{ω} is the direct consequence of non-equilibrium electron and lattice temperatures occurring in femtosecond regime as supported by the temporal variations of electron- gas and lattice temperatures calculated from the coupled equations [4] for respective Au and W experimental conditions.

S_{ω} depends also on the angle of incidence θ and on the polarization (pol.) of the laser beam. Our data, although refuting recent conclusions [5, 6], are in good agreement with the vectorial photoelectric emission theory [7]. They point out that maximum p-pol, sensitivity would be observed for θ close to the Brewster-angle. In every case, we verify that s-pol. quantum efficiencies are much lower that ones corresponding to p-pol. incident laser pulse.

We would like now to emphasize, for the first time, a new fundamental property of metals, in the femtosecond range, resulting from the comparison of their single- and 2-photon photoelectric sensitivities. For relatively low laser intensity, typically up to the laser damage threshold of metals in the picosecond regime, we observe that the 2- photon sensitivity is always lower that the single- photon photoelectric sensitivity. That is a property currently observed from any solid material according to the probability of N-photon photo-electric emission is always lower than the one of any lower -order photoelectric process. On the contrary, when I_{ω} is sufficiently high, but such situation is only possible in femtosecond range to do not damage the surface, the 2- photon sensitivity is higher that the single- photon one. A metal can offer, thanks to such property in femtosecond regime, a photoelectric efficiency for visible radiation so high as this one of a semi-conductor. The photoelectric sensitivity inversion represents therefore a significant step towards the application of metallic photocathodes to the production of very high-

Springer Series in Chemical Physics, Vol. 55 Ultrafast Phenomena VIII
Editors: J.-L. Martin · A. Migus · G.A. Mourou · A.H. Zewail

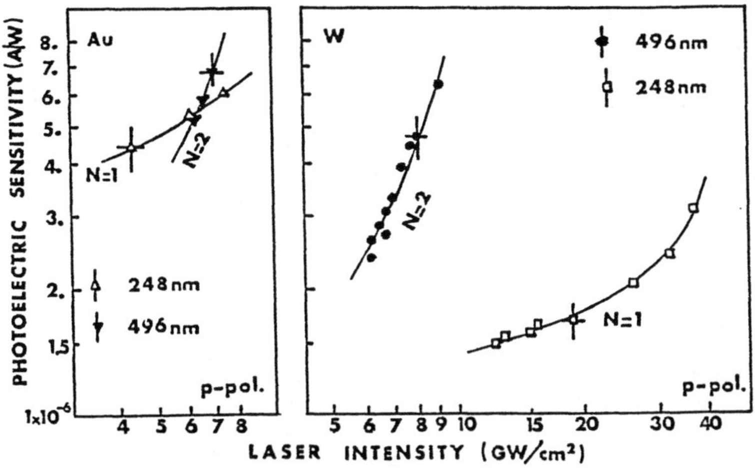

Figure 1 - Crossing and inversion of non-linear femtosecond photoelectric sensitivities of Au and W at 248 nm and 496 nm versus p-pol. laser intensity at θ = 82°.

brightness electron- beam with very short duration, i.e., 1kA/cm^2 and more during less than 1 ps.

References

(1) LAMBDA-PHYZIK Femtosecond Excimer Laser, installed in the ultraviolet Laser Facility, FORTH/IESL, Heraklion, Crete, Greece.

(2) J. P. GIRARDEAU-MONTAUT and C. GIRARDEAU-MONTAUT, "Modeling of electron density produced by femtosecond laser on metallic photocathodes", Proceed. on Indust. and Scientific Uses of High Power Lasers, J.P. Billon and E. Fabre, edit. SPIE, vol. 1502 (1991).

(3) T. SRINIVASAN RAO, T. TSANG and J. FISCHER, Photoemission using femtosecond laser pulses, 13th Int. Free-Electron Laser Conf. Los Alamos (U.S.A.), August 25-30 (1991).

(4) S.I.ANISIMOV, B.L.KAPELIOVICH and T.L.PEREL'MAN, Sov.Phys.JETP, 39 (1974) 375.

(5) Gy.FARKAS, Z.Gy.HORZATH, Cs.TOTH, C.FOTAKIS and E.HONTZOPOULOS, J.Appl.Phys. 62 (11) (1987) 4545.

(6) T.SRINIVASAN-RAO, J.FISHER and T.TSANG, "Femtosecond Photoemission from metals", in Proceedings of the Workshop on short Pulse High Current Cathodes, J.Le Duff.ed., Bendor (Fr) june 18-22 (1990).

(7) R.M.BROUDY, Phys.Rev. B3 (1971) 3641.

Femtosecond Relaxation of Plasma Excitations in Silver Films

R.A. Höpfel[1], *D. Steinmüller-Nethl*[1], *F.R. Aussenegg*[2], and *A. Leitner*[2]

[1]Institut für Experimentalphysik, Universität Innsbruck, Technikerstrasse 25, A-6020 Innsbruck, Austria

[2]Institut für Experimentalphysik, Universität Graz, A-8010 Graz, Austria

Abstract. We report the direct measurement of the relaxation of plasma excitations in silver island films. The lifetime is determined by comparing the second-order autocorrelation of femtosecond laser pulses in nonabsorbing KDP with the second-order-autocorrelation in resonantly excited Ag particles. Applying a forced oscillator model we obtain relaxation times of 40 ± 7 fs for particles with diameters around 12 nm. The damping of the plasma excitations is interpreted as single-particle momentum relaxation through phonon and surface scattering.

Nanometer-scale metal particles reveal strong absorption resonances due to the excitation of localized plasma oscillations [1]. The lifetime and damping mechanisms of these excitations represent interesting physical problems. The linewidths are inhomogeneously broadened in most cases due the nonuniformity and size distribution of the films.

We determine the relaxation times of the plasma excitations in the following way: The second-order autocorrelation of femtosecond laser pulses is measured first with a KDP crystal, where second-harmonic generation occurs quasi-instantaneously, i.e., the nonlinear polarization at time t is a function of the local electric field, $\mathbf{P}(t) \propto \chi^{(2)}:\mathbf{E}(t)\cdot\mathbf{E}(t)$, and does not depend on the polarization at earlier times. Subsequently with the same laser pulse shapes the second-order autocorrelation is measured on Ag island films. The resonant excitation of plasma excitations leads to a local electric polarization and field enhancement also for a certain time *after* the external laser pulse. The resulting broadening of the autocorrelation signal on the Ag islands in comparison to KDP is observed and gives direct information on the damping of the plasma excitations.

The experiments are carried out at room temperature using a CPM dye laser (λ = 610 nm, FWHM = 80 fs to 150 fs, average output power 20 mW). The samples are prepared by evaporating thin inhomogeneous silver films on silicate glass substrates with 200 nm indium-tin-oxide surface layers. The silver film thickness varies between 3 nm and 6 nm, the diameter of the silver islands is determined to be between 10 and 30 nm. The plasma oscillations are localized within the islands, propagation and coupling among different islands is weak.

In Fig. 1 a typical autocorrelation signal of the silver islands (solid curve) via second-harmonic generation as a function of the delay time between the two beams is plotted. The dashed line shows the autocorrelation of the laser pulses which is produced by quasi-instantaneous SHG in KDP. The autocorrelation signal from the silver islands is considerably broader than from KDP.

Measurements for various pulse widths of the CPM laser are shown in Fig. 2, where the FWHM of the correlation signals from silver islands are plotted in comparison to the FWHM of the autocorrelations from KDP. The FWHM of the Ag correlation signal is always larger than the autocorrelation from KDP.

For the interpretation we use the following model: By the incident light field the conduction electrons in the particles, which fulfill the resonance condition [1], are driven

Springer Series in Chemical Physics, Vol. 55 Ultrafast Phenomena VIII
Editors: J.-L. Martin · A. Migus · G.A. Mourou · A.H. Zewail

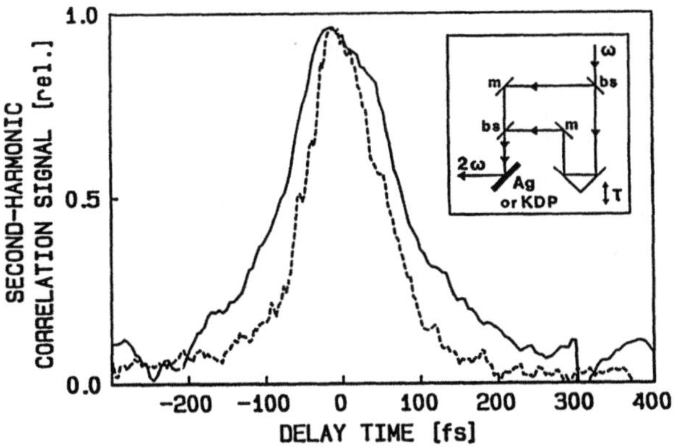

Fig. 1: Second-harmonic correlation signal of the laser pulses on a silver island film (solid) and KDP (dashed).The inset shows schematically the experimental set-up.

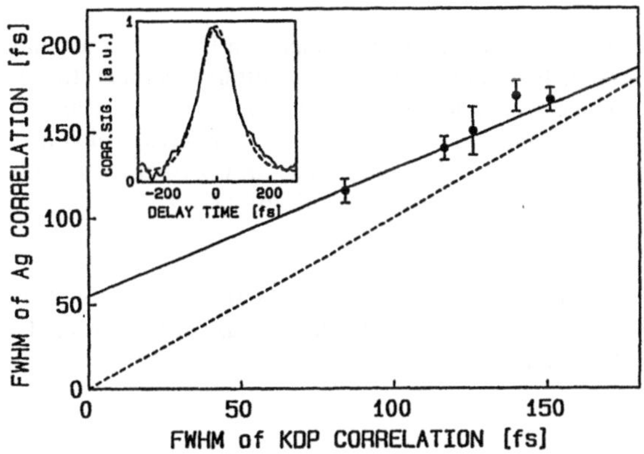

Fig. 2: FWHM of the autocorrelation signal measured on silver islands versus FWHM of the autocorrelation signal of KDP. The dashed line shows a 1:1 relation. Inset: Experimental correlation of silver (solid line) in comparison with the calculated curve (dashed line) using a lifetime of 40 fs.

coherently, forming localized surface plasmons. Due to the broken inversion symmetry by the presence of the surface, the local electric field with frequency ω can excite a 2ω-wave via the local second-order susceptibility of the Ag/ITO/glass system. The local electric field is strongly increased in comparison to the incident light field, due to the excitation of the plasma oscillations. The enhancement factor for our sample parameters is calculated as $f \approx 16$ for cw excitation [2,3]. Therefore, during the complete time range of the excitation and relaxation, the second-harmonic signal is generated mainly by the depolarization field of the plasma excitations. This depolarization field generating the second harmonic is present for the duration of the plasma oscillation and not only for the duration of the light pulse [4]. Thus the second-order autocorrelation is determined by the duration of the plasma oscillation.

Using a driven oscillator model [4] with the light field as the time-dependent excitation, we can calculate the autocorrelation shape from the known laser pulse shape, using the lifetime τ of the plasma excitation as the only variable parameter. Comparison with the whole set of experiments (Fig. 2) gives - independent of the excitation pulse duration - a quantitative value of 40 ± 7 fs for the energy relaxation time of the plasma excitation.

The physical processes that determine the relaxation time of the plasma oscillations are mainly single particle momentum relaxation processes, in the present case electron-phonon scattering (bulk Ag: 37 fs) and carrier-surface interaction (particle size < mean free path). Our result of 40 fs, which is longer than the calculated total momentum relaxation time in the particles, can be understood via the field contribution outside of the metal particles, where no

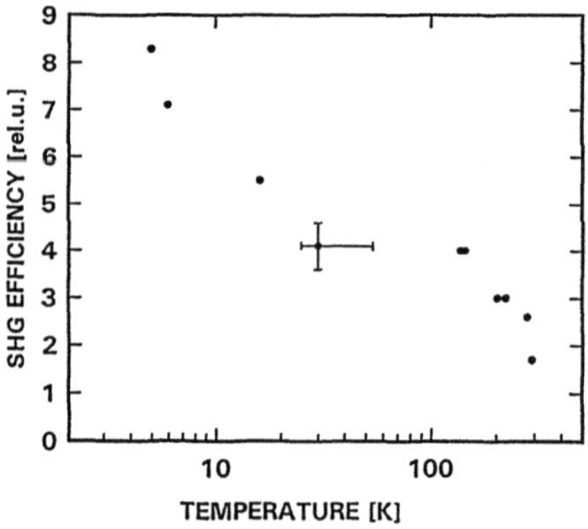

Fig. 3: Temperature dependence of the SHG efficiency on the Ag island film.

damping occurs. Radiative damping plays a minor role for the relaxation [5]. Measurements of the plasmon propagation length in Ag [6] yield values for the lifetime similar to our results. The temperature-dependence of the second-harmonic generation efficiency (Fig. 3.) indicates an increase of the lifetime of the plasma excitations at low temperatures, which may result from weaker electron-phonon scattering. Time-resolved measurements at low temperatures should confirm this interpretation of the increased SHG efficiency.

The authors acknowledge the collaboration with Prof. E. Gornik, S. Juen, K.F. Lamprecht, and R. Rodrigues. This work has been supported by the "Fonds zur Förderung der wissenschaftlichen Forschung" (P 7558 and P 8704).

[1] *Absorption and Scattering of Light by Small Particles*, edited by C. G. Bohren, D. R. Huffman, (John Wiley & Sons, 1983).
[2] P. W. Barber, R. K. Chang, H. Massoudi, Phys. Rev. Lett. **50**, 997 (1983).
[3] A. Leitner, Mol. Phys. **70**, 197 (1990).
[4] D. Steinmüller-Nethl, R.A. Höpfel, E. Gornik, A. Leitner, F.R. Aussenegg, Phys. Rev. Lett. **68**, 389 (1992).
[5] J. Crowell, R. H. Ritchie, Phys. Rev. **172**, 436 (1968).
[6] M. van Exter, A. Lagendijk, Phys. Rev. Lett. **60**, 49 (1988).

Femtosecond Free Induction Decay of Metal Surface Adsorbate Vibrations

J.C. Owrutsky[2], *J.P. Culver*[1], *M. Li*[2], *Y.R. Kim*[2], *M.J. Sarisky*[2], *M.S. Yeganeh*[1], *R.M. Hochstrasser*[2], *and A.G. Yodh*[1]

[1]Department of Physics, [2]Department of Chemistry and Laboratory for Research on the Structure of Matter, University of Pennsylvania, Philadelphia, PA 19104, USA

Recently there has been intense interest in the vibrational dynamics of molecular adsorbates on simple surfaces [1]. In this contribution we present free induction decay (FID) measurements of the vibrational dephasing of CO adsorbed on Cu(111). To our knowledge this represents the first measurement of a FID signal from an adsorbate on a metal, and the fastest coherent transient IR measurement performed on surface adsorbates thus far. The experiments described are the first in a series of ultrafast surface dynamical studies that are underway in our laboratory.

In the experiment we coherently excited molecular CO with an infrared pulse, and then probed the temporal evolution of this coherence by surface induced upconversion using a time-delayed visible probe pulse. The FID measurements were performed using a Nd-YLF regenerative amplifier based laser apparatus. The system can provide 380 fsec, 5 nJ, ~5 μm, IR pulses at a 1 kHz repetition rate by difference frequency generation in $LiIO_3$. Upconversion of the vibrational coherence was accomplished with 270 fsec, 10 μJ, visible pulses that were also mixed with 527 nm pulses to generate the IR.

The sample consisted of approximately 1/3 monolayer of CO adsorbed on the atop sites of Cu(111) in ultrahigh vacuum (1 x 10^{-10} torr) at 93 K. The Cu(111) was prepared by standard UHV procedures and then dosed with 0.45 L of CO. A spectrum of the dosed CO surface was taken by IR-visible sum frequency generation after narrowing the frequency spectrum of the IR pulses. The data yielded a central frequency of 2074 cm^{-1} and a FWHM of 4.2($\pm$.6) cm^{-1} in agreement with reference [2]. The data also enabled us to determine the relative amplitudes and phase of the non-resonant and resonant metal-adsorbate second-order susceptibilities.

The time domain measurement, along with the IR-visible cross-correlation, is shown in Fig. 1. Interestingly we found that the temporal profile of the infrared ($E_{ir}(t)$) and visible ($E_{vis}(t)$) input electric fields could significantly influence the shape of the curve in Fig. 1, and therefore conclusions concerning the total dephasing time. In essence, the resonant and non-resonant contributions of the upconverted field interfere. To account for the interference it is convenient to consider the macroscopic polarization giving rise to the sum frequency signal in the Born-Oppenheimer approximation [3]. In that case the response is given in terms of the macroscopic observables of the surface-adsorbate system. For field envelopes $\mathcal{E}_v^{(j)}(t)$ and $\mathcal{E}_{ir}^{(j)}(t)$ corresponding to polarization j, the complex macroscopic dipole $P_i(t,\tau)$ oscillating at the sum frequency $(\omega_v + \omega_{ir})$ is given by:

$$P_i(t,\tau)e^{-i(\omega_{ir}+\omega_v)t} = \langle\beta_{ijk}\rangle\{\mathcal{E}_v^{(j)}(t-\tau)\mathcal{E}_{ir}^{(k)}(t) + \mathcal{E}_{ir}^{(j)}(t)\mathcal{E}_v^{(k)}(t-\tau)\}$$
$$+ i\mathcal{E}_v^{(j)}(t-\tau)\int_{-\infty}^{t} dt'\mathcal{E}_{ir}^{(k)}(t')\langle[\alpha_{ij},\mu_k(t'-t)]\rangle e^{i\omega_{ir}(t'-t)} \,. \qquad (1)$$

Here μ_k, α_{ij} and β_{ijk} are the dipole, polarizability and hyperpolarizability operators respectively, and τ the time delay between the visible and the IR pulses. The first term corresponds to the instantaneous (electronic) response from the surface-adsorbate hyperpolarizability. The second term is the polarization resulting from the visible field coupling with the polarizability induced by the infrared field driving the nuclear vibrational motions. If there is no angular motion the nuclear term may be written as:

Springer Series in Chemical Physics, Vol. 55 Ultrafast Phenomena VIII
Editors: J.-L. Martin · A. Migus · G.A. Mourou · A.H. Zewail

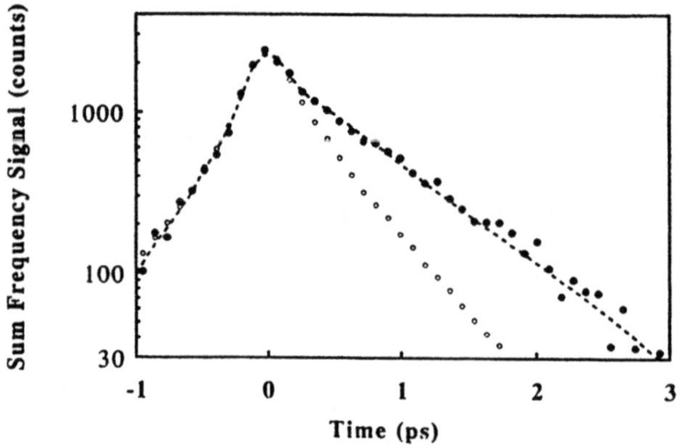

Fig. 1: Measured FID signal (solid circles); IR-visible pulse cross-correlation (open circles); dashed line is our best fit to the data using Eq. 3.

$$-i\left(\frac{d\alpha_{ij}}{dq}\right)_0\left(\frac{d\mu_k}{dq}\right)_0\mathcal{E}_v^{(j)}(t-\tau)\int_{-\infty}^{t} dt'\mathcal{E}_{ir}^{(k)}(t')\langle qq(t'-t)\rangle e^{i\omega_{ir}(t'-t)} \tag{2}$$

and is seen to depend on the polarizability and dipole derivatives of the surface adsorbate system and the convolution of the correlation function of the nuclear displacements, q(t), with the IR field. For a homogeneous system the correlation, or coherence, of the nuclear displacements decays exponentially with a the characteristic decay of T_2 . The signal observed for a time delay τ is the time integral of the square of the sum frequency polarization magnitude, $S(\tau) = \int_{-\infty}^{\infty} dt|P(t,\tau)|^2$. To fit our data we used an equation of the form:

$$S(\tau) = \int_{-\infty}^{\infty} dt I_{vis}(t-\tau)|A\mathcal{E}_{ir}(t) + BG(t)|^2 \,. \tag{3}$$

Here $G(t) = \int_{-\infty}^{t} dt'\mathcal{E}_{ir}(t')e^{-(t-t')/T_2}$, and the coefficients A and B are functions of β_{ijk}, $\left(\frac{d\alpha_{ij}}{dq}\right)_0$, $\left(\frac{d\mu_k}{dq}\right)_0$ and the IR pulse time-bandwidth product. Using this analysis of the decay and modeling the pulses with gaussian shapes, yields a CO vibrational dephasing time $T_2 = 2 \pm .3$ psec which is in good agreement with our frequency domain results.

Acknowledgements: This work was supported by the NSF through the MRL program (RH,AY) and the PYI program (AY), and by the NIH (RH).

References

[1] P. Guyot-Sionnest, Phys. Rev. Lett. **66**, 1489 (1991) ; J. D. Beckerle, M. P. Casassa, R. R. Cavanaugh, E. J. Heilweil, and J. C. Stephenson, Phys. Rev. Lett. **64**, 2090 (1990); A. L. Harris, L. Rothberg, L. H. Dubois, N. J. Levinos, and L. Dhar, Phys. Rev. Lett. **64**, 2156 (1990).

[2] R. Ravel, S. F. Parker, M. E. Pemble, P. Hollins, J. Pritchard and M. A. Chester, Surf. Sci. **203**, 353 (1988).

[3] R.W. Hellwarth, Prog. Quant. Elect. **5**, 1 (1977).

Observation of Laser-Induced Desorption of CO from Cu(111) with 100 fs Time-Resolution

J.A. Prybyla, H.W.K. Tom, and G.D. Aumiller

AT&T Bell Laboratories, Room 4B-421, Holmdel, NJ 07733, USA

Abstract. CO molecules laser-desorb from a Cu(111) surface in $\leq$325 fsec after a 100 fsec, 2 eV photon energy pump pulse. The only mechanism consistent with the observed abrupt, efficient desorption is a novel one: hot electrons ($T_e \sim$ 3600K, while $T_{latt} \sim$ 160K) drive *multiple* electronic excitation-deexcitation cycles of the CO-Cu complex. This vibrationally pumps the molecules, enhancing desorption on subsequent excitations.

A most active pursuit in the field of surface physics has been to develop techniques by which dynamics at surfaces can be resolved on the time-scale of energy transfer within the adsorbate-substrate complex, i.e., subpicosecond. Such techniques would provide a means for directly observing the microscopic mechanisms which make up dynamic processes. Recently, there has been some progress in ultrafast time-resolved studies of energy relaxation at surfaces.[1] Here, we report the first subpicosecond measurement of a surface chemical *reaction*. Specifically, we report that CO molecules desorb from a Cu(111) surface within 325 fsec after irradiation with 100 fsec, 2 eV photon energy laser pulses.

In all previous studies of laser-induced desorption [2,3], the properties of *desorbed* molecules were measured. While informative, such studies can not directly measure the time that the molecules actually desorb from the surface. Here, we use second-harmonic generation (SHG) as a time-resolved probe of the coverage of *adsorbed* molecules.[4] We are thus able to measure the time of desorption with a resolution limited only by the accompanying SH response due to substrate heating. The observation of a surprisingly short desorption time allows us to identify the desorption mechanism. This 100 fsec laser-induced desorption mechanism is distinct from all conventional ones.

The experiments were performed on the CO-saturated Cu(111) held in UHV at a temperature of 95K. The 100 fsec pulses used for this experiment were generated by amplifying the output of a colliding-pulse mode-locked (cpm) dye laser operating at ~620nm. The amplified output was split into a strong pump and a weak probe beam. The probe beam was time-delayed with respect to the pump and then overlapped with the pump on the Cu(111) surface. Both beams were p-polarized. The intensity profile of the beams at the sample was made nearly top-hat by relaying the image from the final prism cell amplifier to the sample. The SH signal generated by the probe was detected with a photomultiplier tube and gated electronics.

Figure 1 (left side) shows the dependence of the SH probe signal on CO exposure. The coverages indicated are those assigned by Raval et al. [5]. On the right-side of this figure, we show the change in SH signal due to desorption from the CO-saturated surface by a *single* 100 fsec laser shot of 4.5 mJ/cm^2 absorbed fluence. Figure 2 shows the result of the pump-probe experiment at the same pump and probe laser intensities as in figure 1. We also show the pump-probe autocorrelation generated by sum-frequency generation (SFG) *on* the Cu surface. This establishes the zero of time delay and the time-resolution of the experiment. The inset shows the same data on an expanded scale together with results using two lower fluences. The spike in the SH signal near zero time-delay is due to a real SH response to electronic excitation which is enhanced by ~2 at zero delay because of the coherent artifact. Analysis of the three curves taken at different fluences (inset) shows that this excitation spike is over in <350 fsec. The constant persisting SH level after that, which has the same magnitude as that due to desorption from a single laser shot shown in figure 1, means that the desorption event is also over by 350 fsec. Deconvolution of this data with the laser pulse gives that >90% of the desorption is complete in <325 fsec.

The desorption yield was also measured (with a mass spectrometer) and was found to be a nonlinear increasing function of the intensity. It can be fit by the power-law ~

Springer Series in Chemical Physics, Vol. 55 Ultrafast Phenomena VIII
Editors: J.-L. Martin · A. Migus · G.A. Mourou · A.H. Zewail

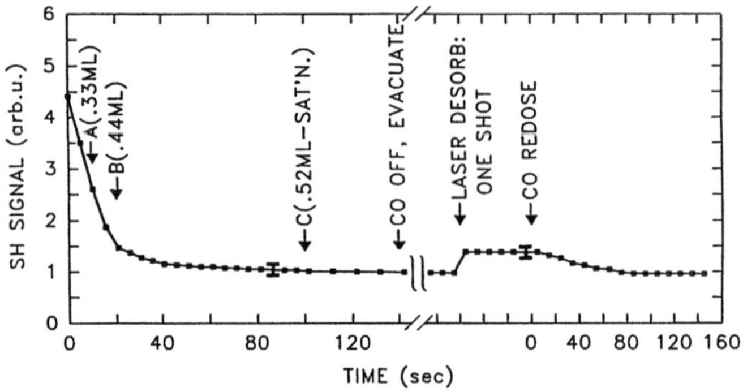

1. Left side: SH signal versus exposure ($P_{CO} \sim 5 \times 10^{-8}$ torr). A,B,C refer to LEED patterns observed (see [5]). Right side: Desorption due to a single laser shot of 4.5 mJ/cm^2, then redose as above.

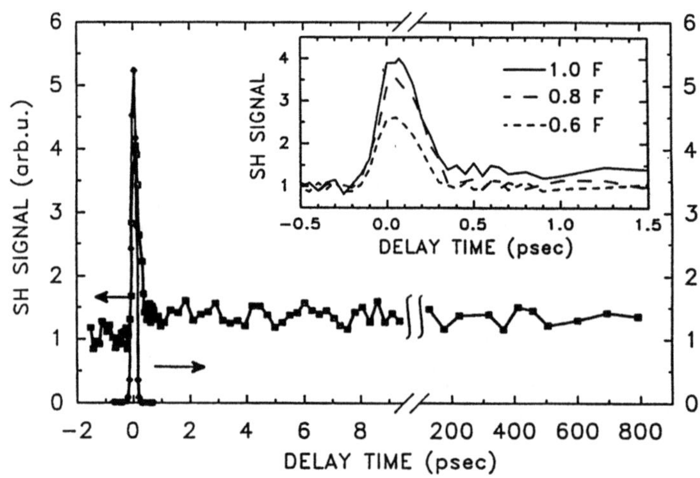

2. SH probe signal versus delay time after the desorption-inducing pump beam. Fluence(F) of pump = 4.5 mJ/cm^2. Intensity autocorrelation also shown (SFG). Inset: Same data (solid line) on expanded time-scale, shown with two other fluences.

$I^{3.7}$ (solid line). For the case of 4.5 mJ/cm^2 absorbed energy, the yield is $\sim$ .05 ML, giving a quantum efficiency of $\sim 10^{-3}$/absorbed photon (cross-section $\sim 10^{-18}$cm^2). This cross-section is much higher (by $\sim 10^3$) than that observed in previous molecule/metal photochemistry studies done using conventional visible sources [3].

The abrupt departure of the molecules can only be caused by hot-electron processes. Modelling (following Ref. 6) of the subpicosecond substrate excitation shows that the electron temperature rises very abruptly ($\sim$100 fsec) to a maximum of $\sim$3600 K, while the lattice temperature increases by only about 100 K over a timescale of several picoseconds. We can immediately rule out all thermal and thermally-assisted processes because the turn-on time for this class of mechanism is too slow. We can also rule out hot electrons *dipole*-coupling to local vibrations [CO–Cu] as the mechanism of desorption because for this case the turn-*off* time is too slow. Additionally, single excitation DIET (desorption induced by electronic transitions) [7] schemes can be ruled out by our observed nonlinear dependence of yield on absorbed fluence. Finally, multiphoton processes are found to be too inefficient to explain our data.

The only mechanism consistent with both the ultrafast desorption time and the superlinear dependence of yield on absorbed fluence is that the hot *substrate* electrons thermalized at $T_e \sim 3600$K, electronically excite the 2π level of the adsorbed CO molecule so that it desorbs. Because the CO-Cu coupling is strong, the CO 2π level is occupied at the Fermi-Dirac temperature of the substrate. At ~3600K, 10% of the CO molecules are excited. To maintain a 10% occupation, 10% of the molecules must be excited every lifetime ~1–10 fsec. Thus in our experiment, the 2π excitation rate is extraordinarily high: >20 times higher than the direct photoexcitation rate. Because of this high rate, a novel mechanism for desorption becomes possible. Multiple excitation-deexcitation cycles within a time short compared to the vibrational relaxation time on the ground state potential can pump the CO molecule to higher vibrational energies than possible with single excitation – this leads to increased desorption probability on subsequent excitation. Note that the additional nonlinearity provided by multiple excitations must be invoked to explain our experimental results. The nonlinear dependence of excitation rate alone on fluence is insufficient.

To examine the relative importance of multiple excitation, we developed a simple mathematical model. This model relies on an excited state potential which provides acceleration to the CO species. For one half of the CO-metal vibrational period, some average vibrational energy, ΔE, is gained (+) for each excitation to 2π, corresponding to an average change in vibrational energy band, Δn=1; for the other half of this cycle, this energy is lost (–). Desorption occurs with probability $\exp[-\tau_{des}/n\tau_{exc}]$, where τ_{des} is the time required for the CO species to evolve into the configuration of no return in the excited state. The model is embodied in two equations (+, for acceleration, and –, for deceleration): $(df_n/dt)^{\pm}=[f_{n\pm1}(1-\exp[-\tau_{des}/n\ \tau_{exc}]) - f_n]G(t)$, where f_n is the fraction of molecules populating the vibrational energy band n (=1 to ∞), and G(t) is the excitation rate to the 2π level. To solve this model we average over all initial CO positions. We find that for reasonable parameters (τ_{exc}=3 fsec, τ_{des}=12 fsec, and E_{thr}=.8eV) we can reproduce all of the experimental results quite well. In particular, we find that ~ 70% of the yield comes from multiple excitations (≥ 2). A more detailed discussion of this model will be given in a subsequent publication.

In summary, we observe that CO molecules desorb from a Cu(111) surface in $\leq$325 fsec after irradiation by a 100 fsec, 2 eV photon energy laser pulse. We find the desorption yield is high and has a super-linear dependence on absorbed fluence. Only a novel mechanism is consistent with our results: the very high electron temperature of ~3600K gives rise to multiple excitation–de-excitation cycles of an adsorbed CO molecule on a timescale short compared to vibrational relaxation. This cumulatively pumps the CO-metal vibration leading to enhanced desorption.

References

1. J. Bokor, Science, *246*, 1130 (1989) and refs. therein; P. Guyot-Sionnest, P. Dumas, Y.J. Chabal, and G.S. Higashi, Phys. Rev. Lett., *64*, 2156 (1990), and refs. therein.
2. J.A. Prybyla, T.F. Heinz, J.A. Misewich, M.M.T. Loy, and J.H. Glownia, Phys. Rev. Lett., *64*, 1537 (1990).
3. a) D.S. King and R.R. Cavanagh, Adv. Chem Phys., *76*, 45 (1989) and references therein. b) S.A. Buntin, L.J. Richter, R.R. Cavanagh and D.S. King, J.Chem. Phys., *91*, 6429 (1989). c) F. Budde, A.V Hamza, P.M. Ferm, G. Ertl, D. Weide, P. Andersen, and H.J. Freund, Phys. Rev. Lett., *60*, 1518 (1988).
4. J.A. Prybyla, H.W.K. Tom and G.D. Aumiller, Phys. Rev. Lett., *68*, 503 (1992).
5. R. Raval, S.F. Parker, M.E. Pemble, P. Hollins, J. Pritchard and M.A. Chesters, Surf. Sci., *203*, (1988).
6. a) R.W. Schoenlein, W.Z Lin, J.G. Fujimoto and G.L. Eesley, Phys Rev. Lett., *58*, 1680 (1987); b) H.E. Elsayed-Ali, T.B. Norris, M.A. Pessot and G.A Mourou, Phys. Rev. Lett., *58*, 1212 (1987); c) P.B. Corkum, F. Brunel, N.K. Sherman, T. Srinivasan-Rao, Phys Rev. Lett., *61*, 2886 (1988).
7. Desorption Induced by Electronic Transitions – DIET III, eds: M.L. Knotek and R.H. Stulen (Springer-Verlag, 1987).

Femtosecond Desorption of Molecular Oxygen from Pt(111)

F.-J. Kao, D.G. Busch, D. Gomes da Costa, D. Cohen, and W. Ho

Laboratory of Atomic and Solid State Physics and Materials Science Center, Cornell University, Ithaca, New York, NY 14853-2501, USA

Abstract. The desorption of chemisorbed O_2 from Pt(111) is shown to have a subpicosecond response time upon excitation with a femtosecond laser pulse.

1. Introduction

Our desorption studies of O_2 from O_2/Pt(111) by irradiation with a femtosecond laser pulse have been motivated by previous results in which by low intensity continuous wave irradiation it has been found that a direct dipole transition of the O_2-Pt complex is responsible for both photodesorption of O_2 and photodissociation[1]. In this case there is no measurable desorption or dissociation at 620 nm while at 300 nm there is a linear dependence of photodesorption yield on absorbed light intensity. In contrast, with the femtosecond excitation, there is a large desorption at both 620 nm and at 310 nm, and at both wavelengths the yield is strongly nonlinearly dependent on the laser fluence, thus pointing to related but alternative mechanisms in the two intensity regimes.

Within the last two years, two related systems have been studied by pump-probe methods with femtosecond resolution, the NO/Pd(111) [2,3] and the CO/Cu(111) [4]. In both of these cases and in the O_2/Pt(111) which we present here the desorption data are incompatible with a phonon assisted desorption process (thermal), or with the Menzel-Gomer-Redhead (MGR) mechanism for desorption induced by an electronic transition [5] which predicts minute desorption rates. A generalization of this mechanism has been introduced to describe desorption from surfaces under conditions of strong electronic excitation which effectively increases the desorption rate by allowing multiple re-excitations of the adsorbate [6]. In all of these three cases of femtosecond excitation there is both a superlinear dependence of the yield on absorbed fluence and an ultrafast turn-on and turn-off time for the desorption event.

It has been demonstrated that an intense femtosecond laser pulse induces nonequilibrium electron heating in a metal [7]; that is, following the optical excitation, carrier-carrier scattering produces a thermalized hot

Springer Series in Chemical Physics, Vol. 55 Ultrafast Phenomena VIII
Editors: J.-L. Martin · A. Migus · G.A. Mourou · A.H. Zewail

electron distribution which is several thousand degrees out of equilibrium with the lattice temperature. Subsequently these hot electrons lose their excess energy to the lattice on a time scale of several picoseconds. In addition, the presence of nonequilibrated carriers in a metal within the first 800 fs after excitation, that is, electrons which have not yet undergone extensive carrier-carrier scattering, has also been demonstrated [8].

2. Experimental

The experiments were performed in an ultrahigh-vacuum system equipped for Electron Energy Loss Spectroscopy, Low Energy Electron Diffraction, Temperature Programmed Desorption, and Time-of-flight measurements. The preparation and characterization of the single crystal Pt(111) was done using standard procedures. The sample was held at a base temperature of 80 K, and was dosed by a microcapillary array. Desorption was caused by the output of an amplified colliding-pulse modelocked laser operating at 10 Hz, 620 nm. Autocorrelation of the amplified laser pulse on the sample surface typically gives us a pulse width (FWHM) of 150-180 fs. The amplified fundamental was doubled in a BBO crystal.

The desorption signal was measured with a quadrupole mass spectrometer operated in the pulse counting mode with a mutichannel scaler. The laser beam entered and exited the UHV chamber at either 28° or 60° from the surface sample normal. The desorption correlation measurements were performed using two orthogonally polarized pulses of equal intensity spatially overlaping onto the crystal in a pump-probe manner. Each data point in Figure 1 corresponds to a single pair of laser pulses on a fresh spot of the surface.

3. Results and Discussion

We have measured the desorption yield to be dependent on the laser fluence to the power of 5.5 at 620 nm in the 3 mJ/cm^2 absorbed fluence region and to the power of 3.0 at 310 nm 1 mJ/cm^2 absorbed fluence region. Within the uncertainity of our measurement there is no appreciable difference between the desorption yields at the two wavelengths. This is interpreted as indication of a different mechanism of desorption from the low intensity regime.

Figure 1 shows the two pulse desorption correlation data as a function of delay between the equal incident fluence pulses of 2 mJ/cm^2 each at 620 nm. The width of the correlation is subpicosecond, indicating a very short lived excitation is responsible for the desorption event. At zero delay we calculate a peak electronic temperature of 1400 K, while

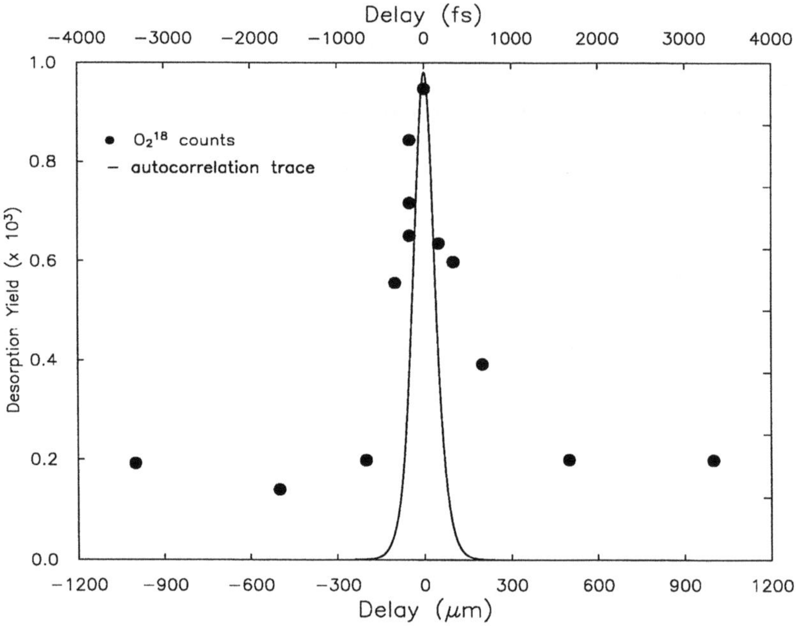

Figure 1 Two pulse desorption correlation data as a function of delay between the equal incident fluence pulses of 2 mJ/cm^2 each at 620 nm. The solid line is the autocorrelation trace of the laser pulse width as measured at the surface.

the lattice temperature reaches a peak value of 180 K ca. 10 ps later. The data at both 310 nm and 620 nm for O_2 desorption from $O_2/Pt(111)$ is in close parallel to the other femtosecond data. However, the thermalized electron temperatures reached in Pt at the fluence levels which we use are not quite as high as those reported for Cu or Pd, there is still a large nonequilibrium between the thermalized electrons and the lattice. In conclusion, our data at both 310 nm and 620 nm are consistent with a multiple electronic excitation mechanism as proposed by Misewich *et.al.* [6], however we can not, at this time, distinguish wheather the process involves nonequilibrated carriers or thermalized carriers, or even a direct multiphoton process.

Support of this work by the Office of Naval Research grant No. N00014-90-J-1214 and the Swiss National Science Foundation is gratefully acknowledged.

References

[1] W. D. Mieher and W. Ho, *J. Chem. Phys.* **91** (1989) 2755, and to be published.

[2] J. A. Prybyla, T. F. Heinz, J. A. Misewich, M. M. T. Loy, and J. H. Glownia, *Phys. Rev. Lett.* **64** (1990) 1537.

[3] F. Budde,T. F. Heinz, M. M. T. Loy, J. A. Misewich, F. de Rougemont, and H. Zacharias, *Phys. Rev. Lett.* **66** (1991) 3024.

[4] J. A. Prybyla, H. W. K. Tom, and G. D. Aumiller, *Phys. Rev. Lett.* **68** (1992) 503.

[5] D. Menzel and R. Gomer, *J. Chem. Phys.* **41** (1964) 3311, P. A. Readhead, *Can. J. Phys.* **41** (1964) 886.

[6] J. A. Misewich, T. F. Heinz, and D. M. Newns, *Phys. Rev. Lett.* in press (1992) .

[7] R. W. Schoenlein, W. Z. Lin, J. G. Fujimoto, and G. L. Eesley, *Phys. Rev. Lett.* **58** (1987) 1680.

[8] W. S. Fann, R. Storz, H. W. K. Tom, and J. Bokor, *Phys. Rev. Lett.* **68** (1992) 2834 .

Femtosecond Carrier Dynamics in Solid C_{60} Films

S.D. Brorson, M.K. Kelly, U. Wenschuh, R. Buhleier, and J. Kuhl

Max-Planck-Institut für Festkörperforschung,
W-7000 Stuttgart, Fed. Rep. of Germany

We have carried out a femtosecond pump-probe study of carrier dynamics in solid films of C_{60}. The experiments were performed using 100 fs pulses of laser light obtained from an amplified CPM laser system. Both undoped as well as Rb-doped films were investigated as a function of pump intensity and sample temperature in the range T = 300 K − 10 K. Also, degenerate four wave mixing experiments on one of the films have been performed to measure the dephasing time.

Samples were made by evaporating C_{60} powder from a Ta boat onto transparent quartz substrates. The films were generally about 5000 Å thick. After deposition, the samples were transferred to a vacuum cryostat fitted with optical ports for measurement. All samples remained under vacuum for the duration of the pump-probe measurements. A Rb source was built into the cryostat to permit *in situ* doping. For the doped samples, the average doping level was determined to be ≈ 2 Rb atoms/C_{60} molecule via ellipsometric measurement and comparison to published spectra.[1]

Typical pump-probe data are given in Fig. 1, which shows ΔT *vs.* t for both an undoped (a), and a doped sample (b). The different curves correspond to different amounts of energy deposited by the pump pulse (E_{pu}). We find that ΔT is negative for all cases. This information coupled with the measured optical constants of C_{60}[1] lead us to conclude that induced absorption is occurring as a result of laser pumping.

As can be seen in Fig. 1, a fast relaxation is evident for times on the order of ≈ 1 ps. This fast relaxation signal is superimposed upon a slower signal, which relaxes on a time scale of the order of hundreds of ps. In the undoped samples the magnitude of the fast decay signal relative to the slow decay signal becomes more pronounced as the pump energy is increased. In the doped samples it remains constant for increasing pump energy, increasing only at the very highest pump power.

In Fig. 2 is plotted τ *vs.* E_{pu}, the energy density of the pump laser pulse. As can be seen, for the undoped samples, τ decreases with increasing E_{pu} for all pump intensities measured,

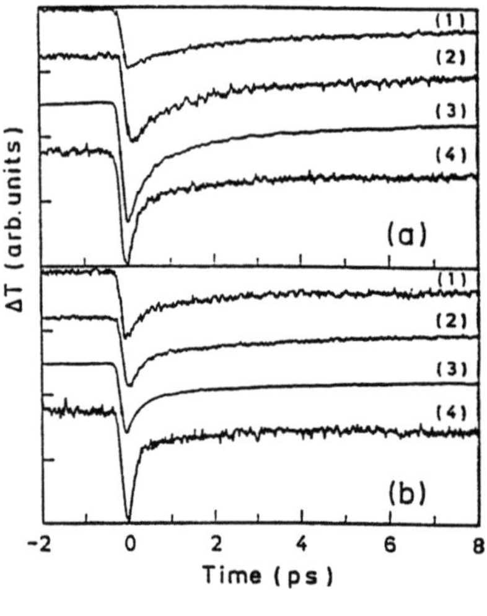

Figure 1. Representative pump-probe data. Shown in (a) is ΔT *vs.* t for undoped C_{60} for E_{pu} = (1) 1.4, (2) 7.2, (3) 29, and (4) 90 mJ/cm^2. In (b), E_{pu} = (1) 4.4, (2) 11, (3) 44, and (4) 73 mJ/cm^2.

Springer Series in Chemical Physics, Vol. 55 Ultrafast Phenomena VIII
Editors: J.-L. Martin · A. Migus · G.A. Mourou · A.H. Zewail

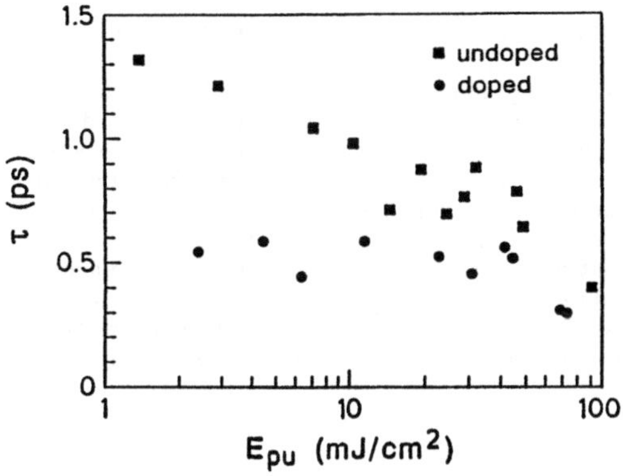

Figure 2. Measured time constant τ *vs.* E_{pu} for doped and undoped samples.

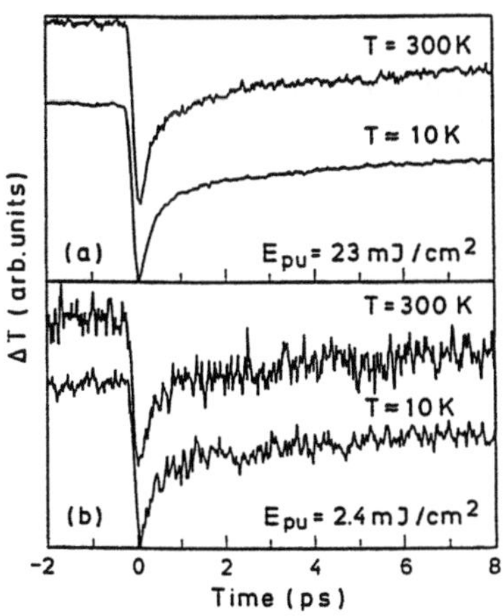

Figure 3. ΔT *vs.* t data for doped C_{60} at T = 300 K and ≈ 10 K. In (a), E_{pu} = 23 mJ/cm²; in (b), E_{pu} = 2.4 mJ/cm².

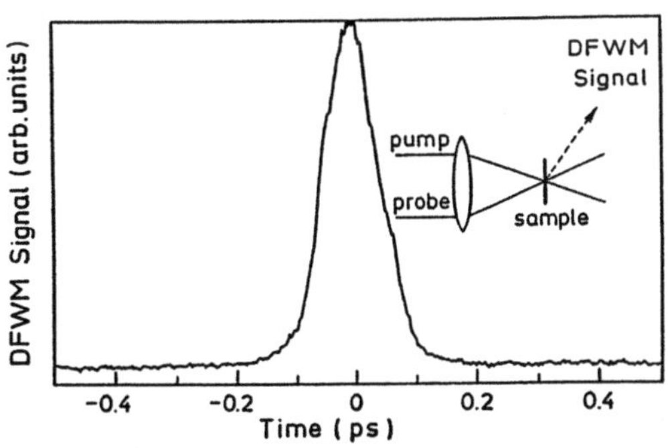

Figure 4. Time dependent diffracted signal observed in undoped C_{60} in the two-beam self-diffraction geometry (see insert).

whereas in the doped samples τ remains constant at about 0.5 ps up to $E_{pu} \approx 40 - 70$ mJ/cm². For $E_{pu} > 50$ mJ/cm², we noticed that irreversible changes occurred in the sample during the course of a run. Therefore, data taken at these high pump energies must be regarded with some suspicion.

In order to learn more about the physical processes at work causing the relaxation, we performed a pump-probe measurement on a doped sample at liquid helium temperature. Upon cooling, the temperature of the cold finger to which the sample was attached was measured to be ≈ 10 K. Relaxation curves were measured before and after cooling for pump fluences of 23 mJ/cm² and 2.4 mJ/cm². The measured relaxation curves are shown in Fig. 3. Identical data were obtained at 300 K and ≈ 10 K for both pump fluences suggesting that the relaxation process is temperature independent in the range ≈ 10 K – 300 K.

Finally, we have also performed a time-resolved degenerate four wave mixing (DFWM) experiment on a film of undoped C_{60}. The experiment was performed using the two-beam (self-diffraction) forward geometry. The magnitude of the diffracted signal *vs.* time is shown in Fig. 4. The diffracted signal shows no obvious decay due to sample response, indicating that the dephasing time (T_2) is shorter than the laser pulse (100 fs). Decreasing the intensity in each beam by a factor of 14 failed to change the time-profile of the diffracted signal, indicating that T_2 remained less than 100 fs.

The occurrence of induced absorption can be explained using a three band model based on the calculated[2] and measured[1,3] electronic level structure of C_{60}. Less readily explained is the physical origin of the observed fast relaxation. Using Allen's theory[4] applicable to electron-phonon relaxation in metals, and assuming a model electron-phonon spectral function which successfully explains the superconducting properties of doped C_{60}[5], we have calculated the relaxation time as a function of sample temperature and pump energy. At 300 K, we obtain a relaxation time of 0.8 ps which is independent of E_{pu}, in agreement with our measurements of doped samples. However, at 30 K the calculated relaxation times depend strongly upon E_{pu}, disagreeing with our results. Indeed, the observed dependence of τ upon excitation density and independence from sample temperature is more symptomatic of a relaxation process mediated by purely Coulomb interactions.

References.

1. M. K. Kelly et al., Phys. Rev. B **46**, 4963 (1992).
2. S. Satpathy, Chem. Phys. Lett. **130**, 545 (1986); S. Saito and A. Oshiyama, Phys. Rev. Lett. **66**, 2637 (1991); W. Y. Ching et al., Phys. Rev. Lett. **67**, 2045 (1991).
3. T. Takahashi et al., Phys. Rev. Lett. **68**, 1232 (1992).
4. P. B. Allen, Phys. Rev. Lett. **59**, 1460 (1987).
5. I. I. Mazin et al., *Submitted to Phys. Rev. B.*

The Role of Covalency in Femtosecond Time-Resolved Reflectivity of Hydrodynamically Expanding Solid Surfaces

X.Y. Wang, H.Y. Ahn, and M.C. Downer

Physics Department, University of Texas at Austin, Austin, TX 78712, USA

Abstract. The ultrafast surface expansion of intensely excited free electron metals and covalently bonded solids are contrasted via time-resolved reflectivity measurements and a numerical wave propagation model.

Intense femtosecond excitation of solid targets has been used to produce and measure the properties of solid density plasmas[1] and pressurized liquids[2]. In several recent studies[1] the ultrafast reflectivity response was observed with a single intense pulse serving both as excitation and optical probe. Consequently, the competing processes of electron heating and hydrodynamic surface expansion, which influence the reflectivity to different degrees and at different times, could be resolved only by a numerical deconvolution of single pulse data. In addition free electron metals were strongly emphasized[1].

In this paper we report femtosecond time-resolved experiments in which a 90 fs, 620 nm, p-polarized pump pulse 45° incidence excites either free electron metal (Ag, Al), or a covalently bonded (W, C) target surfaces at intensities up to 5×10^{13} W/cm^2. The reflectivity of a weak, 90 fs, incident at $\Theta = 20°$, 50°, or 70°, s- or p-polarized probe pulse (620 nm) focused to a spot centered on the pump, is then monitored during the pump-probe time delay $-1\text{ps} < \Delta t < +15\text{ps}$. The sample is rastered during data collection. The results show that 1) electron heating and hydrodynamic expansion can be temporally distinguished, and 2) covalency, as described by non-Drude background dielectric constants $\varepsilon_{10} > 1$ and $\varepsilon_{20} > 0$, is necessary to explain the time-resolved reflectivity of the expanding W and C targets.

Time-resolved p-polarized reflectivity $R_p(t)$ data at $\Theta_{probe} = 70°$ for Al excited at 10^{13} W/cm^2 is shown in Fig.1a. The reflectivity decreases with pulsewidth-limited fall time from an initial value of ~0.8, reaching a minimum of about 0.35 in Δt ~ 1ps, followed by a much slower partial recovery during $1\text{ps} < \Delta t < 15\text{ps}$. We analyze the reflectivity changes by numerically solving the Helmholtz wave equations[3] at each time delay with a model that includes 1) instantaneous heating and subsequent cooling of the electrons, and 2) development of a plasma gradient which expands at a velocity $v_{exp} \sim c_{sound}$ normal to the metal surface. The general Drude dielectric function is $\varepsilon(z,t) = \varepsilon_{10} + i\,\varepsilon_{20} - \omega_p^2/\omega(\omega+i\nu)$. For free electron metals (Al, Ag), $\varepsilon_{10} = 1$ and $\varepsilon_{20} = 0$. The electron collision frequency $\nu(t, n_e(z))$, plasma frequency $\omega_p(t, n_e(z))$ and plasma gradient scale length $L(t)$ evolve with time t; in addition ω_p, ν, and the ε_{10} and ε_{20}, change with the plasma density $n_e(z)$. The dotted curve in Fig.1a was calculated based on electron heating only ($L(t) = 0$), yielding a pulsewidth-limited fall time as observed. However, an unreasonably long electron relaxation time (~10 ps) is required to fit the recovery. Thus a shorter, more realistic recovery is shown. Alternately, the dashed curves in Fig.1a show the calculated reflectivity response of Al considering plasma expansion alone (no heating) at $v_{exp} = 1.2\times10^6$ cm/s ~ c_{sound}. But this model alone cannot account for the rapid fall time without invoking an unphysical $v_{exp} > 10^7$cm/s. Nevertheless, when both effects taken into account, a very satisfying fit for Al is obtained, as shown by the solid curves in Fig. 1a. Very similar data, fitting curves and analysis are obtained for Ag.

Springer Series in Chemical Physics, Vol. 55 **Ultrafast Phenomena VIII**
Editors: J.-L. Martin · A. Migus · G.A. Mourou · A.H. Zewail

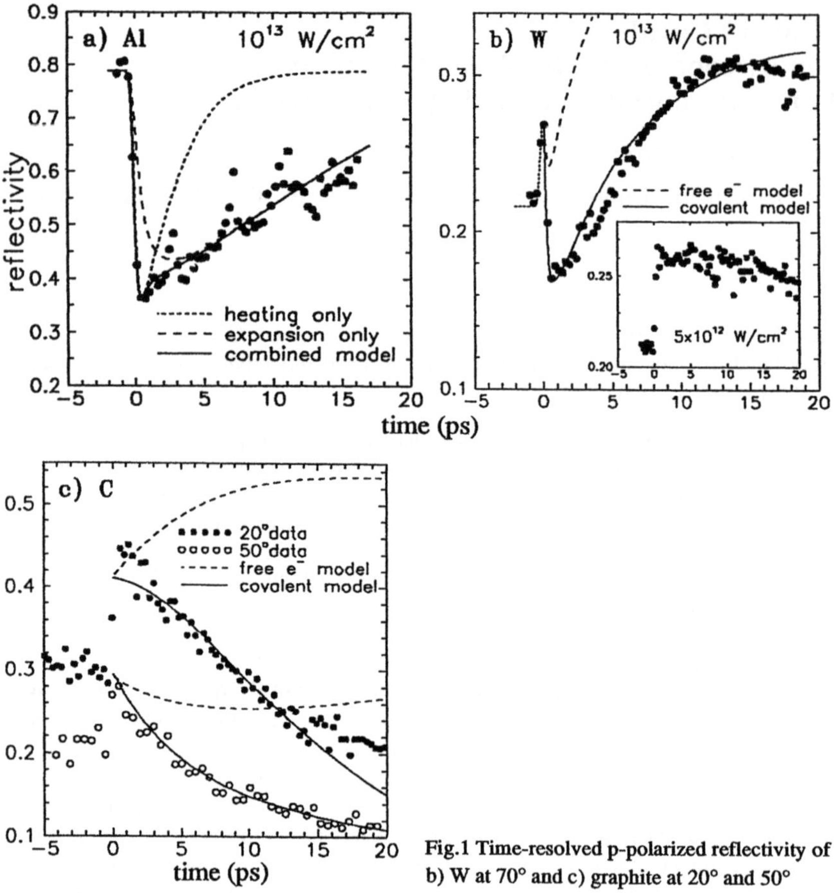

Fig.1 Time-resolved p-polarized reflectivity of a) Al, b) W at 70° and c) graphite at 20° and 50°

Time-resolved data for the non-free electron metal W are shown in Fig.1b. The inset shows $R_p(t)$ for excitation at 5×10^{12} W/cm², $\lambda_{probe} = 620$ nm., $\Theta_{probe} = 70°$. In this case solid W simply melts in < 1 ps into a liquid which is more reflective (~0.26) than the solid (~0.2), without hydrodynamic expansion. Hence the liquid reflectivity persists for > 10 ps. At 10^{13} W/cm² excitation (Fig.1b, main panel), however, an initial reflectivity increase to the liquid value(~0.26), is followed immediately by a reflectivity drop and recovery, which qualitatively resemble the pattern observed in Al and Ag. If, however, we assume liquid W and the expanding material to be a free electron metal (i.e. $\varepsilon_{10}=1$, $\varepsilon_{20}=0$), the reflectivity drop and recovery cannot be similarly explained. The Helmholtz equation yields the dashed line, which disagrees strongly with the data. On the other hand, if we introduce $\varepsilon_{10} = 7.65$, $\varepsilon_{20} = 12.3$, and re-adjust ω_p, ν to fit the initial liquid W reflectivity, the very satisfying solid curve for the W data after melting is achieved using $v_{exp} = 3.3 \times 10^5$ cm/s. In the expansion region, ε_{10} and ε_{20} are assumed to decrease linearly with density down to the vacuum values ($\varepsilon_{10}=1$, $\varepsilon_{20}=0$). These large ε_{10} and ε_{20} values indicate a strong influence of covalency both in the liquid and the denser part of the expansion region.

Covalency plays a similar role in the hydrodynamic expansion of carbon, as shown by $R_p(t)$ data in Fig. 1c for a graphite target excited at 5×10^{12} W/cm², $\Theta_{probe} = 20°$ and 50°.

Melting again induces an initial increase in Rp. The dashed curves show a free electron model sharply deviating from the data, while the solid curves have incorporated finite ε_{10} and ε_{20} values. Note that both models explain the initial liquid W or C reflectivity R_p (t ~ 1 ps) equally well, but only the covalent model is consistent with the delayed reflectivity drop accompanying surface expansion R_p (1ps < t < 15 ps). For all of the best R_p(t) fits, the data begins to deviate slightly from the model for t > 15 ps, probably because of the onset of additional absorption/scattering mechanisms as non-uniform expansion develops. Smaller s-polarized reflectivity changes are also consistent with the model presented. Research supported by the National Science Foundation(Grant DMR-8858388), Air Force Office of Scientific Research (Contract F49620-89-C-0044), and R. A. Welch Foundation (Grant F-1038).

References

1. H.M. Milchberg, R.R. Freeman, S.C. Davey, and R.M. More, Phys. Rev. Lett. **61**, 2364 (1988); R. Fedosejev, R.Ottmann, R. Sigel, G. Kuhnle, S. Szatmari, and F.R. Schafer, Appl.Phys.B**50**, 79(1990).
2. D.H. Reitze, H. Ahn, and M.C. Downer, Phys. Rev. B **45**, 2677 (1992).
3. H. M. Milchberg and R. R. Freeman, J. Opt. Soc. Am. B**6**, 1351 (1989).

Ultrafast Formation Processes of Self-Trapped Excitons in Alkali Iodide Crystals under Band-to-Band Excitation

T. Tokizaki[1], *S. Iwai*[1], *T. Shibata*[2], *A. Nakamura*[1], *K. Tanimura*[2], *and N. Itoh*[2]

[1]Department of Applied Physics, Faculty of Engineering, Nagoya University, Chikusa-ku, Nagoya 464-01, Japan
[2]Department of Physics, Faculty of Science, Nagoya University, Chikusa-ku, Nagoya 464-01, Japan

In alkali halide crystals, an electron-hole pair is localized at a lattice site to form a self-trapped exciton (STE), which consists of an halogen molecule ion (X_2^-) and a trapped electron. The detailed luminescence studies of STEs in many kinds of alkali halide crystals have shown that STEs are classified into three types (I , II and III), according to the configuration of an X_2^- and an electron occupying lattice sites [1,2].

Recently, we have observed the dynamic relaxation processes of the STE of the 2p-excited state pumped by femtosecond pulses from the lowest 1s state which are generated by a nanosecond excimer laser in NaCl [3]. The time evolution of the spectrum revealed the appearance of the shortly-lived STE being type-III, although the stable configuration of the STE in NaCl is type-II. Moreover, the oscillatory behavior of the population between type-II and -III STEs on the adiabatic potential-energy surface (APES) was observed.

In this paper, the formation process of STEs associated with free carriers created by the band-to-band excitation in alkali iodide crystals was investigated to find general feature of the STE dynamics on the APES. The specimens were excited by the two-photon absorption with ultraviolet pulses, whose wavelength and pulse duration were 302 nm and ~200 fs, respectively.

Figure 1 shows transient absorption spectra of KI at 4.2 K in the spectral region of the transition from the 1s to higher excited state of the STE. After the band- to-band excitation, the spectra grow up and the spectrum at τ_d = 95 ps (a solid line) in Fig.1(e) almost coincides with the well-established spectrum of the type-II STE (a dashed line) under the electron-beam excitation except for the higher energy region. The difference seen above 1.5 eV indicates an induced absorption band (we refer to it as Y band). This band may be due to the transient defect pair comprised of an F center and an H center, because the energy is close to the F center band and the absorption for longer delay time disappears at 80 K.

The different spectral behavior for τ_d < 15 ps compared to that observed for 95 ps suggests that the spectrum at early stage consists of the STE band, Y band and another absorption band (X band). The detailed analysis using rate equations allows us to decompose into three bands as shown in Fig.1(d) [4]. The origin of the X band can be discussed on the basis of "Mollwo-Ivey plot" which gives the relationship between a lattice constant and the 1s-2p transition energies of the type-II and -III STE [6]. The transition energy of type-III STE in KI is estimated to be 1.37 eV, which is very close to the peak energy of X band, ~ 1.4 eV.

Moreover, in the time evolution shown in Fig.2, the oscillatory structures with the period of 1.3 ps are obviously noticed, and the absorption oscillates in the anti-correlated manner for 1.33 eV and 1.59 eV. This indicates that the wave

Springer Series in Chemical Physics, Vol. 55 Ultrafast Phenomena VIII
Editors: J.-L. Martin · A. Migus · G.A. Mourou · A.H. Zewail

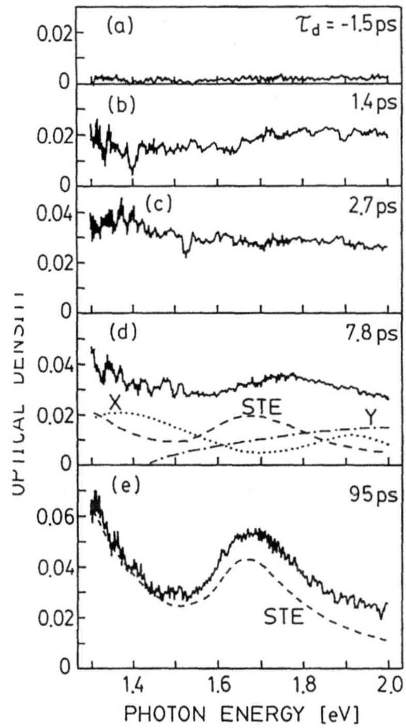

Fig.1 Transient absorption spectra of KI at 4.2 K for the several delay times

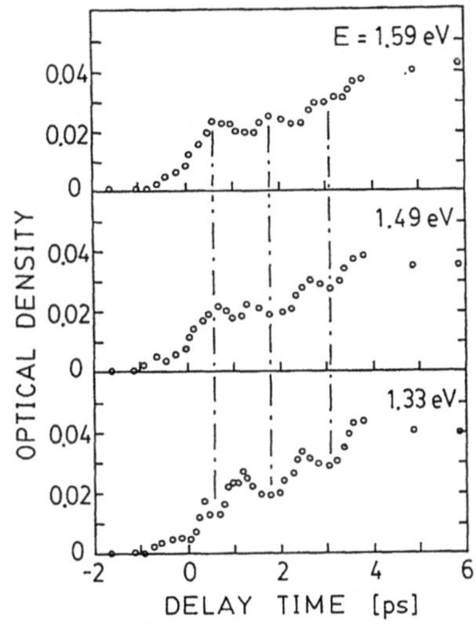

Fig.2 Time evolution of absorption at the several energies

packet of STE oscillates between two minima on the APES after the large de-excitation of band-to-band excitation of the electron-hole pair.

Therefore, the type-II STEs in KI are formed after the relaxation of the type-III STEs within 20 ps accompanying the transient F-H pair formation. This fast formation process was revealed thanks to the observation of the transient absorption in femtosecond time region, which is in contrast to the earlier works in the picosecond region [5,7]. Considering the results of NaCl, we have found that the such dynamical relaxation on the APES is general feature of the STEs at least in alkali halides where the type-II STE is stable.

References

[1] K.Kan'no, K.Tanaka, T.Hayashi, Rev. of Solid State Science, 4, 383 (1990)

[2] R.T.Williams, K.S.Song, J. Phys. Chem. Solid, 51, 679 (1990)

[3] T.Tokizaki, T.Makimura, H.Akiyama, A.Nakamura, K.Tanimura, N.Itoh, Phys. Rev. Lett., 67, 2701 (1991)

[4] S.Iwai, T.Tokizaki, T.Shibata, A.Nakamura, K.Tanimura, N.Itoh, to be published

[5] K.Tanimura, N.Itoh, to be published

[6] Y.Suzuki and M.Hirai, J. Phys. Soc. Jpn., 43, 1679 (1977)

[7] J.D'hertoghe, G.Jacobs, phys. stat. sol. (b) 95, 291 (1979)

Femtosecond Self-Trapping of Interacting Electron–Hole Pairs in α-SiO_2

W. Joosen[1], *S. Guizard*[1], *P. Martin*[1], *G. Petite*[1], *P. Agostini*[1], *A. Dos Santos*[2], *G. Grillon*[2], *J.P. Chambaret*[2], *D. Hulin*[2], *A. Migus*[2], *and A. Antonetti*[2]

[1]Service de Recherches sur les Surfaces et l'Irradiation de la Matière, Centre d'Etudes de Saclay, F-91191 Gif-sur-Yvette, France

[2]Laboratoire d'Optique Appliquée, Ecole Polytechnique, Ecole Nationale Supérieure de Technique Avancées, F-91120 Palaiseau, France

Abstract The 2.8-eV recombination luminescence of the self-trapped exciton (STE) in α-$Si0_2$ is observed under above-bandgap multiphoton excitation with intense femtosecond UV pulses. For electron-hole pair densities up to 10^{18} cm^{-3} its lifetime, polarization, and temperature dependence are characteristic for the isolated STE. Higher-density excitation yields electron-hole pair interactions, enhancing the STE production.

1. Introduction

In principle, electron-hole (e-h) pairs in wide-bandgap insulators such as crystalline quartz can be created with electron pulses, synchrotron radiation, optically induced soft X-rays, and UV pulses. Ultrafast, high-density excitation can only be achieved by multiphoton absorption of intense, femtosecond UV pulses. This is relevant to investigate bi- or multiexciton dynamics[1] and to perform a real-time study of intrinsic defect formation in α-SiO_2 and fused silica[2]. Possible excitation sources include recent excimer laser schemes,[3,4] an amplified, frequency-tripled titanium-sapphire system,[5] and a frequency-doubled, amplified colliding-pulse mode-locked (CPM) dye-laser.

2. Excitation source

We developed a femtosecond UV source, tunable between 4.0 eV and 4.5 eV starting from an amplified Rhodamine 6G - DODCI CPM dye laser system (Fig.1). Second-harmonic generation after five amplification stages yields 80-μJ pulses of 200 fs duration at 4.0 eV. Tunability is achieved by using a fraction of the 2.0-eV beam for continuum generation and spectral selection of a portion of this continuum in the yellow for further amplification. After frequency-doubling we obtain 4.4-eV pulses of 250 fs duration limited by the group velocity dispersion in the 0.5-mm β-BaB_2O_4 crystal. The pulse energy goes up to 10 μJ permitting excitation intensities in the TW/cm^2 regime and e-h pair densities up to 10^{19} cm^{-3} under two-photon absorption.

3. Multiphoton kinetics of the isolated self-trapped exciton

At low temperatures the onset of the excitation spectrum of the 2.8-eV luminescence lies at 8.4 eV.[6] Two 4.4-eV and three 4.0-eV photons are required to achieve above bandgap excitation. Multiphoton absorption in the UV yields broadband emission at 2.8 eV with a FWHM of 0.75 eV (Fig.2). This luminescence is detected with a

Springer Series in Chemical Physics, Vol. 55 Ultrafast Phenomena VIII
Editors: J.-L. Martin · A. Migus · G.A. Mourou · A.H. Zewail

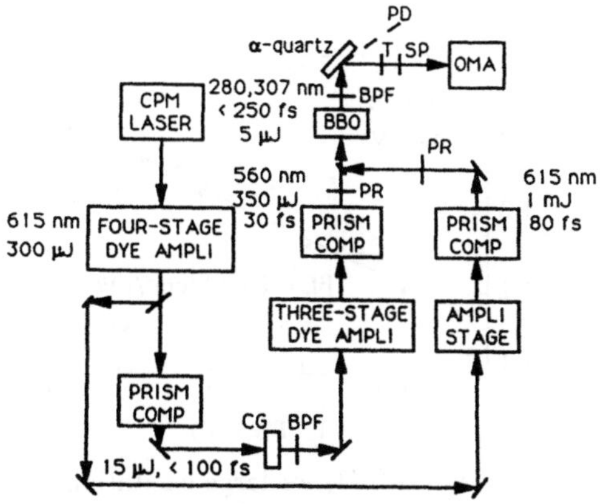

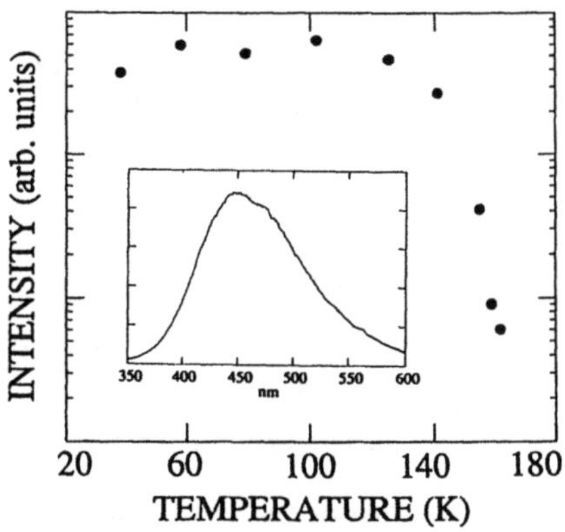

Fig.1 Dye-amplifier for the generation of femtosecond UV pulses and gated OMA detection for the 2.8-eV luminescence. CG = continuum generation, BPF = bandpass filter, T = suprasil collection telescope, PR = polarization rotator, SP = sheet polarizer, PD = photo diode. **Fig.2** Temperature dependence of the 2.8-eV luminescence intensity. The spectrum at 20K is shown in the inset.

gated optical multichannel analyzer (OMA) (Fig.1). It is polarized for 70% along the crystal's z-axis and its lifetime is 1.1 ms. At elevated temperatures the 2.8-eV emission is quenched by radiationless recombination (Fig.2), yielding the formation of E' centers.[2] The spectroscopic properties we find under femtosecond UV excitation are in excellent agreement with the results on electron-irradiated α-SiO_2 by Itoh *et al.*[6,7]

We investigated the multiphoton kinetics of the 2.8-eV luminescence as a function of the excitation intensity, which was varied by a polarizer-rotator combination at the fundamental wavelength. The UV reflected from the crystal surface was used to average the peak-to-peak fluctuations. Absolute energy measurements were performed with a pyroelectric power meter. For excitation intensities typically below 0.5 TW/cm^2, we verified slope values in the logarithmic plots of 1.8 and 2.8 for photon energies of 4.4 eV and 4.0 eV, respectively.

We have checked for possible systematic errors, which could mask the slope values, i.e., (i) self-focusing, (ii) filamentation, or (iii) transient linear absorption. A calculation for the case of resonant self-focusing (n_2 taken equal to 1.10^{-11} esu) shows that the increased excitation density due to beam waist narrowing is compensated by a reduction of the effective interaction volume. Filamentation is kept within reasonable proportions by restricting the sample thickness (< 500 μm): The B-integral is retrieved from the self-phase modulated UV spectrum and is smaller than 2.3. The absorption constant at 4.4 eV for an excitation density of 10^{18} cm^{-3} is 1.5 cm^{-1} (Ref.7), and implies a smaller than 10% correction to the slope value. Therefore, one can attribute physical significance to the slopes for excitation intensities up to a few TW/cm^2.

4. Electron-hole pair interactions

For intensities around 1 TW/cm^2 we establish a slope of four under 4.4-eV excitation (Fig.3). The recombination luminescence intensity is proportional to the square of

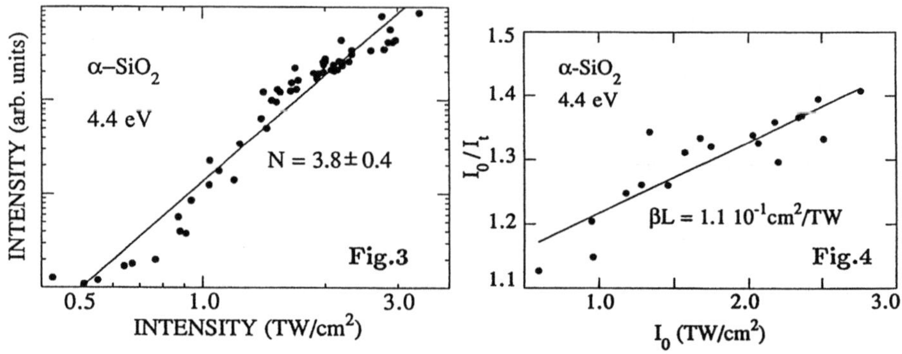

Fig.3 Logarithmic plot of the 2.8-eV luminescence in α-SiO_2 for large excitation intensities. **Fig.4** Inverse UV transmission intensity as a function of the UV excitation intensity in a $L = 0.7$ mm sample. The slope value is βL.

the e-h pair density. We ruled out the possibility of superradiance by verifying (i) the same slope for different delays and (ii) the same decay time in the regimes of moderate and high e-h pair densities. The reliability of the measurement is shown by determining the two-photon absorption coefficient β for the same range of excitation intensities. Its value of (1.8 ± 0.2) cm/TW is consistent with previous experimental findings[8] and the linear behavior of βL in Fig.4 confirms the negligible role of nonlinear propagation effects in Fig.3.

The observed slope value of four evidences interaction between mobile and/or trapped e-h pairs in α-SiO_2. A biexcitonic complex in this indirect bandgap insulator exhibits an instability, which enhances the probability for self-trapping of *both* the e-h pairs. Moreover, the transient loss of 4.4-eV pump pulse photons[7] under 3 TW/cm^2 excitation (not shown in Fig.4) indicates that the STE formation is faster than 250 fs.

Acknowledgement

The authors wish to thank G. Robelin (Quartz & Silice, Pithiviers - France) for supplying some samples and J. Etchepare, J.P. Duraud, and C. Boiziau for interesting discussions. The financial support of the European Community (E.E.C. Contract SC1000103) is kindly acknowledged.

References

1. K. Tanimura and N. Itoh, *Phys. Rev. Lett.* **60**, 2753 (1988).
2. T.E. Tsai and D.L. Griscom, *Phys. Rev. Lett.* **67**, 2517 (1991).
3. J.H. Glownia, M. Kaschke, and P.P. Sorokin, *Opt. Lett.* **17**, 337 (1992).
4. F.P. Schäfer, S. Szatmári, P. Simon, in *Proceedings of the 8th International Conference on Ultrafast Phenomena*, Antibes, France, June 8-12, 1992.
5. A. Nebel and R. Beigang, *Opt. Lett.* **16**, 1730 (1991).
6. C. Itoh, K. Tanimura, N. Itoh, and M. Itoh, *RC Phys. Rev. B* **39**, 11183 (1989).
7. C. Itoh, K. Tanimura, and N. Itoh, *J. Phys. C* **21**, 4693 (1988).
8. P. Liu, W.L. Smith, H. Lotem, J.H. Bechtel, N. Bloembergen, and R.S. Adhav, *Phys. Rev. B* **17**, 4620 (1978).

Ultrafast Soft Mode Dynamics in Ferroelectric Crystals

G.P. Wiederrecht, T.P. Dougherty, and K.A. Nelson

Department of Chemistry, Massachusetts Institute of Technology,
77 Massachusetts Avenue, Cambridge, MA 02139, USA

We report the first femtosecond time-resolved study of soft optic phonon modes near structural phase transitions. The experiments were performed on the perovskites $KNbO_3$ and $BaTiO_3$, and quantitative support for a microscopic model of the ferroelectric phase transition is found.

Introduction

The dynamics of cooperative ordering in crystals near structural phase transitions have long been elusive. Most picosecond and femtosecond time-resolved measurements have involved rapid laser heating of a sample whose phase transition (e.g. melting) dynamics are then monitored. In this case many lattice degrees of freedom are excited, and little information is provided about the roles or dynamics of particular lattice motions. In structural phase transitions, the order parameter is described by one or a very few specific ("soft") lattice modes whose motions bring crystalline constituents from their initial positions into their positions in a new crystalline phase. Experimental characterization of the dynamics is possible in some cases through Raman spectroscopy, but often the highly damped character of the soft mode leads to a central peak in the Raman spectrum which cannot be analyzed accurately. In many cases it is not even possible to tell whether the ordering motion is a heavily damped vibration, Debye relaxation, or some combination of the two. This information is essential for understanding the microscopic mechanism of a transition since vibrational or relaxational character is an indication of a displacive or order-disorder transition (involving motion of ions or molecules within a single potential energy minimum or hopping between different sites) respectively.

$KNbO_3$ and $BaTiO_3$, prototype members of the perovskite crystal family, exhibit a series of phase transitions from a high temperature cubic paraelectric phase to ferroelectric tetragonal, orthorhombic, and rhombohedral phases. Raman spectra have left doubt about the dynamical character (vibrational or relaxational) of the order parameter, and quantitative determination of low phonon frequencies, damping rates, and relaxation times for Debye modes has been frustrated [1,2]. A useful check on these quantities is provided through the Lyddane-Sachs-Teller relation to the low-frequency dielectric constant which is determined by the polar ionic motions. Capacitance measurements have consistently produced a larger value for ε_0 than Raman experiments, leading to the proposal of additional polar relaxational features which have eluded detection in the Raman spectra. This bears directly on microscopic models of the transitions which make predictions about the number and dynamic character of lattice modes of different symmetries.

Here we report direct femtosecond time-resolved observations of soft mode dynamics in $KNbO_3$ and $BaTiO_3$. Impulsive stimulated Raman scattering (ISRS) is used to initiate coherent motion along Raman-active modes, and the ordering dynamics are monitored. Our results distinguish unambiguously the contributions of vibrational and relaxational components with different symmetries. We are able to resolve previous inconsistencies which led to conflicting models for the phase transitions.

Results and Discussion

Figure 1 shows ISRS data for the soft mode of B_2 symmetry in the orthorhombic phase of $KNbO_3$ along with a fit to a single damped harmonic oscillator response. No evidence for relaxational ionic motion is seen, in contrast to the conclusions reached by a recent Raman study and to the common assumption that there are additional slow modes [2]. In order to further

Springer Series in Chemical Physics, Vol. 55 Ultrafast Phenomena VIII
Editors: J.-L. Martin · A. Migus · G.A. Mourou · A.H. Zewail

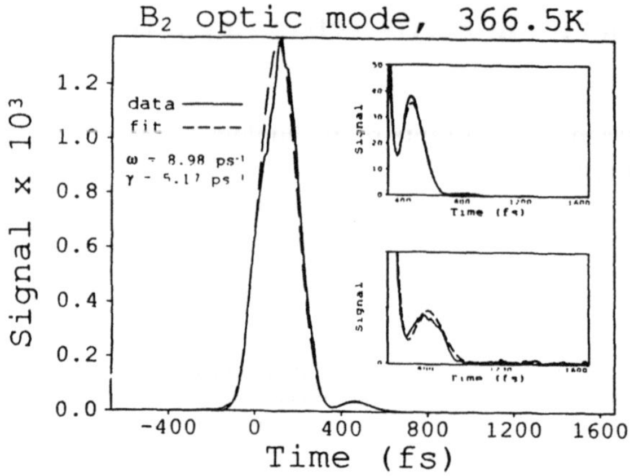

Figure 1. B_2 symmetry ISRS data from $KNbO_3$ at 366.5K showing the dynamics of the soft phonon-polariton mode. The nuclear response is fit well with a single damped harmonic oscillator response.

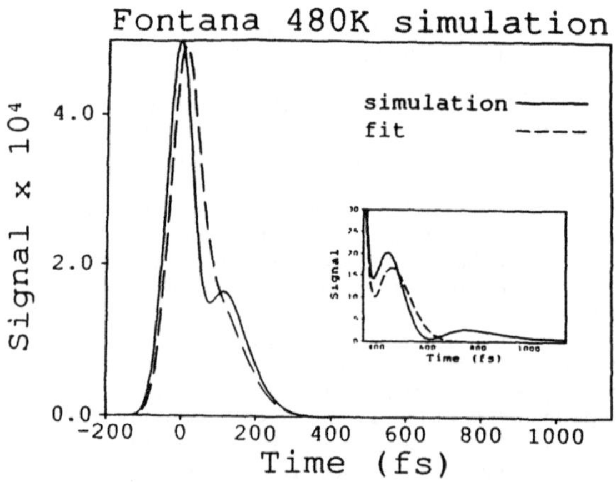

Figure 2. Simulated ISRS data using the parameters given in reference 2 where a relaxational mode of the same symmetry as the soft mode was reported. The attempted fit to the data utilizes only a single damped harmonic oscillator response, which clearly illustrates the extent to which we can eliminate the possibility of a relaxational contribution to the data.

illustrate the extent to which we can eliminate the presence of a relaxational response with the same symmetry as the soft mode, simulated ISRS data is presented in Figure 2 which is based on the results of previous frequency domain light scattering results that report the presence of a relaxational mode of B_2 symmetry. The presence of such an additional mode results in a highly nonexponential decay of the oscillatory signal, even if the relaxational mode has a scattering cross section five orders of magnitude or more weaker than the soft mode.

Due to the polar nature of the soft optic mode, the material excitation displays the dispersive character of a polariton at the small scattering angles typically utilized in a ISRS experiment [3]. The dispersive phonon frequencies and damping rates determined at several stimulated scattering wave vectors allow for accurate determination of the soft mode phonon frequencies. This leads to good quantitative agreement, through the LST relation, with capacitance measurements of the low frequency dielectric constant as illustrated in Figure 3 [4, 5]. These data and data taken at other scattering symmetries (where relaxational motion is observed) strongly support an "eight-site" model for the microscopic nature of the phase transitions [1,2,4,5]. The eight site model is an order-disorder model, which assumes that the central ion (Nb or Ti) occupies one of eight local potential energy minima which are displaced from the center of the unit cell along the diagonal directions. The soft mode is the result of oscillatory motion between "allowed" lowest energy sites separated by very low-lying potential energy barriers. Similar results are found for $BaTiO_3$.

A ramification of the eight site model is that there should be relaxational modes present along the polar axis of the crystal, i.e. perpendicular to the soft mode coordinate. This is the

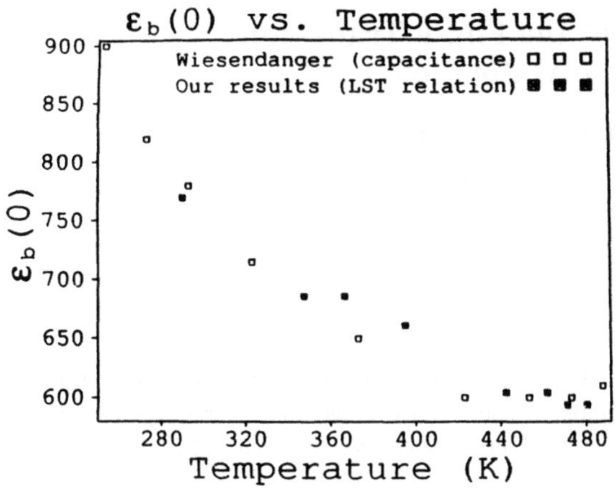

Figure 3. A plot of the low frequency dielectric constant as determined from ISRS data and capacitance measurements. Excellent agreement is found.

result of hopping dynamics between allowed and differing disallowed sites of the central ion (Nb). We have observed two relaxational responses of A_1 symmetry in $KNbO_3$ [3-5] which further supports the eight site model.

In conclusion, we have made femtosecond time-resolved observations of soft modes in the structural phase transitions of $KNbO_3$ and $BaTiO_3$. The results have resolved long-standing inconsistencies and yielded accord with a microscopic model for the transitions.

Acknowledgements

This work was supported in part by NSF Grant No. DMR-9002279, ONR Grant No. N00014-90-J-4070, and contributions from DuPont and Perkin-Elmer.

References

1. J.P. Sokoloff, L.L. Chase, and D. Rytz, Phys. Rev. B **38**, 597 (1988).
2. M.D. Fontana, A. Ridah, G.E. Kugel, and C. Carabatos-Nedelec, J. Phys. C. **21**, 5853 (1988).
3. T.P. Dougherty, G.P. Wiederrecht, and K.A. Nelson, J. Opt. Soc. Am. B, in press.
4. T.P. Dougherty, G.P. Wiederrecht, and K.A. Nelson, Ferroelectrics **120**, 79 (1991).
5. T.P. Dougherty, G.P. Wiederrecht, and K.A. Nelson, Phys. Rev. B, submitted.
6. E. Wiesendanger, Ferroelectrics **6**, 263 (1974).

Temporal Domain Study of the Phase Transition in $PbTiO_3$: A_1 Symmetry Investigation

D.P. Kien, J.C. Loulergue, and J. Etchepare*

Laboratoire d'Optique Appliquée, E.N.S. Techniques Avancées,
Ecole Polytechnique, F-91120 Palaiseau Cedex, France
*Also with: CLOES Université de Metz, Supélec,
2, Rue Belin, F-57078 Metz, France

Abstract. The evolution as a function of temperature of the lowest completely symmetric A_1(TO) polariton is reported, up to the quadratic to cubic phase transition. No relaxational effect, superimposed on the soft mode behavior, has been detected. We deduce that the symmetric part of the transition is of displacive origin.

Spectroscopic studies have been stressed for a long time as adequate tools for the description of phase transition mechanisms. Furthermore, a temporal domain analysis is a priori well suited to elucidate the problem of order-disorder and/or displacive character of a specific transition, as their signature is potentially distinguishable from each other [1].

Ferroelectric $PbTiO_3$ has a tetragonal C^1_{4v} space symmetry below the phase transition temperature T_c, located at 495°C [2]. Above T_c, the crystal becomes paraelectric and its structure cubic. We have used the transient grating technique to describe this phase transition, for which debate is still open [3], as is indeed the case for other ferroelectric crystals [1].

The results reported here correspond to a parallel polarization configuration in the direction perpendicular to c-axis. As shown previously [4], this geometrical arrangement gives access to A_1(TO) polariton modes. We reproduce in figure 1 the temporal evolution of the signal diffracted by a 0.7mm thick $PbTiO_3$ single crystal for three different temperatures.

The discussion will be limited to its behavior in the temporal range greater than ≈ 250 fs; there, an interaction of the polariton with a free carrier grating promoted by two photon absorption, enhances the Raman signal, thus allowing an easier determination of the phonon frequency. As the temperature approaches the quadratic to cubic transition, Ω tends towards zero value and

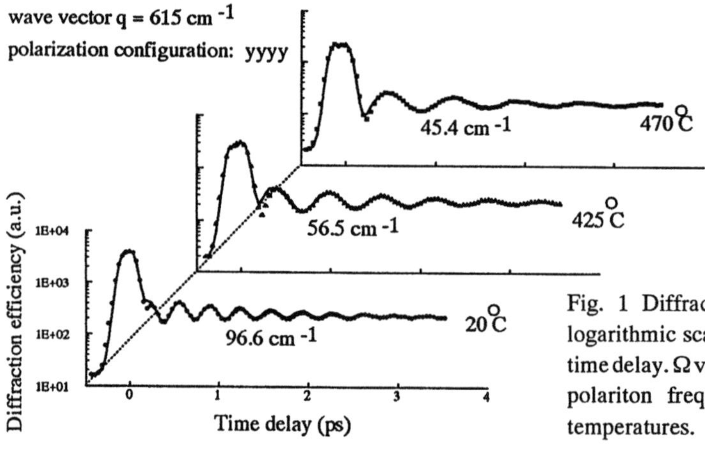

Fig. 1 Diffraction efficiency, in logarithmic scale, as a function of time delay. Ω values stand for fitted polariton frequencies at various temperatures.

Springer Series in Chemical Physics, Vol. 55 Ultrafast Phenomena VIII
Editors: J.-L. Martin · A. Migus · G.A. Mourou · A.H. Zewail

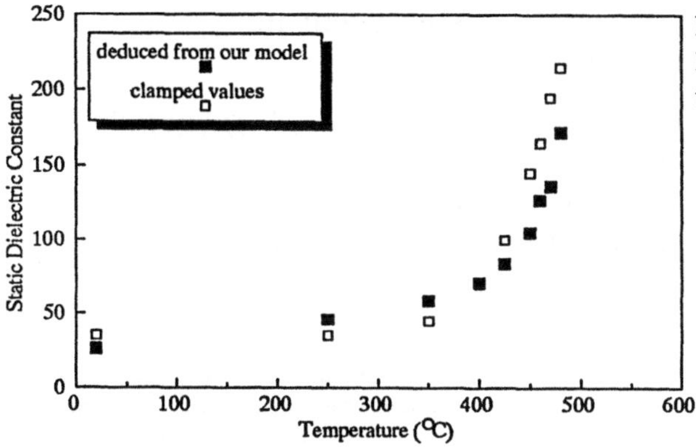

Fig. 2 Evolution of the static dielectric constant as a function of temperature.

any relaxational mode which would contribute to the phase transition for this symmetry can be detected.

We have deduced the evolution of the static dielectric constant $\varepsilon_{||}(0)$ as the function of temperature (Fig. 2), by using an undamped single oscillator model:

$$\frac{c^2q^2}{\Omega^2} = \varepsilon_{||\infty} + [\varepsilon_{||}(0) - \varepsilon_{||\infty}] \frac{\omega_{TO}^2}{\omega_{TO}^2 - \Omega^2}$$

using ω_{TO} and $\varepsilon_{||\infty}$ values given in the literature [3,5]. Comparing our values to measurements of the clamped dielectric constant [3], good agreement leads to ascertain that the lowest frequency phonon-polariton mode contribution to $\varepsilon_{||}(0)$ prevails over others, along the whole temperature range probed.

The measured lowest frequency phonon-polariton describes correctly the static dielectric constant $\varepsilon_{||}(0)$ in the investigated temperature range. Moreover the behavior of the temporal response can be accounted for without any contribution from a relaxational mode. So, we conclude that the A_1 symmetry part of the phase transition is of purely displacive origin, in concordance with previously published work conducted in the spectral domain [3].

References.

[1] T. P. Dougherty, G. P. Wiederrecht and K. A. Nelson. Ferroelectrics 120, 79, (1991).
[2] G. Burns and B. A. Scott. Phys. Rev. B 7, 3088, (1973).
[3] M. D. Fontana, H. Idrissi, G. E. Kugel and K. Wojcik. J. Phys. Condens. Matter 3, 8695, (1991).
[4] J. Etchepare, G. Grillon, A. Antonetti, J. C. Loulergue, M. D. Fontana and G. E. Kugel. Phys. Rev. B 41, 12362, (1990).
[5] W. Kleemann, F. J. Schäfer and D. Rytz. Phys. Rev. B 34, 7873, (1986).

Femtosecond Transient Absorption Measurements on Low Band Gap Thiophene Polymers

A. Cybo-Ottoné[1], *M. Nisoli*[1], *V. Magni*[1], *S. De Silvestri*[1], *O. Svelto*[1], *G. Zerbi*[2], *and R. Tubino*[2,3]

[1]Dipartimento di Fisica, Politecnico, Milano, Italy
[2]Dipartimento di Chimica Industriale, Politecnico, Milano, Italy
[3]Istituto di Fisica e Matematica, Università Sassari, Italy

The excited state kinetics of low band gap thiophene-based polymers has been investigated by femtosecond pump and probe. The transient saturated absorption data show different temporal evolutions that may be related to exciton dynamics.

The growing interest in all optical switching and high speed optoelectronics has promoted the search for new organic materials with large non-linear optical susceptibilities and fast time response. A promising class of organic materials is represented by thiophene-based polymers since they are chemically stable and easily processable. The optical properties of these compounds have not yet been fully investigated and a few measurements of excited state relaxation dynamics after photoexcitation have been so far reported [1-3].

In this work we report the first measurement of ultrafast excited state kinetics of thiophene polymers, characterized by a low energy band gap. The chemical structures of the polymers poly(3-pentoxythiophene) (PPT), poly(3-decylthiophene-vinylene) ($PTVC_{10}$) and poly(3,4-dibutylthiophene-vinylene) ($PTVC_4$) are shown in Fig.1. The electronic absorbtion spectra of these polymers in chloroform solution show an absorption peak around 550 nm. The low band gap in PPT is due to the presence of oxygen in the substituent, while in $PTVC_{10}$ and $PTVC_4$ is due to the vinyle group. The samples were studied in chloroform solution by femtosecond pump and probe. The excitation source consists of a cavity dumped hybrydly mode-locked Rhodamine 6G dye laser, which generates pulses of 60 fs at 1.9 MHz with highly stable temporal and spectral shape. The transient saturated absorption curves for PPT and $PTVC_{10}$, are shown in Fig.2a and b. For $PTVC_{10}$ the data can be fitted by a two exponential curve with a very fast initial decay of about 30 fs followed by a decay of 0.9 ps; for PPT the data show a rather complex non exponential decay, which reaches a plateau after 80 ps. The data for $PTVC_4$ are similar to those for $PTVC_{10}$.

Recent results on transient absorption saturation of poly(3-methylthiophene) [2] have shown the presence of a two exponential decay,

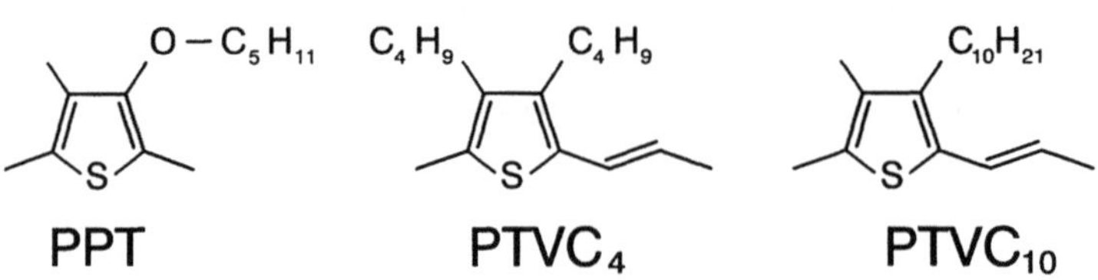

Fig.1 Chemical structures of the low band gap polymers

Springer Series in Chemical Physics, Vol. 55 Ultrafast Phenomena VIII
Editors: J.-L. Martin · A. Migus · G.A. Mourou · A.H. Zewail

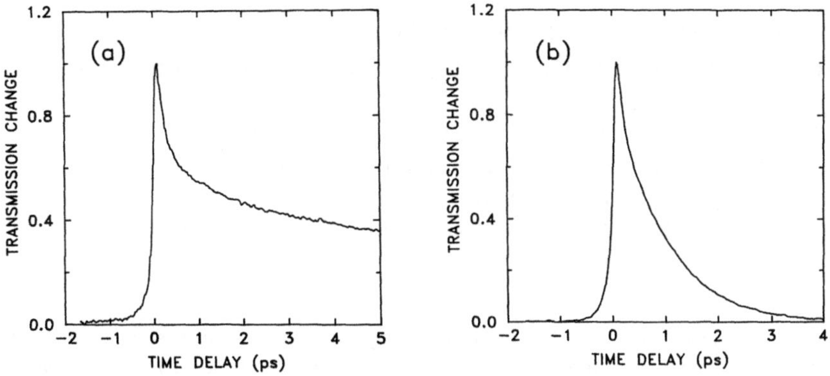

Fig.2 Transient saturated absorption data for PPT (a) and $PTVC_{10}$ (b) as a function of probe delay in chloroform solution

which have been attributed to the formation and decay of self-trapped exciton. We believe our measurements for $PTVC_{10}$ can be interpreted in the framework of this model. The fast initial decay can be attributed to the relaxation process of free excitons to self-trapped excitons, while the slow decay to the ground state recovery of self-trapped excitons. A model based on a random walk process on a finite chain can account for the recombination mechanism of the self trapped excitons. In the case of PPT, the presence of long lived species seems to be dominant. The formation of a neutral exciton state may be responsible for the decay kinetics. As final comment we point out that the backbone structure for $PTVC_{10}$ is more rigid than that of PPT, due to larger extent of delocalization of π electrons. The data obtained with $PTVC_4$ show that changes in the side groups are not relevant for the relaxation process.

In conclusion, we have measured the ultrafast excited state relaxation dynamics of low energy band gap thiophene polymers. The data may be consistent with self-trapped exciton formation and decay.

This work has been partially supported by C.N.R. under the "Progetto Finalizzato" on Telecommunications.

References

[1] Z.Vardeny, H.T.Grahn, A.J.Heeger, F.Wudl, Synth. Metals 28, C299 (1989).

[2] T.Kobayashi, M.Yoshizawa, U.Stamm, M.Taiji, M.Hasegawa, J. Opt. Soc. Am B7, 1558 (1990).

[3] I.D.W.Samuel, K.E.Meyer, S.G. Graham, R.H.Friend, Phys. Rev. B 44, 9731 (1991).

Effects of Crosslinking in Host Polymer on Picosecond Optical Dephasing of Doped Dye Molecules

S. Nakanishi[1], *S. Fujiwara*[1], *M. Kawase*[2], *and H. Itoh*[1]

[1]Department of Physics, Faculty of Education, Kagawa University, Takamatsu City, Kagawa 760, Japan

[2]Department of Chemistry, Faculty of Education, Kagawa University, Takamatsu City, Kagawa 760, Japan

Optical dephasing in polymer host has been attracting much interest because of its distinct behaviors. In this paper, we present the effects of crosslinking in host polymer (polyvinyl alcohol, PVA) on picosecond optical dephasing of the zero-phonon line (ZPL) for two doped dye molecules measured by incoherent photon echoes. We show that the introduction of crosslinking increases the dephasing time of ZPL, and that the effects can be explained as the decrease of a local density of the two-level tunneling system (TLS) in PVA caused by the crosslinking.

Two dyes, Nile blue (NB) and Rhodamine 640 (Rh640), were used as a probe of picosecond optical dephasing in three crosslinked PVA hosts. Crosslinking in PVA was performed by a chemical or photochemical reaction [1]. The crosslinker introduced in PVA was acrolein, glutaraldehyde and stilbazolium group. These crosslinked PVAs are referred to as PVA-A, PVA-G and PVA-SbQ, respectively. The content of crosslinking group was in the range of $1 \sim 2$ mol%. The length of crosslinking was estimated from the standard bond length as 5.6, 6.5 and 14.3 Åfor PVA-A, PVA-G and PVA-SbQ, respectively. The usual PVA without the crosslink, referred to as PVA, was also employed as a standard in evaluating the effect of crosslinking. The molecular structure of these host polymers is shown in Fig. 1. To get a picosecond time resolution in photon echo measurement, a broadband dye laser was employed [2], which produced light pulses of about 8 ns with a bandwidth of 1.6 nm corresponding to the time resolution of approximately 0.5 ps. Two laser dyes, DCM and Rhodamine B, were used to excite the ZPL of NB and Rh640 doped in the crosslinked PVAs at approximately 670 and 590 nm, respectively.

The incoherent photon echo decays obtained at 10 K for NB and Rh640 doped in three crosslinked PVAs and PVA are displayed in Fig. 2 on a semi-log scale. The observed photon echo decay consists of an approximately exponential decay due to the ZPL preceded by an ultrafast decay component resulting from the phonon sideband [3]. As is evident in Fig. 2, the dephasing time of the ZPL for NB and Rh640 increases in the order of PVA, PVA-A, PVA-G and PVA-SbQ, which corresponds to the order of the crosslink length. These results suggest that the introduction of the crosslink to the PVA backbone reduces the effect of TLS on picosecond optical dephasing of ZPL for the doped dye molecule. We consider that, by introducing the crosslink, a large void space is created in PVA and a doped dye is preferentially entrapped into the void space. As the void space probably prevents TLSs from being near the void space, the local TLS density around the dye is expected to decrease compared with PVA. This our model leads to a perturbation theory incorporated with the two-domain model for the distribution of TLS density [4] ; one domain is a sphere with a radius r_s including a void space at the center, where the effective TLS density, ρ_1, is decreased, and the other domain is the outer bulk region of the sphere with the TLS density of ρ_2. The third-order density matrix corresponding to the incoherent photon echoes at the delay time τ between two exciting beams, is described in the form

Springer Series in Chemical Physics, Vol. 55 Ultrafast Phenomena VIII
Editors: J.-L. Martin · A. Migus · G.A. Mourou · A.H. Zewail

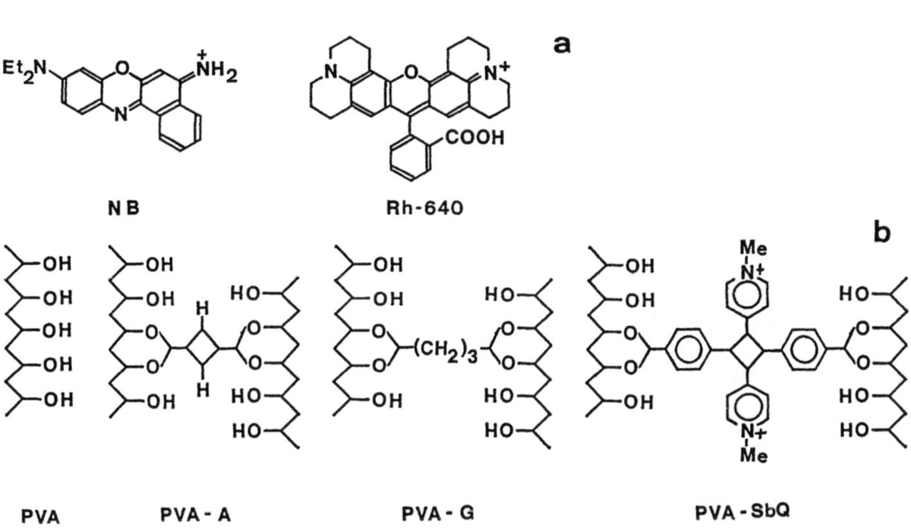

Fig. 1. (a) Molecular structure of two dyes, NB and Rh640, used as a probe of picosecond optical dephasing. (b) Molecular structure of crosslinked PVA hosts used in the dephasing measurement.

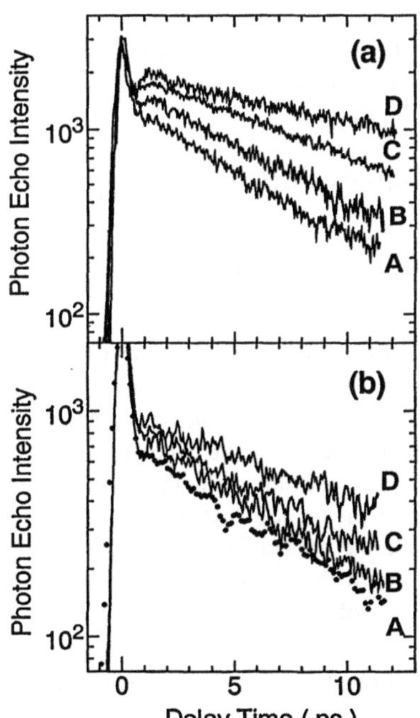

Fig. 2. Incoherent photon echo decays of NB (a) and Rh640 (b) observed at 10 K. The host polymers employed were PVA (A), PVA-A (B), PVA-G (C) and PVA-SbQ (D).

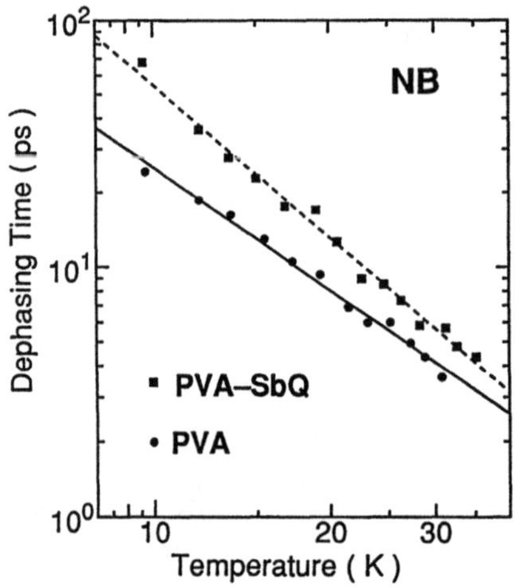

Fig. 3. Temperature dependence of picosecond dephasing time of ZPL for NB doped in PVA (solid circle) and PVA-SbQ (solid square). Solid and broken lines show the power law fitting with $T^{-1.7}$ and $T^{-2.1}$, respectively.

$$\rho_{12}^{(3)}(\tau) \propto \int_0^\infty dt_1 \int_0^\infty dt_2\, E(t-t_1-\tau)G(t_1-\tau)e^{-\gamma_2 t_1}\left(e^{-\gamma_1 t_2}+e^{-\gamma_2 t_2}\right)D(t_1,t_2,t_1) \quad (1)$$

with γ_1 and γ_2 being the population relaxation rates of the ground and excited state for the ZPL, $G(t)$ the field correlation function of laser light, $D(t_1,t_2,t_1)$ the four-point correlation function. The $D(t_1,t_2,t_1)$ can be divided into two parts as [5]

$$\begin{aligned} D(t_1,t_2,t_1) = \exp\Big[&-4\pi\rho_1 \int_0^{r_s} dr\, r^2 \langle 1-\exp[-(\eta/r^3)^2 t_1^2 Q(t_1,t_2,R,E)]\rangle_{R,E} \\ &-4\pi\rho_2 \int_{r_s}^{\infty} dr\, r^2 \langle 1-\exp[-(\eta/r^3)^2 t_1^2 Q(t_1,t_2,R,E)]\rangle_{R,E}\Big], \end{aligned} \quad (2)$$

where $Q(t_1,t_2,R,E) = \text{sech}^2(E/2kT)f^2(Rt_1,Rt_2)$ and the phase fluctuation of the ZPL is assumed to occur through the dipolar-type interaction, η/r^3, with the surrounding TLSs. We also assume that the phase fluctuation caused by a TLS, $\delta\phi(t)$ is described by Gauss-Markoffian stochastic process, $\langle\delta\phi(t)\delta\phi(t')\rangle = (\eta/r^3)^2 e^{-R|t-t'|}$, where R means the fluctuation rate. As photon echoes primarily detect the phase fluctuation on the fast time scale, the first exponent in eq. (2) dominates in photon echo decay. Consequently, the reduced local density ρ_1, caused by the crosslinking, can explain the dependence of dephasing time of ZPL on the crosslinking. The dependence of dephasing time on the crosslink length can be understood by assuming that the longer crosslink creates the larger void space, probably resulting in the lower local density of TLSs.

We also investigated the temperature dependence of the dephasing time in the crosslinked PVA hosts, which is closely related to the distribution $P(E)$ of the energy splitting E of TLSs that make a primary contribution to picosecond optical dephasing. We could not find the apparently different temperature dependence in PVA-A and PVA-G hosts from that in PVA, but we found the difference in PVA-SbQ. Figure 3 compares the temperature dependence of dephasing time for NB doped in PVA-SbQ and PVA. The temperature dependence in PVA-SbQ and PVA was found to obey a power law of $T^{-2.1}$ and $T^{-1.7}$, respectively. This result indicates that the crosslink introduced in PVA-SbQ has changed the distribution $P(E)$ from $E^{0.7}$ to $E^{1.1}$. Similar change of the temperature dependence was also observed for Rh640 in PVA and PVA-SbQ. We consider that the change of $P(E)$ in PVA-SbQ host provides another evidence for the alternation of the local TLS distribution around a doped dye caused by the introduced crosslinker.

References

[1] K. Ichimura, J. Polym. Sci. : Polym. Chem. Ed. 22, 2817 (1984).

[2] S. Asaka, H. Nakatsuka, M. Fujiwara and M. Matsuoka, Phys. Rev. A 29, 2286 (1984), N. Morita and T. Yajima, Phys. Rev. A 30, 2525 (1984).

[3] S. Saikan, T. Nakabayashi, K. Kanematsu and N. Tato, Phys. Rev. B 38, 7777 (1988).

[4] D. W. Pack, L. R. Narasimhan and M. D. Fayer, J. Chem. Phys. 92, 4125 (1990).

[5] M. Kawase, S. Fujiwara, S. Nakanishi and H. Itoh, to be published in Chem. Phys. Lett.

Ultrafast Relaxation of Exciton and Soliton–Antisoliton Pair in One-Dimensional Conjugated Polymers

T. Kobayashi, M. Yoshizawa, S. Takeuchi, and A. Yasuda

Department of Physics, Faculty of Science, University of Tokyo, Hongo 7-3-1, Bunkyo-ku, Tokyo 113, Japan

Abstract. Ultrafast nonlinear optical response in photoexcitations in polydiacetylenes (PDAs), polythiophenes (PTs), and a substituted polyacetylene (PA) were investigated. The decay kinetics in PDAs and PTs are discussed in terms of self-trapped excitons (STE), while the relaxation in the substituted PA is explained by the confinement effect of soliton–antisoliton pairs.

1. Relaxation in Polydiacetylenes and Polythiophenes

Transient photoinduced absorption (ΔA) spectra in blue-phase PDA-4BCMU (butoxycarbonylmethylurethane) film at 290 K observed by femtosecond absorption spectroscopy are shown in Fig. 1. When the pump and probe pulses temporally overlap each other at the sample, ΔA due to several nonlinear optical processes are observed. Two small minima at 1.80 and 1.72 eV are due to Raman gain. They are assigned to the stretching vibrations of the C=C bond and C≡C bond, respectively. Two small maxima observed at 2.18 and 2.25 eV are due to the inverse Raman scattering. Around the pump photon energy (1.98 eV), the spectral changes due to hole burning, optical Stark shift and induced-phase modulation are also observed.[1–3]

In addition to the ultrafast nonlinear optical response the bleaching due to ground-state depletion and transient broad-band absorption below the

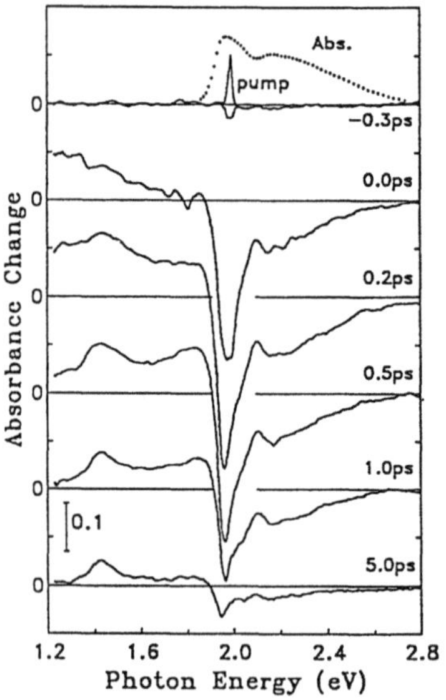

Fig.1 Transient absorption spectra in blue-phase PDA-4BCMU film at 290 K. The absorption (dotted curve) and pump spectra are shown together.

Springer Series in Chemical Physics, Vol. 55 Ultrafast Phenomena VIII
Editors: J.-L. Martin · A. Migus · G.A. Mourou · A.H. Zewail

band-gap, which is attributed to STE, were observed. The formation time of STE was estimated from the rise time of ΔA around 1.8 eV as 150±50 fs. The decay kinetics of STE was explained in terms of the tunneling in configuration space from STE to the ground state, and the time constants were determined as 1.5±0.1 ps at 290 K and 2.0±0.1 ps at 10 K. The decay kinetics in red-phase PDAs and PTs cannot be fitted to single exponential functions. The decay kinetics becomes slower at longer delay time. The decay kinetics can be fitted to biexponential functions with time constants of slightly shorter than 1 ps and about 5 ps. These two decay time constants correspond to relaxations to the ground state, respectively, from the free exciton and hot STE and from the thermalized STE.

2. Confinement Effect of a Soliton-Antisoliton Pair

The photoinduced bleaching at 2.3 eV due to saturation of the interband transition and the photoinduced absorption just below the optical gap around 1.8 eV were observed also in a substituted polyacetylene, poly[o-(trimethylsilyl)phenylacetylene] (PMSPA).[5] The decay kinetics at the photoinduced absorption peak (shown in Fig. 2(b)) was found to follow the power-law behavior over almost three orders of magnitude of the signal intensity with power n=0.65±0.05. This strongly implies the geminate recombination process between the soliton and antisoliton generated by intrachain photoexcitation . By taking into account of the significant reduction in the conjugation length of PMSPA caused by the steric hindrance due to bulky substituents, the features of absorbance change have been concluded to result from the intrachain soliton-antisoliton pairs, which are confined in a conjugation-length-limited segments of the main chain. The distance between the confined soliton and antisoliton was estimated to be about 6 repeat unit lengths. It is fairly consistent with the size of some regular structure indicated to exist in PMSPA by an X-ray diffraction measurement [4]. The photoinduced bleaching also followed the same power-law behavior after the initial fast decay (shown in Fig.2(c)) associated with excitations like intrachain exciton-polarons. They are

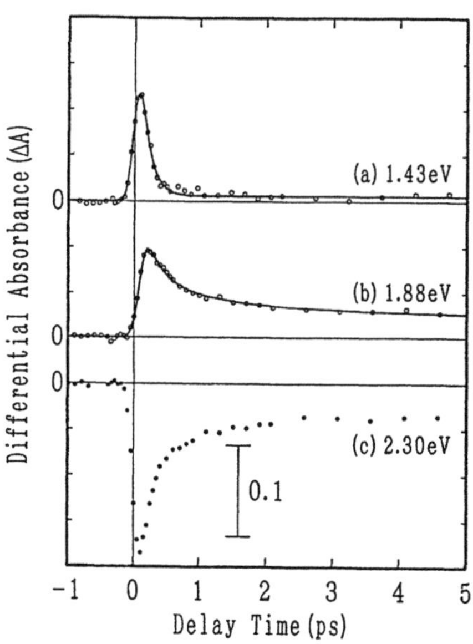

Fig.2 Decay kinetics in PMSPA. Open circles represent the experimental data and solid lines show the fitting curves with exponential function (a) and power-law function (b).

followed by the separation of both charges with spectral blue shift of photoinduced absorption clearly observed in transient ΔA spectra.

Long-lived baseline signals were also observed in both photoinduced absorption and bleaching with the same spectral profiles as the above intrachain features. Since the baseline signals had a quadratic dependence on the pump intensity, they have been concluded to result from pairs of likely charged solitons converted from the interchain polarons by the reaction $P^{\pm} + P^{\pm} \rightarrow S^{\pm} + S^{\pm}$.

References

1. M.Yoshizawa, M.Taiji, and T.Kobayashi, IEEE J. Quantum Electron. QE-25, (1989) 2532.
2. T.Kobayashi, M.Yoshizawa, M.Hasegawa, and M.Taiji, J. Opt. Soc. Am. B7 (1990) 1558.
3. M. Yoshizawa, A. Yasuda, and T. Kobayashi, Appl. Phys. B53 (1991) 296.
4. T.Masuda, T.Hamano, K.Tsuchihara, and T.Higashimura, Macromolecules 23 (1990) 1374.
5. S. Takeuchi, M. Yoshizawa, T. Masuda, T. Higashimura, and T. Kobayashi, IEEE J. Quantum Electron., Special Issue on Ultrafast Phenomena (1992), to be published.

Polarization-Dependent Femtosecond Dynamics of MBE-Grown Phthalocyanine Organic Thin Films

Sandalphon[1], *V.S. Williams*[1,2], *K. Meissner*[1], *N.R. Armstrong*[2], *and N. Peyghambarian*[1]

[1]Optical Sciences Center, University of Arizona, Tucson, AZ 85721, USA
[2]Department of Chemistry, University of Arizona, Tucson, AZ 85721, USA

Ultrafast bleaching of the lowest energy π-π^* state and photoinduced absorption from a sub-gap state is reported in chloro-indium-phthalocyanine grown by molecular beam epitaxy.

Experiment

We have observed ultrafast bleaching and induced absorption in MBE-grown chloro-indium-phthalocyanine (InPcCl) thin films. A 200-Å thick film was deposited using van der Waals molecular beam epitaxy onto a transparent tin disulfide substrate. RHEED measurements indicated epitaxy was maintained throughout sample growth. A comparison of linear absorption spectra for epitaxial versus polycrystalline InPcCl is shown in Figure 1. The Q-band absorption is significantly narrower for the epitaxial sample of Figure 1(a) with peaks at 760 nm and 680 nm compared to the extremely broad Q-band absorption for the 400-Å thick polycrystalline sample of Figure 1(b). The ultrafast measurements were performed using a CPM femtosecond laser system with a 620-nm pump and pulse widths of 80fs. The differential transmission ($-\Delta\alpha l$) spectra were measured at time delays of -300 fs to +1000 fs in 50 fs intervals between the probe and pump pulses.

A comparison of the differential transmission spectra for parallel and perpendicular polarization orientations is shown in Figures 2 (b) and (c) respectively. In both cases we observe

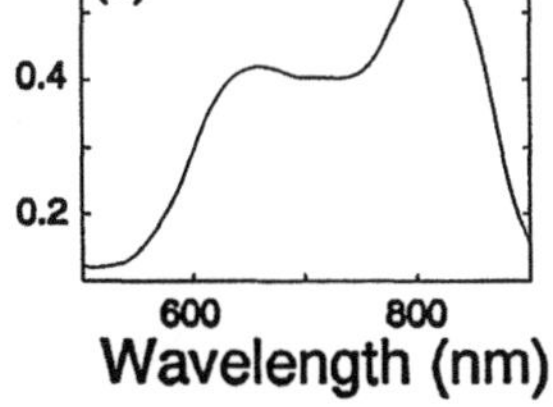

Figure 1. (a) Linear absorbance of epitaxial InPcC
(b) Linear absorbance of polycrystalline InPcC

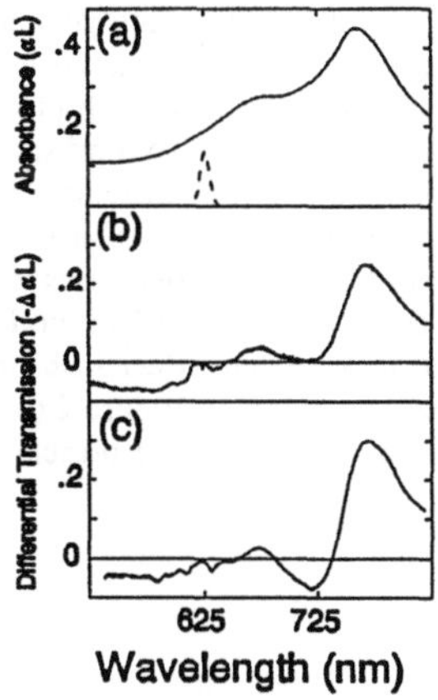

Figure 2. (a) Linear absorbance (solid) of epitaxial InPcCl and pump spectrum (dashed),
(b) Differential transmission spectrum of parallel pump and probe polarization,
(c) Differential transmission spectrum of perpendicular pump and probe polarization.

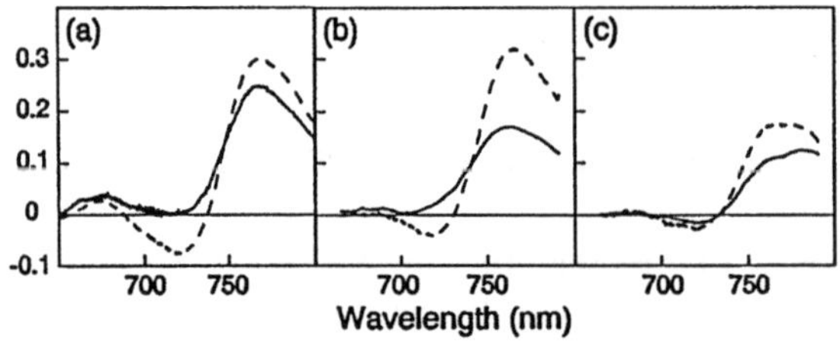

Figure 3. Differential transmission spectra for parallel (solid curves) and perpendicular (dashed curves) pump and probe polarizations. Pump power densities: (a) 7.8 GW/cm², (b) 4.6 GW/cm², and (c) 1.4 GW/cm².

bleaching on the low energy side of the Q-band linear absorption and photoinduced absorption on the high energy side. The linear absorption spectrum is shown for reference along with the spectral position of the pump in Figure 2(a).

For either case of pump and probe polarization the induced absorption signal at higher energy develops ≈ 300 fs before the maximum bleaching signal. The decay dynamics for the two signals differ considerably. A relaxation time of ≈ 1 ps was measured for the bleaching signal while a relaxation time of ≈ 1.7 ps was measured for the induced absorption signal. This seems to indicate the existence of a sub-gap state from which the induced absorption occurs. The rapid decay of the induced absorption appears to rule out a triplet state for its origin. We believe that a one-photon forbidden singlet excited state is a likely candidate for the sub-gap state.

One distinguishing feature between the differential transmission spectra for the parallel versus the perpendicular pump and probe polarizations shown in Figure 2 is a region of induced absorption that lies between the peaks of the bleaching signal , at 720 nm, for the perpendicular case but is absent from the parallel case. This result is surprising since it is not obvious why the two spectra should differ in this way. It is difficult to explain the existence of a transition that is allowed for perpendicular pump and probe polarizations yet not allowed for parallel pump and probe polarizations. Since we were using 7.8 GW/cm^2 for the pump pulse it was possible that we were saturating the induced absorption at 720 nm for the parallel case due to a power broadening of the bleaching peak at 760 nm. Subsequently we repeated the experiment for the spectral region around 720 nm using reduced pump pulse power densities. Figure 3 compares the differential transmission spectra for the two cases at pump pulse power densities of 7.8 GW/cm^2, 4.6 GW/cm^2, and 1.4 GW/cm^2 shown in parts (a), (b), and (c) respectively. In each of the three parts of Figure 3 the bleaching peak at 760 nm is lower and broader for the parallel case compared to the perpendicular case. However, as the pump pulse power density decreases the difference between the two cases in the region around 720 nm also diminishes. For a pump pulse power density of 1.4 GW/cm^2 both cases show induced absorption at 720 nm. Hence it appears that the lack of an induced absorption signal at 720 nm for the parallel pump and probe polarizations shown in Figure 2 (b) and Figure 3 (a) and (b) is not due to a disallowed transition but to a broadening of the bleaching signal.

Detection of a New Strongly-Coupled Vibration Mode During the Exciton Bleaching of Polydiacetylene

J.M. Nunzi[1], *C. Hirlimann*[2], *and J.F. Morhange*[3]

[1]CEA/DTA/LETI/DEIN/SPE, Laboratoire de Physique Electronique des Matériaux, Centre d'Etudes Nucléaires de Saclay, F-91191 Gif-sur-Yvette, France

[2]IPCMS/GONL, Université Louis Pasteur, 5, rue de l'Université, F-67084 Strasbourg, France

[3]Laboratoire de Physique des Solides, Université P. et M. Curie, 4, Place Jussieu, F-75252 Paris Cedex 05, France

Abstract. The femtosecond exciton bleaching dynamics of a polydiacetylene thin film is studied using 2eV resonant pumping. A hole is burnt 20meV below the pump energy. It induces two symmetric phonon replica located 35meV beside the hole.

Summary. Polydiacetylenes have large optical nonlinearities due to the extension of their π-electron wave-functions which are delocalized over many polymer repeat-units. They also exhibit sub-picosecond excited-state relaxation-times due to the efficient non-radiative decay processes which are mediated by the large electron-phonon coupling in those one-dimensional conjugated systems [*1*]. Some vibration modes such as the single, double and triple carbon-carbon bond-stretching ones, which are strongly coupled to the electronic excited-states, have been characterized using various resonant Raman effects [*1,2*]. In order to gain insight into those relaxation processes, we have performed sub-picosecond time-resolved absorption spectroscopy of the exciton bleaching of a polydiacetylene thin film.

The polydiacetylene sample [*3*] was a polycrystalline 5000Å-thick film of the poly-(4BCMU) *blue*-form which was vacuum-deposited on a 1mm-thick fuzed-quartz slide. The spectrum of its largest visible (excitonic) absorption is shown in the upper part of figure 1. The experiments [*4*] were performed using a transmission pump-probe technique. The pulses where orthogonally polarized in order to eliminate pump-light scattered by the sample before the spectral analysis of the probe. The pump which was derived from a 20Hz-amplified CPM-laser had ≈120fsec-duration, 200nJ-average energy per pulse and it was focused in a $4\cdot10^{-5}cm^2$ spot on the sample ($40GW/cm^2$). Its spectrum which peaks at the 2eV-absorption maximum of the sample is shown in the upper part of figure 1. The probe which was generated in a 2cm-long monomode-fiber and was dispersion-recompressed with prisms had ≈20fsec-duration, 150meV-full-width at half height and ≈5nJ-energy.

The pump-induced absorption variations $\delta\alpha l$ are shown at various time delays in the lower part of figure 1. The transient bleaching decays with an exponential time constant of 0.9±0.1psec. The maximum possible bleaching (saturation) is achieved in figure 1. It reaches a limit of 40% of the excitonic absorption. This limit is due to photoinduced absorptions on higher excited states (bi-excitons are possible states). In the studied non-degenerate configuration, with orthogonal polarizations, we could get a $14\pm2GW/cm^2$-saturation intensity for the bleached transition. It is consistent with other measurements performed on similar polymers [*1,5*]. At zero-delay, the 75meV-hole burnt into the 130meV-inhomogeneous line has a bottom which is 15meV-red shifted with respect to the pump energy. The shift may be due to the fact that we pump above the bottom of the exciton band.

The most relevant feature in figure 1 is the appearance at positive delays of two replica (side holes) adjacent to the central hole. They are spaced of 35meV from it. The holes deepen at positive delays when the polymer relaxes. The effect

Editors: J.-L. Martin · A. Migus · G.A. Mourou · A.H. Zewail

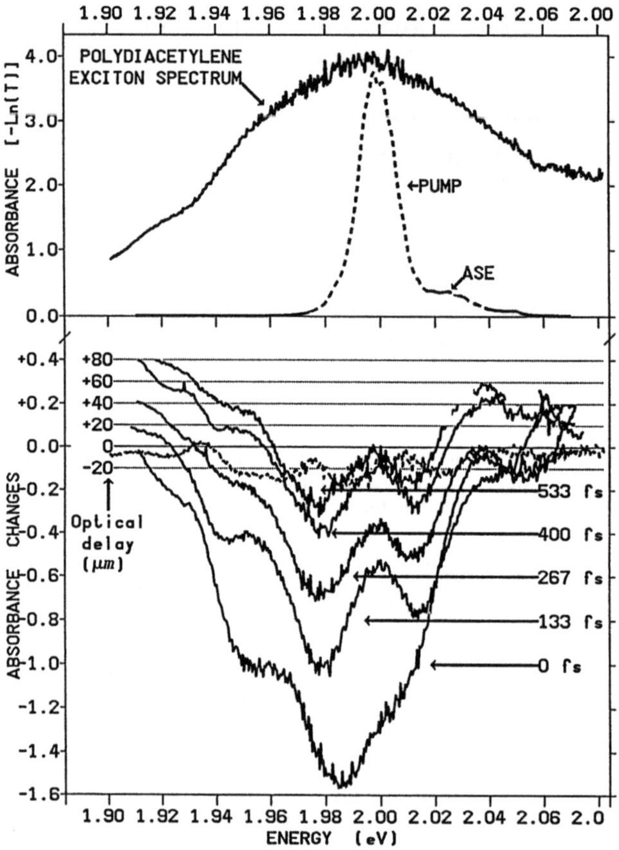

Fig.1 Excitonic absorption of the polydiacetylene film. In the upper part, the continuous line is the absorbance (αl=-Ln[T]) of the probed region and the broken line is the pump spectrum (arbitrary units). ASE is the amplified stimulated emission. The lower part depicts the absorbance changes $\delta\alpha l$ (average on 300 laser shots) induced by the pump at various delays. Its maximum corresponds to a 40% absorption bleaching. Negative and positive delays correspond to broken and continuous lines respectively. The vertical scale of each curve measured for 133fsec steps has been shifted upwards by $\delta\alpha l$=0.1.

may be pictured in terms of the Frank-Condon coupling with vibrations observed in similar experiments performed in dyes [6]. We indeed observe a weak 280cm^{-1}-vibration mode using resonant Raman scattering in the sample. It is in the frequency range of the carbon-skeleton planar shear-modes. Such a weakly coupled mode has been observed (and not considered) in similar polymer forms [2]. The excited vibrational levels have weak population at room-temperature. The central hole is due to both the depletion of the 0-vibrational level in the ground state and the increase in the population of the 0-vibrational level in the excited state. The depletion of the 0-vibrational level in the ground state causes a bleaching of the transition to the 1-vibrational level in the excited state. It is the blue-shifted replica. The transient excess population of the 0-vibrational level in the excited state stimulates light emission at the energy of the transition to the 1-vibrational level in the ground state. It is the red-shifted replica. From the experiment, we estimate the ratio of the Franck-Condon factors [7] to $\chi_{10,01}/\chi_{00} \approx 0.5$. It corresponds to a ≈0.1Å well displacement in the excited state. The faster relaxation of the low energy side-hole reflects the faster depopulation of the exciton with respect to the ground state recovery time. In conclusion, we have detected a new motion linked with the polymer deformations which are typical of the exciton relaxation.

References:

1. B.I. Greene et al., *Phys. Rev.Lett.* **61**, 325 (1988).
2. M.L. Shand et al., *Phys. Rev. B* **25**, 4431 (1982).
3. J.M. Nunzi and F. Charra, *Appl. Phys. Lett.* **59**, 13 (1991).
4. C. Hirlimann et al., *Appl. Phys. Lett.* **55**, 2307 (1989).
5. T. Kobayashi et Al., *J. Opt. Soc. Am. B* **7**, 1558 (1990).
6. C.H. Brito Cruz et al., *Chem. Phys. Lett.* **132**, 341 (1986).
7. D.N. Batchelder and D.Bloor, *J. Phys. C* **15**, 3005 (1982).

Pressure-Induced Vibrational Relaxation and Electronic Dephasing in Molecular Crystals

E.L. Chronister and R.A. Crowell

Department of Chemistry, University of California, Riverside, CA 92521, USA

Abstract. Picosecond coherent anti-Stokes Raman (CARS) measurements on strain free naphthalene crystals at low temperature (≤15K) in a high pressure diamond anvil cell are presented. The pressure induced vibrational relaxation (VR) results are interpreted in terms of pressure induced changes in anharmonic coupling and the density of two phonon states. Photon echo measurements of electronic dephasing at high pressure are also discussed.

1. Introduction

Conventional Raman spectroscopy under high pressure conditions has been a useful probe of molecular solids[1]. More recently, low temperature time resolved coherent Raman and photon echo experiments have been shown to be useful probes of pressure induced VR[2-4] and electronic dephasing[4].

In this study pressure dependent VR is used to probe the anharmonicity of the intermolecular potential. Furthermore, such measurements provide an experimental tool with which to evaluate solid state potential energy functions. A "golden rule" description of cubic relaxation (i.e. fission into two lower energy modes), yields a relaxation rate determined by the product of the square of an anharmonic coupling matrix element and a density of two phonon states (2PDOS). In this description, pressure effects will be due to either changes in the magnitude of the anharmonic coupling coefficient and/or changes in the 2PDOS. Although the average density of acoustic and optical phonon states per volume at a given frequency changes little over the pressure range investigated, spectral expansion results in higher frequency phonon modes which can open up new relaxation pathways.

2. Results and Discussion

To determine the effect of crystal strain on psCARS measurements, we performed several pressure *release* experiments to induce a large reproducible crystal strain, the results of which are shown in Figure 1. For the ν_8 mode of naphthalene a pressure release of $\Delta P = -6.8$kbar resulted in a decrease in the dephasing rate to a value equal to the homogeneous low pressure relaxation rate[5]. In Figure 2 we also show several semilog psCARS decays for the ν_8 mode in strain free naphthlene crystals at various pressures.

The exponential CARS decays from Figure 2 yield dephasing rates, $1/T_2$, which are attributed to vibrational relaxation. Figure 3 shows the relative pressure induced change in the VR rate compared with the relative change in the 2PDOS for both the ν_8 and ν_9 vibrons of naphthlene[6]. From Figure 3 we see that the pressure dependence of the VR rate of ν_8 follows closely the changes in the 2PDOS for this mode. However, the situation is much different for the ν_9 mode for which the pressure induced VR rate increases dramatically while the 2PDOS is nearly pressure independent. Thus, for the ν_9 mode, pressure induced anharmonicity appears to be the dominant cause of the relaxation [4,6].

Springer Series in Chemical Physics, Vol. 55 **Ultrafast Phenomena VIII**
Editors: J.-L. Martin · A. Migus · G.A. Mourou · A.H. Zewail

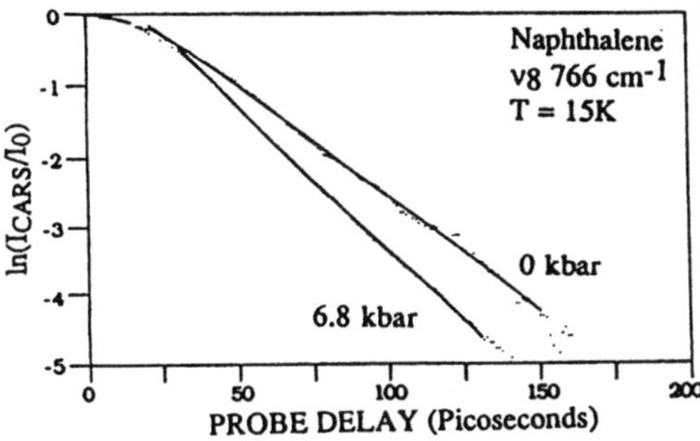

Figure 1. Semilog psCARS decays at 15K for the ν_8 vibron of naphthalene at 6.8kbars and following a pressure release to 0kbar.

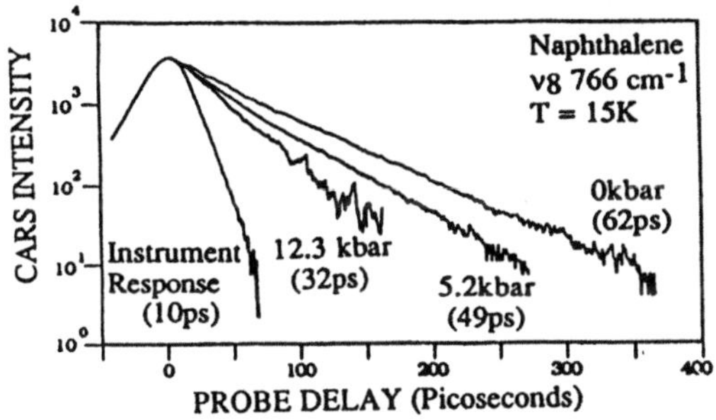

Figure 2. Semilog psCARS decays at 15K for the ν_8 vibron of naphthalene at various pressures.

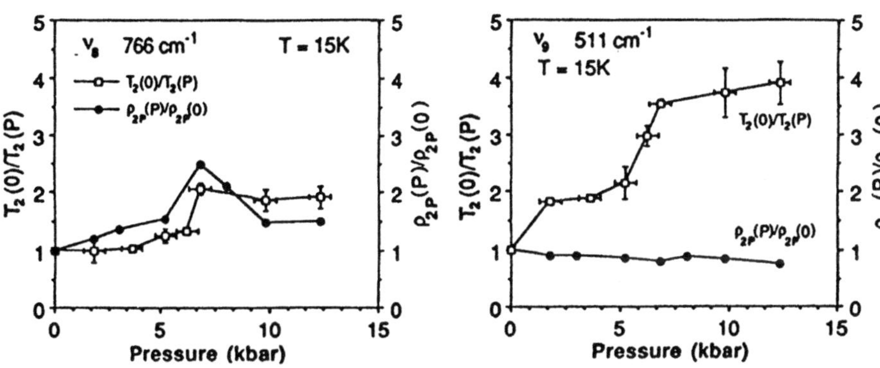

Figure 3. The relative change in the low temperature dephasing rates of the ν_8 and ν_9 modes of crystalline naphthalene as a function of pressure, i.e. $T_2(0)/T_2(P)$. Also shown is the relative pressure induced change in the 2PDOS, $\rho_{2p}(P)/\rho_{2p}(0)$.

We have also performed photon echo measurements of the homogeneous electronic dephasing rate of pentacene doped p-terphenyl at high pressure. Measurements on the O_1 photosite origin at a temperature of 15K and a pressure of 1kbar yield a dephasing time, T_2 = 280ps which is *slower* than the 65ps value obtained for the ambient pressure solid at the same temperature (i.e. pressure induced line narrowing)[4]. At 1.1K the dephasing rate is pressure independent. The line narrowing observed at higher temperatures can be explained by a pressure induced increase in the pseudolocal phonon frequency resulting in a lower phonon occupation at that temperature.

3. Acknowledgements

We acknowledge the National Science Foundation (CHE#9008551) and the Los Alamos Center for Nonlinear Studies INCOR Program, for financial support of this research.

4. References

1. J. Ferraro: Vibrational Spectroscopy at High External Pressure (Acad., New York, 1984)
2. M. Baggen ,A. Lagendijk Chem. Phys. Lett. **177**, 361 (1991).
3. R. A. Crowell, E. L. Chronister, J. Phys. Chem. **96,** 1045 (1992);
   E. L. Chronister, R. A. Crowell, Chem. Phys. Lett., **182,** 27 (1991).
4. E. L. Chronister, R. A. Crowell, Mol. Cryst. Liq. Cryst. **211**, 361 (1992).
5. R. A. Crowell, E. L. Chronister, Chem. Phys. Lett., in press.
6. R. A. Crowell, E. L. Chronister, J. Phys. Chem., submitted.

Ultrafast Reversible Phase Changes for Optical Recording

J. Solís, C.N. Afonso[1], *F. Catalina*[1], *and C. Kalpouzos*[2]

[1]Insituto de Optica, CSIC, S/Serrano 121, E-28006 Madrid, Spain
[2]FORTH-IESL, P.O. Box 1527, GR-7110 Heraklion, Crete, Greece

Abstract. Reversible phase changes with optical contrast are induced by ps and fs laser pulse irradiation in GeSb thin films. The feasibility of phase change optical recording at ultrafast rates is therefore demonstrated for the first time. The phase reversal induced by ultrashort laser pulses seems to be related to the dependence of the degree of undercooling achieved prior to solidification on the irradiation energy density.

1.Introduction

Optical storage systems offer higher storage densities than magnetics, magneto-optics being the most mature technology [1]. Phase change storage media take advantage of the different optical properties which may present amorphous and crystalline phases of a given material. The erasing process is usually referred as the slow process since crystallization requires laser pulses long enough to trigger both nucleation and growth of crystals [2]. Current trends in the field include the development of materials which crystallize under ns laser pulses [2,3]. However, to our knowledge no attempt has been made up to date to perform reversible phase change optical storage using subnanosecond laser pulses. The aim of this work is to study the structural and optical transformations induced in Sb-rich GeSb thin films by ps and fs laser pulses, both in the amorphous and crystalline material. It aims also to demonstrate for the first time the reversibility of the crystallization process by using ps and fs laser pulses.

2.Experimental

The samples are amorphous $Sb_{0.9}Ge_{0.1}$ films grown by DC magnetron sputtering (Fig.1a). Crystalline material is obtained by UV laser irradiation of the amorphous as-deposited samples with 12 ns ArF excimer laser pulses (Fig.1b) being crystallization accompanied by a reflectivity increase of 15-20 %. Both amorphous and crystalline films are then irradiated with 500 fs and 5 ps (FWHM) laser pulses at 248 nm. The pulses are obtained from a frequency doubled distributed feedback dye laser seed pulse in a two pass amplification through a KrF amplifier cavity. After irradiation the samples are analyzed by optical microdensitometry, transmission electron microscopy, and selected area diffraction.

3.Results and Discussion

The irradiation of the crystalline films with ps and fs laser pulses produces the formation of an amorphous phase (Fig.1d). It exhibits similar optical properties and diffraction patterns than the initial as-deposited material and therefore it is demonstrated that the crystallization process can be reversed by using subnanosecond laser pulses. Due to inhomogeneities in the ps, fs laser beam, crystalline and amorphous phases are formed together being the latter

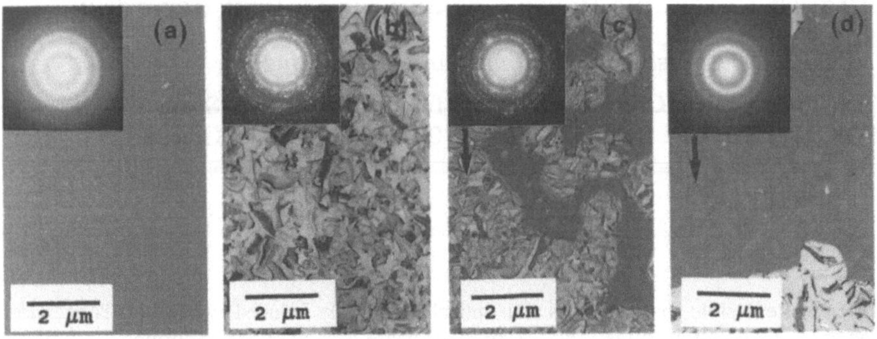

Figure 1: Transmission electron micrographs and difraction patterns of as-deposited (a) and ns pulse (56 mJ/cm^2) crystallized (b) thin films; (c) and (d) show images of (a) and (b) upon irradiation with 500 fs pulses with fluences of 15 mJ/cm^2 and 10 mJ/cm^2 respectively. The morphology inhomogeneities in (c) and (d) are due to the non homogenous profile of the laser caused by diffraction effects in the KrF amplifier cavity.

induced in those regions where the energy density is higher. The same effect is observed by irradiating the amorphous as-deposited material with 500 fs laser pulses (Fig.1c).

The real time reflectivity measurements obtained during the nanosecond pulse induced crystallization [4] allow us to establish that crystallization is induced for quenching rates as large as 10^9 K/s. The quenching rates expected for irradiation with 500 fs and 5 ps laser pulses are in both cases no less than 10^{12} K/s [5] and therefore only the formation of amorphous phases is expected in absence of other factors affecting the nature of the induced microstructures like undercooling. Experimental evidence of the influence of the energy density on the undercooling achieved in thin films in laser induced rapid solidification processes has been provided previously [6]. Basically, when the energy deposition is large enough, the substrate can be effectively heated reducing the thermal gradient along the surface normal and therefore causing solidification to occur at higher temperatures (undercooling decrease). A beautiful example of this is the comparison of the crystalline microstructures induced by irradiation with fs laser pulses (Fig.1c,1d) and with nanosecond laser pulses (Fig.1b). Both types of crystalline microstructures are essentially the same indicating that solidification has occurred at similar temperatures in spite of the expected differences in the cooling rate. As a general conclusion, our results show that irradiation of $Sb_{0.9}Ge_{0.1}$ thin films with 500 fs or 5 ps laser pulses may yield the formation of either a high reflectivity crystalline phase or a low reflectivity amorphous phase depending on the pulse energy density (undercooling achieved). An implication of this fact is the existence of a real candidate material to perform subnanosecond reversible phase change optical recording

4.References

[1] E.M.Engler; Adv.Mater.**2** (1990) 166, [2] K.A.Rubin, M.Chen; Thin Solid Films **181** (1989) 129, [3] D.J.Gravesteijn, H.M.van Tongeren, M.Sans, T.Bertness, C.J.van der Poel; Appl.Optics **26** (1987) 4772, [4] J.Solis, J.F.Trull and C.N.Afonso; (to be published), [5] F.Spaepen in "Ultrafast Phenomena V", Ed. G.R.Fleming and A.E.Siegman, Springer (Berlin, 1986), [6] C.N.Afonso, F.Catalina, J.Solis; Mat.Sci.Eng **B7** (1990) 169

Picosecond Transient Absorption and Fluorescence Emission Studies of C_{60} and C_{70} in Solution

D. Kim[1], *Y.D. Suh*[2], *S.K. Kim*[2], *and M. Lee*[1]

[1]Korea Research Institute of Standards and Science (KRISS), Taedok Science Town, Taejon 305-606, Republic of Korea
[2]Department of Chemistry, Seoul National University, Seoul 151-742, Republic of Korea

Abstract. The excited state photophysics of C_{60} and C_{70} in solution were studied by picosecond transient absorption spectroscopy and time-correlated single photon counting technique.

The recent development of the production of C_{60} and other fullerenes in high yields made it possible to study the basic properties of the newly discovered molecules, employing various experimental techniques. The excited state photophysical properties of C_{60} and C_{70} have also been investigated extensively, using time-resolved spectroscopy by many groups. In this study, we carried out transient absorption and fluorescence lifetime measurements on those molecules and present new experimental results which would provide useful information on understanding more fully the excited state properties of C_{60} and C_{70}.

The fullerenes were produced by the contact-arc method and purified by now-standard procedures. FT-IR, Raman, ^{13}C NMR spectroscopy as well as HPLC and mass spectrometry were employed to identify the species. We confirmed no presence of impurities within the detection limits of the above techniques. In Fig. 1, we report the transient absorption spectra of C_{60} and C_{70} in the wavelength region of 600-1000 nm, partially covering the near infrared region that was not seen in the previous studies. The exciting and probing continuum pulses of the 800 fs pulse width at 20 Hz were obtained from a combination system of CW mode-locked Nd:YAG laser, hybridly mode-locked synchronously pumped dye laser, Nd:YAG regenerative amplifier, and tunable picosecond dye amplifier. The transient absorption plotted in Fig. 1 was measured at 20 ps, 153 ps, 0.8 ns, 1.8 ns, and 3.1 ns after the picoseocond laser pulse excitation of C_{60} and C_{70} in toluene at 587 nm. Since these trasient absorption spectra were most comprehensive among the experimental data reported so far, it gives new information especially at the short time delay. The excited singlet lifetime was drawn from the decrease in S_1-S_n absorption and the increase in T_1-T_n. They appear to be 1.3±0.2 ns for C_{60} and 0.7±0.05 ns for C_{70}.

The C_{60} and C_{70} show very weak fluorescence emission in solution, having maximum at 710 nm and 670 nm respectively. The emission intensity of C_{60} at the peak is ca. five times weaker compared with C_{70} fluorescence. Fig. 2 shows the fluorescence lifetime of C_{60} and C_{70} in toluene measured by the time-correlated single photon counting system. The sample was degassed. The excitation wavelength is 587 nm and the

Springer Series in Chemical Physics, Vol. 55 Ultrafast Phenomena VIII
Editors: J.-L. Martin · A. Migus · G.A. Mourou · A.H. Zewail

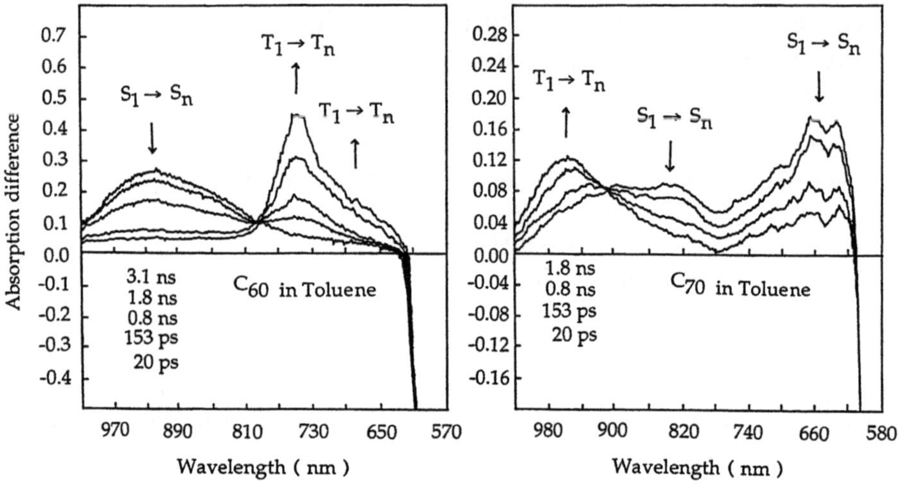

Fig.1 Transient absorption spectra of C_{60} and C_{70} in toluene at different delay times. The arrows ↑ and ↓ indicate the increase and decrease of the transient absorption intensity.

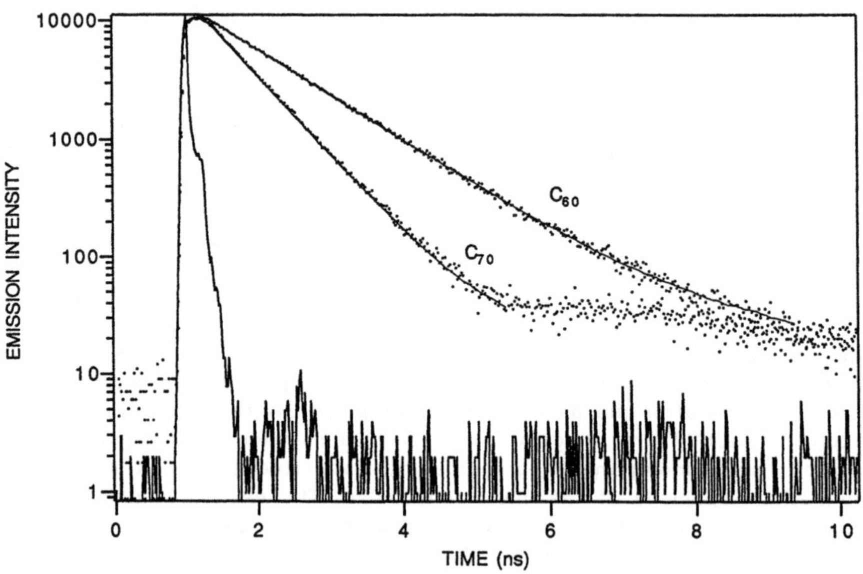

Fig. 2 Emission decay curves for C_{60} and C_{70} in toluene at 22 ℃. Shown in the short time region is the instrument response function of our TCSPC. Excitation wavelength is 587 nm. The fitting gives acceptable χ^2 of 1.22 for C_{60} and 1.18 for C_{70} data.

instrument function is 25 ps. The decay curves fit a single exponential with the lifetime of 1.17±0.02 ns for C_{60} and 0.66±0.02 ns for C_{70}, respectively.

References

1. R. J. Sension, C. M. Phillips, A. Z. Szarka, W. J. Romanow, A.R. McGhie, J.P. McCauley, Jr., A.B. Smith III, and R. M. Hochstrasser, J. Phys. Chem. 95, 6075 (1991).
2. K. Tanigaki, T. W. Ebbesen, and S. Kuroshima, Chem. Phys. Lett. 185, 189 (1991).
3. M. lee, O.-K Song, J. -C Seo, D. Kim, Y. D. Suh, and S. K. Kim, submitted to Chem. Phys. Lett..
4. D. Kim, M. Lee, Y. D. Suh, and S. K. Kim, J. Am. Chem. Soc., in press.

Part VII

Semiconductors, Confinement and Opto-Electronics

Transient Absorption-Edge Singularities in GaAs

D. Hulin[1,2], *J.-P. Foing*[1,*], *M. Joffre*[1,2], *M.K. Jackson*[1,*], *J.-L. Oudar*[3], *C. Tanguy*[1,3], *and M. Combescot*[2]

[1]Laboratoire d'Optique Appliquée, Ecole Nationale Supérieure des Techniques Avancées, Ecole Polytechnique, Centre de l'Yvette, F-91120 Palaiseau, France

[2]Groupe de Physique des Solides, Universités Pairs VI et Paris VII, 2, place Jussieu, F-75005 Paris, France

[3]Centre National d'Etude des Télécommunications, 196, avenue Henri Ravera, F-92220 Bagneux, France

*J.P. Foing is now at ETCA, Arcueil, France. M.K. Jackson is now at the Department of Electrical Engineering, University of British Columbia, Vancouver, Canada.

The induced absorption observed on the high energy side of spectral hole-burning in GaAs is discussed in terms of coherence effects with a chirped pump pulse and in terms of absorption edge singularities at the limits of sharp carrier distributions.

Nonequilibrium carrier distributions can be observed in semiconductors by monitoring the corresponding induced transparency on a femtosecond timescale. This bleaching occurs because optical transitions cannot occur for final states occupied by previously created carriers due to Pauli exclusion. Just after the excitation by a short pump pulse, the nonthermal carrier distribution reflects the excitation energy profile. This appears experimentally as a spectral hole-burning in the interband absorption. The bleaching does not remain peaked at the excitation energy since different interactions tend to thermalize the carrier distributions.

Pump-probe experiments are performed in GaAs by using femtosecond pulses amplified by a copper vapor laser at 6.5 kHz. In order to take full advantage of the spectral informations, carriers have to be generated with a well-defined energy. Therefore a trade-off between spectral width and temporal length has been chosen leading to a 120 fs pump pulse duration. The chirp due to the group velocity dispersion of the continuum used as the probe is compensated with a pair of prisms and a pair of gratings in order to correct terms in the phase up to third order. The temporal resolution is better than 30 fs. For the pump beam, a pair of prisms can be adjusted to obtain positive or negative chirp or no chirp at all. The sample is a 0.75-μm GaAs layer grown by molecular beam epitaxy and cooled down to 20 K.

By recording the induced changes in the absorption spectra, we expect to monitor the evolution of the electron and hole occupations. Figure 1 shows the induced transparency ($-\Delta\alpha d$) in the very early times for a 120-fs pump pulse. Spectra are recorded 40 fs apart and show the build-up of the induced transparency. The most obvious feature is the fact that three maxima appear in the bleaching, one approximately at the spectral position of the pump as expected in a hole-burning experiment, and two others on each side with a smaller amplitude. When the wavelength of the pump is varied, the energy differences between the peaks change, excluding phonon emission or absorption. They go to zero when the pump is close to the absorption edge. These three different maxima are due to the presence of heavy-hole (h-h) and light-hole (l-h) valence bands. For the same pump photon energy, two different electron populations are created corresponding to transitions from the h-h and l-h valence bands. The "white" probe beam experiences then an increased transmission at those wavelengths corresponding to transitions for which states in the conduction band or in the valence bands or in both bands are occupied. In the following, we focus our attention on the main peak around the pump wavelength.

Springer Series in Chemical Physics, Vol. 55 Ultrafast Phenomena VIII
Editors: J.-L. Martin · A. Migus · G.A. Mourou · A.H. Zewail

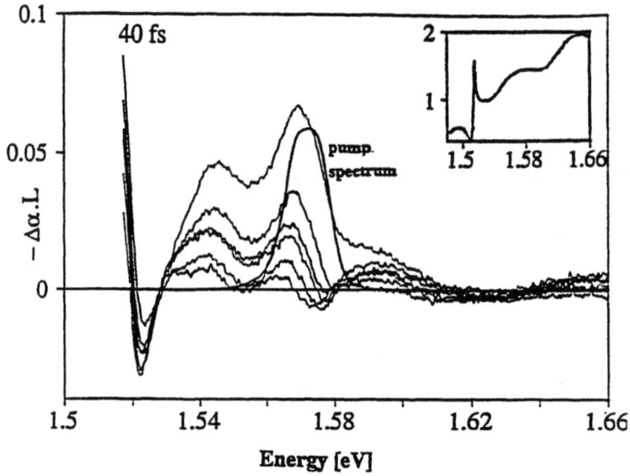

Figure 1
Experimental differential spectra recorded for various pump-probe time delays 40 fs apart. The large signal around 1.51 eV is due to the bleaching of the excitonic absorption. The pump pulse has a positive chirp and the correction of the probe chirp is not perfect around 1.53 eV.

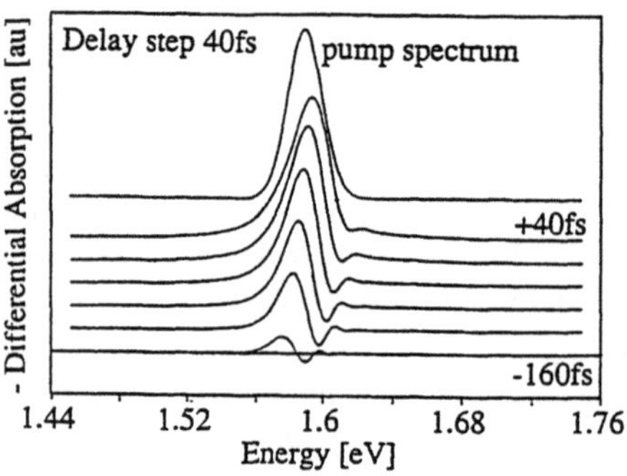

Figure 2
Differential absorption spectra for an assembly of two-level systems excited with a linearly chirped pump pulse. Both the coherent grating term and the hole-burning term are considered

We have extended the calculations for a nonzero chirp of the pump pulse and found asymmetric coherence oscillations (Figure 2). This results mainly from the diffraction term on the induced grating and leads to a sizeable negative signal on the wing of the hole-burning. This feature appears on one side or the other depending upon the sign of the chirp. To know how these coherence effects should contribute to our experimental results, we purposely introduced a chirp in the excitation. We indeed observed a pronounced asymmetry in the hole-burning spectra, appearing mainly as a negative signal on the high- or low-energy side of the hole burning. The maximum of the observed bleaching shifts as a function of time delay. However we still observe in every case the unexpected induced absorption on the high-energy side of the hole-burning. Figure 3 shows the differential transmission spectra for an unchirped pump pulse.

During the build-up of the induced transparency, a negative signal appears on the high energy side of the bleaching, corresponding to a transmission decrease. This feature shows up only at the beginning of excitation, the redistribution of carriers leading rapidly to a large positive signal over the spectral range of interest which overwhelms any negative contribution. Furthermore, the maximum of the induced transparency does not coincide with the pump wavelength.

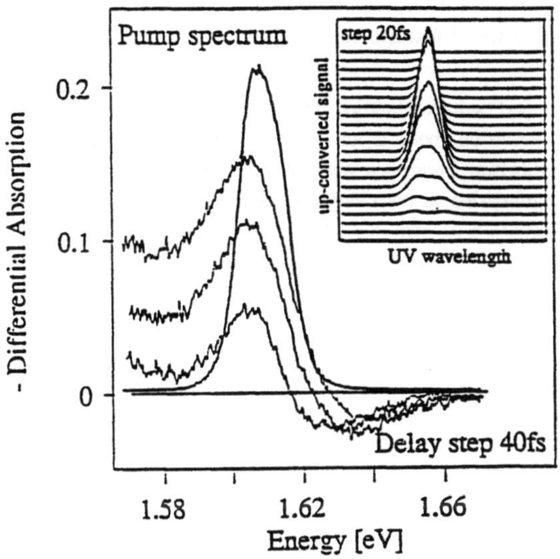

Figure 3
Experimental differential spectra recorded at various delays 40 fs apart. The chirps of both the pump pulse and the probe pulse have been corrected, as can be seen in the inset where cross-correlation signals have been recorded with steps of 20 fs. The induced bleaching seen at 1.56 eV corresponds to occupied states in the conduction band seen by an optical transition from the heavy holes.

Coherent effects can be very important for spectra observed during the temporal overlap of the pump and probe pulses. Spectral oscillations may provide the appearance of a negative signal such as the one we have observed. The situation has been analyzed [1] for hole-burning experiments with an unchirped pump pulse. In addition to the absorption saturation which provides information on the carrier populations, two other factors may change the probe transmission. The first one originates from the diffraction of both beams on the induced grating created in the material through the interference of the incident beams. The second one is the so-called perturbed free-induction decay. Both terms give rise to oscillatory structures in the transmission spectrum. However they are symmetrical with respect to the pump central wavelength. Furthermore, in the semiconductor where the density of states is almost flat, the calculated oscillations are weak, particularly the perturbed free-induction decay term.

Since coherence effects do not explain our experimental results, we turn now to interpretations linked to solid state physics. Optical transitions can be affected by the presence of a strong electromagnetic field (optical Stark effect) in resonance with band to band transitions. The opening of a gap has been predicted [2,3]; this leads to a decreased absorption at the spectral position of the pump which cannot be differentiated from state filling effects. In addition, this gap opening gives rise to an increased absorption on the low energy side of the gap and a decreased absorption on the other side [3]. This is opposite to our experimental observation, thus model cannot be used to describe our results.

Band gap renormalization shows up as soon as the carrier density is high enough. However a rigid band gap shift, taking into account the actual shape of the absorption spectrum, should provide an almost flat negative signal over all the spectral range of observation in differential transmission, and this is not observed. Furthermore, the magnitude of the observed signal would correspond to a large band gap renormalization appearing below the energy of the unperturbed exciton and this is not seen in our experiments.

The increased absorption we observe has a strong resemblance with the edge singularities found in luminescence and absorption spectra of highly doped semiconductors [4,5]. They originate from many-body effects between electrons and the optically created hole [6]. For an infinite hole mass and an abrupt Fermi cutoff, a divergent absorption is predicted at the Fermi level. In GaAs, this singularity reduces to a broad feature, but still persists, the hole mass being quite large.

In contrast with the case of doped semiconductors, the carrier distribution in our experiment has, just after the laser excitation, both a low- and a high-energy cutoff. This situation can be modelled by a nonequilibrium Fermi sea extending between (μ_1,μ_2) instead of $(0,\mu)$. These two sharp edges induce singularities on both sides of the induced transparency. In the limit of infinite hole mass, the absorption spectra close to threshold are found to have a power law behavior [7]

$$I_{12}(\omega) = 1 / \ |\omega - \mu_{1,2}|^{\alpha_{1,2}}$$

The critical exponents $\alpha_{1,2}$ being linked to the phase shift $\delta_{1,2}$ induced by the attracting potential of the optically created hole at the $\mu_{1,2}$ energy level.

$$\alpha_{1,2} = \gamma_{1,2} - \sigma$$

$$\gamma_1 = -2\delta_1/\pi \qquad \gamma_2 = +2\delta_2/\pi$$

$$\sigma = (\delta_1^2 + \delta_2^2)/\pi^2$$

It can be shown that in systems with a large amount of quite heavy holes such singularities should persist as an increased absorption at the upper threshold and a reduced absorption at the lower threshold. This would lead, in the differential transmittance spectra, to a negative signal on the high energy side of the hole-burning and a positive one on the low energy side. The first feature corresponds well to our experimental observation whereas the second one cannot be distinguished from state filling occurring at the same energy.

The above theoretical results are in good qualitative agreement with the experimental data and provides a satisfying physical explanation for the observed induced absorption [8]. This explanation emphasizes the fact that hole-burning spectra reflect not only the distribution of photoexcited carriers, but also the change of the oscillator strength due to many-body effects between the photo-created electron-hole pair and the previously created electron-hole plasma. These effects are strongly linked to non-equilibrium distributions with sharp edges as created by ultrashort pump pulses.

1) C.H. Brito-Cruz, J.P. Gordon, P.C. Becker, R.L. Fork and C.V. Shank
IEEE J. Quantum Electr. **24**, 261 (1989)

2) V.M. Galitskii, S.P. Goreslavskii and V.F. Elesin
Sov. Phys. JETP **30**, 117 (1970)

3) S. Schmitt-Rink, D.S. Chemla and H. Haug
Phys. Rev. B **37**, 941 (1988)

4) A.E. Ruckenstein, S. Schmitt-Rink and R.C. Miller
Phys. Rev. Lett. **56**, 504 (1961)

5) M.S. Skolnick, J.M. Rorison, K.J. Nash, D.J. Mowbray, R. Tapster, S. Bass and A.D. Pitt
Phys. Rev. Lett. **58**, 2130 (1987)

6) For a good recent review, see K. Ohtaka and Y. Tanabe
Rev. Mod. Phys. **62**, 929 (1990) and references therein

7) C. Tanguy and M. Combescot
Phys. Rev. Lett. **68**, 1935 (1992)

8) J.P. Foing, D. Hulin, M. Joffre, M.K. Jackson, J.-L. Oudar, C. Tanguy and M. Combescot
Phys. Rev. Lett. **68**, 110 (1992)

Nonthermal Distribution of Electrons in GaAs

D. Snoke and W.W. Rühle

Max-Planck-Institut für Festkörperforschung, Heisenbergstr. 1,
W-7000 Stuttgart 80, Fed. Rep. of Germany

Abstract. We have observed the evolution of the energy distribution of free conduction-band electrons from a nonthermal to a Maxwell-Boltzmann distribution. This allows us to estimate the electron-electron scattering rate and to test various models of dynamic screening in simulating the experimental results with a numerical solution of the Coulomb-scattering Boltzmann equation.

The question of how fast free electrons scatter with each other as a function of carrier density is a fundamental question of solid state physics. The most direct method of finding the answer to this question is to experimentally observe the energy distribution of an electron plasma on timescales short enough to see a nonequilibrium distribution. The amount of time required to evolve to a thermal Maxwell-Boltzmann or Fermi-Dirac distribution would then provide a measure of the electron-electron scattering rate.

Several experimental groups have attempted to perform such a measurement using femtosecond techniques which require relatively high excitation density. Most recently, Elsaesser et al.[1] used luminescence up-conversion to observe band-to-band recombination on 100-fs timescale at pair densities around 10^{17} cm^{-3} . They observed a relatively thermalized distribution at all times.

In our experiments we have used a streak camera to observe free electrons at extremely low ($10^{12} - 10^{14}$ cm^{-3}) densities on a picosecond timescale. The band-to-acceptor recombination spectrum in bulk GasAs:Ge, with roughly $8 \cdot 10^{16}$ cm^{-3} doping level, yields the energy distribution of the electrons directly.[2]

Figure 1 shows the band-to-acceptor luminescence at two densities as a function of time, with the long-lifetime donor-acceptor luminescence subtracted and an $E^{1/2}$ density-of-states factor divided out. A Maxwell-Boltzmann distribution gives a straight line on a log scale. At the lower density, the luminescence is hotter and far from a thermal distribution at early times. The fact that the spectrum at lower density is hotter also indicates the lack of thermalization, because less LO-phonon emission occurs at low density because the electrons are not scattered into states above the LO phonon energy.

The primary unanswered question remaining in this work is why the thermalization rate at densities below $4 \cdot 10^{13}$ cm^{-3} remains constant. The time for thermalization remains about 50 ps over an order of magnitude of density variation below this density.

Springer Series in Chemical Physics, Vol. 55 Ultrafast Phenomena VIII
Editors: J.-L. Martin · A. Migus · G.A. Mourou · A.H. Zewail

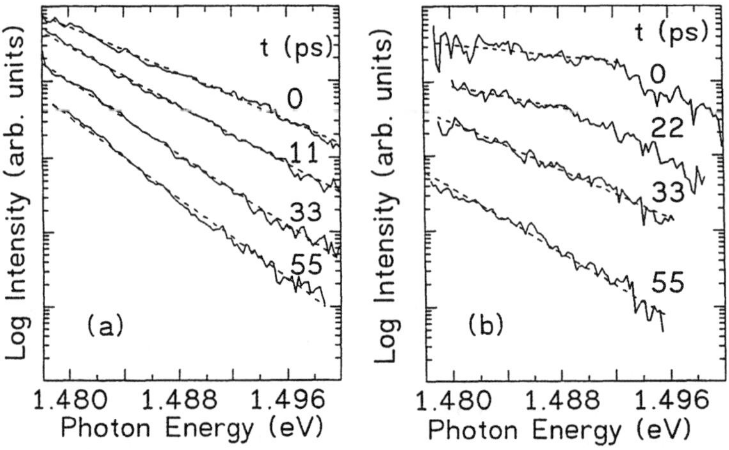

Fig. 1. a) Log $I(E)/(E-E_0)^{1/2}$ for the band-to-acceptor luminescence at a pair density of $4 \cdot 10^{14}$ cm^{-3}, at several delays relative to a 5-ps laser pulse at 1.536 eV. b) Spectra at the same delays for a pair density of $4 \cdot 10^{13}$ cm^{-3}. Spectra remain the same at lower densities.

We have previously examined the density-independent mechanisms of phonon scattering, impurity scattering, and spatially-correlated electron- hole scattering.[3] The first two are not adequate to explain the broad distribution seen within 10 ps over this range of density, while the last mechanism should depend strongly on elastic scattering mechanisms to confine a generated electron-hole pair in a small region of space. Another possibility is an increase in electron-electron scattering cross section as density decreases, due to the rapid increase of screening length in this nearly-unscreened density regime.

Numerical modeling of the evolution of the electron distribution by P. Kocevar[4] and ourselves, taking into account the electrons generated from both the heavy-hole and light-hole valence bands and weak electron-plasma dynamic screening, has shown that the electrons can obtain the early broad distribution within the experimental time resolution via Coulomb scattering. So far, these models fail to explain the constant rate of thermalization at densities below 10^{14} cm^{-3}. In treating Coulomb scattering in this regime, however, extreme care must be taken with the numerical integration with very low screening because of the near-singularities. We view these experiments as the first accurate test of models of electron-plasma screening because of the extreme sensitivity of the electron thermalization rate to the exact model chosen for the screening length, which appears to the fourth power in the scattering rate.

Acknowledgement. We thank E. Bauser and Y.-C. Lu for high-quality LPE GaAs epilayers and heterostructures and collaboration at an early stage in these experiments.

References

1. T. Elsaesser et al., Phys. Rev. Lett. **66**, 1757 (1991).
2. See, e.g. R.G. Ulbrich et al., Phys. Rev. Lett. **62**, 949 (1989).
3. D.W. Snoke et al., Phys. Rev. B **45**, 10979 (1992).
4. P. Kocevar, Proceedings of the 1992 European Physical Society, and private communication.

Femtosecond Carrier–Carrier Interaction in GaAs

T. Gong[1], *K.B. Ucer*[1], *L.X. Zheng*[1], *G.W. Wicks*[2], *J.F. Young*[3], *P.J. Kelly*[3], *and P.M. Fauchet*[1]

[1]Laboratory for Laser Energetics, University of Rochester, Rochester, NY 14623, USA
[2]The Institute of Optics, University of Rochester, Rochester, NY 14627, USA
[3]Institute for Microstructural Sciences, National Research Council, Ottawa, Canada K1A OR6

Abstract. We present a series of pump-probe measurements on GaAs for carrier densities from 6×10^{15} to 10^{19} cm^{-3} excited at 2 eV. The role of electron-electron, electron-plasmon, and electron-hole interactions in hot-electron relaxation is established. The time scales, carrier-density, and temperature dependence of these processes are also revealed.

Because of its intrinsic many-body nature, Coulomb-mediated carrier-carrier scattering is the most complex, and therefore one of the least understood processes. Even though recent studies using femtosecond spectroscopy have improved our quantitative understanding of many interaction mechanisms, such as dephasing, internal thermalization, screening, renormalization, etc., a comprehensive picture of carrier-carrier interaction has not yet emerged.

We have made a series of femtosecond spectroscopic measurements on intrinsic, n-type and p-type GaAs using the usual pump-probe arrangement and a novel three-pulse-pump-probe technique. By investigating the initial scattering and the cooling rate of hot carriers injected by 2-eV photons for densities (N) varied from 6×10^{15} to 10^{19} cm^{-3}, we have acquired quantitative knowledge on electron-electron, electron-plasmon, and electron-hole interactions.

The changes of absorption coefficient [$\Delta\alpha(t)$] at 2 eV are measured for N from 8×10^{15} cm^{-3} up to 7×10^{18} cm^{-3} on intrinsic GaAs, as shown in Fig. 1. The fast recovery of the bleaching gives a measure of the *dwell time* of the hot carriers, which is the effective time the carriers take to leave the initial excited Bloch states. This time appears relatively insensitive (within ±30%) to N, even when N changes by nearly 3 orders of magnitude. However, there is a clear sublinear relation between $|\Delta\alpha_{max}|$ and N, in particular for N > 10^{17} cm^{-3}, as shown in the insert of Fig. 1. We have distinguished three regimes of carrier density, for which electron-electron scattering influences the Γ-valley electron distribution in different ways, and discuss them with the help of additional measurements.

For $N < 8 \times 10^{16}$ cm^{-3}, the very weak sublinear relation between $|\Delta\alpha_{max}|$ and N indicates that electron-electron scattering plays a secondary role in comparison with intervalley scattering. Therefore, the initial recovery of the bleaching is mainly caused by two quasi-monoenergetic electron distributions excited from the heavy- and light-hole valence bands. The role of electron-electron scattering in this low-density regime is further revealed by a three-pulse-pump-probe measurement (Fig. 2). When the probe pulse arrives at $t \sim 0^+$ with respect to the pump pulse (which injects $N \sim 3 \times 10^{16}$ cm^{-3}), it measures the maximum bleaching ($\Delta\alpha_{max}$).

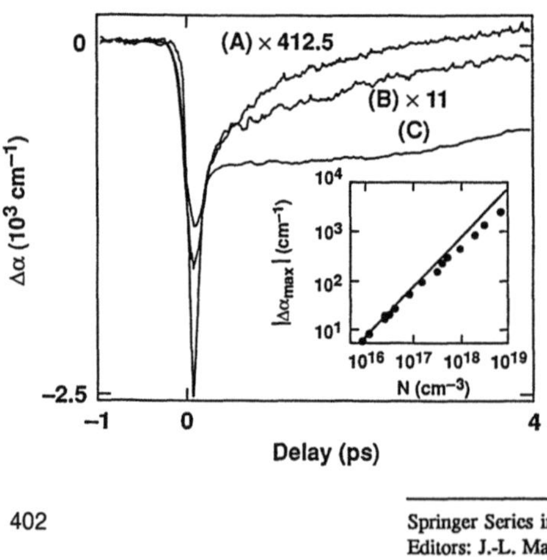

Fig. 1 Measured $\Delta\alpha(t)$ at 2 eV on an intrinsic film for N~ (A) 8×10^{15} cm^{-3}, (B) 3×10^{17} cm^{-3}, and (C) 3.3×10^{18} cm^{-3}. The curve (A) and (B) are magnified by a factor of 412.5 and 11, respectively. The insert shows measured $|\Delta\alpha_{max}|$ as a function of N. The dashed line represents a linear relation between $|\Delta\alpha_{max}|$ and N.

Springer Series in Chemical Physics, Vol. 55 Ultrafast Phenomena VIII
Editors: J.-L. Martin · A. Migus · G.A. Mourou · A.H. Zewail

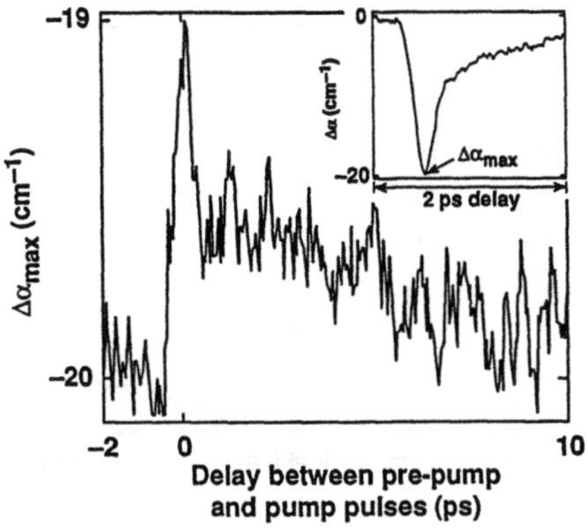

Fig. 2 Measured $\Delta\alpha_{max}$ at 2 eV on an intrinsic film as a function of the delay-time between the pre-pump pulse ($N_{pre} \sim 3 \times 10^{16}$ cm^{-3}) and the pump pulse ($N \sim 3 \times 10^{16}$ cm^{-3}). The insert shows measured $\Delta\alpha(t)$ when the pre-pump pulse is not present. The arrow indicates when $\Delta\alpha_{max}$ is measured.

Injection of an additional $N_{pre} \sim 3 \times 10^{16}$ cm^{-3} by a pre-pump pulse at $t \sim 0^+$ causes a reduction of $\Delta\alpha_{max}$ of ~5%, due to enhanced electron-electron scattering. By monitoring this reduction as a function of the time delay between pump and pre-pump pulses, as illustrated in Fig. 2, we conclude that single-particle scattering is most efficient when all the electrons are in Bloch states with the same (large) kinetic energy. This result can be explained by a simple theory [1,2], which shows that the short-ranged collision between two electrons is more efficient when momentum exchange is smaller.

For $N > 10^{17}$ cm^{-3}, the enhanced sublinear relation between $|\Delta\alpha_{max}|$ and N indicates that a larger portion of the electrons has left the initial excited states via electron-electron scattering. A recent photoluminescence measurement [3] and our pump-continuum-probe results [2] have shown that, even for N in the low-10^{17} cm^{-3}, the Γ-electron distribution has broadened significantly via efficient electron-electron scattering on a 100-fs time scale. However, $\Delta\alpha(t)$ measured around the band-edge exhibits a similar slow (>1 ps) dynamics for $N \sim 10^{17}$ to 10^{18} cm^{-3} [2,4]. This result suggests that complete internal Γ-valley thermalization is not established on a time shorter than the average intervalley scattering time. Thus, only the small fraction of electrons excited from the split-off valence band contributes to band-edge filling on a subpicosecond time scale, whereas the electrons excited from heavy- and light-hole valence bands first transfer to the X and L valleys and then return to the Γ valley on a picosecond time scale. The recovery of the bleaching measured at 2 eV for intermediate N is thus attributed to intervalley scattering of a significantly broadened Γ-electron distribution.

For $N > 3 \times 10^{18}$ cm^{-3}, the observation of subpicosecond gain around the band-edge [5] indicates that intra-Γ-valley thermalization via strong electron-electron scattering takes place on a time that is shorter than or comparable to the average intervalley scattering time, and a much greater fraction of electrons reaches the band-edge on a subpicosecond time scale. Since the initial electron temperature is very high (>2000 K), many electrons are energetically allowed to scatter to the X and L valleys. The recovery of the bleaching measured at 2 eV for high N is thus caused by intervalley cooling of a quasi-thermalized Γ-electron distribution. Calculations show that more than 40% of the total excited electrons remain in the Γ-valley, and the corresponding bleaching maximum due to *thermal* state filling at 2 eV is also close to the experimentally observed bleaching peak.

In addition to establishing the efficiency of electron-electron scattering, hole-hole scattering, and intervalley cooling, subpicosecond gain measurements also provide quantitative information on electron-hole scattering. The subpicosecond transient heating and subsequent cooling of holes indicates that at high-plasma density, when the Frohlich interaction (LO-phonon) is strongly screened for electrons, the efficient electron-hole and hole-lattice interactions can supply an efficient energy-loss channel for hot electrons on a subpicosecond time scale [6].

A more quantitative understanding of electron-hole and electron-plasmon interactions can be established by measuring $\Delta\alpha(t)$ at 2 eV for $N \sim 8 \times 10^{16}$ cm^{-3} in p-type ($N_a \sim 2.1 \times 10^{18}$ cm^{-3}) and n-type ($N_d \sim 3.3 \times 10^{18}$ cm^{-3}) GaAs, as shown in Fig. 3. The fastest recovery time of the bleaching changes little when a dense hole plasma is present. There is no evidence of hole-plasmon emission in the initial scattering process because of the heavily damped nature of the hole plasma [7]. This result suggests that electron-hole scattering does not play a dominant role on a sub-100 fs time scale. In contrast, no strong bleaching is observed when a dense electron plasma is present. Thus, on a time scale much shorter than the pulse width (~75 fs), the injected hot electrons have left the initial excited states. A comparison of this measurement to that on intrinsic GaAs at $N \sim 3.3 \times 10^{18}$ cm^{-3} (all hot electrons) indicates that the *dwell time* of the hot electrons is reduced by more than a factor of 3 in the presence of a dense and cold electron plasma. This plasma provides an efficient energy-loss channel for hot electrons through plasmon emission.

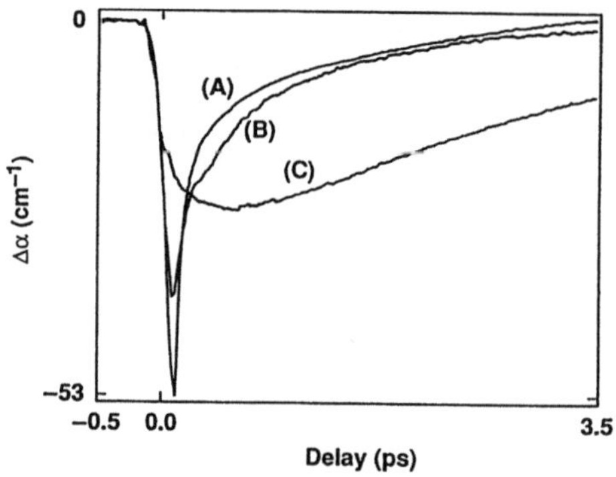

Fig. 3 Measured $\Delta\alpha(t)$ at 2 eV for N ~ 8 × 10^{16} cm^{-3} on thin films of (A) intrinsic, (B) p-type (N_a ~ 2.1 × 10^{18} cm^{-3}), and (C) n-type (N_d ~ 3.3 × 10^{18} cm^{-3}) GaAs.

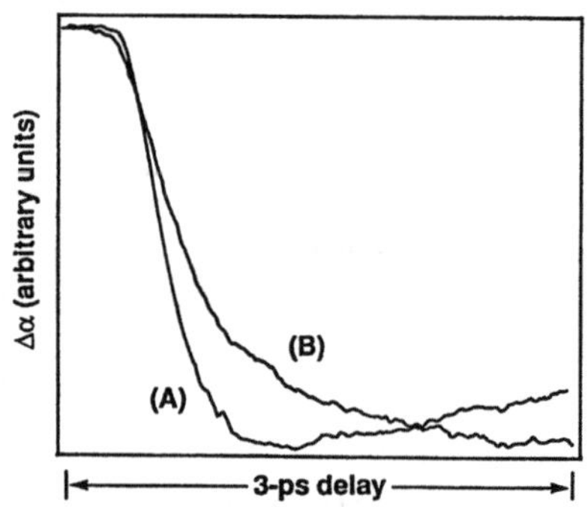

Fig. 4 Measured $\Delta\alpha$ at 2 eV on n-type (N_d ~ 3.3 × 10^{18} cm^{-3}) GaAs at T_L ~ 14 K for N~ (A) 6 × 10^{16} cm^{-3} and (B) 6 × 10^{15} cm^{-3}. Each curve is normalized by its own maximum value.

In n-type GaAs, the signal of Fig. 3 rises relatively slowly and reaches a maximum ~500 fs after the weak bleaching that follows the laser pulse. This effect results from heating of the majority electrons by the minority holes. Energy exchange between 8 × 10^{16} cm^{-3} hot electrons and 3.3 × 10^{18} cm^{-3} cold electrons produces a thermalized distribution with an electron temperature <400 K, which is lower than the initial hole temperature of ~600 K. Energy can therefore be transferred from holes to electrons through electron-hole scattering until thermal equilibrium between electrons and holes is established in ~500 fs. It is worth noting that this time should be also very close to the overall cooling time of holes, a result consistent with recent time-resolved photoluminescence [8] and our band-edge gain measurements [6]. This effect is further magnified when measurements are performed at 14 K (Fig. 4), when the initial electron temperature (<100 K) is much lower than the hole temperature (~600 K). It is also clear that the time to reach thermal equilibrium between electrons and holes is quite long at very low N (6 × 10^{15} cm^{-3}), because weak electron-hole interaction and inefficient hole-lattice interaction at T_h <100 K leave the holes warmer than the electrons for several picoseconds.

For n-type GaAs, when the injected N exceeds 7.5±2.5 × 10^{17} cm^{-3}, a strong bleaching reappears, followed by a fast recovery. The dwell time of the hot electrons thus becomes longer as their density increases. Three-pulse-pump-probe measurements show that this unexpected effect has little dependence on the temperature of plasma and the density of electrons, but instead appears to be related to the presence of a dense *hole* plasma. Damping of electron-plasmon interactions due to inter-valence band transitions [9], phase space filling, screening, and kinematic blocking [10] may explain this result.

This work was supported by the U.S. Office of Naval Research (N00014-91-J-1488) and the U.S. Army Research Office (DAAL03-91-G-0173).

References

[1] J. M. Ziman, *Electrons and Phonons* (Clarendon Press, Oxford, 1960), pp. 170–172.
[2] T. Gong, L. X. Zheng, K. B. Ucer, J. F. Young, P. J. Kelly, G. W. Wicks, and P. M. Fauchet, submitted to Phys. Rev. B.
[3] T. Elsaesser, J. Shah, L. Rota, and P. Lugli, Phys. Rev. Lett. **66**, 1757 (1991).
[4] T. Gong, W. L. Nighan, Jr., and P. M. Fauchet, Appl. Phys. Lett. **57**, 2713 (1990).
[5] T. Gong, P. M. Fauchet, J. F. Young, and P. J. Kelly, Phys. Rev. B **44**, 6542 (1991).
[6] T. Gong, P. M. Fauchet, J. F. Young, and P. J. Kelly, submitted to Appl. Phys. Lett.
[7] K. Wan and J. F. Young, Phys. Rev. B **41**, 10,772 (1990).
[8] X. Q. Zhou, K. Leo, and H. Kurz, Phys. Rev. B **45**, 3886 (1992).
[9] A. Pinczuk, J. Shah, and P. A. Wolff, Phys. Rev. Lett. **47**, 1487 (1981).
[10] J. F. Young and P. J. Kelly, to appear in SPIE Proc. **1679** (1992).

Quantum Beats versus Polarization Interference: An Experimental Distinction

M. Koch[1], *J. Feldmann*[1], *G. von Plessen*[1], *E.O. Göbel*[1], *P. Thomas*,[1], *and K. Köhler*[2]

[1]Department of Physics and Materials Sciences Center, Philipps-University, W-3550 Marburg, Fed. Rep. of Germany
[2]Fraunhofer Institute of Applied Solid State Physics, W-7800 Freiburg, Fed. Rep. of Germany

Abstract: We report the first experimental distinction between quantum beats and polarization interference of independent quantum mechanical systems by time-resolving the four-wave mixing signal emitted from specially designed GaAs/AlAs heterostructures.

In recent years, optical quantum beat spectroscopy has become a powerful tool also in the field of semiconductor physics. This technique allows to determine the energy splitting between different coherently excited transitions, even in the case of strong inhomogeneous broadening. In particular, transient four-wave mixing (FWM) experiments with time-integrated detection have been widely used for this purpose [1, 2]. However, this method does not provide information on the details of the electronic level scheme involved in the coherent excitation.

In this paper, we demonstrate that a three-level system, where two different transitions have one state in common, can be distinguished unambiguously from a set of two independent two-level transitions by time-resolving the diffracted FWM signal. In case of a three-level system, quantum beats (QB) appear in the FWM signal, which are due to the quantum mechanical interference of the two transitions via the common state. In contrast, the observed beating of the FWM signal in case of two independent two-level systems is due to the interference of the emitted electric fields driven by the coherent third-order polarizations of the two independent optical transitions (polarization interference: PI).

In order to realize the respective model systems, we have designed two different GaAs/AlAs multiple quantum well (MQW) structures. Sample A consists of one MQW structure with a GaAs layer-width of 150 Å leading to an energetic splitting of $\Delta E = 5.4$ meV between the heavy-hole (hh) and light-hole (lh) excitonic transitions. Sample B consists of two successive and thus spatially separated MQW structures with GaAs layer-widths of 80 Å and 90 Å, respectively. The energetic splitting between the heavy and light hole excitonic transitions is sufficiently large; the two hh-excitons, however, are energetically close together ($\Delta E = 12.9$ meV). Thus, sample A provides a three-level system in the form of hh and lh excitons, which are coupled via the common electron state (QB-system), whereas two independent two-level systems can be studied in case of sample B in the form of two spatially separated hh excitons (PI-system).

We have performed two-beam FWM experiments for both samples with time-resolved detection (via up-conversion) [3, 4, 5, 6] of the diffracted FWM signal using short laser pulses (typically 100 fs) with a spectral width of 21 meV (FWHM). The sample temperature was 5 K.

The three dimensional plots in Fig.1 show the real-time (t) evolution of the FWM signal for different time-delays τ between the excitation laser pulses. Pronounced modulations of the FWM signals are observed in real-time in case of sample A as well as for sample B. The observed beat periods T_b correspond to the above mentioned energetic splittings of hh-lh excitons (sample A), and of the two independent hh excitons (sample B), respectively.

Springer Series in Chemical Physics, Vol. 55 Ultrafast Phenomena VIII
Editors: J.-L. Martin · A. Migus · G.A. Mourou · A.H. Zewail

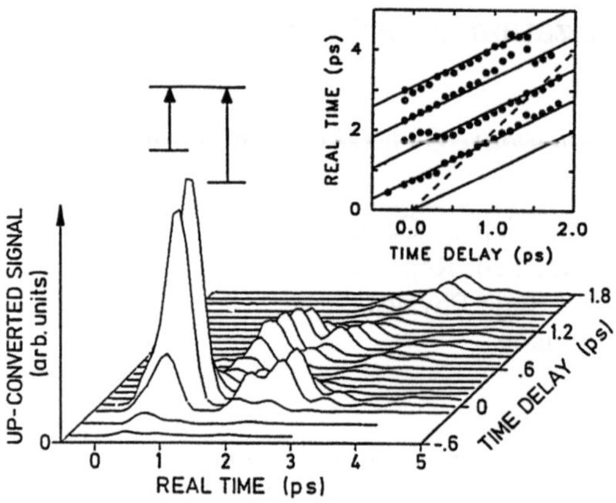

Figure 1: Sample A (QB)

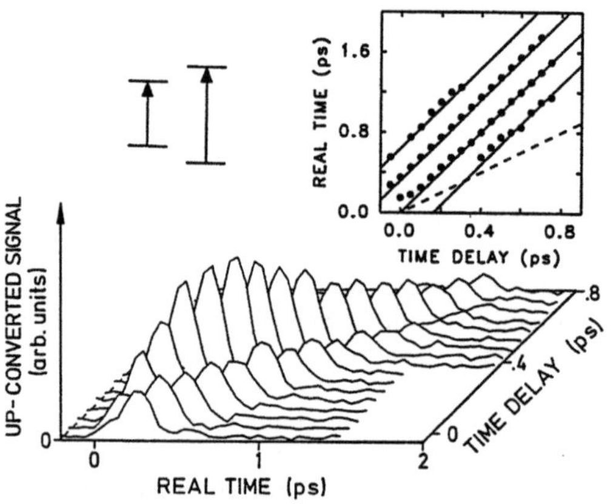

Sample B (PI)

Yet, the (t, τ)-positions of the beat maxima are different for the two cases. This can be seen in the insets of Fig.1, where the experimentally determined maxima are depicted as dots. For the case of QBs (sample A), the peak positions follow the condition $t = \tau + nT_b$ (n being an integer) represented by the solid lines. The situation is obviously different for PI (sample B) were the solid lines correspond to the condition $t = 2\tau \pm nT_b$. For comparison the opposite situations for $t = 2\tau$ and $t = \tau$, respectively, are shown as dashed lines. Indeed, the characteristic differences in the real-time behaviour of QBs and PI are predicted theoretically by the optical Bloch equations [6].
We therefore conclude that performing FWM experiments with time resolved detection on a system involving two different optical transitions, and plotting the positions of the signal maxima vs. t and τ allows an unambiguous distinction between QBs and PI. The experimental approach described here for the case of a system with two different transition frequencies can be generalized to cover problems involving a greater number of transitions. This new experimental method should thus provide a powerful diagnostic tool in the field of coherent laser spectroscopy.

References

[1] E.O. Göbel, K. Leo, T.C. Damen, J. Shah, S. Schmitt-Rink, W. Schäfer, J.F. Müller, and K. Köhler, Phys. Rev. Lett. **64**, 1801 (1990).
[2] K. Leo, T.C. Damen, J. Shah, E.O. Göbel, and K. Köhler, Appl. Phys. Lett. **57**, 19 (1990).
[3] M.D. Webb, S.T. Cundiff, and D.G. Steel, Phys. Rev. Lett. **66**, 934 (1991).
[4] D.S. Kim, J. Shah, J.E. Cunningham, T.C. Damen, S. Schmitt-Rink, and W. Schäfer, Phys. Rev. Lett. **68**, 2838 (1992).
[5] M.-A. Mycek, S. Weiss, J.-Y. Bigot, S. Schmitt-Rink, D.S. Chemla, and W. Schäfer, Appl. Phys. Lett. **60**, 2666 (1992).
[6] M. Koch, J. Feldmann, G. von Plessen, E.O. Göbel, P. Thomas, and K. Köhler, submitted for publication.

Plasmon–Phonon Coupling and Hot Carrier Relaxation in GaAs and Low-Temperature-Grown GaAs

R.I. Devlen, J. Kuhl, and K. Ploog

Max-Planck-Institut für Festkörperforschung, Heisenbergstr. 1, W-7000 Stuttgart 80, Fed. Rep. of Germany

Abstract. Time-resolved Raman experiments in low temperature grown GaAs and conventional GaAs reveal the strong dependence of the phonon occupation number on the plasma density.

1. Introduction

Within the first picoseconds after photogeneration, the initially hot carrier distribution in bulk GaAs cools predominantly by emitting optical phonons. The emitted phonons exist as a nonthermal phonon population until they decay anharmonically. Previously, the dynamics of the nonthermal phonon population was studied using time-resolved anti-Stokes Raman spectroscopy [1]. For a carrier density larger than 10^{17} cm^{-3} the plasma frequency is comparable to the phonon frequency and coupling of the two excitations becomes nonnegligible. In GaAs plasmon-phonon coupled modes have been observed in both single component plasmas (via doping)[2] and two component plasmas (via optical excitation)[3]. To our knowledge there have been no measurements of the plasmon-phonon coupled mode population dynamics, however, these nonequilibrium modes play an important role in theories of hot carrier relaxation at high densities. In particular the existence of a large nonthermal phonon population and its reabsorption by carriers has been used to explain a reduction in the carrier energy loss rate[4].

2. Results and Discussion

We have performed Raman scattering experiments on high quality GaAs and on GaAs grown at a very low substrate temperature (200°C) and subsequently anneallled at 600°C for 10 minutes (LT-GaAs). Both samples have ⟨100⟩ surfaces and all the measurements were done in a cryostat held at 80K. Using a two pulse cross-correlation luminescence technique, we have measured the carrier lifetime in the LT-GaAs sample to be approximately 1ps. In the same sample we have measured the phonon population lifetime using the technique described in Ref[1]. It is approximately 5ps at 80K. This agrees with a measurement of the Raman linewidth in the frequency domain and is less than a factor of two smaller than the lifetime in GaAs (7ps).

To measure the dynamics of a nonthermal phonon population in the presence of a large electron-hole plasma we have performed dichroic pump and probe measurements. The pulse from a YLF laser was split into two parts. One half was used as the pump pulse for the experiment (pulse duration=40ps; photon energy=2.35eV) and the rest synchronously pumped a R6G dye laser (pulse duration=2.3ps; photon energy=2.15eV). Both pulses were focused to the same spot on the sample and yielded per pulse total absorbed photon densities of 9x10^{18}cm^{-3} (pump) and 4x10^{17}cm^{-3} (probe). Backscattered light was detected using a monochromator (spectral resolution=30cm^{-1}) and a standard photon counting scheme. Both time-dependent and time-independent hot luminescence signals were subtracted from the Stokes (S) and anti-Stokes (AS) signals. In Fig.1 we show the time dependent S and AS signals originating from the probe pulse for which the incident light was polarized along ⟨010⟩ and the detected light along ⟨001⟩ (LO-phonons allowed). For the GaAs sample both S and AS signals increase concurrent with the onset of the pump pulse followed by a loss of Raman signal which is maximum at the end of the pump. At later times the signal starts to recover. The S and AS signals for LT-GaAs show no loss of signal at any time, but an increase which approximately follows the pump pulse profile.

The intensity of the probe Raman signals may be expressed as S $\propto \sigma(N+1)$ and AS $\propto \sigma N$ where σ is the scattering cross-section and N is the phonon occupation number. At 80K the

Springer Series in Chemical Physics, Vol. 55 Ultrafast Phenomena VIII
Editors: J.-L. Martin · A. Migus · G.A. Mourou · A.H. Zewail

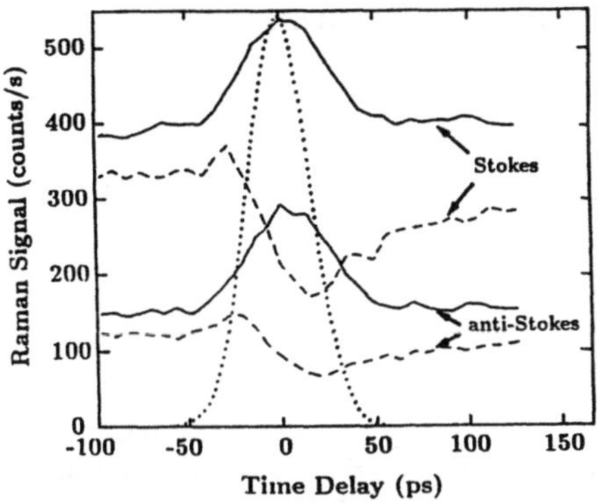

Fig. 1. Time dependent S and AS signals for GaAs (dashed line) and LT-GaAs (solid line); the YLF pulse (dots).

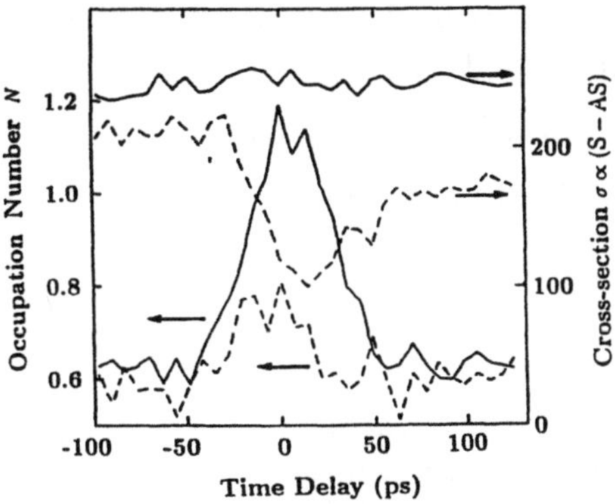

Fig. 2. Time-dependent phonon occupation numbers and Raman cross-sections for GaAs (dashed) and LT-GaAs (solid).

contribution to N from thermal phonons is approximately zero [1] and the AS signal observed at negative time delays is due to nonthermal phonons generated and detected by the probe pulse. For probe times coincident with and following the pump pulse, the S and AS intensities will reflect changes in both σ and N. To separate these two contributions we have calculated N from the relation $N = [S/AS - 1]^{-1}$ and σ from the proportionality $\sigma \propto$(S-AS) (see Fig.2).

In LT-GaAs the short carrier lifetime implies that the maximum instantaneous pump induced carrier density is about $2x10^{17}cm^{-3}$. At this density the coupling between plasmon and phonon modes is weak. This was confirmed by a spectral measurement of the pump pulse LO phonon line from which no broadening was observed. From inspection of Fig.2 it is seen that the AS and S transients shown in Fig.1 are dominated by changes in the nonthermal phonon population. The phonons are emitted during the relaxation of pump-excited carriers. The curve for N in Fig.2 is shifted slightly toward positive time delays with respect to the pump pulse due to the phonon lifetime of 5ps.

By contrast the Raman signals associated with GaAs contain components associated with changes in both N and σ. Fig.2 shows that σ decreases continuously during the pump pulse and then starts to recover. It is known that the presence of a two component plasma broadens the Raman line and reduces the peak signal [3]. At the time delay corresponding to the minimum value of σ the Raman linewidth was measured and found to have broadened which accounts for part, if not all, of the change in σ. The GaAs shows only a small increase in the nonthermal population, and unlike the LT-GaAs the peak of the signal is shifted slightly towards negative time delays. Both the magnitude and shift of this curve as compared with LT-GaAs suggest two effects which modify the phonon population in the presence of a high density plasma: (a) The emission of Raman detectable coupled modes by relaxing carriers is suppressed (either due to carrier-carrier scattering which dominates the energy relaxation [5], or because the Fröhlich interaction is screened by the plasma [6]); (b) The energy relaxation time of these coupled modes is greatly reduced (via either hot phonon reabsorption [4] or plasma damping). At present we cannot distinguish between these two possibilities.

References

[1] D.von der Linde, J.Kuhl and H.Klingenberg, Phys. Rev. Lett. **44**, 1505 (1980).
[2] A.Mooradian and A.L.McWhorter, in *Light Scattering Spectra of Solids*, ed. by G.B.Wright (Springer, New York, 1968) p. 297.
[3] C.L.Collins and P.Y.Yu, Solid State Commun. **51**, 123 (1984).
[4] W.Pötz and P.Kocevar, Phys. Rev. B **28**, 7040 (1983).
[5] H.Sato and Y.Hori, Phys. Rev. B **36**, 6033 (1987).
[6] W.Graudszus and E.O.Göbel, Physica **117B & 118B**, 555 (1983).

Femtosecond Carrier–Carrier Interaction Dynamics in Doped GaAs

T. Furuta and A. Yoshii

NTT LSI Laboratories, 3-1, Morinosato Wakamiya,
Atsugi-shi, Kanagawa 243-01, Japan

Carrier-carrier interaction is a very important energy-relaxation mechanism for carriers in semiconductors. In our previous investigation of carrier transport properties, we discussed the energy relaxation process in connection with transport parameters [1-3]. Through these works, we concluded that carrier-carrier interaction plays an important role in the energy relaxation process in highly-doped samples. Although the conclusion derived from analyzing the transport parameters is plausible, it has not been verified by direct measurement of the relaxation time. Recent advances in ultrafast optical pulse generation and measurement techniques make it possible to study the ultrafast phenomena in semiconductors that determines the ultimate high-speed-performance of electric and optical devices [4]. In this contribution, we report systematic studies of femtosecond carrier-carrier interaction dynamics in *p*-doped and *n*-doped GaAs samples with doping concentration as a parameter.

We employed the up-conversion technique to perform time-resolved photoluminescence (PL). This method provides very high time resolution measurements. Femtosecond optical pulses of a mode-locked dye laser (repetition rate : 76 MHz, pulse duration : 130 fs) with a photon energy of 1.90 eV were used to generate electron-hole pairs in doped GaAs. The PL from the sample was collected by a dispersion free optics and focused on a 0.5-mm-thick $LiIO_3$ nonlinear optical crystal. The delayed laser beam was also focused non-collinearly at the same point in the nonlinear crystal. The up-converted signal was dispersed by a monochrometer, and then detected by a photon counting technique. From the cross-correlation trace, the time resolution of the total experimental system was 130 fs. All experiments were performed at 300 K.

Figures 1 (a) and (b) show the time-evolution of PL intensity at a photon energy of 1.45 eV for *p*-GaAs and *n*-GaAs samples, respectively. The excited carrier density is about 10^{16} cm^{-3}. The data reveal two features of special interest. First is the observation that the rise in PL intensity, which corresponds to the time for photoexcited electrons to relax from an excited energy state to the bottom of the conduction band edge, becomes shorter with increasing doping concentration. Generally, the initial energy relaxation of photoexcited carriers is determined by a competition between carrier-phonon and carrier-carrier interactions. The effect on the energy relaxation due to carrier-carrier interaction becomes stronger as the doping concentration increases and thus, the rise time becomes shorter. Furthermore, it is found that the rise time in *p*-GaAs is shorter than that in *n*-GaAs with the same doping concentration. This result is attributed to the different energy relaxation mechanism in *p*-GaAs and *n*-GaAs. In *p*-GaAs, the electron-hole interaction is the dominant carrier-carrier interaction. Because of the large effective mass of hole, the large energy transfer from photoexcited electrons to majority holes leads to a faster rise of PL intensity in *p*-GaAs. On the other hand, the hole-electron interaction affects the relaxation process in *n*-GaAs. This difference is more clearly observed in the time-evolution curves at higher photon energies.

The second point of interest is that the rises of PL for the most highly doped samples in *p*-GaAs and *n*-GaAs consist of two time constants, while the samples with low carrier concentration have a single time constant. One is a fast time constant which appears at the initial time stage and the other is a slow one that characterizes the curve near saturation. The two time components of rise in PL intensity can be understood as

Springer Series in Chemical Physics, Vol. 55 Ultrafast Phenomena VIII
Editors: J.-L. Martin · A. Migus · G.A. Mourou · A.H. Zewail

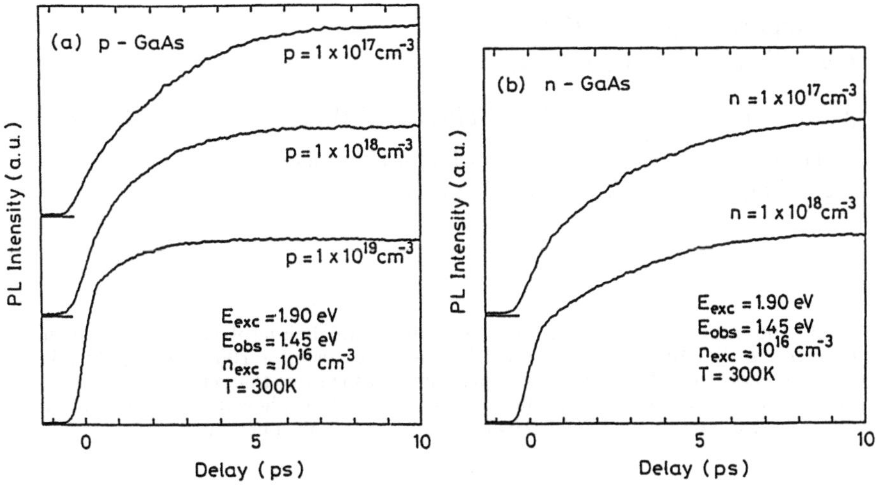

Fig.1. Time evolution curves of near-band-edge photoluminescence intensity in *p*-GaAs (a), and in *n*-GaAs (b) samples with various doping concentrations.

follows: In these samples, the Fermi energies are approximately 40 meV and 58 meV for p=1 $\times$ 10^{19} cm^{-3} and n=1 $\times$ 10^{18} cm^{-3} samples, respectively. The chemical potential of the photoexcited carrier system is much higher than that of majority carriers at the initial time stage after excitation. Hence, the energy of the photoexcited carrier system is transfered to the majority carrier system through carrier-carrier interaction according to their large chemical potential difference. This energy transfer eventually ceases when the chemical potential coincides with that of majority carriers after several scatterings. After this time, the photoexcited carriers hardly interact with majority carriers; and as a consequence, the carriers relax by emitting optical phonons and approach their distribution function to lattice temperature. Thus, the presence of Fermi energy strongly affects the total energy relaxation process of photoexcited carriers in highly-doped GaAs. Similar results are also observed in the cooling curves of carrier temperature and these results are supported by the calculated ones, which include the energy loss rates of carrier-phonon and carrier-carrier interactions.

In summary, we investigated the energy relaxation process of photoexcited carriers in doped-GaAs by time-resolved photoluminescence measurement with up-conversion technique. The rise of PL intensity strongly depends on carrier concentration, which suggests the important role of carrier-carrier interaction in the energy relaxation process. Furthermore, it is found that the presence of Fermi energy strongly affects the total energy relaxation process in highly-doped GaAs.

References

[1] T. Furuta, H. Taniyama, and M. Tomizawa, J. Appl. Phys. **67**, 293 (1990)
[2] T. Furuta and M. Tomizawa, Appl. Phys. Lett. **56**, 824 (1990)
[3] H. Taniyama, M. Tomizawa, T. Furuta, and A. Yoshii, J Appl. Phys. **68**, 621 (1990)
[4] J. Shah, Solid State Electron. **32** 1051 (1989)

Femtosecond Carrier Kinetics in Low-Temperature-Grown GaAs

X.Q. Zhou, H.M. van Driel, A.P. Heberle, W.W. Rühle, and K. Ploog

Max-Planck-Institut für Festkörperforschung, Heisenbergstr. 1, W-7000 Stuttgart 80, Fed. Rep. of Germany

Abstract. The carrier energy and recombination kinetics are time-resolved in low-temperature-grown GaAs at 295 K from upconverted luminescence using a 100 fs Ti:sapphire laser. Electrons and holes are observed to have approximately the same trapping time of about 1 ps. Because the carrier lifetime is less than the lifetime of emitted LO phonons, the energy relaxation rate after 1 ps is more strongly inhibited by hot phonon effects than in normal GaAs.

1. Introduction

Crystalline GaAs grown at low temperatures (LT-GaAs) by MBE and annealed at 900K has been shown to have an excess of 1% As in the form of microclusters and point defects with density approximately 10^{18} cm^{-3}. As a semi-insulating material with reasonably high carrier mobility and a carrier lifetime of about one picosecond [1], it has been used to improve the performance of picosecond photoconductive switches and MESFETs. LT-GaAs has also proven to be an interesting material in which to study non-equilibrium carrier and phonon effects because the carrier lifetime is comparable to the carrier cooling time and less than the lifetime of emitted LO phonons[2].

2. Femtosecond Luminescence Upconversion Experiments

Here we report studies of the carrier kinetics in LT-GaAs using femtosecond time-resolved photoluminescence upconversion [3]. An 80 MHz, 80 fs, $\lambda = 720$ nm pulse train with average power up to 2W is simultaneously used to excite a LT-GaAs specimen at T= 295 K and upconvert/gate the photoluminescence; the upconverted beam is dispersed with a monochromator and detected with a photomultiplier. Fig. 1 shows typical spectra for LT-GaAs and normal GaAs at the peak of the excitation pulse when the carrier density near the surface is estimated to be $2\text{x}10^{18}$ cm^{-3}. The comparable spectrally integrated intensities observed in the two materials indicates that the radiative recombination coefficient is essentially the same and the reported low luminescence efficiency for longer pulse excitation is undoubtedly related to a short carrier trapping time[2]. The luminescence in LT-GaAs at energy 1.63 eV (200 meV above the band edge) decays with a time constant of 550 fs while that at an energy as low as 1.1 eV, significantly below the band-edge, decays with a time constant of about 900 fs. The difference is attributed to carrier energy relaxation which removes carriers from states at higher energy and adds them to states at lower energy. The similarity of the luminescence onset and decay times at energies above and below the band-gap indicates that recombination below the band-gap is through delocalized states which most likely exist due to a smearing of the band-edge by the microclusters or defects. To determine the individual electron and hole lifetimes we used a luminescence autocorrelation technique [4], and from the single exponential decay of the luminescence at the band-edge we conclude that the electron and hole trapping times are *both* approximately equal to 1 ps.

Fig. 2 shows the behavior of the characteristic carrier temperature in LT-GaAs and GaAs for an initial excitation density of $3\text{x}10^{18}$ cm^{-3}. The higher initial carrier temperature for the LT sample is consistent with 5% of the carriers being activated directly from mid-gap As defect states with a higher initial excess energy (1 eV) than is produced through band-to-band

Springer Series in Chemical Physics, Vol. 55 Ultrafast Phenomena VIII
Editors: J.-L. Martin · A. Migus · G.A. Mourou · A.H. Zewail

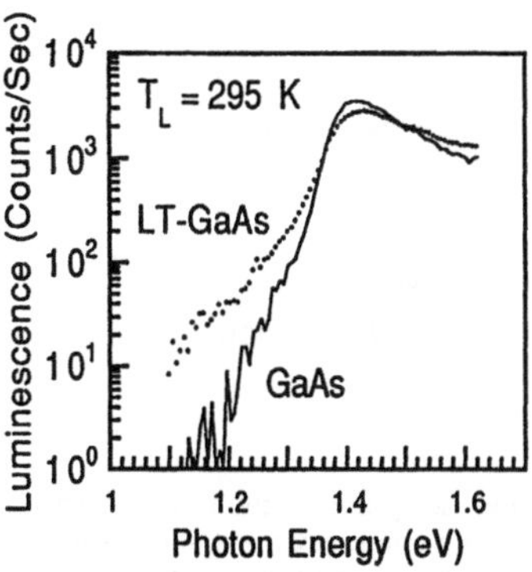

Fig. 1 Luminescence spectra of LT-GaAs and GaAs at peak of excitation pulse.

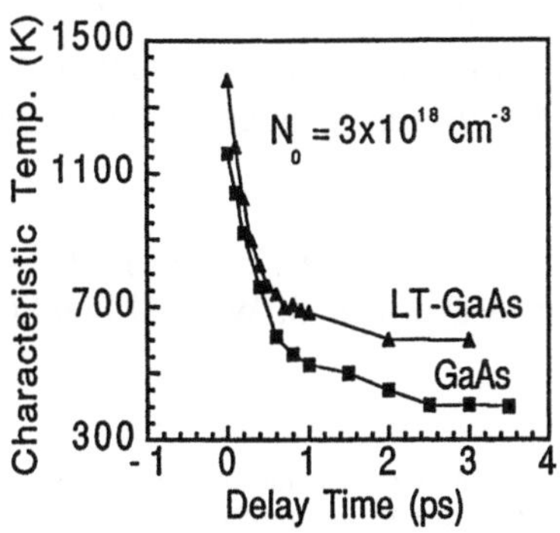

Fig. 2 Characteristic carrier temperature in LT-GaAs and GaAs as a function of time.

excitation (which gives about 300 meV to electrons and holes together). The carriers in both samples cool at the same rate for about 500 fs but thereafter carriers in the LT sample cool more slowly. This is thought to be due to the fact that the carrier lifetime is shorter than the approximately 3 ps LO phonon lifetime at 295 K. The ratio of the nonequilibrium LO phonon population to carrier density is therefore expected to increase for the first few picoseconds following excitation, thereby leading to significant phonon reabsorption by the carriers and a reduced cooling rate. When a 5x lower initial excitation density is used we observe that the carrier temperature in the two samples never differ by more than 50 K after 1 ps, confirming our conclusions about the influence of hot phonon effects at the higher excitation level.

3. Conclusions

In conclusion we have shown that the presence of excess As produces a significant sub-band-gap luminescence in LT-GaAs. It also give the carriers a picosecond trapping time and alters their initial carrier temperature and cooling kinetics from what is observed in GaAs.

4. References

1. S. Gupta, M. Y. Frankel, J. A. Valdmanis, J. F. Whitaker, G. A. Mourou, F. W. Smith, and A. R. Calawa, Appl. Phys. Lett. **59**, 3276 (1991).

2. H. M. van Driel, X.-Q. Zhou, W. W. Rühle, J. Kuhl, and K. Ploog , Appl. Phys. Lett., **60,** 2246 (1992).

3. J. Shah, I.E.E.E. J. Quant. Elec., **24**, 276 (1988).

4. D. von der Linde, J. Kuhl, and E. Rosengart, J. Lumin. **24/25,** 675 (1981).

Transient Anisotropic Luminescence and Long-Living Polarization of an Optically Excited Dense Electron–Hole Plasma

A.L. Ivanov and H. Haug

Institut für Theoretische Physik, Unversität Frankfurt a.M., Robert-Mayer-Str. 8, W-6000 Frankfurt a.M., Fed. Rep. of Germany

Abstract. A strongly anisotropic transient luminescence and long-living induced interband polarization due to the unrelaxed total momentum transfered to the dense electron-hole plasma of a semiconductor by a femtosecond optical excitation pulse are predicted and described.

Currently transient optical phenomena in direct-gap semiconductors are receiving extensive experimental [1-2] and theoretical [3] attention in connection with recent advances in ultrashort (femtosecond) pulse spectroscopy. Such investigations mainly deal with an optical generation, real or virtual, of hot carriers and their following relaxation in a subpicosecond time-scale. The very first stage of the optical exitation of carriers and their thermalization are in the focus of the current experimental and theoretical interest.

For the theoretical description of the evolution of a dense electron-hole plasma in a semiconductor previously excited by a strong femtosecond optical pulse one has to consider two different time scales given by τ_s and τ_R . The first one characterizes the relaxation into quasi-equilibrium by the carrier-scattering processes (typically 20 fs - 500 fs), the second one is the recombination lifetime of the electron-hole plasma (typically 0,1 ns - 1 ns). The carrier-scattering processes are determined both by the Coulomb mechanism which conserves the total momentum of an electron-hole (e-h) system and dominates at high plasma-densities and by the interaction with crystal imperfections, i.e. phonons or impurities. Moreover, such two scattering mechanisms can be separated easily in time, if the carrier frequency Ω of the initial pump pulse lies close to the bandgap E_g of a semiconductor, i.e. if $\Omega - E_g \leq \Omega_{opt}$, where Ω_{opt} is the frequency of an optical phonon [2] .

Anisotropic transient luminescence occurs because during the time $t < \tau_b$ an optically excited dense e-h plasma can exist as a momentum-conserving system, if the Coulomb interaction is the dominant scattering mechanism. In this regime the e-h system preserves the total momentum $\hbar N\mathbf{k}$ of the initially exciting femtosecond pulse, where N is the number of the excited e-h pairs, and $\mathbf{k}$ is the carrier photon wavenumber of the optical pulse. The created electron-hole cloud moves ballistically with a drift velocity $v_b = \hbar\mathbf{k}/(m_e + m_h)$ inside the crystal. Here, m_e and m_h are the effective electron and hole masses, respectively. Only through the interaction with LA-phonons and impurities the drift slows down during a characteristic time $\tau_b >> \tau_s$. As long as the ballistic propagation persists, strongly anisotropic transient photon emission has to appear due to conservation of the total momentum. The anisotropic luminescence has a clear mechanical analog. After the explosion of a flying artillery shell, the space diagram of it fragments is directed along the vector of a shell velocity.

A ballistic velocity v_b of an optically excited electron-hole cloud is small in comparison with a background light velocity $c/\sqrt{\epsilon_g}$ in a crystal (for GaAs, e.g., $v_b \simeq 0.61\times 10^6$cm/s) . Nevertheless, transient anisotropic luminescence has a macroscopic character because a photon emission process is governed just by the momentum conservation law, rather than by the ballistic velocity v_b . This situation resembles the usual polariton picture, where an exciton-like polariton with a small group velocity can exist. In most theoretical treatments the finite photon momentum $\hbar s$ has been neglected because the electron and hole momenta $\mathbf{p}_e$ and $\mathbf{p}_h$ are cosiderably large than that of the photon. But just the optical transition with $\mathbf{p}_e - \mathbf{p}_h = \hbar s$ retains the information about the initial momentum $N\hbar\mathbf{k}$ transfered from the optical pump pulse.

If the LO and TO-phonon scattering processes are forbidden, an optically excited e-h plasma can reach a quasi-equilibrium in 200-500 fs [1] , i.e. still during the ballistic motion. In this case the nondegenerate electron and hole distributions with a common temperature T produce luminescence intensity I_s in a photon mode s :

Springer Series in Chemical Physics, Vol. 55 Ultrafast Phenomena VIII
Editors: J.-L. Martin · A. Migus · G.A. Mourou · A.H. Zewail

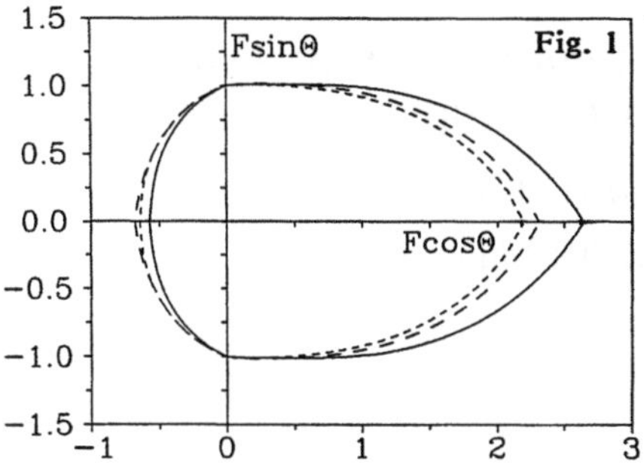

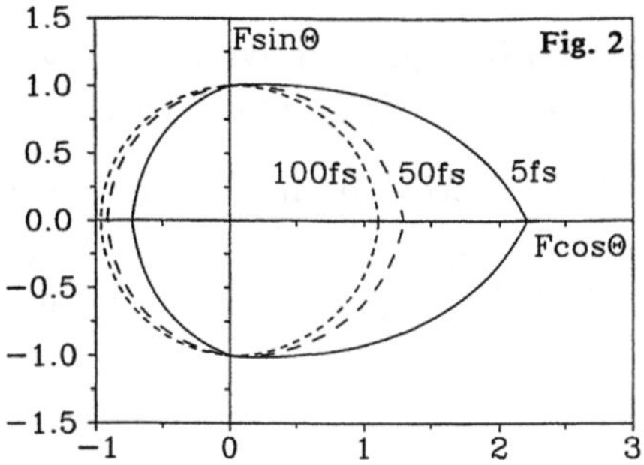

Fig. 1 Space diagram of the transient luminescence 5 fs after the beginning of the excitation, $\omega_{\ell um} = \Omega = E_g+10$ meV (solid line), $\omega_{\ell um} = E_g-10$ meV (long dashes) and $\omega_{\ell um} = E_g+20$ meV .

Fig. 2 Time evolution of the space diagram of the transient luminescence at $\omega_{\ell um} = \Omega$.

$$I_s = I_0 \exp\left[-\frac{\hbar^2(\mathbf{s}-\mathbf{k})^2}{2(m_e+m_h)T}\right] \simeq I_0\left[1 - \frac{2\hbar^2 k^2}{(m_e+m_h)T}\sin^2\frac{\Theta}{2}\right] ,$$

where Θ is the angle between the wave-vectors **k** and **s** of the absorbed and emitted photons.

Even for a completely quasi-equilibrium e-h plasma the anisotropy degree $\delta = (I_f - I_b)/I_f$ can reach observable values. Here, I_f and I_b are the intensities of the forward- and backward-luminescence, respectively. For GaAs plasma at $T \simeq 100$ K the parameter δ is about 4% . Moreover, in all stages before a complete thermalization the degree of the anisotropy of a transient luminescence is much greater.

For a detailed investigation of the appearance and evolution of an anisotropic transient luminescence, the self-consistent quantum kinetic equations for the e-h-photon system of a direct-gap semiconductor have been derived [4] on the basis of the microscopic approach. The computer simulations of this set of equations have been performed for a 2D e-h plasma in a GaAs QW. The calculated diagrams of the time-integrated transient luminescence signals $F_s(t) = \int I_s(t)dt$ at the different frequencies are presented in Fig. 1 . Although such signals represent the luminescence during only 5 fs after the initial stage of the optical excitation of a GaAs QW by 100 fs Gaussian pump pulse, their intensities may be detectable with a single-photon counting technique of femtosecond luminescence spectroscopy [2] . The space diagrams in Fig. 1 reflect the hot luminescence at the initial stage of the thermalization process. An anisotropy ratio F_f/F_b of 3-4 can be reached. The time evolution of the luminescence space diagram at the given frequency $\omega_{\ell um} = \Omega$ is shown in Fig. 2 . After approximately 100 fs the space diagram becomes quasi-stationary on the subpicosecond time scale. In this regime $\delta \simeq 15$-20% .

During the ballistic regime, carrier-carrier Coulomb scattering processes preserve the phase memory of the preceding optical excitation act and do not destroy the initial interband polarization induced by the coherent light pulse. As in the usual polariton picture, the decay of the interband polarization is determined by scattering processes with phonons or impurities. Our treatment connects the propagation of the initial light pulse with the concept of the inter-band polariton wave [5] for the case of a dense electron-hole plasma.

One of us (A.L.I.) thanks the Alexander von Humboldt Stiftung for support. This work has also been supported by the Deutsche Forschungsgemeinschaft and the Volkswagen Stiftung.

References

[1] J. Shah, Solid State Electron. 32, 1051 (1989) .
[2] T. Elsaesser, J. Shah, L. Rota, and P. Lugli, Phys. Rev. Lett. **66**, 1757 (1991) .
[3] H. Haug and S.W. Koch, *Quantum Theory of the Optical and Electronic Properties of Semiconductors* (World Scientific, Singapore, 1990) .
[4] A.L. Ivanov and H. Haug, Phys. Stat. Sol.(b), **173** (in print) .
[5] A.L. Ivanov, in *Proceedings of the XVII International Conference on Quantum Electronics*, USA, Technical Digest Series, **8**, 88 (1990).

Hot Hole Capture by Shallow Acceptors in p-Type GaAs Studied by Picosecond Infrared Spectroscopy

A. Lohner, M. Woerner, T. Elsaesser, and W. Kaiser

Physik Department E 11, Technische Universität München, W-8046 Garching, Fed. Rep. of Germany

Abstract. Capture of free holes by shallow acceptors monitored with picosecond infrared pulses occurs on a time scale of 50 ps via emission of longitudinal-optical phonons.

The capture of free carriers by shallow impurities in doped crystalline semiconductors represents an important relaxation process which is of particular relevance for hot carrier transport. Until now, those phenomena have been investigated by stationary transport or noise measurements, giving approximate cross sections of electron or hole trapping for a number of shallow donors or acceptors, respectively [1]. However, very limited information exists on the underlying scattering dynamics by which carriers are transferred from continuum states in the respective bands to localized states of the impurity atoms. In this paper, we discuss the first time-resolved observation of the ultrafast recombination of free holes with ionized shallow acceptors in a III-V semiconductor.

Pump-probe experiments were performed with zinc-doped GaAs crystals (acceptor binding energy 30 meV) at a lattice temperature of 10 K. The excitation pulse in the mid-infrared promotes holes which are initially bound to the (neutral) acceptor atoms, to the valence bands, i.e. free carriers are created by photoionization of the impurity atoms. The relevant absorption band and the spectral position of the pump pulses are shown in Fig. 1a [2]. After excitation, the ionized acceptors give rise to a new transient absorp-

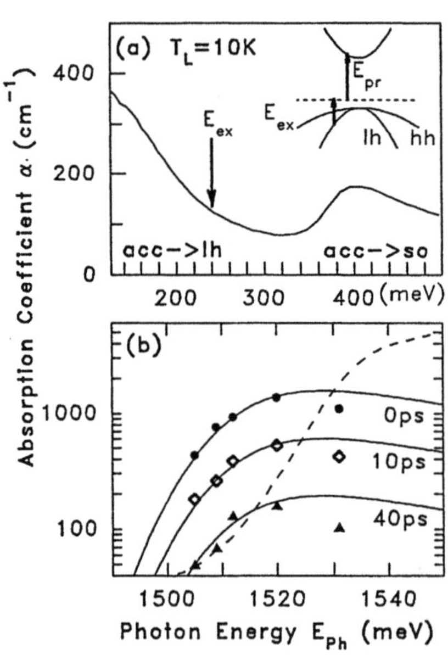

Figure 1 (a) Mid-infrared absorption of p-type GaAs (lattice temperature T_L=10K, doping concentration $N_A = 7 \times 10^{17}$ cm^{-3}). The arrow marks the photon energy of the picosecond excitation pulses.
(b) Absorption edge of the p-type sample (dashed line). The transient absorption band due to transitions from the photoionized acceptors to the conduction band is plotted for delay times of 0ps, 10ps, and 40ps (solid lines: calculated spectra). The decay of the band is due to the capture of free holes. The different optical transitions are depicted in the insert.

Springer Series in Chemical Physics, Vol. 55 Ultrafast Phenomena VIII
Editors: J.-L. Martin · A. Migus · G.A. Mourou · A.H. Zewail

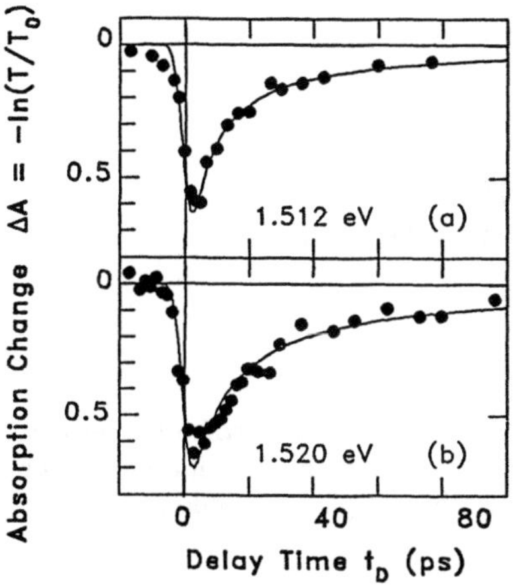

Figure 2 Transient interband absorption of negatively charged acceptors occurring after picosecond infrared ionization at E_{ex}= 0.24 eV. The change of absorption ΔA=-ln(T/T_0) is plotted vs delay time for probe energies of (a) 1.512 eV , and (b) 1.520 eV (points; T,T_0: transmission of the sample with and without excitation). The decay of the signal follows a $1/t_D$-behavior (solid lines).

tion in the near infrared (Fig. 1 b), the deionization band which is due to transitions from the acceptor ground state to the conduction band [3]. The strength of this absorption is determined by the momentary density of ionized acceptors that is identical to the density of free holes. Thus the decay of the deionization band monitored by near-infrared probe pulses directly reveals the kinetics of hole-acceptor recombination. Pump and probe pulses were derived from a mode-locked Nd:glass laser and have durations of 2 ps and 4 ps, respectively.

In Fig. 2, the time evolution of the transient absorption ΔA=-ln(T/T_0) is presented for two photon energies of the probe pulses. We find a rise within the time resolution of the apparatus and a subsequent relaxation on a time scale of several tens of picoseconds which follows a nonexponential kinetics. A more detailed inspection of the data demonstrates a decay of the enhanced absorption proportional to $1/t_D$. The symbols in Fig. 1 b give the transient deionization band derived from the time-resolved measurements. This band decays without any reshaping, i.e. the time resolved data taken at different photon energies give the same time evolution of the density of free holes. A hole capture rate per ionized acceptor of $C_{trap} \simeq 1.5 \times 10^{-6}$ cm^3/s is derived from the experimental results.

The data were quantitatively analyzed by theoretical calculations. The model gives - in agreement with the experiment - a $1/t_D$ behavior of the density of free carriers. Different trapping mechanisms, e.g. radiative recombination of hole and acceptor, Auger processes and scattering with phonons, are considered in the simulation. The numerical evaluation gives evidence that hole and acceptor predominantly recombine via the emission of single longitudinal optical phonons (E_{ph}=36 meV). The capture rate $C_{trap} \simeq 2 \times 10^{-6}$ cm^3/s calculated for this mechanism compares very well with the experimental results.

References

[1] V.N. Abakumov, V.I. Perel, I.N. Yassievich, Sov. Phys. Semicond. **12**, 1 (1978)

[2] E.J. Mayer, A. Lohner, M. Woerner, T. Elsaesser, Phys. Rev. **B**, in press

[3] W.P. Dumke, Phys. Rev. **B 2**, 1998 (1963)

Ultrafast Dephasing and Interference of Coherent Phonons in GaAs

W. Kütt, T. Pfeifer, T. Dekorsy, and H. Kurz

Institute of Semiconductor Electronics, RWTH Aachen, Sommerfeldstr. 24, W-5100 Aachen, Fed. Rep. of Germany

Abstract. Impulsive excitation of coherent phonons in GaAs via ultrashort laserpulses and time-resolved detection of amplitude and phase of coherent lattice vibrations allows to study details of the dynamics and interference of coherent phonon modes.

Optical excitation of coherent LO–phonons in GaAs by irradiation with femtosecond laser pulses is based on the ultrafast screening of the surface–space–charge field [1]. Both amplitude *and* phase of the coherent vibration can be determined in time domain by using highly sensitive electro–optic detection schemes in reflection mode. In this paper we report for the first time the application of this method to study phonon coupling and interference phenomena.

At low excitation densities ($<5*10^{17}$ carriers/cm^3) only LO–phonons are generated, which decay with an exponential time constant of 4 ps, indicating a "dephasing" through population relaxation via anharmonic phonon–phonon coupling [2]. This decay channel is verified by the temperature dependence of the time constant. If the excitation density is increased above $4*10^{18}$ cm^{-3} LO– and TO–modes appear, as shown in the Fourier spectrum in the insert of Fig.1. In time domain a clear mode beating becomes visible.

Numerical evaluation of the amplitude and phase reveals that the LO–mode decays with two time constants: In addition to the 4 ps decay a second fast decay with 440 fs appears. The TO–mode decays within 780 fs. At higher excitation densities ($>5*10^{18}$ cm^{-3}) the TO–mode with the fast decay time dominates.

The TO/LO interference occurs at the density, where the plasma frequency comes in resonance with the LO–mode. The plasma frequency, however, is strongly sensitive to side valley occupation within the first picosecond, giving rise to strong temporal variation of the plasma resonance. No plasmon–phonon–coupled modes well above and below the

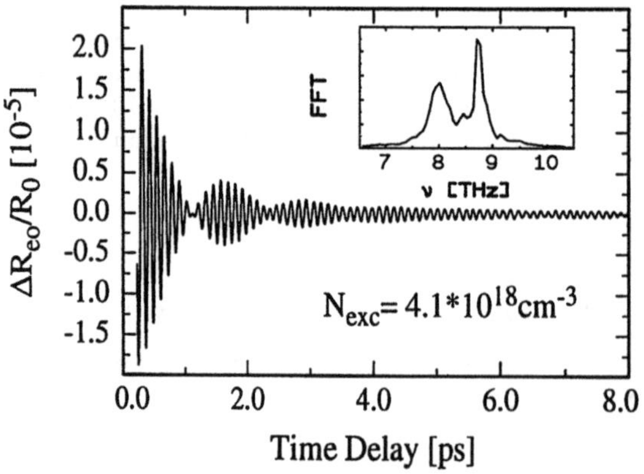

Fig. 1: Time–resolved interference of coherent LO– and TO–phonon modes in GaAs at high carrier density. The insert shows the fast Fourier transform of the oscillation signal. The data are taken at room temperature on a (100) orientated n–type GaAs sample (N_D=$4*10^{17}$ cm^{-3}).

Springer Series in Chemical Physics, Vol. 55 Ultrafast Phenomena VIII
Editors: J.-L. Martin · A. Migus · G.A. Mourou · A.H. Zewail

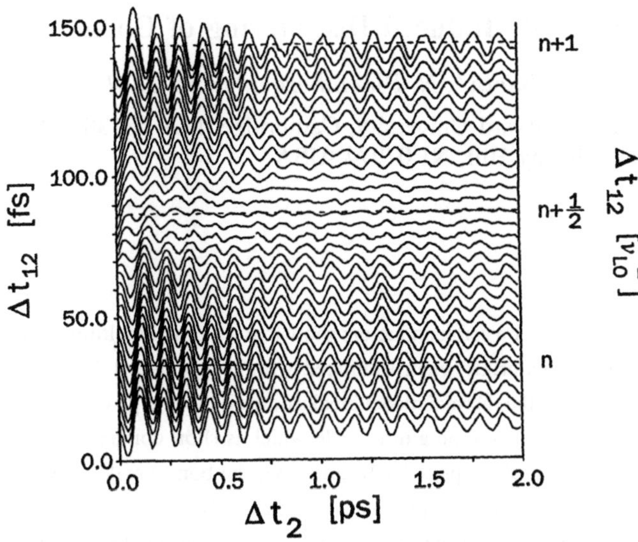

Fig. 2: Two pulse experiment demonstrating constructive and destructive interference of two coherent LO–phonon modes. Δt_2 denotes the time delay between the second pump and the probe pulse. Δt_{12} is the time between the two pump pulses, covering the range of one complete phonon vibration period ν_{LO}^{-1}.

LO– and TO–frequencies are observed due to the strong overdamped nature of the hole plasma [3] and large spatial gradients in the carrier distributions [4].

Coherent phonon interferences are demonstrated in a two-pulse experiment. At sufficiently low excitation densities, two successive pump pulses with a time delay of Δt_{12} excite two coherent lattice displacements consecutively, which are displayed in time domain in Fig.2. At a time delay $\Delta t_{12} = n\nu_{LO}^{-1}$ (n integer, ν_{LO} LO–phonon frequency) constructive interference occurs, while the oscillations vanish due to destructive interference at $(n+\frac{1}{2})\nu_{LO}^{-1}$. Although this experiment at the moment adds no physical insight to the theory of lattice vibrations, the demonstration of their interference in time domain is pure fun for any experimentalist.

References

[1] G.C. Cho, W. Kütt and H. Kurz, Phys. Rev. Lett. 65, 764 (1990)

[2] F. Vallee and F. Bogani, Phys. Rev. B43, 12049 (1991)

[3] A. Mlayah, R. Carles, G.Landa, E. Bedel and A. Munoz–Yague, J. Appl. Phys 69, 4096 (1991)

[4] C. L. Collins and P. Y. Yu, Solid State Commun. 51, 123 (1984)

Femtosecond, Electronically-Induced Disordering of GaAs

J.-K. Wang[1], *Y. Siegal*[1], *P.N. Saeta*[2], *N. Bloembergen*[1], and *E. Mazur*[1]
[1]Gordon McKay Laboratory, Harvard University, Cambridge, MA 02138, USA
[2]AT&T Bell Laboratories, Murray Hill, NJ 07974, USA

Picosecond laser pulse experiments on semiconductors show that a thermal melting model can explain the behavior of these materials when irradiated with such pulses.[1] However, femtosecond experiments suggest that the thermal model breaks down if the energy of the laser pulse is deposited into the electronic system on a time scale which is short compared to the electron-lattice energy relaxation time (2-3 ps).[2,4-7] In this case, the bond breaking which occurs through the excitation of an electron-hole plasma can lead directly to ionic disordering while the lattice is still cold.[3] We have performed a series of measurements on GaAs using 160-fs, 620-nm pulses which show that above a threshold pump fluence of 0.1 J/cm^2, the material takes on a centrosymmetric structure within the pulse duration in a process which is directly driven by the electronic excitation.

We chose GaAs for this experiment because its lack of inversion symmetry results in a non-zero bulk, dipole, second-order susceptibility. Because disordering leads to a loss of this second-order susceptibility, measuring the reflection second-harmonic (SH) signal generated by a probe beam as a function of pump-probe delay time monitors the symmetry of the sample. The sample was aligned such that SH generation from the pump beam was suppressed, and the polarization of the probe beam was orthogonal to that of the probe beam. The sample was translated between shots to assure a fresh spot for each measurement.

Figure 1 summarizes the immediate response of the material for an incident fluence of 0.36 J/cm^2, which is about 3.6 times the damage threshold, on the (110) GaAs sample. Both the reflectivity and the SH signal are normalized to unity for no pump excitation. The SH efficiency drops to zero with a decay time of 90 fs while the linear reflectivity increases to a plateau 44% greater than that of the unpumped material with a rise time of 170 fs. The sudden rise in the reflectivity is evidence for a transformation to a metallic state. The vanishing of the SH suggests that the electronic system takes on a centrosymmetric character consistent with disordering of the lattice. This transition to a disordered phase occurs an order of magnitude faster than the electron-phonon energy relaxation time of 2 ps.[2,5] From the reflectivity results,[5-7] the dielectric constant of the transient disordered phase is determined to be $-0.31+i0.34$, which is inconsistent with a Drude model. This suggests a phase which is not an equilibrium liquid.

Information on the carrier density upon disordering can be derived from the results on the (100) GaAs sample. Figure 2 shows the results on the (100) GaAs sample at a pump fluence of 0.58 times the damage threshold. The SH efficiency rises to 2.4 while the reflectivity decreases to 0.85 by about 130 fs after the pump pulse peak. This rise in the SH is not due to changes in

Springer Series in Chemical Physics, Vol. 55 Ultrafast Phenomena VIII
Editors: J.-L. Martin · A. Migus · G.A. Mourou · A.H. Zewail

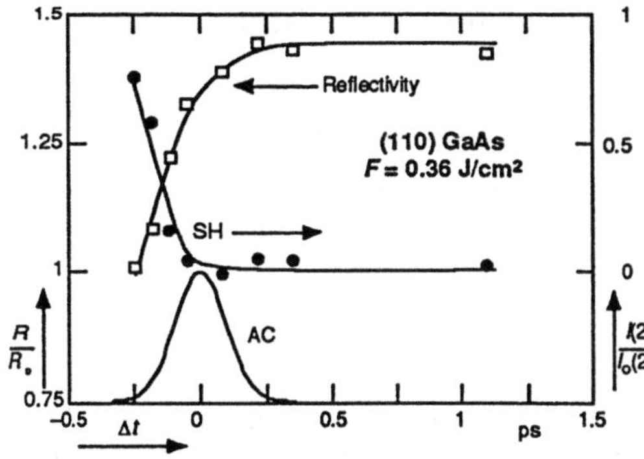

Fig. 1 Time response of second harmonic and linear reflectivity in (110) GaAs to 160-fs laser pulse. Second harmonic vanishes with 90-fs decay time while reflectivity rises to a plateau with 170-fs rise time.

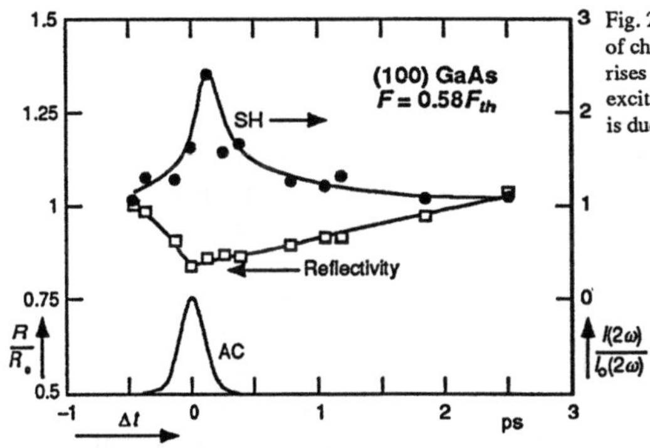

Fig. 2 Excitation at 0.58 times damage threshold shows effects of changes in linear dielectric constant. Second harmonic rises to peak of 2.4 while reflectivity drops to 0.85 due to excitation of free carriers. Quick recovery of second harmonic is due to fast Auger recombination.

the second-order susceptibility, but to changes in the linear dielectric constant caused by the excited free carriers, providing a very sensitive probe of the free carrier density. In this regime, we can use a Drude model for the dielectric constant, and fits to both the SH and linear reflectivity results yield a free carrier density of about 1×10^{22} cm^{-3}. The fast decay of the rise in the SH signal indicates an Auger recombination time of 400 fs. This rapid recombination puts a cap on strong nonlinear absorption, dominated by impact ionization driven by free-carrier absorption. Thus, the carrier density saturates at about 1×10^{22} cm^{-3} before the second-order susceptibility begins to drop due to the onset of disordering as the incident fluence approaches the damage threshold. This experimentally determined critical free carrier density agrees with that predicted by Van Vechten for electronically-induced disordering.[3]

This work was supported by the Joint Services Electronics Program and the Material Research Laboratory Program under contracts with Harvard University. YS acknowledges a Ph.D. fellowship from the National Science Foundation.

1. Malvezzi, A. M. in *Excited-state spectroscopy in Solids* , edited by U. M. Grassano and N. Terzi (North-Holland Phsyics Publishing, Amsterdam, 1987), 335.
2. J. S. Kash, J. C. Tsang, and J. M. Hvam, Phys. Rev. Lett. **54**, 2151 (1985).
3. J. A. Van Vechten, R. Tsu, and F. W. Saris, Phys. Lett. **74A**, 422 (1979).
4. H. W. K. Tom, G. D. Aumiller, and C. H. Brito-Cruz, Phys. Rev. Lett. **60**, 1438 (1988).
5. P. Saeta, J.-K. Wang, Y. Siegal, N. Bloembergen, and E. Mazur, Phys. Rev. Lett. **67**, 1023 (1991).
6. Sokolowski-Tinten, K., H. Schulz, J. Bialkowski and D. von der Linde, Appl. Phys. A, **53**, 227 (1991).
7. Govorkov, S. V., I. L. Shumay, W. Rudolph and T. Schroder, Opt. Lett., **16**, 1013 (1991).

Laser-Induced Ultrafast Order–Disorder Transitions in Semiconductors

K. Sokolowski-Tinten, J. Bialkowski, and D. von der Linde

Institut für Laser- und Plasmaphysik, Universität-GHS-Essen, W-4300 Essen 1, Fed. Rep. of Germany

Abstract. Using fs-time-resolved linear and non-linear optical spectroscopy we characterize the different stages of the transformation from the the crystalline state to the metallic liquid state.

During the past several years a number of reports have been published [1–5] suggesting that ultrafast melting occurs when semiconductors (typically Si and GaAs) are irradiated with laser pulses 100 fs or less in duration. The evidence for such transitions was based on the observed changes of the optical reflectivity which were shown to be consistent with a transition from a covalent crystalline to a metallic liquid state. Measurements of the reflected second harmonic (SH) supported this picture. In particular, for non-centrosymmetric GaAs a sharp drop of the SH is expected when a transition to a centrosymmetric disordered state takes place. However, it must be noted that the changes of reflectivity and SH resulting from a possible structural transition are obscured by the laser-induced changes of the linear and non-linear optical properties due to the creation of a dense e-h-plasma.

Here we present new data and a detailed analysis of the situation: Careful measurement of the reflectivity (Si and GaAs) and of the reflected SH (GaAs) for different polarizations and over a wide range of times and fluences allows us to monitor the complex evolution of the laser-induced phase-transition.

We find that two fundamentally different types of phase transitions can be produced by femtosecond laser pulses. There is a distinct range of laser fluences where a relatively slow transition to the liquid state is observed which takes typically 100 ps to develop [3]. The results in this fluence regime are very similar to the observed melting behaviour in the case of ps–laser–induced phase–transitions [6].

For higher fluences, however, we observe a dramatic change of the time evolution suggesting a different type of transition which develops on a sub-ps time scale. Figure 1 gives an example of the ultrafast phase transformation in GaAs. From the measured time dependencies of the s– and p–polarized reflectivity at 620 nm we were able to determine the changes of the optical constants. Using standard formula of nonlinear optics we then calculate the expected SH assuming $\chi^{(2)}$ is constant (dotted curve in Fig. 1). It can be seen that suprisingly large changes of the SH are caused simply by changes of the optical constants alone. The dashed curve in Fig. 1 shows the measured SH. From these data and the measured optical constants one can deduce the actual changes of $\chi^{(2)}$ (solid curve). It turns out that there is a very large and extremely fast drop of $\chi^{(2)}$ within about 100 fs followed by a more gradual decrease over approximately several hundred femtoseconds.

At least two explanations can be offered for the fast component of $\chi^{(2)}$: (i) collaps of the non-centrosymmetric crystal lattice, and (ii) bond charge depletion associated with the intense optical pump pulse. At this point there is not sufficient evidence to

Springer Series in Chemical Physics, Vol. 55 Ultrafast Phenomena VIII
Editors: J.-L. Martin · A. Migus · G.A. Mourou · A.H. Zewail

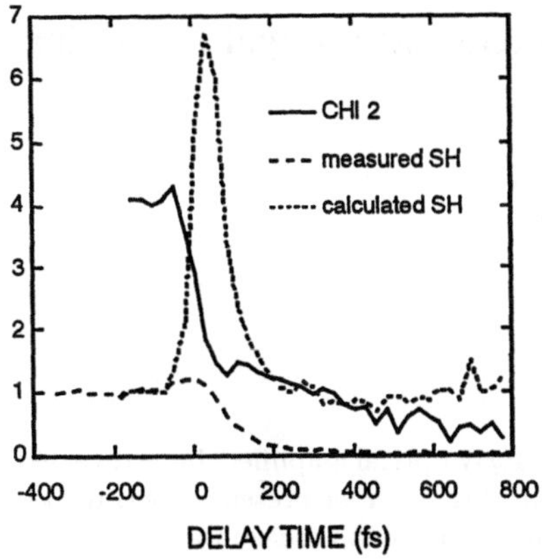

Figure 1: Measured reflected second harmonic of the probe pulse (dashed curve), calculated second harmonic (dotted curve, see text), and the measured nonlinear susceptibility $\chi^{(2)}$ of GaAs as a function of time. Pump fluence: $0.3\,J/cm^2$ (approximately two times the threshold fluence).

determine unambiguously the contribution of the various processes to the measured changes of $\chi^{(2)}$ during the first hundred femtoseconds.

On the other hand, rather firm conclusions can be drawn concerning the evolution during the several hundred femtoseconds following the initial sharp drop of $\chi^{(2)}$. During this time the optical reflectivity continues to increase, approaching values that are consistent with the metallic liquid state having an electronic density about one order of magnitude greater than the estimated initial density of photoexcited free carriers. In fact our time-resolved measurements of the optical reflectivity spectra indicate that only 300 fs after the pump pulse the spectra are Drude-like over a wide spectral regime (380 nm - 850 nm).

This research was financially supported by the Deutsche Forschungsgemeinschaft.

[1] C. V. Shank, R. Yen, and C. Hirlimann, Phys. Rev. Lett. **50**, 454 (1983)

[2] H. W. K. Tom, G. D. Aumiller, and C. H. Brito–Cruz, Phys. Rev. Lett. **60**, 1438 (1988)

[3] K. Sokolowski–Tinten, J. Bialkowski, and D. von der Linde, Appl. Phys. **A 53**, 227 (1991)

[4] P. Saeta, J.-K. Wang, Y. Sigal, N. Bloembergen, and E. Mazur, Phys. Rev. Lett. **67**, 1023 (1991)

[5] S. V. Govorkov, I. L. Shumay, W. Rudolph, and T. Schröder, Opt. Lett. **16** 1013 (1991)

[6] D. von der Linde, B. Danielzik, K. Sokolowski–Tinten, and P. Harten: In *Ultrafast Phenomena VI*, ed. by T. Yajima, K. Yoshihara, C. B. Harris, S. Shionoya, Springer Ser. Chem. Phys., Vol. 48 (Springer, Berlin, Heidelberg 1988)

Femtosecond Carrier Dynamics in InGaAsP Optical Amplifiers

J. Mark and J. Mørk

TFL Telecommunications Research Laboratory,
Lyngsø Allé 2, DK-2970 Hørsholm, Denmark

Pump-probe measurements on InGaAsP bulk and MQW optical amplifiers have revealed ultrafast dynamics not yet fully understood [1,2]. In this paper we present experimental and theoretical results which indicate that the dynamics is due to the combined effect of spectral holeburning (SHB), carrier heating (CH), and two-photon absorption (TPA).

The experimental setup is a conventional pump-probe setup with TM-polarized pump and TE-polarized probe beams, which are coupled in and out of the active region of the amplifier by short pieces of tapered single-mode fiber. The second output from the polarization beamsplitter, which combines the pump and probe beams, is lowpass-filtered after detection and used as a reference signal for the zero delay position. Transform-limited pulses with a typical width of 125 fsec, increasing to 150 fsec at the tuning range boundaries, are generated by an F-center laser using additive pulse modelocking. The center wavelength of the 20 nm wide spectrum can be tuned from 1485 nm to 1540 nm.

Measured pump-probe signals for an InGaAsP DC-PBH bulk amplifier are shown in Fig. 1a for a bias current of 10 mA and increasing center wavelength. At negative delay times the probe pulse is travelling ahead of the much stronger pump pulse (the probe power is $\sim$ 100 times smaller than the pump power). The transparency point at this particular bias current is $\lambda \simeq 1515$ nm. At wavelengths below (above) this point the pump pulse increases (decreases) the carrier density, and hence the gain, seen by the probe pulse at large positive delay times. The "slow" ($\sim$ 0.7 ps) transient is attributed to relaxation of the carrier temperatures to the lattice temperature, following carrier heating induced by the pump pulse [1,2]. The physical mechanism responsible for the fast positive dynamic seen in the absorption region was not yet identified.

To help the interpretation of the experimental data we have set up a model for the gain dynamics based on the density matrix equations (within the relaxation time approach). The model includes the dynamics of CH [3] and SHB. The non-adiabatic treatment of SHB was essential for the results. Futhermore, TPA was taken into account [1,2].

In Fig. 1b we show the calculated response at different wavelengths (for the same wavelength and gain change as in Fig. 1a, keeping all other parameters fixed). Pulse-propagation effects, including group-velocity mismatch between pump and probe pulse, were also included. The qualitative agreement between measured and calculated responses leads us to attribute the positive peak in the absorption region to SHB. Such a component changes sign at the transparency point, which is a very clear feature in the experimental data.

At low bias currents and high pump powers, experimental (and theoretical) results show that the SHB induced peak becomes very large, corresponding to an on-off ratio of more than 10 and a switching time of less than 200 fsec. This physical effect might therefore serve as the basis for an ultrafast optical switch or a fast saturable absorber for applications in laser diode modelocking.

Springer Series in Chemical Physics, Vol. 55 Ultrafast Phenomena VIII
Editors: J.-L. Martin · A. Migus · G.A. Mourou · A.H. Zewail

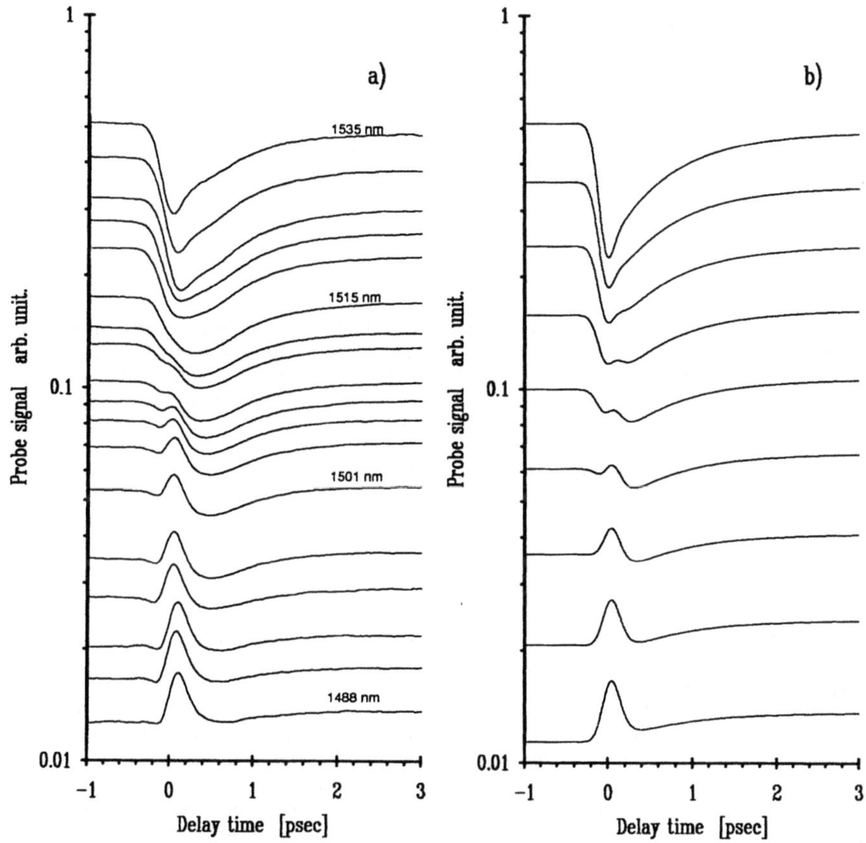

Figure 1: a) Measured and b) calculated pump-probe response for an InGaAsP optical amplifier at fixed pump and probe power levels, a bias current of 10 mA, and different wavelengths. The wavelength covers a range of 44 nm in b).

In conclusion, we have shown that sub-picosecond gain dynamics in InGaAsP laser amplifiers can be understood in terms of carrier heating, spectral holeburning and two-photon absorption.

[1] K. L. Hall, J. Mark, E. P. Ippen, and G. Eisenstein, Appl. Phys. Lett, vol. 56, 1740–1742 (1990).

[2] K. L. Hall, Y. Lai, E. P. Ippen, G. Eisenstein, and U. Koren. Appl. Phys. Lett, vol. 57, 2888–2890 (1990).

[3] M. Willatzen, A. Uskov, J. Mørk, H. Olesen, B. Tromborg, and A.-P. Jauho, IEEE Photon. Technol. Lett., vol. 3, 606–609 (1991).

Ultrafast Nonlinear Refraction in Semiconductor Laser Amplifiers

M. Sheik-Bahae and E.W. Van Stryland

Center for Research in Electro-Optics and Lasers, University of Central Florida, Orlando, FL 32816, USA

Ultrafast large nonlinear refraction in semiconductor diode lasers at wavelengths near their transparency point ($\hbar\omega > E_g$) has been recently reported.[1,2] This has generated much interest due to its potential use for all optical switching. We have extended an earlier model [3] which successfully described the dispersion and band-gap scaling of the electronic n_2 in passive semiconductors in their transparency range ($\hbar\omega < E_g$) to explain such nonlinearities in semiconductor amplifiers.

We use a nonlinear Kramers-Kronig transformation to obtain the change in refractive index from the calculated change in the absorption spectrum $\Delta\alpha$:

$$\Delta n(\omega\,;\,\omega) = \frac{c}{\pi}\int_0^{\infty}\frac{\Delta\alpha(\omega'\,;\,\omega)}{\omega'^2-\omega^2}d\omega'. \tag{1}$$

The change in absorption $\Delta\alpha$ due to virtual excitation is calculated using a "dressed state" formalism accounting for the effects of two-photon absorption, electronic Raman, and the a.c. Stark shift (light-induced gaps) of the bands.[3]

In extending this model to active semiconductors, we consider the general case in which the excitation photon energy can be above the bandgap energy. In practice, this leads to additional real excitation processes such as carrier heating, band-filling and gain-saturation effects that are distinguished from ultrafast virtual processes due to their longer recovery times.[1] In modeling, we account for the virtual processes by considering only the states with energy E_{cv} whose virtual carrier liftime ($|\omega-E_{cv}/\hbar|^{-1}$) is shorter than the phenomenological dephasing time (T_2) of that state. The gain (or loss) characteristics of the medium are obtained using Fermi-Dirac (FD) distribution functions for electrons and holes with quasi-Fermi levels $E_{cf}=E_{vf}$. Fig.1 (solid line) shows the calculated $\Delta\alpha(\omega';\omega)$ with $\hbar\omega/E_g=1.03$ denoting the transparency point.

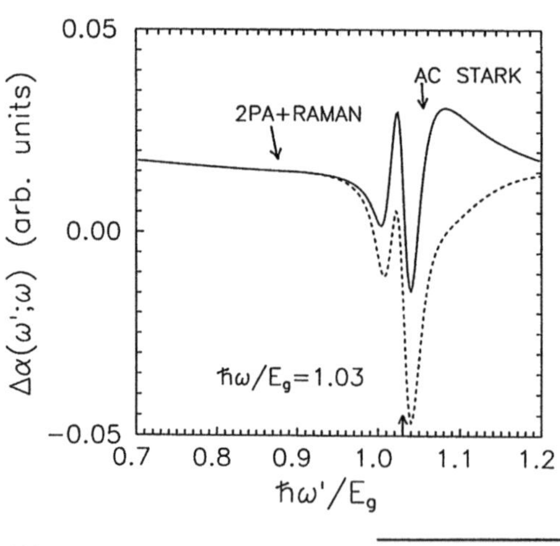

Fig.1 The calculated change in absorption due to virtual excitation at the transparency wavelength ($\hbar\omega/E_g$=1.03). The solid and dashed lines correspond to before and after a.c. Stark shift of the quasi-Fermi levels respectively.

Springer Series in Chemical Physics, Vol. 55 **Ultrafast Phenomena VIII**
Editors: J.-L. Martin · A. Migus · G.A. Mourou · A.H. Zewail

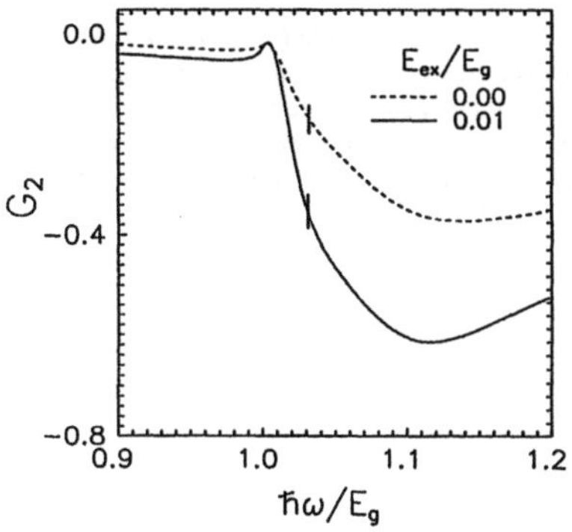

Fig.2 The calculated G_2 dispersion function for various quasi-Fermi levels. The dashed line represents the G_2 for a passive semiconductor. The transparency point of each curve (at $\hbar\omega/E_g$=1.0, 1.03, and 1.05) is marked.

The effects of two-photon absorption and electronic Raman are shown as a net increase in the absorption. This effect increases as the probe frequency approaches the infrared. The quadratic a.c. Stark effect (QSE), on the other hand, is more localized in energy since the optical field induces a gap that alters the density of states only in the vicinity of the excitation photon energy. An important consequence of this induced change in the density of states is the self-adjustment of the quasi-Fermi levels in order to conserve the total carrier density. This adjustment is not instantaneous and has a characteristic time determined by the carrier-carrier scattering ($\simeq$100 fs). For an excitation at the transparency point, the E_{fc} (or E_{fv}) is shifted to higher (or lower) energies in **k** space. The overall change in the absorption spectrum, including the quasi-Fermi shift, is depicted in Fig.1 (dashed line). Therefore, for a probe beam having the same wavelength as the pump, this is manifested by an initial decrease in the transmission due to two-photon absorption followed by an increase due to the QSE and subsequent blue shift of the quasi-Fermi levels. This description is consistant with the experimentally observed ultrafast gain dynamics in AlGaAs amplifier waveguides.[5] It must be noted again that here we are not concerned with the effect of carrier heating. It has been shown that, with a longer time-scale ($\simeq$0.6 ps), this effect would tend to red-shift the quasi-Fermi levels (i.e. opposing the effect of the QSE) and hence decrease the transmittance of the probe near the transparency point.[1,5]

Following Eq.(1), the induced change in the refracive index is now obatined using the KK transformation of the calculated $\Delta\alpha$. Similar to the case of passive materials [3] we derive the following formula for the nonlinear index n_2:

$$n_2\ (\mathrm{x}10^{-11}\ \mathrm{cm^2/W}) = K\,\frac{\sqrt{E_p(\mathrm{eV})}}{E_g^4(\mathrm{eV})\ n_0^{\,2}}\,G_2(\hbar\omega/E_g)\ , \tag{2}$$

where K is a constant, E_p is the Kane energy ($\simeq$21 eV for most semiconductors [3]) , n_0 is the linear index and G_2 is the dispersion function. Fig.2 depicts the calculated G_2 function for various quasi-Fermi levels.

It is seen that n_2 is negative and dispersive at the transparency point of each curve. The transparency values of n_2, however, do not vary strongly as the quasi-Fermi levels are changed. In calculating the dispersion curves of Fig.2, we assumed the parmeters E_g/kT=50 for the carrier temperature and $E_gT_2/\hbar$=50 for the broadening. The value of $|G_2|$ at the transparency point increases almost linearly as a function of these two parameters. The dispersion curves in Fig. 2 contain the contributions from 2PA and Raman in addition to the QSE of the quasi-Fermi levels, all having a negative sign at the transparency point. However, as pointed out for $\Delta\alpha$, the latter effect has a characteristic risetime determined by carrier-carrier scattering.

For bulk passive semiconductors, the constant K was evaluated in ref.[3] to be $\simeq$6 (units of Eq.(2)). However, this underestimated (up to a factor of 4) the n_2 for AlGaAs samples measured at wavelengths very near the band-edge.[4] This may be partially attributed to the electron-hole Coulomb interaction that was not included in the calculation.[3] In a simple approximation, we account for the Coulomb interaction by using a generalized (screened) Elliott enhancement factor for the continum of excitons.[6] Applying this enhancement factor

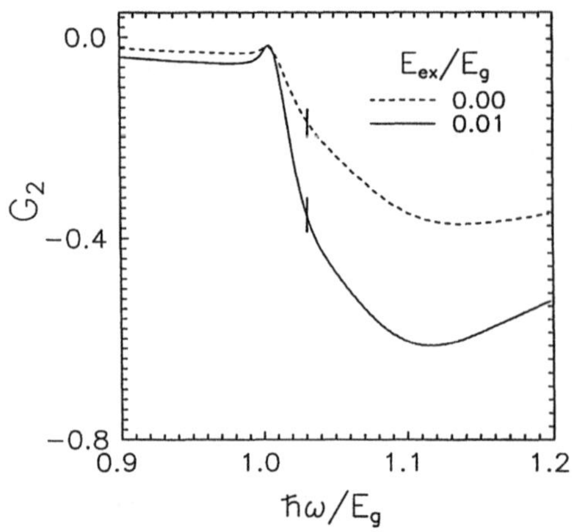

Fig.3 The calculated n_2 dispersion function including the screened exciton enhancement (solid line) in comparison with the no exciton case (dashed line). The transparency point is at $\hbar\omega/E_g=1.03$.

to all the processes involved in the transition rate, we re-evaluate $\Delta\alpha$. The exciton-enhanced G_2 function obtained using this approximation is shown in Fig.3 for a typical exciton binding energy of $E_{ex}=0.01\text{x}E_g$. Although no appreciable qualitative difference is seen, it is evident that, in this specific case, more than a 2 times enhancement of the transparency value of G_2 is predicted.

Using the enhanced G_2 function, we find that the n_2 formula given by Eq.(1) gives a relatively good agreement with the sign and magnitude of the measured n_2 values in AlGaAs [1] and InGaAsP [2] diode lasers. For instance, using $E_g=1.5$ and $n_0=3.4$ for AlGaAs, we obtain $n_2\simeq-2\text{x}10^{-12}$ cm²/W compared to the measured value of $-5\text{x}10^{-12}$ cm²/W reported in [1].

In conclusion, we have obtained a simple expression (giving the dispersion and band-gap scaling) for the ultrafast n_2 in semiconductor lasers near their transparency point. We have accounted for the effects of two-photon absorption, electronic Raman and a.c. Stark shift, including the self-adjustment of the quasi-Fermi levels.

References

[1] C.T. Hultgren and E.P. Ippen, Appl. Phys. Lett. **59**, 635(1991).
[2] R.S. Grant and W. Sibbett, Appl. Phys. Lett. **58**,1119(1991).
[3] M. Sheik-Bahae *et al*, IEEE J. Quantum Electron. *QE-27*, 1296(1991).
[4] M.J. LaGasse, *et al*, Appl. Phys. Lett. **56**, 417(1990).
[5] K.L. Hall, *et al*, Opt. Lett. **17**, 874(1992).
[6] L. Banyai, and S.W. Koch, Z. Phys. B -Condenssed Matter, *63*, 283(1986).

Femtosecond Luminescence Spectroscopy of Indium Phosphide

E. Fazio[1] *and G.M. Gale*[2]

[1]Dipartimento di Energetica, Università "La Sapienza",
Via Scarpa, I-00161 Roma, Italy
[2]Laboratoire d'Optique Quantique du CNRS, Ecole Polytechnique,
F-91128 Palaiseau, France

Abstract. **Femtosecond luminescence spectroscopy of intrinsic and p-doped bulk indium phosphide at 77 K is reported. Fast carrier relaxation and complex dynamics including intervalley transfer are observed. The results obtained as a function of electron energy are interpreted in the framework of a simple model.**

1. Experimental

Intrinsic (IN), moderately (MD $2x10^{18}$) and heavily p-doped (HD 10^{19}) bulk samples of InP cooled to 77K are photoexcited with a 60 fs, 2 eV laser pulse. Luminescence from the samples is detected using an up-conversion system with 90 fs time-resolution, at energies up to 250 meV above the band-gap of InP, depending on doping.

2. Results

The luminescence signal from all three samples rises very rapidly (100 fs to 2 ps - see below) but on a longer time-scale we observe a second, previously unnoticed, slowly rising luminescence component (see Fig. 1) which we attribute to intervalley transfer of electrons from the Γ to the L valley in InP at 77K. Our data indicate that about 70% of the initially excited electrons transfer rapidly, independent of doping, to the L valley and return only very slowly with an exponential time constant of 45 ps. This observation is consistent with the hypothesis that essentially only electrons excited from the split-off valence band remain in the Γ valley.

With our experimental technique we can resolve luminescence rise-time constants well below 100 fs. The ensemble of luminescence rise-times, obtained from exponential fits, in all three samples is reported in Fig. 2(a) as a function of luminescence energy above the band gap.

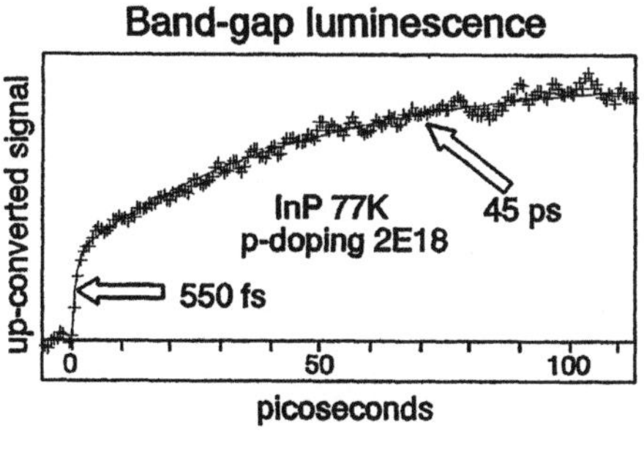

Fig. 1. Band-gap luminescence from the MD sample over 110 ps. After an initial fast rise, luminescence increases very slowly with an exponential time-constant of 45 ps. This rise, which is independent of excitation intensity, is attributed to intervalley return from the L-valley to the Γ-valley.

Springer Series in Chemical Physics, Vol. 55 Ultrafast Phenomena VIII
Editors: J.-L. Martin · A. Migus · G.A. Mourou · A.H. Zewail

3. Interpretation

In weakly photo-excited p-doped InP samples the observed luminescence is determined uniquely by electron dynamics. In view of the relatively long relaxation times obtained in the intrinsic sample we may assume hole relaxation to be instantaneous. We accordingly interpret these results in the framework of a model which includes re-injection of electrons high in the Γ valley by intervalley return, relaxation of electrons by LO phonon emission and interaction of electrons with a thermalised hole bath. Electron thermalization is introduced into the model phenomenologically.

In the case of the intrinsic sample only LO-phonon emission is of primordial importance, although saturation of this channel must be taken into account. LO-phonon saturation is suppressed in the doped samples because of LO-phonon lifetime reduction [1]. Finally, it is necessary to include the interaction of the hot electrons with the cold hole bath produced by p-doping, which becomes the dominant relaxation channel in the HD sample. Fig. 2(b) shows the calculated luminescence rise-times obtained from the model (to be compared with Fig. 2(a)).

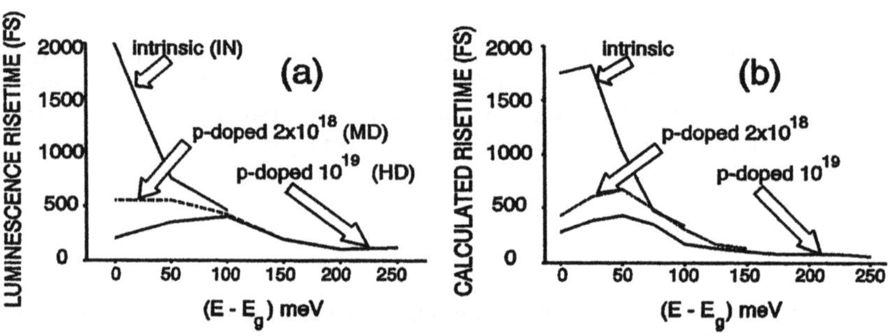

Fig. 2. (a) Measured luminescence rise-times (obtained from an exponential fit to the data) for the samples IN, MD and HD as a function of luminescence energy above the gap . (b) Calculated luminescence rise-times using the model described in this paper.

References

1] J. A. Kash and J. C Tsang
Solid State Electronics **31**, 419 (1988)

Dynamics of Excitons Probed by Accumulated Photon Echo

T. Bouma, P. Vledder, and J.I. Dijkhuis

Faculty of Physics and Astronomy, and Debye Research Institute,
University of Utrecht, P.O. Box 80000, NL-3508 TA Utrecht, The Netherlands

Accumulated photon echo[1] (AccPE) is an ultra-sensitive tool for studying optical dephasing of excitations in solids. The essential prerequisite for the succesful application of this technique is the presence of a bottleneck in the optical pumping cycle, preventing the system from a quick return to the ground state. A frequency grating in the ground state may then be formed and serve to create a stimulated echo. The technique has been applied to study *long*-lived optical impurity centers and bound excitons in solids. Here we present the first exploratory measurements on *short*-lived excitons in semi-insulating GaAs, and we will discuss the accumulation mechanism.

In Fig. 1 we present a typical echo trace taken at 2 K in the standard AccPE geometry on a thin layer of LEC-grown GaAs, rendered semi-insulating by the EL2 defect. The picosecond Ti-sapphire laser was tuned through frequencies in the near-band-edge region. At the low energy side where the neutral-acceptor-bound excitons (A^0,X) reside[2], we observe pronounced quantum beats. We attribute these to the internal level structure of the bound-exciton complex. At energies close to the band edge a marked decrease in the dephasing time is found, possibly due to delocalization effects of excitons in a disordered surrounding. The

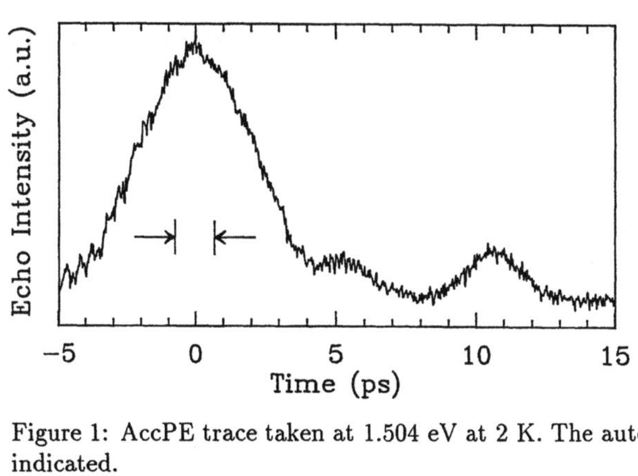

Figure 1: AccPE trace taken at 1.504 eV at 2 K. The autocorrelation time of the laser is indicated.

accumulation time of the echo was measured to be 1 μs, three orders of magnitude longer then the exciton lifetime.

It has been shown by electronic Raman experiments[2] that at low temperatures a non-equilibrium population of holes and ionized acceptors A^- can be maintained in the presence of EL2-defects in the metastable state. We suggest the following accumulation mechanism: A fraction of the excitons bound to A^0 decays by Auger recombination and creates hot holes and ionized acceptors. The remaining population of A^- serves as the bottleneck and gives rise to a frequency-selective bleaching of the (A^0,X) transition, because the A^--bound excitons absorption is at a different energy (~1.490 eV) *well beyond* the laser line, and a frequency grating may be formed. The accumulation time (i.e. the lifetime of the frequency grating) is thus limited by the capture of the holes by A^- and *not* by the exciton lifetime. The lifetime of 1 μs observed, together with the capture crossection $\sigma = 8.7.10^{-14}$ cm^2 for holes [3] leads to an estimated A^--concentration of 10^{13} cm^{-3}.

In conclusion we have observed for the first time AccPE of *short*-lived excitons. This opens new perspectives to apply this ultra-sensitive technique to study low-density excitons and charge carriers in semiconductors with sufficiently high defect densities, in particular amorphous materials.

We acknowledge C. R. de Kok and F. J. M. Wollenberg for invaluable technical assistance, J. E. M. Haverkort and C. M. van Es for providing the sample and R. P. Tijburg for the etching. Special thanks to Spectra Physics for giving us the Tsunami picosecond Ti:sapphire laser on loan. This work has been partly supported by the Netherlands organizations 'Fundamenteel Onderzoek der Materie (FOM)' and 'Nederlandse Organisatie voor Wetenschappelijk Onderzoek (NWO)'.

[1] W. H. Hesselink and D. A. Wiersma, Phys. Rev. Lett. **43** (1979) 1991.
[2] R. Bray, K. Wan, and J. C. Parker, Phys. Rev. Lett. **57** (1986) 2434.
[3] D. Bimberg, H. Münzel, A. Steckenborn, and J. Christen, Phys. Rev. B **31** (1985) 7788.

Time-Resolved Measurement of Hot Carrier Cooling Rates in a-Si:H and a-Ge:H

M. Wraback and J. Tauc

Department of Physics and Division of Engineering, Brown University, Providence, RI 02912, USA

Abstract. Femtosecond pump and probe technique with 4eV pulses was employed in the direct measurement of hot carrier cooling rates in a-Si:H and a-Ge:H.

While great progress has been made in the measurement and understanding of hot carrier relaxation processes in crystalline semiconductors, little is known about the hot carrier cooling rates necessary for an understanding of carrier-phonon interactions in the extended states of amorphous semiconductors. Most experiments have been performed on the technologically important amorphous hydrogenated silicon (a-Si:H), a material which, because of its large optical gap (1.7eV), does not lend itself easily to the study of hot carrier dynamics with conventional femtosecond laser systems operating at or below photon energies in the spectral region surrounding 2eV. As a result, previous experiments [1] employing the pump-probe technique with 2eV excitation pulses as short as 100fsec have not revealed any evidence of hot carrier relaxation in a-Si:H, although indirect measurements [2] suggest that the carrier cooling rate is greater than 1eV/psec.

In this paper we present the first direct measurements of hot carrier cooling rates in a-Si:H and high quality amorphous hydrogenated germanium (a-Ge:H) [3]. Femtosecond ultraviolet pulses of 4eV photon energy were produced by frequency doubling of an amplified CPM ring dye laser, and along with 2eV pulses were employed in the excitation and probing of carriers with considerable excess energy in extended states required for the study of the ultrafast cooling processes in amorphous materials. Figure 1 shows the change in the imaginary part of the dielectric function $\Delta\epsilon_2$ in a-Si:H as measured with a 4eV probe after excitation with either a 2eV (a) or 4eV (b) pump pulse. The negative onset at $t=0^+$ represents photoinduced bleaching associated with the saturation of interband transitions [3]. Theoretical calculations incorporating the effects of both interband and intraband contributions to $\Delta\epsilon_2$ indicate that this bleaching remains essentially unchanged during the dissipation of ~1eV of excess energy by carriers in the bands. The ~0.5psec delay in the $\Delta\epsilon_2$ decay for 4eV excitation (curve b) therefore provides a direct measure of the time evolution of the hot carriers in extended states, with their subsequent dynamics governed by trapping in shallow localized states. The best fits to the data yield an average carrier cooling rate and trapping time of 2±0.2eV/psec and 0.4±0.1psec, respectively.

Similar measurements have been performed on a-Ge:H. Figure 2 shows the time evolution of $\Delta\epsilon_2$ in this material as obtained from 4eV pump and 2eV probe

Springer Series in Chemical Physics, Vol. 55 Ultrafast Phenomena VIII
Editors: J.-L. Martin · A. Migus · G.A. Mourou · A.H. Zewail

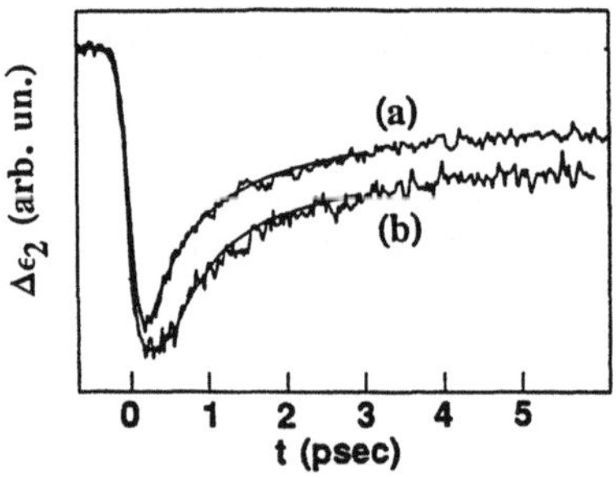

Figure 1 $\Delta\epsilon_2$ decays in a-Si:H for 4eV probe and a) 2eV pump; b)4eV pump.

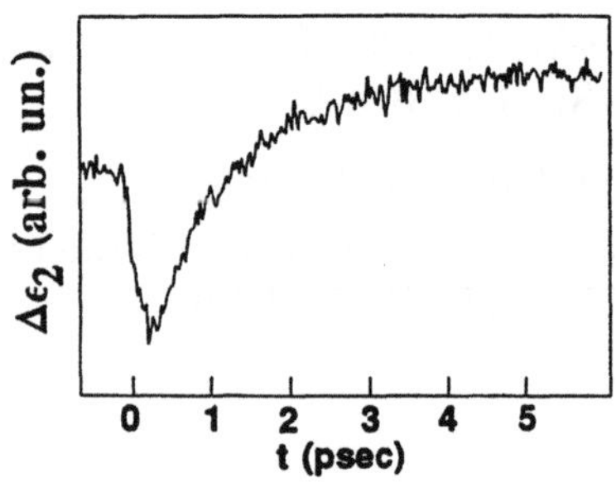

Figure 2 $\Delta\epsilon_2$ decay in a-Ge:H for 4eV pump and 2eV probe experiments.

experiments. The negative onset resulting from the saturation of interband transitions decays to a positive plateau at longer times due to both ultrafast heat production associated with the cooling of the hot carriers and the subsequent trapping of these carriers. Employing a similar model as for a-Si:H, we calculate an average cooling rate of ~1.5eV/psec. This value and the one obtained for a-Si:H are significantly larger than the ones predicted in the weak coupling limit from perturbation theory, indicating that carrier interaction with high frequency silicon- and germanium-hydrogen vibrational modes as well as disorder-enhanced carrier-phonon interactions may play an important role in the cooling process in these materials.

The authors thank J.-K. Lee and E.A. Schiff for the a-Si:H samples and D. Pang and W. Paul for the a-Ge:H samples. This work was supported by NSF grant DMR9014977. M. Wraback acknowledges the additional support of a National Research Council-ETDL Research Associateship.

[1] For a review see J. Tauc and Z. Vardeny in Critical Reviews in Solid State and Materials Sciences, Vol. 16, (CRC Press, 1990), p.403.
[2] D. Hulin, A. Mourchid, P.M. Fauchet, W.L. Nighan Jr., and R. Vanderhaghen, J. Non-Cryst. Solids, 137&138, 527 (1991).
[3] M. Wraback, J. Tauc, D. Pang, W. Paul, J.-K. Lee, and E.A. Schiff, J. Non-Cryst. Solids, 137&138, 531 (1991).

Dephasing of the Short Exciton–Polariton Pulses in Polar Semiconductors: The Cuprous Chloride Case

F. Vallée, F. Bogani, and C. Flytzanis*

Laboratoire d'Optique Quantique du C.N.R.S., Ecole Polytechnique, F-91128 Palaiseau Cedex, France

*Permanent address: Dipartimento di Fisica, Universita di Firenze, I-50125 Firenze, Italy

Abstract. We demonstrate a non-linear optical technique that allows the direct observation of the spatio-temporal evolution of the coherent short exciton pulses in polar semiconductors. The detailed intrinsic dephasing of the Z_3 longitudinal and transverse exciton in CuCl is obtained by this technique.

It has recently been shown that the dynamics of an exciton-polariton can be directly accessed in the time domain using time resolved two-photon excitation and probing[1].This technique, which allows the investigation of excitons in the bulk of crystals, avoids in particular the high excitation regime where polariton dephasing is dominated by exciton-exciton interactions. Exploiting the non-local possibility of the non-linear techniques, the spatial propagation of the polariton pulse can also be followed, as demonstrated for phonon-polariton[2]. The first measurements were performed on the Z_3 exciton-polariton in CuCl for which the relaxation was found to occur much faster than the propagation allowing a local probing of the polariton wave packet. The same experimental method can thus be used to investigate the nonpropagating longitudinal exciton, allowing a comparison of the relaxation mechanism of the transverse and longitudinal modes.

The principle of the two-photon excitation and detection of a coherent exciton wavepacket is depicted in Fig. 1 [2]. First, a coherent exciton mode of frequency ω_e and wave vector k_e is realized in the bulk of the crystal by two-photon absorption of two synchronized picosecond pulses with frequency ω_1 and ω_2 and wavevector $\mathbf{k}_1$ and $\mathbf{k}_2$ such that $\omega_1 + \omega_2 = \omega_e$ and $\mathbf{k}_1 + \mathbf{k}_2 = \mathbf{k}_e$; this restricts the applicability of the technique only to the upper polariton branch and longitudinal components. The evolution of the exciton coherence, subsequent to the local excitation process , is followed by the parametric emission at $\omega_s = \omega_e - \omega_p$ stimulated by a third picosecond pulse of frequency ω_p delayed by t_D and spatially separated with respect to the excitation stage.

The experiments were performed using a passively mode-locked Nd^3 glass delivering a single 5-ps pulse in the infrared (λ_1 = 1054 nm) which is partially frequency converted in order to create three independent pulses ω_1 ω_2 and ω_p. The one at ω_1 is a small

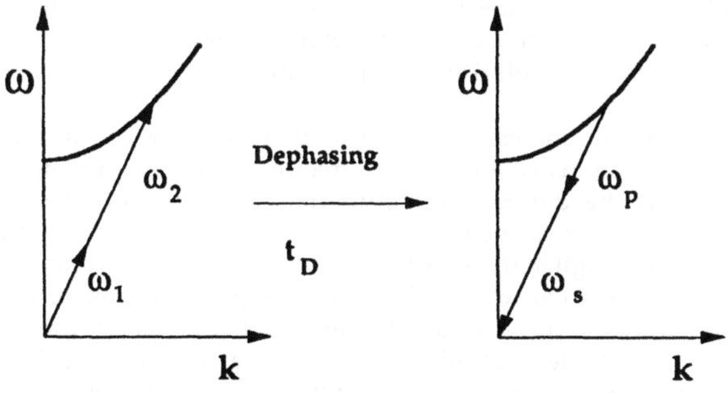

Fig. 1. Principle of two-photon excitation and probing.

Springer Series in Chemical Physics, Vol. 55 **Ultrafast Phenomena VIII**
Editors: J.-L. Martin · A. Migus · G.A. Mourou · A.H. Zewail

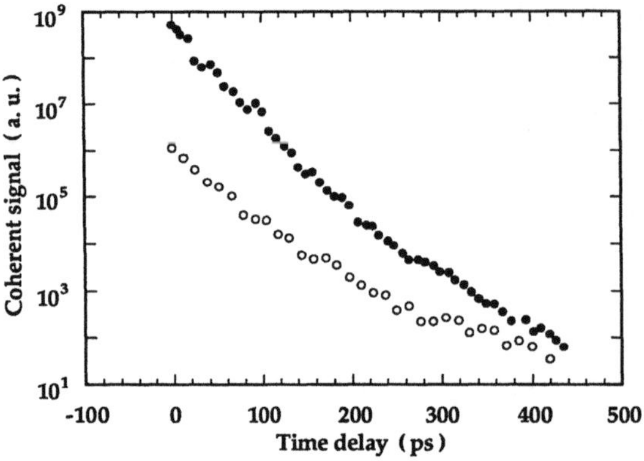

Fig. 2. Coherent signal, as a function of the probe delay, measured at 6K in a phase-matched (dots) and a non phase-matched (circles) geometry.

fraction of the initial infrared pulse while ω_2 is tunable ($\lambda_2 \sim 611$ nm); the probe beam frequency ω_p ,in the infrared, is adapted to phase matching requirements. The experimental set-up is described in [1]. Several samples have been used in the experiments with (111) surface for the study of the longitudinal exciton and with (110) surface for the transverse exciton-polariton.

The signal observed on the longitudinal exciton exhibits an exponential decay for all crystal temperatures between 20 and 70K, allowing the first determination of the dephasing time of a longitudinal exciton. For lower temperature, en exponential decay is also observed over almost five orders of magnitude, followed by a slower nonexponential evolution over two orders of magnitude (Fig.2). Such a deviation has not been observed for the transverse exciton, suggesting that the spatial dispersion of the longitudinal mode plays a role. In order to check the wave vector sensitivity of the signal, we have taken advantage of the high dynamics of the technique to perform measurements in a non phase-matched geometry ($\Delta k \sim 800$ cm^{-1}). The results show a very different behavior of the temporal evolution of the much weaker signal which exhibts a longer nonexponential decay. This suggests that the deviation observed in the phase matched situation originates from the excitation of different longitudinal excitons with different relaxation times due to spatial dispersion. In the phase matched geometry, excitons created with a wavevector satisfying phase-matching give the dominant contribution to the small delay (<250 ps) signal leading to the observed exponential decay.

The temperature dependence of the dephasing rate measured in the phase matched geometry is plotted in Fig.3. The overall temperature dependence is similar to the one observed for the transverse mode and can be interpreted on the same basis. It has been shown [1] that the relaxation of an upper branch exciton-polariton is mainly due to down and up-conversion into a lower branch polariton with respectively emission or absorption of a phonon.

At low temperature the relaxation rate increases linearly with T, as observed in Fig. 3. This is a signature of a mechanism dominated by the low frequency acoustic phonons interacting with the exciton through the deformation potential and the piezoelectric effect.Including only acoustic phonons the dashed line of Fig. 3 is obtained. A strong deviation is observed for temperatures higher than 40 K ,for which a strong increase of the relaxation rate is measured. This is a consequence of the strong interaction of the exciton with the longitudinal optical (LO) phonon mediated by the Fröhlich hamiltonian. Only up-conversion processes are observed because of the very low probability of the down-conversion mechanism associated with a very low density of final states. Considering the interaction with both optical and acoustical phonons, a good

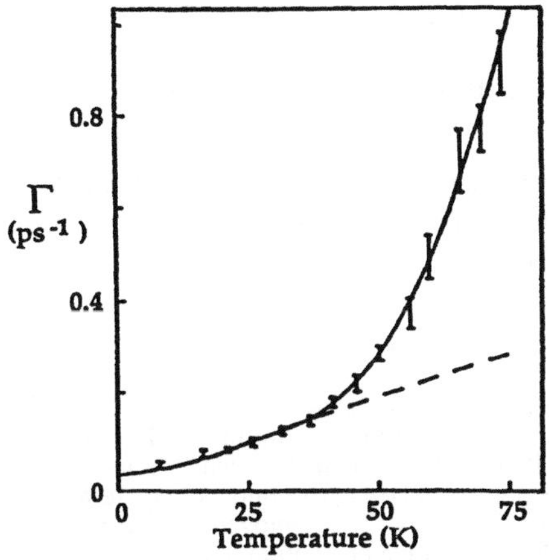

Fig. 3. Temperature dependence of the longitudinal optical exciton dephasing rate. The continous line is the best fit obtained taking into account the relaxation mechanisms due to acoustic phonons (dashed line) and to LO phonons.

description of the experimental results is obtained (Fig. 3). In fact, the absolute values of the measured dephasing rates of the longitudinal component are identical with the one previously determined for the exciton polariton of almost the same energy [1]. This similarity is expected for the investigated modes for which the interaction coefficients calculated from the Raman and Brillouin tensors were found to be identical.

Acknowledgements

We wish to thank Prof. D. Fröhlich of the University of Dortmund, Germany, for providing samples and for valuable information; we thank J. Godard of the Laboratoire de Physique des solides, Université de Paris-Sud, France, for preparing the samples.

References

1. F. Vallée, F. Bogani and C. Flytzanis, Phys. Rev. Lett. **66**,1509 (1991).
2. F. Vallée ,G.M. Gale and C. Flytzanis,Phys .Rev. Lett. **61** ,2102 (1988).

Femtosecond Electronic Dynamics of CdSe Nanocrystals

C.V. Shank[1,2], R.W. Schoenlein[3], D.M. Mittleman[1], J.J. Shiang[2], and A.P. Alivisatos[2]

[1]Department of Physics, University of California, Berkeley, CA 94720, USA
[2]Department of Chemistry, University of California, Berkeley, CA 94720, USA
[3]Lawrence Berkeley Laboratory, MS 70-193A, 1 Cyclotron Rd., Berkeley, CA 94720, USA

Abstract. We report the first direct measurements of electronic dephasing in CdSe nanocrystals using two-pulse and three-pulse photon echo techniques. Photon echo measurements reveal that the electronic dephasing is strongly modulated by coherent vibrational excitation of the nanocrystals. Using three-pulse photon echo measurements, we are able to separate the vibrational dynamics from the electronic dynamics. The homogeneous linewidth of these materials is shown to result from electronic dephasing on a one hundred femtosecond time scale with significant contributions from both LO and acoustic phonon modes.

Semiconductor nanocrystals of a size comparable to or smaller than the volume of the bulk exciton exhibit a number of novel properties resulting from the three-dimensional quantum confinement. In such systems, the quasi-continuum of electronic states in the bulk material is reduced to a series of essentially discrete electronic transitions. This change in the density of states gives rise to large optical nonlinearities and a blue-shift of the excitonic optical absorption. Due to the strong dependence of band-gap (Homo-Lumo transition) on size, $E_{gap}\sim R^{-2}$[1], even highly monodisperse nanocrystals are substantially inhomogeneously broadened. In addition, the electron dynamics of these materials are strongly influenced by size. This results from the confinement of the electron and hole wavefunctions, as well as from the influence of surface states and phonon coupling, which also vary with size.

The electronic properties of CdSe nanocrystals have been previously investigated using a variety of techniques. Raman measurements have been used to investigate electron-phonon coupling in these materials, and have shown significant coupling to the dominant LO phonon mode at 205 cm^{-1}[2]. However, this coupling is considerably reduced relative to electron-LO phonon coupling in bulk CdSe, as a consequence of the size quantization[3]. The confinement enhances the overlap between the electron and hole wavefunctions which results in a corresponding reduction in the polar interaction with LO phonons. Photoluminescence excitation experiments have also shown evidence of electron-phonon coupling in the absorption spectra of individual CdSe clusters[4]. Time resolved luminescence measurements as well as spectral hole burning on time scales of nanoseconds[4,5] and hundreds of femtoseconds[4,6,7] have been used to investigate carrier relaxation dynamics, including surface trapping and dephasing. Pump-probe measurements have shown clear evidence of spectral hole burning at low temperatures[5] with hole widths of several hundred wavenumbers. However, such spectral holes may in principle include contributions from electronic dephasing as well as phonon broadening. At room temperature, these effects have prevented the observation of spectral holes that are narrower than the inhomogeneous distribution. Thus such measurements are unable to determine the various contributions to the homogeneous linewidth.

The recent development of an ultrafast laser source in the blue-green spectral regime[8], now enables the direct investigation of the transient electronic properties of semiconductor nanocrystals with 10 fs time resolution. In this paper we demonstrate the use of photon echo techniques to investigate the various processes which contribute to the dephasing of electron-hole pairs on such short time scales. In particular, using the three-pulse photon echo and a novel technique for suppressing the effects of quantum beats, we are able to clearly distinguish between electronic dephasing due to dynamic processes and linewidth broadening due to phonon line structure.

The laser system consists of a colliding-pulse modelocked (CPM) dye laser producing 50 fs pulses at 620 nm. These pulses are amplified to several microjoules in a multipass dye amplifier pumped at 400 Hz by a XeF excimer laser. The amplified pulses generate a white-light continuum in a 1-mm jet of ethylene glycol, and the blue-green (490 nm) portion of the continuum is re-amplified in a second dye amplifier pumped by the same excimer laser. A

Springer Series in Chemical Physics, Vol. 55 Ultrafast Phenomena VIII
Editors: J.-L. Martin · A. Migus · G.A. Mourou · A.H. Zewail

sequence of gratings and prisms are used to directly compress the amplified portion of the continuum, resulting in nearly transform-limited pulses of 15 fs duration. Using appropriate dyes and dye mixtures, we are able to generate ~15 fs pulses over a broad spectral range, enabling us to investigate a variety of particle sizes.

The CdSe nanocrystals used in this study are grown in solution using the inverse micelle technique and are capped with Se[5]. Following the growth process, the particles are annealed at 170 °C, and the crystalline quality and size distribution are characterized by X-ray powder diffraction, TEM, and linear optical absorption. The variation in particle diameter (~5%) results in an inhomogeneous half-width of ~800 cm^{-1}. The ~42 Å CdSe nanocrystals used in these experiments are uniformly dispersed in a 300 μm free-standing polymer film. Linear absorption measurements at room temperature show a distinct excitonic (homo-lumo) absorption peak at ~490 nm, which merges with an absorption continuum at shorter wavelengths. In all experiments the excitation fluence is ~0.1 mW/cm^{-2} which is less than one photon per nanocrystal at our sample densities. The ~25 nm bandwith of the pulses is centered on the excitonic absorption peak to avoid exciting carriers to the continuum states.

Our initial studies of the electronic dephasing in these materials were performed using the two-pulse photon echo in which the integrated echo intensity in the phase-matched direction, $2k_2-k_1$, is measured as a function of delay (t_{12}) between the first pulse (k_1) and second pulse (k_2). For an ideal (inhomogeneously broadened) two-level system, the integrated echo signal is predicted to decay as $\exp[-4t_{12}/T_2]$. The measured two-pulse echo decay times range from ~9 fs at a temperature of 30 K to ~6 fs at 235 K. The extremely rapid echo decay and the temperature dependence suggest that the polarization dephasing may be strongly modulated by the vibrational modes of the system, resulting in significant deviation from the ideal $T_2/4$ behavior. This was confirmed by numerical simulations using a two-level system model with offset harmonic potentials to describe the effects of the 205 cm^{-1} LO phonon mode. The presence of such a mode results in a periodic modulation of the echo, with a rapid echo decay at early times due to the quantum beat. Although it is expected that the two-pulse echo signal will reach a second maximum at 163 fs (one vibrational period), this recurrence was not observed experimentally since the overall echo signal decays nearly two orders of magnitude in that time.

Using three-pulse photon echo measurements, we are able to separate the effects of the quantum beat from the underlying dephasing dynamics. The three-pulse photon echo is simply a generalization of the two-pulse echo whereby the two field interactions with the second pulse ($2k_2$) are replaced with one field interaction with the second pulse and one field interaction with a third pulse (k_3+k_2). The first two pulses (k_1 and k_2) create a population grating with an amplitude which is determined by the time delay between the pulses and the dephasing time of the system (T_2). The interaction of the third pulse with this grating results in the formation of an echo in the phase-matched ($k_3+k_2-k_1$) direction. As a result, the delay of the third pulse determines when the echo is formed. By appropriately delaying the third pulse, it is possible to measure the echo at a particular phase of the quantum beat. In particular, if the third pulse is delayed exactly one vibrational period, then the nuclei have returned to their original position (corresponding to the Franck-Condon region of the harmonic potential surface) and the quantum-beat modulation of the photon echo can be suppressed. In contrast, if the arrival of the third pulse is out of phase with the vibrational period, then the nuclei are maximally displaced and the echo decay is strongly modulated by the quantum beat.

Figure 1 illustrates the predicted three-pulse photon echo response in the presence of quantum beats resulting from coherent vibrational excitation. This illustrative calculation includes a dephasing time of 80 fs, as well as the (damped) vibrational dynamics of the 205 cm^{-1} LO phonon mode. The figure shows the integrated echo signal (in the $k_3+k_2-k_1$ direction) as a function of the delay between the first and second pulse (t_{12}) and as a function of the delay between the first and third pulse (t_{13}). The solid line (A) indicates the expected echo response for a fixed delay of the second pulse (t_{12}=50 fs) as a function of the third pulse delay. The oscillations in (A) result from the vibrational coherence of the LO phonon mode.

The experimental result corresponding to the prediction of Fig. 1 line (A) is shown in Fig. 2. This is a measurement of the three-pulse echo response of a 42 Å diameter CdSe sample at 15 K as a function of the third pulse delay (t_{13}) with the delay of the second pulse, (t_{12}) fixed at 33 fs. We observe oscillations resulting from the coherent vibrational excitation, and the measured period is consistent with the 205 cm^{-1} LO-phonon mode. In addition, we observe that the phonon oscillations are damped on a time scale of ~4 ps. This damping is likely due to decay of optical phonons into acoustic phonons, and is in good agreement with the

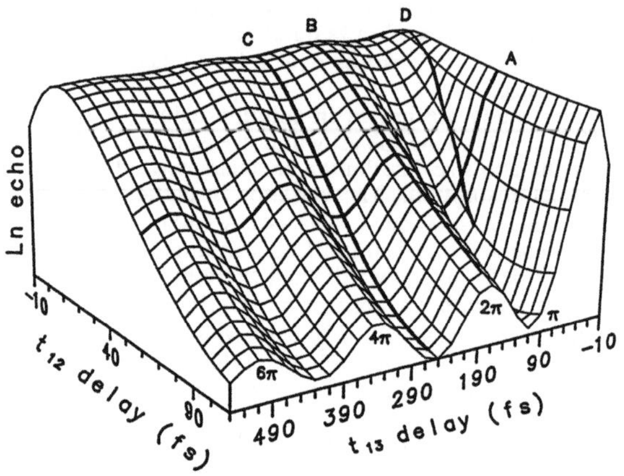

Figure 1 Calculated three-pulse photon echo signal in the presence of a 200 cm^{-1} phonon mode as a function of delay between pulse one and two (t_{12}) and betweeen pulse one and three (t_{13}). Solid lines A, B and C correspond to t_{12}=50 fs, t_{13}=170 fs, and t_{13}=250 fs. Solid line D corresponds to the two-pulse echo in which t_{12}=t_{13}.

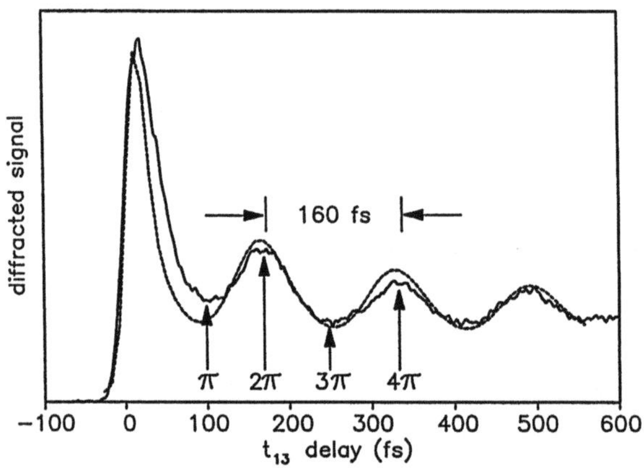

Figure 2 Three-pulse photon echo (42 Å particles, 15 K) as a function of t_{13} with t_{12}=33 fs. Dashed line is a theoretical fit including 205 cm^{-1} phonon mode (Δ=0.7), 4.4 ps vibrational damping time, and T_2=85 fs.

~15 cm^{-1} Raman linewidth measured by Alivisatos et al [2]. Such measurements enable the direct investigation of the vibrational dynamics of these nanocrystals.

In order to directly investigate the electronic dephasing, we can use the results of Fig. 2 to determine the optimal delay for the third pulse. By positioning the third pulse at exactly one LO-phonon period (t_{13}=2π) and scanning the delay of pulse two (t_{12}) the effects of the vibrational dynamics can be effectively suppressed (Fig. 1 curve B). Conversely, if the third pulse is positioned out of phase with the LO-phonon (t_{13}=3π), then the echo decay will be strongly modulated by the quantum beat (Fig. 1 curve C). The corresponding experimental measurements in the 42 Å sample at 15 K are shown in Fig. 3. With the third pulse in the 2π position we measure an echo decay time of ~20 fs. With the third pulse in the 3π position, the echo decay is strongly modulated by the quantum beat, and we measure a much faster echo decay (~9 fs). Finally, with the third pulse in the 4π position, the quantum beat modulation is again suppressed and the echo decay becomes longer. Note that at successively longer delays of the third pulse (t_{13}=4π, 6π, 8π...) the suppression of the vibrational dynamics is less effective. This is a result of the vibrational damping which is evident in the measurements of Fig. 2. This damping of the nuclear motion is an additional contribution to the electronic dephasing which can only be suppressed for $t_{13} \ll T_{damping}$.

In order to better understand these results, we have theoretically modeled both the electronic and vibrational dynamics of the system. The electronic dynamics are described as an ideal two-level system using a standard density-matrix formalism to calculate the third-order polarization response. The vibrational dynamics of the 200 cm^{-1} mode are described by

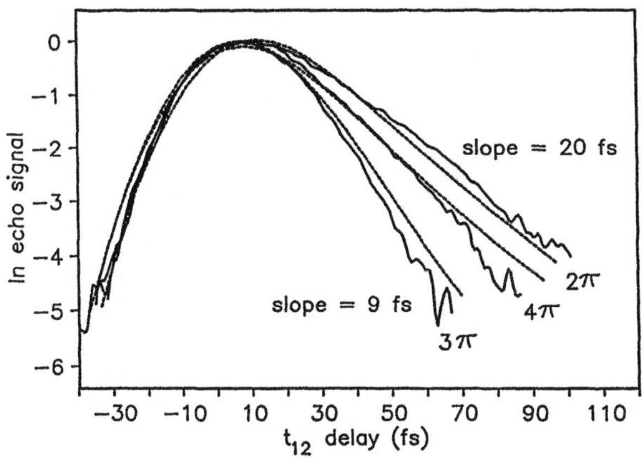

Figure 3 Three pulse echo (42 Å particles, 15 K) as a function of t_{12}, with the third pulse in phase ($t_{13}=2\pi$, and $t_{13}=4\pi$), and out of phase. ($t_{13}=3\pi$), with the quantum beat (162 fs period). Dashed lines are theoretical fits including the 205 cm^{-1} phonon mode ($\Delta=0.7$), 4.4 ps vibrational damping time, and $T_2=85$ fs.

displaced harmonic potentials in the ground and excited states. Such a description is readily incorporated into the density matrix formalism using a semi-classical Liouville-space formalism [9]. Finally, the vibrational damping is accounted for by introducing frictional damping, exp[-γt], following the approach of Yan et al. [10]. The dashed lines in Figs. 2 and 3 show the results of such a calculation, which includes a single vibrational mode (205 cm^{-1} LO-phonon) with a displacement parameter, $\Delta=0.7$, consistent with Raman measurements of the electron-phonon coupling[2]. We assume a vibrational damping time of 4.4 ps and an electronic dephasing time $T_2=85$ fs. With only these parameters we obtain remarkably good fits to the measurements shown in Fig. 2, including the period, amplitude, and vibrational damping of the oscillations. The same calculation also gives a good fit to the results in Fig. 3 including the variation of the echo decay times for $t_{13}=2\pi$, 3π, and 4π. Note that the ~20 fs echo decay for the $t_{13}=2\pi$ measurement in Fig.3 is approximately $T_2/4$ ($T_2=85$ fs). The discrepancy is again the result of the vibrational damping which has some small contribution to the electronic dephasing even at $t_{13}=2\pi$, and of course increases for longer delays of the third pulse.

These results provide a good picture of the dynamic evolution of the electron-hole coherence in the 42 Å CdSe particles at low temperatures. There are two important components which contribute to the polarization dephasing in very different ways. First, the measurements of Fig. 3 indicate a "stochastic" polarization dephasing (T_2) on a time scale of 85 fs. In addition, significant electronic coupling to the LO-phonon mode at 205 cm^{-1} results in a periodic modulation of the polarization. This damped quantum beat modulation is clearly observed in the three-pulse echo measurements of Fig. 2. The very rapid echo decay observed in the two-pulse echo measurements ($t_{12}=t_{13}$, corresponding to curve D in Fig. 1) results from the simultaneous effects of both processes.

We can obtain some insight into the nature of the "stochastic" component of the polarization dephasing by suppressing the quantum beat modulation, and investigating the temperature dependence of the echo decay. Figure 4 shows the echo decay rate (measured with $t_{13}=2\pi$) for the 42 Å particles as a function of temperature. The decay rate scales roughly linearly with temperature from ~0.05 fs^{-1} at 15 K to ~0.16 fs^{-1} at room temperature. This rather strong temperature dependence of the echo decay can not be accounted for by the increase in the LO phonon population. This is illustrated by the dashed line in Fig. 4 which shows the calculated temperature dependence predicted by the same model used to fit the data in Figs. 2 and 3. The weak temperature dependence predicted by the dashed line is easily understood since the effects of the LO phonon mode have been largely suppressed by the position of the third pulse. In addition, since the LO phonon energy is 25 meV, the population does not change significantly at room temperature.

One possible explanation for the temperature dependence of the dephasing is enhancedcoupling to acoustic phonons. In bulk CdSe, coupling to the acoustic phonons via the deformation potential is rather weak. In nanocrystals however, coupling to acoustic modes is significantly enhanced due to the quantum confinement[3,11]. Such compression and expansion modes of the nanocrystal effectively modulate the band gap, thereby contributing to

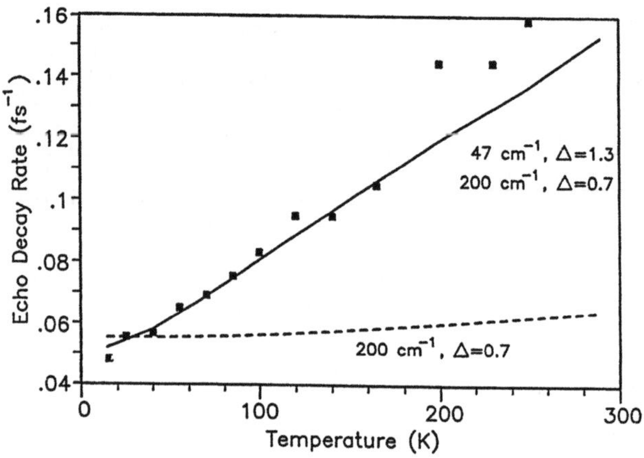

Figure 4 Temperature dependence of the echo decay rate (42 Å particles) with $t_{13}=2\pi$ for suppression of the quantum beat. Dashed line is the temperature dependence predicted using the fit parameters from Figs 2 and 3. Solid line is a theoretical fit using the parameters from Figs 2 and 3 with a single additional mode (47 cm^{-1}, $\Delta=1.3$) to approximate the coupling to the continuum of acoustic phonons.

the polarization dephasing. If we model the quasi-continuum of acoustic modes by a single mode at $\hbar\omega=47$ cm^{-1}, $\Delta=1.3$, following [5], then we can fit the temperature dependence of Fig. 4 with surprising accuracy (solid line). This strongly suggests that coupling to acoustic phonons is a major contribution to the electronic dephasing at room temperature.

In summary, the three-pulse photon echo is a powerful technique for investigating the dephasing dynamics in quantum confined semiconductor nanocrystals. In 42 Å CdSe particles, we observe a coherent phonon mode at 205 cm^{-1} which is damped on a picosecond time scale, and strongly modulates the polarization dephasing even at low temperatures. By appropriate positioning of the third pulse, the effects of the quantum beat can be suppressed in the three-pulse echo measurements, enabling us to directly investigate the "stochastic" dephasing process. Our results indicate that the underlying polarization dephasing occurs on a time scale of ~85 fs at 15 K. The temperature dependence of this dephasing suggests that enhanced coupling to a continuum of acoustic modes may account for a significant part of this dephasing. Additional studies over a range of particle sizes as well as varied surface passivation should allow us to separate the effects of surface trapping and acoustic phonon coupling on the fundamental dephasing dynamics.

This work was supported by the U.S. Department of Energy under contract DE-AC03-76SF00098.

References

1. L. Brus, *J. Chem. Phys.* **80**, 4403 (1984).
2. A.P. Alivisatos, T.D. Harris, P.J. Carroll, M.L. Steigerwald, and L.E. Brus, *J. Chem. Phys.* **90**, 3463 (1989).
3. S. Schmitt-Rink, D.A.B. Miller, and D.S. Chemla, *Phys. Rev. B* **35**, 8113 (1987).
4. M.G. Bawendi, W.L. Wilson, L. Rothberg, P.J. Carol, T.M. Jedju, M.L. Steigerwald and L.E. Brus, *Phys. Rev. Lett.* **65**, 1623 (1990).
5. A.P. Alivisatos, A.L. Harris, N.J. Levinos, M.L. Steigerwald and L.E. Brus, *J. Chem. Phys.* **89**, 4001 (1988).
6. P. Roussignol, D. Ricard, C. Flytzanis and N. Neuroth, *Phys. Rev. Lett.* **62**, 312 (1989).
7. N. Peyghambarian, B. Fluegel, D. Hulin, A. Migus, M. Joffre, A. Antonetti, S.W. Koch, and M. Lindberg, *IEEE J. Quant. Elect.* **25**, 2516 (1989).
8. R.W. Schoenlein, J.-Y. Bigot, M.T. Portella and C.V. Shank, *Appl. Phys. Lett.* **58**, 801 (1990).
9. Y.J. Yan and S. Mukamel, *J. Chem. Phys.* **94**, 179 (1991).
10. Y.J. Yan and S. Mukamel, *J. Chem. Phys.* **89**, 5160 (1988).
11. S. Nomura and T Kobayashi *Phys. Rev. B* **45** 1305 (1992).

Quantum Beats Spectroscopy of Exciton Spin Dynamics in GaAs Heterostructures

S. Bar-Ad and I. Bar-Joseph

Department of Physics, The Weizmann Institute of Science,
Rehovot 76100, Israel

Abstract. Absorption quantum beats between Zeeman split levels and between excitons and bi-excitons are observed in system of spin oriented excitons in GaAs quantum wells.

The study of the spin dynamics in semiconductor quantum wells (QW) is of growing interest in recent years [1-6]. In this paper we describe a powerful novel technique, absorption quantum beats, used to investigate this process. Controlling the initial spin orientation by the polarization of the pump, and separately varying the polarization of the probe, we observe two kinds of absorption quantum beats: beats between Zeeman-split levels and beats between excitons and bi-excitons. The frequency of these beats corresponds to the small Zeeman splitting and bi-exciton binding energy, respectively. Our experiments are performed at 4K and in magnetic fields up to 7T, in a standard pump-probe configuration with a time resolution of ~ 1 ps, on several narrow QW samples which show evidence of exciton localization.

The induced absorption in a magnetic field of 5T, with *orthogonal linearly polarized* pump and probe, is shown in Fig. 1 for the case of a step-MQW sample, with adjacent layers of 30 Å GaAs and 100 Å $Al_{0.3}Ga_{0.7}As$. In this case the initial spin orientation of the excitons is a coherent superposition of σ^- ($|-3/2, -1/2>$) and σ^+ ($|+3/2, +1/2>$) states. The observed oscillations have a period of 9 ps. Oscillations appear *only* when magnetic field is applied, and their period is inversely proportional to its strength. We interpret these oscillations as beats between the Zeeman split levels [4]. This interpretation is supported by the fact that when parallel polarizations are used the phase of the beats is shifted by π.

In order to observe these quantum beats spin orientation must be conserved for the duration of the oscillations. It follows that spin relaxation for holes, which is believed to be instantaneous in bulk GaAs, is slowed down in QW. To validate this we have performed experiments with *circularly polarized* pump and probe, in which we prepare the system in a spin eigen-state, and probe both the decay of this state and the build-up of population

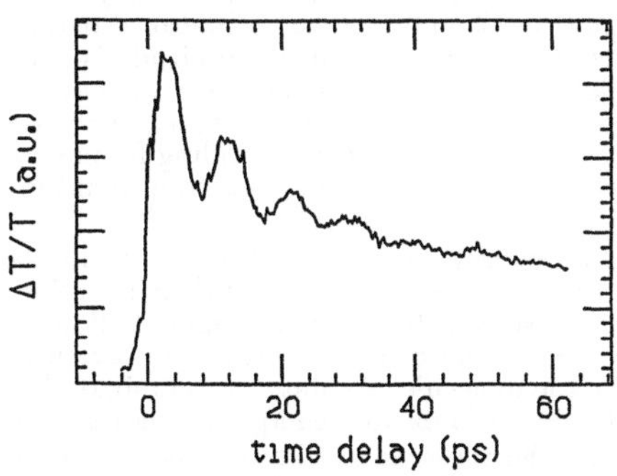

Fig. 1. *The differential absorption signal for linearly polarized pump and probe at 5T, for a step-MQW sample with adjacent layers of 30 Å GaAs and 100 Å* $Al_{0.1}Ga_{0.9}As$ *between* $Al_{0.3}Ga_{0.7}As$ *barriers.*

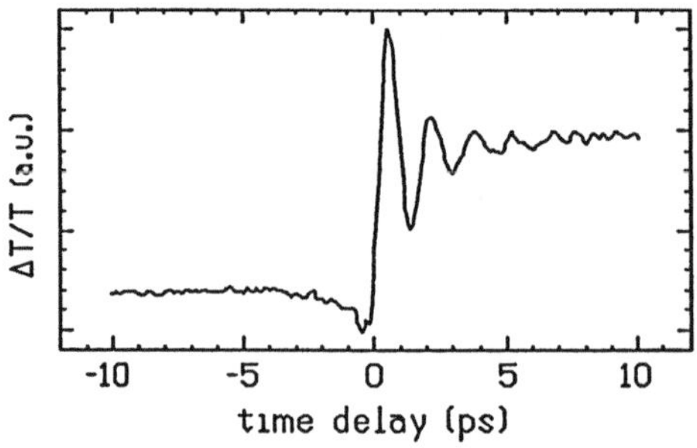

Fig. 2. *The differential absorption signal, measured with a σ^+ pump and a σ^- probe, for a step-MQW sample with adjacent layers of 30 Å GaAs and 100 Å $Al_{0.1}Ga_{0.9}As$ between $Al_{0.3}Ga_{0.7}As$ barriers.*

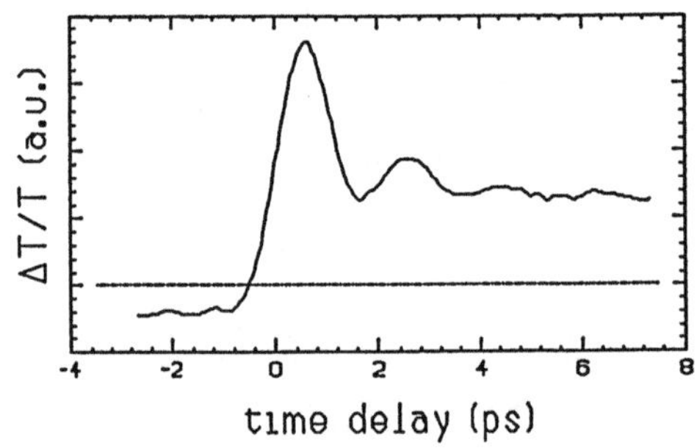

Fig. 3. *The differential absorption signal, measured with a σ^+ pump and a σ^- probe, for a 30 Å GaAs / 30 Å $Al_{0.3}Ga_{0.7}As$ superlattice.*

in the opposite spin state. These experiments confirmed that spin orientation is indeed maintained for the duration of the beats [5]. This observation is in agreement with recent theories, which predict long spin relaxation times for holes in narrow QW [6].

A striking phenomenon is observed when the system is pumped and probed by oppositely handed circular polarizations. An immediate response to the excitation is observed, accompanied by deep damped oscillations. The period of the oscillations is sample dependent, with typical values from 1.6 ps to 2 ps (Fig. 2 and 3). The oscillations' period is independent of both the intensity and the excitation energy. Their amplitude is suppressed with temperature, and they are completely washed out at 60K. As a magnetic field is applied a change of the oscillations' period is observed. The period increases at one polarity of the magnetic field and decreases at the opposite. The corresponding frequency change is within ten percent of the Zeeman splitting, as measured by Zeeman quantum beats (Fig. 1).

These oscillations indicate the existence of an energy gap of 2 - 2.5 meV. While we can rule out explanations related to removal of spin degeneracy and optical nutation, we find strong evidences that the oscillatory response is of bi-excitonic origin. Following excitation by a σ^+ pump, excitons created by a σ^- probe can bind with the previously created σ^+ pump excitons to form a singlet bi-exciton molecule. As states are eliminated at the exciton energy, other states are created at the lower bi-exciton energy, giving rise to a change in the absorption spectrum for a σ^- probe. The observed oscillations, as well as the small dip observed at short negative delays (Fig. 2), can be explained as an interference effect between several terms of the nonlinear optical susceptibility, which couple the two excitonic states through the bi-excitonic state. This implies that oscillations are expected as long as there is partial overlap between the pump and probe pulses, and that the decay of the oscillations reflects the autocorrelation of the pulse rather than a dephasing process. Indeed, the decay time of the beats was strongly influenced by changing the pulse shape.

In conclusion, systems in which spin orientation is maintained over long periods allow the observation of coherent quantum beats between the excited spin states. These in turn allow the measurement of small energy gaps, such as the Zeeman splitting in a magnetic field and the bi-exciton binding energy. Our measurements suggest that the role of bi-excitons in GaAs QW is more important than previously believed.

REFERENCES

1. M. Potemski, J.C. Maan, A. Fasolino, K. Ploog and G. Weimann, Phys. Rev. Lett. **63**, 2409 (1989).
2. M. Kohl, M.R. Freeman, D.D. Awschalom, and J.M. Hong, Phys. Rev. **B 44**, 5923 (1991).
3. T.C. Damen, L. Vina, J.E. Cunningham, Jagdeep Shah and L.J. Sham, Phys. Rev. Lett. **67**, 3432 (1991).
4. S. Bar-Ad and I. Bar-Joseph, Phys. Rev. Lett. **66**, 2491 (1991).
5. S. Bar-Ad and I. Bar-Joseph, Phys. Rev. Lett. **68**, 349 (1992).
6. R. Ferreira and G. Bastard, Phys. Rev. **B 43**, 9687 (1991).

Evidence of Slow Hole Spin Relaxation in n-Modulation Doped GaAs/AlGaAs Quantum Well Structures

Ph. Roussignol[1], *P. Rolland*[1], *R. Ferreira*[1], *C. Delalande*[1], *G. Bastard*[1], *A. Vinattieri*[2], *J. Martinez-Pastor*[2,*], *L. Carraresi*[2], *M. Colocci*[2], *J.F. Palmier*[3], and *B. Etienne*[3]

[1]Laboratoire de Physique de la Matière Condensée de l'Ecole Normale Supérieure, 24, rue Lhomond, F-75005 Paris, France
[2]Dipartimento di Fisica and LENS, Universita' di Firenze, Largo E. Fermi 2, I-50125 Firenze, Italy
[3]Centre de Bagneux, 196 Avenue Henri Ravera, F-92220 Bagneux, France
*On leave from Dept. Fisica Aplicada, Universitat de València, 46100 Burjassot, València, Spain

Abstract. The hole spin relaxation in a n-modulation doped quantum well has been investigated as a function of the electron density N_s. A strong decrease of the initial polarization when increasing N_s from 10^{11} cm^{-2} to 10^{12} cm^{-2} and a slow spin relaxation around 1 ns are found at low temperature.

Increasing interest is being currently addressed to carrier spin relaxation in quantum wells. In fact, on one hand, relaxation and transport phenomena in these systems are strongly affected by the actual processes that relax the carrier spin alignement; on the other, a substantial slowing down in the hole spin relaxation is expected, compared to bulk semiconductors, as a consequence of lifting the degeneracy of the valence band top due to the confinement potential. However, hole spin relaxation times scattered over one order of magnitude and not far from the bulk values have been reported [1,2].
In this work we present CW and time-resolved PL measurements in a n-modulation doped 75Å GaAs/$Al_{.3}Ga_{.7}As$ QW, embedded in a p-i-n diode, as a function of the electron density N_s after excitation with circularly polarized light. Details of the sample growth and geometry have been reported elsewhere [3]. The electron density has been estimated from the Stokes shift between the PL and PLE peaks [4] and varies between 10^{11} cm^{-2} to $1.2\,10^{12}$ cm^{-2}. All measurements have been performed at liquid helium temperature.
The dependence of the polarization $P=(I^+-I^-)/(I^++I^-)$ on the electron density N_s is reported in Fig.1a. A decrease of the polarization with increasing N_s is clearly observed for CW excitation at the fundamental transition CB1-HH1 (•); a similar decrease of the inital polarization P_0 after picosecond excitation is found (♦), while no dependence on N_s has been observed for CW excitation at CB2-HH1 (o). We report in Fig.1b the dependence on N_s of the recombination time and the spin relaxation time as obtained from the time dependence of the quantities $(I^+(t)+I^-(t))$ and $(I^+(t)-I^-(t))$, respectively [3]. We find that τ_R is approximately constant at 300 ps for N_s in the interval 0.6-1.2 10^{12} cm^{-2} and starts decreasing when N_s is decreased below 0.6 10^{12} cm^{-2} towards a value around 100 ps for $N_s = 0.5\,10^{12}$ cm^{-2}. No significant variation of τ_S from the surprisingly long value of 1 ns is instead measured for the hole spin relaxation time. We believe that the decrease of τ_R is very likely due to an increase of the non radiative recombination probability connected with the appearance of a reverse current in the p-i-n diode. More challenging, indeed, is the explanation of the long time measured for τ_S, in striking contrast with the bulk value around 1 ps. In fact, since the hole population, created at $K_{||}\neq 0$, is detected at $K_{||}=0$, the initial polarization can be lost during relaxation towards the band edge by emission of acoustic phonons, or elastic scattering by ionized impurities

Springer Series in Chemical Physics, Vol. 55 Ultrafast Phenomena VIII
Editors: J.-L. Martin · A. Migus · G.A. Mourou · A.H. Zewail

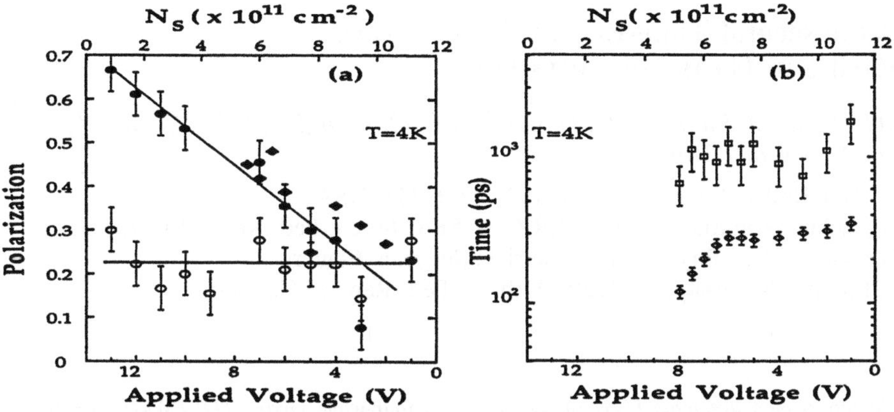

Fig.1: (a) CW polarizations for excitation energies corresponding to CB1-HH1 (•) and CB2-HH1 (o) as a function of the applied voltage; the initial polarization in time-resolved measurements is also reported (♦). (b) Recombination time τ_R(◊) and spin relaxation time τ_S(❑) as a function of the applied bias.

or interface roughness. By a calculation in the first Born approximation, we find that while the emission time of an acoustic phonon is fairly constant with $K_{||}$ for hole-spin conserving processes, very long times (infinite at band edge) are predicted by the theory for $K_{||} \rightarrow 0$ in the case of spin-flipping processes. Note that the same theory when applied to the bulk semiconductors predicts the correct value around 1 ps. We believe that the measured value (1 ns) should be retreived in the framework of a complete theory along these lines with the inclusion of parity breaking terms and band bending potentials neglected at this stage.

References

[1] T.C. Damen, L. Vina, J.E. Cunningham, J. Shah, L.J. Sham, Phys. Rev. Lett. **67**, 3432 (1991)

[2] S. Bar-Ad and I. Bar-Joseph, Phys. Rev. Lett. **68**, 349 (1992)

[3] Ph.Roussignol, P.Rolland, R.Ferreira, C.Delalande, G.Bastard, A.Vinattieri, J.Martinez-Pastor, L.Carraresi, M.Colocci, J.F.Palmier and B.Etienne: submitted to Phys.Rev.Lett. (1992).

[4] C. Delalande, G. Bastard, J. Orgonasi, J.A. Brum, H.W. Liu, and M. Voos, Phys. Rev. Lett. **59**, 2690 (1987)

Femtosecond Time-Resolved Four-Wave Mixing in GaAs Quantum Wells

D.S. Kim[1], *J. Shah*[1], *T.C. Damen*[1], *J.E. Cunningham*[1], *W. Schäfer*[2], *and S. Schmitt-Rink*[3]

[1]AT&T Bell Laboratories, Holmdel, NJ 07733, USA
[2]Forschungszentrum Jülich, HLRZ, W-5170 Jülich, Fed. Rep. of Germany
[3]Department of Physics and Materials Sciences Center, Philipps University, W-3550 Marburg, Fed. Rep. of Germany

We report measurements [1-3] of time-resolved four-wave mixing (TR-FWM) signals in intrinsic and modulation-doped GaAs quantum wells with femtosecond time resolution. These measurements, as a function of excitation and modulation-doping densities, temperature and incident photon energy, show a number of unusual features and provide new insights into the nature of excitonic and continuum states in semiconductors.

I. Introduction.

In this paper we concentrate on *femtosecond time evolution* of the FWM signal in high quality GaAs quantum wells. We use the standard two-beam FWM geometry with two incident beams at the directions of $\mathbf{k}_1$ and $\mathbf{k}_2$. We have investigated three different quantum well samples: sample A, a high quality intrinsic quantum well sample with an exciton FWHM of only 0.8 meV when T_2=1.5 ps (i.e. nearly homogeneously broadened exciton), sample B, a modulation-doped sample with a doping density of 2×10^{11} cm^{-2} (Fermi energy 7 meV), and sample C, with a doping density of 6×10^{11} cm^{-2} (Fermi energy 20 meV). We up-convert the FWM signal along $\mathbf{k}_d = 2\mathbf{k}_2 - \mathbf{k}_1$ with a suitably delayed third beam in a nonlinear crystal to obtain the necessary time resolution. A passively mode-locked Ti:Sapphire laser with tunable 100 fs pulses was used for these experiments. We have made extensive measurements as a function of excitation density, temperature etc. but we discuss here mostly the results at 5 K with an excitation density of 1×10^{10} cm^{-2}. We consider time delay T between the two pulses 1 and 2 as positive when pulse 1 precedes pulse 2; time t, which is the "real time", is measured from pulse 2.

II. Intrinsic Sample: Dominance of Exciton Interactions in coherent regime

The TR-FWM signal from the intrinsic sample A (Fig. 1) is the most interesting. It shows an unusually slow rise of the signal with the peak at t > 1000 fs, ten times the pulsewidth. Furthermore, the temperature dependence at fixed T (Fig. 2) shows that the peak becomes narrower and shifts to shorter t with increasing temperature. We conclude from this observation that the position of the peak is related to the dephasing time T_2.

The generally accepted independent two-level model predicts that the peak should occur within 1 or 2 pulsewidths of the second pulse. This model is clearly completely inadequate to explain our results. It has been shown recently that interaction between excitons leads to the diffraction of polarization, in addition to the usual diffraction of fields, which leads to the observation of the time-integrated FWM signal at negative time delays [4]. However, the strength of the interaction-induced (I-I) signal could not be determined.

Springer Series in Chemical Physics, Vol. 55 Ultrafast Phenomena VIII
Editors: J.-L. Martin · A. Migus · G.A. Mourou · A.H. Zewail

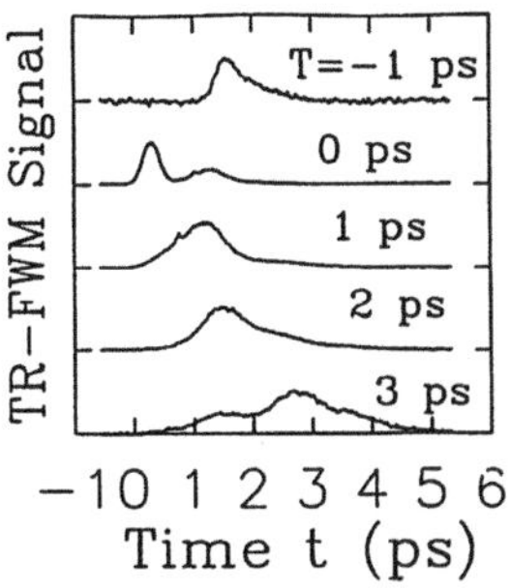

Fig. 1. TR-FWM signal at various time delays T with linear co-polarized beams at 10 K for sample A.

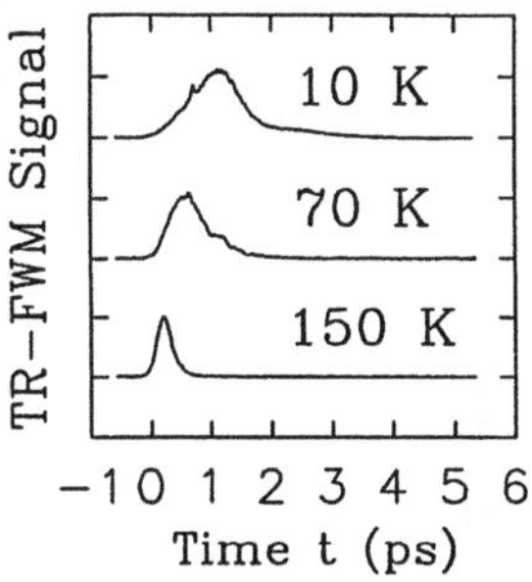

Fig. 2 Temperature dependence of TR-FWM for linear copolarized beams at T=1 ps for sample A.

Our measurements show that the generally-neglected I-I effects in fact dominate the FWM signal. The observations that the TR-FWM signal peaks at approximately T_2, and that there is an intrinsic photon-echo-like behavior, are in agreement with recent theories [5]. The experiments also show interesting polarization dependence which will be discussed elsewhere [2].

III. Heavily Modulation-doped Sample: A case of intrinsic inhomogeneous broadening in momentum space

The heavily-doped Sample C shows the simplest behavior. Time-integrated FWM signals provide the first **direct** measurement of scattering rates in a degenerate electron system, and show a strong inhibition of the carrier-carrier scattering rates near the Fermi edge [1]. The results agree with the predictions of Landau Fermi liquid theory and provide the first direct measurements of the electron and hole scattering times in a semiconductor. TR-FWM signals with circularly co-polarized beams (Fig. 3) show a clear photon echo and no other features, demonstrating the inhomogeneous character of this **intrinsic** system in momentum space.

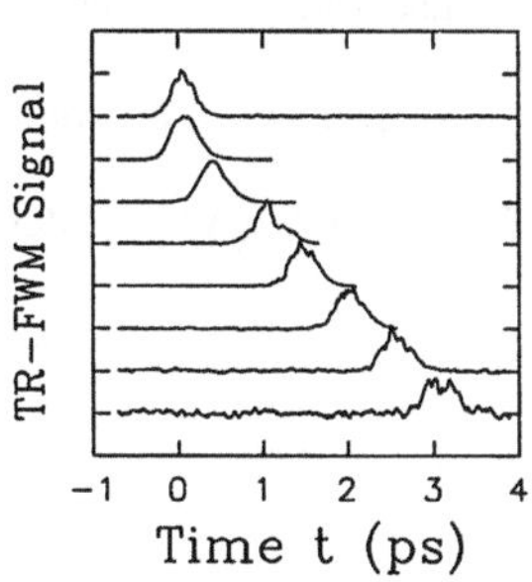

Fig. 3. TR-FWM signal at 10 K for co-circular polarized beams at varopis T for sample C. T is, from top to bottom, -0.2, 0, 0.5, 1, 1.5, 2, 2.5, and 3 ps.

IV. Lightly Modulation-doped Sample: Persistence of some excitonic features

Comparison of time-integrated FWM results from sample B and C allow us to verify the scaling behavior expected for degenerate Fermi liquids [1]. The TR-FWM signal with co-

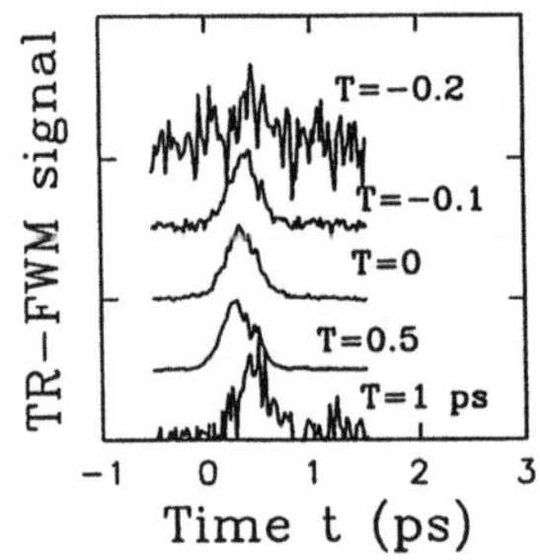

Fig. 4. TR-FWM signal with cross-circular polarized beams for sample B at various time delays at 10 K.

circularly polarized beams shows the photon echo which broadens with increasing T. This is because at larger time delays, only those states very close to Fermi energy, with long dephasing times, can contribute to the photon echo. Equally interesting is the observation of only a **prompt** TR-FWM signal in the cross-polarized case (Fig. 4). This is an indication that intrinsic exciton-like effects persist at this relatively low doping concentration.

V. Conclusions.

In conclusion, we have shown that investigating the time-evolution of FWM signal is a very powerful tool in understanding intrinsic natures of excitons and free electron-hole pairs in semiconductors.

References.

[1] D. S. Kim, J. Shah, J. E. Cunningham, T. C. Damen, S. Schmitt-Rink, and Schäfer, Phys. Rev. Lett. **68**, 2838, 1992.

[2] D. S. Kim, J. Shah, T. C. Damen, W. Schäfer, F. Jahnke, S. Schmitt-Rink, and K. Köhler, submitted to Phys. Rev. Lett., 1992.

[3] A femtosecond TR-FWM experiment on high density excitons at room temperature has been reported: M. -A Mycek, S. Weiss, J. -Y. Bigot, S. Schmitt-Rink, and D. S. Chemla, Appl. Phys. Lett., **60**, 2666, 1992.

[4] K. Leo, M. Wegener, J. Shah, D. S. Chemla, E. O. Göbel, T. C. Damen, S. Schmitt-Rink, and W. Schäfer, Phys. Rev. Lett., **65**, 1340.

[5] M. Lindberg, R. Binder, and S. W. Koch, Phys. Rev. B, **45**, 1865, 1992; W. Schäfer and S. Schmitt-Rink, unpublished.

Exciton Radiative Lifetimes in GaAs Quantum Wells

R. Eccleston, J. Kuhl, W.W. Rühle, and K. Ploog

Max-Planck-Institut für Festkörperforschung, Heisenbergstr. 1,
W-7000 Stuttgart 80, Fed. Rep. of Germany

Abstract Complementary time-resolved photoluminescence (TRPL) and dephasing measurements support the theoretically predicted relationship between the exciton radiative lifetime and exciton homogeneous linewidth.

1. Introduction

The radiative lifetime τ_x of delocalized 2D excitons in quantum wells has been a subject of wide theoretical and experimental interest over recent years [1][2]. A macroscopic theory has proposed a strong link between τ_x and the homogeneous linewidth, Γ_h [1]:

$$\tau_x = C\frac{\Gamma_h}{(1-\exp(-\Gamma_h/kT_x))} \tag{1}$$

The term in the denominator increases τ_x at finite exciton temperature, T_x, by taking into account only the fraction of the exciton distribution within one homogeneous linewidth of zero centre-of-mass kinetic energy (and therefore momentum-matched for recombination). The validity of this thermal average is well-accepted experimentally [3]. The proportionality of τ_x to Γ_h in the numerator arises from limited spatial coherence of the ideal $\mathbf{K}_{||}$ states describing the exciton centre of mass motion. In a real crystal this coherence is limited by impurity, defect, carrier or phonon scattering which occurs on a faster timescale than the exciton radiative decay: thus a centre-of-mass motion term in the exciton wavefunction of well-defined $\mathbf{K}_{||}$, extending coherently throughout the quantum well layer never actually exists. This $\mathbf{K}_{||}$ broadening, (which gives Γ_h=0.2-1meV, typically) efficiently mixes radiatively inactive higher $\mathbf{K}_{||}$ states into the small (≈0.1meV) range of radiatively active $\mathbf{K}_{||} \approx 0$ exciton states below the photon branch [2], leading to an increase in τ_x as Γ_h increases.

This picture, that scattering leads to a mixed $\mathbf{K}_{||}$ exciton state which then interacts with the photon field, has however escaped clear experimental verification due to the difficulty of comparative τ_x and Γ_h measurements on delocalized excitons [3]. In this paper we measure the radiative lifetime of resonantly excited delocalized excitons in a very narrow (0.3meV) linewidth [4] 27nm GaAs/$Al_{0.3}Ga_{0.7}As$ single quantum well and correlate this with the increase in Γ_h with incoherent exciton density obtained by degenerate four-wave-mixing (DFWM). According to Equation (1) a correlated decrease in the instantaneous radiative recombination rate with increasing exciton density is expected.

2. Experimental Results

Γ_h was investigated using DFWM at 6K in the 'reflection' self-diffraction configuration (Fig. 1 inset). The exciton density dependence was obtained by pre-injection of excitons at various densities via a third pulse (beam 3) timed to arrive 60ps before the grating pulses. Because 60ps> T_2, we obtain the dephasing efficiency of a predominantly incoherent exciton ensemble, the situation which is relevant to the TRPL measurements. Fig. 1 shows the values of Γ_h obtained as a function of total exciton density, n_x. TRPL measurements using a streak camera were performed at 1.8K. The sample was excited by 5ps pulses from a dye laser tuned to resonance with the heavy-hole exciton line. The rapidly decaying signal around t=0 is due to elastically scattered laser light. At low exciton density (Fig.2(e)), an exponential decay of the TRPL signal

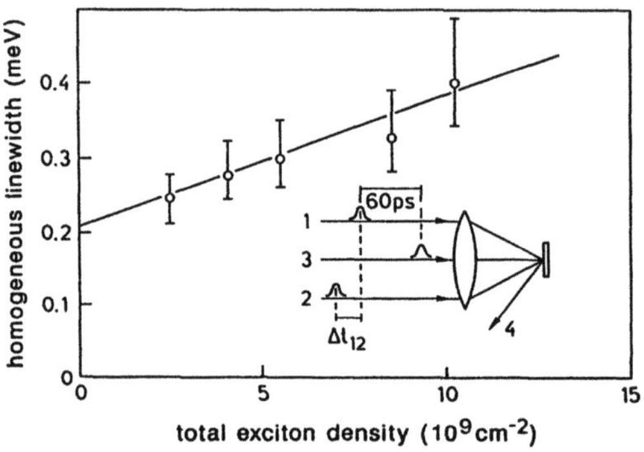

Fig. 1. Γ_h, vs. total exciton density, showing a linear dependence: $\Gamma(n_x) = 0.226 + 0.154 \times 10^{-10} n_x$. Inset: DFWM configuration.

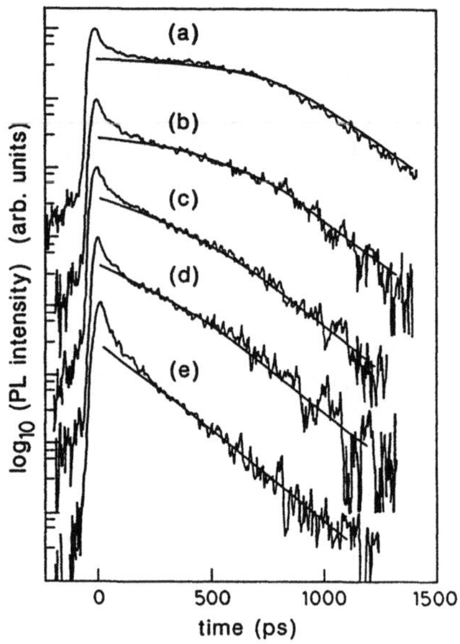

Fig. 2. PL intensity vs. time. (a) $n_x = 8 \times 10^{10} cm^{-2}$, (b) $n_x = 5.2 \times 10^{10} cm^{-2}$, (c) $n_x = 2.5 \times 10^{10} cm^{-2}$, (d) $n_x = 1.5 \times 10^{10} cm^{-2}$, (e) $n_x = 1.8 \times 10^{9} cm^{-2}$

with a time constant of 170ps is obtained. However, as the density is raised (Figs. 2(a)-(d)), a progressively slower non-exponential decay is observed with the 170ps exponential decay only recovered at later times when the exciton population has predominantly decayed away. Over this density regime we observe no change in time-integrated PL efficiency. The increase in initial PL lifetime with n_x therefore cannot be related to nonradiative saturation effects. We conclude that the instantaneous radiative recombination rate is reduced by increasing exciton-exciton scattering as implied by Equation (1) and the $\Gamma_h(n_x)$ data of Fig.1. This conclusion is supported by the recovery of the low density fast radiative decay in the high density measurements at later time when the majority of the exciton population has decayed away.

We have used (1) to model the TRPL profiles using the extrapolated n_x dependence of Γ_h shown in Fig. 1. The constant C is adjusted to set the time constant of the calculated decay for $n_x \rightarrow 0$ to the measured value of 170ps. Good quantitative agreement with the experimental data is obtained. The results clearly show the dependence of the exciton radiative lifetime on the homogeneous linewidth of the exciton state. This effect represents a breakdown of the familiar assumption of calculating each interaction rate from a single interaction Hamiltonian and well-defined $\Psi^{\mathbf{K}_\parallel}$ extracted from the entire exciton Hamiltonian for a real crystal. Interactions with phonons, impurities and other excitons always lead to a broadening of exciton $\mathbf{K}_\parallel$ states. But since these interactions are only weak functions of $\mathbf{K}_\parallel$, each individual interaction rate is basically unchanged by the presence of the others. The above assumption remains valid. Radiative decay is an exception because the critical range of single $\mathbf{K}_\parallel$ exciton states which may couple to the photon field is extremely small [2], and is almost always exceeded by $\mathbf{K}_\parallel$ broadening.

3. Conclusions

We have validated experimentally the fundamental relationship between exciton radiative lifetime and exciton homogeneous linewidth in quantum wells, and shown that time- and density-dependent exciton-exciton scattering can significantly influence exciton recombination dynamics in high-quality quantum well structures. Measured lifetimes in good quality samples are therefore likely to be strongly influenced by exciton density, as well as temperature and sample parameters.

[1] J.Feldmann, G.Peter, E.O.Göbel, P.Dawson, K.Moore, C.Foxon, and R.J.Elliot, Phys. Rev. Lett. **59**, 2337 (1987).

[2] E. Hanamura, Phys. Rev. **B 38**, 1228, (1988), L.Andreani and F.Bassani, Solid State Comm. **B 38**, 1228, (1988)

[3] See: R.Eccleston, B.F.Feuerbacher, J.Kuhl, W.W.Rühle, and K.Ploog, Phys. Rev. **B 45**, 11403 (1992), for a more detailed discussion of this, and other aspects of this work.

[4] K.Ploog, A.Fischer, L.Tapfer, and B.F.Feuerbacher, Appl. Phys. **A 52**, 135 (1991).

Optical Investigation of Bloch Oscillations in a Semiconductor Superlattice

J. Feldmann[1,3], *K. Leo*[2,3], *J. Shah*[3], *D.A.B. Miller*[3], *J.E. Cunningham*[3], *T. Meier*[1], *G. von Plessen*[1], *P. Thomas*[1], *and S. Schmitt-Rink*[1]

[1]Department of Physics and Materials Sciences Center, Philipps-University, W-3550 Marburg, Fed. Rep. of Germany
[2]Institute of Semiconductor Electronics, Techn. University, W-5100 Aachen, Fed. Rep. of Germany
[3]AT&T Bell Laboratories, Holmdel, NJ 07733, USA

Abstract: We report transient four-wave mixing experiments on a GaAs/AlGaAs superlattice in the presence of an electric field. The diffracted non-linear signal exhibits pronounced modulations with a time period varying linearly with the electric field. These modulations are a consequence of Bloch oscillations in the mini Brillouin-zone of the superlattice structure.

1 Introduction

The optical excitation of a multi-level system by short laser pulses allows to construct an electronic wavepacket by the coherent superposition of several eigen-states of the system. Various time resolved methods of coherent optical spectroscopy allow to observe the temporal dynamics of such a wavepacket in atomic systems [1] as well as in semiconductor heterostructures [2]. The work by Leo et al. [2] reported the time resolved detection of the oscillatory motion of an electronic wavepacket within an asymmetric double quantum well.

In this paper, we study the coherent dynamics of optical excitations in electrically biased GaAs/AlGaAs superlattices by performing sub-picosecond four-wave mixing experiments. In such a superlattice, the quantum-mechanical coupling of adjacent electronic quantum well states leads to the formation of so-called minibands, when no electric field is applied. However, the application of an electric field $\vec{F}$ parallel to the growth direction replaces the continuous spectrum of miniband states by a series of equally spaced levels called the Wannier-Stark (WS) ladder [3]. As schematically illustrated in the bandscheme of Fig.1, the equal level spacing amounts to

$$\Delta E = eFd \tag{1}$$

and gives rise to a series of optical WS ladder transitions labelled $S_\nu(\nu = 0, \pm1, \pm2, ...)$. Increasing the electric field F leads to an increasing (Stark) localization of the WS wavefunction and thus diminishes the oscillator strengths of the 'oblique' inter-well transitions ($\nu \neq 0$).

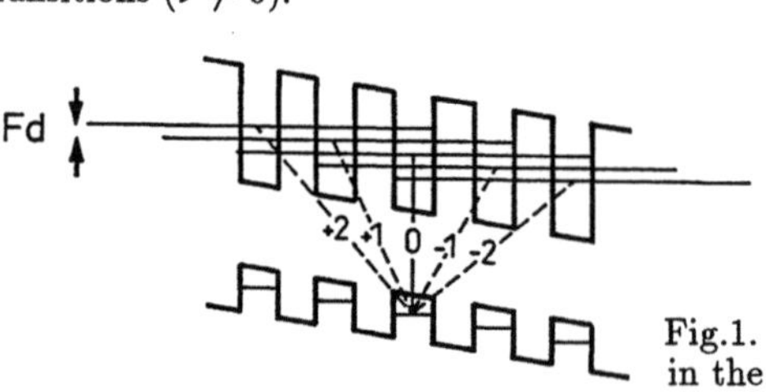

Fig.1. Schematic bandscheme of a superlattice in the presence of an electric field $\vec{F}$.

Springer Series in Chemical Physics, Vol. 55 **Ultrafast Phenomena VIII**
Editors: J.-L. Martin · A. Migus · G.A. Mourou · A.H. Zewail

An alternative description in $\vec{k}$-space relies on the so-called acceleration theorem, which states that the wavevector $\vec{q}$ of a superlattice Bloch electron varies linearly in time:

$$\hbar\dot{\vec{q}} = e\vec{F} \tag{2}$$

However, when the electron reaches the mini Brillouin-zone (BZ) boundary it undergoes a Bragg-reflection, i.e., an 'umklapp' process occurs. As a consequence, the electron periodically traverses the BZ with a time period

$$T_B = \frac{h}{eFd}, \tag{3}$$

where d is the superlattice period.

A theoretical study by von Plessen and Thomas [4] shows that a transient FWM experiment performed on a semiconductor superlattice should be capable of measuring Bloch oscillations directly in the time domain. Their conclusion relies on the fact that, in $\vec{k}$-space, the transition frequencies of the interband two-level systems evolve in time according to the acceleration theorem given by Eq.(2) and thus reflect Bloch oscillations [4, 5].

2 Experimental

As shown in Fig.2, the two-beam FWM experiment is performed on a p-i-n diode with a superlattice structure located in the intrinsic region. The superlattice consists of 91 periods of 95 Å GaAs and 15 Å $Al_{0.3}Ga_{0.7}As$. We use a tandem synchronously pumped dye laser providing pulses of about 500 fs and, alternatively, a mode-locked Ti-sapphire laser emitting pulses of about 110 fs for the low temperature dephasing experiments. The non-linear FWM signal emitted into the phase-matching direction $2\vec{k}_2 - \vec{k}_1$ is then measured time integrated as a function of the time delay τ between the two pump pulses.

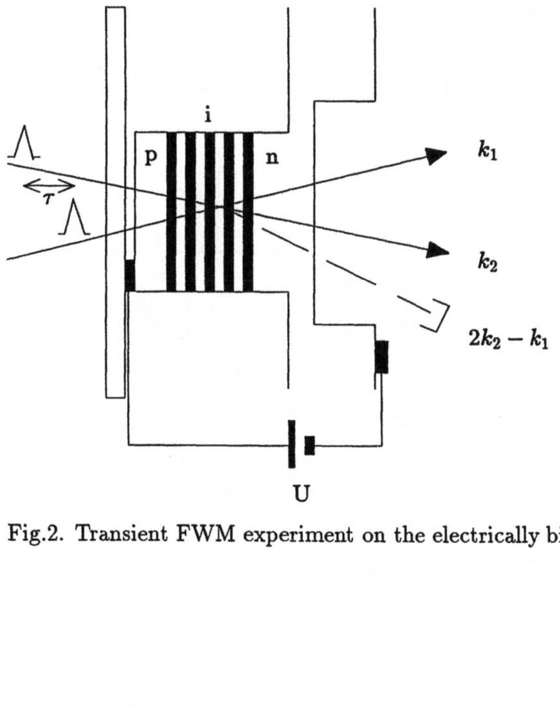

Fig.2. Transient FWM experiment on the electrically biased superlattice.

3 Results

In order to characterize the superlattice sample we first measure the photocurrent spectrum as a function of the applied voltage and of the laser wavelength. The result is shown in Fig.3. The spectrum is taken using the pulsed dye laser assuring that the same conditions hold as in the corresponding FWM experiments. This is important for quantitative comparisons between spectral and temporal features, since screening of the electric field due to photoinduced space charges is unavoidable in the FWM experiments. In Fig.3, WS ladder transitions labelled S_ν are seen at intermediate electric fields. The observed peak positions of the oblique inter-well transitions are drawn as circles in the fan-chart of Fig.4b.

In Fig.4a, the temporal evolution of the diffracted signal is shown for several applied voltages at forward bias. The central photon energy of the dye laser pulses is located in the spectral regime of the oblique WS transitions as indicated by the vertical arrow in Fig.4b. At low electric field (0.95 Volt), the signal decay is approximately exponential. However, for higher fields (0.7 - 0.4 Volt) the signal shows pronounced modulations. The

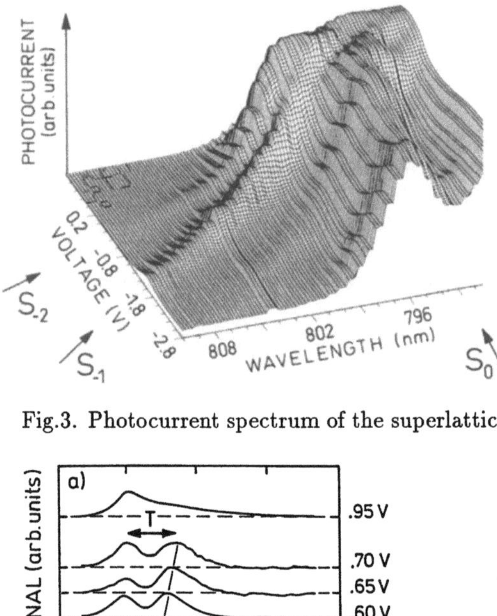

Fig.3. Photocurrent spectrum of the superlattice sample taken with the pulsed dye laser.

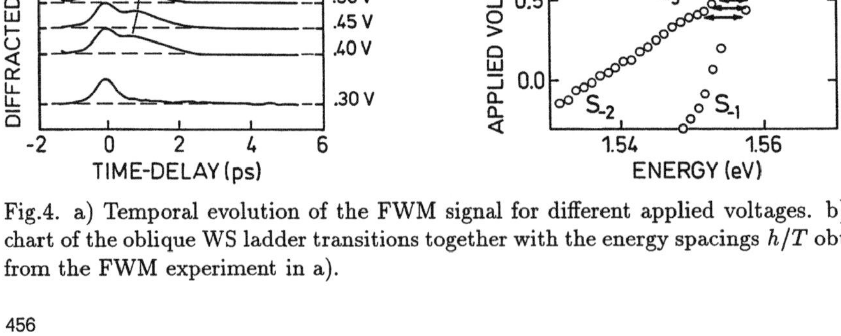

Fig.4. a) Temporal evolution of the FWM signal for different applied voltages. b) Fan-chart of the oblique WS ladder transitions together with the energy spacings h/T obtained from the FWM experiment in a).

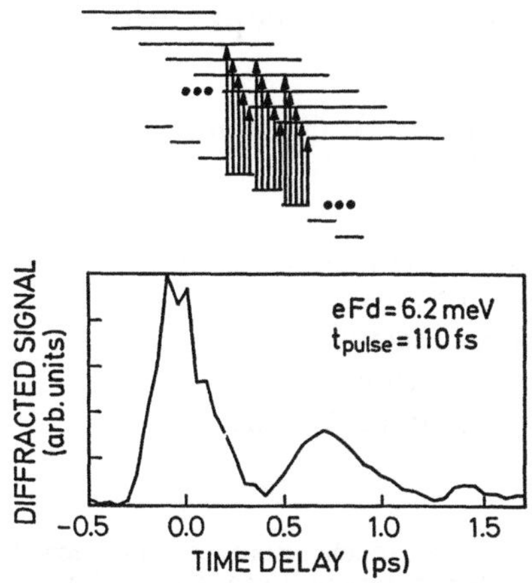

Fig.5. FWM curve using 110 fs laser pulses. The upper part illustrates the simultaneous coherent excitation of various WS ladder transitions.

time duration between the observed peaks, T, decreases with increasing electric field. The horizontal arrows in Fig.4b indicate the energy splitting $\Delta E = h/T$ calculated from the period T of the signal modulations seen in Fig.4a. We find reasonable agreement with the energy spacing between the S_{-1} and S_{-2} transitions.

The result of a transient FWM experiment using 110 fs pulses of the mode-locked Ti-sapphire laser is shown in Fig.5. As indicated by the arrows in the upper part of Fig.5, the spectrum of the pulse now encompasses about five WS transitions. For the experimentally determined WS ladder spacing of about 6.2 meV, which is comparable to the situation at 0.4 V in Fig.4, the FWM curve now exhibits three peaks for time delays of 0 ps, 0.7 ps and 1.4 ps. With $eFd = 6.2$ meV, the time period T_B expected for Bloch oscillations amounts to 0.67 ps (Eq.3), again in good agreement with the observed oscillation period of $T = 0.7$ ps.

On the basis of the theoretical results obtained in Ref.[4], we can definitely conclude that the modulations of the FWM curves shown in Fig.4 and 5 are associated with Bloch oscillations in the superlattice mini Brillouin-zone.

References

[1] A. ten Wolde, L.D. Noordam, A. Lagendijk, and H.B. van Linden, Phys. Rev. Lett. **61**, 2099 (1988).

[2] K. Leo, J. Shah, E.O. Göbel, T.C. Damen, S. Schmitt-Rink, W. Schäfer, and K. Köhler, Phys. Rev. Lett. **66**, 201 (1991).

[3] E.E. Mendez, F. Agullo–Rueda and J.M. Hong, Phys. Rev. Lett. **60**, 2426 (1988).

[4] G. von Plessen and P. Thomas, Phys. Rev. **B45**, 9185 (1992).

[5] J. Feldmann, in: Festkörperprobleme/Advances in Solid State Physics, Vol. 32, ed. by U. Rössler, (Vieweg, Braunschweig 1992).

Coherent Pulse Breakup in Femtosecond Pulse Propagation in Semiconductors

P.A. Harten, A. Knorr, S.G. Lee, R. Jin, F. Brown de Colstoun, E.M. Wright**, G. Khitrova, H.M. Gibbs, S.W. Koch**, and N. Peyghambarian*

Optical Sciences Center, University of Arizona, Tucson, AZ 85721, USA
*Also with Friedrich-Schiller-University, Faculty of Physics and Astronomy, Institute of Optics and Quantum Electronics, Max-Wien-Platz 1, O-6900 Jena, Fed. Rep. of Germany
**Also with the Physics Department of the University of Arizona

Abstract. Coherent pulse breakup is observed for below-exciton resonance femtosecond excitation in a room-temperature GaAs/AlGaAs multiple quantum well waveguide using the cross-correlation technique. Calculations based on the coupled semiconductor Maxwell-Bloch equations show that coherent self-phase modulation during propagation drives the system out of the initial adiabatic following regime into oscillatory excitation density, and eventually, pulse shape modulations.

Coherent pulse propagation and pulse breakup in atomic media [1] have been extensively studied. The long dephasing time in atomic media facilitates the observation of these phenomena using relatively long pulses. However, such coherent processes have only recently been considered for semiconductors where the phase relaxation times are very rapid and the band structure often prevents the study of a single isolated electronic transition. Here we report on our experimental and theoretical investigations showing coherent pulse breakup in a semiconductor waveguide with photon energies below the 1s exciton. In this spectral region, the optical Stark effect should produce an ultrafast change in the nonlinear refractive index with a response time limited only by the polarization dephasing time. For the femtosecond experiments reported here, however, the input pulses are shorter than the polarization dephasing time so that the coherent nature of the light-semiconductor interaction becomes of paramount importance. Coherent effects are shown to be responsible for the observed pulse breakup of large area femtosecond pulses by comparison with calculations based on the coupled semiconductor Maxwell-Bloch equations. We show that in spite of the fact that the input pulse satisfies the adiabatic following condition [2], upon propagation in the semiconductor, an intense input pulse can develop oscillations which violate the adiabatic approximation, and pulse breakup follows for high-intensity, large area pulses, while no such breakup is observed for low-intensity, small area pulses.

The experiments were performed using the cross-correlation technique. Figures 1(a) and 1(b) show the experimentally measured cross correlations of the transmitted pulses for both low and high intensity, respectively, in the 0.37-mm-long waveguide. The peak intensities were 8.5 GW/cm^2 (Fig. 1(b)) and 0.8 GW/cm^2 (Fig. 1(a)). The transmitted low-intensity pulse in Fig. 1(a) shows no significant distortion compared with the input pulse. In contrast, pulse breakup is clearly seen in Fig. 1(b) for the high-intensity pulse. The main peak of the transmitted pulse has a secondary peak on the trailing edge of the pulse.

The experiments were analyzed using theoretical calculations that rely on the semiconductor Bloch equations [2] to model the medium, resulting in a nonlinear polarization that was then inserted into the appropriate wave equation. The two band semiconductor Bloch equations in the low density limit and the wave equation for the electro-magnetic field were solved self-consistently for each small slice of material, one after the other, thus, propagating through the waveguide. The simulations took into account temporal asymmetry of the input pulse, group velocity, and group velocity dispersion. Figures 1(c) and 1(d) show the calculated cross-correlations of the transmitted pulses for comparison with the experimental results of Figs. 1(a) and 1(b). Good agreement between experiment and theory is obtained, with the pulse breakup clearly evident for the 7.5-π pulse in Fig. 1(d).

To obtain physical insight into what is responsible for the observed pulse breakup, we have performed an extensive series of diagnostic tests involving the calculated excitation density

Springer Series in Chemical Physics, Vol. 55 **Ultrafast Phenomena VIII**
Editors: J.-L. Martin · A. Migus · G.A. Mourou · A.H. Zewail

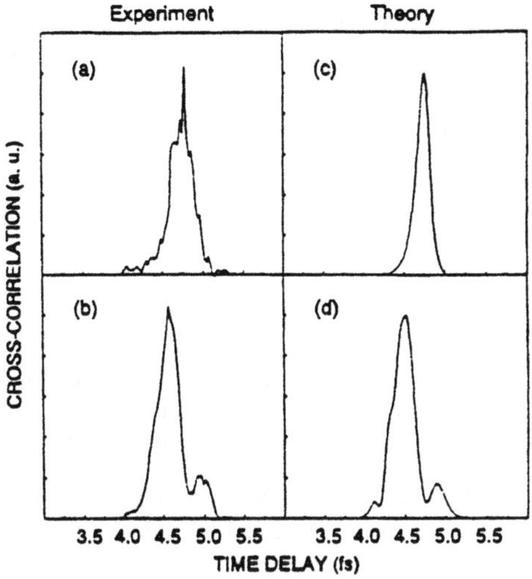

Fig. 1. (a) and (b) are measured cross correlations of transmitted low-intensity and high-intensity femtosecond pulses, respectively. (c) and (d) are the corresponding calculated cross-correlation pulses with areas of 2π and 7.5π. The time delay zero point is defined by the peak signal without waveguide.

and the pulse shape at different locations within the waveguide. [3] We found that at the input face of the sample, the excitation density follows the field without showing any oscillatory structure as expected in the adiabatic following regime. Therefore, the pulse breakup can not be simply ascribed to off-resonance Rabi oscillations in the excitation density, as found in self-induced transparency. However, the calculations show that oscillations in the excitation density precede the pulse breakup, and we therefore trace the source of the pulse breakup to the excitation density oscillations.

Our further analysis showed that for the coherent case, when the pulse duration is less than the dephasing time, the propagating field is sufficiently frequency chirped by self-phase modulation, the phase of the electric field changes rapidly enough to escape from the adiabatic following regime and produce interference terms that drive oscillations in the excitation density, which in turn cause pulse breakup.

In conclusion, we have observed below-resonance, coherent pulse breakup in a room-temperature semiconductor waveguide for the first time. Simulations of the semiconductor Maxwell-Bloch equations identify the effect as propagation-induced escape from adiabatic following. This coherent effect, which is distinctly different from self-induced transparency, a temporal soliton, or self-steepening, should be ubiquitous for pulse propagation in the transparency region of semiconductors with pulse durations less than the polarization dephasing time.

The authors would like to acknowledge support from ARO, NSF, JSOP, NATO travel grant, the Optical Circuitry Cooperative of the University of Arizona, and NEDO. One of us, A. K., thanks the Studienstiftung des deutschen Volkes, Germany, for support.

References

1. S. L. McCall and E. L. Hahn, Phys. Rev. Lett. **18**, 908 (1967); Phys. Rev. **183**, 457 (1969); see also R. E. Slusher and H. M. Gibbs, Phys. Rev. A **5**, 1634 (1972).
2. R. Binder, S. W. Koch, M. Lindberg, N. Peyghambarian, W. Schäfer, and F. Jahnke, Phys. Rev. Lett. **65**, 899 (1990).
3. P. Harten et al., submitted for publication.

Absorption Saturation of the Urbach's Tail in Multiple Quantum Wells

R. Raj, B.G. Sfez, D. Pellat, and J.L. Oudar

Centre National d'Etudes des Télécommunications,
196, Avenue Henri Ravera, F-92220 Bagneux, France

Pico-second pump and probe measurement of the nonlinear transmission identifies saturation of the photo-excited carrier density in the Urbach's tail as the origin of the saturating behaviour of the nonlinear index δn_{NL}observed in the band tail of GaAs/GaAlAs Multiple Quantum Wells

It has been dicovered recently that the dispersive nonlinearity of MQW exhibits a strong saturating behavior [1], which constiutes a serious limitation in the design of optical bistable etalons.

This saturation can be discussed in terms of the dependence of the nonlinear index δn_{NL} on the carrier density N and N's dependence on the incident light intensity I (i.e. $\delta n_{NL}(I) = \delta n_{NL}\{N(I)\}$). One possible mechanism, arising from the $\delta n_{NL}(N)$ is the saturation of the excitonic and band-filling contribution. As the limiting values thus obtained are incompatible with the measured values the N(I) dependence is a more likely candidate.

A deeper analysis[2] shows that δn_{NL} saturates when either the absorption $\alpha(N)$ or the carrier lifetime $\tau(N)$ vanishes for a given N. Although τ decreases with increasing N, due to bimolecular or Auger recombination, it cannot vanish for finite values of N.

Therefore the most probable mechanism is the saturation of absorption. In order to independantly check this hypothesis, we performed pico-second pump and probe experiments which allow measurements independant of the carrier lifetime. As the bistable operation occurs in a region of low absorption, we made use of excitonic saturation as a sensitive measure of the photo-excited carrier density,to measure directly the nonlinearity of band-tail absorption similar to Von Lehmen et al's[3] measurement of linear absorption in the QW band-tail. We extended the measurements to the nonlinear regime

The sample studied consists of AlAs/$Al_{0.1}Ga_{0.9}Al$ quater-wave thick layers as the back mirror and a 130 period GaAs/$Al_{0.3}Ga_{0.7}As$ MQW active layers with 10nm/10nm nominal thickness . It is anti-reflection coated to prevent Fabry-Perot resonance and to allow double-pass transmission through the MQW's .The pump-probe experiments were performed at room temperature with two independently tunable beams from Nd:YAG pumped $LiIO_3$ optical parametric generators giving 15 psec, 1nm and 1μJ light pulses. The probe energy was low enough to avoid saturating the exciton absorption. The pump wavelength was scanned from band-to-band (834nm) through the exciton (842nm) up to the transparency region.

A series of nonlinear transmission spectra were recorded for pump intensities ranging from 0.05 mJ/cm^2 to 42 mJ/cm^2. In Figure1 the normalised differential nonlinear transmission $\Delta T/T = \frac{T(I_{pump})-T(0)}{T(0)}$ is plotted against excitation intensity. We fitted the data with a saturation model $\frac{\Delta T}{T} = \delta T_{limit}\frac{I/Is}{1+I/I_s}$ where I_s is the saturation intensity, δT_{limit} the limiting value of the normalised differential nonlinear transmission reached asymptotically when the sample becomes trans-

Springer Series in Chemical Physics, Vol. 55 Ultrafast Phenomena VIII
Editors: J.-L. Martin · A. Migus · G.A. Mourou · A.H. Zewail

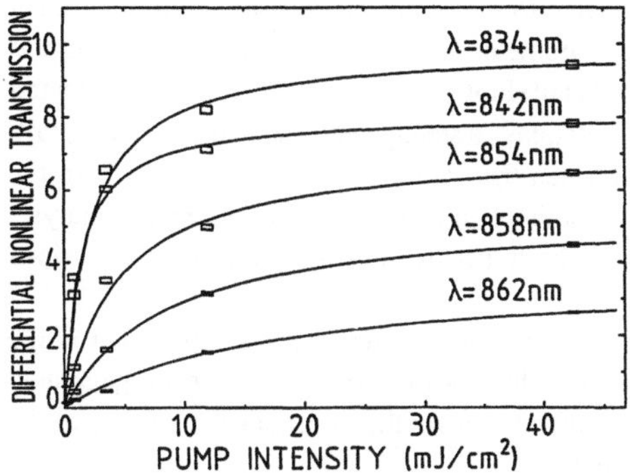

Fig.1 Differential nonlinear transmission for five chosen pump wavelengths in the band (834nm), on the exciton (842nm), in the Urbach's tail (854nm,858nm and 862nm), as a function of the pump intensity. The continuous lines are the fit using a saturation model described in the text.

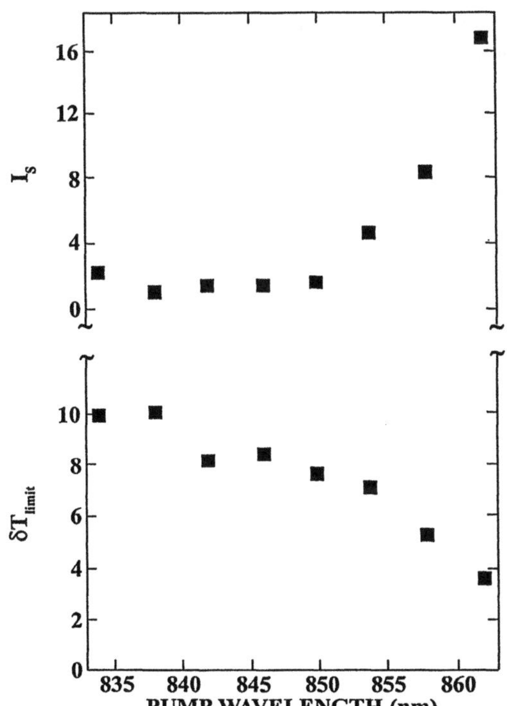

Fig.2 Spectral variation of the saturating intensity I_s (top) and of δT_{limit} (bottom) deduced from the fitted experimental points.

parent at the pump wavelength. It is directly related to the N_t carrier density required to reach transparency at the excitation wavelength. The parameter I_s decreases slowly from 834nm to 842nm then increases very rapidly in the Urbach's tail (Figure 2), consistent with the decreasing pump power absorption. More strikingly, δT_{limit} displays a steady decrease in the Urbach's tail. This unambiguously shows that N_t is smaller when the carriers are photo-excited from the Urbach's tail than from the excitonic and band-to-band spectral regions. A more detailed analysis allows us to obtain a quantitative evaluation of the transparency carrier densities at various wavelengths.

In conclusion this specific measurement of absorption saturation shows that it is the mechanism which effectively limits the maximum density of carriers that can be generated through Urbach's tail absorption and identifies unambigously this mechanism as being at the origin of the saturation of nonlinear refractive index.

REFERENCES

[1] B.G. Sfez, J.L. Oudar, J.C. Michel, R. Kuszelewicz, and R. Azoulay, Appl. Phy. Lett., 1, 324 (1990)

[2] B.G. Sfez, J.L. Oudar, R. Kuszelewicz and D. Pellat, submitted to Appl. Phys. Lett.

[3] A. Von Lehmen, J.E. Zucker, J.P. Heritage, D.S. Chemla and A.C. Gossard, Appl. Phys. Lett.,48, 1479 (1986).

Photon Echo Polarisation Rules in GaAs Quantum Wells

R. Eccleston[1], D. Bennhardt[2], J. Kuhl[1], P. Thomas[2], and K. Ploog[1]

[1]Max-Planck-Institut für Festkörperforschung, Heisenbergstr. 1, W-7000 Stuttgart 80, Fed. Rep. of Germany

[2]Phillips-Universität Marburg, Renthof 5, W-3550 Marburg, Fed. Rep. of Germany

Abstract Measurements of the photon echo in GaAs quantum wells as a function of the polarization of the incident fields can be used to show coupling of the σ^+ and σ^- exciton transitions.

1. Introduction

Two-pulse self-diffraction or spontaneous photon echo measurements [1] have been increasingly widely used to investigate dephasing processes and coherent phenomena in low-dimensional semiconductors. Such measurements are usually performed with the excitation pulses either parallel or cross polarized. However, little attention has been given to detailed dependence of the measured signal on arbitary linear polarization orientation of the incident pulses. We show here that the polarization of the emitted signal, and also the phase of light-hole (lh) and heavy-hole (hh) exciton quantum beats, can be used to investigate the degree of coupling between different exciton angular momentum states. Recent experiments have suggested this coupling is much larger than previously believed [2].

2. Theory

The optical dipole transitions in a quantum well for light propagating perpendicular to the plane of the layers are given by a pair of *uncoupled* 3 level systems (Fig.1). σ^+ and σ^- represent the allowed transitions due to clockwise and counter-clockwise circularly polarized light respectively. To treat different orientations of the linear polarization vectors of the two incident pulses $\vec{E}_1$ and $\vec{E}_2$ (directions $\vec{k}_1$ and $\vec{k}_2$), we define an $(\hat{x}, \hat{y})$ coordinate system (Fig. 1) in the plane of the quantum well with $\vec{E}_2$ always lying along $\hat{x}$. We obtain $P^{(3)}$ in the $2\vec{k}_2 - \vec{k}_1$ direction from the generalized photon echo result of a N $\times$ M level system [3] for arbitary orientation of $\vec{E}_2$. In the following, we state only the principle results obtained. First, for selective excitation of either the hh or lh exciton transition alone, the orientation θ_{sig} of the self-diffracted signal obeys the relation $\theta_{sig} = -\theta_{12}$, as expected for two independent two level

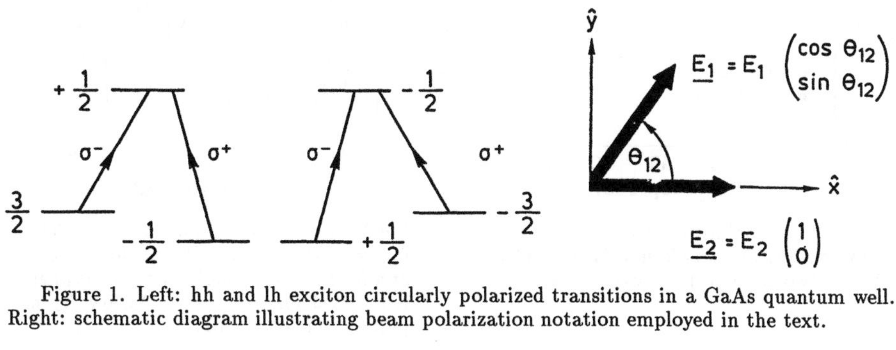

Figure 1. Left: hh and lh exciton circularly polarized transitions in a GaAs quantum well. Right: schematic diagram illustrating beam polarization notation employed in the text.

Springer Series in Chemical Physics, Vol. 55 Ultrafast Phenomena VIII
Editors: J.-L. Martin · A. Migus · G.A. Mourou · A.H. Zewail

systems with opposite circular polarization selection rule [4]. Second, the photon echo intensity $|P^{(3)}|^2$ is independent of θ_{12}. Third, if both transitions are excited, quantum beats between lh and hh polarizations are obtained. However, the phase of the quantum beats relative to zero delay changes from 0° at $\theta_{12} = 0^o$ (parallel polarized fields), to 180° at $\theta_{12} = 90^o$ (cross-polarized fields), due to the phase shift induced between the constituent circular polarizations of the incident light field when the linear polarizations are crossed. We now consider how these rules would be influenced in the presence of a coherent coupling of the circularly polarized transitions by either the conduction band or valence band levels. In the extreme case, where only the bonding state of the coupled levels is excited, we find quite different behavior, because the transitions now have mixed circularly polarized character. The signal polarization is now always aligned along $\vec{E}_2$ (i.e. it follows the polarization of the delayed pulse) for any θ_{12}, leading to the rule $\theta_{sig} = 0^o$. The echo intensity follows a $\cos^2\theta_{12}$ dependence which goes to zero at $\theta_{12} = 90^o$, as has been noted previously [4]. No phase shift in the quantum beat signal is obtained as a function of θ_{12}. These predictions differ greatly from those given above for no coupling of the upper or lower levels. Polarization studies of the photon echo signal are therefore a sensitive measure of the coupling between the σ^+ and σ^- exciton transitions.

3. Experimental

Experiments were performed at 8K on a 27nm single GaAs quantum well, with a 0.2meV PL linewidth. Fig. 2 shows θ_{sig} vs. θ_{12} for two different photoexcited exciton densities, N_x. At $N_x = 4 \times 10^9\text{cm}^{-2}$, we obtain good agreement with the $\theta_{sig} = -\theta_{12}$ rule predicted earlier for decoupled circularly polarized transitions. However, at $N_x = 4 \times 10^8\text{cm}^{-2}$ we observe a different dependence. For $\theta_{12} < 60^o$, we find $\theta_{sig} = 0^o$. For $60^o < \theta_{12} < 70^o$ the polarization of the diffracted signal becomes undefined, as represented by the oblique dashed line in Fig. 2. At large angles the $\theta_{sig} = -\theta_{12}$ rule is recovered. Fig. 2, therefore, suggests that at low exciton density, strong coupling of the circularly polarized exciton transitions occurs. At large θ_{12}, the uncoupled case appears dominant because the coupled contribution is suppressed at large angle by its $\cos^2\theta_{12}$ dependence (which we have also verified). The unpolarized signal in

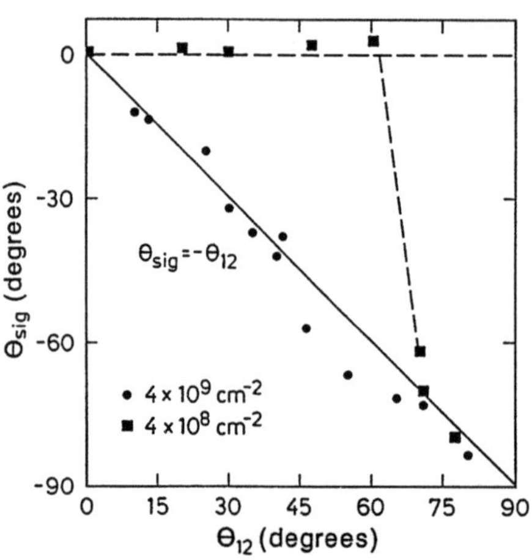

Fig. 2. The polarization of the photon echo signal θ_{sig} vs. θ_{12} from a 27nm GaAs QW for two different photoexcited exciton densities.

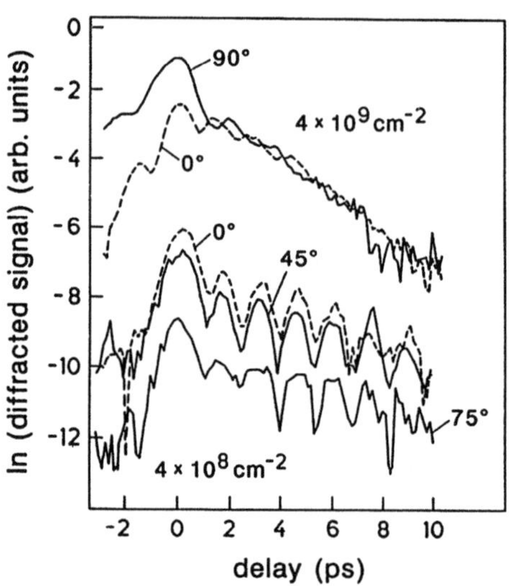

Fig.3. Self-diffracted Intensity vs. delay for excitation of both lh and hh excitons at different θ_{12}. At high N_x the beat phase is inverted for $\theta_{12} > 45^o$. At low N_x no phase inversion occurs.

the transitional regime, results from roughly equal contribution of two nonorthogonal signals. Fig. 3 shows the quantum beat signal for simultaneous excitation of lh and hh excitons in these high and low density regimes. At $N_x = 4 \times 10^9 \mathrm{cm}^{-2}$, we observe the phase shift in the quantum beat signal with θ_{12} predicted for uncoupled transitions. However, at $N_x = 4 \times 10^8 \mathrm{cm}^{-2}$ no phase shift is observed beyond 45°, indicative of coupled transition behavior.

4. Conclusions

Analysis of the polarization rules for the spontaneous photon echo in a GaAs quantum well shows that linear polarization dependent measurements are strongly sensitive to the degree of coupling between circularly polarized exciton transitions. Experiments show unexpectedly strong coupling at low exciton density.

[1] E.g. see J.Kuhl, L.Schultheis, A.Honold, and C.W.Tu, Festkörperprobleme **29**, 157, (1989).
[2] S.Bar-Ad and I.Bar-Joseph, Phys.Rev. Lett. **68**, 349, (1992)
[3] W. Huhn, A. Stahl, Phys. Stat. Sol. (b) **124**, 167 (1984), a fuller discussion of our analysis is the subject of a larger paper, R.Eccleston et al. (submitted to Phys. Rev. **B**).
[4] 'Optical Resonance and the Two-level Atom', L.Allen and J.H.Eberly, (Wiley 1975).

Observation of Many-Body Effects in the Femtosecond Temporal Profile of Quasi-2D Exciton Free-Induction Decay

S. Weiss, M.-A. Mycek, J.-Y. Bigot, S. Schmitt-Rink, and D.S. Chemla*

Physics Department, University of California at Berkeley, and Materials Sciences Division, Lawrence Berkeley Laboratory, 1 Cyclotron Road, Berkeley, CA 94720, USA
*Permanent address: Groupe d'Optique Nonlineaire, CNRS, F-67084 Strasbourg, France

Abstract: We demonstrate that the temporal lineshape of coherent emission from semiconductor quantum wells is dominated by many-body interactions among excitons. We present investigations of the density dependence of the processes involved.

1. Introduction

Time resolved coherent wave-mixing techniques have been used very successfully to investigate nonlinear optical processes and relaxation mechanisms in semiconductors [1-4]. These experiments have been analysed often using isolated-level models [5] which were developed for the case of atomic systems [6]. Recent experiments in which the ultrafast coherent emission from semiconductor heterostructures [7,8] was integrated by slow detectors, have given results qualitatively different from those predicted by Ref.[5]. These effects were explained by theories which account for the Coulomb mediated many-body correlation, which governs the optical response of media with delocalized electronic elementary excitations [8-10]. In semiconductors, the coupling of the interband dipole μ_k with the electromagnetic field (EMF) is expressed by the Rabi-frequency $\mu_k E$, and is renormalized by the Coulomb force [11]. In k-space, where this is most naturally expressed, the renormalized coupling is written:

$$\mu_k E \rightarrow \mu_k E + \sum_{k'} V_{k,k'} \psi_{k'}, \tag{1}$$

where $V_{k,k'}$ is the Coulomb potential and ψ_k is the k-component of the pair amplitude. As a consequence, excitonic polarization waves, $P = \sum_k \mu_k^* \langle\psi_k\rangle$, interact not only with the applied EMF, E(t), but also with other excitonic polarization waves [8-12]. In the case of excitation by ultrashort pulses, the temporal profile of the polarization waves is step-like. Hence, processes due to the Coulomb mediated interaction between polarization waves have a temporal dependence fundamentally different from that due to EMF interactions [8-12].

The simplest coherent wave-mixing configuration is that of two beam four-wave-mixing (FWM). In such a configuration two ultrashort laser pulses, labeled pulse-2 and pulse-1, separated by a time delay $\Delta t = t_2 - t_1$, and propagating in the directions k_2 and k_1, interfere in a sample to generate a transient grating, which diffracts photons into the background-free direction $k_s = 2k_2 - k_1$. In the case of homogeneously broadened resonances the FWM-signal is emitted immediately after the second pulse. It originates from the natural decay of the component of the nonlinear polarization P(t), which emits in the direction k_s, and corresponds to free induction decay (FID) [6]. For inhomogeneously broadened lines, the FWM-signal is delayed by Δt after the second pulse and corresponds to a "photon echo" [6]. As mentioned above, this signal is usually measured as a function of Δt by using a slow detector to integrate the light emitted in the direction k_s. The many-body theories of Ref. [8,10] have predicted that, for homogeneously broadened resonances, important information on exciton-exciton interactions is contained in the **temporal lineshape of the FWM-signal itself.** In this paper, we present results of experiments in which this temporal-profile has been resolved with a resolution of ≈ 100fs for the case of FWM resonant with quasi-two dimensional excitons in GaAs quantum wells (QW) [13,14]. These first observations of many-body mediated polarization wave interactions demonstrate that they dominate the low excitation nonlinear response of semiconductors.

Springer Series in Chemical Physics, Vol. 55 **Ultrafast Phenomena VIII**
Editors: J.-L. Martin · A. Migus · G.A. Mourou · A.H. Zewail

2. Experimental Technique and Results

Excitonic resonances in semiconductors are usually inhomogeneously broadened at low temperatures. In QW-structures, however, the quantum confinement in ultrathin layers, narrower than the bulk Bohr radius, stabilizes the quasi-2D excitons up to room temperature [15]. Collisions with the large population of thermal-phonons homogenize the resonances and shorten their dephasing time. We have investigated a sample consisting of 47 periods of 98Å GaAs QWs and 96Å $Al_{0.3}Ga_{0.7}As$ barrier layers. The substrate was removed by selective etching and anti-reflection coated to allow for transmission experiments without Fabry-Perot interference. The output of a mode-locked Ti:Sapphire laser, delivering extremely stable ≈ 70-100fs transform-limited Gaussian pulses at 88 MHz, was tuned to the exciton resonance and split into three beams. Two of these beams were used to generate the FWM-signal, which, if desired, could be detected directly as a function of Δt using a slow detector in the conventional way. The signal obtained in such a measurement is called the time-integrated-signal (TIS). In order to **time-resolve the FWM signal,** for each Δt the light emitted in the direction k_s was cross-correlated with the third laser beam by sum frequency generation in a highly transparent nonlinear crystal. This cross-correlation determined the temporal-profile of the FWM-signal vs **the absolute time** t. The signal obtained in such a measurement is called the time-resolved-signal (TRS). The overall time resolution of the experiment is better than 100fs.

In Fig. (1a) we present TRS vs absolute time t for various values of the time delay Δt, measured with a laser intensity such that the total (generated by both pulses) exciton density is $N_x \approx 3\times10^{11}cm^{-2}$. The laser pulse duration was 78±3fs. The weaker pulse-1 acts at t = 0 and the stronger pulse-2 acts at t = Δt, so that for Δt<0 pulse-2 acts first and pulse-1 acts second, whereas for Δt>0 pulse-1 acts first and pulse-2 second. Clearly, the time traces are asymmetric both in t and Δt. In Fig. (1b) we have plotted in the Δt-t plane, the position of the TRS maximum vs Δt (solid circle). Superimposed on this plot are a line of slope 1 (short-dashed line) and a line of slope 2 (long-dashed line), which correspond to the case of FID and photon echo, respectively, in the limit of infinitely narrow (δ-function) laser pulses. Clearly the experimental data follow the line of slope 1, showing that that the FWM-signal is emitted immediately after the second pulse (the position of the maximum for Δt<0 and the variations of this position relative to the line of slope 1 for Δt>0 are due to finite pulse duration effects). This behavior has been checked for the numerous sets of experiments which we performed at various laser intensities. It confirms that the exciton transition is predominantly homogeneously broadened at room temperature and therefore, that the FWM-signal

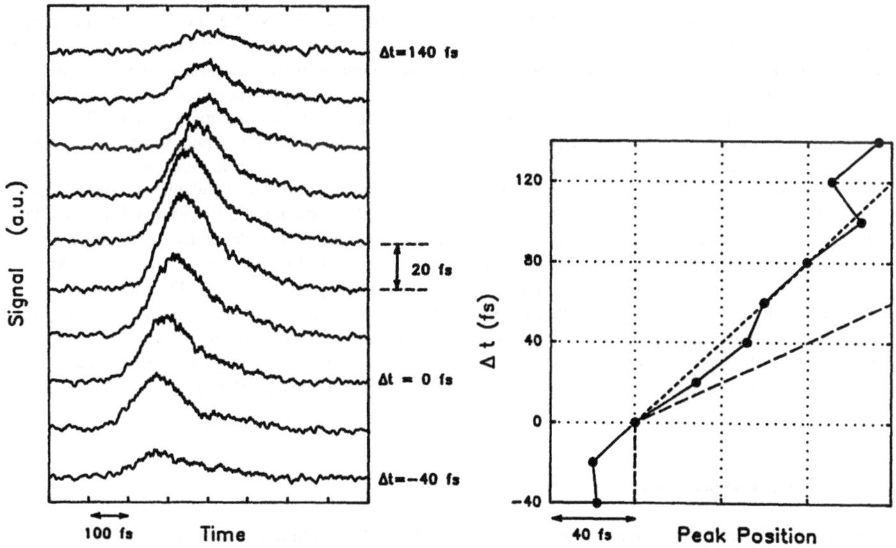

Figure (1): (a) Time resolved signal vs absolute time t, for a series of time delays Δt and a total exciton density $N_x \approx 3\times10^{11}cm^{-2}$. (b) Absolute time position of the maximum of time resolved signal vs Δt (solid circle). The short-dashed line and the long-dashed lines have respectively slope 1 and slope 2.

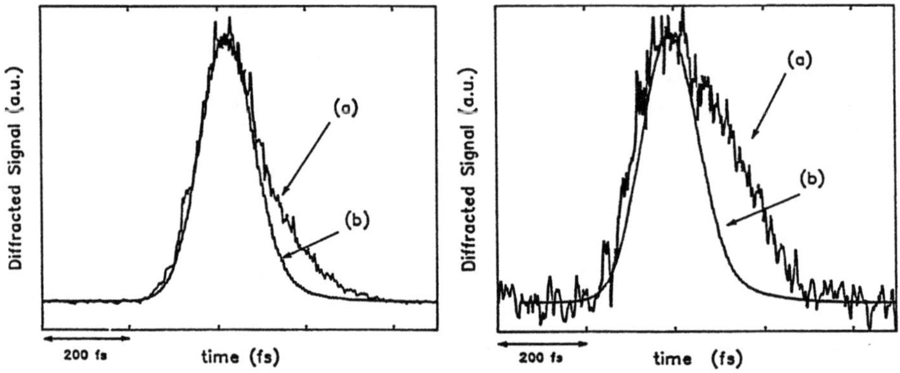

Figure 2: Comparison of the temporal profile of (a) the time resolved signal at $\Delta t = 0$ and (b) the time integrated signal, for two total exciton densities: $N_x \approx 4\times 10^{11}cm^{-2}$ (left traces), and $N_x \approx 10^{11}cm^{-2}$ (right traces).

corresponds to a free-induction decay. Furthermore, the self-consistency of the data was checked by numerically integrating for each Δt the TRS vs t, and comparing the result to the TIS vs Δt measured with a slow detector. Again, the agreement is excellent in our experiments.

In Fig. (2) we display the temporal profile of (a) the TRS vs the absolute time t at $\Delta t = 0$ and (b) the TIS vs Δt, for two total exciton densities, $N_x \approx 10^{11}cm^{-2}$ and $N_x \approx 4\times 10^{11}cm^{-2}$. For this set of experiments, the laser pulse duration was (98±2)fs. Since the relevant information is contained in the lineshape, the two sets of curves have been normalized to unity and the unrelated time-axes have been shifted to bring the maxima into coincidence. The difference between the TRS and TIS profiles is evident: the former is clearly broader than the latter, with a slower rising edge and a significantly non-exponential trailing edge. This difference is density dependent and shows up specially on the trailing side of the profiles. Within a Δt series the total exciton density is constant and, therefore, all the TRS traces are expected to have similar lineshapes (although their height may depend on Δt) [13,14]. This is indeed what is observed at low densities, $N_x \approx 10^{11}cm^{-2}$. At moderate density, $N_x \approx 2\text{–}4\times 10^{11}cm^{-2}$, noticeable changes in the temporal profile are seen within a Δt series as in Fig. (1). In particular the sign of Δt is found to influence the temporal lineshape. As the exciton density increases further, the TRS and TIS profiles become more similar (compare Fig. 2a & 2b). Finally, at very high densities, $N_x \approx 10^{12}\text{–}10^{13}cm^{-2}$, where the band gap renormalization has completely washed out the exciton resonances, the TRS and the TIS become of the order of the laser pulse and are beyond our time-resolution. Then, the profile of all the traces of a Δt series become similar again. It is worth noting that in the experimental situation where the TRS is measured, the sample is in a steady state (fixed Δt) and the "time-resolved measurement" occurs after and independent of the sample. This is in contrast with the conventional situation where the TIS is measured as Δt is varied, and thus the density, nature, and dynamics of the elementary excitations evolve during the measurement. For example, if the low density TIS displayed in Fig. (2) is analysed according to the conventional method [5], one obtains a dephasing time $T_2 = (135 \pm 5)$fs. This interpretation is incorrect since i) it neglects the fact that as Δt varies the excitation of the sample is not constant, ii) it completely ignores the contribution of the polarization wave scattering. Instead, if one uses the correct analysis of the TRS profile (described below) one obtains $T_2 = (190 \pm 5)$fs, for a constant excitation of the sample ($\Delta t = 0$ fixed) and a dominant contribution from polarization wave scattering.

3. Model and Interpretation

In order to be more quantitative, we must account for the finite duration of the laser pulses and compare the experimental results to a theoretical model. In a semiconductor excited by an optical field, E(t), the exciton amplitude, $\psi(t)$, satisfies the nonlinear Schrödinger equation [9]

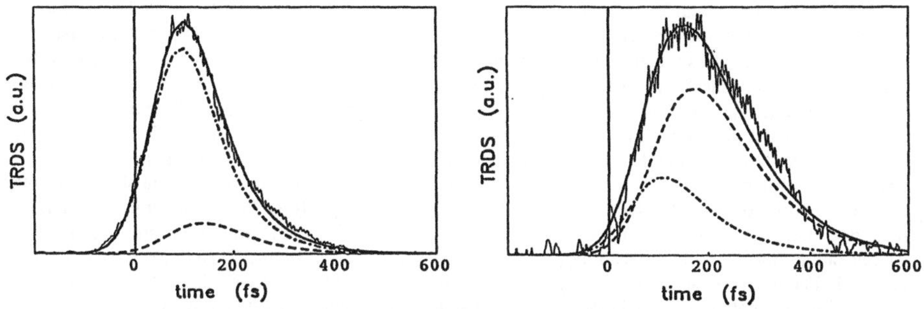

Figure (3): Fit of the temporal profiles shown in Figure (2), using the model discussed in the text. The dashed-dotted lines give the (isolated-level-like) phase space filling contribution, the dashed lines give the exciton-exciton interaction contribution, and the smooth solid lines give their sum.

$$\frac{\partial}{\partial t}\psi(t) = -i(\Omega - i\Gamma)\,\psi(t) + i\mu E(t) \qquad (2)$$
$$- i\mu E(t)\,\frac{|\psi(t)|^2}{\psi_s^2} - iV|\psi'(t)|^2\psi(t),$$

where only terms up to third order in the field have been retained. In this equation, the interband dipole element, μ, is assumed k-independent, and Ω and $\Gamma(=T_2^{-1})$ are, respectively, the exciton frequency and half-width at half maximum. The first line describes the linear response, while the two terms on the second line account the nonlinear response. The first nonlinear term originates from the Pauli-exclusion reduction of the EMF-exciton coupling, $\mu E(t) \rightarrow [1-(|\psi(t)|/\psi_s)^2]\mu E(t)$. The saturation parameter, ψ_s, is a material parameter that depends on the exciton characteristics [15]. This term is, of course, always present and accounts for the isolated-level-type nonlinearities [5,6]. The second nonlinear term is specific to dense media semiconductors [8-10] and molecular crystals [16,17]. In semiconductors, it describes the Coulomb mediated exciton-exciton interaction [9]. Its physical interpretation is that, via the potential V, excitons at one site, $\psi(t)$, interact with excitons at other sites, $\psi'(t)$. It is this last term which describes an interaction between polarization waves within the medium. We call these two nonlinear terms the Phase Space Filling (PSF) and the exciton-exciton interaction (XXI) term, respectively.

We have solved Eq. (2) numerically, in the mean field approximation, for the nonlinear polarization, $P^{(3)}(t,k_s) = \mu^* \langle\psi^{(3)}(t,k_s)\rangle$ radiating in the direction k_s, using Gaussian laser pulses with a duration corresponding to that of our laser and accounting for the effect of upconversion in the time-resolved (TRS) measurement. Using this model, we fit the TRS by imposing the severe constraint that all TRS curves in a Δt series must have the same origin of the absolute time t and a constant calibration. In this model there are only two fitting parameters: the exciton dephasing time, T_2, and the ratio of the two nonlinear terms, $v = V\psi_s^2$. We find that it is impossible to fit the data with such a constraint if we retain only the PSF term in Eq. (2). An excellent fit is obtained, however, if both the XXI and PSF contributions are considered. This is shown in Fig. (3), where we fit the TRS temporal profiles of Fig. (2). The dashed-dotted lines give the PSF contribution, the dashed lines give the XXI contribution, and the smooth solid lines give their sum.

At low total exciton densities, $N_x \approx 10^{11}\,cm^{-2}$ (right traces in Fig. 2 & 3), the dephasing time, $T_2 = (190 \pm 5)$fs, and the ratio of nonlinearities, $v = (0.01 \pm 0.001)$, remain essentially constant for all Δt. The contribution of the XXI to the total energy of the pulse emitted by the sample (which is proportional to the TRS area) dominates the emission. It is ≈ 2.5 larger than that due to PSF. As the total exciton density is increased, both T_2 and v become smaller. For example for moderate density, left traces of Fig. (2&3), $T_2 = (130 \pm 5)$fs and the XXI contributes to only 14% of the emission. Interestingly, we found that the XXI contribution dominates as long as the exciton density does not exceed the exciton saturation density in the sample, $N_s \approx 3\times10^{11}\,cm^{-2}$ [15]. When this critical density is surpassed the XXI contribution decreases very rapidly and becomes negligible at high densities.

As mentioned above, at low density both T_2 and v remained approximately constant for all the TRS fits within a single Δt series. At moderate densities, however, this was not found to be the case within a single

Δt series (i.e. fixed N_x) both T_2 and ν had to be varied to fit the data. These observations can be understood in terms of the well established dynamics of excitons in quantum wells at room temperature [15]. Consider first the case of a total excitation density low enough that photo-generated excitons are in bound states (binding energy $\approx$ 10meV for $\approx$ 100Å QW) and are spatially well separated. They interact effectively via the Coulomb potential and their dephasing time is determined by phonon collisions. Their environment is independent of the instantaneous density determined by the order in which the laser pulses arrive in the sample (i.e. Δt). At moderate densities, however, when the strongest pulse-2 arrives first in the sample, it generates a substantial number of excitons in the bound states. They are ionized by collisions with the energetic thermal phonons (phonon energy $\approx$ 36meV for GaAs), in $\approx$ 100–200fs, generating e-h pairs in scattering states with a significant excess energy ($\approx$ 25meV). The charged carriers, in turn, both shorten the relaxation time T_2, owing to their larger effects on the neutral bound states, [2] and screen the Coulomb potential. For the reverse time ordering, i.e. when the weaker pulse-1 arrives first, less e-h pairs are generated by the first pulse, the effects described above are less pronounced, and the sample remains closer to steady state during the FID emission. Finally, at very high densities, the band gap renormalizes so much during the laser pulses that the excitons are generated in scattering states, giving free e-h pairs immediately. They, of course, shorten the relaxation time below the experimental resolution, but, more importantly, they screen the Coulomb potential to the point that the XXI contribution to the emission is eliminated. This description of the exciton dynamics also demonstrates that the conventional experimental technique, TIS, does not correspond to a sample in a steady state and therefore provides unreliable data, unless the excitation is extremely weak (which is not the case for most of the published experiments).

The fits also show that the maximum of the TRS is delayed with respect to the second pulse. We found that the delay is of the order T_2. This indicate that the time required to **establish** the coherent polarization wave interaction within the sample (rise-time of the TRS) is directly related to the dephasing time T_2 (usually associated with the signal decay-time).

4. Conclusion

We have shown that, because of exciton-exciton interactions, semiconductors emit coherent light in a way fundamentally different from isolated atoms. We have found that the polarization wave scattering originating from this interaction is, in fact, dominant at low excitation densities. At low and moderate densities it produces non-trivial effects that must be accounted for in a correct description of many phenomena occurring in these materials.

This work is supported by the Director, Office of Energy Research, Office of Basic Energy Sciences, Division of Materials Sciences of the US Department of Energy, under Contract No. DE-AC03-76SF00098.

REFERENCES

[1] L. Schultheis, M.D. Sturge, and J. Hegarty, Appl. Phys. Lett. 47, 995 (1985).

[2] L. Schultheis, J. Kuhl, A. Honold, C.W. Tu, Phys. Lett. 55, 1635, (1986)

[3] G. Noll, U. Siegner, S. Shevel, and E.O. Göbel, Phys. Rev. Lett. 64, 792 (1990).

[4] M.D. Webb, S.T. Cundiff, and D.G. Steel, Phys. Rev. Lett. 66, 934 (1991).

[5] T. Yajima,and Yoichi Taira, J. Phys. Soc. Jap. 47, 1620 (1979)

[6] See for example: L. Allen and J.H. Eberly, "Optical Resonances and Two Level Atoms" (Wiley, New York 1975)

[7] K. Leo, M. Wegener, J. Shah, D.S. Chemla, E.O. Göbel, T.C. Damen, S. Schmitt-Rink, and W. Schäfer, Phys. Rev. Lett. 65, 1340 (1990);

[8] M. Wegener, D.S. Chemla, S. Schmitt-Rink, and W. Schäefer, Phys. Rev. A 42, 5675 (1990).

[9] S. Schmitt-Rink, S. Mukamel, K. Leo, J. Shah, and D.S. Chemla, Phys. Rev. A 44, 2124 (1991).

[10] W. Schäfer, F. Jahnke, and S. Schmitt-Rink, Proceeding of the International Meeting on Optics of Excitons in Confined Systems, Gardini Naxos, Italy 1991, Inst. Phys. Conf. Ser. No. 123, and submitted to Phys. Rev. Lett.

[11] S. Schmitt-Rink, D.S. Chemla, Phys. Rev. Lett. 57, 2752 (1986). and S. Schmitt-Rink, D.S. Chemla, H. Haug, Phys. Rev. Lett. B37, 941 (1988).

[12] A.V. Kuznetsov, Phys. Rev. B44, 8721 and 13381 1991-II

[13] M.-A. Mycek, S. Weiss, J.-Y. Bigot, S. Schmitt-Rink, D.S. Chemla Appl. Phys. Lett. 60, (1992)

[14] S. Weiss, M.-A. Mycek, J.-Y. Bigot, S. Schmitt-Rink, D.S. Chemla Submitted to Phys. Lett.

[15] S. Schmitt-Rink, D.S Chemla, D.A.B. Miller, Adv. in Phys. (1989)

[16] J.R. Kuklinski and S. Mukamel, Phys. Rev. B42, 2959 (1990), and B42, 11938 (1990)

[17] J.A. Leegwater and S. Mukamel, to be published in Phys. Rev. A46, (1992)

Radiative Recombination of Free Excitons in GaAs Quantum Wells

B. Sermage[1], *K. Satzke*[1], *C. Dumas*[1], *N. Roy*[1], *B. Deveaud*[2], *F. Clerot*[2], and *D.S. Katzer*[3]

[1]Centre National d'Etudes des Telecommunications, B.P. 107, F-92225 Bagneux Cedex, France
[2]Centre National d'Etudes des Télécommunications, F-22300 Lannion, France
[3]Naval Research Laboratory, Washington, DC 20375, USA

Abstract. We have measured the radiative lifetime of excitons in GaAs quantum wells under resonant excitation at 10 K using time resolved luminescence with 6 ps time resolution. The luminescence decay has two components: a fast one with a time constant τ_1 (~17-40 ps) and a slow one with a time constant τ_2 (~200-300 ps). We interprete the first decay as due to the radiative recombination of excitons with $k_{\parallel} < k_0$.

The radiative lifetime of excitons in quantum wells pose a problem concerning the volume on which the elementary excitonic oscillator strength participate to the recombination [1-3]. The excitons lifetime can be measured by time resolved luminescence. Till now most measurements have been done with an excitation energy E_l higher than the free exciton energy E_0 and the luminescence decay is perturbed by the cooling of the hot excitons. In our measurements [4], the quantum wells are excited resonantly.

We have studied three kind of wells: AlAs-GaAs 4.8 nm thick single well (S1) [5], AlAs-GaAs 6.3 nm thick single well (S2) [5] and $Al_{0.3}Ga_{0.7}As$-GaAs 6 nm thick triple well (DB07). Samples S_1 and S_2 had very thin luminescence peaks (about 2 meV wide) presenting monolayer splittings. The samples were cooled at 10 K and excited by a pyridine or a Titanium Saphire mode locked laser. The luminescence was spectrally dispersed by a 32 cm focal length monochromator and dispersed in time by a streak camera with a 2 dimensions detection. The time resolution is 6 ps and the energy resolution is 1 meV. Most of the time, the measurements were done at low excitation power (~ 1mW/cm^2 i.e. 10^{15} excitons/cm^3). For each experiment, we select a position in the sample where the scattering of the laser is minimum out of resonance. Pratically the intensity of the diffusion of the laser out of resonance is smaller than a tenth of the luminescence intensity at resonance.

The following results were observed on the three samples. Fig. 1 shows the luminescence decay curves obtained with resonant excitation and out of resonance on sample DB07. The

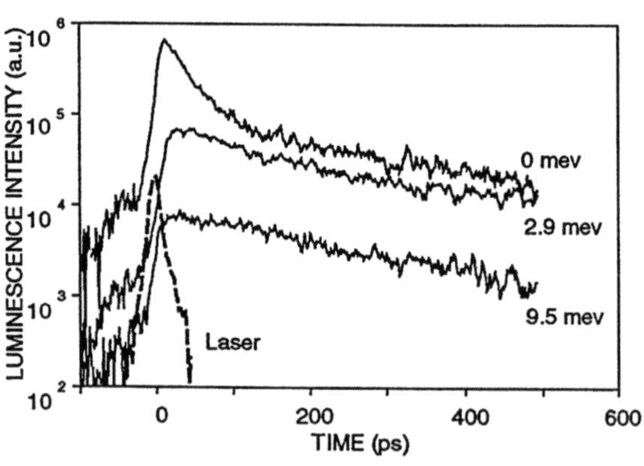

Fig. 1 Excitonic luminescence decay curves for different shifts of the excitation energy on sample DB07 at 10 K. The dashed curve shows the laser pulse out of resonance.

Springer Series in Chemical Physics, Vol. 55 Ultrafast Phenomena VIII
Editors: J.-L. Martin · A. Migus · G.A. Mourou · A.H. Zewail

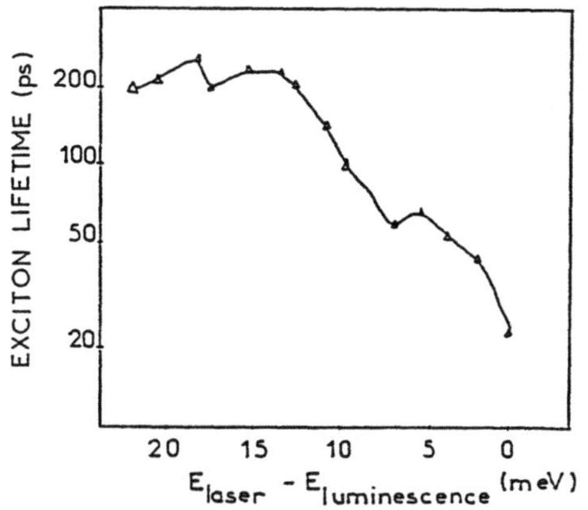

Fig. 2 Variaton of τ_1 with the shift between the laser and the luminescence on sample S_1.

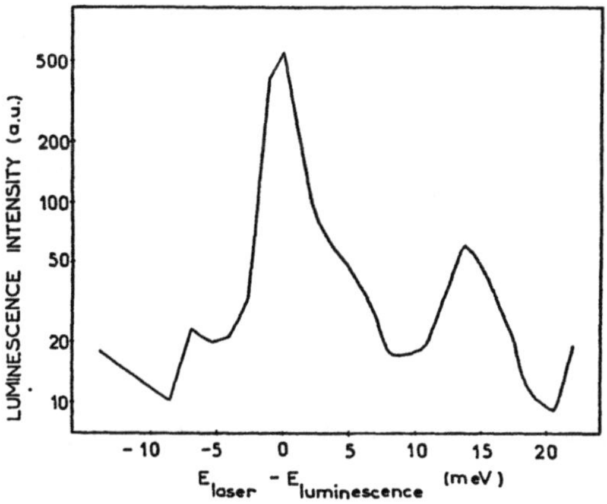

Fig. 3 Variation of the luminescence intensity with the shift between the laser and the luminescence energy on sample S_1.

luminescence intensity increases by more than two orders of magnitude at resonance and the decay time decreases.

At resonance, the luminescence decay is the sum of two exponentials: a first one with a decay time τ_1 and a second one with a decay time τ_2. τ_1 vary between 17 and 40 ps depending on the samples and τ_2 vary between 200 and 300 ps. Out of resonance, we only observe the decay time τ_2. Figures 2 and 3 show the variation of τ_1 and of the luminescence intensity as a function of the energy difference between the laser and the exciton. The large increase in the luminescence intensity at resonance is due for one part (~5x) to the increase of the absorption coefficient and for the other part (~10x), to the decrease of the radiative exciton lifetime τ_r. Indeed, when the laser beam has the energy of the free excitons, it creates excitons with $k_{\|} < k_0$ which can radiate easily with a lifetime τ_0. After a time τ_s, they are scattered by accoustical phonons to states with $k_{\|} > k_0$ where they can not radiate. τ_1 is then given by the following equation:

$$\frac{1}{\tau_1} = \frac{1}{\tau_0} + \frac{1}{\tau_s}$$

τ_2 is the radiative lifetime of the excitons thermalized at the temperature of the crystal. This appears clearly on fig. 4 showing that the variation of τ_2 with temperature is nearly linear:

τ_2 (ps) = 37. x T (K).

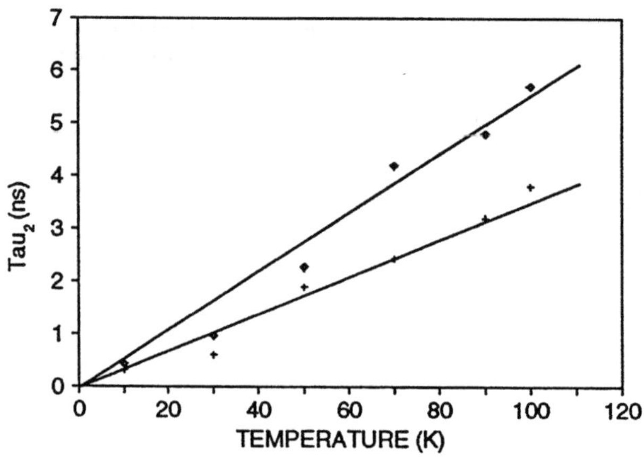

Fig. 4 τ_2 at resonance (+) and the decay time with an excitation at E_0 + 20 meV (◆) as a function of temperature in sample DB07.

This expression is in good agreement with Andreani's calculations [3] τ (ps) = 34. x T (K).
The decay curve at resonance at 10 K shows that there has been loss of excitons during the first exponential so that τ_0 is close to τ_1 in agreement with our value of τ_2 and with ref. 3.

Acknowledgements: We wish to thank M. Berz, J. Worlock, I. Abram and C. Tanguy for helpful discussions. One of the samples was grown by F. Mollot to whom we are debtful.

References.

1. J. Feldman, G. Peter, E.O. Göbel, P. Dawson, K. Moore, C. Foxon and R.J. Elliott, Phys. Rev. Lett. 59, 2337 (1987).
2. E. Hanamura, Phys. Rev. B 38, 1228 (1988)
3. L.C. Andreani, Solid State Comm., 77, 641 (1991).
4. B. Deveaud, F. Clérot, N. Roy, K. Satzke, B. Sermage and D.S. Katzer, Phys. Rev. Lett. 67, 2355 (1991).
5. D. Gammon, B.V. Shanabrook and D.S. Katzer, Appl. Phys. Lett., 57, 2710 (1990).

Field-Enhanced GaAs/AlGaAs Waveguide Saturable Absorbers

J.R. Karin[1], *D.J. Derickson*[1], *R.J. Helkey*[1], *J.E. Bowers*[1], *and R.L. Thornton*[2]

[1]Ultrafast Optoelectronics Laboratory, Department of Electrical & Computer Engineering, University of California at Santa Barbara, Santa Barbara, CA 93106, USA
[2]Xerox Palo Alto Research Center, Palo Alto, CA 94304, USA

Bulk and quantum well semiconductors are important saturable absorbers and have been used to generate picosecond and femtosecond pulses in color center[1], erbium-doped fiber[2], dye[3] and semiconductor[4] laser systems. These applications typically use damage sites or traps to decrease the absorption recovery time. In semiconductor diode lasers, monolithic structures incorporating waveguide saturable absorber sections for passive modelocking have produced pulses of 0.6 picoseconds, at repetition rates up to 350 GHz. However, the exact mechanism for such high frequency modelocking is not well understood. In order to produce a stable train of short pulses at these millimeter-wave repetition rates, the absorption recovery must be fast enough to recover on each round trip in the cavity, e.g., 10 ps for a 100 GHz repetition rate. The saturable absorber sections are typically p-i-n structures operated under reverse bias to remove photogenerated carriers and achieve this fast recovery, as in a p-i-n photodiode. Knowledge of the dynamics of the saturable absorbers is necessary for design optimization, as well as insight into the potential of these devices for producing short optical and electrical pulses. Absorption recovery is also thought to be responsible for elimination of multiple pulses in external cavity lasers[5]. In this paper we present a study of the bleaching and recovery dynamics of these waveguide saturable absorbers for different bias conditions, and discuss the implications of the results for modelocked laser systems.

Figure 1 shows a schematic drawing of the experimental setup and the layer structure of a device with a bulk $Al_{0.04}Ga_{0.96}As$ active region used in these experiments. Devices with four GaAs quantum wells and with a single AlGaAs quantum well were also studied. The lateral waveguide is formed by impurity-induced disordering[6]. The end facets are anti-reflection coated with SiN_x, with a residual facet reflectivity of 10^{-3}. The device is biased through a bias tee and the high frequency port is terminated in a 50 Ω sampling head. We use a modelocked Ti:Sapphire laser which emits pulses of ~100 fs, with the center wavelength adjusted to the lasing wavelengths of the devices when they are used in modelocking experiments. A pump-and-probe technique is used to measure the change in transmission of a weak probe beam in the presence of a strong pump.

Figure 2 shows data from bulk multiple quantum well devices, for different bias values. The overall signal amplitudes decrease with increasing bias. This could be due to the Franz-Keldysh shift of the absorption edge of the material with bias. For all values of bias the 1/e recovery time is quite short, typically less than 15 ps, and as short as 4 ps at high bias levels. The recovery time decreases with increasing bias, due to reduced emission time across the heterojunction and shorter transit times

Springer Series in Chemical Physics, Vol. 55 Ultrafast Phenomena VIII
Editors: J.-L. Martin · A. Migus · G.A. Mourou · A.H. Zewail

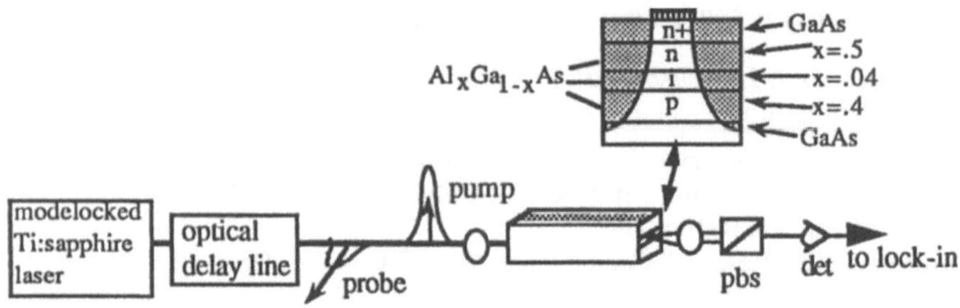

Figure 1: Experimental configuration and device structure for pump-probe measurements of waveguide saturable absorbers formed by impurity-induced disordering.

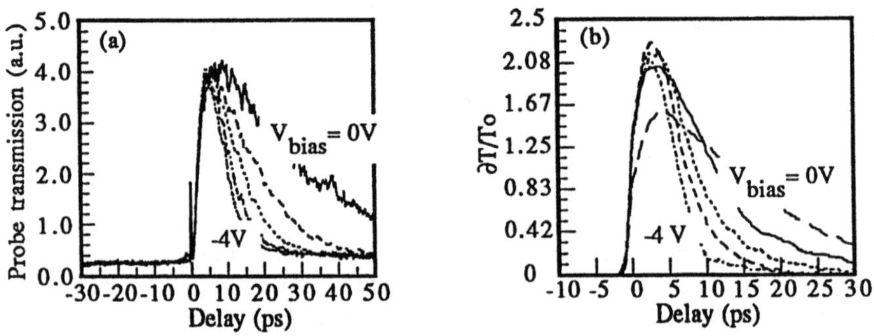

Figure 2: Differential transmission data for (a) bulk and (b) multiple quantum well waveguide saturable absorbers, for different bias values.

across the undoped region. The effect of space charge created by the photogenerated carriers is also important in these experiments. As the carriers are swept out by the reverse bias, the space charge field will act to oppose the applied field, reducing the effective field across the junction and affecting the transit time of the carriers and thus the absorption recovery time, particularly when the applied bias is small. In addition, the effect of space-charge fields on intervalley scattering must be taken into account for a full understanding of the physics. The very high fields due to the built-in potential and applied bias accelerate the photogenerated carriers and give them enough energy to scatter into satellite valleys. This process happens in tens of femtoseconds, and we expect to see a very fast transient in the differential transmission due to the initial bleaching and subsequent scattering of electrons from the Γ valley into the L valley, limited in time resolution to the width of the laser pulse. The data show an unexpected slower (<~5ps) rise time before the final recovery due to carrier sweepout. This slow risetime is likely due to scattering of carriers back to the Γ valley from the L valley, thus repopulating the states in the Γ valley and rebleaching the absorption. In addition, the absorption edge will shift via electroabsorption as the fields in the region change due to the space charge.

The short recovery times measured indicate that high frequency modelocking with these devices is possible. A recovery time of 4 ps translates to a maximum repetition rate in excess of 250 GHz. We have successfully fabricated monolithic and external cavity modelocked lasers containing saturable absorber sections from this material. The best results were obtained with an absorber length of 16μm, and pulsewidths of 2.8 picoseconds were achieved [5].

Using the pump-probe technique to study the dynamics of these devices is necessary in order to tailor the device designs for specific applications. It is also important to understand the mechanisms of the non-linear absorption properties of waveguide structures and the inherent limits to their recovery times. It should be possible to design a structure to exploit the subpicosecond response measured here, and make even faster saturable absorbers using semiconductor structures.

References:

1. C.E. Soccolich et al., Appl. Phys. Lett. **56**, 2177 (1990)
2. M. Zirngibl et al.,OSA Topical Meeting on Optical Amplifiers,# PDP9 (1991)
3. J. Feldmann et al., Optics Letters **16**, 241 (1991)
4. C. Harder et al., Appl. Phys. Lett. **42**, 772 (1983)
5. D.J. Derickson et al., Photon. Tech. Lett. **4**, 333 (1992)
6. R.L. Thornton et al., IEEE J. Lightwave Technology **6**, 786 (1988)

Picosecond Excitonic Nonlinearities in the Presence of Disorder

S.T. Cundiff and D.G. Steel

Harrison M. Randall Laboratory of Physics, University of Michigan,
Ann Arbor, MI 48109, USA

Abstract. The polarization dependence of the low temperature excitonic nonlinear optical susceptibility in GaAs/$Al_{0.3}Ga_{0.7}As$ multiple quantum wells (MQW) exhibits unexpected properties. In a three pulse transient four wave mixing experiment (TFWM), the emitted signal becomes prompt when the first field is cross-linearly-polarized (with respect to the second and third fields), in contrast to the delayed signal observed for co-polarized fields. This demonstrates the presence of both a homogeneously broadened and an inhomogeneously broadened resonance. The characteristics of the nonlinear response indicate that the inhomogeneous resonance is due disorder localized excitons and is suppressed by cross-polarizing the first field, and that the homogeneously broadened resonance is due to extended excitons. Circularly polarized TFWM and transient absorption measurements indicate that this suppression occurs due to coupling between opposite spin excitonic transitions.

The optical and electronic properties of semiconductor heterostructures depend on the energy level structure and dynamics of the electronic excitations. These in turn are sensitive to the presence of disorder, which arises in heterostructures from non-ideal growth conditions. The inhomogeneous broadening of the excitonic resonance reflects the modification of the energy level structure due to the presence of disorder induced localization of the states [1]. Migration between the localization sites, an entirely new dynamical process resulting from disorder, is responsible for relaxation of the excitation at a given energy [2]. To probe the effects of disorder in GaAs MQW we use resonant TFWM and non-degenerate circularly polarized transient absorption at low excitation density. Examination of the polarization dependences yields results which are unexpected for an ideal semiconductor (i.e. no disorder), but which can be explained by the presence of disorder.

Recent results have shown that in a two-pulse TFWM experiment the decay of the signal for increasing interval between pulses changes dramatically in time scale and strength when the pulses are cross-polarized [3]. However a more dramatic effect is observed if the time resolved signal is examined in a three pulse experiment for all fields co-polarized (Fig. 1a) and for $\mathbf{E}_1 \perp \mathbf{E}_2 \| \mathbf{E}_3$, $\mathbf{E}_s \| \mathbf{E}_1$ (Fig. 1b). The shifting of the signal with increasing delay between $\mathbf{E}_1$ and $\mathbf{E}_2$ for co-polarized fields makes it apparent that it is a stimulated photon echo (SPE), arising from an inhomogeneously broadened resonance. However for $\mathbf{E}_1 \perp \mathbf{E}_2$, the emission time of the primary signal is independent of delay between $\mathbf{E}_1$ and $\mathbf{E}_2$, and hence is a free polarization decay (FPD) and arises from a homogeneously broadened resonance (there is a weak SPE present in Fig 1b, however it arises from a resonance which is 2.5 meV Stokes shifted from linear absorption peak, we tentatively assign this to an impurity bound exciton). The signal saturation characteristics for $\mathbf{E}_1 \| \mathbf{E}_2$ and $\mathbf{E}_1 \perp \mathbf{E}_2$ (Fig. 1b inset) are also quite different. For $\mathbf{E}_1 \| \mathbf{E}_2$ the signal exhibits strong saturation at fluxes above 3×10^{10}, whereas no saturation occurs for $\mathbf{E}_1 \perp \mathbf{E}_2$ at densities up to 2×10^{11}.

These differences indicate that the signals must be arising from separate resonances. The disparity in strength suggests that if both resonances are contributing for co-polarized fields only the signal from the inhomogeneously broadened one would be observable. This is confirmed by the presence of both a prompt and delayed contribution at high intensity where the inhomogeneously broadened resonance is strongly saturated.

The SPE for co-polarized fields has been attributed to localized excitons, where the dephasing and population relaxation are due to migration of excitons amongst localization sites [4]. The rapid dephasing and homogeneous nature of the resonance for cross-polarized fields is characteristic of the response from extended excitons. The observed dephasing rate of 5 ps is comparable to that observed for homogeneously broadened excitons in a high quality single quantum well [5]. Comparison of the temperature dependence of the dephasing

Springer Series in Chemical Physics, Vol. 55 **Ultrafast Phenomena VIII**
Editors: J.-L. Martin · A. Migus · G.A. Mourou · A.H. Zewail

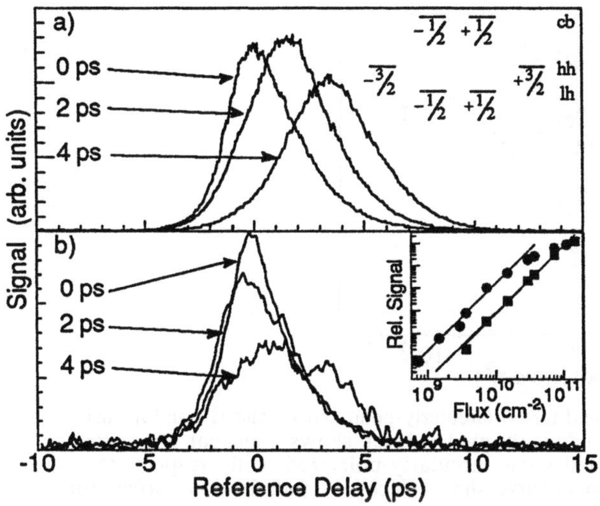

Figure 1. Cross correlation between the TFWM signal and a reference pulse. Upper panel is for co-polarized fields, the lower for $E_1 \perp E_2 \| E_3$, $E_s \| E_1$. In each panel the curves are for increasing E_1-E_2 delay. Upper inset is energy level structure for a GaAs quantum well, lower inset is saturation behavior of the two signals (circles for $E_1 \| E_2$, squares for $E_1 \perp E_2$, straight lines shows cubic behavior).

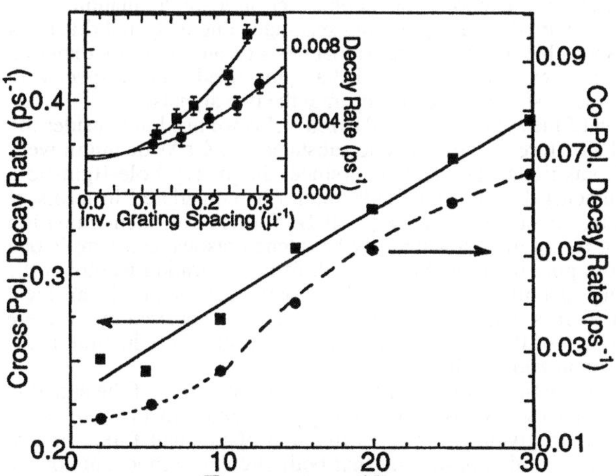

Figure 2. Temperature dependence of dephasing rates for co-polarized (circles, right axis) and cross-polarized (squares, left axis) signals. Lines are fit to theory. Note difference in axis scales. Inset shows grating spacing dependence of grating decay (same legend). The quadratic dependence shows that the diffusion coefficient, D, is larger for the prompt signal (D=10) than the echo (D=5).

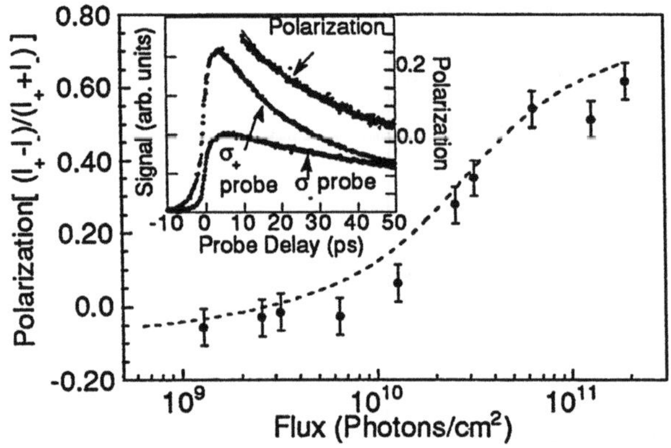

Figure 3. Polarization of signal field for co-circularly-polarized $\mathbf{E}_1$ and $\mathbf{E}_2$, and linearly polarized $\mathbf{E}_3$ (dashed lined is a guide to the eye). Inset shows transient absorption response for probe fields co- and cross-circularly-polarized with respect to a circularly (σ_+) polarized pump, upper curve shows decay of polarization (offset for clarity)

rates (Fig. 2) and the relative mobilities (Fig. 2 inset) are consistent with this distinction. The temperature dependence of the co-polarized signal dephasing rate has the same functional form, as observed earlier, and matches theoretical predictions for phonon-assisted-migration of excitons at low temperature and thermal activation at slightly higher ones [4]. The dephasing rate for the cross-polarized signal exhibits a linear dependence on temperature, as expected for single phonon scattering of excitons and as observed in a single quantum well [5]. The increase in relaxation rate with decreasing grating spacing is due to spatial diffusion and shows that the excitation responsible for the prompt signal has a greater mobility than that responsible for the SPE, as would be expected for extended and localized states respectively (here the differing saturation characteristics are used to separate the two signals).

To understand the suppression of the signal for co-polarized fields we need to consider the origin of the TFWM signal and structure of the magnetic substates in a GaAs quantum well. Because the only allowed "transitions" are $\Delta m=\pm 1$ (since the heavy-hole-light-hole degeneracy is lifted by confinement) the incident fields should be considered in terms of circularly polarized components. A separate grating will be formed for each transition (although there is no net grating for an inhomogeneously broadened resonance, there is one for each frequency group). Cross-polarizing the fields which form the grating results in the grating for each transition being spatially 180° out of phase with each other. In an ideal semiconductor this will result in a cross-polarized signal. However, it is apparent that if there is a coupling between the excitation of the two transitions it will result in a reduction of the grating contrast for both, and a suppression of the signal.

To probe for such a coupling of the transitions we examine the polarization of the signal in a TFWM experiment with the first two fields co-circularly-polarized and the third field linearly polarized. These results are given in Fig. 3 as function of incident flux. At low density the signal is linearly polarized, demonstrating that both circular components of the third field are strongly scattered. Hence, there exists a coupling of the two transitions. At increasing density the inhomogeneous resonance saturates, and the signal is more circularly polarized (in the same sense as the first two fields) as would be expected in an ideal semiconductor.

Further evidence of a coupling is observed in transient absorption experiments using circularly polarized fields. Using a circularly polarized pump and a linearly polarized probe, which is resolved into circular components after passing through the sample, we are able to simultaneously probe the response of both transitions. The inset in Fig. 3 shows that both transitions respond essentially simultaneously. This has been confirmed with 200 fsec pulses.

The observed coupling is unexpected for an ideal semiconductor, however the presence of disorder modifies the response. Although spin relaxation results in a coupling of the transitions, it occurs on slower time scales [6, 7]. The relaxation of the polarization in Fig. 3

inset shows that it occurs on a 40 ps time scale in our sample. Based on circularly polarized transient absorption measurements in a narrow stepped quantum well it has been suggested that biexcitonic effects are important [7]. The coupling arises because the presence of an exciton in a localization site modifies the eigenstates for the creation of a second exciton in the vicinity, i.e. a biexciton is created. This has the effect of shifting the effective oscillator frequency by the biexciton binding energy and results in an induced transmission resonant with the pump, and an induced absorption Stokes shifted by the biexciton binding energy. This effect has been observed in our non-degenerate circularly polarized transient absorption measurements.

In summary, we have shown that there are two resonances present in GaAs MQW, and that their response can be separated by their polarization properties. The properties of these resonances are consistent with the simultaneous presence of both localized and extended excitons. The apparent simultaneous presence of both localized and extended states is not predicted by localization theory, especially in two dimensions where all states are thought to be localized by any disorder [8]. However, the interface disorder is complicated and recent work has suggested that the roughness spectrum may be bimodal where some regions may permit extended states [9]. Our results also clearly show that there is a strong coupling between opposite spin exciton transitions.

This work was supported by ARO and AFOSR, samples provide by P.K. Bhattacharya.

References

1. C. Weisbuch, R. Dingle, A.C. Gossard, and W. Wiegmann, Solid State Comm. **38**, 709 (1981).
2. J. Hegarty and M.D. Sturge, J. Opt. Soc. Am. B. **2**, 1143 (1985).
3. H.H. Yaffe, Y. Prior, J.P. Harbison, and L.T. Florez, Quant. Electr. Laser Sci., 1991; K. Leo, J. Shah, S. Schmitt-Rink, and K. Köhler, VIIth Int. Sym. Ultrafast Pro. Spec., 1991.
4. T. Takagahara, J. Lumin. **44**, 347 (1989); M.D. Webb, S.T. Cundiff, and D.G. Steel, Phys. Rev. B **43**, 12658 (1991).
5. L. Schultheis, A. Honold, J. Kuhl, K. Köhler, and C.W. Tu., Phys. Rev. B **34**, 9027 (1986).
6. T.C. Damen, K. Leo, J. Shah, and J.E. Cunningham, App. Phys. Lett. **58**, 1902 (1991).
7. S. Bar-Ad and I. Bar-Joseph, Phys. Rev. Lett. **68**, 349 (1992).
8. E. Abrahams, P.W. Anderson, D.C. Licciardello, and T.V. Ramakrishnan, Phys. Rev. Lett. **42**, 673 (1979).
9. D. Gammon, B.V. Shanabrook, and D.S. Katzer, Phys. Rev. Lett. **67**, 1547 (1991).

Fast Optical Nonlinearities in Semiconductor Quantum Dots

G. Tamulaitis[1], *R. Baltramiejũnas*[1], *S. Pakalnis*[1], *and A.I. Ekimov*[2]

[1]Institute of Physics, Lithuanian Academy of Sciences,
G. Goštauto 12, 2600 Vilnius, Lithuania
[2]Physical-Technical Institute, Russian Academy of Sciences,
198904 St. Petersburg, Russia

In comparison with bulk crystals, semiconductor-doped glasses (SDG) attract a specific interest particularly due to short relaxation times of their optical nonlinearities. The investigation of the fastest room-temperature nonlinearities of SDG by means of excite-and-probe technique is reported on in the paper. We used the third harmonic of $KGd(WO_4)_2:Nd^{3+}$ laser radiation (FWHM duration 5 ps, quantum energy 3.05 eV) for band-to-band excitation of semiconductor microcrystals, and the output of parametric generator, pumped by the second harmonic of the same laser, for probing.

The most interesting results were obtained with microcrystals, their average radius exceeding the exciton Bohr radius, i.e. when quantum confinement effect was rather week. Nevertheless, from the point of view of temporal characteristics of nonlinear response, the SDG with large microcrystals are rather specific objects due to the rapid surface recombination of electron-hole pairs, and the spatial confinement of electron-hole plasma (EHP) and excitons by natural boundaries of a microcrystal. The latter effect prevents diffusion and enables to create a high density of quasiparticles. Both, exciton system and EHP, were investigated. The former was generated in CuBr microcrystals. The linear absorption spectrum of this material in the region of a band gap is dominated by $Z_{1,2}$ exciton peak. The nonlinear absorption under the excitation is properly described by the broadening of Lorentzian-shaped exciton band. Both, temperature and exciton-density dependence of the damping parameter, were studied in order to elucidate the mechanisms of the broadening. The relaxation of nonlinear absorption is strongly effected by the size of the microcrystals (Fig. 1). The absorption recovery time, as short as 30 ps in the smallest microcrystals, was interpreted to be governed mainly by two effects: radiative exciton-exciton interaction and surface recombination. The absence of long-lasting components of nonlinear absorption was proved by the experiments with two sets of pulses (5 ps FWHM) separated by 120 ps.

The EHP was created in CdS microcrystals of 12 nm average radius. Alongside with the bleaching of the sample in the spectral region near the band edge, the competition of this effect with band-gap renormalization, causing the induced absorption, was observed when probe-quantum energy significantly exceeded the band gap. The dependence of nonlinear absorption on excitation intensity was investigated and is consistent with the results of investigations in time domain. Due to the competence of the two opposite-sign effects, the time of switching from the bleaching of the sample to the state of induced absorption can be as short as 10 ps (see Fig.2.).

Springer Series in Chemical Physics, Vol. 55 **Ultrafast Phenomena VIII**
Editors: J.-L. Martin · A. Migus · G.A. Mourou · A.H. Zewail

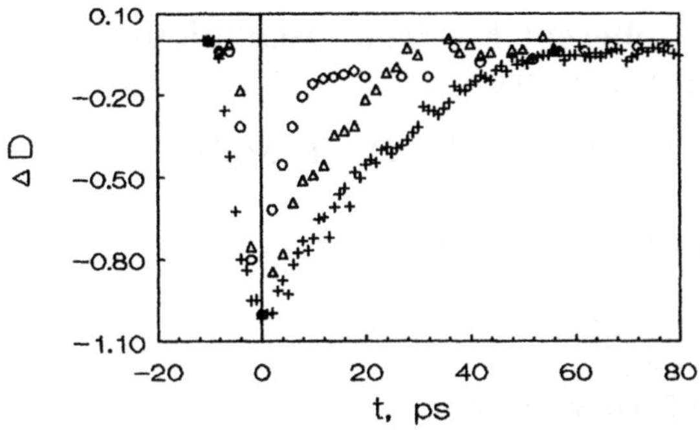

Fig.1. Time evolution of excitation-induced changes of optical density at $Z_{1,2}$ exciton peak for three average sizes of CuBr microcrystals: 6 (o), 14 (Δ) and 21 nm (+). All curves correspond to the same excitation intensity and are normalized to peak ΔD of the last sample.

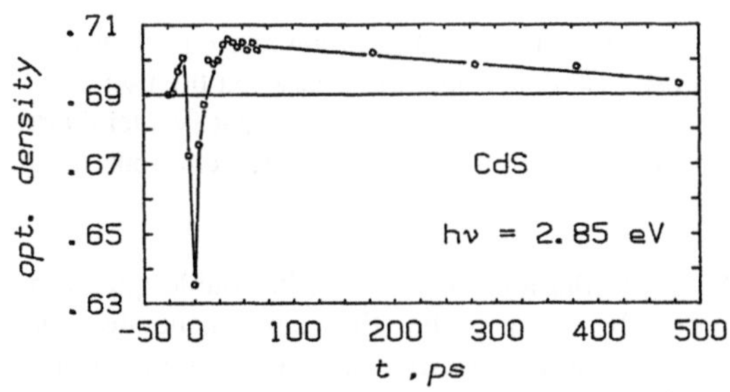

Fig.2. Time evolution of optical density of photoexcited CdS microcrystals at 2.85 eV.

Terahertz Radiation from Coherent Electron Oscillations in a Double-Quantum-Well Structure

H.G. Roskos[1,*], *M.C. Nuss*[1], *J. Shah*[1], *K. Leo*[1,*], *D.A.B. Miller*[1], *S. Schmitt-Rink*[2], *and K. Köhler*[3]

[1]AT&T Bell Laboratories, Crawfords Corner Road, Holmdel, NJ 07733-3030, USA

[2]Dept. of Physics & Mat. Science Center, Philipps-University, W-3550 Marburg, Fed. Rep. of Germany

[3]Fraunhofer-Institute for Applied Solid State Physics, W-7800 Freiburg, Fed. Rep. of Germany

*Present address: Institute of Semiconductor Electronics II, Technical University of Aachen, Sommerfeldstr. 24, W-5100 Aachen, Fed. Rep. of Germany

Abstract. We detect the submillimeter-wave radiation from excitonic wave packets that, after excitation by an ultrashort laser pulse, oscillate coherently between the two wells of a semiconductor coupled-quantum-well structure. In addition, we identify a terahertz signal originating from the instantaneous polarization of photo-excited excitons in an electrical bias field.

The dynamics of coherent spatial charge oscillations of excitons in double-quantum-well structures has recently been studied by pump-probe (PP) and degenerate four-wave-mixing (DFWM) measurements [1]. Here, we describe an experiment that probes the coherent charge oscillations more directly by detecting the electromagnetic field transient emitted by the oscillating excitonic wave packets [2].

Our asymmetric GaAs double-quantum-well structure (well widths: 145 Å / 100 Å), with a thin (25 Å) $Al_{0.2}Ga_{0.8}As$ tunneling barrier between the wells, allows tuning of the energy levels of the two wells by an external electrical bias field. The bias is applied between the n-doped substrate and a semi-transparent Cr Schottky top contact. At a certain bias, the lowest conduction band levels are aligned and the electron wave functions extend over both wells. A 250 fs optical pulse (bandwidth: 7.5 meV) resonant in energy to the lowest light-hole excitonic interband transition ($\hbar\omega$ = 1.541 eV) then generates a wave packet in the wide well only (see Fig. 1 (a)) [2, 4]. From this non-stationary state, the electrons oscillate between the two wells at a frequency that is determined by the probability for tunneling through the $Al_{0.2}Ga_{0.8}As$ barrier. The holes remain localized because the energy levels in the valence band are not aligned at this bias. Electrons and holes together form an oscillating excitonic dipole that emits electromagnetic radiation in the THz frequency range until the phase coherence of the oscillating wave packets is destroyed.

We detect the transient radiation field with the help of a photoconductively driven subpicosecond dipole antenna (see Fig. 1 (b) for the geometry of the experiment) [3]. Fig. 2 shows measured THz waveforms at a sample temperature of 10 K for internal reverse bias fields from close to flatband up to 16.3 kV/cm. Fig. 3 presents the Fourier spectra of these data. We observe two distinct contributions to the signal: (i) a spectrally broad single-cycle contribution that for low bias fields depends linearly on the bias and disappears at flatband, and (ii) an oscillatory component at roughly 1.6 THz.

Springer Series in Chemical Physics, Vol. 55 Ultrafast Phenomena VIII
Editors: J.-L. Martin · A. Migus · G.A. Mourou · A.H. Zewail

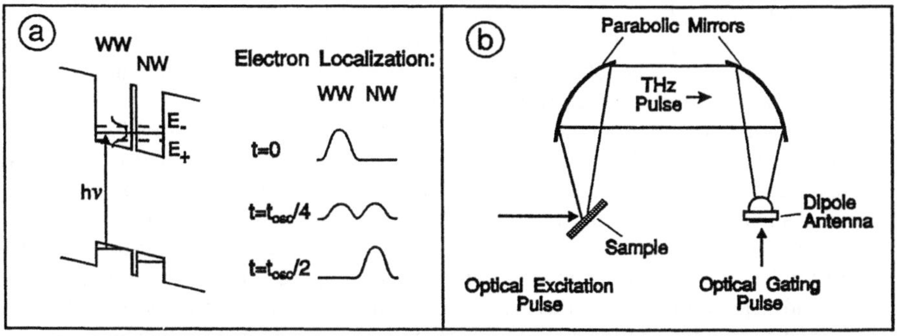

FIG. 1. (a) Schematic band diagram of the asymmetric coupled GaAs quantum wells at a bias that brings the lowest conduction band levels of the two wells into resonance. The aligned levels split into a bonding and an antibonding state (E_+ and E_-), separated by an energy gap proportional to the tunneling probability of the carriers through the AlGaAs barrier. A short laser pulse resonant to the lowest excitonic transition excites both split levels simultaneously, i.e. generates exciton wave packets in the wide well (WW) only. The electrons begin to oscillate between the wells while the holes remain confined in the WW. (b) Experimental setup to observe the terahertz transient emitted by the photo-created excitons.

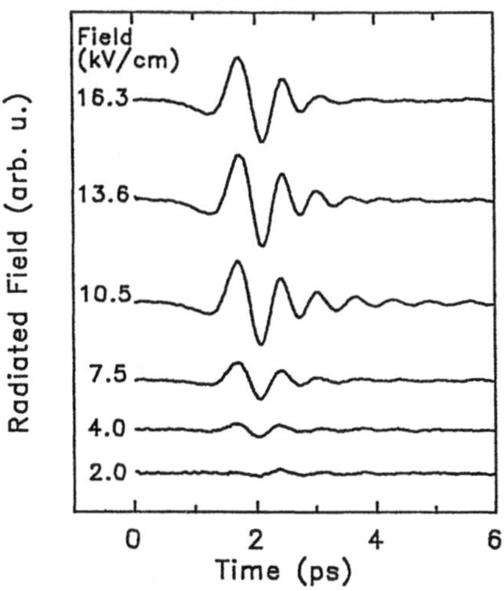

FIG. 2. Detected electromagnetic transients from the coupled quantum well for different internal bias fields and at low temperature (10 K).

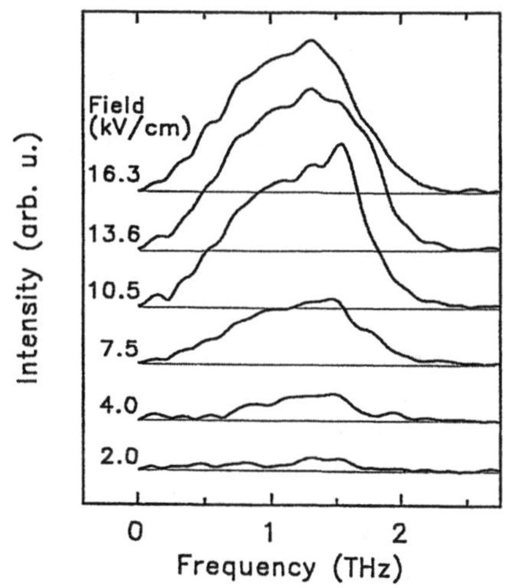

FIG. 3. Fourier spectra of the waveforms of FIG. 2. The spectra are composed of a narrow-band contribution from the coherent exciton oscillations and a broad-band signal from the impulsive generation of electric-field-polarized excitons.

We attribute the oscillatory signal contribution to the sinusoidal polarization change from electrons tunneling through the barrier between the quantum wells. This signal is strongest at a bias field of 10.5 kV/cm and decreases in amplitude and lifetime for higher and lower fields. Photocurrent measurements show that the lowest wide-well and narrow-well conduction band levels are aligned at 10.5 kV/cm and charge oscillations should be possible. The observed THz signal decays with a time constant of 3 ps because of dephasing of the oscillating exciton wave packets. The oscillation frequency is 1.56 THz. For higher fields, it increases up to 1.84 THz (at 15 kV/cm), for lower fields it does not vary significantly from the resonance value. The bias dependence is weaker than expected, an observation that has also been made in PP and DFWM experiments [1, 5] and requires further theoretical analysis.

We identify the initial single-cycle contribution to the detected waveforms as the fingerprint of the impulsive creation of polarized excitons by the laser pulse. Under the influence of the electrical bias field, the electrons and holes are generated spatially displaced. As long as the excitation pulse feeds the instantaneous polarization, an electromagnetic transient is emitted. This optical rectification process has recently been predicted for the case of bulk semiconductors [6]. The amplitude of the instantaneous signal should depend linearly on the bias field. In bulk materials, however, it is difficult to distinguish the rectification signal from the radiation resulting from transport currents of the photogenerated charge carriers accelerated in the surface field. Both possible radiation sources are instantaneous to the excitation on a time scale of 100 fs and may contribute to the observed THz mono-cycle transients. In our experiment, the confinement of the charge carriers in the quantum wells inhibits ultrafast carrier transport across the quantum well barriers along the bias field. The only possible fast transport current is that associated with the coherent oscillation of the electrons between the quantum wells. Since it occurs on a time-scale slower than optical rectification, our experiment temporally separates the THz transient resulting from optical rectification and the THz signal from the oscillating wave packets.

In conclusion, we have for the first time directly observed the electromangentic radiation emitted by coherently oscillating exciton wavepackets. We also identify for the first time a contribution to a THz transient resulting from an optical rectification process in a quantum well structure.

REFERENCES:

[1] K. Leo, E. O. Göbel, T. C. Damen, S. Schmitt-Rink, W. Schäfer, and K. Köhler, Phys. Rev. Lett. **66** (1991) 201.

[2] H. G. Roskos, M. C. Nuss, J. Shah, K. Leo, D. A. B. Miller, A. M. Fox, S. Schmitt-Rink, and K. Köhler, Phys. Rev. Lett. **68** (1992) 2216.

[3] P. R. Smith, D. H. Auston, and M. C. Nuss, IEEE **JQE-24** (1988) 255.

[4] in preparation for submission to Phys. Rev. B.

[5] K. Leo, J. Shah, T. C. Damen, A. Schulze, T. Meier, S. Schmitt-Rink, P. Thomas, E. O. Göbel, S. L. Chuang, M. S. C. Luo, W. Schäfer, K. Köhler, and P. Ganser, to be published in IEEE J. Quantum Electron.

[6] S. L. Chuang, S. Schmitt-Rink, B. J. Greene, P. N. Saeta, and A. F. J. Levi, Phys. Rev. Lett. **68** (9192) 102.

Optical Generation of Terahertz Pulses from Polarized Excitons in Quantum Wells

P.C.M. Planken and M.C. Nuss

AT&T Bell Laboratories, Holmdel, NJ 07733-3030, USA

Abstract. We generate terahertz pulses in quantum wells proving that the optical creation of polarized excitons is an important mechanism for the generation of terahertz radiation. In a wide well sample, we observe oscillatory terahertz emission from the coherent excitation of both light-hole and heavy-hole excitons. The radiation is tunable from 1.4 to 2.6 THz, with a dephasing time of 3-5 ps.

After the original discovery of optical rectification at semiconductor surfaces [1], significant attention has focused on the origin of this radiation. Initial interpretations concentrated on transient transport currents in the surface fields of the semiconductors. Recently however, an additional explanation was given [2] based on the optical creation of e-h pairs in states in which they are already polarized. This leads to a time-dependent total polarization P that grows with the integrated pulse energy and hence radiates an electrical transient according to $E \sim \partial^2 P / \partial t^2$. In quantum wells, transport current is significantly inhibited, allowing us to finally prove the importance of the creation of polarized e-h pairs for the generation of THz pulses. This is demonstrated by the experiment sketched in Fig. 1.

A MQW sample consisting of 82 periods of 145 Å thick GaAs wells, separated by 102 Å $Al_{0.3}Ga_{0.7}As$ barriers (Fig. 1), is illuminated by femtosecond pulses from a mode-locked Ti:sapphire laser. The laser is tuned roughly 4 nm above the bandedge, near the lh and hh excitons. Because of the near-bandgap excitation, we create excitons (rather than uncorrelated e-h pairs) in our experiment. Although the lh-hh splitting is within our excitation bandwidth, we do not observe any effects from the simultaneous excitation of the lh and hh excitons as discussed below, because of the broad exciton linewidths in the present sample. By applying a voltage to a chromium Schottky contact on top of the sample, we can control the electric field in the structure. Fig. 2 shows the THz transients coherently detected with a 50μm photoconductive dipole antenna [3] as a function of electric field in the structure at 77 K. The THz amplitude at a field of 15 kV/cm is reduced by a factor of two compared with the typical sig-

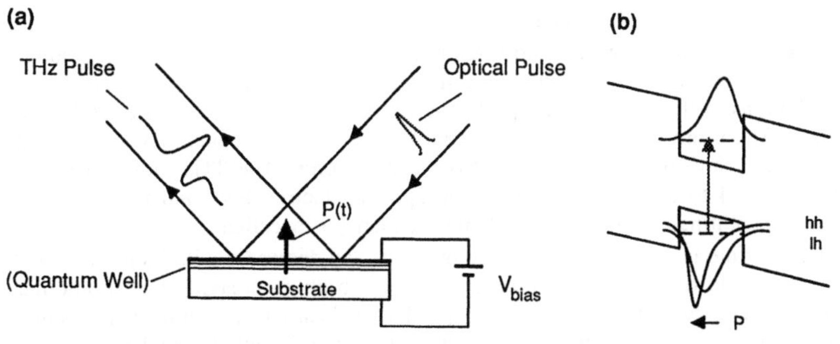

Fig. 1: (a) Schematic of experimental setup, and (b) band diagram and envelope functions of hh- and lh-excitons.

Springer Series in Chemical Physics, Vol. 55 Ultrafast Phenomena VIII
Editors: J.-L. Martin · A. Migus · G.A. Mourou · A.H. Zewail

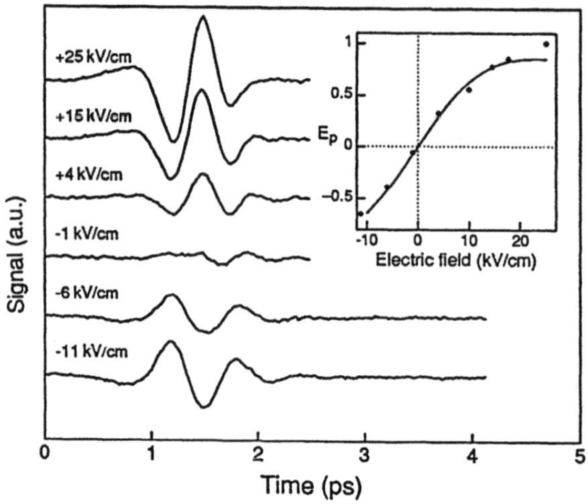

Fig. 2:
THz-transients for different electric fields. The insert shows the measured THz amplitude as function of electric field. The solid line is based on a calculation, as described in the text.

nal from the surface of bulk GaAs at the same temperature, whereas we estimate that in this sample the transport current is at least a factor of 100 smaller. We therefore conclude that the THz pulse is generated by the creation of polarized excitons and not by transient transport currents.[4]

The insert shows the amplitude of the THz pulses as a function of electric field. Initially the amplitude rises linearly with field, but shows saturation for higher fields. This can be understood from the decreasing oscillator strength of the transition as the electron and hole are pulled apart and from the presence of the barriers which prevent the dipole from growing linearly with electric field. We have modeled both these effects by calculating the field-induced polarization $P \sim |\langle \phi_e | \phi_h \rangle|^2 \cdot [\langle \phi_h | z | \phi_h \rangle - \langle \phi_e | z | \phi_e \rangle]$. The solid line in the insert is the result of this calculation with the proportionality constant adjusted to give the best fit to the data.

In another experiment we resonantly excite both hh- and lh-excitons in a 15-period 175 Å GaAs, 150 Å $Al_{0.3}Ga_{0.7}As$ MQW-structure. Fig. 3 shows the detected THz transients as a function of bias voltage applied across the structure at 10 K. The first cycle of radiation results from the excitons already being in a polarized state upon creation [2] as discussed above. The origin of the later, oscillatory radiation is that in an electric field, the hh-exciton is polarized more than the lh-exciton (Fig. 1). When a coherent superposition of both lh- and hh-excitons is excited, these two polarizations will beat against each other, leading to a dipole moment μ that oscillates at the hh-lh splitting frequency. The size of this dipole moment corresponds to the difference of the field-induced dipole moments of both types of excitons. (This should be compared to the oscillatory radiation that is generated in a coupled double quantum well [5], in which it is the quantum mechanical coherent tunneling process that leads to spatial oscillations of the electron wavepacket between both wells and hence to oscillatory THz radiation.) The insert compares the observed THz oscillation frequency with the splitting of the hh- and lh-excitons as observed from PLE-spectra. The THz radiation can be tuned easily from 1.4 to 2.6 THz by varying the applied electric field, and the agreement between the hh-lh splitting and the THz oscillation frequency is excellent.

The dephasing time in Fig. 3 is in the range of 3-5 ps and decreases with increasing field. We attribute the origin of the dephasing process in part to phase breaking due to interface roughness of the quantum wells. Intensity-dependent measurements show that exciton-exciton and exciton-free carrier scattering do not contribute significantly to the dephasing process in the considered excitation density range ($n = 10^9 - 10^{10}\ cm^{-2}$).

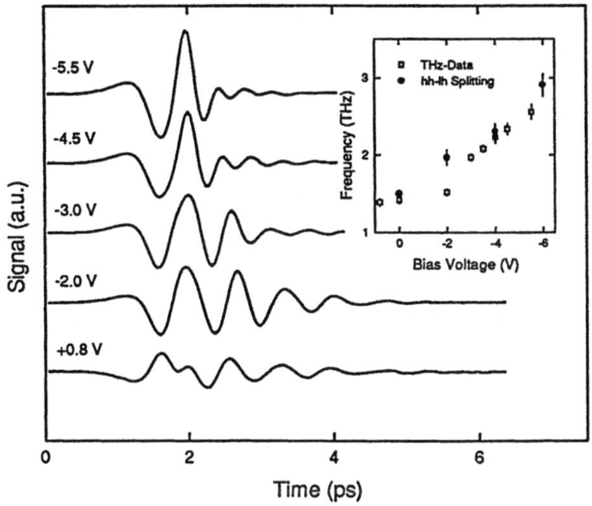

Fig. 3:
Oscillatory THz-transients as a function of applied voltage. The insert shows the measured oscillation frequency compared to the hh-lh splitting.

Unlike in degenerate four-wave-mixing experiments, we do not expect any THz radiation at flatband (0.8 V), when neither hh- nor lh-excitons are polarized. The observation of THz radiation even at flatband indicates that the picture outlined above is oversimplified. We attribute this to the valence-band mixing in GaAs quantum wells, leading to a nonvanishing transition dipole moment even at flatband. Unlike the field-induced part, this dipole moment also has components in the plane of the quantum well.

1. X.-C. Zhang, B. B. Hu, J. T. Darrow, and D. H. Auston; Appl. Phys. Lett. **56**, 1011 (1990)
2. S. L. Chuang, S. Scmitt-Rink, B. I. Greene, P. N. Saeta, and A. F. J. Levi, Phys. Rev. Lett. **68**, 102 (1992).
3. P. R. Smith, D. H. Auston, and M. C. Nuss; IEEE J. Quant. Electr. **24**, 255 (1988)
4. P. C. M. Planken, and M. C. Nuss; submitted to Appl. Phys. Lett.
5. H. G. Roskos, M. C. Nuss, J. Shah, D. A. B. Miller, S. Schmitt-Rink, and K. Köhler; Phys. Rev. Lett. **68**, 2216 (1992)

Generation of High-Power Single-Cycle Picosecond Radiation

D.R. Dykaar[2], *R.R. Jones*[1], *D. You*[1], *D. Schumacher*[1], *and P.H. Bucksbaum*[1]

[1]Department of Physics, University of Michigan, Ann Arbor, MI 48109-1120, USA
[2]AT&T Bell Laboratories, Holmdel, NJ 07733-3030, USA

Abstract. We have created single-cycle picosecond (THz bandwidth) pulses with pulse energies of up to 1 μJ. Such pulses are unique new sources for the study of atoms and molecules in intense coherent elecromagnetic radiation fields.

1. Introduction

The production of single-cycle transverse electromagnetic radiation has opened up a new field of spectroscopy[1]. A significant issue is the scaling of these pulses to high peak electric fields, where nonlinear effects may be studied. We report the creation of single cycle picosecond pulses with energies of $\simeq 1\mu$J. The efficiency and the dependence on experimental parameters suggests that scaling to 1 mJ or more should be straightforward.

2. Making single-cycle radiation

The single cycle pulses are made by illuminating a biased GaAs wafer with 100 fs, 770 nm laser pulses. The semiconductor mounting details such as crystalligraphic orientation of the bias field and laser polarization, or angle of incidence, make little or no difference in performance. We have seen strong evidence for sample-dependence that may suggest a preferred doping or growth method, but this has not yet been investigated systematically.

In a typical setup a 20 mil (.05 cm) wafer is illuminated at normal incidence while a 40 kV/cm electric field is applied in the wafer plane. The wafer will not support this field dc, but a slow (50 μs) pulsed bias works well. We have applied up to 70 kV/cm without damaging the wafer. Thz radiation is generated in transmitted and reflected directions with comparable amplitudes, although second pulses, presumably due to multiple reflections from the wafer surfaces, are seen in transmission (see fig. 1). The energy in the pulse is largest for short (~100 fsec) applied laser pulses. The THz energy is a complicated nonmonotonic function of

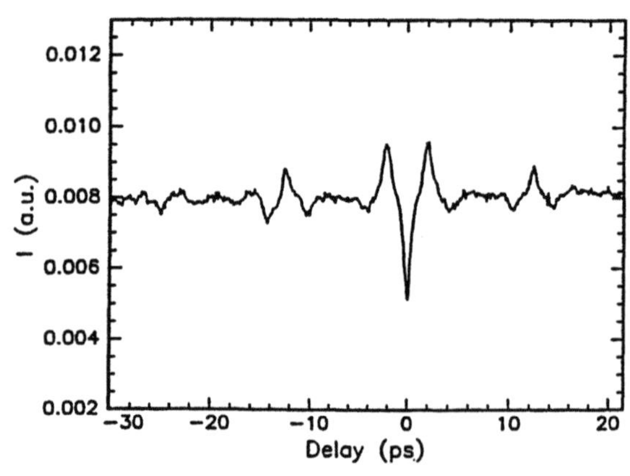

Fig.1 Interferometric measurement of the THz pulse shape.

Springer Series in Chemical Physics, Vol. 55 **Ultrafast Phenomena VIII**
Editors: J.-L. Martin · A. Migus · G.A. Mourou · A.H. Zewail

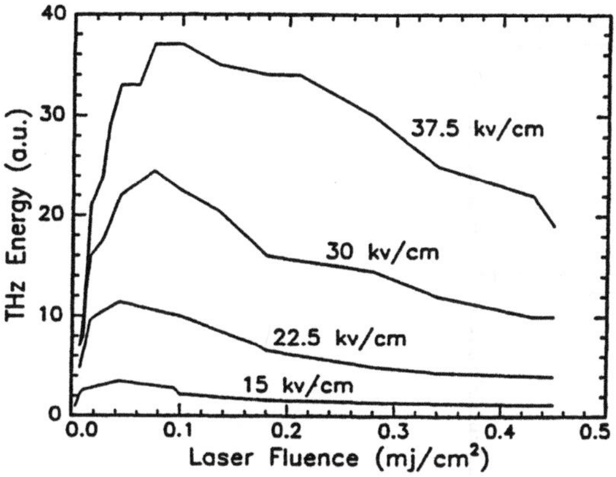

Fig.2 Dependence of THz pulse energy on laser fluence.

laser fluence(fig. 2). The function may depend on the pulse shape; this has not yet been fully investigated. Note that the peak energy occurs for rather modest fluence. We need use only a few percent of the 8 mJ supplied by our Ti:Sapphire laser system to illuminate a 2 cm^2 area. Most interesting is the quadratic dependence of the THz energy on the bias field. We have not seen any saturation in this behavior up to the largest bias fields we can create.

3. Pulse detection

The THz pulse shape is measured with a device similar to a Michaelson interferometer, described in reference [2] (see fig. 1). This technique has the advantage of not requiring a fast sampling gate, so that there is no roll-off in resolution at high frequency. The output energy is measured with a pyroelectric detector. The separation between the multiple pulses in figure 1 is equal to the round trip time of the pulse in the GaAs wafer. We have measured autocorrelations of THz pulses generated with different laser intensities and bias fields. No obvious change in pulse length is observed. However, when no bias field is used, significantly shorter pulses may be observed[2].

4. Ionization experiments

In our first atomic physics experiment with single-cycle pulses, Rydberg atoms of Barium were irradiated with the THz pulses from a ~1cm^2 GaAs wafer focused to a 3 mm (FWHM) waist in vacuum. We were able to completely ionize of states with n > 25 (fig. 3). Further work on focusing Thz radiation to high peak fields for excitation and ionization of atoms is in progress.

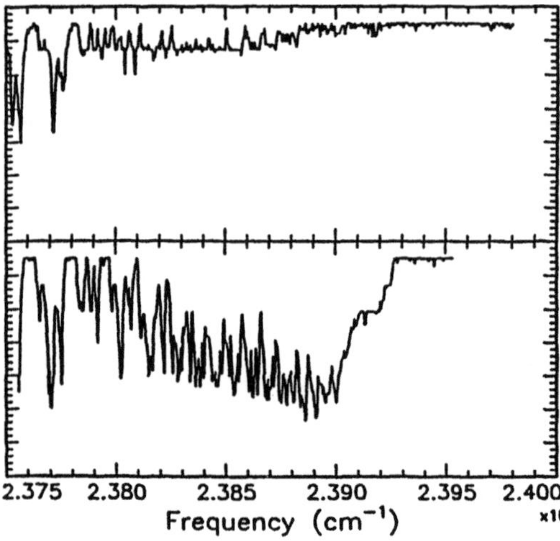

Fig.3 Bottom: Population of Rydberg states produced by laser excitation of 6s6p-6snk transitions in barium in a 100 v/cm static clearing field. Top: Following excitation, the same Rydberg states are depleted by ionizing with a focused single cycle picosecond pulse.

This research is supported by National Science Foundation.

References

[1]B.B. Hu, J.T. Darrow, X.-C. Zhang, and D.H. Auston, Appl. Phys. Lett. 56, 886 (1990); M. van Exter, Ch. Fattinger, and D. Grischkowsky, Opt. Lett. 14, 1128 (1989); B.I. Greene, J.F. Federici, D.R. Dykaar, A.F.J. Levi, and L. Pfeiffer, Opt. Lett. 16, 48 (1991); G. Mourou, C.V. Stancampiano, A. Antonetti, and A. Orzag, Appl. Phys. Lett. 39, 295 (1981).

[2]B.I. Greene, J.F. Federici, D.R. Dykaar, R.R. Jones, and P.H. Bucksbaum, Appl. Phys. Lett. 59, 893, (1991).

Transient Electron Transport in GaAs Quantum Wells: From the Ballistic to the Quasi-Equilibrium Regime

W. Sha, J. Rhee, and T.B. Norris

Center for Ultrafast Optical Science, University of Michigan, 2200 Bonisteel, Ann Arbor, MI 48109-2099, USA

The study of transient electron transport has fundamental importance for the understanding of high-speed semiconductor devices. We have attempted a fairly comprehensive experimental investigation of high-field electron dynamics on femtosecond and picosecond time scales corresponding to the transient regimes of ballistic, overshoot, and quasi-equilibrium transport [1,2].

The basic scheme is to inject electron-hole (e-h) pairs with a 100-fs optical pulse into $GaAs/Al_{0.3}Ga_{0.7}As$ multiple quantum wells, with a *parallel* field (up to 16 kV/cm) applied across a 50-μm gap (using a horizontal p-i-n structure). In fact, carriers are injected at the edge of the conduction band with zero initial kinetic energy (pump wavelength 820nm, see fig.1). The subsequent dynamics are studied by three complementary experiments on the *same* sample:

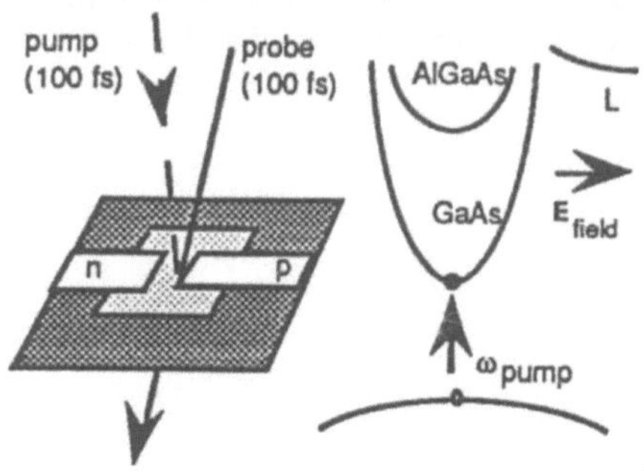

fig.1 experimental setup

1. Ballistic acceleration of electrons on a 150 fs time scale.

Using a femtosecond supercontinuum as our probe, we measure the normalized differential transmission spectra by modulating the electric field, with results shown in fig.2. The electron distribution functions are measured via the absorption saturation due

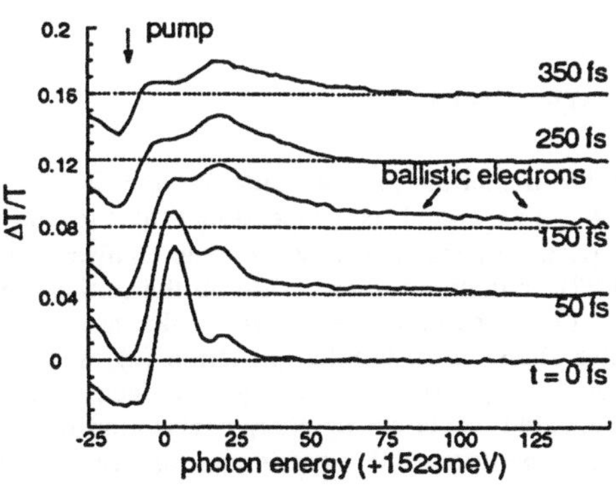

fig.2 probe with fs supercontinuum

Springer Series in Chemical Physics, Vol. 55 Ultrafast Phenomena VIII
Editors: J.-L. Martin · A. Migus · G.A. Mourou · A.H. Zewail

to band filling for transitions above the exciton lines (E > 40meV); the high-energy tail at 150 fs indicates the presence of electrons ballistically accelerated from the band edge by the applied electric field, which is directly observed for the first time. [3]

2. Transient electron velocities obtained by measuring relaxation of the electric field.

The field in the gap switches off due to (1) space-charge screening and (2) THz radiation by accelerating ballistic electrons. We probe the temporal behavior of the field via excitonic electro-absorption [4]. A 76-MHz 100-fs Ti:Sapphire laser is tuned to probe the electric-field-induced absorption change at the hh exciton line (the contribution from exciton saturation due to carriers is small). The differential transmission ΔT_ε is taken by modulating the bias field (fig.3). For t<0, ΔT_ε corresponds to the bias field 16kV/cm. After the pump pulse, the field switches off (apparent as $\Delta T_\varepsilon \rightarrow 0$), displaying both fast and slow components. The fast component is partly due to radiation. The slow component we attribute to space charge screening, and can be modeled simply by noting that as the space charges separate by dx, the field will be screened by $d\varepsilon$ proportional to dx. Two types of field screening are observed in fig.3 ((a) and (c)). In curve (a), for high carrier density, the field relaxes initially by radiation into the low field ($\leq$ 3 kV/cm) regime, and afterwards exponentially which indicates $v = \mu\varepsilon$ (fitting indicates μ=4500 $cm^2/Vsec$). In curve (c), for low density, the initial radiation loss is small, and as the field is still high, the linear screening indicates a saturated electron velocity (v_s = 2.5×10^7cm/sec). The time interval from the initial photoinjection of e-h to the established quasi-equilibrium states (described either by μ or v_s) is less than 1 ps.

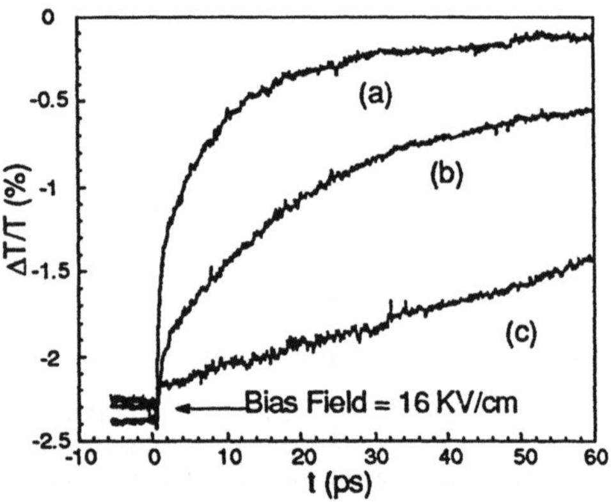

fig.3 transient internal field probe

3. THz radiation as a measure of carrier transport.

By using standard THz photoconductive detection [5], we measure the far-field radiation pulse from the same gap (fig.4). The signal is proportional to the time derivative of the photocurrent. The main positive peak in the two curves confirm that ballistic electrons (fig.2) generate efficient THz radiation which results in the initial fast relaxation of the internal field (fig.3). (The secondary positive peaks are from reflections.) We observe a change of pulse shape from bipolar (in curve (a)) at high carrier density to unipolar (b) at low carrier density. At high density, the electric field collapses rapidly in the gap region; while coupled to the strong energy relaxation of the electrons through LO-phonon emission, the photocurrent experiences an overshoot

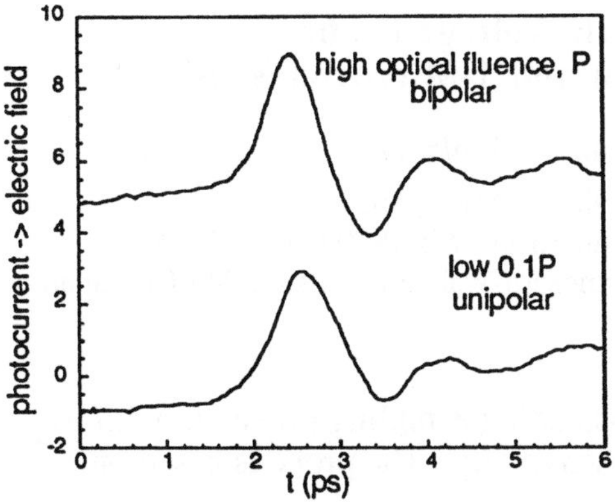

fig.4 observed THz pulse

producing the bipolar radiation pulse. This dramatic overshoot can also be observed in fig.1, as the electron distribution at high energies rises at 150 fs and drops at 250 and 350 fs. At low carrier density, the field stays on; the velocity of the electrons reaches a saturated value due to real-space transfer or intervalley scattering, and so thus the photocurrent. In this case, the radiated pulse is unipolar.

This work was supported by AFOSR grant URI-090-0214 and NSF PHY-8920108.

1. K.Hess and G.J.Iafrate, Proc.IEEE 76, 519 (1988), and references therein.
2. W. Sha, J. Rhee, T. B. Norris, W. J. Schaff, to be published in IEEE JQE, special issue on ultrafast optics and electronics, October, 1992
3. W. Sha, T. B. Norris, W. J. Schaff and K. E. Meyer, Phys. Rev. Lett. 67, 2553 (1991).
4. W. H. Knox *et al.*, IEEE JQE 25, 2586 (1989), and references therein.
5. P. R. Smith, D. H. Auston, and M. C. Nuss, IEEE JQE 24, 255 (1988).

A Novel Free-Standing Absolute-Voltage Probe with 2.3-Picosecond Resolution and 1-Microvolt Sensitivity

J. Kim[1], *S. Williamson*[1], *J. Nees*[1], *and S. Wakana*[2]

[1]Ultrafast Science Laboratory, University of Michigan, 2200 Bonisteel, Room 1006, IST, Ann Arbor, MI 48109-2099, USA

[2]Fujitsu Laboratories Ltd., 10-1 Moorinosato-Wakamia, Atsugi 243-01, Japan

Abstract. We demonstrate a free-standing sampling probe with 2.3 ps temporal resolution and $1\mu V/\sqrt{HZ}$ sensitivity. The probe is based on low-temperature-grown-GaAs

We are reporting on a new optically-based external probe that overcomes limitations of temporal resolution and invasiveness met by current all-electronic instrumentation while maintaining excellent voltage resolution. The probe concept is based on photoconductive (PC) sampling, the most sensitive optically-based sampling technique.[1] The embodiment we have developed has a sensitivity of $1\ \mu V/\sqrt{Hz}$ and is capable of measuring absolute voltage without in-situ calibration. Its dynamic range is $>10^6$, making possible measurements from the microvolt to the 10-volt level. Though contacting in nature, the probe is non-invasive with a load resistance of $>100\ k\Omega$ and a capacitance of <1 fF. It makes contact to the device-under-test using a small tip, defined by conventional photolithography. The sensitivity and speed of the probe are independent of the tip-to-device contact resistance up to $10\ k\Omega$.

The high-speed, high-sensitivity, and non-invasive features of this probe are realized by the use of low-temperature-grown-GaAs (LT-GaAs). This material's subpicosecond carrier lifetime, high mobility ($>200\ cm^2/V\cdot s$), and high breakdown field strength (>500 kV/cm), make it an ideal material for electrical pulse switching and gating. Last year, we reported on a high-sensitivity LT-GaAs-based metal-semiconductor-metal photoconductive detector (MSM) that takes advantage of these features.[2] By decreasing the carrier transit time across the semiconductor gap in the MSM to a value comparable to the carrier lifetime we increase the photocurrent gain to a value near unity. With unity gain, the responsivity of an LT-GaAs MSM becomes comparable to that of an ideal detector while the speed is still dominated by the (sub)picosecond intrinsic carrier lifetime.

As it is not practical to fabricate a photoconductive gate on every circuit we wish to test, we designed an external probe with a sampling gate which resides on an independent substrate. The probe consists of an LT-GaAs MSM switch which is integrated with a small tip that can be used to make contact with the circuit-under-study.

Springer Series in Chemical Physics, Vol. 55 Ultrafast Phenomena VIII
Editors: J.-L. Martin · A. Migus · G.A. Mourou · A.H. Zewail

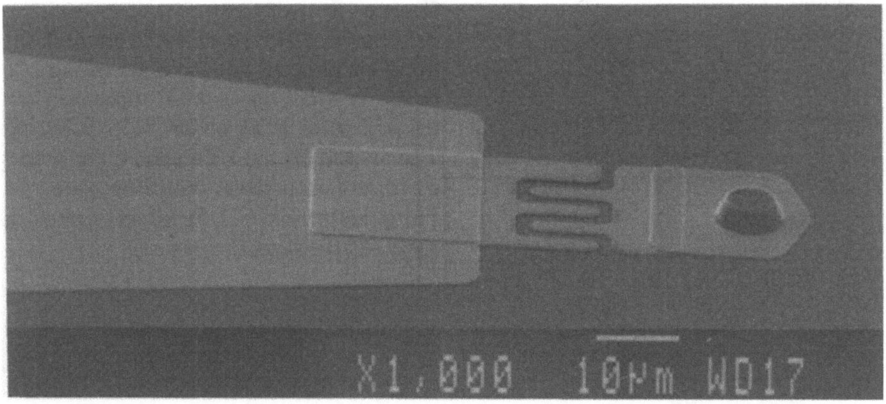

figure 1
Scanning-electron-micrograph of photoconductive sampling probe. It consists of 2-µm interdigital photoconductive switch and titanium contacting tip of 8-µm diameter. The switch and tip metalizations were photo-lithographically defined and deposited by e-beam evaporation.

To demonstrate the concept of external photoconductive sampling we fabricated a picosecond sampling probe on an LT-GaAs layer. Using standard lithographic methods, we defined a 2-µm finger-width interdigital photoconductive switch with one electrode extending to a 20-µm square pad with a triangular extension (See figure 1). A 3-µm-high titanium tip of 8-µm diameter was placed near the apex of this shape, again, by standard lift-off techniques, but using a thick phohotoresist layer.[3] The other electrode connectes the switch to lock-in detection electronics. No ground plane or transmission line structure is necessary on the probe because the high-speed sampling occurs at the gate in immediate proximity to the circuit-under-test.

We used this photoconductive sampling probe to measure picosecond electrical waveforms on coplanar transmission lines. Figure 2 shows a schematic diagram of the experimental set-up. An electrical signal is generated by the photovoltiaic effect at the edge of the 20-µm coplanar striplines fabricated on damaged silicon-on-sapphire (SOS). A 120-fs Ti:sapphire laser generates an electrical signal and is also used to activate the probe gate. We controlled the contact resistance between the probe tip and transmission line electrode by monitoring both the dc electrical conductivity and optical interference. Contact was stable and repeatable, resulting in measurement of absolute voltage. Figure 3 is the electrical pulse measured by the photoconductive sampling probe. The rise time of the pulse is 2.3 ps and the signal-to-noise ratio 350:1. By moving the probe we were able to measure picosecond transients at any point along the transmission lines, demonstrating the versatility of the probe for

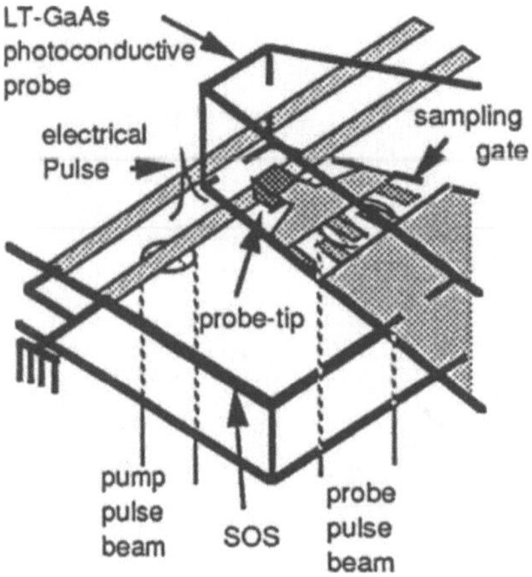

figure 2
Schematic diagram of experimental set-up. The photoconductive sampling probe is touching one of coplanar transmission lines on an SOS substrate. The switch on the LT-GaAs layer works as photoconductive sampling gate. Pump and probe beams are generated by a 120-fs Ti-sapphire laser.

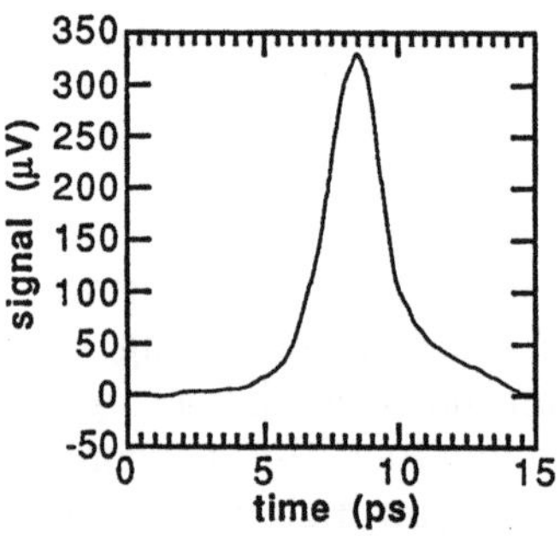

figure 3
Photovoltaic pulse generated on SOS. The transmission lines have 20-µm width and spacing. The signal has 1-µV RMS-noise.

sensitive, high-speed, non-destructive measurements. Owing to the low capacitance of the sampling tip (less than 1 fF) and its high off-state resistivity (>10 MΩ), the invasiveness of the probe is minimal.

In conclusion, we fabricated and tested a contacting photoconductive sampling probe with 2.3-ps temporal resolution, 1-µV sensitivity, and 8-µm spatial resolution. It will be applicable to testing of high speed electronic devices and circuits.

The Authors wish to acknowledge helpful discussions with John Whitaker. This work is supported by AFOSR-90-0214, ARO-DAAL-03-89-k-0071, and by Fujitsu Laboratories Ltd.

References

1. D.R.Grischkowsky, M.B.Ketchen, C-C Chi, I.N.Dulling III, N.J.Halas, and J-M. Halbout, *IEEE J. Quantum Electron.* **24**, 221(1988)
2. Y. Chen, S. L. Williamson, T.Brock.(1991). *Appl. Phys. Lett* **59**(16), 14.
3. C. Spindt, I. Brodie, L. Humphrey, and E. Westerberg, "Physical Properties of Thin Film Field Emission Cathodes with Molybdenum Cones," J. Appl. Phys., Vol. **37**(12), December 1976.

Picosecond Pseudomorphic AlGaAs/InGaAs MODFET Large-Signal Switching Measured by Electro-Optic Sampling

M.K. Jackson[1,4], *M.Y. Frankel*[2], *J.F. Whitaker*[3], *G.A. Mourou*[3], *D. Hulin*[4], *A. Antonetti*[4], *M. Van Hove*[5], *W. De Raedt*[5], *P. Crozat*[6], and *H. Hafdallah*[6]

[1]Department of Electrical Engineering, University of British Columbia, 2356 Main Mall, Vancouver, BC V6T-1Z4, Canada
[2]Naval Research Laboratory, Washington, DC 20375, USA
[3]Center for Ultrafast Optical Science, University of Michigan, Ann Arbor, MI 48109, USA
[4]Laboratoire d'Optique Appliquée, ENSTA, Ecole Polytechnique, Centre de l'Yvette, F-91120 Palaiseau, France
[5]Interuniversity Microelectronics Center, Kapeldreef 75, B-3001 Leuven, Belgium
[6]Institut d'Electronique Fondamentale, Université Paris Sud, F-91405 Orsay, France

ABSTRACT: We present a study of the large-signal switching characteristics on a picosecond time scale of an AlGaAs/InGaAs modulation-doped field-effect transistor (MODFET) using electro-optic sampling.

The MODFET has been of great interest for high-speed digital and analog applications since its demonstration. Recent progress in the fabrication of such devices has resulted in extrinsic cutoff frequencies exceeding 250GHz, corresponding to response times on the order of a picosecond. The rapid increase in the speed of MODFET devices has outstripped the ability to measure their performance with all-electronic equipment. Devices are most often characterized at lower frequencies and the results extrapolated to much higher frequencies. These difficulties can be avoided using very high-bandwidth techniques based on ultrafast lasers, where it is routine to generate and detect electrical signals with supicosecond rise and fall times. These optoelectronic techniques have been applied to small-signal characterization of passive coplanar striplines to 1THz[1] and MODFET devices to 100GHz.[2] In addition to the very high bandwidth of optoelectronic methods, the ability to generate large electrical signal amplitudes has been employed in the study of switching of active devices[3]

The AlGaAs/InGaAs MODFET is grown by molecular beam epitaxy, and a 0.35μm gate is defined using electron-beam lithography. For characterization the wafer is lapped to 100μm, and a single device is diced out and wire bonded into a coplanar stripline test fixture, as shown in Fig.1. A colliding pulse mode-locked (CPM) dye laser provides 100fs optical pulses. One of the CPM output beams illuminates a photoconductive switch located 3mm from the gate, and the second delayed beam measures the electrical transient on the coplanar stripline via electro-optic sampling.

The measured rise time of the input signal at the gate is 2.8ps, due mainly to dispersion on the coplanar stripline between the photoconductive switch and the gate input. The subsequent decay of the gate input is much slower, because of the long lifetime of the semi-insulating GaAs used in the photoconductive switch.

In Fig.2 we show the drain output voltage, measured about 50μm from the drain wire bond. The dc operating point of V_{ds}=1V and V_{gs}=0.5V is chosen to make the transistor

Springer Series in Chemical Physics, Vol. 55 Ultrafast Phenomena VIII
Editors: J.-L. Martin · A. Migus · G.A. Mourou · A.H. Zewail

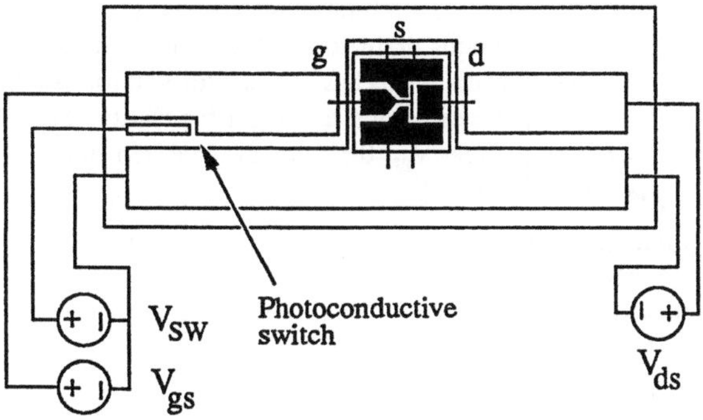

Fig.1 Coplanar stripline test fixture on semi-insulating GaAs. The photoconductive switch sends a step signal to the gate input. Voltage is measured using an external electro-optic probe (not shown).

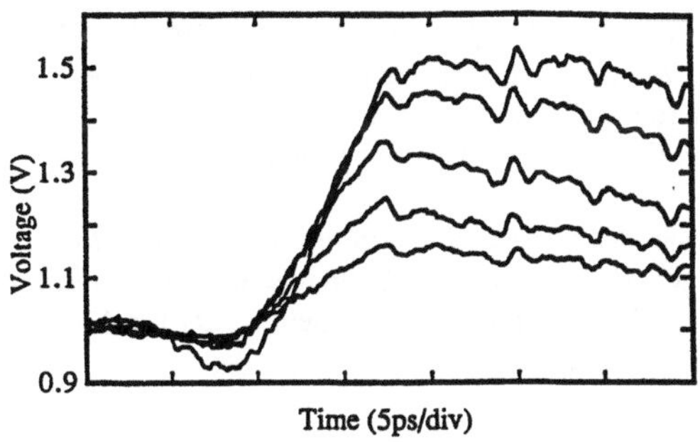

Fig.2 Drain output for negative-going gate transient (switch-off) for increasing gate step amplitudes of 1.0, 1.9, 2.7, 3.6, and 4.4 times ΔV_g, where ΔV_g is approximately –0.17V.

conducting. The gate transient in Fig.2 was negative, which leads to a positive-going drain signal. The drain output is shown for increasing gate transient amplitudes (see caption). From Fig.2 we see two main features in the drain response: a small, negative-going transient followed by switching of the transistor. After switching, the drain voltage decreases slowly, due to slow decrease of the gate input signal. The switching time in Fig.2 is 6.5ps, which corresponds well with cutoff frequency measurements of other devices from the same wafer.

In Fig.3 we show the drain voltage for transistor turn-on (positive gate transient) for various drain-source biases. From Fig.3 we see that for V_{ds}=1.0V, the drain transient has a very small precursor, no overshoot, and a switching time of 6.9ps. In contrast, for V_{ds}=0.4V, the precursor is much larger. For V_{ds}=1.5V, a significant overshoot is observed. It is interesting to note that in this case the switching time is reduced to 6.2ps. We can attribute the precursor signal seen in Figs.2 and 3 to feedthrough of the gate input signal, which is coupled directly from the gate to the drain by the capacitance C_{gd}. The variation of the precursor seen in Fig.3 can be explained by the variation of C_{gd} with operating point: C_{gd} changes as the form of the depletion region under the gate changes.

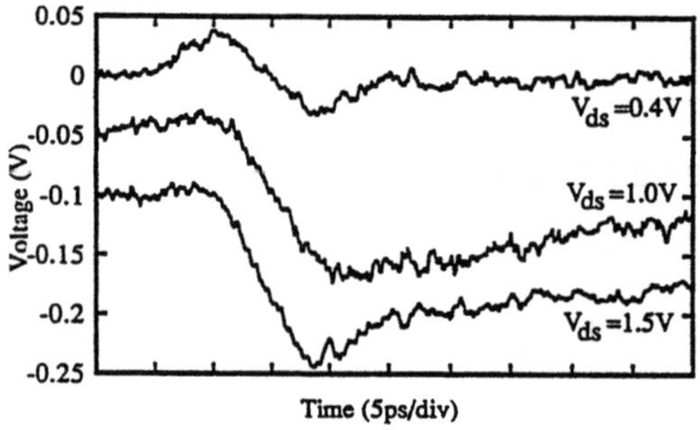

Fig.3 Switch-on drain voltage transients for various drain-source biases as labelled. The dc drain bias has been subtracted from all the curves, which are offset for clarity. For all curves V_{gs}=0.5V and the gate step amplitude is 0.16V.

This work was supported by ESPRIT basic research action 3042, "Nanofet."

References

[1] D.R. Dykaar, A.F.J. Levi and M. Anzlowar, Appl. Phys. Lett. **57**, 1123(1990); M.Y. Frankel, S. Gupta, J.A. Valdmanis and G.A. Mourour, I.E.E.E. Trans. M.T.T.**39**, 910(1991).

[2] M.Y. Frankel, J.F. Whitaker, G.A. Mourou, J.A. Valdmanis and P.M. Smith, in *OSA Proceedings on Picosecond Electronics and Optoelectronics*, Sollner and Shah, eds. (OSA, Washington, DC, 1991), Vol.9, p146.

[3] K.E. Meyer, D.R. Dykaar and G.A. Mourou, in *Picosecond Electronics and Optoelectronics*, Mourou, Bloom, and Lee, eds. (Springer, 1985), p54; J.A. Valdmanis, in *Measurement of High-Speed Signals in Solid State Devices*, R.B. Marcus, ed. (Academic, 1990) p.136.

Ultrafast Decay of Photodiffractive Gratings in Hetero n-i-p-i's by Enhanced In-Plane Transport

A.L. Smirl[1], *D.S. McCallum*[1], *A.N. Cartwright*[1], *X.R. Huang*[1], *T.F. Boggess*[1], *and T.C. Hasenberg*[2]

[1]Center for Laser Science & Engineering, University of Iowa, 124 AMRF Oakdale, Iowa City, IA 52242, USA

[2]Hughes Research Laboratories, 3011 Malibu Canyon Road, Malibu, CA 90265, USA

Charge separation is shown to cause an order-of-magnitude enhancement of the in-plane diffusive decay of photorefractive and photoabsorptive gratings in a semiconducting hetero n-i-p-i.

Hetero n-i-p-i's are one of several semiconductor quantum-well devices that use second-order electro-optic effects and that continue to attract interest for possible applications to low-power two-dimensional switching arrays and all-optical spatial light modulators. The turn-on time for a hetero n-i-p-i device is usually determined by transport perpendicular to the quantum wells. More specifically, it is determined by the time required for carriers generated in the quantum wells to escape the wells and to move to screen the built-in electric field, thus shifting the exciton. Consequently, typical turn-on times are of the order of a few ps. By contrast, when used in the conventional single-beam geometry, the recovery (or turn-off) time of hetero n-i-p-i's is determined by the slow recombination of the spatially-separated charges in the doped regions and is typically in the range of μs-ms. If instead, however, we use a two-beam mixing geometry for the device (where gratings are written in the material by the interference of the two beams), then the decay or turn off of the signal is determined by the decay of the gratings by in-plane transport over micron dimensions. Here, we use transient grating techniques to measure the recovery of such photorefractive and photoabsorptive gratings written in all-binary hetero n-i-p-i's. In this geometry, we show that the separation of photo-generated charge actually speeds the recovery by enhancing the effective in-plane ambipolar diffusion coefficient by roughly an order of magnitude (in contrast to the single beam geometry where charge separation prolongs the recombination and recovery time).

The structure we will focus on here is a hetero n-i-p-i containing quantum wells in the intrinsic regions which are composed of all-binary InAs/GaAs short-period strained-layer superlattices. The n and p regions are 20 nm thick layers of GaAs doped to a concentration of $2x10^{17}$ and $1.2x10^{18}$ cm^{-3}, respectively. Each intrinsic region is a 120 nm wide layer of undoped GaAs, which contains three quantum wells. Each quantum well consists of 6 periods of 2 monolayers of InAs alternating with 5 monolayers of GaAs. The growth and the dynamics of single-beam excitation of this sample are described elsewhere [1].

Here, we measure the in-plane ambipolar diffusive transport at room temperature by using a conventional transient grating technique. That is, a ~1 ps optical pulse produced by a cavity-dumped, synchronously-modelocked dye laser, which can be tuned across the excitonic resonance, is split into three parts. Two parts are spatially and temporally overlapped in the sample. The absorption of these two interfering pulses creates a periodic modulation of the carrier density in the plane of the quantum wells. These carriers then escape the wells and move perpendicular to the wells to produce a periodic screening of the built-in field and, thus, a periodic modulation of the quantum-confined Stark shift in the plane of the wells. This photorefractive and photoabsorptive grating, associated with spatially-separated carriers, will decay by ambipolar diffusive transport of the carriers in the plane of the wells. The decay of this grating was monitored by measuring the diffraction efficiency of the third probe pulse as a function of time delay. These measurements were repeated as a function of wavelength and fluence.

The decays of the gratings written in the hetero n-i-p-i at several fluences are compared in Fig. 1 to the decay in an undoped sample containing identical quantum wells (i.e. under flat band conditions). Notice that the decay rate is strongly fluence dependent. Also notice that at the lowest fluence the decay rate in the hetero n-i-p-i is ~7 times faster than that in the undoped sample and ~7 times faster than that predicted using the conventional ambipolar diffusion coefficient. These points are more apparent in Fig. 2 where we summarize the effective ambipolar diffusion coefficients, D_a, extracted from measurements at several fluences and three wavelengths. These wavelengths correspond to

Springer Series in Chemical Physics, Vol. 55 **Ultrafast Phenomena VIII**
Editors: J.-L. Martin · A. Migus · G.A. Mourou · A.H. Zewail

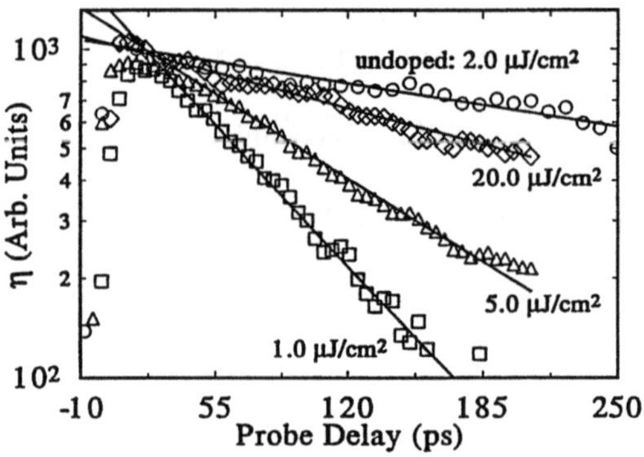

FIG. 1 Diffraction efficiency, η, vs probe delay for the hetero n-i-p-i for total peak pump fluences 1.0 $\mu J/cm^2$ (squares), 5.0 $\mu J/cm^2$ (triangles) and 20 $\mu J/cm^2$ (diamonds), and for the undoped sample at total peak pump fluence 2.0 $\mu J/cm^2$ (circles).

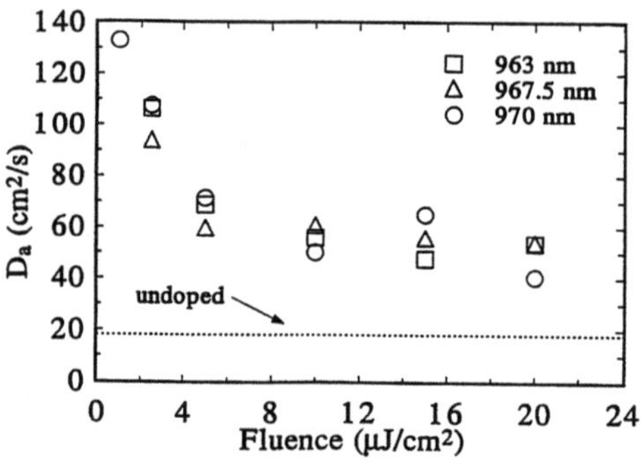

FIG. 2 Effective ambipolar diffusion coefficient, D_a, of the hetero n-i-p-i for total peak pump fluences in the range 1.0 $\mu J/cm^2$ to 20 $\mu J/cm^2$, at the excitonic peak (triangles), 2.5 nm to the red (circles) and 4.5 nm to the blue (squares). D_a for the undoped sample is indicated by the broken line.

exciting on the excitonic peak, 2.5 nm to the red and 4.5 nm to the blue. The effective D_a decreases smoothly from a maximum of 133 cm^2/s observed at a total peak pump fluence of 1.0 $\mu J/cm^2$ monotonically toward the conventional ambipolar value of 18 cm^2/s measured in the undoped sample and shown as the dotted line.

This enhanced in-plane transport in a semiconducting hetero n-i-p-i is a direct consequence of the charge separation dictated by the built-in field. That is, the photogenerated electrons and holes escape the wells in the intrinsic region and move perpendicular to the wells to screen the built-in field. Since the photogenerated charge is periodically modulated in the plane of the sample, the screening will be similarly modulated. The gradient of the potential (i.e., the field) associated with this periodic screening is the source of an additional force driving the grating decay. Since this in-plane drift field (associated with the screening) is proportional to the carrier density, the transport appears diffusive in nature. The importance of charge separation in this process is confirmed and illustrated by examining the dependence of D_a on fluence. That is, as the fluence is increased, the screening becomes more complete and charge separation (and hence D_a) is reduced. At the highest fluences, we approach complete flat band conditions, there is little charge separation, and D_a approaches the conventional value. Consequently, we see that the same space charge separation that slows the recombination can significantly enhance the diffusive decay. Although it differs in detail, this process is similar to the high-speed diffusive conduction recently reported in p-i-n structures [2] and to the giant ambipolar diffusion reported in semi-metallic p-n junctions [3].

References

1. D.S. McCallum, X.R. Huang, M.D. Dawson, T.F. Boggess, A.L. Smirl, T.C. Hasenberg and A. Kost, J. Appl. Phys. **70**, 6891 (1991).
2. G. Livescu, D.A.B. Miller, T. Sizer, D.J. Burrows, J. E. Cunningham, A.C. Gossard and J.H. English, Appl. Phys. Lett. **54**, 748 (1989).
3. K.H. Gulden, H. Lin, P. Kiesel, P. Riel, G.H. Dohler and K.J. Ebeling, Phys. Rev. Lett. **66**, 373 (1991).

Picosecond High-Sensitivity $In_xGa_{1-x}As$ Photodetectors

S. Gupta[1], *J.F. Whitaker*[1], *S.L. Williamson*[1], *P. Ho*[2], *J.S. Mazurowski*[2], *and J.M. Ballingall*[2]

[1]Center for Ultrafast Optical Science, University of Michigan, Ann Arbor, MI 48109, USA

[2]Electronics Laboratory, General Electric, Syracuse, NY 13221, USA

For ultrafast optoelectronic applications, photoconductors with sub-picosecond carrier lifetime and high-resistivity are desired. Previous work has demonstrated sub-picosecond carrier lifetime by lowering the growth temperature to ~200°C in MBE grown GaAs epilayers [1]. In addition, the semi-insulating nature and high breakdown-voltage of these epilayers has enabled the fabrication of sub-micron interdigitated photodetectors having an unprecedented bandwidth of 375 GHz and responsivity of ~0.1A/W [2]. The excess-As incorporation during the low-temperature MBE growth is believed to lead to the above properties [3]. Hence this approach of low-temperature(LT) MBE growth should be applicable to other As-based materials, for achieving ultrashort carrier lifetimes. We investigate the ultrafast carrier dynamics in low-temperature MBE grown $In_xGa_{1-x}As$ epilayers, for applications to the near infra-red wavelength detection.

Lattice-matched $In_xGa_{1-x}As$ (x=0.53) epilayers grown on semi-insulating InP at reduced growth temperatures (200°C to 450°C) exhibited a very low resistivity of ~10^{-2} Ω-cm. Therefore we have used lattice-mismatched growth of $In_xGa_{1-x}As$ on GaAs to give higher resistivity epilayers. Table I summarizes the Hall resistivity and mobility of 1.0μm thick $In_xGa_{1-x}As$ epilayers (x=0.25,0.35 and λ_g=1.1,1.35 μm, respectively) on semi-insulating GaAs substrate (lattice mismatch of 2% and 3%, respectively), grown by MBE at lower substrate temperatures. Similar epilayers were also grown, which were subsequently annealed at 600°C for 10 min. under an As-overpressure.

Table I. Properties of LT-$In_xGa_{1-x}As$ (T=300 K)

Growth temp. (°C)	Anneal	Resistivity (Ω-cm)	Mobility (cm^2/V-s)	Lifetime (ps)	Responsivity (10^{-5} A/W)
x = 0.25					
250	no	5.54×10^2	1110	8.0	3.0
200	no	1.63×10^2	20	3.0	0.6
160	no	78	47	- *unmeasurable* -	
250	yes	1.43×10^3	2730	40.0	3.0
200	yes	8.78×10^3	2260	7.3	5.5
160	yes	4.50×10^4	3271	2.0	0.4
x = 0.35					
200	no	72	44	- *unmeasurable* -	
160	no	268	91	4.5	0.2
120	no	2.1×10^4	2180	<1.0	0.03
200	yes	1.32×10^3	303	4.0	3.0
160	yes	260	894	5.5	2.1
120	yes	78	642	7.0	3.1

Springer Series in Chemical Physics, Vol. 55 Ultrafast Phenomena VIII
Editors: J.-L. Martin · A. Migus · G.A. Mourou · A.H. Zewail

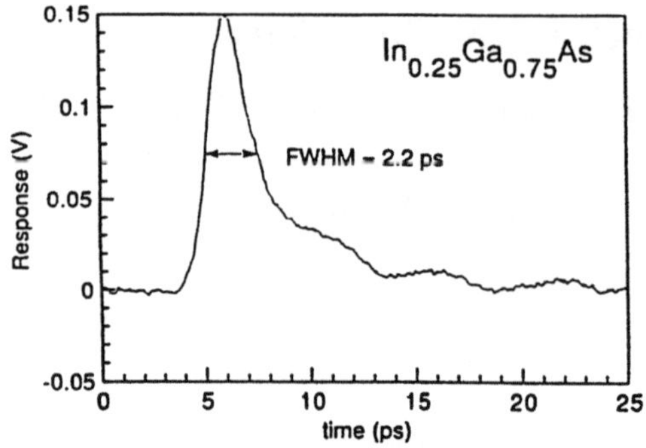

Fig.1 Photoconductive response of annealed LT-$In_{0.25}Ga_{0.75}As$ grown at 160°C.

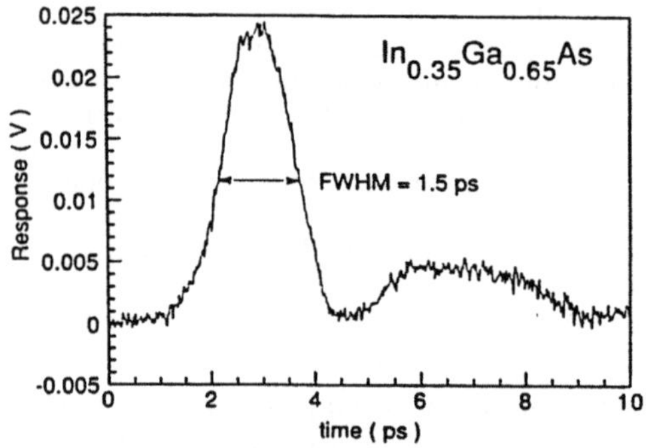

Fig.2 Photoconductive response of unannealed LT-$In_{0.35}Ga_{0.65}As$ grown at 120°C.

High-resolution transmission-electron-micrograph (TEM) shows all the as-grown epilayers to be heavily dislocated due to the large lattice-mismatch. When the layers are annealed, in addition to the dislocations, precipitates are also observed, which have been tentatively identified as being due to excess arsenic.

To measure the response of the material for high-speed photoconductive detectors, high temporal resolution switching measurements were performed. A 15-μm gap switch (biased at 15 V) was defined in a coplanar strip structure of wave impedance 73 Ω. External electro-optic sampling using a $LiTaO_3$ micro-probe, and a CPM dye laser producing 100 fs optical pulses is used to measure the ultrafast switch response [4]. Fig.1 and Fig.2 show the measured response for the lowest growth temperatures, for the x=0.25 and x=0.35 material, respectively. Table I also shows the carrier lifetimes and responsivities of the various epilayers, as obtained from the above measurement. For growth temperatures in the range of 120-160°C, single picosecond carrier lifetimes are obtained. Annealing, in general, improves the responsivity while degrading the lifetime. (The oscillations in the measured data are due to the reflections within the micro-probe.)

To improve the responsivity as a photoconductive detector the electrode spacing can be reduced, provided the resistivity of the epilayer is high enough to sustain practical voltage levels [2]. For a carrier lifetime of a few picoseconds as measured above, and the high resistivity of the epilayers, a 1-2 μm finger spacing of the interdigitated detectors is possible, leading to a near unity gain for the detector. The x=0.25,0.35 material is suited for applications requiring detection of upto 1.1 and 1.3 μm in wavelength. In addition, this scheme of lattice-mismatch low-temperatue growth can be easily extended to the important wavelength region of 1.55 μm, by increasing the mole fraction of In (x>0.45) in the epilayer. This material therefore shows great promise for application as an ultrafast and sensitive longer-wavelength photodetector.

[1] S. Gupta, M. Frankel, J. A. Valdmanis, J. Whitaker, G. Mourou, F. W. Smith, and A. R. Calawa, *Appl. Phys. Lett.* **59** (25), 3276 (1991).
[2] Y. Chen, S. Williamson, T. Brock, F. W. Smith, and A. R. Calawa, *Appl. Phys. Lett.* **59** (16), 1984 (1991).
[3] F. W. Smith, Ph.D. thesis, Massachusetts Institute of Technology, MIT (1990).
[4] J. A. Valdmanis, *Electron. Lett.* **23**, 1310 (1987).

An Ultrafast Polarization-Independent All-Optical Demultiplexer Utilizing Induced-Frequency Shift

T. Morioka, K. Mori, and M. Saruwatari

NTT Transmission Systems Laboratories, 1-2356 Take, Yokosuka, Kanagawa 238-03, Japan

Abstract. A novel polarization-independent all-optical demultiplexer is proposed utilizing induced-frequency shift in Kerr media. Experimental results of 4-port, 60 Gb/s operation at 1.307 μm are presented.

1. Introduction

All-optical switching using intensity-dependent refractive index in Kerr media is promising to perform ultrafast optical demultiplexing, optical routing, and logic operations that can handle ultrafast time-division-multiplexed(TDM) signals. In this paper, we propose the use of induced optical carrier frequency shift through cross-phase modulation(XPM)[1] combined with a "polarization rotation mirror"[2] to construct a novel polarization-independent optical demultiplexer and present its stable 4-port, 60 Gb/s operation. The proposed switch features multi-outputs with one operation compared with two outputs given by many other all-optical switching schemes based upon induced-phase shift.

2. Principle of Operation and Experimental Results

Fig.1 illustrates the experimental setup of the proposed polarization-independent all-optical demultiplexer. An ultrafast 60 Gb/s, 10 ps TDM signal pulse train (t_1,t_2,t_3,t_4) from a 1.307μm gain-switched(GS) laser-diode(LD) and a broad 70 ps pump pulse from a 1.313 μm Nd:YLF laser are coupled by a WDM coupler and directed to a "polarization rotation mirror" that is composed of a looped fiber-pigtailed polarization beam splitter(PBS). The two output ports of the PBS are connected so that the polarization of the returning pulse from the loop is flipped to its crossed state. A 3.4 km polarization-maintaining(PM) fiber that operates as a Kerr medium is inserted in the loop. Note that the two counter-propagating lights in the loop propagate along the same principal axis of the PM fiber and experience no path difference.

The pump polarization is set at 45 degrees with respect to one of the principal axes of the PBS while the signal polarization is arbitrary. Therefore, in the PM fiber, the pump is divided into two counter-propagating pulses of equal intensity and as a result, two counter-propagating signal pulse trains experience the same optical carrier frequency shift(f_1,f_2,f_3,f_4) produced by their copropagating pump pulses through XPM, regardless of the input signal polarizations. Here, the frequency shift is determined by a temporal location with respect to the pump pulse as illustrated in the inset of Fig. 1. The signal and the pump polarizations are flipped by 90 degrees in the "polarization rotation mirror" and they are directed back to the WDM coupler. At the exit, signal pulses are spatially separated from one another using a dispersion element such as a grating or a Fabry-Perot interferometer as shown in Fig.1.

Fig. 2 shows the experimental results of 4-port, 60 Gb/s switching observed with a streak camera of 6 ps temporal resolution. The switched waveforms kept unchanged for various input signal polarizations as predicted from a theory. The required switching pump power was 7.5W. The induced optical carrier frequency difference between the two adjacent pulses was 98 GHz. The signal pulse had a 93 GHz spectral width. The relative walk-off between the pump pulse and the CW signal was 40 ps and the temporal duration between the maximum and the minimum was measured to be 62 ps. One of the advantages of this device is that it is capable of having more than 2 outputs and demultiplexing N-time-multiplexed signals into N original

Springer Series in Chemical Physics, Vol. 55 **Ultrafast Phenomena VIII**
Editors: J.-L. Martin · A. Migus · G.A. Mourou · A.H. Zewail

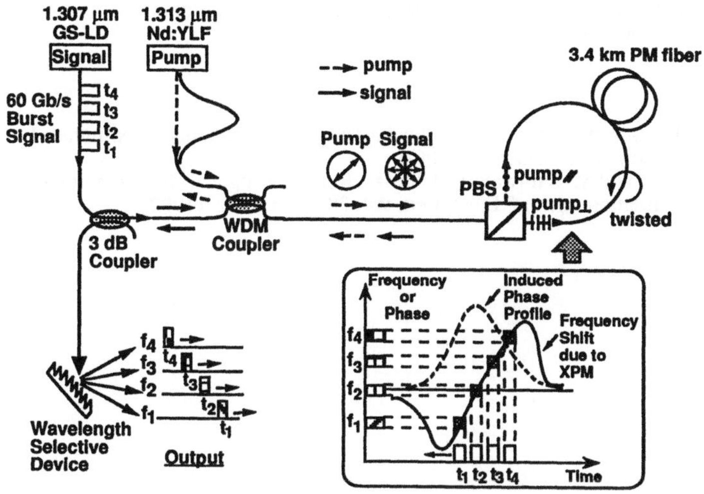

Fig. 1 Schematic of Polarization-independent All-optical Demultiplexer

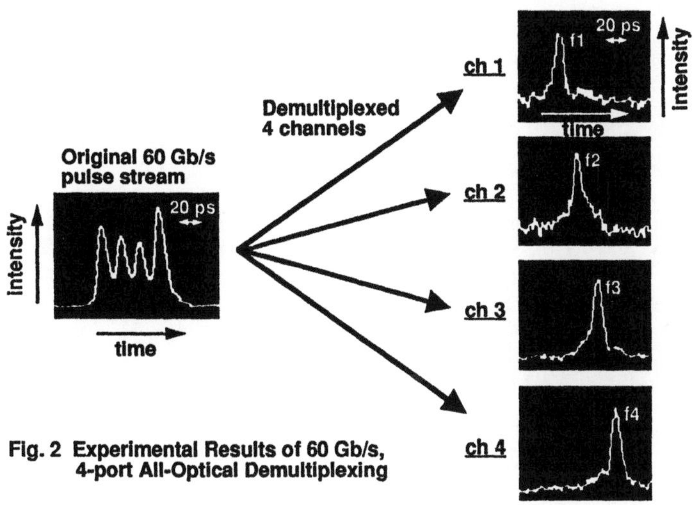

Fig. 2 Experimental Results of 60 Gb/s, 4-port All-Optical Demultiplexing

signal pulse streams with one operation. Moreover, the signal can be carrier frequency coded and can be used in WDM optical routing applications.

3. Conclusion

A novel polarization-independent all-optical demultiplexer has been proposed utilizing induced-frequency shift through cross-phase modulation in Kerr media. Stable experimental results of 4-port, 60 Gb/s operation at 1.307 μm has been demonstrated using a polarization-maintaining fiber combined with a polarization rotation mirror. The proposed scheme is also applicable to all-optical time-frequency converter, optical pulse frequency shifter, chirping manipulator, and logic devices, especially in ATM (Asynchronous Transfer Mode) photonic networks applications where optical burst signals are utilized.

References

[1] T. Morioka and M. Saruwatari, "Novel all-optical routing switch utilizing pump-induced optical carrier frequency shift," CLEO'90 CFC6

[2] T. Morioka, H. Takara and M. Saruwatari, "Ultrafast Reflective optical Kerr demultiplexer using a polarisation rotation mirror," Elect. Lett. vol. 28, pp. 521-522(1992)

Electrical Soliton Devices as >100 GHz Signal Sources

E. Carman[1], *M. Case*[1], *M. Kamegawa*[2], *R. Yu*[1], *K. Giboney*[1], *and M. Rodwell*[1]

[1]ECE Department, University of California, Santa Barbara, CA 93106, USA
[2]Shimadzu Corporation, Kyoto, Japan

We report a GaAs device which with 39 GHz and 490 mW input, generates 33 mW at 117 GHz, and 2.8 mW at 156 GHz using electrical soliton propagation. This device will serve as a signal source for ultrafast electronics.

Solitons are pulses which propagate without distortion in media having both nonlinearity and dispersion [1,2]. A nonlinear transmission line (NLTL), a ladder network of high-impedance transmission line sections periodically loaded with reverse-biased diodes (Fig. 1), exhibits nonlinearity from the diodes voltage-variable capacitance and dispersion from the periodic network. Our soliton frequency multiplier, as a NLTL, can therefore support soliton propagation.

Soliton propagation in optical fibers [6] results from the interaction between linear group velocity dispersion and the cubic term in the material polarization-field characteristics. Nonlinear propagation in optical fibers is described by the nonlinear Schrödinger Equation [2], which predicts the propagation of ***envelope solitons***, e.g. a sinusoidal optical carrier modulated by a hyperbolic-secant-squared intensity envelope. In contrast, soliton propagation on the NLTL results from the interaction between linear dispersion and the quadratic term in the diode charge-voltage characteristics. Nonlinear propagation on the NLTL is described by the Toda lattice equations [1], to which the Korteweg de Vries equation is a first approximation. The KdV equation also describes surface waves in shallow water. In both cases, ***baseband solitons*** of hyperbolic-secant-squared profile result [2].

The velocity at which solitons propagate is a function of their amplitude, with large solitons traveling faster than small solitons. Because of this velocity difference, solitons traveling the same direction will collide. The pulse waveform during collision corresponds to a nonlinear superposition of solitons and has smaller amplitude and larger pulse width than the largest of the colliding solitons. In driving the device with a 39 GHz sinusoid each negative lobe of the sine wave corresponds to a superposition of solitons. These separate during propagation into three solitons per cycle, generating at the output a waveform which has substantial third harmonic power at 117 GHz.

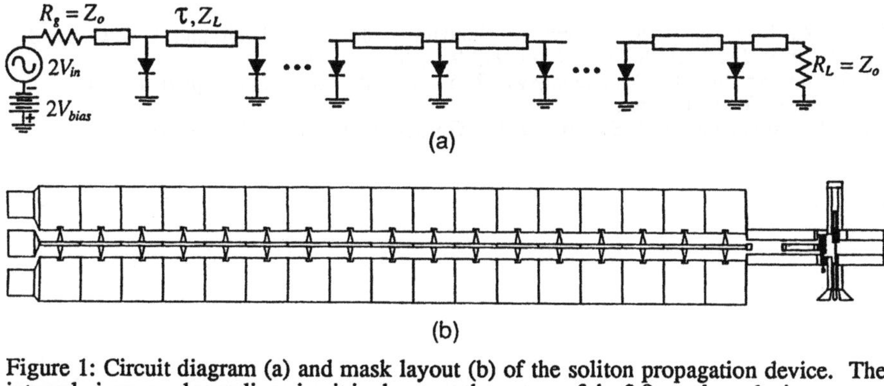

Figure 1: Circuit diagram (a) and mask layout (b) of the soliton propagation device. The integral picosecond sampling circuit is shown at the output of the 2.9 mm long device.

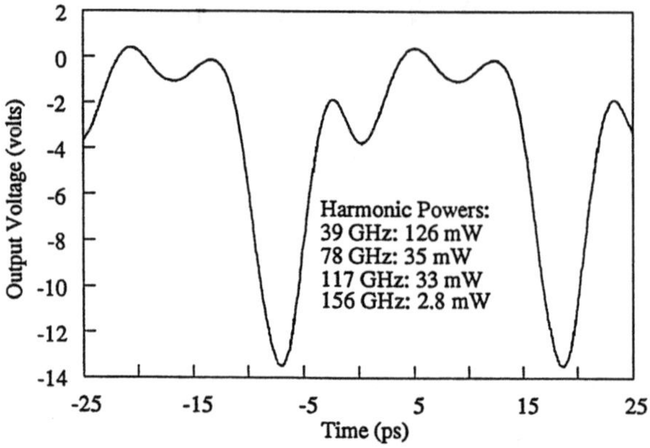

Figure 2: Measured output waveform for soliton frequency multiplier. With 490 mW input power at 39 GHz, 33 mW are generated at 117 GHz.

We have fabricated a soliton frequency multiplier which has a diode epitaxial design and fabrication process as described in [3], with the addition of airbridges to eliminate unbalanced modes in the coplanar waveguide structure. The multiplier output waveform was measured using an integrated on-wafer picosecond electrical sampling circuit [4], and harmonic powers were obtained by numerical Fourier transformations.

Figure 2 shows the measured output waveform. The output waveform also has substantial pulse characteristics with the largest soliton having 12.9 volts amplitude at the 90Ω load. This device is an improvement on earlier, lower frequency devices reported elsewhere [5], generating both higher output power and higher output frequency, at greater than 100 GHz. This work was supported by the U.S. Air Force Office of Scientific Research under grant number AFOSR-89-0394.

References

[1] R. Hirota and K. Suzuki, *Proc. IEEE*, vol. 61, no. 10, pp. 1483-1491, Oct. 1973.

[2] A. Scott, F. Y. F. Chu, and D. W. McLaughlin, *Proc. IEEE*, vol. 61, pp. 1443-1482, Oct. 1973.

[3] E. Carman, K. Giboney, M. Case, M. Kamegawa, R. Yu, K. Abe, M. Rodwell, and J. Franklin, *IEEE Microwave and Guided Wave Letters,* vol. 1, no. 2, pp. 28-31, Feb. 1991.

[4] R. Yu, M. Case, M. Kamegawa, M. Sundaram, M.J.W. Rodwell, and A. C. Gossard, *Electron. Lett.*,vol. 26, no. 13, pp. 949-951, June 21, 1990.

[5] E. Carman, M. Case, M. Kamegawa, R. Yu, and M. Rodwell, accepted at the 1992 IEEE MTT Symposium.

[6] L.F. Mollenauer, R.H. Stolen, and J.P. Gorden, *Opt. Lett* 8, 289 (1983).

Determination of Photonic Band Gaps and Dispersion in Two-Dimensional Dielectric Arrays with Ultrafast Electromagnetic Transients

W.M. Robertson[1,*], *G. Arjavalingam*[1], *R.D. Meade*[2], *K.D. Brommer*[2], *A.M. Rappe*[2], *and J.D. Joannopoulos*[2]

[1]IBM Research Division, T.J. Watson Research Center, P.O. Box 218, Yorktown Heights, NY 10598, USA
[2]Department of Physics, MIT, Cambridge, MA 02319, USA
*Present Adress: National Research Council Canada, Institute for Information Technology, M-50 Montreal Road, Ottawa, Ontaria, Canada K1A OR6

Coherent scattering of radiation in periodic dielectric structures results in an electromagnetic dispersion relation which exhibits novel features such as new propagation modes and photonic band gaps. The latter topic has received much current attention for a variety of fundamental and practical reasons [1,2]. Furthermore, the theoretical calculation of the electromagnetic dispersion relation in such structures has become relatively sophisticated. However, experimental verification using traditional microwave techniques has been limited to determining the frequencies which define the gaps and to study localized defect modes. Here, we present the dispersion relation of a periodic dielectric array measured explicitly using the coherent microwave transient spectroscopy (COMITS) technique [3,4]. The experimental results are compared with theoretical predictions obtained using the plane-wave expansion technique [5].

The experimental setup is shown schematically in figure 1. The exponentially tapered coplanar stripline antennas are fabricated on implanted silicon-on-saphhire substrates. Optical pulses from a mode-locked, pulse-compressed, and frequency-doubled, Nd:YLF laser are used in a standard pump/probe configuration to excite the dc-biased transmitter and photoconductively sample the received waveforms. Hemispherical fused silica lenses collimate the radiation diverging from the transmitter and focus it onto the receiver. The freely-propagating ultrashort electromagnetic transient is polarized with the E-field in the plane of the antennas. The measured signal has a 7 ps wide central peak, and frequency components from 0

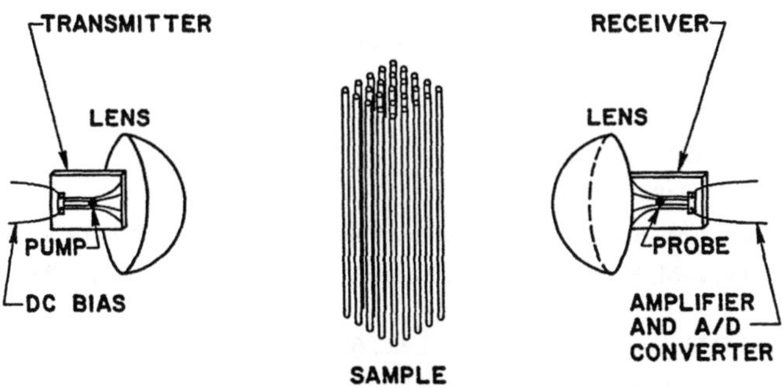

Figure 1. Schematic of the coherent microwave transient spectroscopy experimental set up. The photonic crystal is made of 100 mm long 0.74mm diameter alumina ceramic rods arranged in a square array with 1.87 mm lattice constant.

Springer Series in Chemical Physics, Vol. 55 Ultrafast Phenomena VIII
Editors: J.-L. Martin · A. Migus · G.A. Mourou · A.H. Zewail

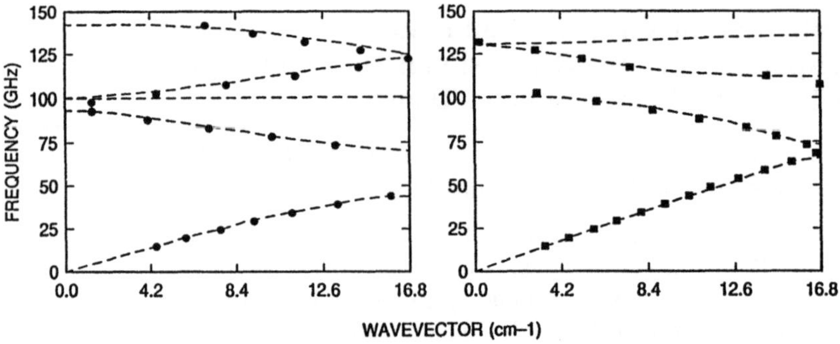

Figure 2. The electromagnetic dispersion along the <10> direction with polarization (left) parallel and (right) perpendicular to the rod axis. The points are experimental values and the lines are theoretical predictions.

GHz to 150 GHz. Since it is proportional to the time-dependent field incident on the receiver, phase information is preserved [4].

The two-dimensional photonic crystal consists of 0.74 mm diameter alumina ceramic rods arranged in a square array of lattice constant 1.87 mm. Time-domain waveforms are recorded with and without the array in the beam path. The waveforms are numerically Fourier transformed and the corresponding spectra are divided to obtain the frequency-dependent complex transmission function of the sample. Much like results obtained with traditional microwave techniques, the amplitude transmission (not shown) clearly shows the frequency regions in which electromagnetic-wave propagation is inhibited. In addition, using the phase data and the known thickness of the array we derive an effective dielectric constant at each frequency point [4]. Hence, the frequency-dependent dispersion relation is determined explicitly. The dispersion relation for propagation along the <10> direction of the array, with the E-field oriented parallel and perpendicular to the rod axis, is shown in figure 2. The points are experimental results while the lines are predicted theoretically using the plane-wave expansion technique [6]. The agreement between theory and experiment is generally good. However, the theoretical calculations also predict modes which are not observed experimentally (Fig 2). Further investigations reveal that the unobserved modes have opposite symmetry to that of the incoming plane wave. Hence, even though these modes may exist in the photonic crystal they cannot be observed with plane-wave radiation. Similar results were obtained for other propagation directions and crystal structures &lbt7].

Acknowledgements
G.A. thanks Jeff White (Hughes Research Lab) for making this presentation at Antibes.

REFERENCES

1. E. Yablonovitch, Phys. Rev. Lett., 58, 2059 (1987).
2. S. John, Phys. Rev. Lett., 58, 2486 (1987).
3. Y. Pastol, G. Arjavalingam, J-M. Halbout, G.V. Kopcsay, Appl. Phys. Lett. 54, 307 (1989).
4. G. Arjavalingam, Y. Pastol, J.-M. Halbout, and G.V. Kopcsay, IEEE Trans. Microwave Theory Tech., 38, 615 (1990).
5. K.M. Ho, C.T. Chan, and C.M. Soukoulis, Phys. Rev. Lett., 65, 3152 (1990).
6. W.M. Robertson, G. Arjavalingam, R.D. Meade, K.D. Brommer, A.M. Rappe, and J.D. Joannopoulos, Phys. Rev. Lett, 68, 2023, (1992).
7. W.M. Robertson, G. Arjavalingam, R.D. Meade, K.D. Brommer, A.M. Rappe, and J.D. Joannopoulos, to be published.

Part VIII

Biology: Primary Dynamics, Electron and Energy Transfer

Ultrafast Infrared Spectroscopy of Protein Dynamics

R.M. Hochstrasser[1], *R. Diller*[1], *S. Maiti*[2], *T. Lian*[1], *B. Locke*[1], *C. Moser*[2], *P.L. Dutton*[2], *B.R. Cowen*[1], *and G.C. Walker*[1]

[1]Department of Chemistry, University of Pennsylvania, Philadelphia, PA 19104, USA

[2]Department of Biochemistry and Biophysics, University of Pennsylvania, Philadelphia, PA 19104, USA

1. Introduction

The focus of this paper is the examination of the responses of proteins to optically triggered events. A primary goal is to step beyond studies of protein dynamics based on chromophore responses and make direct observations of the protein itself. It is hoped to discover whether or not there are ultrafast responses of functional interest and to develop a molecular level rationale for such responses. The experimental method is transient infrared spectroscopy.

Infrared methods have some advantages for the study of proteins. The method is nondestructive and essentially all polyatomic molecules and ions can be accessed. An extensive study of spectra-structure relationships exist because IR was once a principal structural director in synthetic organic chemistry. In addition the orientations and correlation functions of the dipole derivatives available from polarized IR measurements are useful, especially in the time regime 100fs to 5ps for comparison with MD simulations. Two hurdles to overcome in each application are the absorption by H_2O and the spectral complexity from overlapping transitions.

In order to study the time evolution of the protein response it is necessary to have a triggering mechanism that is faster than the timescale of interest. This is the case even if the object is to determine equilibrium correlation functions - though in that case the mid-infrared pulses can be used as triggers. In the present case when picosecond and femtosecond responses are desired, the necessity for optical triggering at present seems inevitable. The various optical triggers that are already well-known involve: (a) radiationless decay; (b) photodissociation; (c) isomerism; and (d) electron transfer. The processes or responses that can be examined in each of these cases are: (a) Energy transduction: Examples include cooling of Mb, Hb, Cyt-C which can be studied by observing Kerr effects, thermal lensing or changes in the IR spectrum of water. (b) Protein influence on geminate dynamics: Examples include HbNO and HbCO where IR spectroscopy can be used to evaluate the protein influence on the vibrational spectra of NO or CO diatomic molecules in process of being buffeted by the protein. (c) Protein response to sudden structural change: Examples include protein responses to retinal isomerization in bacteriorhodopsin and rhodopsin where IR spectroscopy is used to observe peptide vibrations. (d) Protein response to charge separation: An example is the reaction center where IR methods can identify the intermediates by time gating and seek protein structure relaxation. The study of protein dynamics in systems where there is no natural optical trigger (e.g. protein folding) presents significant challenges to ultrafast technology and to synthetic chemistry.

2. Methods

Ultrafast IR spectroscopy is a method of choice for direct observations of protein responses. Probing in the mid-IR with pulses as short as 180fs was recently accomplished in this

Springer Series in Chemical Physics, Vol. 55 Ultrafast Phenomena VIII
Editors: J.-L. Martin · A. Migus · G.A. Mourou · A.H. Zewail

laboratory [1]. However, in the work presented here the quasi-CW method [2,3] is used exclusively. Certainly this is a useful method when the signals are small and the time resolution is not critical (i.e. > ca. 1ps).

In the quasi-CW method a carbon monoxide (CO) laser is used to interrogate an optically triggered sample. The generated IR field interferes with the input CW field as the result of the square law detection of the sum frequency pulse generated in a nonlinear crystal by mixing the total IR field with a suitably delayed optical pulse. This technique measures not only changes in population induced by the pump pulse but all other birefringent and dichroic responses that arise in the third order. Kerr type nonlinearities for many liquids and coherent transient beats in vapors have been measured in this way. Such effects were present but unrecognized in some of the early work on photoreactions employing this technique[4]. The presence of the IR field demands that terms not normally considered in the third order response to the Born-Oppenheimer approximation be considered, such as those involving the hyperpolarizability-dipole correlation function[5]. It is important to recognize that the up-conversion step in which the transmitted IR field is focused inside the optical spot on a nonlinear crystal is effectively a slit. Slight changes in the IR beam diameter (as a result of the nonlinear index $n_2(t)$ or thermal lenses) can be very accurately determined as changes in the sum frequency signal. Hence this approach may be configured into a rather precise method of determining such nonlinear transient responses.

In a recent experiment a titanium sapphire (TiS) laser and associated Q-switched YAG pumped regenerative amplifier was used to carry out IR spectroscopy at ca. 200fs time resolution in conjunction with a CO laser (see below under reaction centers). The YAG delivered 650μJ in 80ns green pulses at 1.5kHz. The regen was seeded by Bragg diffracting 150fs pulses from a Mira oscillator. The output ca. 10μJ was grating compressed. Clearly the TiS laser presents broad possibilities for IR spectroscopy in the future[6].

3. Bacteriorhodopsin: (BR; R.Diller, S.Maiti, B.R.Cowen)

The IR spectra of the K intermediate of the BR photocycle [7] is shown in Figure 1 and the kinetics over a wide range in Figure 2. This study presents the first ultrafast examples of protein responses to a photoinduced process. Three different responses were deduced [8]; (a) An intermediate response of a peptide carbonyl group indicated by the presence of an amide-I band (1660cm^{-1}) in the IR difference spectrum; (b) A slow ns response most likely involving a carboxylic group of the protein in the 1730-1740cm^{-1} region which may also correspond to Asp115; and (c) A rapidly relaxing peptide (amide-I) state (1556cm^{-1}) that is present in low temperature FTIR but absent in the 300K difference spectrum. Detailed structural conclusions are still limited by lack of assignments, especially for the amide-I and amide-II bands of the protein backbone. However, the prognosis for significant new knowledge of

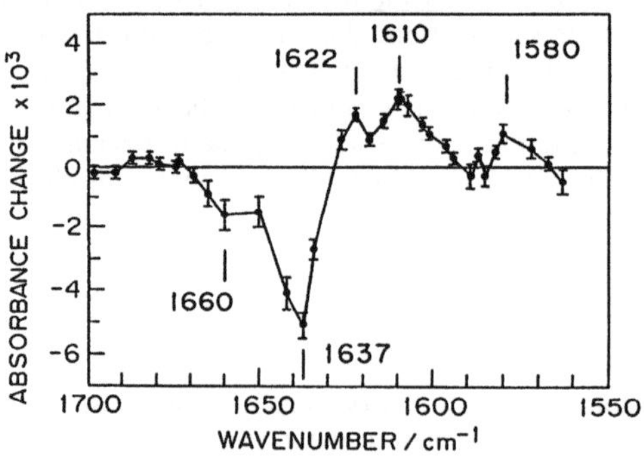

Figure 1. Infrared BR_{570}-K (KL) difference spectrum. Excitation at 550nm. The spectrum is representative for between 100ps and 14ns [after reference 8].

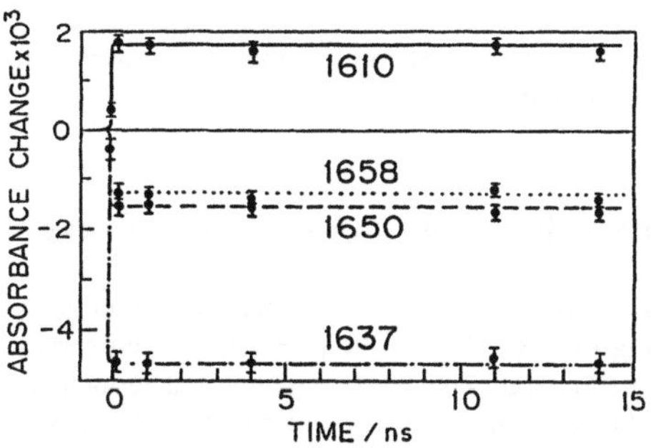

Figure 2. Infrared absorbance kinetics following excitation of BR_{570} at 550nm. Probed as indicated at 1658, 1650, 1637, and 1610 cm^{-1} [after reference 8].

the operation of BR with ultrafast IR seems outstanding. The J states of BR that exist prior to 5ps can now be examined directly in the IR using the TiS laser described above.

4. Nitrosohemoglobin: (HbNO; T.Lian and R.B.Locke)

It is inferred from transient optical spectroscopy that photodissociated NO recombines on the ps timescale to form HbNO [9,10]. Direct IR studies of the released NO may provide information on the protein dynamics occurring within the pocket region near the heme and provide opportunities for comparisons to molecular dynamics theory. In recent experiments we have observed directly the spectra of NO released by photodissociation in less than 300fs. The assignment to NO is confirmed by isotopic studies that show the peak absorption at 1827cm^{-1} (±2) for $N^{15}O^{16}$ having a linewidth (FWHM) of ca. 15cm^{-1}. The free spectrum NO disappears in ca. 10ps consistent with optical studies [8,9]. This work confirms that the optical work is measuring geminate reactions of NO on the ps timescale.

In the present experiments the NO is superimposed on a thermally induced background and an electronic response that tracks the pump pulse, as shown in Figure 3 (compare HbNO with Hb). The thermal signal appears to arise from changes in the IR transmission due to the heating of water which occurs in a diffusive manner with a Gaussian time constant of 11±2ps. This response of the protein and solvent shows that the heme cools faster than 4-5ps, consistent with other work [11].This background is present in all the heme proteins studied to date (Mb, MbCO, Hb, HbCO and HbNO) and the growth time is clearly thermal. A systematic study of these systems allows the background to be subtracted to yield the ligand

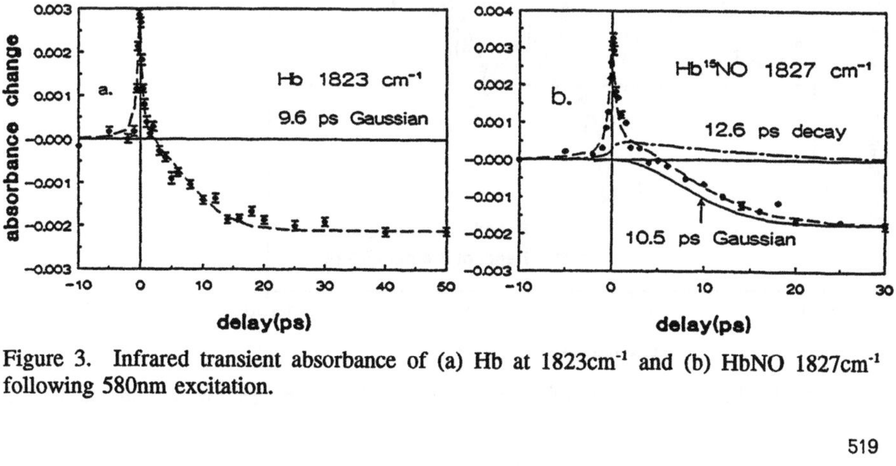

Figure 3. Infrared transient absorbance of (a) Hb at 1823cm^{-1} and (b) HbNO 1827cm^{-1} following 580nm excitation.

spectrum. The sharp spike at t=0 appears to be partly caused by the simultaneous absorption of an IR and visible photon to produce a heme excited state.

5. Reaction Centers: (RC; P.L. Dutton, C. Moser, S. Maiti, R. Diller, B. Cowen, and G.C. Walker)

In the operation of RC's (in the present work we use *Rhodobacter sphaeroides*) charge is shifted over large distances in time steps ranging from ca. 1ps to 300μs. The protein serves as the scaffolding that maintains the proper spatial arrangement of oxidants and reductants during the electron transfer and as the ideal dielectric medium for optimization of charge separation and recombination. There is therefore expected a protein response to charge separation, but whether this involves a small contribution from many protein constituents and/or the crucial involvement of a few residues is not yet known. We have recently been successful in obtaining transient IR spectra of RC's in various time regimes. The 70ps spectra capture the difference between the initial and $BChl_2^+BPh^-$ states whereas the spectrum at 600ps contains $BChl_2^+Q_a^-$. As a result of kinetic measurements we were able to identify $BChl_2^+Q_a^-$ components in the ~1660cm^{-1} region. It is clear from these spectra that the differences in the IR spectra at 70ps and 600ps are small: Perhaps this implies that the protein dielectric responses are not highly local or that the protein is dielectrically "mushy". On the other hand we have seen some differences in the 1550-1670cm^{-1} region between our spectra in solution and low temperature FTIR difference spectra, but these have not yet been interpreted.

In a recent advance we have directly observed the 3.5ps formation time of $BChl_2^+$ through a study of the growth of its IR absorption at 1702cm^{-1}. The pump pulse in this case was obtained from a regeneratively amplified TiS laser operating at 868nm. The first result obtained with this system and shown in Figure 5 can have greatly improved signal:noise in the near future. The time resolution of 300fs is already adequate for these experiments. Clearly these methods will permit the acquisition of the IR spectra and kinetics of chromophore and protein regions at times short enough to intercept intermediates, such as $BChl^-$ proposed recently [12].

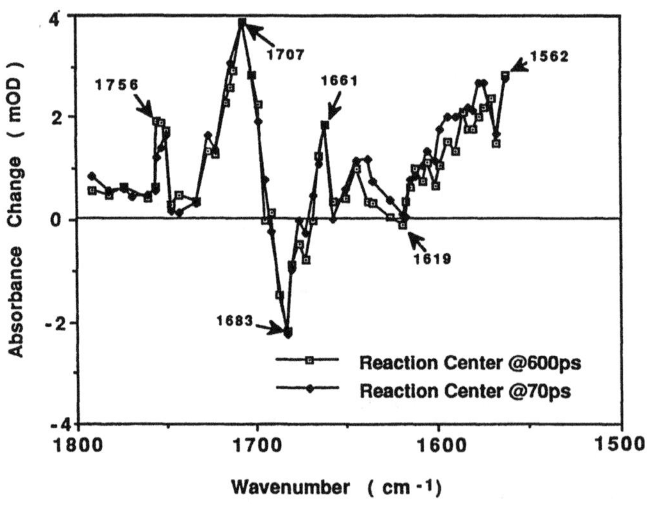

Figure 4. IR difference spectra of *R. sphaeroides* at 70 and 600ps after 810nm excitation.

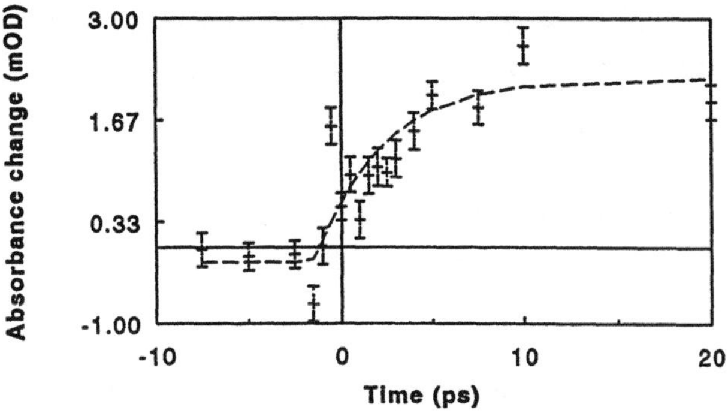

Figure 5. IR transient absorbance of $BChl_2^+$ in *R. spheroides* at 1702cm^{-1} following excitation at 868nm.

References

1. J.C.Owrutsky, J.P.Culver, M.Li, Y.R.Kim, M.J.Sarisky, M.S.Yeganeh, A.G.Yodh, and R.M.Hochstrasser, J. Chem. Phys. (in press).
2. J.N.Moore, P.A.Hansen and R.M.Hochstrasser, Chem. Phys. Lett. 138, 110 (1987).
3. P.A.Anfinrud, C.Han and R.M.Hochstrasser, Proc. Natl. Acad. Sci. USA 86, 8387 (1989).
4. P.A.Anfinrud, C.H.Han, T.Lian and R.M.Hochstrasser J. Phys. Chem. 94, 1180 (1990).
5. R.W.Hellwarth, Prog. Quant. Electr. 5, 1 (1977).
6. F.Salin, J.Squier, and G.Mourou, Optics Letters 16, 1964 (1991).
7. T.Kouyama, K.Kinosito and A.Ikegami, Adv. Biophys. 24, 121 (1988).
8. R.Diller, M.Ianonne, R.Bogomolni and R.M.Hochstrasser, Biophys. J. 60, 286 (1991); R.Diller, M.Ianonne, B.R.Cowen, S.Maiti, R.Bogomolni and R.M.Hochstrasser, Biochem. (in press).
9. P.Cornelius, R.M.Hochstrasser, and W.A.Steele, J. Mol. Biol. 163, 119 (1983).
10. J.W.Petrich, J.-C.Lambry, K.Kuczera, M.Karplus, C.Poyart and J.-L.Martin, Biochem. 30, 3975 (1991).
11. R.Lingle, Jr., X.Xu, H.Zhu, S.Yu and J.B.Hopkins, J. Am. Chem. Soc. 113, 3992 (1991); R.Lingle, Jr., X.Xu, H.Zhu, S.Yu and J.B.Hopkins, J. Phys. Chem. 95, 9320 (1991).
12. W.Holzapfel, U.Finkele, W.Kaiser, D.Oesterhelt, H.Scheer, H.UStilz and W.Zinth, Chem. Phys. Lett. 160(1), 1 (1989).

Ultrafast Near-IR Spectroscopy of Carbonmonoxymyoglobin: The Dynamics of Protein Relaxation

M. Lim, T.A. Jackson, and P.A. Anfinrud

Department of Chemistry, Harvard University,
12 Oxford Street, Cambrige, MA 02138, USA

Abstract. The dynamics of heme and protein relaxation following photolysis of carbonmonoxymyoglobin have been measured using time-resolved near-IR spectroscopy.

1. Introduction

The conformation of a protein can influence its activity, yielding a structure-function relationship. X-ray diffraction studies have shown that the tertiary structures of ligated and deligated myoglobin (Mb) are somewhat different: the iron is located in the plane of the porphyrin in MbCO [1] but is displaced 0.5 Å to the proximal side of the plane in deoxy Mb [2]. Consequently, photodissociation of MbCO generates Mb* (Mb in the MbCO conformation) and initiates an out-of-plane displacement of the iron. Motion of the iron is coupled to motion of the protein through the proximal histidine, yielding a change in the tertiary conformation of the protein.

We have probed the dynamics of conformational relaxation by monitoring band III, a weak ($\epsilon_{max} \approx 100\ M^{-1}\ cm^{-1}$ at room temperature) $\pi \rightarrow d$ charge transfer transition [3] whose center frequency is near 13120 cm^{-1} (762 nm). Since both porphyrin-π and iron-d orbitals are involved in this transition, the center frequency of band III is sensitive to the position of the iron relative to the plane of the porphyrin. The time-dependent shift of band III therefore reveals the time evolution of the iron-heme displacement and the correlated motion of the protein.

2. Experimental Technique

Tunable femtosecond pulses are generated with a single-jet rhodamine 6G dye laser that is pumped by the compressed and frequency-doubled output from a mode-locked Nd:YLF laser. The femtosecond pulses are amplified to 40 μJ at 1.5 kHz in a dye amplifier that is pumped by the frequency-doubled output from a Nd:YLF regenerative amplifier. One portion of the amplified pulse is used to photolyze the sample while the remainder is used to generate a white-light continuum by focussing into a 1 cm cell that contains water. The continuum is spatially and spectrally filtered to yield a broadband near-IR probe pulse. The probe is split into two portions: one portion passes through the photolyzed volume of the sample and the other portion, a reference, bypasses the sample. The sample and reference pulses are routed through a monochromator with a bandpass of $\approx$ 10 cm^{-1} and detected with amplified photodiodes. The two signals are digitized and their ratio is computed on a shot-by-shot basis in order to recover the transmission of the sample at the selected wavelength and at a delay time established by a computer-controlled optical delay. The pump-induced change in the absorbance of the sample is computed to a precision of $< 10^{-3}$ after one second of signal averaging. The instrument response function, as measured by the cross-correlation of the pump and probe pulses, is typically 350 fs.

A 10 mM MbCO sample was prepared in a pH 7 phosphate buffer and loaded into a 1 mm path-length rotating sample cell. The wavelength of the photolyzing pulse was tuned to 597

Springer Series in Chemical Physics, Vol. 55 Ultrafast Phenomena VIII
Editors: J.-L. Martin · A. Migus · G.A. Mourou · A.H. Zewail

nm, where the absorbance of the sample was ≈ 1. The photolyzing pulse contained 3 μJ of energy and was focussed to 120 μm, yielding on average 0.15 photons per heme group.

3. Results and Discussion

The time-resolved near-IR spectra of Fig. 1 represent the average of data accumulated in less than 6 hours. The time-resolved absorbance at 13410 cm^{-1}, shown in Fig. 2, is well-described by a convolution of the function $\Delta A(t) \propto 0.82 \exp(-t/0.24\,ps) + 0.07 \exp(-t/1.5\,ps) + 0.11$ with the cross-correlation of the pump and probe pulses. The 0.24 ps component may correspond to relaxation from an excited electronic state of the heme. This state may be responsible for the short-lived feature observed near 11800 cm^{-1}. The 1.5 ps component, which is much smaller in amplitude, may be due to thermal relaxation of the heme.

Time-resolved spectra in the vicinity of band III were collected over a broad range of times ranging from 1.78 ps to 3.16 ns (excited-state absorptions complicated the characterization of band III at earlier times). Representative spectra are shown in Figure 3. The center frequency of band III was determined by nonlinear least-squares fitting of these spectra with a Gaussian function plus a cubic polynomial background. The time-dependent shift of band III relative to equilibrium Mb is plotted (squares) in Fig. 4a.

The center frequency of band III is temperature dependent [4] so the spectral evolution of band III at early times must be interpreted carefully. The thermal contribution to the shift of band III was estimated by measuring time-resolved near-IR absorption spectra of photoexcited Mb [5]. The results are plotted (triangles) in Fig. 4a. Photoexcitation of Mb with a 597 nm photon deposits 47.9 kcal/mole of energy in the heme, causing the temperature of the heme to rise by ≈ 390 °K [6]. This rise in temperature causes a red-shift of the center frequency of band III. Vibrational energy redistribution between the heme and the surrounding protein cools the heme causing band III to shift toward its room-temperature position. The characteristic time constant for this process is ≈ 6 ps, which is similar to that predicted by molecular dynamics simulations [6] and measured by transient anti-stokes Raman spectroscopy [7].

The data of Fig. 4b reveal the time-dependent shift of band III after correction for the thermal contribution. The relaxation is well-described by a sum of three exponential terms, $\Delta\bar{\nu}(t) = 61 \exp(-t/7\,ps) + 20 \exp(-t/90\,ps) + 24 \exp(-t/4700\,ps)$, where $\Delta\bar{\nu}(t)$ is the magnitude of the red-shift in wavenumbers. The fast and slow relaxations differ by nearly three orders of magnitude, illustrating the highly nonexponential character of the relaxation.

A recent molecular dynamics simulation [8] probed the time-dependent iron-heme displacement over a 100 ps period following ligand dissociation. They reported that 80% of

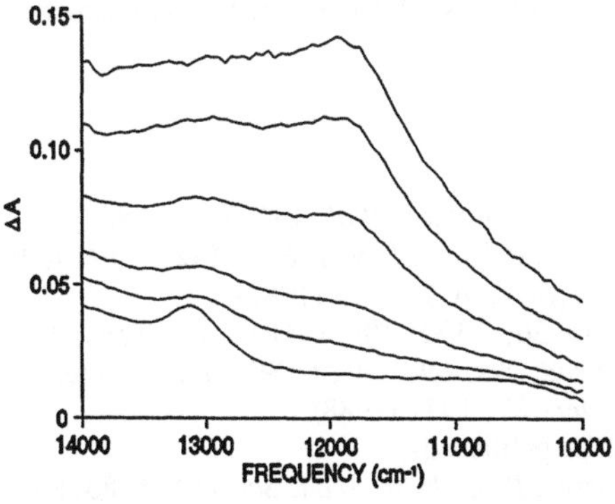

Figure 1. Time-resolved near-IR spectra of photolyzed MbCO at 100 fs, 316 fs, 562 fs, 1.00 ps, 1.78 ps, and 100 ps (top to bottom).

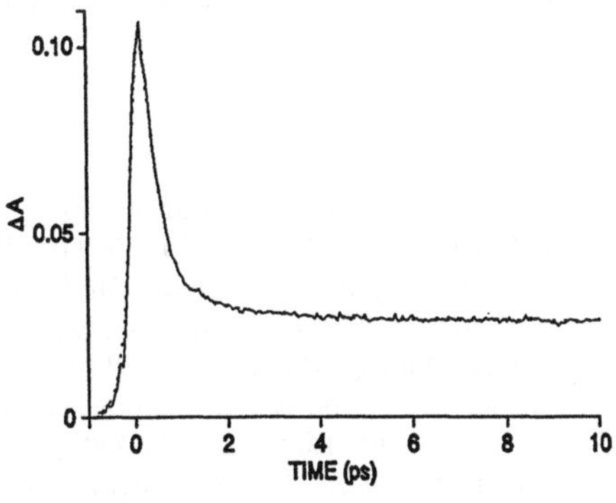

Figure 2. Transient absorption at 13410 cm^{-1}. The relaxation is modeled (dashed line) with a multiexponential function.

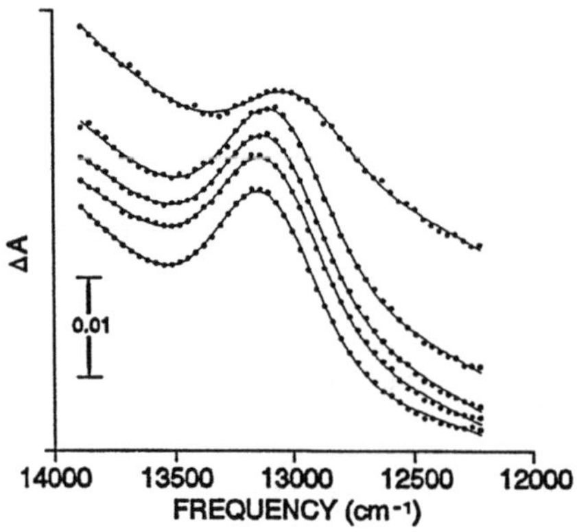

Figure 3. Time-resolved spectra of band III at 1.78 ps, 17.8 ps, 178 ps, 1.78 ns, and scaled spectrum of Mb (top to bottom; offset vertically to avoid overlap). The data are modeled (line) with a Gaussian function plus a cubic polynomial background.

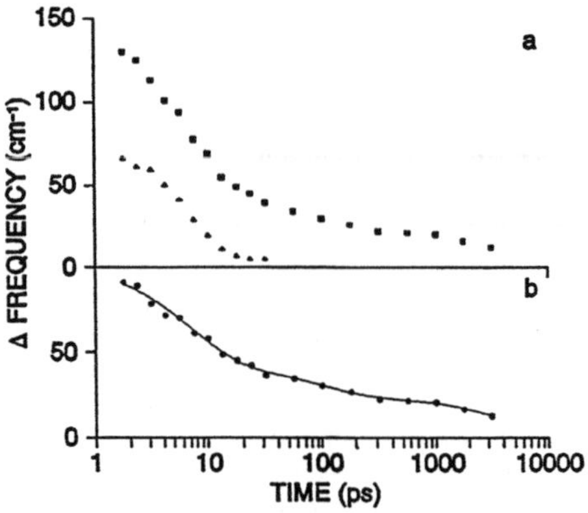

Figure 4. Time-dependent shift of band III: a) Photolyzed MbCO (squares) and photoexcited Mb (triangles); b) Photolyzed MbCO data corrected for thermal contribution to the shift (circles) and modeled (line) with a sum of three exponentials.

the iron-heme displacement occurred within 50 fs, which is less than half of a vibrational period of the iron-histidine stretch. The displacement at subsequent times involved conformational relaxation of the protein. The shift of band III, as reported in Fig. 4b, likely correlates with the last 20% of the iron-heme displacement. The relationship between the position of band III and the iron-heme displacement would be nearly linear over such a small excursion, facilitating a comparison between the results. The data of Fig. 4b, when scaled appropriately, are virtually superimposable over the curve defined by the molecular dynamics simulation. However, the fluctuations of the simulation are significantly larger than the scatter of the data in Fig. 4b, inhibiting a detailed comparison.

4. Acknowledgements

We are grateful to the Petroleum Research Fund and Mitsubishi Kasei Corporation for support.

5. References

1. J. Kuriyan, S. Wilz, M. Karplus and G.A. Petsko, *J. Mol. Biol.* **192**, 133 (1986).
2. T. Takano, *J. Mol. Biol.* **110**, 569 (1977); S.E.V. Phillips, *J. Mol. Biol.* **142**, 531 (1980).
3. W.A. Eaton and J. Hofrichter, *Methods In Enzymology* **76**, 175 (1981).
4. V. Srajer and P.M. Champion, *Biochemistry* **30**, 7390 (1991); G.U. Nienhaus, J.R. Mourant and H. Frauenfelder, *Proc. Natl. Acad. Sci.* **89**, 2902 (1992); A. Cupane, M. Leone, E. Vitrano and L. Cordone, *Biopolymers* **27**, 1977 (1988).
5. T. Jackson, M. Lim and P.A. Anfinrud, (to be published).
6. E.R. Henry, W.A. Eaton and R.M. Hochstrasser, *Proc. Natl. Acad. Sci. U.S.A.* **83**, 8982 (1986).
7. R. Lingle, Jr., X. Xu, H. Zhu, S. Yu and J.B. Hopkins, *J. Phys. Chem.* **95**, 9320 (1991).
8. K. Kuczera and M. Karplus, (to be published).

Energetics and Dynamics of Global Protein Motion

R.J.D. Miller, J. Deak, S. Palese, M. Pereira, L. Richard, and L. Schilling

Department of Chemistry and Institute of Optics,
University of Rochester, Rochester, NY 14627, USA

Functionally important motions in biological systems require the correlated displacement of thousands of atoms. The exact mechanism of the protein response to a stimulus for the functionally relevant motions is poorly understood. The classic models for deterministic protein motion rely on potential energy gradients, created by the interaction with a stimulus, to provide the forces that orchestrate the molecular motion. The question is over what length scale are these forces distributed and what are the magnitudes of the driving forces, i.e., the energetics for the structural changes. In this regard, heme proteins provide ideal systems for addressing these issues. Photodissociation of the axial ligand at the heme site selectively triggers the functionally important structural changes involved in oxygen binding and allosteric regulation of oxygen transport in heme proteins.

Picosecond phase grating studies of heme proteins have discovered that the protein motion to the deoxy tertiary structure launches coherent acoustic waves which match the grating wave vector.[1] The generated acoustics are nonthermal and satisfy t=0 strain boundary conditions, as shown by previous studies at the zero thermal expansion point of water[1] and absolute diffraction efficiency studies as a function of wavelength.[2] The protein motion following CO photodissociation leads to material displacement which is holographically recorded as a density induced change in the index of refraction. The coupling of this protein motion to the hydrodynamics requires a global change in protein structure (i.e., acoustic volume change). Thus, the grating formation dynamics give direct measurement of the global relaxation of the protein.

A four pulse grating geometry has been used to study this relaxation process for carboxymyoglobin (MbCO). The laser system used for these studies was a hybridly modelocked dye laser synchronously amplified by a Nd^{+3}:YLF regenerative amplifier. The grating excitation pulses were 1-2 picoseconds in duration, centered at 575 nm. The grating dynamics were probed off resonance at 1.053 μm by diffracting the 50 picosecond fundamental of the regenerative amplifier and summing the diffracted IR with a residual dye pulse to conserve time resolution. The results from this study are shown in figure 1 for MbCO. The grating rise time is related to the structural changes leading to changes in protein strain. There appears to be a pulse width limited rise and decay time in the diffracted signal that arises from the optically induced Kerr effects. This component to the signal is observed in the absence of the protein. There may also be a contribution from the short lived excited state of deoxy-Mb. The exact contribution from either source is not well defined at the present time. This spike on the data makes it difficult to extract the grating rise time associated with the protein structural relaxation. Regardless, it is clear from this study that the global relaxation is found to have an effective time constant of less than 10 picoseconds. There appears to be no evidence for significant slower relaxation components, indicative of conformation substates along the relaxation coordinate, except at low temperature.

The grating studies probe the global or long length scale of protein motion. Time resolved Raman studies of the proximal histidine have also found that the tertiary structural changes occur on a ten picosecond time scale.[3] The Raman studies provide a probe of the short length scale motion of the protein, at what is considered the focal point for the forces driving the relaxation. The close correspondence of protein relaxation over two length scales provides evidence that the structural changes are propagated by collective modes. This is the most efficient mechanism possible for the correlated displacement of the atomic coordinates.

Springer Series in Chemical Physics, Vol. 55 Ultrafast Phenomena VIII
Editors: J.-L. Martin · A. Migus · G.A. Mourou · A.H. Zewail

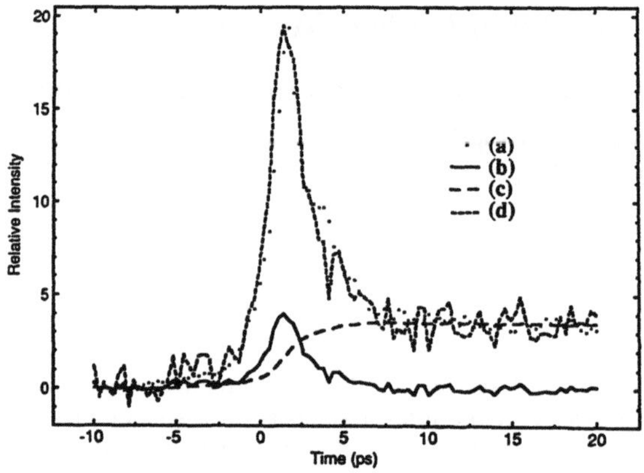

Figure 1. a) MbCO transient grating signal. b) Instrument function derived from a Kerr grating component. c) Convolution of the instrument function with a infinitely fast system response (step function), scaled to match the long time signal. d) Conveniently scaled sum of the last two plots showing that the rise time of the protein driven density changes occur in less than 10 ps.

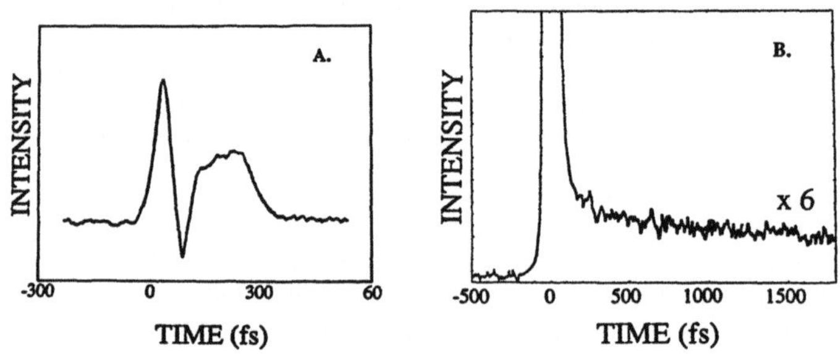

Figure 2.

The grating studies provide information on the protein's structural relaxation processes that lead to changes in volume or strain. There is also the possibility that there are structural relaxation processes that lead to changes in protein shape rather than volume. These dynamics would lead to changes in symmetry of the protein and would modify the material birefringence. Femtosecond studies of the changes in material birefringence following CO photodissociation of MbCO were conducted using a synchronously amplified CPM operating at 627 nm which is on the red edge of the Q-band of MbCO. The excitation and probe pulses were 50 femtoseconds in duration with the probe polarized 45° relative to the excitation. The probe intensity was monitored through an analyzer polarizer which was adjusted to be either orthogonal to the probe or 45° off from this position to enable differential heterodyne detection, as described previously.[4] Results from this study are shown in figure 2. Figure 2A shows the non-resonant laser induced birefringence (Optical Kerr Effect) for water which serves as a control for the system without the protein present. These results are in relatively good agreement with that expected for water.[5] The important feature is that this nonresonant contribution from the water does not contribute to the observed signal for $t > 300$ fs. With MbCO present (figure 2B.), there is observed an additional component of a few picoseconds. This component is seen as a small amplitude decay in the induced birefringence. This relaxation component is comparable in time scale to that observed in the grating studies which implies the structural relaxation processes leading to changes in strain and symmetry are related. However, the dynamic range needs to be extended, along with other controls to directly compare the two studies, as there may be other explanations for the transient birefrigence.

The grating studies give information on structural relaxation components which lead to volume changes. These structural changes are analogous to collective acoustic modes in solid state systems. On the other hand, the studies of the dynamic birefringence gives information on structural relaxation processes that include both acoustic and nonacoustic modes. In both studies, the observed dynamics are consistent with an initially rapid phase for the structural relaxation in which the doming of the heme, upon ligand dissociation, couples to spatially extended modes of the surrounding protein on a picosecond time scale. Other studies have given evidence for multiexponential processes occuring on much longer time scales which are indicative of conformational substates along the reaction coordinate.[6] The energetics associated with each phase of the structural relaxation can be used to determine the relative importance of each phase of the protein relaxation. In addition, the energetics determine the amplitudes of the driving forces and the time scale under which the protein accesses the stored potential energy in its structure.

The energetics for the protein relaxation have been determined using thermal phase grating methods and changing to a glycerol/water medium in which the thermal mechanism dominates the acoustics. Under these conditions, there is minimal interference between the protein driven and thermally driven acoustics. The amplitude and dynamics of the thermal grating are directly related to the energy relaxation processes. The results to date show that the photodissociation of CO from carboxymyoglobin is endothermic by 21 ± 2kcal/mole in which the energy relaxation is complete in less than 200 picoseconds. In contrast, carboxyhemoglobin is endothermic by only 11 ± 3kcal/mole, i.e., $\sim$ 10 kcal/mole of excess energy is released. This energy difference between monomeric myoglobin and tetrameric hemoglobin is related to structural differences in the two proteins. The important point is that for both myoglobin and hemoglobin, the measured energetics are fairly close to those obtained by thermodynamic measurements (i. e., steady state). This observation and the lack of any significant slow relaxation components ($>$ 200ps) suggest that the protein relaxation is **energetically** complete on a picosecond time scale.

Both the time scale for the observed global relaxation and energetics are consistent with a collective mode mechanism as the dominant mechanism for the tertiary structural changes in the heme proteins.

This research was supported by the NIH (1 R01 GM41909-01A1).

1. L. Genberg, L. Richard, G. McLendon, and R. J. D. Miller, (1991) Science **251**, 1051.
2. L. Richard, L. Genberg, J. Deak, H. Chiu, and R. J. D. Miller, Biochemistry, submitted.
3. a) E. W. Findsen, J. M. Freidman,, M. R. Ondrias, and J. R. Simon,(1985) Science, **229**, 661.
   b) J. W. Petrich and J. L. Martin, (1989) Chem. Phys., **131**, 31.
4. D. McMorrow and W. T. Lotshaw,(1991) Chem. Phys. Lett. **174**, 85-94.
5. S. Palese, W. Bosma, L. Schilling, S. Mukamel and R. J. D. Miller, submitted for publication.
6. X. L. Xie and J. D. Simon, (1991) Biochemistry **30**, 3682.

Investigation of the Reaction Coordinate for Ligand Rebinding in Photoexcited Hemeproteins Using Transient Raman Spectroscopy

H. Zhu, R. Lingle, Jr., X. Xu, and J.B. Hopkins

Department of Chemistry, Louisiana State University,
Baton Rouge, LA 70803, USA

ABSTRACT. Transient picosecond Raman spectroscopy is capable of differentiating vibrational relaxation from conformational changes by comparing the Stokes and anti-Stokes dynamics. We report pump-probe picosecond Raman experiments on oxy- and deoxyhemoglobin (oxyHb and deoxyHb, respectively) using 8 ps 532 nm pump pulses and 8 ps 355 nm probe pulses. Heme-to-protein vibrational cooling has been directly observed in deoxyHb for the first time, and the deconvolved cooling time constant is measured to be 2 - 5 ps.[1-2] By applying our mode-specific Stokes and anti-Stokes technique to oxyHb, we find that any geminate recombination of photodeligated O_2 must occur in either less than two picoseconds or longer than a nanosecond.

1. INTRODUCTION

The advent of ultrafast spectroscopy has raised hopes that the ligational and conformational dynamics of heme proteins can be directly observed in the vibrational spectrum. However, vibrational dynamics, conformational dynamics, and ligation dynamics complicate the interpretation of transient spectra. Several attempts have been made to characterize the ultrafast dynamics of vibrational energy dissipation in photoexcited heme proteins.[3-4] Previous Raman experiments were unable to probe the transient anti-Stokes spectrum and were interpreted using indirect analyses. An accurate measurement of the vibrational energy relaxation (VER) which follows photoexcitation is necessary to correctly interpret ultrafast electronic absorption data. Our experiment is unique in that it *directly* and *unambiguously* detects hot vibrations in both low- and high-frequency porphyrin modes. We are also able to estimate the vibrational temperature from anti-Stokes transient data alone. Confusion also exists in the literature over the time scale for ligand recombination following photoexcitation. We have applied our mode-specific picosecond Raman technique to oxyhemoglobin in order to separate some of the complex vibrational and conformational dynamics.

2. VIBRATIONAL RELAXATION IN DEOXYHEMOGLOBIN

We have studied deoxyHb by exciting at 532nm and probing at 355nm. A pure transient Raman spectrum is produced by subtracting the background

Springer Series in Chemical Physics, Vol. 55 Ultrafast Phenomena VIII
Editors: J.-L. Martin · A. Migus · G.A. Mourou · A.H. Zewail

probe-only components from the spectrum obtained with the pump-probe sequence. The resonant Raman enhancement for excited vibrations turns out to be lower than that of the ground vibrational level. Negative Stokes bands appear in the transient spectrum as a result of the removal of population (by the 532 nm pump photons) from those levels thermally populated at room temperature. Positive transient anti-Stokes bands indicate that there is more population in the low-lying excited vibrational modes of the photoexcited heme than in unexcited heme at room temperature. Notice that the frequency-resolved transient data show opposite and complementary behavior for the Stokes and anti-Stokes cases.

We have observed the VER dynamics of photoexcited deoxyHb by monitoring the *Stokes* and *anti-Stokes* transient signals for specific deoxyHb vibrational modes as the time delay between pump and probe pulses is varied.[2] Deconvolution of the 8 ps laser pulse width yields a VER exponential time constant of 2-5 ps for heme vibrations at 671, 1424, and 1563 cm^{-1}. The complementary behavior of the Stokes and anti-Stokes transients is illustrated by the dynamics observed for the heme ν_2 band (not shown here). That is, the anti-Stokes transient signal rises promptly and decays with 2-5 ps time constant, while the Stokes transient signal drops negative promptly and recovers to the baseline with a 2-5 ps time constant.[2] We can estimate the heme vibrational temperature by comparing the magnitudes of the anti-Stokes transient signals at the ν_7 and ν_4 band positions.[1] Our temperature jump estimate for the photoexcited deoxy heme, averaged over the convolution of our 8 ps pump and probe pulses, is $\approx$36 K above room temperature. A 532 nm photon can produce a heme internal temperature jump in excess of 400 K. Therefore the 2-5 ps VER time constant represents the loss of most of the heme internal energy.

3. HEME-LIGAND DYNAMICS IN OXYHEMOGLOBIN

The dynamics of photoexcited oxyHb are more complicated due to presence of the O_2 ligand. Following photoexcitation, heme-oxygen deligation and geminate recombination may occur in addition to vibrational relaxation. We can separate the heme-ligand dynamics from the vibrational relaxation dynamics by comparing *Stokes* and *anti-Stokes* Raman dynamics. VER with $\tau \approx 5$ ps is observed for both hot oxyHb (which could result from photoexcited oxyHb which did not lose its ligand or from geminate recombination that occurs faster than VER) and hot deoxyHb (resulting from photodeligated oxyHb). Our results indicate that O_2 recombination occurs either faster than 2 ps or slower than 1 ns.

Photodeligation of oxyHb yields a deoxy form of hemoglobin which has a vibrational spectrum very similar to that of oxyHb. The differences in the resonant Raman spectra for the ligated and unligated forms consist of band frequency shifts and/or band intensity changes. These differences in the resonant Raman of oxyHb and deoxyHb will dramatically affect the dynamics observed for the photodeligation reaction of oxyHb. For example, the oxyHb ν_4 band appears at 1376 cm^{-1}while the deoxyHb band is shifted

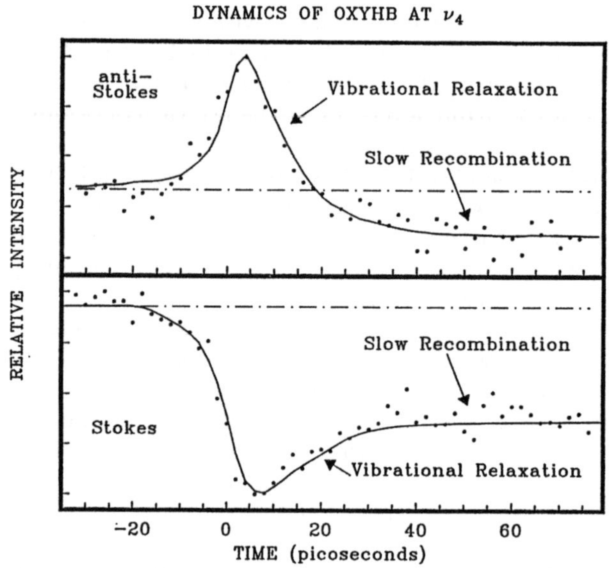

Figure 1
Dynamics of oxyHb for the ν_4 vibrational mode (1376 cm^{-1}). (Top) anti-Stokes dynamics; (Bottom) Stokes dynamics. The circles are data points; the solid line is the fit.

to 1359 cm^{-1}. The oxyHb and deoxyHb ν_{16} bands are at roughly the same Raman shift. For Stokes Raman scattering, the ν_4 band is almost a factor of three more intense for oxyHb than for deoxyHb, while ν_{16} is about a factor of three stronger for deoxyHb than for oxyHb. In the anti-Stokes spectrum, both ν_4 and ν_{16} are more intense for deoxyHb than for oxyHb.

Figure 1 shows both Stokes and anti-Stokes dynamics of the ν_4 vibrational mode at 1376 cm^{-1}. Following a pulse-width limited bleaching of the Stokes signal in Figure 1, we clearly observe two dynamics as the transient returns toward the baseline (established by the signal at negative time). The anti-Stokes dynamics of Figure 1 rises promptly with the laser pulse width, and then decays rapidly to a level *below* the baseline. Taken together, these dynamics indicate both vibrational cooling and slow recombination of the O_2 ligand. Photodeligation will remove a certain number of oxyHb molecules from the sample volume, leaving deoxy-like Hb in their place. Because the deoxyHb signal is weak at 1376 cm^{-1} (especially on the Stokes side), photodeligation will result in negative transient signals for both the Stokes and anti-Stokes bands. This is clearly observed in the long time dynamics, which represents heme-ligand recombination on the nanosecond time scale. Because the deoxyHb ν_4 band is shifted 20 cm^{-1} off of the position we are observing, *the oxyHb signal dominates the observed dynamics*.

The early time dynamics of Figure 1 is assigned to VER. The complementary negative/positive behavior in the Stokes/anti-Stokes early time

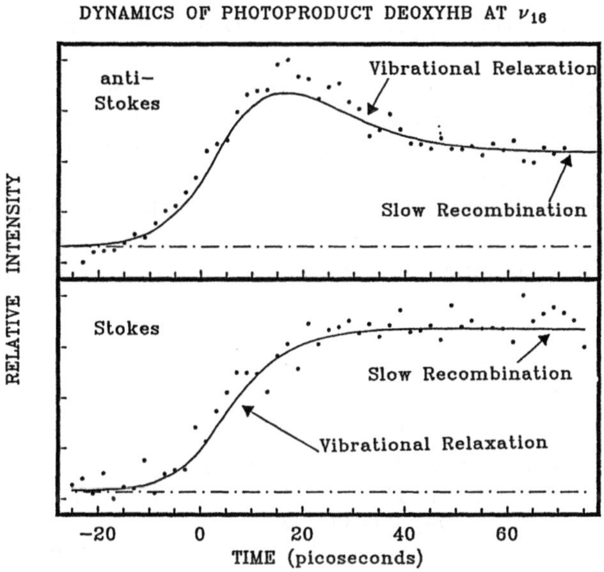

Figure 2
Dynamics of deoxyHb for the ν_{16} vibrational mode (755 cm^{-1}). (Top) anti-Stokes dynamics; (Bottom) Stokes dynamics. The circles are data points; the solid line is the fit.

dynamics is exactly the same as for the earlier study of deoxyHb. Because the oxyHb resonant Raman signal dominates the spectrum of Figure 1, the early-time fast dynamics is believed to result from VER in hot *oxyHb* which either never lost its ligand or else underwent geminate recombination faster than 2 ps. Deconvolution of both Stokes and anti-Stokes dynamics gives same VER rate of $\tau \approx 5$ ps. On the other hand, fast recombination would appear as a transient moving in the positive direction in both the Stokes and anti-Stokes dynamics. This is not observed.

Both the Stokes and anti-Stokes dynamics of the ν_{16} band in Figures 2 show *positive* transient signal at long times, in contrast to the ν_4 dynamics in Figure 1. The transient is positive at long times because the resonant Raman enhancement for the ν_{16} Stokes and anti-Stokes bands is larger for deoxyHb than for oxyHb, so *the observed dynamics of ν_{16} are dominated by the deoxy photoproduct*. These deligated hemes will be vibrationally hot, as shown in the early time dynamics. The anti-Stokes transient signal, shown in Figure 2, rises as the deoxyHb low-lying excited levels become populated; it then decays to the long time offset as the deoxyHb population cools to thermal equilibrium at room temperature. The Stokes transient for this band, shown in Figure 2, serves as a window to observe the cold vibrationless level of deoxyHb; thus the dynamics rises as the deoxyHb vibrationless level is populated. We see no evidence for fast geminate recombination in the Stokes dynamics, which would appear as a negative-going feature in Figure 2.

Geminate recombination for photolyzed oxyHb has been observed on the nanosecond time scale using transient absorption.[5] Petrich et al.[3] observed a 2.5 ps dynamics which they propose to be a deligated but highly reactive electronic state that relaxes back to the ligated form. However, we see no evidence of ligand recombination on a 2.5 ps time scale. A 2.5 ps geminate recombination is on roughly the same time scale as the VER we observe in deoxyHb, so it should appear in the dynamics of Figures 1 and 2 if it exists.

4. REFERENCES

[1.] Robert Lingle, Jr., Xiaobing Xu, Huiping Zhu, Soo-Chang Yu, J. B. Hopkins, J. Am. Chem. Soc., **113**, 3992 (1991).

[2.] Robert Lingle, Jr., Xiaobing Xu, Huiping Zhu, Soo-Chang Yu, J. B. Hopkins, J. Phys. Chem., **95**,, 9320 (1991).

[3.] J. W. Petrich and J. L. Martin, Chemical Physics, **131**, 31 (1989).

[4.] R. G. Alden, et al., J. Am. Chem. Soc., **112**, 3241 (1990).

[5.] J. M. Friedman et al., Science, **229**, 187 (1985).

Resonance Raman Studies of Electronic and Vibrational Relaxation Dynamics in Heme Proteins

P.M. Champion, J.T. Sage, and P. Li

Department of Physics, Northeastern University, Boston, MA 02115, USA

Abstract. Resonant Stokes and anti-Stokes scattering as a function of laser excitation rate is used to determine the electronic relaxation times and vibrational temperatures of heme proteins. A simple thermal transport model is used to extract a value for the product of the heme area and the coefficient of surface heat transfer between the heme and the surrounding protein. This leads to a ~4ps time constant for the short-time exponential phase of heme cooling.

1. Introduction

The spontaneous Raman scattering intensity observed under resonance conditions is reduced at high photon fluxes, due to light driven depopulation of the electronic ground state. Increases in the anti-Stokes scattering intensities are also observed at high laser flux and attributed to rapid redistribution of vibrational energy and heating of the heme chromophore. Separation of the electronic and vibrational relaxation using resonance Raman scattering should prove to be a useful complement to time-resolved absorption studies, which in general involve a convolution of time scales due to electronic (bleaching recovery) and vibrational (thermal broadening) relaxations.

2. Results

We have previously discussed the technique of resonance Raman saturation spectroscopy [1] and summarize here some experimental results that probe relaxation processes in heme proteins following electronic excitation in the Soret band. Analysis of the Stokes intensity saturation, using a model that allows for parallel relaxation channels [1] leads to effective electronic relaxation times for deoxymyoglobin (4.8±1.5ps), ferrocytochrome c (6.4±2.0ps), and deoxyhemoglobin (2.0±0.7ps). The more rapid relaxation times found for hemoglobin and its isolated chains are not currently understood.

Studies of the asymmetric broadening of the ν_4 mode of cytochrome c at high laser flux, reveal that the lineshapes in the Stokes and anti-Stokes region are inequivalent. Time-reversal symmetry dictates that this broadening is due to an underlying Raman band associated with an excited electronic state that is populated at high laser flux. Similar linebroadening effects, observed in hemoglobin and myoglobin samples, are also shown [1] to arise from Raman scattering of excited electronic states rather than Rabi broadening [2] or anharmonic coupling to vibrationally hot low frequency modes [3].

Analysis of the Stokes/anti-Stokes Raman scattering intensity [1] leads to a determination of the heme vibrational temperature as a function of the incident laser flux (averaged over the 10ns laser pulse). This analysis, which includes scattering within a complete set of harmonic basis states, should be contrasted with the approach of Lingle et al. [4], which is based on the scattering properties of a two level system. As can be seen from the Bose-Einstein factors in Eqs. 1 and 2 below, consideration of the vibrationally excited states leads to an increase in **both** the Stokes and anti-Stokes

Springer Series in Chemical Physics, Vol. 55 Ultrafast Phenomena VIII
Editors: J.-L. Martin · A. Migus · G.A. Mourou · A.H. Zewail

scattering as the vibrational temperature is increased. Thus, the claim by Lingle et al. [4], that vibrational heating is signaled by a decrease of the Stokes scattering along with a simultaneous increase of the anti-Stokes scattering, is not consistent with theory.

The Stokes (anti-Stokes) Raman intensity observed at high laser flux can be normalized by the intensity obtained in a low flux (cw) experiment in order to minimize the resonance effects and expose the thermal effects through the Bose-Einstein factors ($\langle n_i \rangle = [exp(\tilde{\nu}_i/kT) - 1]^{-1}$). For example, we can express the high to low flux ratio for the Stokes (+) and anti-Stokes (-) scattering cross-sections of the ith totally symmetric fundamental in the harmonic limit as:

$$I_r^+ \equiv \frac{I_{tot}^+}{I_{tot}^+(k_L \to 0)} = \frac{\langle N_0 E_s^2 \rangle}{\langle E_s^2 \rangle} \frac{\langle n_i^* \rangle + 1}{\langle n_i \rangle + 1} R_{T,T^*}^+(\Phi) \tag{1}$$

$$I_r^- \equiv \frac{I_{tot}^-}{I_{tot}^-(k_L \to 0)} = \frac{\langle N_0 E_s^2 \rangle}{\langle E_s^2 \rangle} \frac{\langle n_i^* \rangle}{\langle n_i \rangle} R_{T,T^*}^-(\Phi) \tag{2}$$

$$R_{T,T^*}^{\pm}(\Phi) = \frac{| \Phi_{T^*}(\tilde{\nu}_L) - \Phi_{T^*}(\tilde{\nu}_L \mp \tilde{\nu}_i) |^2}{| \Phi_T(\tilde{\nu}_L) - \Phi_T(\tilde{\nu}_L \mp \tilde{\nu}_i) |^2} \tag{3}$$

where $\Phi_T(\tilde{\nu}_L)$ is the temperature dependent complex lineshape function, derived from either a Kramers-Kronig transform of the resonance absorption lineshape or a half Fourier transform of a generalized time-correlator expression [5], and $\tilde{\nu}_L$ is the laser excitation frequency. The orientationally averaged electronic ground state population factor, $\langle N_0 E_s^2 \rangle / \langle E_s^2 \rangle$, is a function of the laser excitation rate (k_L) and the electronic relaxation time (τ). Careful analysis of the terms in Eq 1 shows that, if ν_L is chosen appropriately, τ can be directly extracted from measurements of I_r^+ vs. k_L. Taking the ratio of Eqs. 2 and 1 cancels the electronic ground state saturation effects and leads to a method for determining the laser excited heme temperature, T^*.

Analysis of T^* as a function of the laser flux J leads to

$$T^* - T_0 = \frac{1 + aH/K}{4\pi a^2 H} \frac{J\sigma_A}{1 + \tau\sigma_A \tilde{J}} \tag{4}$$

where a is the heme length scale, H is the coefficient of surface heat transport between the heme and its surroundings, and K is the thermal conductivity of the protein ($\tilde{J} = J\lambda/hc$ is the photon flux). The factor $4\pi a^2 H$ can be extracted from the data and used to calculate the initial heme cooling rate, $k_c = 4\pi a^2 H/C_h$, where $C_h \sim 10^{-21}$J/K is the heme vibrational specific heat. This leads to a time constant for heme cooling of ~4ps which is in good agreement with molecular dynamics simulations [6].

References

1. P. Li, J.T. Sage and P.M. Champion, J. Chem. Phys. (in press).
2. R. Alden et al., J. Phys. Chem. **94**, 85 (1990).
3. J. Petrich et al., Biochemistry **26**, 7914 (1987).
4. R. Lingle et al., J. Phys. Chem. **95**, 9320 (1991).
5. K.T. Schomacker and P.M. Champion, J. Chem. Phys. **90**, 5982 (1989).
6. E. Henry, W. Eaton and R.M. Hochstrasser, Proc. Natl. Acad. Sci. 83, 8982 (1986).

Molecular Processes in the Primary Reaction of Photosynthetic Reaction Centers

W. Zinth[1], C. Lauterwasser[1], U. Finkele[2], P. Hamm[1], S. Schmidt[2], and W. Kaiser[2]

[1]Institut für Medizinische Optik der Universität München, Barbarastr. 16/IV, W-8000 München 40, Fed. Rep. of Germany

[2]Physik Department E 11 der Technischen Universität München, W-8046 Garching, Fed. Rep. of Germany

Abstract: The primary electron transfer is investigated for wildtype reaction centers from Rhodobacter sphaeroides, mutant reaction centers and reaction centers with modified bacteriochlorophyll . Experimental results are presented which strongly support the idea that the electron transfer is stepwise involving the accessory bacteriochlorophyll as a real intermediate electron carrier.

1. Introduction

The energy converting processes of bacterial photosynthesis start in a large chromoprotein called reaction center (RC). At first light energy leads to electronic excitation of a chromophore. Subsequently the electronic excitation initiates an electron transfer (ET) via several acceptor molecules. Finally the electron reaches the quinone Q_B. Further secondary transfer processes involve diffusive motion of protons and large organic molecules. Schematically the structural arrangement of the electron carrying pigments in the RC's is shown in Fig. 1 /1, 2/. The pigments are incorporated in two branches (A and B) related by an approximate C_2 symmetry. The two branches cross at two strongly interacting bacteriochlorophyll molecules forming the special pair P which acts as a primary electron donor. Subsequently each branch contains a monomeric bacteriochlorophyll molecule B_A and B_B, a bacteriopheophytin (H_A, H_B) and a quinone (Q_A, Q_B). Various experiments have revealed that the two pigment branches are not equivalent and that the reactive electron transfer uses the A branch.

The very first reaction processes include the initial charge separation and the ET until the bacteriopheophytin H_A. These events take several picoseconds /3-9/. In the next step the electron reaches within approximately 200 ps the quinone Q_A. The secondary quinone Q_B is reduced only much slower on the 10^{-6} s time scale. Transient absorption spectroscopy of the

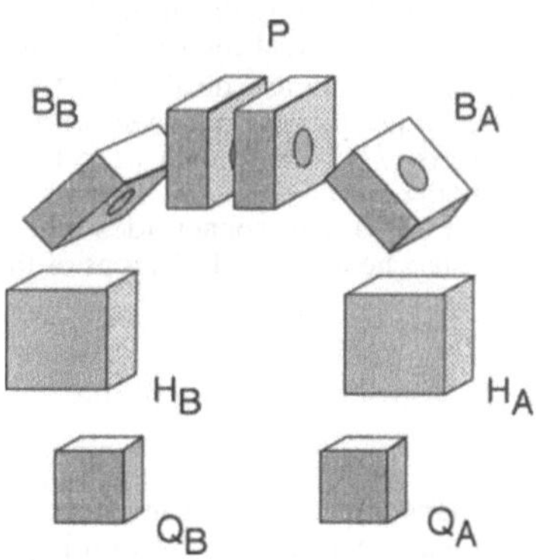

Fig. 1: Schematic of the molecular arrangement of the four bacteriochlorophylls (P, B_A, B_B), the two bacteriopheophytins (H_A, H_B) and the two quinones (Q_A, Q_B) in reaction centers

Springer Series in Chemical Physics, Vol. 55 Ultrafast Phenomena VIII
Editors: J.-L. Martin · A. Migus · G.A. Mourou · A.H. Zewail

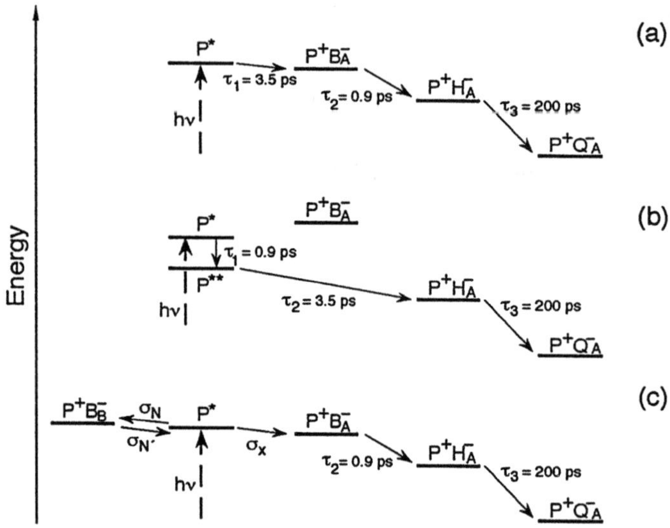

Fig. 2: Reaction models for the primary electron transfer

primary reaction dynamics at room temperature exhibit one time constant around 3.5 ps (in the literature values are reported between 2.8 ps and 4 ps /3-9/) related to the decay of the excited electronic state P* and a weak subpicosecond kinetic (0.9 ps in Rb. sphaeroides) best seen in the absorption range of the BChl and the BChl anion /7-9/. Until now no generally accepted molecular interpretation of the fast kinetic component exists /10-12/. Two reaction models are currently discussed (see Fig. 2a and b):

The structural arrangement of the reaction centers strongly suggests the stepwise electron transfer model of Fig. 2a: According to this model the electronically excited state P* of the special pair decays with the time constant of 3.5 ps. Simultaneously an electron is transferred from the special pair to the monomeric bacteriochlorophyll B_A. The second electron transfer is faster and carries the electron with a time constant of 0.9 ps to the bacteriopheophytin H_A. Finally the 200 ps process generates the radical pair $P^+Q_A^-$ where the electron has reached the first quinone.

In model 2b the fast time constant of 0.9 ps is assigned to an excited state relaxation process of the special pair. This process should be vibrational relaxation from the initially populated Franck-Condon-state. In this reaction model the first electron transfer drives the electron with a time constant of 3.5 ps directly to the bacteriopheophytin H_A. This fast long-distance electron transfer is only possible if the monomeric BChl B_A is involved as a virtual intermediate in a superexchange interaction. In this case the energy level of the corresponding radical pair $P^+B_A^-$ is higher than the energy of P*. From extensive experimental studies the absorption spectra of the different intermediates could be calculated for both models /7-9/. These data were fully consistent with the molecular interpretation of the respective model.

In this paper we will present transient absorption data on RC of Rb. sphaeroides where additional information on the reaction model is obtained by: (i) biochemical modifications of the RC and (ii) changing the temperature of the sample.

2. Results and Discussion

Measurements on low temperature RC are performed on quinone depleted RC from the carotenoid free strain R26.1. The preparation of RC where BChl a at the accessory position B_A is exchanged to approximately 70 % by [3-vinyl]-13^2-OH-BChl a is described in Ref. 13. The time resolved absorption experiments are performed using the excite and probe technique with weak subpicosecond pulses (pulse duration ≃ 150 fs) generated by a laser-amplifier-system

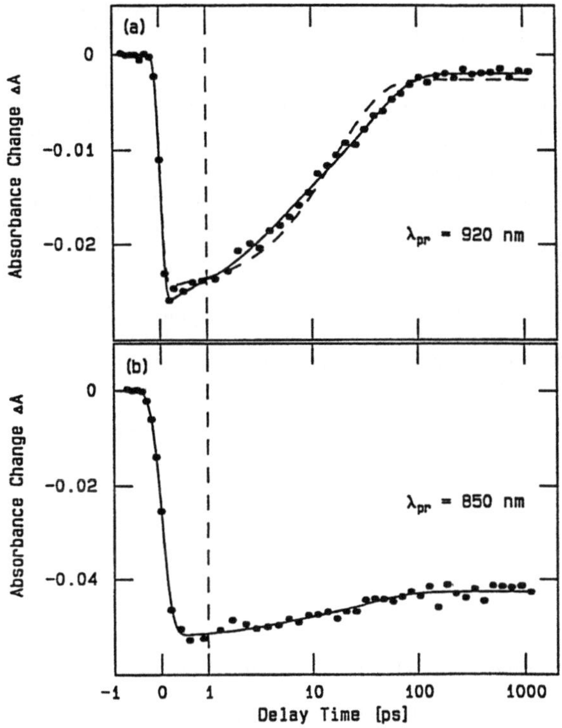

Fig 3: Transient absorption data on [3-vinyl]-13^2-OH-RC-preparations. Probing in the gain region at $\lambda = 920$ nm (a) and in special pair absorption band $\lambda = 850$ nm (b)

with a repetition rate of 10 Hz. The temporal width of the instrumental response function is below 300 fs. Details of the experimental system are described in reference /14/.

Upon lowering the temperature of the RC preparation the picosecond kinetics become faster /14, 15/. The decay of the gain having a time constant of around 3.5 ps at room temperature speeds up to 1.4 ps at 25 K. The 0.9 ps kinetic component also accelerates at lower temperatures. At 25 K a value of 0.3 ps is reached. Throughout the whole temperature range from 300 K to 25 K the spectral signature of the fast component does not change.

Experiments on [3-vinyl]-13^2-OH-BChl a containing RC's are shown in Fig. 3. The modification due to the 3 vinyl group is expected to change the redox potential of the BChl and the energy of the radical pair state P^+B^-. This change should have pronounced consequences on the ET when the accessory BChl B_A is involved. Indeed, one finds a strong slowing down of the decay of P* (see Fig. 3a for a measurement in the gain region). The experimental data indicate that the RC's containing [3-vinyl]-13^2-OH-BChl a have a decay time of P* of 32 ps. Fig. 3b, measured at a wavelength where the P absorption is strong, shows a long-lasting bleaching of P. These observations prove that the exchange leads to RC's which are still photochemically active but where the ET-step out of P* is slowed down by a factor of ten. In the [3-vinyl]-13^2-OH-BChl a containing RC's the 0.9 ps component is not visible. Experimental indications exist that a related process appears with a longer time constant in the 5 ps domain. The disappearance of the 0.9 ps component can be taken as a strong indication that the 0.9 picosecond process in wildtype RC is not related to vibrational relaxation of P* since P* is unchanged upon modifying B. This finding argues against the superexchange electron transfer of model 2b.

Additional support of the stepwise model of Fig. 2a comes from the low temperature hole-burning experiments where narrow holes related to a $\tau >$ 1ps process are observed. According to these experiments the 0.3 ps process observed in transient absorption spectroscopy can not be the first reaction. As a consequence the 0.3 ps time constant (and due to the smooth temperature dependence the 0.9 ps time at room temperature) must be related to the second reaction step.

Other important experimental information is obtained from transient absorption spectroscopy of an antenna deficient mutant (strain U43, pTXA6-10) of Rb. capsulatus. Here whole chromatophores are investigated instead of isolated RC's. The experiments exhibit the same subpicosecond component as observed in the RC preparation. This finding clearly indicates that the subpicosecond component is not an artefact due to the preparation procedure. Recently emission experiments on RC have shown that the 3.5 ps kinetic component of wildtype RC has to be split into a 2.3 ps and a 7 ps process/17/. Within the scope of the stepwise ET model this biexponentiality can be explained by a transient ET to the acccessory bacteriochlorophyll on the B branch as indicated in scheme C of Fig. 2. According to this interpretation the ET on the B branch would be blocked efficiently only between B_B and H_B.

In conclusion, we have shown that the primary processes in photosynthetic RC can be explained by a stepwise electron transfer from the special pair over the different electron carrying pigments. The distance between the pigments is small enough to allow fast and efficient electron transfer in the stepwise model.

Acknowledgement: The author thank H. Scheer and G. Drews for providing us with high quality preparations.

References

1. J.Deisenhofer, H.Michel, EMBO J. 8, 2149 (1989).
2. C.H.Chang, D.Tiede, J.Tang, U.Smith, J.Norris, M.Schiffer, FEBS Letters 205, 82 (1986).
3. N.W.Woodbury, M.Becker, D.Middendorf, W.W.Parson, Biochem. 24, 7516 (1985).
4. J.Breton, J.L.Martin, A.Migus, A.Antonetti, A.Orszag, Proc. Natl. Acad. Sci. US 83, 5121 (1986).
5. J.L.Martin, J.Breton, A.J.Hoff, A.Migus, A.Antonetti, Proc. Natl. Acad. Sci. US 83, 957 (1986).
6. C.Kirmaier, D.Holten, W.W.Parson, Biochim. et Biophys. Acta 810, 49 (1985).
7. W.Holzapfel, U.Finkele, W.Kaiser, D.Oesterhelt, H.Scheer, H.U.Stilz, W.Zinth, Chem. Phys. Letters 160, 1 (1989).
8. W.Holzapfel, U.Finkele, W.Kaiser, D.Oesterhelt, H.Scheer, H.U.Stilz, W.Zinth, Proc. Natl. Acad. Sci. US 87, 5168 (1990).
9. K.Dressler, E.Umlauf, S.Schmidt, P.Hamm, W.Zinth, S.Buchanan, H.Michel, Chem. Phys. Letters 183, 270 (1991).
10. M.Plato, K.Möbius, M.E.Michel-Beyerle, M.Bixon, J.Jortner, J. Am. Chem. Soc. 110, 7279 (1988).
11. R.A.Marcus, Chem. Phys. Letters 133, 471 (1987).
12. M.Bixon, J.Jortner, M.E.Michel-Beyerle, Biochim. et Biophys. Acta 1056, 301(1991).
13. A. Struck, E. Cmiel, I. Katheder, H. Scheer, FEBS Lett. 268, 180 (1990).
14. C.Lauterwasser, U. Finkele, H. Scheer, W. Zinth; Chem. Phys. Letters 183, 471 (1991).
15. G.R.Fleming, J.L.Martin, J.Breton, Nature 333, 190 (1988).
16. S.G. Johnson, D. Tang, R. Jankowiak, J.M. Hayes, G. J. Small, J. Phys. Chem. 93, 5953 (1989) .
17. P. Hamm, W. Zinth, (1992) same proceedings.

Femtosecond Spontaneous Emission Studies of Photosynthetic Bacterial Reaction Centers

S.J. Rosenthal[1], *M. Du*[1], *X. Xie*[1], *T.J. DiMagno*[1], *M.E. Schmidt*[1], *J.R. Norris*[1,2], *and G.R. Fleming*[1]

[1]Department of Chemistry, The University of Chicago, 5735 So. Ellis Ave., Chicago, IL 60637, USA
[2]Chemistry Division, Argonne National Laboratory, Argonne, IL 60439, USA

Spontaneous fluorescence decays of photosynthetic reaction centers were measured by fluorescence upconversion with a time resolution of 50 fs. In all cases non-exponential decays were observed.

1. Introduction

The mechanism of the initial electron transfer step in photosynthetic bacterial reaction centers has been the subject of intense study over the past ten years. This initial step is ultrafast, occuring in about 3 ps at room temperature[1] . Previous studies have relied mostly on transient absorption or stimulated emmision techniques. As the understanding of the reaction center improves more precise kinetic data is required. Particular questions concern the exponentiality of the observed kinetic signals[2], the possibility of differing behavior at different wavelengths[3] , the existence of oscillatory components[2] and of spectral shifts accompanying the excitation and subsequent electron transfer process. We have measured the spontaneous emission decay of the primary donor (P*) in Rb. Sphaeroidis R26, deuterated R26, Rb. Capsulatus and mutations of Rb. Capsulatus at sites L181 and M208 using the fluorescence upconversion technique with 50 fs time resolution[4]. The R26 measurements were performed with both direct excitation of P at 850nm and indirect excitation through internal conversion from the Q_x band of P and energy transfer from the accessory bacterial chlorophyll excited at 608nm. All other samples were excited at 850nm.

2. Results and Discussion

All of the samples displayed nonexponential fluoresence decays which were well fit as biexponential decays. The short component of the decays compares very well with the single decay components preveiously determined by stimulated emission measurements[5]. However, the improved dynamic range and absence of other components (excited state absorption and ground state bleaching) make the nonexponentiality quite clear. The time constants determined in the Capsulatus measurements vary as the redox potential of P* is varied by the mutations. Substitution of phenylalanine L181 with tyrosine increases the rate of primary electron transfer while substitution of tyrosine M208 with phenylalanine or reversal of phenylalanine L181 and tyrosine M208 decreases the electron transfer rate. As figure 1 shows deuteration of Rb. Sphaerodis R26 does not , within the current error of the measurements, yield a different rate than the undeuterated R26. Finally, with 608 nm excitation a risetime of 200 fs is apparent while at 850 nm excitation we are unable to detect a risetime at 940 nm emission. The decay times observed at both excitation wavelengths are the same within experimental error.

We propose three mechanisms which may give rise to the observed nonexponential behavior: 1) Protein inhomogeneity leading to a distribution in parameters affecting the electron transfer rate; 2) The inherent mechansim of electron transfer, i.e. the competition between sequential with a reversbiel first step, and one step (super exchange) electron transfer; and 3) slow or incomplete vibrational relaxation on the electron transfer time scale. The third mechanism seems to be ruled out by the comparison of 608 nm and 850 nm excitation. We cannot yet distinguish between the other two possibilities, although in view of hole burning studies it seems likely that inhomogeneity must contribute to at least part of the non-exponentiality. Detailed temperature-dependent and wavelength resolved studies should help to reveal the source of non-

Springer Series in Chemical Physics, Vol. 55 Ultrafast Phenomena VIII
Editors: J.-L. Martin · A. Migus · G.A. Mourou · A.H. Zewail

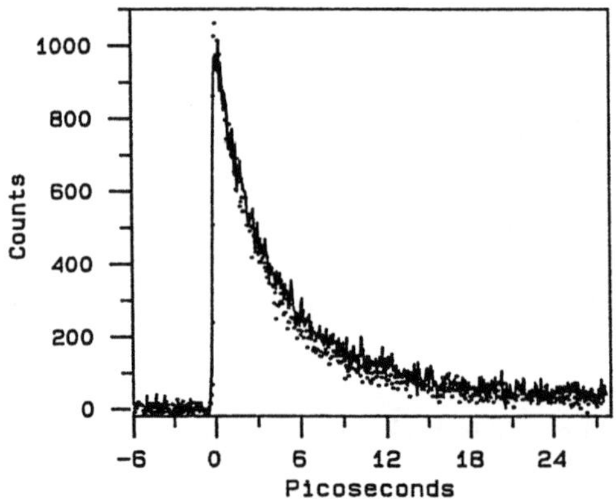

Figure 1. Spontaneous fluorescence decay of Rb. Sphaerodis R26 reaction centers excited at 850 nm and detected at 940 nm. Solid line: fully deuterated R26; dotted line: normal R26. Best fits to double exponential decays give 2.7 ps (81%) and 12.1 ps (19%) for R26, and 3.2 ps (83%) and 16.5 ps (17%) for deuterated R26.

exponentiality, an understanding of which is crucial for an understanding of the role of mutations in the primary charge separation process.

This work was supported by an NSF grant (GRF) and by the Department of Energy under contract W-31-109-Eng-38 (JRN).

3. References

1. For a recent review and references therein see D. Hoten and C. Kirmaier, Photosynthesis Research 13, 225 (1987).
2 (a) M.H. Vos, J.C. Lambry, S.J. Robles, D.C. Youvan, J. Breton and J.-L. Martin, (1991) Proc. Natl. Acad. Sci. USA 88, 8885; (b) PNAS 89, 613 (1992). (c) C. Lauterwasser, U. Finkele, H. Scheer and W. Zinth, Chem. Phys. Lett. 183 (1991) 471.
3 (a) D. Holten and C. Kirmaier, Proc. Natl. Acad. Sci. USA 87 (1990) 3552; (b) Biochem. 30, (1991) 609.
4 M. Du, S. J. Rosenthal, X. Xie, T. J. DiMagno, M. Schmidt, J. R. Norris, and G.R. Fleming Proc. Natl. Acad. Sci. USA in press.
5. C.-K. Chan,L. X.-Q. Chen, T.J. DiMagno, D. K. Hanson, S. L. Nance, M. Schiffer, J.R. Norris and G.R. Fleming, Chem. Phys. Lett. 176 (1991) 366.

Subpicosecond Emission Studies of Bacterial Reaction Centers

P. Hamm and W. Zinth

Institut für medizinische Optik der Ludwig-Maximilians-Universität München, Barbarastr. 16, W-8000 München, Fed. Rep. of Germany

The emission of the electronically excited state of the special pair in the reaction centers of the photosynthetic bacterium Rhodobacter (Rb.) sphaeroides R26.1 and several mutants was observed by upconversion. In all samples, a biexponential decay of the fluorescence emission was observed.

In the primary photosynthetic reaction energy is stored by a charge transfer across a membrane bound pigment protein complex called reaction center (RC). The x-ray structure analysis of RC's from Rb. sphaeroides has shown that the electron carrying pigments are arranged in two branches (A and B) related by an approximate C_2-symmetry. It is known that the electron uses predominantly the A branch. By transient absorbance measurements 3 kinetic components were found which could be associated with a sequential reaction model. After electronical excitation of the special pair (P) by a photon the electron is transfered via a bacteriochlorophyll monomer (B_A) in 3.5 ps to a bacteriopheophytin (H_A) in 0.9 ps, and finally to a quinone (Q_A) in 200 ps.

In order to get new and independent information on the first reaction step we investigated fluorescence emission of the electronically excited state P^*. We present data on native RC's from Rb. sphaeroides and from mutants where the tyrosine at position M210 and the phenylalanine L181 were exchanged. In detail, we have investigated (YM210→F), (FL181→Y), and (YM210→F,FL181→Y). The mutations were at symmetry related positions between the special pair and chlorophyll monomer on each branch.

The experiments were based on a femtosecond near infrared laser-amplifier system operating at a repetition rate of 10 Hz. Exitation was performed at a wavelength of 865nm. The fluorescence light at a wavelength of 920 nm was upconverted in a 1mm thick BBO crystal. Uncritical collinear Type II phasematching was used. The time resolution was approximately 400 fs.

The transient absorbance measurements on native RC's in the spectral region of stimulated emission indicated that P^* decays monoexponentially with a time constant of 3.5ps. The results of the emission experiments are plotted in Fig. 1. No data trace follows a monoexponential decay. However, the measurements can be explained well by a biexponential model function with time constants of 2ps-6ps for a faster component and 7ps-26ps for the slower component. The values of the time constants (τ_1, τ_2) and the amplitude ratios (A_2/A_1) are summarized in Table 1.

No other fast time constant is required to explain the experimental data; e.g. the 0.9 ps kinetic component which was observed by transient absorbance measurements in the spectral region of the absorption bands of B_A is not observed in the emission experiment. On the other hand one would expect the 0.9 ps kinetic if it reflects an energy relaxation process within P^*. Therefore our finding supports the stepwise electron transfer model where the 0.9 ps kinetic is related to the transfer step from B_A to H_A.

Within the scope of nonadiabatic electron transfer theory the observation of an additional time constant (the 3.5ps kinetic splits into two kinetics) indicates that a new intermediate state has to be considered. Although we can not finally rule out other

Springer Series in Chemical Physics, Vol. 55 Ultrafast Phenomena VIII
Editors: J.-L. Martin · A. Migus · G.A. Mourou · A.H. Zewail

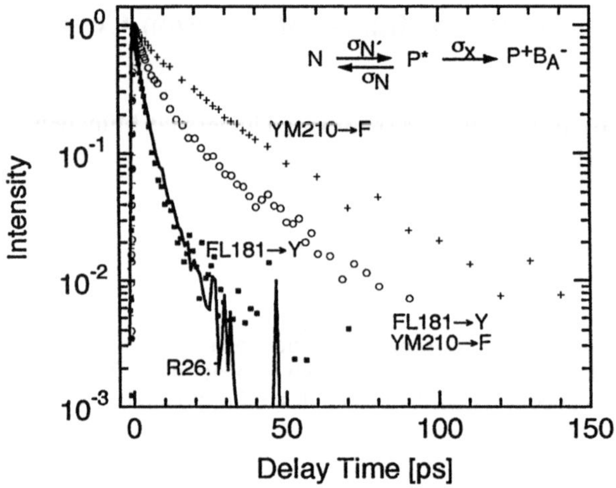

Fig.1: The emission experiments on RC's of R26.1 and of the mutants (YM210→F), (FL181→Y), and (YM210→F,FL181→Y).

Table 1:

	τ_1 [ps]	τ_2 [ps]	A_2/A_1	$1/\sigma_x$ [ps]	$1/\sigma_N$ [ps]	$1/\sigma_N'$ [ps]
R26.1	2.3	7	0.25	3.2	13	4.8
YM210→F	6.1	26	1.37	18	29	9.0
FL181→Y	2.1	15	0.06	2.8	10	11
YM210→F,FL181→Y	3.5	18	0.59	8.8	11	7.1

reaction models we want to focus on the model which is shown as an insert in Fig 1. The model contains a "parking state" N coupled to P^*. The reaction rates σ_x, σ_N, and σ_N' can be evaluated directly from the measured time constants and amplitude ratios and are summarized in Table 1. It can be seen, that a replacement of a phenylalanine by a tyrosine tends to speed up the reaction into the direction of the branch where the tyrosine is introduced.

A reevaluation of our former absorbance measurements taking into account the time constants given above allows us to compute the spectral properties of state N to be similar to those expected for a P^+B^- state. This leads to the conclusion, that the "parking state" N may be $P^+B_B^-$ were the electron is transiently brought on the inactive, the B branch.

Acknowledgement: The experiments were performed in collaboration with W.Kaiser, H.Scheer, D.Oesterhelt and K.Gray

Picosecond Fluorescence Kinetics of Purple Bacterial Reaction Centers

M.G. Müller, K. Griebenow, and A.R. Holzwarth

Max-Planck-Institut für Strahlenchemie,
W-4330 Mülheim a.d. Ruhr, Fed. Rep. of Germany

Abstract. A new component of 12 ps lifetime has been found in the decay kinetics of the primary donor P* of isolated reaction centers from *Rb. sphaeroides* by fluorescence. Several kinetic models of the primary electron transfer processes are discussed which could possibly explain this new component.

1. Introduction

The sequence and kinetics of the primary electron transfer processes in isolated purple bacterial reaction centers (RCs) are still discussed controversially. The model which has the strongest experimental support at present is the so-called sequential two-step model [1]. It involves a reduced accessory bacteriochlorophyll (B_L^-) as a short-lived intermediate which is formed in 3.5 ps and is reoxidized in 0.9 ps. All kinetic data so far are based on transient absorption techniques only [1,2]. We have recently reported the first measurements of the fluorescence kinetics for RCs with high time resolution in the picosecond range [3,4]. For RCs from *Rb. sphaeroides* a new kinetic feature, not observed in transient absorption measurements so far, was a 12 ps component of substantial amplitude, in addition to the expected main 3 ps decay component [3]. The 12 ps component has a substantial amplitude. The observation of this new component poses new questions as to the nature of the primary processes in these RCs. It will be necessary to include this component in a coherent mechanistic framework which is consistent with the existing transient absorption data. It is the purpose of this contribution to present a detailed kinetic analysis of the fluorescence data from *Rb. sphaeroides* RCs by testing various kinetic models.

2. Materials and Methods

RCs were isolated from *Rb. sphaeroides* w.t. and the measurements were carried out by the single-photon-counting technique as described [3]. The full width at half maximum of the system temporal response function was about 30 ps [3]. All measurements were carried out at room temperature (22°C) up to about 20000 counts in the peak channel with a time resolution per channel of 2.0 ps. The sample was excited at a wavelength of 800 nm which corresponds to the maximum absorption of the accessory bacteriochlorophylls (B).

3. Results and Discussion

Fluorescence decays were recorded at various emission wavelengths across the whole emission region. The data were analyzed both by the global lifetime and the global target analysis procedures. The global lifetime analysis requires four decay components for good fits. The results, presented as decay-associated spectra (DAS), are shown in Fig. 1. The two shortest-lived components show decay-associated spectra with a maximum wavelength of 900 nm indicating that their origin is fluorescence from the special pair (P*). In contrast the 930 ps component reflects the fluorescence decay of a small amount of decoupled BChl with an emission maximum at about 860 nm. The 3 ps component has a dominant amplitude and is related to the primary charge separation process. The 12 ps

Springer Series in Chemical Physics, Vol. 55 Ultrafast Phenomena VIII
Editors: J.-L. Martin · A. Migus · G.A. Mourou · A.H. Zewail

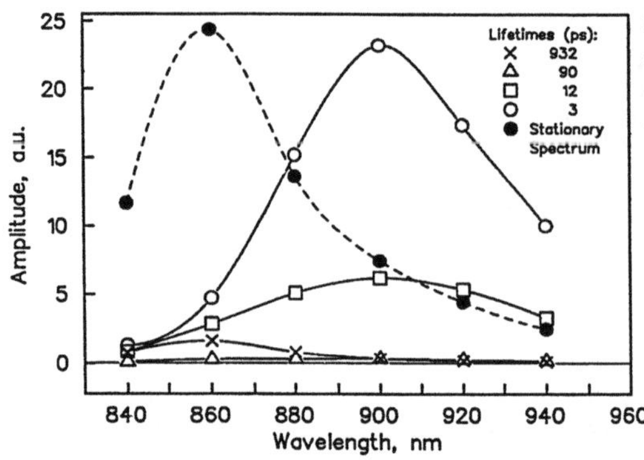

Fig. 1: Decay-associated fluorescence spectra of RCs from *Rb. sphaeroides* as calculated by the global lifetime analysis. $\lambda_{exc} = 800$ nm.

Model	χ^2, Lifetimes
Sequential (X^- could be B_L^-)	**1.124** 4 ps, 13 ps, 97 ps
Branching (X^- could be B_M^- or H_M^-)	**1.122** 2 ps, 12 ps, 92 ps
Heterogeneity (X^- could be H_L^-)	**1.114** 3 ps, 250 ps, 12 ps, 2.3 ns

Rate constants (ns^{-1})

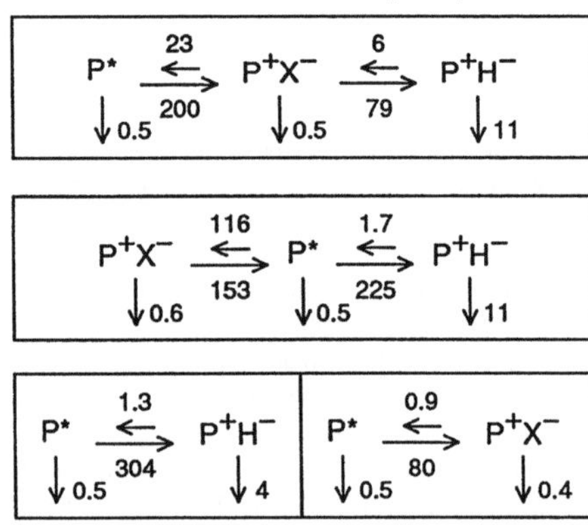

Fig. 2: Rate constants, lifetimes, χ^2-values for three different kinetic models as calculated by the global target analysis. Possible assignments for X^- are given.

component shows about one third of the amplitude of the 3 ps component. It seems also to be associated with a charge separation process in the reaction center. Also the other components have very similar spectra except for the 970 ps component, which has a blue-shifted spectrum and results from a very small amount of uncoupled bacteriochlorophyll.

Various kinetic models were tested by global target analysis. Three different kinetic models are presented here which formally explain the data about equally well. All models assume P* as the only radiative excited state and X^- as an a priori unknown intermediate.

The models are shown in Fig. 2 along with the results of the analysis: i) A sequential model where an additional charge transfer or relaxation step after the much faster primary charge separation process is assumed. ii) A branching model which describes the 12 ps component in terms of a parallel electron transfer to another chromophore, presumably located in the so far considered 'inactive' M-branch. iii) An inherent heterogeneity in the RCs giving rise to a heterogeneity in the rate of primary electron transfer for different RCs. The sequential and the branching models describe the fluorescence kinetics about equally well as judged by the χ^2-values and the residuals while the heterogeneity model appears to have a slightly lower χ^2-value. At the present level of S/N ratio it is not possible to distinguish between these models

4. References

[1] Holzapfel, W., Finkele, U., Kaiser, W., Oesterhelt, D., Scheer, H., Stilz, H.U. and Zinth, W. (1990). *Proc. Natl. Acad. Sci. USA* 87: 5168-5172.

[2] Breton, J., Martin, J.-L., Fleming, G.R. and Lambry, J.-C. (1988). *Biochemistry* 27: 8276-8284.

[3] Müller, M.G., Griebenow, K. and Holzwarth, A.R. (1991). *Biochim. Biophys. Acta* 1098: 1-12.

[4] These data have been presented in part also at the "International Congress for Photobiology", Amsterdam, September 1991.

Primary Radical Pair Formation in Photosystem-Two Reaction Centres

D.R. Klug, J.R. Durrant, G. Hastings, Q. Hong, D.M. Joseph, J. Barber, and G. Porter

Centre for Photomolecular Sciences, Imperial College of Science Technology and Medicine, Prince Consort Road, London SW7 2BB, UK

1. Introduction

Photosystem Two (PS2) produces our oxygenic atmosphere by splitting water to release protons electrons and oxygen. This ability allows higher plants to use water as a supply of electrons for their PS2 reaction centres. Despite the homology and presumed structural similarity between the proteins of the PS2 reaction centre and those of purple bacteria [1], the latter cannot produce a high enough oxidising potential to drive water splitting reactions. This means that the purple bacteria are limited to more easily oxidised compounds such as hydrogen sulphide as substrates for their reaction centres.

PS2 reaction centres (RC's) are relatively unstable even *in vivo*. The plant continually strips out and rebuilds the D1 protein of the RC in response to damaging side reactions which are thought to be a consequence of the modifications required to split water [2]. Much effort has been directed towards characterising and stabilising the isolated PS2 RC to make it suitable for spectroscopic studies. Reasonable stability and high activity have now been achieved [3].

Low temperature measurements demonstrate that the reaction centre chlorins fall approximately into two pools of molecules. One of these pools has its spectrum centred at 670nm (C670) while the other has its spectrum centred at 680nm (C680)[4]. The C680 pool is dominated by P680 which is the primary electron donor in PS2 and is postulated to be a special pair of chlorophyll-a molecules. This means that relatively good photoselectivity of P680 can be achieved in femtosecond transient absorption experiments such as those discussed below.

Previous femtosecond studies of PS2 RC's have reported a (3±0.6)ps rate of formation of the primary radical pair $P680^+Ph^-$ [5]. The results we report here and elsewhere [6,7] appear to contradict these previous studies.

2. Materials and Methods

Experimental details can be found elsewhere [6,7]. The samples were exposed to light in the spectrometer for approximately one hour at 295K during which less than 5% activity was lost. The femtosecond spectrometer had a instrument response of ≈150fs (FWHM), with a white light probe and tunable excitation beam. Two forms of excitation pulse were used in the experiments described here. One pulse was centred at 665nm which selectively excites C670, and one at 694nm which selectively excites C680. Data was analysed globally taking account of components monitored over a range of timescales and wavelengths.

Only 5-10% of the reaction centres in the interogated volume were excited by each pulse. It is critical to use low excitation intensities in order to avoid multiphoton effects. These are easily generated as each reaction centre contains 8 chlorins. Previous femtosecond studies [5] of PS2 reaction centres used excitation intensities 18 times greater than those used in our work.

3. Results

The radical pair spectrum was measured at time delays of 100ps and 9ns, and is shown at 100ps in figure 1. The major features of this spectrum are essentially the same as determined by Danielius et al.[8]. Moreover the radical pair spectrum is essentially invariant in shape from 100ps to 9ns using both 665nm and 694nm excitation. The radical pair state is much longer lived (mean lifetime of 37ns [3,8]) than chlorin singlet states (maximum lifetime of 6ns). Our observation of similar spectra at both 100ps and 9ns indicates that the only excited states which contribute significantly to these spectra are the P680 cation and pheophytin anion states.

Springer Series in Chemical Physics, Vol. 55 Ultrafast Phenomena VIII
Editors: J.-L. Martin · A. Migus · G.A. Mourou · A.H. Zewail

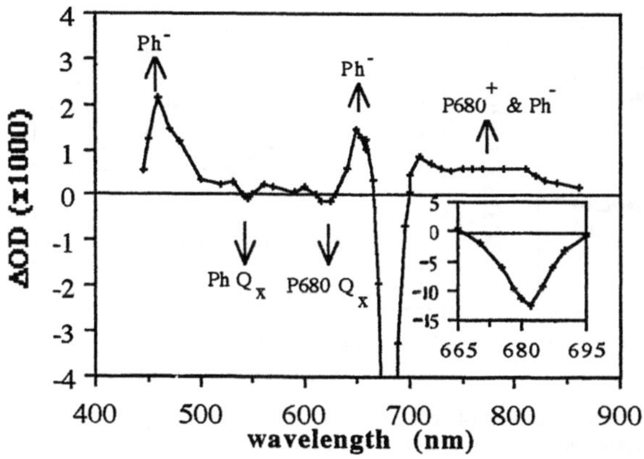

Figure 1. Transient absorption spectrum 100ps after excitation of PS2 reaction centres at 694nm, this spectrum is assigned to formation of the state $P680^{+}Ph^{-}$

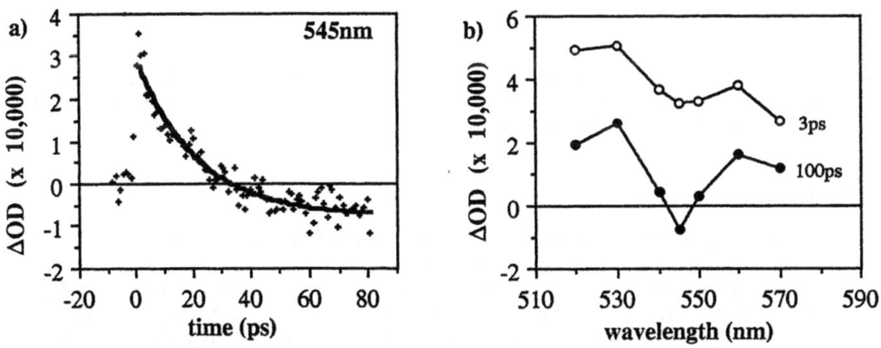

Fig. 2. a) Kinetics of the transient absorption changes (+) observed between 0-80ps at 545nm, following excitation of PS2 reaction centres at 694nm. The solid line is a fitted function.
b) Transient absorption spectra at 3ps and 100ps after excitation at 694nm.

The radical pair spectrum is formed with a $(21ps)^{-1}$ rate when P680 is directly excited using 694nm pulses [6,7]. A lifetime of (21±3)ps was obtained for the beaching of pheophytin ground state Q_X-absorption band at 545nm (figure 2a). We have also shown that the pheophytin anion band at ≈460nm grows in at this rate [7]. The pheophytin Q_X-band does not overlap with any other absorption bands, which makes observation of the bleaching of this band unambiguous. The transient spectrum of the pheophytin Q_X band, before and after the action of the 21ps component, is shown in figure 2b. The 21ps component accounts for 60% of the total pheophytin bleach present in the final radical pair spectrum. The pheophytin which is already bleached prior to the action of the 21ps component is present at times as early as 300fs, and is probably directly excited by the excitation pulse or is bleached by fast energy transfer (see below). However, some fast electron transfer cannot be ruled out at this stage in our studies.

There are a number of other kinetic components [6] which can be observed following the direct excitation of P680, with lifetimes between 100fs and 3.5ps. Not all of these components have yet been assigned. The fastest component, with a lifetime of ≈100fs, is easily distinguished from the slower components by its dependence upon excitation wavelength. If 665nm rather than 694nm excitation is used then this component becomes inverted (shown at ≈670nm in figure 3a), whilst the slower components are relatively unchanged. The spectrum of the 100fs component (figure 3b) following 665nm excitation is consistent with energy transfer from the initially excited 670nm pool to the 680nm chlorin pool, where the chromophores in the 680nm pool have a higher oscillator strength on average. This energy transfer process can be reversed by exciting the 680nm pool. In this case the kinetic spectrum would be expected to

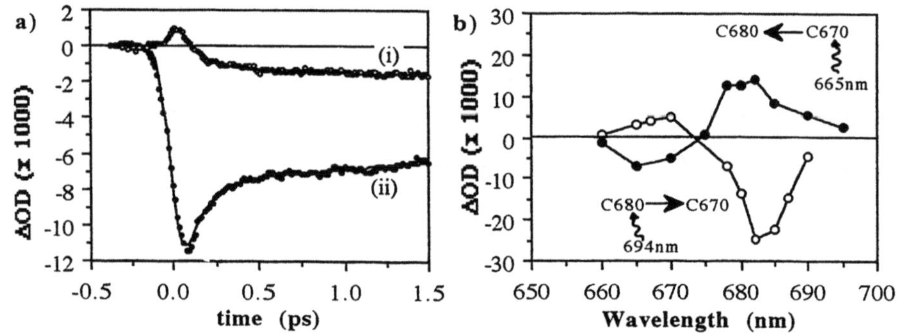

Fig. 3. a) Kinetics of transient absorption changes observed at (i) 667nm following excitation at 694nm and (ii) 670nm following excitation at 665nm. The solid lines are fitted functions.
b) Spectra of the (100±50)fs component following excitation at 665nm (•) and 694nm (o).

invert completely and this can indeed be seen in figure 3b. It can be concluded that significant equilibration of excitation energy between the 670nm and 680nm chlorin pools occurs on a time scale which is much faster than radical pair formation.

4.Discussion

The RC's of ***Rb.Sphaeroides, R.Viridis*** and ***R.Capsulatus,*** all show an approximately $(3.5ps)^{-1}$ rate of pheophytin reduction at room temperature. In these RC's the amino acid M208 (M210 for ***Rb. Sphaeroides***) is a tyrosine while the corresponding amino acid on the L branch (L181) is a phenylalanine. In one mutant of ***Rb.Sphaeroides*** L181 is preserved as a phenylalanine while M210 is changed to leucine. In these mutated RC's the rate of bacteriopheophytin reduction is lengthened considerably to 22±8ps [9]. The sequences of the D1 and D2 polypeptides line up with sequences of the L and M subunits such that D1-206 corresponds to L-181 and D2-206 to M-208 [1]. In pea PS2 RC's D1-206 is a phenylalanine while D2-206 is a leucine. One might therefore surmise a relationship between the 21ps rate of pheophytin reduction in higher plants and the phenomenological model for the influence of the identities of L-181 and M-208 on the rate of pheophytin reduction in purple bacteria [10].

The bi-directional energy transfer which we observe in PS2 is not found in purple bacteria. The reason for this is that the lowest excited singlet state of the primary donor in purple bacteria (P870 in ***Rb.Sphaeroides***)is significantly lower in energy than the other chlorin singlet states, whereas in PS2 RC's these states are within k_bT of each other at room temperature. The 100fs equilibration of excitation energy between the C680 and C670 pools of chromophores might limit the observed rate of pheophytin reduction in PS2. We estimate the equilibrium constant to be between 1:2 and 2:1, which suggests an 'intrinsic' rate of pheophytin reduction of between 7ps and 14ps. It has not yet proved possible to determine how many electron transfer steps are involved in pheophytin reduction in PS2 but our future work will attempt to address this issue.

References

1) Michel, H. and Deisenhofer, J. (1988) Biochemistry 27, 1-7.
2) Barber, J. and Andersson, B. (1991) Trends in Biochem. Sci., 17, 61-66.
3) Booth, P. J. et al. (1991) Biochemistry 30, 7573-7586.
4) van Kan, P. J. M. et al. (1991) Biochim. Biophys. Acta 1020, 146-152.
5) Wasielewski, M. R. et al. (1989a) Proc. Natl. Acad. Sci. USA 86, 524-528.
6) Durrant, J. R. et al. (1992) Chem. Phys. Letts. 188, 54-60.
7) Hastings, G. et al. (1992) Biochemistry, in press.
8) Danielius, R. V. et al. (1987) FEBS Lett. 213, 241-244.
9) Finkele U. et al. (1990) Biochemistry 29, 8517-8521.
10) Chan, C-K. et al. (1991) Chem. Phys. Letts. 176, 366-372.

Energy Transfer and Primary Charge Separation in Heliobacteria by Picosecond Transient Absorption Spectroscopy

P.I. van Noort, T.J. Aartsma, and J. Amesz

University of Leiden, Department of Biophysics, Huygens Laboratory, P.O. Box 9504, NL-2300 RA Leiden, The Netherlands

Introduction

In photosynthetic bacteria the excitation energy of excited antenna pigments is transferred to the reaction center within tens of picoseconds. After trapping of the excitation energy by the reaction center, charge separation is induced as the first step in a series of electron transfer reactions.

The recently discovered heliobacteria contain Bchl *g* as their major pigment in relatively small pigment pools, consisting of 35-50 Bchls. In *Heliobacterium chlorum* three different Bchls *g* complexes can be distinguished at low temperature: Bchl *g* 778, Bchl *g* 793 and Bchl *g* 808. In this paper we will discuss experimental results concerning the decay of excitation energy and the rate of charge separation in membranes of *Heliobacterium chlorum* at room temperature with a time resolution of 2 ps (one-color pump-probe) or 10 ps (two-color pump-probe).

Results and discussion

Figure 1 shows the absorbance difference spectrum of *H. chlorum* at room temperature in the spectral region from 770 nm to 820 nm upon excitation at 585 nm with reduced ("open") reaction centers. The solid circles (measured at a delay of 5 ps) show the spectrum due to excitation of antenna Bchls (mainly Bchl-808). At a delay of 100 ps (open circles) the excitations in the antenna have disappeared and the spectrum is due to oxidation of the primary donor P-798 in the reaction center, i.e. the formation of P-798$^+$. This result is in agreement with the fact that the absorption of the primary donor occurs at shorter wavelength than that of BChl-808. The relative shift of the spectra in fig. 1 shows that the excitations are transferred to the long-wavelength antenna-component, BChl-808, prior to energy transfer to the reaction center.

Figure 2 shows the time resolved bleaching of membranes of *H. chlorum* upon excitation at 585 nm in the presence of open reaction centers, measured at two different wavelengths: 793 nm (fig. 2A) and 808 nm (fig. 2B). The 808 nm transient in fig. 2B reflects the energy transfer from *Bchl* 808 to the reaction center. Below 808 nm the differential spectral bleaching (that is, the difference between open and solid circles in fig. 1) is near zero so that the bleaching in time is virtually constant. The bleaching due to P-798$^+$ persists for about 6 ms. The characteristic time constant of energy transfer to P-798 is 24 ± 2 ps, in accordance with time resolved fluorescence measurements [2].

Similar experiments were performed with closed reaction centers, in which case the transient absorbance difference spectra only show features associated with the decay of excited BChl-808: the spectral shift in fig. 1, and the constant component in fig. 2 are virtually absent. The absorption

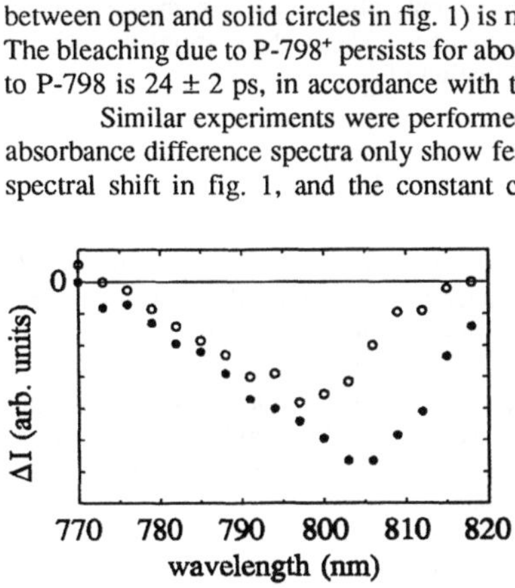

Figure 1 *Absorbance difference spectrum of H. chlorum with excitation at 585 nm. Solid circles: 5 ps delay; open circles: 100 ps delay.*

Springer Series in Chemical Physics, Vol. 55 Ultrafast Phenomena VIII
Editors: J.-L. Martin · A. Migus · G.A. Mourou · A.H. Zewail

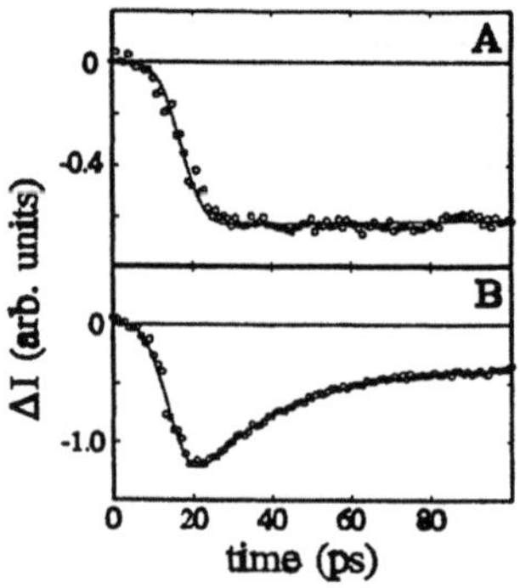

Figure 2. *Absorbance changes of H. chlorum around with excitation at 585 nm. Detected at A: 793 nm, B: 808 nm.*

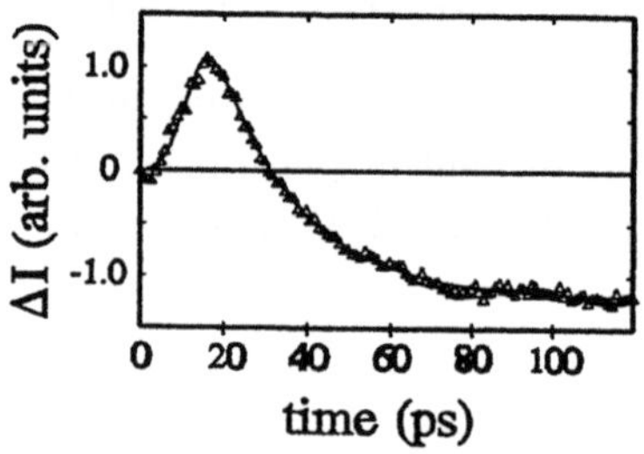

Figure 3. *Absorbance changes of H. chlorum at 670 nm. Excitation wavelength: 793 nm.*

recovery under these conditions is still very fast, with a time constant of 18 ± 2 ps, which is attributed to efficient quenching of excitations by P-798$^+$. The somewhat shorter lifetime in this case is consistent with a decrease of the fluorescence quantum yield observed in closed reaction centers, as compared to open reaction centers. One-color pump-probe experiments with a time-resolution of 2 ps in the 800 nm region show a mono-exponential recovery of the absorption with the same time constant as in two-color experiments. This indicates that the energy is equilibrated amongst the different antenna components within 2 ps.

Positive evidence for electron transfer in the reaction center was obtained by examining the absorbance transients around 670 nm, where a bleaching can be observed due to oxidation of the primary acceptor I [3]. Figure 3 shows a typical result of the absorbance transient, in open reaction centers, at 670 nm with excitation at 793 nm. The solid line represents a fit with an instantaneous absorbance increase, and an exponentially increasing bleaching. The former reflects excited state absorption of antenna pigments in this wavelength region, while the latter is due to the oxidation of the primary acceptor. The decay of the absorbance increase and the growth of the bleaching are expected to have the same time constants, as indeed was assumed in fitting the data. This time constant was found to be 19 ± 2 ps, in agreement with the time constant of absorption recovery around 800 nm.

The amplitude of the bleaching is strongly reduced when the reaction centers are closed. Furthermore, we also obtained the absorbance difference spectrum around 670 nm at a time delay of 120 ps and at 1.3 ns. They resemble the P^+I^- difference spectra reported in ref. [3] and [4]. At the isosbestic point of these absorbance difference spectra we only observe the absorbance increase associated with the excited state absorption of antenna pigments. These observations corroborate the assignment of the bleaching around 670 nm to the formation of I.

Conclusions

We have investigated the energy transfer and charge separation in membranes of *H. chlorum*. It was found that the excitation energy in the antenna is equilibrated within less than 2 ps, and that energy transfer to the reaction center occurs with a time constant of 23 ± 2 ps. The bleaching at 670 nm is unambiguously identified as due to reduction of the primary acceptor. Closed reaction centers are more effective in quenching excitations in the antenna than open reaction centers, consistent with fluorescence yield measurements [5]. Similar experiments as described above are currently being performed at low temperature in order to examine energy transfer in this inhomogeneous antenna system in more detail.

REFERENCES

1. E.J. van de Meent, F.A.M. Kleinherenbrink, and J. Amesz, Biochim. Biophys. Acta, **1015** (1019), 223.
2. J.T. Trost, R.E. Blankenship, Biochemistry **28** (1989), 9898.
3. A.M. Nuys, R.J. Van Dorssen, L.N.M.Duysens and J. Amesz, Proc.Natl.Acad.Sci. USA **82** (1985), 6865.
4. F.A.M. Kleinherenbrink, T.J. Aartsma, J. Amesz, Biochim. Biophys. Acta, **1057** (1991), 346.
5. G. Deinum, T.J.Aartsma, F.A.M.Kleinherenbrink, and J. Amesz, Biochim. Biopys. Acta, **1058** (1991), 339.

Excitation Energy Transfer in Mutants of *Rb. sphaeroides*: The Effects of Changes in the Core Antenna Size

L.M.P. Beekman[1], *R.W. Visschers*[1], *K.J. Visscher*[3], *B. Althuis*[1], *W. Barz*[2], *D. Oesterhelt*[2], *V. Sundström*[3], and *R. van Grondelle*[1]

[1]Biophysics Department, Physics Laboratory,
Free University of Amsterdam, Boelelaan 1081, Amsterdam, The Netherlands
[2]Max-Planck-Institut für Biochemie,
W-8033 Martinsried bei München, Fed. Rep. of Germany
[3]Department of Physical Chemistry, University of Umea,
S-901 87 Umea, Sweden

Picosecond absorption and steady-state fluorescence measurements were made on five mutant strains from the photosynthetic bacterium *Rb. sphaeroides* strain ATCC17023, displaying varying core antenna sizes due to mutations in the *puf*-operon. For Förster type of energy transfer a linear dependence of the mean-lifetime of excitations in the antenna with the antenna size is to be expected. It was found that both at room temperature and at 77K this linearity seems to hold. For one mutant strong deviations were observed at 77K which are ascribed to a broad distribution in core antenna sizes combined with spectral inhomogeneity at 77K giving rise to multi-exponential and strongly temperature sensitive excited state decays.

Introduction

A model of the exciton migration in a photosynthetic antenna has been presented that treats the antenna as a two-dimensional lattice with all but one sites occupied by antenna pigments and the one remaining site occupied by the RC [1]. An absorbed photon creates a mobile excited state, which migrates through the antenna by hopping to neighboring pigments. Eventually the exciton hops onto the RC, and charge separation may occur. By making assumptions on spectral and spatial homogeneity, on the probability of a pigment being initially excited and on the charge separation time [1,2], a linear relationship between the mean-lifetime of an exciton and the lattice size can be obtained (Eq. 1).

$$M_0 = k_p^{-1} + (N-1)\left(\frac{F_D}{F_T}k_p^{-1}\right) + (N-1)\left(\left[\frac{1}{qF_T} - \frac{1}{qF_A}\right] + \frac{\alpha}{F_A}\right)$$

Eq. 1: M_0 is the mean lifetime of an exciton in an antenna lattice. (N-1) is the relative lattice size. F_A, F_T and F_D are respectively the hopping rate between the antenna pigments, the trapping and de-trapping rate constants of the reaction center. α is a lattice constant and k_p is the photoconversion rate.

The lattice size *(N-1)* corresponds to the number of antenna pigments per RC. The mutant system we used [3,4] displayed varying core antenna sizes, so this system gives the opportunity to test the theory on the antenna system of a purple bacterium.

Materials and Methods

All experiments were performed on chromatophores in a 50 mM Potassium Phosphate buffer (pH7.5) and 65% v/v Glycerol. The RC were kept open during the measurements. The experimental pump-probe setup consisted of two dye-laser/cavity dumper combinations synchronously pumped by a Nd-YAG laser with cross-correlation pulselength of about 8ps[7], FWHM, with excitation wavelength at 890nm and detection at 905nm. Decay-times were estimated by computer analysis [5].

Springer Series in Chemical Physics, Vol. 55 Ultrafast Phenomena VIII
Editors: J.-L. Martin · A. Migus · G.A. Mourou · A.H. Zewail

Results

Table 1 shows the LHI:RC ratio's determined by flash induced absorbance difference measurements. The fluorescence quantum yield was obtained by integrating the total fluorescence spectrum. Figure 1 shows that a linear dependence between the fluorescence quantum yield and the LHI:RC ratio is observed for strains 2-5 at 77K. Strain 1 however shows a large deviation from linearity, the obtained fluorescence quantum yield and therefore mean exciton lifetime at 77K is twice the expected value. At room temperature however linearity seems to hold for all strains (not shown). In purified antenna complexes and RC-less strains[6,7] long mean-lifetimes were observed and this should also be seen in the decay of strain 1 if part of the LHI is not connected to the core. A complication in the analysis of the decays is the transient formation of the state P^+I^-. This produces a long-lived bleaching which interferes with the contribution of long-lived excited states. Figure 2 shows the picosecond absorption profiles observed for the WT (strain 5), the RC-less mutant (strain 6) and strain 1.The initial bleaching in all measurements were almost pulse limited indicating fast equilibration within the antenna (< 0.1ps).

The long lifetime for strain 6 is in good agreement with previous time-resolved measurements on RC-less mutants and B880 complexes[6,7]. The wild-type sample (strain 5), gives an almost mono-exponential decay with a major component of 24 ps and a constant bleaching in the tail due to P^+I^-[5,7] The traces of strains 1 and 4 clearly exhibit multi-exponential behavior, with fast components of 34 and 37ps respectively, and a strong slow phase in the 200-300ps range. For strain 1 and the WT the ratio of the integrated decays (excluding components due to P^+I^-) corresponds very well to the ratio of their low-intensity fluorescence yields. This shows that in strain 1 no strong 840 ps is observed reflecting disconnected LHI. For strain 4 the integrated decay does not match the relative fluorescence yield, indicating that in this particular sample a significant fraction of the traps was closed by the laser light.

Table 1: List of strains used. with official name used in references 3 and 4. The column LHI:RC shows the LHI:RC ratio estimated with flash induced absorption measurements.

Name	*Official name*[3,4]	*LHI:RC*
Strain 1	*PUFΔLMX(pRKXmut2T)*	*1:1.6*
Strain 2	*PUFΔLMX(pRKX%3.8)*	*1:1.0*
Strain 3	*PUFΔLMX(pRKX5.3)*	*1:0.9*
Strain 4	*ATCC17023(pRKQABKan)*	*1:1.0*
Strain 5	*ATCC17023 (WT)*	*1:0.6*

Table 2: Decay times (in ps) and relative amplitudes of the traces obtained for the different strains fitted with the conv-fit program. T_3 is fixed to 1 μs, and this accounts for the long lifetime of P^+I^-. Strain 6 does not contain a RC and is therefore not fitted with T_3.

	A_1	T_1(ps)	A_2	T_2(ps)	A_3	T_3
strain 1	0.39	37	0.56	214	0.05	μs
strain 4	0.49	34	0.43	296	0.08	μs
strain 5	0.76	24	0.15	96	0.09	μs
strain 6	0.18	60	0.82	840	--	--

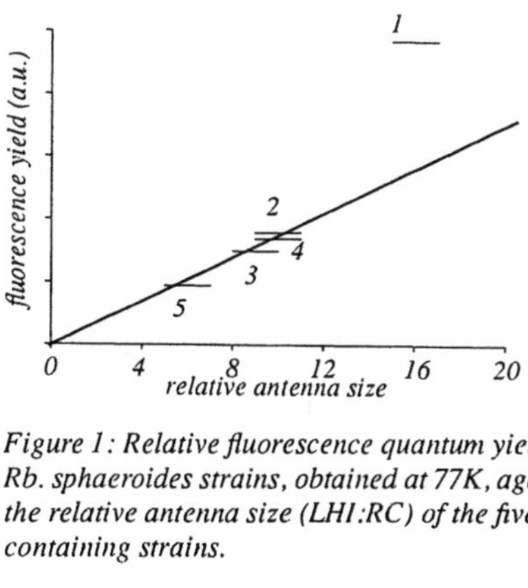

Figure 1: Relative fluorescence quantum yield of Rb. sphaeroides strains, obtained at 77K, against the relative antenna size (LHI:RC) of the five RC containing strains.

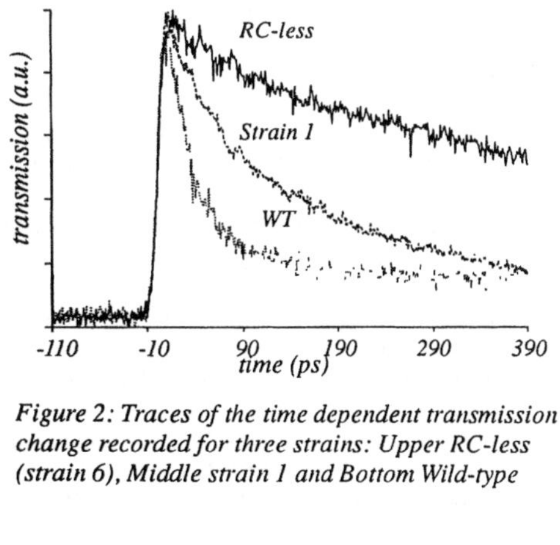

Figure 2: Traces of the time dependent transmission change recorded for three strains: Upper RC-less (strain 6), Middle strain 1 and Bottom Wild-type

Discussion

The measurements show that at room temperature a linear relation exists between the fluorescence emitted by LHI and the LHI:RC ratio, which implies that the mean exciton lifetime, scales linearly with N, the number of antenna pigments per RC, suggesting a well connected light-harvesting antenna. This relation persists for strains 2-5 at 77K. All the differences between these strains can be solely ascribed to the variations in antenna size without changes in any of the molecular rate constants.

The time-resolved measurements clearly show that the 840 ps component, representative for a fully disconnected LHI antenna is absent in this mutant. The observed excited-state decay is multi-exponential with time-constants in the 35-400 ps range, with a dominant contribution from the longer decays. We tentatively conclude that in this strain the LHI:RC ratio exhibits a broad distribution. Some LHI-RC complexes are close to 'normal', while others have a large antenna.

We assume that the fast decay phases observed in WT and strain 4 represent trapping in 'intact' cores with either a 'normal' or a modified LHI:RC ratio. Taking for k_p^{-1}=1.5ps at 77K as experimentally observed for isolated RC's we can find the relation between the observed trapping rate and the relative antenna size and this is illustrated in figure 3. Taking N=12 for the WT (assuming the BChl *a*-dimer as fundamental spectral unit) we can estimate $1/F_A$= 0.8ps, assuming all Förster rates in eq.1 to be equal (which is an oversimplification). Since the fast phase in mutant 1 is close to that of mutant 4, it probably represents trapping in cores of similar size.

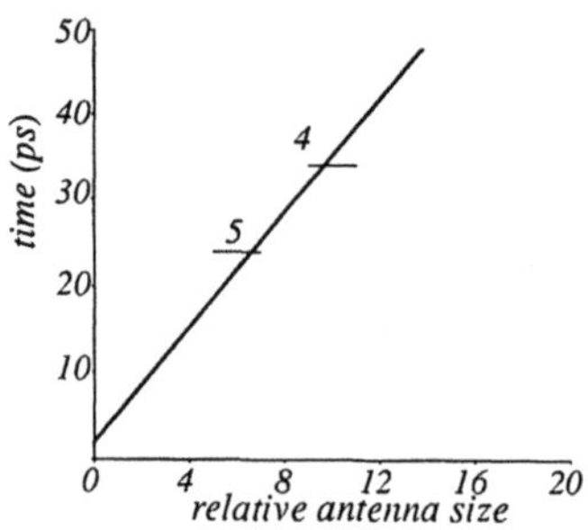

Figure 3: Fast decay times (T_1 in table 2) plotted against the relative antenna size.

Conclusions

If in photosynthetic purple bacteria the LHI:RC ratio is varied over a factor of about 2, the system behaves precisely as predicted theoretically by Pearlstein [1]. The experiments suggest a nearest-neighbor transfer time of 0.8 ps in WT *Rb. sphaeroides*, which is in reasonable agreement with estimates based on singlet-singlet annihilation [8]. The value is slightly higher because we use a dimer as spectral unit and performed the measurements at 77K. A mutant with an anomalously high fluorescence yield probably has a broad distribution of LHI:RC ratios.

References

[1] Pearlstein, R.M.,(1982) Photochem. Photobiol. Vol 35, pp. 835-844.

[2] Owens, T.G., Webb, S.P., Mets, L., Alberte R.S. & Flemming, G.R., (1987) Proc. Natl. Acad. Sci. USA.Vol. 84, pp 1532-1536.

[3] Farchaus, J.W. & Oersterhelt, D. (1989) EMBO-journal vol. 8 no.1 p47-54.

[4] Farchaus, J.W., Grünberg, H. & Oesterhelt, D. (1990) J. of Bacteriology, vol 172, no. 2, pp. 977-985

[5] van Grondelle, R., Bergström, H., Sundström, V.,& Gillbro, T. (1987) Biochimica et Biophysica Acta 894, pp 313-326.

[6] Sebban, P., Jolchine, G. & Moya, I. (1984) Photochemistry and Photobiology. Vol. 39, No. 2, pp. 247-253.

[7] Bergström, H., Westerhuis, W. H. J., Sundström, V., van Grondelle, R., Niederman, R.A. & Gillbro, T. (1988) FEBS-Let. Vol. 233, number 1, pp 12-16

[8] Bakker, J. G. C., v.Grondelle, R. & Den Hollander, W. Th. F. (1983) B.B.A. 725 pp.508-518

Femtosecond Excitation Transfer in Allophycocyanin

A.V. Sharkov[1], *E.V. Khoroshilov*[1], *I.V. Kryukov*[1], *P.G. Kryukov*[1], *T. Gillbro*[2], *R. Fischer*[3], *and H. Scheer*[3]

[1]P.N. Lebedev Physics Institute, Russian Academy of Sciences, 117924 Moscow, Russia

[2]Department of Physical Chemistry, University of Umea, S-901 87 Umea, Sweden

[3]Botanisches Institut der Universität München, W-8000 München, Fed. Rep. of Germany

Allophycocyanin (APC), C-phycocyanin (C-PC) and other biliproteins (phycoerythrin and phycoerythrocyanin) of cyanobacteria and red algae harvest solar energy in regions of the visible spectrum having weak chlorophyll absorption. This excitation energy is then efficiently transferred to chlorophyll in the photosynthetic membrane [1]. Light-harvesting complexes of cyanobacteria (phycobilisomes) consist of several hundred chromophores. The smallest (monomeric) subunit of C-PC consists of three (α-84, β-84 and β-155) chromophores. The APC monomeric subunit consists of two (α-84 and β-84) chromophores. In phycobilisomes APC and C-PC monomeric subunits are organised as trimers. Absorption spectrum of APC monomeric subunit changes drastically upon trimeric formation. Absorption maximum shifts from 620 nm for monomeric preparations to 650 nm with a 620-nm shoulder for trimeric preparations.

Isotropic absorption recovery kinetics with τ=440 fs was recently observed in APC trimers at 615 nm [2-3]. The corresponding anisotropy was constant and close to 0.4 during the first few picoseconds. The results are consistent with a model of APC trimer in which α-84 and β-84 chromophores have different absorption spectra and femtosecond process corresponds to Förster energy transfer from 620-nm chromophore to neighbouring 650-nm one. Now new results from measurements in the 650-nm absorption band is presented here for further understanding of the ultrafast processses in APC trimers.

70-fs pulses at 620 nm from a CPM-laser were amplified at 100 Hz repetition rate in a multipass jet amplifier (pumping at 308 nm) and served as exciting pulses. A part of the amplified light was used for continuum generation. Pump-probe measurements were made with probe pulse polarization parallel or perpendicular to the polarization of the exciting pulse. APC was isolated from *Mastigosladus laminosus*.

Figure 1 shows difference spectra obtained in the 635-690 nm spectral region at parallel polarization of probe relative to the pump pulse. One can see that the bleaching of the 650-nm band is delayed relative to the bleaching of the 620-nm band. The strong red-shifted bleaching at 660 nm, due to stimulated emission, is further delayed relative to the bleaching of the 650-nm band. Fast recovery kinetics (corresponding to the 440-fs process observed at 615 nm [2-3]) was measured at 635 nm after initial bleaching (not shown). In contrast, only rise term was observed at 658 nm within the first picosecond after excitation. Decay of anisotropy from 0.4 to 0.2, however, occurs at that wavelength during this period. Femtosecond processes were not observed in monomeric preparations.

There are two possible explanations for the absorption spectrum of APC trimer. Due to the first one, 650-nm band and 620-nm shoulder both belong to α-84 and β-84 chromophores. One may believe that excitonic coupling between neighbouring α and β chromophores of different monomeric subunits is very strong (620-nm and 650-nm excitonic bands). For C-PC trimers the excitonic coupling (112 cm^{-1}) was calculated on the base of X-ray crystallography data [4]. Förster energy transfer between neighbouring excitonically coupled α and β chromophores with nearly similar absorption spectra was recently observed in C-PC trimers [5].

Editors: J.-L. Martin · A. Migus · G.A. Mourou · A.H. Zewail

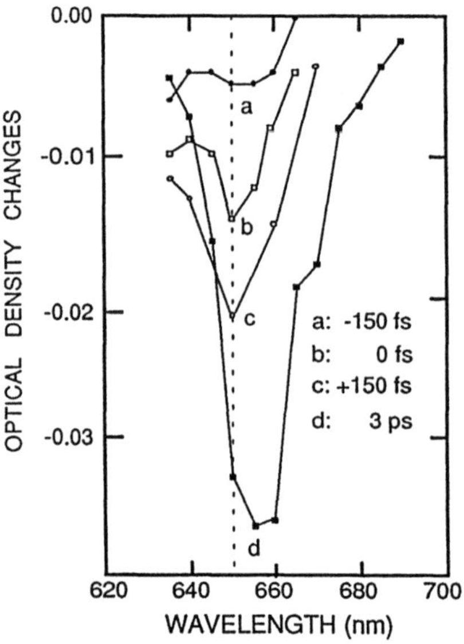

Fig.1 Difference absorption spectra measured at -150 fs (a), 0 fs (b). +150 fs (c) and 3 ps (d) delay after excitation at 620 nm with polarization of probe pulse parallel to the polatization of exciting pulse. Vertical line corresponds to APC absorption maximum

Unfortunately X-ray data are not available for APC trimers, but the excitonic splitting similar to C-PC could be assumed [6]. The 620-nm band in this case is a vibronic band of 650-nm electronic transition of α and β chromophores. An alternative explanation [3] is that one chromophore in a pair absorbs at 620 nm and the other (most likely β-84) at 650 nm due to a conformational change.

Our results strongly support the latter model. If the 650-nm band and the 620-nm shoulder belong to both chromophores, one should observe simultaneous bleaching of both bands due to ground state depletion. The delay in the bleaching of the 650-nm band (0 fs spectrum in comparison to the -150 fs spectrum) corresponds to Förster energy transfer from a 620-nm donor to the 650-nm acceptor. Absorption recovery observed at 615 nm [2-3] and at 635 nm is due to relaxation of donoı molecules to the ground state as a result of excitation energy transfer. Acceptor molecules absorbing mainly at 650 nm participate to a small extend to the isotropic or anisotropic kinetics measured at 615 nm [3]. the strong bleaching appearing near 660 nm is red-shifted relative to the bleaching of the 650-nm band (3 ps spectrum in comparison to 150 fs and 0 fs spectra). This shift is probably due to vibrational relaxation in the acceptor's excited state. Anisotropy decay observed at 658 nm is a result of Förster energy transfer between differently oriented donor and acceptor chromophores (as predicted from C-PC data).

References

1. R.MacColl and D.Guard-Friar, "Phycobiliproteins", CRC Press, Boca Raton (1987)
2. E.V.Khoroshilov, I.V.Kryukov, P.G.Kryukov, A.V.Sharkov and T.Gillbro, Proc. SPIE Vol. 1403, Laser Applications in Life Sciences (1990), pp. 431-433
3. A.V.Sharkov, I.V.Kryukov, E.V.Khoroshilov, P.G.Kryukov, R.Fischer, H.Scheer and T.Gillbro, Chem.Phys.Lett. **191**, 633 (1992)
4. K. Sauer and H.Scheer, Biochim. Biophys. Acta **936**, 157 (1988)
5. T.Gillbro, A.V.Sharkov, I.V.Kryukov, E.V.Khoroshilov, P.G.Kryukov, R.Fischer and H.Scheer, Biochim. Biophys. Acta, in press
6. A.R.Holzwarth, E.Bittersmann, W.Reuter and W.Wehrmeyer, Biophys.J. **57**, 133 (1990)

Femtosecond Förster Energy Transfer over 20 Å in Phycoerythrocyanin (PEC) Trimers

L.O. Palsson[1], T. Gillbro[1], A. Sharkov[1], R. Fischer[2], and H. Scheer[2]

[1]Department of Physical Chemistry, University of Umea, S-901 87 Umea, Sweden

[2]Botanisches Institut der Universität München, W-8000 München, Fed. Rep. of Germany

Energy transfer between violobilin (α-84) and phycocyanin (ß-84) chromophores in phycoerythrocyanin trimers from *Mastigocladus laminosus* occurs in 400 fs indicating a Förster-type energy transfer over 20 Å.

Introduction.

Phycobilisomes constitute the antenna for capture of light enrgy and are active in the subsequent transfer of the excitaion energy to the reaction centers in red algea and cyanobacteria. The phycobilisome contains chromophores of several different types and thus have the capacity to absorbe light of different wavelengths. This enables the organism to harvest light energy and grow in different radiation environments. Thus far only a few photosynthetic antenna complexes have been crystallized and their structure determined. One example of a highly resolved structure is the phycoreythrocyanin antenna complex isolated from phycobilisomes of *Mastigocladus laminosus* [1]. In this work the positions and orientations of the three chrompohores (α-84, β-84 and β-155) were determined. The closest contact is between the α-84 (violobilin) and β-84 (phycocyanin) chromophores with a center-to-center distance of 19 Å. The α-chomophores is supposed to absorbe strongly about 572 nm, while the β-subunit absorbes at longer wavelengths, i.e. about 592 nm [2].

Results and discussion.

Exciting at 575 nm using cavity dumped compressed pulses, FWHM ~ 50 fs we resolve a very fast 400 fs partial recovery of the bleaching signal.This indicates an ultrafast energy transfer from α-84 to a red-shifted β-chromophore. The process we observe is actually an equlibration of the excitation energy between these chromophores. A

Springer Series in Chemical Physics, Vol. 55 **Ultrafast Phenomena VIII**
Editors: J.-L. Martin · A. Migus · G.A. Mourou · A.H. Zewail

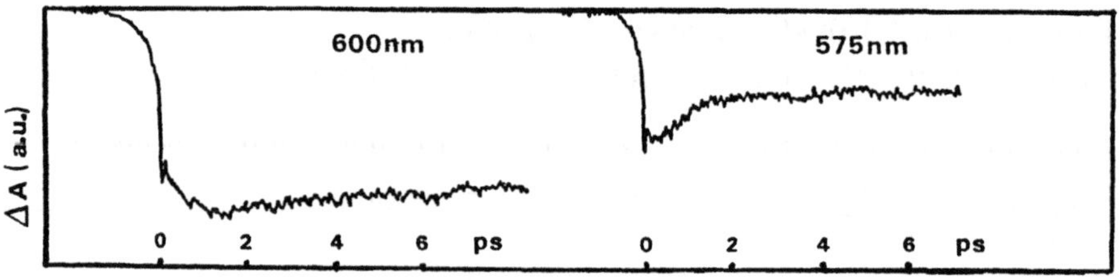

Fig. Absorption recovery at 575 nm and at 600 nm.

similar but reversed 400 fs process is observed if we excite at 600 nm (mainly β-chromophores). In this case a slow recovery of the transient bleaching is estimated to occur in about 50 ps is observed as well. From the crystal structure and an estimated Förster radius (R_0) of 60-70 Å for this sub picosecondenergy transfer we calculate a transfer rate, k, of

$$k=\frac{\kappa^2}{\tau}\left(\frac{R_0}{R}\right)^6 = 978\ \mathrm{ns}^{-1}$$

where τ = 2.0 ns, κ = -1.10, R_o = 65 Å and R = 19 Å. For the equilibration process $\Sigma k \cong 1950\ \mathrm{ns}^{-1}$, giving a lifetime of about 500 fs in fair agreement with the experimental results.The results reported here are also consistent with recent work on allophycocyanin and c-phycocyanin trimers [3,4].

References:

1. M. Duerring, R Huber, W. Bode, R. Ruembeli and H. Zuber J.Mol.Biol. (1990) 211, 623-644.
2. R. Füglistaller, F. Suter and H.Zuber Hoppe-Seyler's Z.Physiol.Chem. (1983) 365, 1085-1096.
3. A. Sharkov et al Chem.Phys Letters (1992) 191, 633-638.
4. T Gillbro et al Biochim. Biophys. Acta submitted

Ultrafast Energy Transfer Within the Light-Harvesting Antenna of Photosynthetic Purple Bacteria

K.J. Visscher[1], *V. Gulbinas*[1], *R.J. Cogdell*[2], *R. van Grondelle*[3], and *V. Sundström*[1]

[1]Department of Physical Chemistry, University of Umea, S-90 187 Umea, Sweden

[2]Department of Botany, University of Glasgow, Scotland

[3]Department of Biophysics, Free University, Amsterdam, The Netherlands

Abstract. Energy transfer within the LH2 antenna of several photosynthetic purple bacteria have been studied using subpicosecond transient absorption measurements, at 300 and 77 K. The results show that the transfer from the high energy pigment B800 to the low energy pigment B850 is very fast ~0.5-2.4 ps at 77K, depending on species and the wavelength of excitation. At 300K the transfer is faster.

1. Introduction

The photosynthetic pigment apparatus serves two functions, to collect light energy and transform it into chemical energy. These two functions are maintained by the light-harvesting and photochemical reaction center pigments, respectively. In photosynthetic purple bacteria both of these functions have been extensively studied using various time-resolved ultrafast methods. It has been shown that after absorption of light the energy is rapidly equilibrated over the entire light-harvesting system within ~10 ps, and concentrated to the low-energy antenna pigments [1]. Here we report on some results obtained for one particular light-harvesting pigment, the LH2 antenna of purple bacteria.

2. Results and Discussion

The energy transfer from B800 to B850 was monitored in a one-color pump and probe experiment with ~500 fs low intensity dye laser pulses (~10^{13} photons cm^{-2} $pulse^{-1}$). Kinetics was measured within the wavelength range 791- 809 nm. Figure 1 illustrates the kinetics observed for *Rb. sphaeroides* chromatophores at 803 nm and 77K. After deconvolution with the autocorrelation response function and removal of the coherent coupling artifact a time constant of 2.4 ps is obtained for the B800-->B850 energy transfer. At room temperature the corresponding transfer time is considerably shorter, 0.7 ps, in agreement with an earlier measurement by Owens et al [2]. These two results clearly shows that the value 2.3 ps [3] for the B800-->B850 transfer time deduced from measurements where the carotenoid is excited and the B800 ground state bleaching monitored, is in error. At 77 K the energy transfer time is clearly wavelength dependent, becoming shorter at shorter wavelengths it reduces to 1.9 ps at 794 nm (see Fig. 2).

For the species *Rps.palustris* a much more pronounced wavelength dependence of the B800-->B850 transfer is observed, decreasing from ~1.5 ps to ≈0.5 ps in going through the B800 absorption band from the red to the blue edge (791-809 nm). Two different strains of *Rps.palustris* were examined, high and low light grown cells, respectively. Within experimental error both types yielded the same results (Fig. 2). The results of Fig. 2 represents a single exponential analysis of the kinetics. For *Rps.palustris* a somewhat better fit to the data in the red and blue extremes of the spectrum could be obtained with two lifetimes, $\tau_{(1)}$=0.6 and $\tau_{(2)}$=1.5 ps. This result appears reasonable if the wavelength variation of the measured B800 decay is a result of two (or more) spectrally different components,

Springer Series in Chemical Physics, Vol. 55 Ultrafast Phenomena VIII
Editors: J.-L. Martin · A. Migus · G.A. Mourou · A.H. Zewail

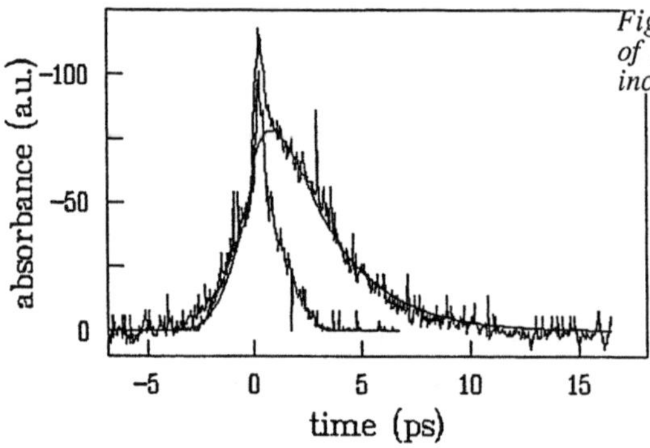

Fig. 1. Measured ground state recovery of Rb.Sphaeroides at 803 nm and 77K, including response function and fit.

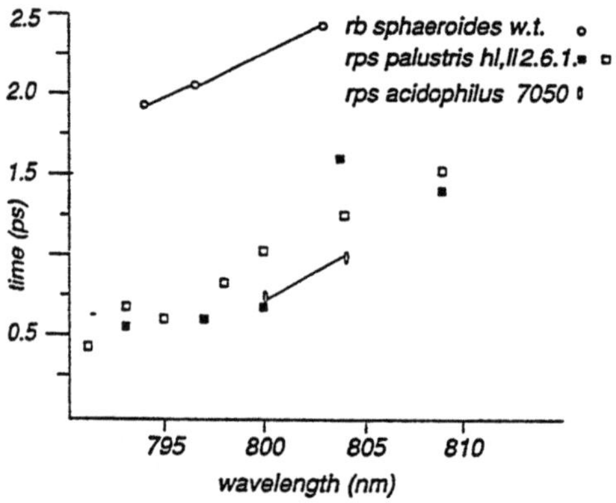

Fig. 2. Excitation wavelength dependence of B800-->B850 energy transfer in several purple bacteria.

because the variation is largest for *Rps.palustris* and a nonexponential decay is expected to be most easily resolved in the wings of the absorption spectrum where the amplitude difference would be largest.

Measurements were also performed on two other species having a shifted B850 band. In the B800-820 complex of *Rps.acidophila* a much faster 0.7-1.0 ps transfer (again depending on wavelength, see Fig. 2) was observed at 77K, in agreement with the better spectral overlap for Förster energy transfer. In the B800-838 complex of *Chr.purpuratum* the energy transfer from B800 was observed to occur with almost the same time constant (2.1 ps) as in B800-850, also this in good agreement with spectral overlap.

The wavelength dependence of the energy transfer from B800 suggests a spectral heterogeneity of the B800 band. At least two different explanations could be envisaged for this. One possibility is that the spectral heterogeneity represents inhomogeneous broadening of the B800 absorption band due to chromophore-protein interactions. In this case part of the transfer rate observed in the blue wing of the B800 band would represent transfer to other B800 chromophores absorbing at lower energy. Another possibility is that the wavelength dependence represents relaxation between two exciton states. At the shortest wavelengths an upper exciton state is predominantly excited, and the fast relaxation time represents decay to the lower exciton state. In the red wing of the B800 band the lower exciton state is mainly excited, and the decay time would in this case represent B800-->B850 energy transfer.

Energy transfer among the B800's was also examined by measuring the time resolved anisotropy within the B800 absorption band. For *Rb.sphaeroides* a high (≈0.35) and time independent (during 0-10 ps) anisotropy was observed. This observation suggests that there is little B800<-->B800 energy transfer, or if there is some transfer the B800 transition dipoles are almost parallel. In

Rps.palustris on the other hand there is a more complex time and wavelength dependence of the anisotropy, which may support the idea that the fast ≈500 fs relaxation observed in the blue part of the B800 absorption band represents relaxation between two different exciton states.

3. References

1. F. Geng Zhang, R. van Grondelle and V. Sundström, Biophys.J. **61**, 911 (1992)
2. A.P. Shreve, J.K. Trautman, H.A. Frank, T.G. Owens and A.C. Albrecht, Biochim. Biophys. Acta, **1058**, 280 (1991)
3. J.K. Trautman, A.P. Shreve, C.A. Violette, H.A. Frank, T.G. Owens and A.C. Albrecht, Proc. Natl. Acad. Sci. USA, **87**, 215 (1990)

Femtosecond Dynamics in Rhodopsin

T. Kobayashi[1], *M. Taiji*[1], *K. Bryl*[1], *M. Nakagawa*[2], *and M. Tsuda*[2]

[1]Department of Physics, University of Tokyo, Bunkyo, Tokyo 113, Japan
[2]Department of Life Science, Himeji Institute of Technology, Hyogo 678-12, Japan

1. Introduction

Many evidences indicate that the primary photoprocess in rhodopsin after absorbing photons is 11-*cis* → all-*trans* isomerization. A number of picosecond experiments performed[1] could not succeed to clarify the dynamics of the *cis-trans* isomerization. In these studies three intermediates (primerhodopsin(photorhodopsin)[1, 2], hypsorhodopsin[1, 3], and bathorhodopsin[1]) are found, but their relationship was still unclear because of their limited time resolution. Very recently, two groups reported femtosecond studies of rhodopsin[4, 5], but their results and interpretations are inconsistent.

In this study, we report femtosecond spectroscopy of octopus rhodopsin over very wide spectral region of 400–1000 nm, which made us possible to determine the time constant of *cis-trans* isomerization to be about 400 fs. The first intermediate primerhodopsin was converted to bathorhodopsin with a time constant of 10–20 ps along with a blue shift of about 10 nm. We found a new absorption from the excited state and two different stimulated emissions, which may be relevant with the two forms of bathorhodopsin reported previously[6].

2. Experimental

The method to isolate microvillar membranes of octopus rhodopsin was similar to that described previously[1, 7]. The green femtosecond pulses of 525 nm were generated by wavelength selection of white contiuum and were amplified by a two-stage dye amplifier pumped by the third harmonic of a Q-switched Nd:YAG laser. The pulse energy, pulse duration, repetition rate and spectral width were 8 μJ, 300 fs, 10 Hz, and 14 nm, respectively.

3. Results and Discussion

Figure 1 shows difference absorption spectra of octopus rhodopsin in H_2O at several delay times. The data were not displayed in 500–534 nm region because of the unavoidable disturbance due to scattered pump light at 525 nm. The obvious features of these spectra were (1) breaching in the region of $\sim$ 500 nm, (2) prominent induced absorption around 580 nm, (3) small gain in the region of 630 nm around time zero, (4) very small but reproducible induced absorption around 700 nm, and (5) gain in the broad spectral region of the longer wavelength than 850 nm. All these features were also observed in the sample in D_2O.

The time dependence of the absorbance change of the samples in H_2O is shown in Fig. 2. First, the induced absorption appeared within 100 fs in the 680–730 nm and 640–

Springer Series in Chemical Physics, Vol. 55 Ultrafast Phenomena VIII
Editors: J.-L. Martin · A. Migus · G.A. Mourou · A.H. Zewail

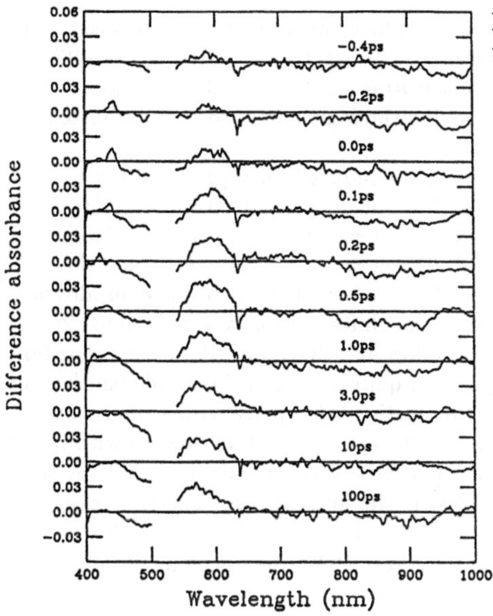

Fig.1 Difference absorption spectra of octopus rhodopsin in H_2O at 8 °C.

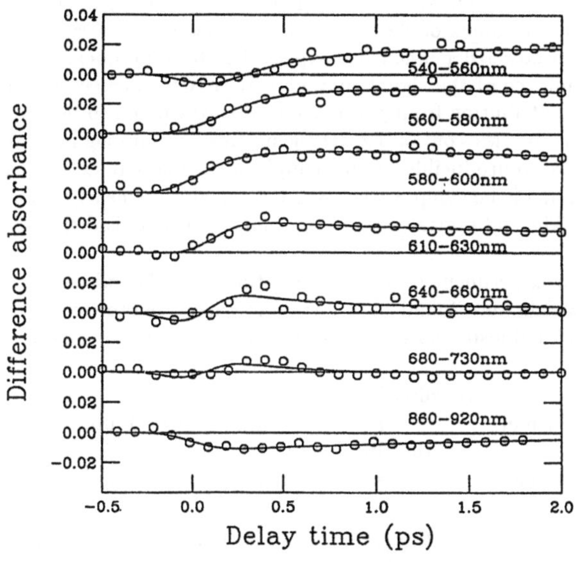

Fig.2 Transient absorption change of octopus rhodopsin in H_2O at 8 °C.

660 nm regions decayed with time constant of 140 ± 70 fs and 360 ± 180 fs, respectively. Therefore the "hot" stimulated emission from the state not much relaxed from the Franck-Condon state could be the origin of such a short-lived signal.

The induced absorption observed in the 640–730 nm region is considered due to the excited-state, since it grows up within 100 fs. The fact that the decay time constant was faster in a longer wavelength region (~ 700 nm) than in a shorter (650 nm) as shown in Table 1 can be explained by the wave packet motion in the excited-state.

In the 540–600 nm region, the most prominent induced absorption was observed. The rise times of the induced absorption are estimated to be in the range of 200–400 fs, which coincide with the decay time of weak absorption in the 640–730 nm region. The similar femtosecond spectroscopic results of bovine rhodopsin were recently reported by Schoenlein et al.[4].

After the 2–3 ps decay in the sample in H_2O, the induced absorption shifted toward blue by about 10 nm with a relaxation time of about 20 ps and no more change was observed till 100 ps. The spectrum at 100 ps coincides with those in previous picosecond experiments of octopus or squid rhodopsin[1, 2], therefore we can safely conclude that the final product at 100 ps is bathorhodopsin, which is well established to have an all-*trans* chromophore. In the previous experiments the similar shifts of spectra were found[1, 2] and the intermediate before shift was called "primerhodopsin". Since the difference absorption spectra of primerhodopsin and bathorhodopsin are quite similar, primerhodopsin has an all-*trans* chromophore. Therefore, *cis-trans* isomerization takes place with a time constant of 400 fs after excitation. Schoenlein et al. reported the isomerization finishes within 200 fs[4]. These seem to be inconsistent with each other, however, we must note that the time constant depends on the wavelengths and 400 fs is the longest one. The time constant of 400 fs may correspond partial thermalization of primerhodopsin. It may also be due to the difference between octopus and bovine rhodopsin. This difference must be studied further.

The origin of the gain in the wavelength region of longer than 850 nm can be explained by considering another pathway. One possibility of the origin of the gain is the stimulated emission from the photoproducts of rhodopsin. In the case of a contribution from photoproducts the excitation power-dependence is expected to be superlinear. The results of excitation power dependence of the signal showed a linear dependence within experimental error. Therefore a contribution from photoproducts is concluded to be negligibly small. We also measured samples containing detergent only, but no signal was detected. The gain signal was also reproducible in the sample with different detergent (SM–1200). The lifetime of the gain of the sample was 1.8 ± 0.6 ps, which is nearly equal to that in L–1690 (2.0 ± 0.8 ps). The fact indicates that the origin of the gain is definitely rhodopsin.

Another pathway found in our experiment may have a relation with the existence of two forms of bathorhodopsin, which were found in a low-temperature photoexcitation study[6]. The primary process of rhodopsin can be then schematically shown as below.

$$\begin{array}{lllllll}
\mathrm{Rh} & \xrightarrow{h\nu} & \mathrm{Rh}^{*} & \xrightarrow{400\mathrm{fs}} & \mathrm{Prime}_1 & \xrightarrow{10\text{–}20\mathrm{ps}\ (2\text{–}3\mathrm{ps})} & \mathrm{Batho}_1 \longrightarrow \mathrm{Lumi}_1 \\
(\mathrm{Rh}_2 & \longrightarrow & \mathrm{Rh}_2^{*}) & \xrightarrow{2\mathrm{ps}} & \mathrm{Prime}_2 & \xrightarrow{10\text{–}20\mathrm{ps}} & \mathrm{Batho}_2 \longrightarrow \mathrm{Lumi}_2
\end{array}$$

There are two possibilities of the origin of the two forms of bathorhodopsin. The first one is that in the ground states (before excitation) rhodopsin already have two form, and the second is that it branches into two in the excited state. Sasaki et al.[6] excluded the possibility of the heterogeneity of rhodopsin itself because of the kinetics of the photoconversion process from rhodopsin to bathorhodopsin at 77 K. They proposed that different bathorhodopsin can be formed by clockwise and counterclockwise isomerizations. But if there are two forms of rhodopsin which are interconvertible each other even at 77 K under illumination because of possible resemblance of the conformations, two forms of rhodopsin can exist also in the ground state. Since protein is a glassy material and it can have many substates, the model is considerable. In this case, a question arises whether the process exists in native membranes under physiological condition.

References

[1] H. Ohtani, T. Kobayashi, M. Tsuda, and T. G. Ebrey, *Biophys. J.* **53**, 17 (1988), and references therein.

[2] Y. Shichida, S. Matuoka, and T. Yoshizawa, *Photobiochem. Photobiophys.* **7**, 221 (1984).

[3] T. Kobayashi, *FEBS Letters* **106**, 313 (1979).

[4] R. W. Schoenlein, L. A. Peteanu, R. A. Mathies, and C. V. Shank, *Science* **254**, 412 (1991).

[5] M. Yan, D. Manor, G. Weng, H. Chao, L. Rothberg, T. M. Jedju, and R. R. Alfano, *Proc. Natl. Acad. Sci. USA* **88**, 9809 (1991).

[6] N. Sasaki, F. Tokunaga, and T. Yoshizawa, *Photochem. Photobiol.* **32**, 433 (1980).

[7] M. Tsuda, *Biochimica et Biophysica Acta* **545**, 537 (1979).

Subpicosecond Time-Resolved Spectroscopy of Halorhodopsin and Comparison with Bacteriorhodopsin

H. Kandori[1], K. Yoshihara[1], H. Tomioka[2], H. Sasabe[2], and Y. Shichida[3]

[1]Institute for Molecular Science, Myodaiji, Okazaki 444, Japan
[2]Frontier Research Progam, RIKEN (The Institute of Physical and Chemical Research), Wako 351-01, Japan
[3]Department of Biophysics, Faculty of Science, Kyoto University, Kyoto 606, Japan

Halorhodopsin (hR) and bacteriorhodopsin (bR) are the typical retinal proteins present in *Halobacterium Halobium* and both possess the same chromophore (all-*trans* retinal) and similar protein moieties. The primary photochemical events are all-*trans* to 13-*cis* isomerizations of the chromophores in both hR and light-adapted bR. The subsequent thermal reactions, however, cause different physiological functions of these proteins, light-driven chloride for hR and proton pump for bR. Although a number of studies have been done with bR with ultrashort pulses [1-6], there have been few on hR [7]. In the present paper, we compare the primary photochemical events of hR and bR under the same experimental conditions.

For time-resolved spectroscopy, hR was solubilized by a detergent (MEGA-9), while the membrane fraction was used for the bR sample. To ensure light-adaptation of bR, the sample reservoir was illuminated during experiments. The apparatus for time-resolved measurements is a double-beam spectrometer linked with an amplified subpicosecond light source described elsewhere [8,9].

Upon excitation of hR with a 600 nm, 0.6 ps laser pulse, excited state absorption and stimulated emission appear immediately in the 420-530 nm and 650-770 nm wavelength regions (Fig. 1a and 1b), respectively, and both decay with a time constant of 2.3 ps. The absorption spectrum of hR^* had a peak at 516 nm and a shoulder at about 460 nm [8]. Accompanying the decay of hR^*, primary ground-state product appears at around 645 nm (Fig. 1c). The quantum yield of product formation was estimated to be 0.27. It should be noted that even when generation of the product is complete, the stimulated emission signal still remains at 720 nm (Fig. 1c). It is clearly illustrated by kinetics (Fig. 2) that the rise time of the product (645 nm) is faster than the decay of hR^* (470, 510, 720 nm). A detailed analysis elucidated the rise time of hR_J to be shorter than 1.0 ps [9]. This indicates that the simple sequential kinetic model, considering three states of all-*trans,* 13-*cis* with a common excited-state, seems not to be appropriate for hR, although it has been commonly applied for bR. Thus a parallel channel model is proposed for the primary process of hR (Fig. 3), in which the cis-trans isomerization and the relaxation to hR^* takes place simultaneously after Franck-Condon excitation. The relaxation in the excited state might involve the other conformational change than C_{13}=C_{14} rotation of the chromophore or an electronic transition

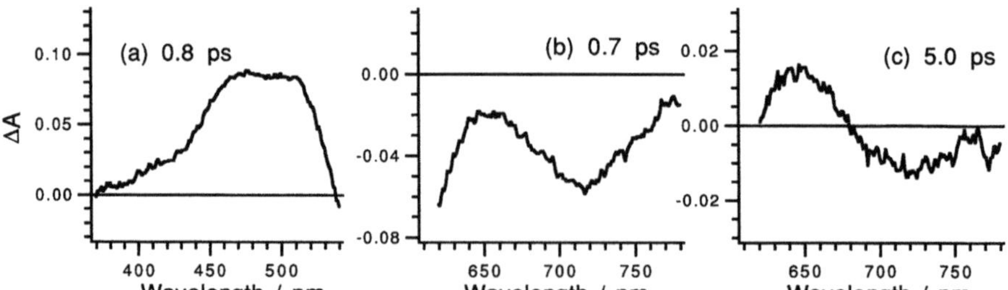

Fig. 1 Transient absorption spectra of halorhodopsin at respective times.

Springer Series in Chemical Physics, Vol. 55 Ultrafast Phenomena VIII
Editors: J.-L. Martin · A. Migus · G.A. Mourou · A.H. Zewail

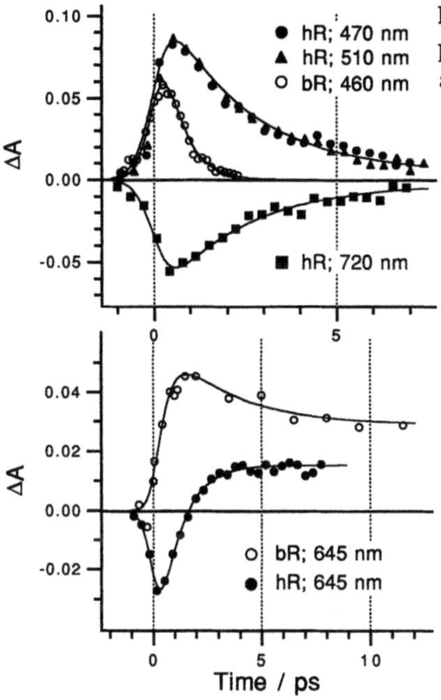

Fig.2

Kinetic data for hR and bR. Both pigments are excited under the same condition.

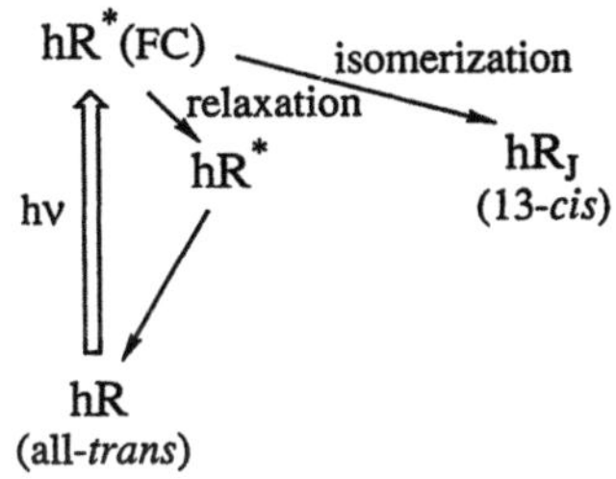

Fig.3

Schematic drawing of primary reaction of hR.

to another low-lying state as is typically observed in linear polyenes. Regardless, the present results suggest that the relaxation process in the excited state proceeds multidimensionally and it is a process in competition with isomerization.

The results obtained for bR under the same experimental condition are essentially similar to those of hR, and reproduce the previous results [1-6]. Moreover, the detailed measurement of bR^* absorption spectrum showed that its spectral behaviour is similar to that of hR [8]. It is noted that the formation yield of the primary ground-state product is twice as large in bR as in hR (Fig. 2), suggesting that cis-trans isomerization is twice as efficient in bR as in hR. On the basis of a three dimensional model for bR, by Henderson et al. [10], Asp85 of bR is proposed to be a key residue for highly efficient cis-trans isomerization of bR [11].

References

1. E.P. Ippen, C.V. Shank, A. Lewis and M.A. Marcus, Science 200, 1279 (1978).
2. A.V. Sharkov, A.V. Pakulev, S.V. Chekalin and Y.A. Matveetz, Biochim. Biophys. Acta 808, 94 (1985).
3. J.W. Petrich, J. Breton, J.L. Martin and A. Antonetti, Chem. Phys. Lett. 137, 369 (1987).
4. J. Dobler, W. Zinth, W. Kaiser, D. Oesterhelt, Chem. Phys. Lett. 144, 215 (1988).
5. R.A. Mathies, C.H. Brito Cruz, W.T. Pollard and C.V. Shank, Science 240, 777 (1988).
6. T. Kobayashi, M. Terauchi, T. Kouyama, M. Yoshizawa and M. Taiji, in Laser Applications in Life Sciences. SPIE 1403, 407 (1990).
7. H.J. Polland, M.A. Franz, W. Zinth, W. Kaiser, P. Hegemann and D. Oesterhelt, Biophys. J. 47, 55 (1985).
8. H. Kandori, K. Yoshihara, H. Tomioka and H. Sasabe, Chem. Phys. Lett. 187, 579 (1991).
9. H. Kandori, K. Yoshihara, H. Tomioka and H. Sasabe, J. Phys. Chem. in press (1992).
10. R. Henderson, J.M. Baldwin, T.A. Ceska, F. Zemlin, E. Beckmann and K.H. Downing, J. Mol. Biol. 213, 899 (1990).
11. H. Kandori, K.Yoshihara, H.Tomioka, H.Sasabe and Y.Shichida, to be published.

Part IX

Chemistry: Electron and Energy Transfer, and Solvation Dynamics

Femtosecond Intermolecular Electron Transfer: Dye in Weakly Polar Electron-Donating Solvent

K. Yoshihara[1], *A. Yartsev*[1], *Y. Nagasawa*[2], *H. Kandori*[1], *A. Douhal*[1], and *K. Kemnitz*[3]

[1]Institute for Molecular Science, Myodaiji, Okazaki 444, Japan
[2]The Graduate University for Advanced Studies, Myodaiji, Okazaki 444, Japan
[3]Government Industrial Research Institute, Midorigaoka Ikeda, Osaka 563, Japan

Abstract. An ultrafast electron transfer (ET) was found in diffusionless weakly polar contact systems of various electron donors and acceptors, namely, xanthene dyes in aniline. ET was measured as nonexponential when it is determined by mutual displacement of the reactants and by the solvent relaxation. ET process becomes single exponential in a case when ET is determined by electronic interaction between donor and acceptor and nuclear motion in reacting systems.

1. Introduction

Intermolecular ET reactions have attracted great attention because of their universal importance in many chemicals and biological processes. Different aspects of the ET process were examined theoretically [1,2], and numerous experimental investigations were made. To obtain direct information on the rate constant of fast intermolecular electron transfer, it is favorable to perform the experiment in systems in which the electron donor and acceptor are in contact, in order to eliminate the effect of translational diffusion. We have reported ultrafast ET (as fast as 10^{13} s^{-1}) between excited dye and weakly polar electron-donating solvent [3]. It is also the aim of present study to investigate the role of intramolecular processes such as vibrational relaxation, nuclear motion, intermolecular interaction, and molecular reorientation in relation to ET.

2. Experimental

Ultrafast intermolecular ET (forward ET) between the excited dye molecule and solvent was investigated by probing femtosecond fluorescence quenching. Synchronously pumped hybridly mode-locked dye laser with group velocity dispersion (GVD) control by a prism pair was used to generate optical pulses at different wavelengths at 76 MHz. Using usual combination of Rh-6G and DODCI/DQOCI and spatial filters between the second prism and high reflecting mirror, 70 fs pulses were generated at 583, 605, and 615 nm with output power of ca. 100 mW. With sulforhodamine 101 and DQTCI generation of 50 fs pulses at 675 nm was achieved. The fluorescence decay was measured at the magic angle by a conventional fluorescence up-conversion system using a 0.3 mm

Springer Series in Chemical Physics, Vol. 55 Ultrafast Phenomena VIII
Editors: J.-L. Martin · A. Migus · G.A. Mourou · A.H. Zewail

BBO crystal. A quartz flow-cell of 1 mm optical path length was used to minimized thermal effects on the fluorescence decay. In order to avoid long term instabilities of the dye laser, fluorescence decay was measured many times to increase accumulation time to several minutes at each delay position.

The transient absorption of reaction products, which provides a direct evidence of ET was investigated with time resolution of 0.7 ps. The reverse ET was observed by both the decay of the reaction products and the ground-state recovery.

Nile Blue perchlorate (NB) and oxazine 1 perchlorate (OX1) were obtained from Exiton and used without purification. Aniline (AN) and N,N-dimethylaniline (DMA) were vacuum distilled. The concentration of NB and OX1 in electron-donating solvent was varied from 0.5 to 2×10^{-4} M. Dye solutions were bobbled by N_2 to eliminate a possible effect of solvent oxidation.

3. Results

In Fig. 1 fluorescence decays measured at different emission wavelengths for OX1 dissolved in AN excited at 605 nm (blue side of the absorption spectra) are presented. It is easy to see that fluorescence decays at different observation wavelengths with experimental accuracy have the same time courses. In Fig. 2(a) one of the decays is plotted on the semilogarithmic scale. Nonexponential character of these decays is obvious. Very similar nonexponential time behavior at all measured fluorescence wavelengths was found for NB in AN. A decay curve of OX1 in DMA upon excitation at 605 nm is presented in Fig. 2(b). Since fluorescence in DMA is quenched very efficiently, the signal level is weaker and accuracy of data is not as good as in the case of AN. Nevertheless, decay was well fitted by a single exponential dependency with the time constant of 280 fs for OX1 and 160 fs for NB in DMA. To analyze the influence of the excess vibrational energy in the excited state, fluorescence decays were measured upon excitation at 675 nm which corresponds to the red slope of the absorption spectra for both dyes in AN and DMA. In AN nonexponential decays were observed and almost the same multiexponential decay components were obtained by deconvolution as in the case of excitation at 605 nm. Some

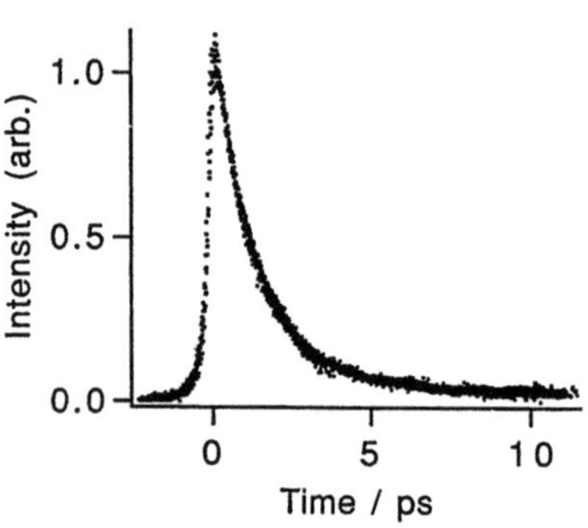

Fig. 1. Superposition of 4 fluorescence (normalized to the ~0.5 ps delay) decays of OX1 in AN observed at 680, 700, 725, and 830 nm. Excitation wavelength is 605 nm.

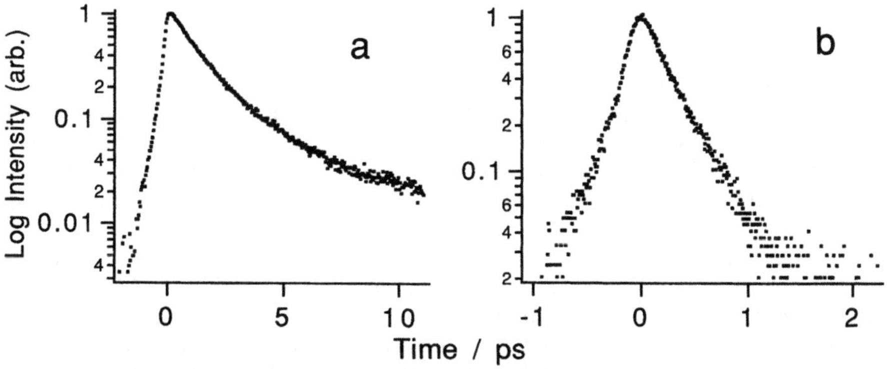

Fig. 2. Nonexponential fluorescence decay of OX1 in AN (a). Single exponential fluorescence decay of OX1 in DMA (b).

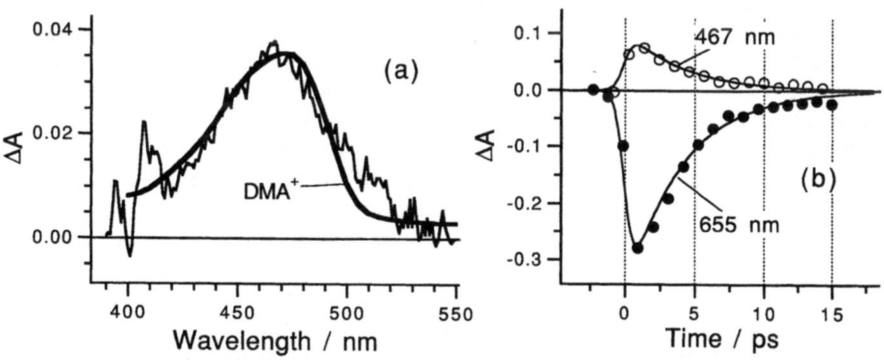

Fig. 3. (a). Transient absorption spectrum NB/DMA at 4.0 ps. The spectrum is composed of those of DMA^+ (smooth line) and NB· (neutral radical). (b). Kinetic changes of absorbances at 467 nm (absorption of DMA^+ and NB·) and at 655 nm (ground-state depletion). Both traces are fitted by a lifetime of 4.0 ps, indicating that the direct decay of the product back to the ground state.

difference was found only for the longest component of deconvolution (longer than 10 ps) which increased from 3% to 7%. The fluorescence of OX1 in DMA also gives single exponential decay with the decay constant of 130 fs which is also close to that obtained at 605 nm excitation.

The ultrafast quenching due to intermolecular ET resulted in formation of solvent cation and neutral radical of dye. Figure 3(a) shows the transient absorption spectrum of NB/DMA, which coincides well with that of DMA^+ [4]. Thus subpicosecond transient absorption spectrum provided the direct proof of ET. Both of the transient species appear rapidly and decrease exponentially with a decay time constant of 4.0 ps, indicating that the reverse ET occurs in 4.0 ps. In the present system, the reverse ET takes place also very rapidly and it performs a full cycle of charge separation and recombination in less than 10 ps.

4. Discussion

As it was clearly shown in transient absorption measurements, the process under investigation is ET. A possibility of proton trasfer is denied by the analogy in fluorescence decays for similar dye NB, where there is a possibility of proton transfer, and OX1, where it is not possible because of its chemical structure. It allows us to consider the fluorescence quenching dynamics as transfer of electron.

There are two main results in the present fluorescence decay measurements. 1) Fluorescence decays for NB/AN and OX1/AN upon excitation at 605 nm showed the same time behavior in wide spectral region from 670 to 830 nm. 2) The time behavior of fluorescence signal upon excitation at 605 nm with much excess of vibrational energy and at 675 nm close to 0-0 transition gave very similar results. These observations allow us to ascribe the measured quenching decays as ET which takes place from the bottom of the excited state potential after ultrafast intramolecular vibrational relaxation. One of the important consequence of this result is the possibility to eliminate the influence of vibrational relaxation on the ET dynamics. Excited state relaxation faster than 100 fs, which should provide the present result, can be inferred from the measured two exponentials with 70 fs and 410 fs time constants with 70 fs resolution [5]. The fluorescence of NB and OX1 in methanol gave instantaneous rise times at different wavelengths with 55 fs resolution [6]. The thermalization of ground and excited states was reported as fast as 50 fs for the same oxazine dye, cresyl violet, in transient spectroscopy with 60 fs pulse excitation and 10 fs pulse probing [7].

Clearly nonexponential fluorescence decays measured for NB and OX1 in AN have to be related with the nature of ET itself. This nonexponentiality can be caused by static and dynamic effects both based on the assumption of the strong interaction between donor-acceptor orientation and ET rate (the case of "narrow window" in Sumi-Marcus theory [1]). Static effect occurs when ET takes place for the different solvent-solute mutual orientation where the different magnitude of electronic interaction gives different ET rate. In the all ensemble of the molecules ET will give nonexponential decay. Dynamic nonexponentiality realizes if the donor-acceptor orientation changes during ET and disturbs its single exponential character. As the measured signal is the sum of all these different disturbed exponential processes for each separated orientation, it becomes nonexponential. To explain observed experimental time behavior we suggest the effect of molecular reorientation with the rate in the order of ET (dynamic nonexponentiality). Upon excitations at 605 and 675 nm due to inhomogeneous broadening it would be possible to obtain different ET rate in the case of ET rate much faster than solvent reorientation time. Same fluorescence decay measured upon 605 nm and 675 nm excitations may be explained as the molecular reorientation rate which is in the order of ET. This conclusion is in agreement with the values of longitudinal reorientation times of AN, which were calculated from Debye relaxation time [8] as 0.4 ps and 4.3 ps.

ET in the order of 100 fs which was measured for NB and OX1 in DMA is due to strong interaction between donor and acceptor molecules. ET rate is increased in these cases because of increasing electron exchange matrix element. Almost single exponential ultrafast fluorescence decays obtained in this case can be explained in two aspects. A single exponential ET can be realized if the ET rates are close for different molecular orientations ("wide window" case) and/or if the molecular reorientation takes place much faster than ET occurs. Assumption of "wide window" means that ET is determined only by nuclear motion and is not affected by solvent reorientation. Nuclear motion in this strongly coupled donor-acceptor system can also reorient the interacting molecules. Generally, molecular vibrations (intramolecular vibration and/or some intermolecular vibrational motion) induce ultrafast ET. One particular motion named libration was recently found to be as fast as 60 fs in a light polar solvent [9]. It is also important to emphasize that for realization of ET with such a fast rate, some mechanism of ET stabilization process is required. It means that relaxation of the ET products dissipates the excess energy above the bottom of the products energy surface to eliminate the back ET reaction. This relaxation is as fast as ET, when the ET rate is determined by stabilization process.

The authors would like to thank Prof. H. Sumi for valuable discussions.

References:

1. H. Sumi and R.A.Marcus: J. Chem. Phys., **84**, 4894 (1986).
2. M. Bixon and J. Jortner: Faraday Discusson Chem. Soc., **74**, 17 (1982).
3. T. Kobayashi, Y. Takagi, H. Kandori, K. Kemnitz, and K. Yoshihara: Chem. Phys. Lett., **180**, 416 (1991).
4. H. Kandori, K. Kemnitz, and K. Yoshihara: J. Phys. Chem., in press (1992).
5. A.M. Weiner and E.P. Ippen: Chem. Phys. Lett., **114**, 456 (1985).
6. A. Mokhtari, A Chebira, and J. Chesnoy: J. Opt. Soc. Am. B., **7**, 1551 (1990).
7. C.H. Brito Cruz, R.L. Fork, W.H. Knox, and C.V. Shank: Chem. Phys. Lett., **132**, 341 (1986).
8. J. Bhattacharyya, A. Hasan, S.B. Roy, and S. Kastha: J. Phys. Soc. Japan, **28**, 204 (1970).
9. S.J. Rosenthal, X. Xie, M. Du, and G.R. Fleming: J. Chem. Phys., **95**, 4715 (1991).

Ultrafast Studies and Simulations on Direct Photoinduced Electron Transfer in the Betaines

A.E. Johnson, N.E. Levinger, G.C. Walker, and P.F. Barbara

Department of Chemistry, University of Minnesota,
Minneapolis, MN 55455, USA

The direct photoinduced electron transfer of betaine-30 has been studied by femtosecond pump-probe spectroscopy in the visible and near infrared. The ultrafast pump-probe data have been analyzed to extract the time scales for the reverse thermal electron transfer and for vibrational cooling processes that occur after the thermal electron transfer process.[1-3] These time scales have been measured in a broad range of solvents, over a large temperature range, in the liquid and solid state. This paper emphasizes new variable wavelength pump-probe measurements and simulations of the spectral dynamics of the betaines. The new data facilitate a detailed analysis of the evolution of the spectral band shape during the electron transfer process.

Figures 1 and 2 display the change in optical density, ΔOD, as a function of time for betaine-30 in acetone probed at the low energy band-edge (800nm), near the band maximum (700nm) and at the high energy band-edge (640nm) of the absorption band. Similar curves were observed for betaine-30 in other quickly relaxing polar solvents, e.g., acetonitrile. The initial bleach component in the absorption transients is followed by complex dynamical behavior that depends strongly on the detection wavelength and environment. In the case of quickly relaxing polar solvents such as acetone, at detection wavelengths near the maximum of the static absorption band, the initial bleach exhibits a recovery which fits well to a single exponential function (Fig. 1b). A dramatically different effect is observed at the low energy band-edge in solvents of this type, shown in Fig. 1a. Here, the decay of the initial bleach is followed by an increased absorption signal which decays back to zero in a few picoseconds. At the high energy band-edge, the decay of the bleach requires a biexponential fit to account for the dynamics (Fig. 1c).

The basic model for the betaines that is most consistent with the data is a sequence of events where population is transferred from an equilibrium position in the ground state (D^+A^-) to a nonequilibrium position in the excited state (DA) by photoexcitation and evolves along the solvent coordinate as summarized pictorially in Fig. 3. The solvent coordinate is a collective degree of freedom involving the response of the solvent polarization to the charge distribution of the solute. A sequence of events accounting for the data are represented by the numbered probability distributions in Fig. 3. Before optical excitation, the solvent coordinate is equilibrated in the ground state, as depicted by distribution 1. This distribution is transferred to the excited state by the excitation pulse yielding distribution 2. Two important dynamical processes affect the population distribution in the DA state, namely transient solvation in DA and $DA \rightarrow D^+A^-$ electron transfer. The rate of electron transfer accelerates as solvation occurs because the vertical energy gap decreases along the solvent coordinate from 1 to 0. Competition due to radiative decay and intersystem crossing processes can be ruled out. Radiative decay is too slow to be significant.[4] Intersystem crossing must be negligibly slow because the bleach in the $DA \rightarrow D^+A^-$ absorption fully recovers in tens of picoseconds indicating that there is no significant build up of population in the triplet state. The consequence of the solvation dynamics and electron transfer is to produce an excitation of the solvent coordinate in the initially excited solute molecules. Eventually distribution 4 evolves back to equilibrium on the ground D^+A^- state by solvation dynamics on the state.

Because the electron transfer in betaine is a direct optical transition, the parameters needed to predict the rate and the dynamics can be obtained directly from the static absorption spectrum. Predictions by various simple theoretical electron transfer models found in the literature were many orders-of-magnitude too slow when compared to experimental results for the betaines.[5.6] The failure of these models stems from the approximate means by which they describe nuclear degrees of freedom of the solute/solvent system. As a result, we have developed a hybrid theoretical model.[3]that is an extension of the Sumi-Marcus approach.[5]

Springer Series in Chemical Physics, Vol. 55 Ultrafast Phenomena VIII
Editors: J.-L. Martin · A. Migus · G.A. Mourou · A.H. Zewail

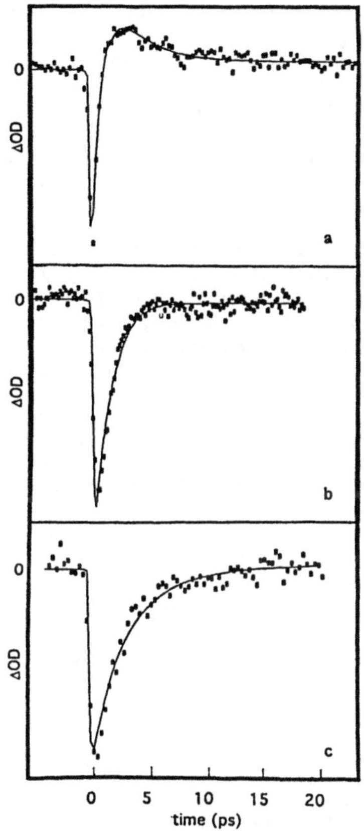

Figure 1. Change in absorbance, ΔOD, as a function of time for betaine-30 in acetone, λ_{pump}= 800nm, a) λ_{probe}= 800nm, b) λ_{probe}= 700nm, and c) λ_{probe}= 640nm. Points are data and solid lines are fits to the data.

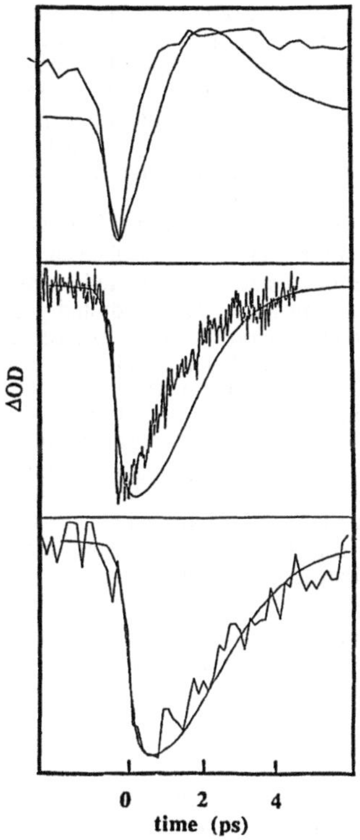

Figure 2. Comparison of the predicted and observed dynamics for betaine-30 in acetone probed at a) 800nm, b)700nm and c) 640nm.

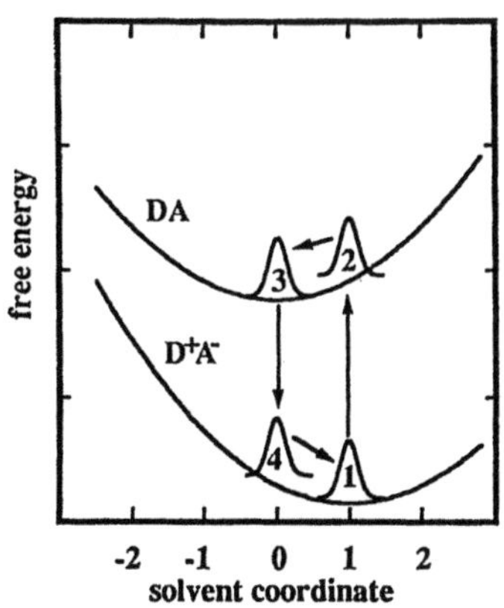

Figure 3. Free energy as a function of solvent coordinate for the lowest energy transition in betaine-30. Distributions labeled 1,2,3, and 4 indicate the sequence of events following laser excitation.

The results demonstrate that an accurate model for electron transfer in the inverted regime in solution minimally requires the following three nuclear degrees of freedom: (i) a solvent mode with a frictional response, (ii) an intramolecular low frequency (classical) mode, and (iii) a high frequency (quantum mechanical) intramolecular mode. Furthermore, the analysis allows for a detailed understanding of the combined effects of solvation dynamics and vibrational excitation in ultrafast electron transfer kinetics. Simulations from this model are in semi-quantitative agreement with various aspects of the time-dependent spectroscopy of the betaines. We present dynamics predictions from the hybrid model in this paper. A detailed description of these calculations will appear in another article.[7]

To simulate the time dependence of the absorption spectra, we must consider the population dynamics in the ground and excited states as a function of time, including motion upon the excited state surfaces as well as transitions between the two surfaces. At any given time, population which is missing from the ground state contributes to a bleach signal while population on the excited state surface can contribute to the signal by undergoing an absorption to a higher excited state or stimulated emission to the ground state. We consider only the case of stimulated emission. The predicted population dynamics are convoluted with a spectral shape for the absorption spectrum for direct comparison to experiment.

We have extended our hybrid model to predict ground *and* excited state population dynamics and electron transfer. Qualitatively, the model outlined is consistent with the experimental observations on betaine-30. This can be seen in Figure 2 which displays the measured and predicted absorption transients for betaine-30 in acetone at 800nm (low energy edge), 700nm (absorption peak) and 640nm (high energy edge). The low energy band-edge observation of an initial bleach followed by a sign change to an increased absorption is consistent with the predictions because the nonequilibrium distribution in the ground state following electron transfer should provide a higher absorption probability on the low energy band edge than the equilibrated distribution. At the high energy band-edge, the model predicts that bleach recovery should appear slower because the population responsible for this signal must first undergo internal conversion and then transient solvation in D^+A^- before the transient absorption signal can decay to zero.

ACKNOWLEDGMENTS

Support of this research by the US. National Science Foundation and the Office of Naval Research is gratefully acknowledged. NEL was supported by a National Science Foundation Postdoctoral Fellowship.

References

1. E. Åkesson, G.C. Walker, and P.F. Barbara, *J. Chem. Phys.*, **95**, 4188 (1991).
2. E. Åkesson, A.E. Johnson, N.E. Levinger, G.C. Walker, T.P. DuBruil, and P.F. Barbara, *J. Chem. Phys.* **96**, 7859 (1992).
3. G. C. Walker, E. Åkesson, A. E. Johnson, N. E. Levinger and P. F. Barbara, *J. Phys. Chem.* **96**, 3728 (1992).
4. Using the Strickler-Berg analysis (S. J. Strickler and R. A. Berg, *J. Chem. Phys.* **37**, 814 (1962)) we estimate that the fluorescence lifetime for betaine-30 in acetone is $7x10^{-9}$s.
5. H. Sumi and R.A. Marcus, *J. Chem. Phys.*, **84**, 4894 (1986).
6. J. Jortner and M. Bixon, *J. Chem. Phys.*, **88**, 167 (1988).
7. A. E. Johnson, N. E. Levinger and P. F. Barbara, in preparation.

Picosecond Infrared Study of Ultrafast Electron Transfer and Vibrational Energy Relaxation in $[(NC)_5RU^{II}CNRu^{III}(NH_3)_5]^{1-}$

P.O. Stoutland[1], *S.K. Doorn*[2], *R.B. Dyer*[1], *and W.H. Woodruff*[2]

[1]Divisions of Chemical and Laser Sciences (Mail Stop J-567), Los Alamos National Laboratory, Los Alamos, NM 87545, USA

[2]Isotope and Nuclear Chemistry (Mail Stop C-346), Los Alamos National Laboratory, Los Alamos, NM 87545, USA

Mixed-valent transition metal dimers have become key systems for understanding many fundamental aspects of electron transfer dynamics.[1] We have been interested in the mixed-valent dimer $[(NC)_5Ru^{II}CNRu^{III}(NH_3)_5]^{1-}$ for which optical excitation into the metal-metal charge transfer (MMCT) band produces a state which is highly unstable with respect to back electron transfer:

$$[(NC)_5Ru^{II}CNRu^{III}(NH_3)_5]^{1-} \underset{ET}{\overset{h\nu}{\rightleftharpoons}} [(NC)_5Ru^{III}CNRu^{II}(NH_3)_5]^{1-}$$

The thermal back electron transfer in such systems may occur on very short timescales, allowing one to observe effects due to ultrafast reorganization processes, to discern excited state/ground state coupling phenomena, and to test modern electron transfer theories. We have studied[2] the dynamics of $[(NC)_5Ru^{II}CNRu^{III}(NH_3)_5]^{1-}$ following MMCT optical excitation with the use of picosecond infrared spectroscopy. We monitor the terminal RuC≡N stretching vibrational frequency which is sensitive to the oxidation state of the ruthenium cyanide center and, because of anharmonicity, to the degree of vibrational excitation of this mode.

Upon excitation at 600 nm we observe an instrumentally limited ($\tau < 0.5$ ps) decrease in the RuC≡N absorbance for the ground state at 2053 cm^{-1} which recovers with $\tau = 6 \pm 1$ ps. Simultaneously, a new absorption centered at 2110 cm^{-1} is observed which decays with $\tau < 0.5$ ps. Subsequently, a broad feature extending from 1950 to 2040 cm^{-1} is seen. While the rise times of the transients at the low energy end of this band are instrumentally limited, they become progressively slower when moving to the high energy side of the band, and in fact correspond to the *decay* times found for the lower energy transients. The decay times of the transients range from < 2 ps at 1954 cm^{-1} to 6 ps at 2034 cm^{-1}, matching the recovery time of the ground state band.

MMCT excitation should lead to a complex in which the ruthenium cyanide center has formally changed from a 2+ to a 3+ oxidation state, resulting in a shift of the RuC≡N stretch from the ground state value of 2053 cm^{-1} to ca. 2120 cm^{-1}. We can thus assign the short-lived transient at 2110 cm^{-1} unambiguously to the MMCT excited state. The features between 1950 and 2040 cm^{-1} are consistent with vibrationally hot ground state molecules formed

Springer Series in Chemical Physics, Vol. 55 Ultrafast Phenomena VIII
Editors: J.-L. Martin · A. Migus · G.A. Mourou · A.H. Zewail

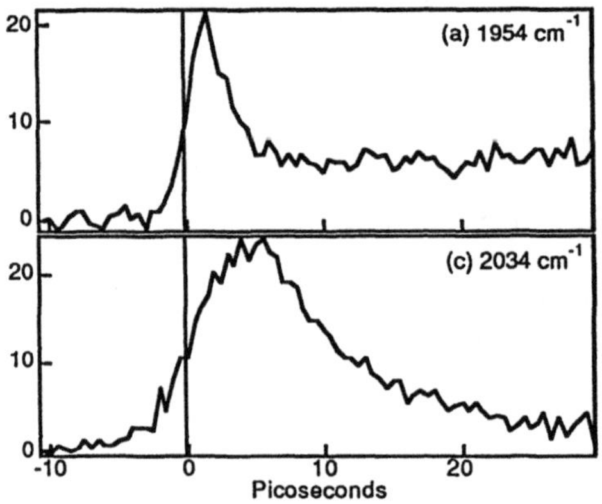

Figure 1. Transient decays obtained at (a) 1954 and (b) 2034 cm^{-1}. Appropriate rise and decay times are shown in the table. Offsets remaining at long times are due to thernal transients and do not affect the fits obtained. Vertical axis is $\Delta A \times 10^2$.

Table I. Rise and decay times for the best fit to data from 1954 to 2060 cm^{-1}.[a]

Wavelength (cm^{-1})	Rise time (ps)	Fall time (ps)
1954	0.1	0.7
1974	0.3	1.9
1994	0.6	3.6
2014	0.7	4.6
2034	1.4	5.9
2060[b]	0.0	6.9

[a] Values given correspond to the best fit obtained when convolving the instrument response (typically a 3-4 ps FWHM $sech^2$ function) with a biexponential function; estimated relative error is ± 0.5 ps. [b]Ground-state bleach.

following the back electron transfer; the frequencies observed nominally correspond to population of the vibrationally excited states v = 1-7. Additional experiments in which the pump pulse was changed to 700 nm (14,285 cm^{-1}) demonstrated that the degree of vibrational excitation was dependent on the frequency of the excitation wavelength; only states up to v = 4 were observed in this case. The nature of the cooling process is presently under investigation; both inter- and intramolecular processes are possible. Comparison of the relaxation rates observed with known *intermolecular* relaxation rates,[3] however, suggests that the cooling is dominated by an intramolecular mechanism. Additionally, preliminary experiments have shown that vibrational relaxation in the osmium derivative is 3-4 times slower, consistent with the notion that IVR rates are affected by the mass of the metal atom.[4]

These findings demonstrate that the back electron transfer from the MMCT state is ultrafast ($\tau < 0.5$ ps); the infrared frequency observed suggests that nearly a full charge transfer has occurred. More intriguing, however, is that upon return to the ground electronic state large amounts of energy (up to 14,000 cm^{-1}) are placed into a single vibration: the terminal RuC≡N stretching mode. The large energy difference (ca. 8000 cm^{-1}) between the Ru^{II}--Ru^{III} and the Ru^{III}--Ru^{II} species potentially allows the deposition of large amounts of energy into selected vibrational modes of the product of the thermal back electron transfer. The efficiency and selectivity of this process in the present case is, however, quite remarkable, and provides a unique opportunity to study the deposition and relaxation of large amounts of vibrational energy in an inorganic molecule. Predictions of the rates of vibrational relaxation processes in highly excited inorganic molecules have never been directly tested. Vibrationally excited products following electron transfer have been observed in the gas phase, but to our knowledge this is the first direct observation in the solution phase.[5] Moreover, the relative rates observed for electronic and vibrational relaxation in $[(NC)_5Ru^{II}CNRu^{III}(NH_3)_5]^{1-}$ suggest that activated modes in the excited state remain so on the timescale of electron transfer, holding important consequences for electron transfer theories and for fast electron transfer processes such as charge separation in photosynthesis.

References.

[1] See, for example: (a) *Mixed-valence Compounds*, D.B. Brown, ed., D. Reidel, Dordrecht: Holland, 1980. (b) Creutz, C. *Prog. Inorg. Chem. 30*, 1 (1983).

[2] Experiments were performed as described previously: Dyer, R.B.; Peterson, K.A.; Stoutland, P.O.; Woodruff, W.H. *J. Am. Chem. Soc. 113*, 6276 (1991).

[3] (a) Heilweil, E.J.; Cavanagh, R.R.; Stephenson, J.C. *J. Chem. Phys. 89*, 230 (1988), (b) Heilweil,E.J.; Cavanagh, R.R.; Stephenson, J.C. *Chem. Phys. Lett. 134*, 181 (1987).

[4] See for example: (a) Rogers, P.J.; Selco, J.I.; Rowland, F.S. *Chem. Phys. Lett. 97*, 313 (1983). (b) Rogers, P.; Montague, D.C.; Frank, J.P.; Tyler, S.C.; Rowland, F.S. *Chem. Phys. Lett. 89*, 9 (1982).

[5] (a) Liao, C.L.; Shao, J.D.; Xu, R.; Flesch, G.D.; Li, Y.G.; Ng, C.Y. *J. Chem. Phys. 85*, 3874 (1986), (b) Nieder, G.; Noll, M.; Toennies, J.P. *J. Chem. Phys. 87*, 2067 (1987) and references therein. For a general discussion of vibrational excitation accompanying electron transfer reactions, see: Jortner, J.; Bixon, M. *J. Chem. Phys. 88*, 167 (1988).

Ultrafast Studies on Intervalence Charge Transfer*

K. Tominaga[1], *D.A.V. Kliner*[1,**], *J.T. Hupp*[2], *and P.F. Barbara*[1]

[1]Department of Chemistry, University of Minnesota, Minneapolis, MN 55455, USA

[2]Department of Chemistry, Northwestern University, Evanston, Illinois 60208, USA

*Supported by the U.S. National Science Foundation

**National Science Foundation Postdoctoral Fellow

Mixed-valence metal complexes have played a central role in the development of the theory of electron-transfer (ET) reactions.[1] These compounds contain two metal atoms in different oxidation states, typically +2 and +3. An absorption band is often observed in the near IR or visible, corresponding to the transfer of an electron between the two metal centers, i.e. intervalence absorption. Much theoretical work has addressed the connection between this photoinduced charge transfer and the analogous thermal ET reaction. Recently, the first direct ultrafast measurements were reported for an intervalence absorption.[2]

In this paper we report the femtosecond transient-absorption spectra for the following metal/metal intervalence photoinduced charge-transfer reaction in a broad range of solvents and temperatures.[3]

$$(NH_3)_5Ru^{III}NCRu^{II}(CN)_5^- \underset{k_{ET}}{\overset{h\nu}{\rightleftharpoons}} (NH_3)_5Ru^{II}NCRu^{III}(CN)_5^- \quad (1)$$

The observed electron-transfer rates (k_{ET}) fall in the range of $\sim 10^{12}$ sec^{-1} to greater than $10^{13} sec^{-1}$. The critical role played by intramolecular vibrational modes in promoting the ET reaction is demonstrated by the relative insensitivity of k_{ET} to the solvent and temperature and by the fact that, in many instances, k_{ET} exceeds $1/\tau_s$ (where τ_s is the time scale for diffusive solvent motion), which is in marked contrast to several recently reported fast ET reactions.[4]

The kinetics have been analyzed using various theoretical descriptions available from the literature[1] to reveal a surprisingly complex interaction of intramolecular and intermolecular modes in these ET reactions. Especially noteworthy is the wealth of spectroscopic data available for these mixed-valence compounds. In particular, resonance Raman and static absorption spectra of the metal-metal charge-transfer transition provide frequencies and displacements (reorganization energies) of the vibrational modes coupled to the ET reaction.[5] These molecular parameters have been used to evaluate multi-mode expressions for the reaction rate. The ET rates are calculated by employing the current ET theories with assumption of the separation of the solvent motion, a fast dissipative (<50 fs) and a slow diffusive (~0.5 ps) motions. The predicted values for $1/k_{ET}$ range from 25 to 160 fs, depending on the fraction of the reorganozation energy of the fast solvent motion. This shows a good agreement with the measured ET rates. This analysis elucidates the important pathways for electron transfer and allows a quantitative evaluation of the theoretical treatments to be performed.

The transient pump-probe apparatus has been described in previous publications.[6] Two ultrafast spectrometers were used, one for single-wavelength measurements ($\lambda_{pump} = \lambda_{probe}$) and one for measurements in which λ_{pump} and λ_{probe} were different. The temperature of the sample could be varied between -100° and 70° C.

Figure 1 shows transients recorded for Reaction (1) in H_2O and NMF at 20° C and glycerol at -100° C. The transients were fit to a sum of exponentials and can generally be characterized by two time scales (Table 1): 1) A pulse-limited (<0.1 ps) transient bleach (except glycerol, see below), and 2) A slower bleach or increased absorption with a decay time that depends on the solvent and an amplitude that is 5-20% of the fast component. Similar results were obtained for other intervalence ET reactions that we have investigated.

The dominant fast response is ascribed to electron transfer. For many of the solvents and temperatures investigated, the ET time ($1/k_{ET}$) is <0.1 ps (see Table 1) despite large variations in τ_s. In glycerol, however, the ET rate for Reaction (1) is not pulse limited and is sensitive to

Springer Series in Chemical Physics, Vol. 55 Ultrafast Phenomena VIII
Editors: J.-L. Martin · A. Migus · G.A. Mourou · A.H. Zewail

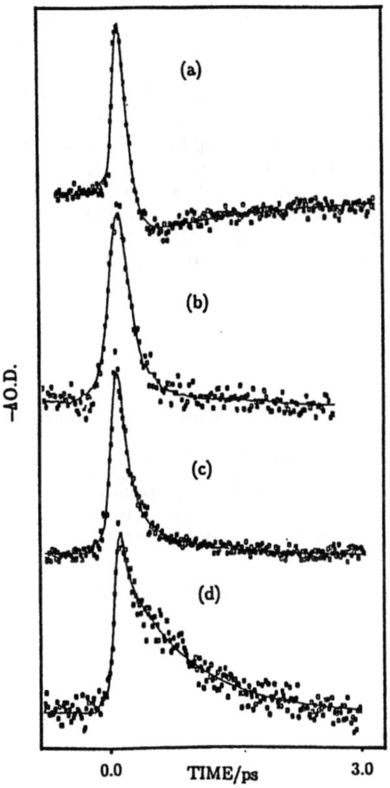

Figure 1. Change in absorption as a function of time delay between the pump and probe pulses for Reaction (1) in H_2O at 20°C ((a) and (b)), NMF at 20°C (c), and glycerol at -100°C (d). The pulse width was 70 fs and $\lambda_{pump} = \lambda_{probe}$ = 792 nm for (a), (c), and (d), and λ_{pump} = 800 nm and λ_{probe} = 700 nm for (b). Positive signal indicates a bleach. The points are the experimental data, and solid curves show biexponential fits to the data.

Table 1
Biexponential Fits to Kinetic Data for Reaction (1)[a]

Solvent	T/°C	τ_1/ps[b] (A_1)	τ_2/ps (A_2)
H_2O	0	≤0.1 (0.88)	1.6 (-0.12)
	20	≤0.1 (0.89)	1.5 (-0.11)
NMF	0	≤0.1 (0.72)	0.5 (0.28)
	20	≤0.1 (0.86)	0.9 (0.14)
formamide	0	≤0.1 (0.85)	0.6 (0.15)
	20	≤0.1 (0.90)	0.7 (0.10)
ethylene glycol	20	0.3 (0.96)	>5 (0.04)
	70	≤0.1 (0.72)	0.4 (0.28)
glycerol	-100	0.74 (1.00)	---
	-5	0.43 (0.90)	~6 (-0.10)
	20	0.33 (0.95)	~5 (-0.05)
	50	0.27 (0.97)	~5 (-0.03)
glycerol-d_3	-100	1.3 (1.00)	---
	20	0.40 (0.95)	~5 (-0.05)

a. Transients were fit to $A_1\exp\{-t/\tau_1\} + A_2\exp\{-t/\tau_2\}$. Uncertainties in τ_1 and τ_2 are typically ±10-15%. Positive (negative) values of A_1 and A_2 refer to a bleach (increased absorption). For glycerol and glycerol-d_3, the fit is improved by inclusion of a small pulse-limited component, which is attributed to a coherent artifact.
b. τ_1 is identified with electron transfer, i.e., $k_{ET} = 1/\tau_1$.

temperature and to deuteration of the solvent OH groups. The ET time in glycerol varies from 0.27 ps at 50° C to 0.74 ps at -100° C. A pronounced, temperature-dependent isotope effect is observed: k_{ET}(glycerol)/k_{ET}(glycerol-d_3) = 1.2 at 20° C and 1.5 at -100° C. Thus, high-frequency solvent fluctuations are clearly coupled to the ET reaction, with H-atom motion involved in the relevant solvent modes. Such modes may include librational or inertial motions.

The behavior of the slower component depends on the relative wavelengths of the analyzing light and the peak of the absorption spectrum. Solvent- and temperature-dependent shifts of the absorption spectrum and the use of both one- and two-color measurements allowed the probe pulse to interrogate different regions of the absorption band. The slower component is tentatively attributed to local heating or ground-state solvation, i.e., to nuclear motions in the solute and solvent. This interpretation is consistent with the measured temperature- and solvent-dependence of the static absorption spectrum.

References

1. N.S. Hush, Prog. Inorg. Chem. **8**, 391 (1967); C. Creutz, *ibid.* **30**, 1 (1983), R.A. Marcus and N. Sutin, Biochim. Phys. Acta **811**, 265 (1985).
2. G. C. Walker, P.F. Barbara, S. K. Doorn, Y. Dong, and J. T. Hupp, *J. Phys Chem.* 95, 5712(1991).
3. Doorn *et al.* recently studied reaction (1) using visible pump/IR probe spectroscopy and concluded that the ET reaction takes place within 0.5 ps after the photon excitation followed by vibrational relaxation in the ground state. S.K. Doorn, P.O. Stoutland , R.B. Dyer, and W.H. Woodruff, *J. Am. Chem. Soc.* 114, 3133 (1992).
4. M. Maroncelli, J. MacInnis, and G.R. Fleming, Science **243**, 1674 (1989); P.F. Barbara and W. Jarzeba, Adv. Photochem. **15**, 1 (1990).
5. S.K. Doorn and J.T. Hupp, J. Am. Chem. Soc. **111**, 1142 (1989).
6. M.A. Kahlow, W. Jarzeba, T.P. DuBruil, and P.F. Barbara, Rev. Sci. Instrum. **59**, 1098 (1988); A.E. Johnson, N.E. Levinger, and P.F. Barbara (to be published).

Picosecond Infrared Study of Intramolecular Energy Transfer in $[(phen)(CO)_3Re^{I}(NC)Ru^{II}(CN)(bpy)_2]^+$

R.B. Dyer[1], *K.A. Peterson*[1], *K.C. Gordon*[1], *W.H. Woodruff*[1], *J.R. Schoonover*[2], *T.J. Meyer*[2], *and C.A. Bignozzi*[3]

[1]Photochemistry and Photophysics Group, CLS-4, MS J567, Los Alamos National Laboratory, Los Alamos, NM 87545, USA

[2]Department of Chemistry, University of North Carolina at Chapel Hill, Chapel Hill, NC 25799-3290, USA

[3]Dipartimento di Chimica, Universita di Ferrara, Ferrara, Italy

Significant advances have been made in the design and characterization of molecular assemblies which, upon photoexcitation, undergo intramolecular electron or energy transfer [1]. These reactions have typically been followed by transient electronic absorption spectroscopy. This technique, however, suffers from the disadvantage that absorption bands tend to be broad and featureless which can lead to ambiguous interpretation in complex assemblies, especially where there is more than one absorbing chromophore. Time-resolved resonance Raman spectroscopy has been applied to the study of excited states [2], but time-resolved infrared spectroscopy (TRIR) is particularly well-suited to probing complexes containing ligands such as CO or CN which are bound to the metal [3]. Unlike the transient Raman experiment, TRIR does not rely on resonance enhancement in the excited state since the metal-CO and -CN stretching vibrations have large oscillator strength providing high sensitivity. We describe here a novel application of TRIR spectroscopy to the elucidation of intramolecular energy transfer in the binuclear, cyano-bridged complex, $[(phen)(CO)_3Re^{I}(NC)Ru^{II}(CN)(bpy)_2]^+$ [4], representing the first application of the technique to a process of this type.

The TRIR difference spectra shown in the Figure are obtained 1 ps (A) and 40 ps (B) after 300 nm excitation of $[(phen)(CO)_3Re^{I}(NC)Ru^{II}(CN)(bpy)_2]^+$ in CH_3CN. Typical of these types of complexes, the broad electronic absorptions of both metal centers overlap considerably such that at this pump wavelength the excitation of both $Re^{I} \rightarrow$ phen and $Ru^{II} \rightarrow$ bpy metal-to-ligand charge transfer (MLCT) transitions occur. The IR frequencies of the CO and CN bands are very sensitive to metal oxidation state however, making it possible to determine unambiguously the nature of the intermediates formed by the excitation pulse. Assignments are based on comparison to ground and excited state IR spectra of monomeric model complexes.

The 1 ps spectrum indicates that the predominant species present at very early times is the $Re^{I} \rightarrow$ phen MLCT state as illustrated in (A). Both the number of CO bands and their frequencies provide direct evidence for oxidation of Re^{I} to Re^{II}. While only two CO bands (1921 and 2028 cm^{-1}) are observed in the ground state, (suggesting effective C_{3v} symmetry for the 3 CO and 3 N ligands), the lowest energy CO band is split (1964 and 2015 cm^{-1}) in the 1 ps spectrum, indicative of a lower effective symmetry. Symmetry reduction in the $Re^{II}(bpy^{\cdot -})$ MLCT state is expected because of the presence of a hole in the $d(\pi)$ orbitals of Re^{II} (d^5). In addition, an increase in $\nu(CO)$ of 50-70 cm^{-1} is expected due to reduced π backbonding.

Subsequently this state rapidly (10 ps) decays to a state which lives for hundreds of nanoseconds and gives the TRIR spectrum shown in (B). This spectrum is identical within experimental error to the spectrum obtained 200 ns after 355 nm excitation. The small shifts of the CO bands and the large shifts of the CN bands are consistent with the assignment of this spectrum to the $Ru^{II} \rightarrow$ bpy MLCT state as illustrated in (B).

The net reaction following $Re^{I} \rightarrow$ phen excitation is $Re^{II}(phen^{\cdot -}) \rightarrow Ru^{III}(bpy^{\cdot -})$ energy transfer which is spontaneous by ~ 0.2 eV and the long-lived intermediate is a mixed-oxidation state complex. The slight increase in $\nu(CO)$ ($\Delta\nu(CO)$ ~ 5 cm^{-1}) for this intermediate is consistent with partial $Re^{I} \rightarrow Ru^{III}$ electronic delocalization by orbital mixing across the cyano bridge. Assuming a linear relationship between the charge density at Re and $\nu(CO)$, we calculate the extent of charge transfer across the bridge as ~10 % based on the

Springer Series in Chemical Physics, Vol. 55 Ultrafast Phenomena VIII
Editors: J.-L. Martin · A. Migus · G.A. Mourou · A.H. Zewail

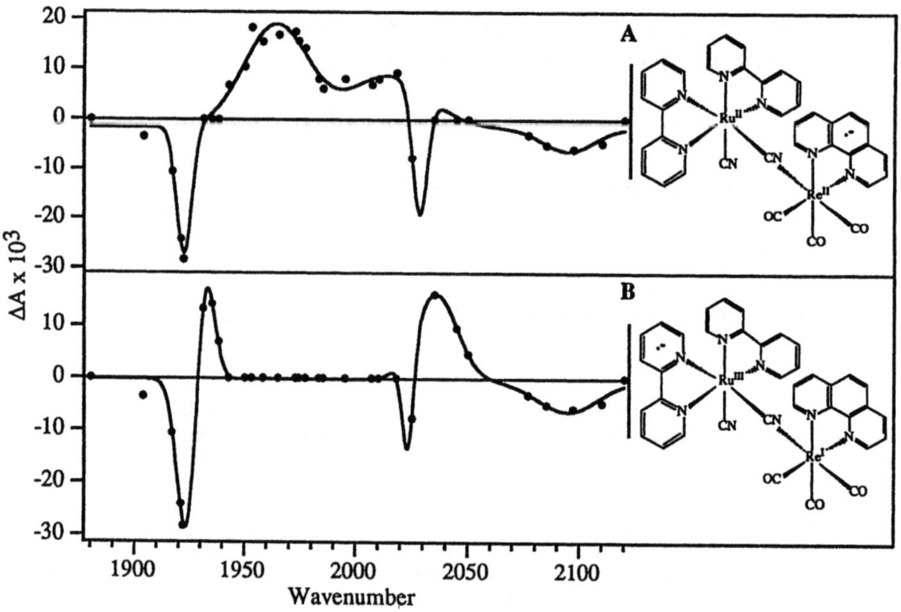

Figure. TRIR spectra and assignments obtained 1 ps (A) and 40 ps (B) after 300 nm excitation of $[(phen)(CO)_3Re^I(NC)Ru^{II}(CN)(bpy)_2]^+$ in CH_3CN.

averaged shifts in ν(CO). This value can be compared to 15 % for the extent of delocalization from Re^I to Ru^{III} across the CN bridge in $[(phen)(CO)_3Re^I(CN)Ru^{III}(CN)(bpy)_2]^+$ as estimated by integration of the $Re^I \rightarrow Ru^{III}$ intervalence transfer absorption band.

The TRIR results enable unambiguous assignment of the intermediates produced by UV excitation and indicate quenching of the Re^{II}phen·⁻ MLCT state via ultrafast energy transfer. These results demonstrate the effectiveness of TRIR spectroscopy at providing detailed information on the nature and dynamics of photo-induced, short-lived intermediates of complex assemblies.

References.

(1) (a) Balzani, V; Scandola, F. Supramolecular Photochemistry ; Ellis Harwood: Chichester, 1990. (b) Fox, M.A.; Chanon, M., eds. Photoinduced Electron Transfer ; Elsevier: New York, 1988.

(2) See for example, Morris, D.E.; Woodruff, W.H. Spectroscopy of Inorganic-Based Materials, Clark, R.J.H. and Hester, R.E. (Eds.); John Wiley & Sons; 1987; pp 285-332.

(3) Dyer, R.B.; Einarsdottir, O; Killough, P.M.; Lopez-Garriga, J.J.; Woodruff, W.H. *J. Am. Chem. Soc.* **1989**, *111*, 7657-7659.

(4) Bignozzi, C.A.; Argazzi, R.; Schoonover, J. R.; Dyer, R. B.; Scandola, F. manuscript in preparation.

Noise-Induced Intramolecular Electron Transfer Processes in Polar Media

P.O.J. Scherer

Physikdepartment T38, Technische Universität München, W-8046 Garching, Fed. Rep. of Germany

Abstract. A stochastic theory for photoinduced intramolecular charge separation in polar media is developed. The fluctuation of the solvent tunes the energy difference of the initial and final electronic state whereas internal motion provides a quasicontinuum of accepting modes.

In two recent papers the photoinduced electron transfer process within a porphyrin quinon cyclophane complex has been studied experimentally [1] and theoretically [2] as a model system for the photosynthetic reaction center. The observed rate of the charge separation process depends strongly on the solvent polarity. This dependence could be described analytically using macroscopic models for the dielectric behavior of the solvent. In general, however, such systems may show nonexponential behavior [3] and the solvent dynamics has to be considered explicitly in a microscopic model. This is the aim of the present paper.

We present a microscopic model for the solvent induced transition from the optically excited state (P*) to the charge separated state (CT). Due to its large permanent dipole moment the CT state energy fluctuates strongly in a polar environment. In the usual treatment the electron transfer takes place whenever the energy difference Δ=E(CT)–E(P*) becomes comparable to or smaller than the electronic coupling matrix element V(P*,CT). However, for large molecules the internal modes provide a quasicontinuum of accepting modes. In analogy to the theory of internal conversion processes [4] the rate for the transition P* $\longrightarrow$ CT becomes

$$k = \frac{2\pi}{\hbar} V^2 \bar{\rho} \int_{\Delta+E(P*)}^{\infty} \delta(E(P*)-E)\, dE \;=\; k_i\, \theta(-\Delta)\ .$$

Here $\bar{\rho}$ is a Franck–Condon weighted density of states and the weak energy dependence of the Franck–Condon factors is neglected. Thus in our model the rate switches between the fixed value k_i and zero due to the motion of the solvent. Molecular dynamics calculations give typical switching rates in the order of 1 $psec^{-1}$ for CH_2Cl_2 at 300 K.

These fluctuations are now modeled by a dichotomic process where the energy difference Δ switches between two values $\Delta^+ > 0$ and $\Delta^- < 0$ according to a master equation for the probabilities P(+)=P(Δ>0) and P(–) =P(Δ<0)

$$\dot{P}(+) = -\alpha P(+) + \beta P(-)$$

$$\dot{P}(-) = \alpha P(+) - \beta P(-)\ .$$

This process is characterized by its correlation time $\tau_c = 1/(\alpha+\beta)$ and the stationary probability $P(-)_0 = \alpha/(\alpha+\beta)$.

The combination of solvent fluctuations plus intramolecular transitions is described by the slightly more complicated system of equations

$$\dot{P}(+,P*) = -\alpha P(+,P*) + \beta P(-,P*)$$

$$\dot{P}(-,P*) = \alpha P(+,P*) - (\beta + k_i) P(-,P*)$$

Springer Series in Chemical Physics, Vol. 55 Ultrafast Phenomena VIII
Editors: J.-L. Martin · A. Migus · G.A. Mourou · A.H. Zewail

$$\dot{P}(+,CT) = -(\alpha+k_{cr})\,P(+,CT) + \beta\,P(-,CT)$$

$$\dot{P}(-,CT) = \alpha\,P(+,CT) - (\beta+k_{cr})\,P(-,CT) + k_i\,P(-,P*)$$

where k_{cr} is the rate for charge recombination. This system of equations can be solved analytically. In general the decay of the optically excited state P*, which is observable in absorption and emission experiments, shows two time constants

$$k_{12} = \left[\, \alpha+\beta+k_i \pm \sqrt{(\alpha+\beta)^2 + 2k_i(\beta-\alpha) + k_i^2}\,\right]/2$$

The occupation of the CT state which may be observed in an absorption experiment shows k_{12} plus the bare recombination rate k_{cr}.

The amplitudes of the exponential functions can be evaluated analytically. As the results are rather clumsy we concentrate on some limiting cases which are of interest for comparison with simpler models:

i) If the fluctuations of the solvent are fast compared to the intramolecular transition rates, P* decays monoexponentially with the average rate $k_{av}=k_iP(-)_0$. The occupation of CT becomes

$$P(CT) = \frac{k_{av}}{k_{av}-k_r}\left(e^{-k_rt} - e^{-k_{av}t}\right)$$

ii) In the opposite case of slow fluctuations $\alpha,\beta << k_i,k_{cr}$ we have

$$P(*) = \frac{\alpha}{\alpha+\beta}\,e^{-k_it} + \frac{\beta}{\alpha+\beta}\,e^{-\alpha t}$$

$$P(CT) = \frac{\alpha}{\alpha+\beta}\frac{k_i}{k_i-k_{cr}}\left(e^{-k_{cr}t} - e^{-k_it}\right) + \frac{\beta}{\alpha+\beta}\frac{\alpha}{k_{cr}}\left(e^{-\alpha t} - e^{-k_{cr}t}\right).$$

The decay of P* shows a fast decaying component with the internal rate k_i and a slower time constant α which is characteristic of the solvent. CT shows mainly a population with k_i followed by depopulation with k_{cr}.

The model can be easily extended to involve more intramolecular transitions, for instance the backreaction CT → P*. Even without inclusion of the backreaction in our model several time constants show up. If the solvent motion becomes slow at lower temperatures or for high viscosity the rate is determined by the solvent dependent parameter α. There is experimental evidence for such a transition to solvent controlled electron transfer at low temperatures [5].

Acknowledgement. The author thanks prof. Sighart Fischer for many stimulating discussions. This work was supported by the DFG SFB143C2.

References:

1 W.Frey, R.Klann,F.Laermer,T.Elsaesser,E.Baumann,M.Futscher and H.A. Staab, Chem.Phys.Letters 190 (1992) 567

2 P.O.J. Scherer and Sighart F.Fischer, Chem.Phys.Letters 190 (1992) 574

3 H.Heitele,F.Pöllinger,K.Kremer and M.E.Michel–Beyerle, M.Futscher,G.Voit, J.Weiser and H.A.Staab,Chem.Phys.Letters 188 (1992) 270

4 C.A.Coulson, F.R.S. and K. Zalewski, proceedings of the royal society Vol. 268 A (1962) 437

5 U. Rempel, B. von Maltzan and C. von Borczyskowski, Zeischrift für physikalische Chemie Bd.170 (1991) 107

Femtosecond Proton Transfer in the Electronic Ground State of Vibrationally Hot Molecules

T. Elsaesser, W. Frey, and M.T. Portella

Physik Department E 11, Technische Universität München,
W-8046 Garching, Fed. Rep. of Germany

Abstract. Intramolecular proton transfer in the ground state of benzotriazoles is monitored with femtosecond ultraviolet pulses and occurs with time constants of about 700 fs. The reaction creates vibrationally hot molecules which cool down within 40 ps by collisional interaction with the surrounding solvent.

Photoinduced intramolecular proton transfer represents a chemical reaction that has mainly been studied in electronically excited molecules. For benzothiazole and -triazole compounds, recent femtosecond experiments gave evidence of excited state proton and deuterium transfer on a time scale of 100 to 200 fs [1,2]. In 2-(2'-hydroxy-5'-methylphenyl)-benzotriazole (trade name: TINUVIN P, TIN), the excited state enol→keto reaction (see structures (I) and (II)) is followed by the radiationless relaxation to the ground state of the keto-type molecules, occurring with a time constant of 150 fs. This species eventually undergoes a proton back-transfer to the ground state of the original enol structure. Until now, there was little information on the ground state reaction of TIN and of other compounds performing a closed reaction cycle. In particular, vibrational redistribution processes randomizing the large amount of energy from the initial photoexcitation ($\simeq$ 4 eV) have not been observed. In this paper, we discuss the first study of those processes with a time resolution of 100 fs, directly revealing the subpicosecond proton transfer in the ground state of TIN and the strong vibrational heating of the reaction product.

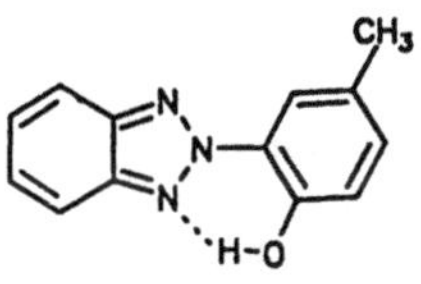

(I) enol

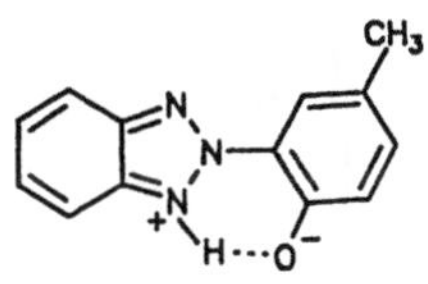

(II) keto-type

In our experiments, we studied TIN dissolved in the nonpolar solvent cyclohexane (concentration c=2 $\times$ 10^{-3} M). Femtosecond pump and probe pulses are derived from the amplified output of a colliding-pulse-modelocked dye laser working at 620 nm. The second harmonic of those pulses serves for excitation of the molecules. Probe pulses tunable between 325 nm and 400 nm were generated by frequency doubling the recompressed output of a femtosecond traveling wave dye laser. Probe light in the wavelength range from 400 to 800 nm was derived from a femtosecond white light continuum. The time resolution of the apparatus has a value of 100 fs.

The three transients in Fig. 1 give direct information on the kinetics of the ground state reaction. After excitation of the enol tautomer at λ_{ex}=310 nm, a strong transient absorption is found between 500 and 600 nm. The transmission change at 570 nm (Fig. 1 a) exhibits a fast rise and a subsequent decay with a time constant of 700±100 fs. The absorption found at delay times longer than 1 ps when the excited enol and keto states

Springer Series in Chemical Physics, Vol. 55 **Ultrafast Phenomena VIII**
Editors: J.-L. Martin · A. Migus · G.A. Mourou · A.H. Zewail

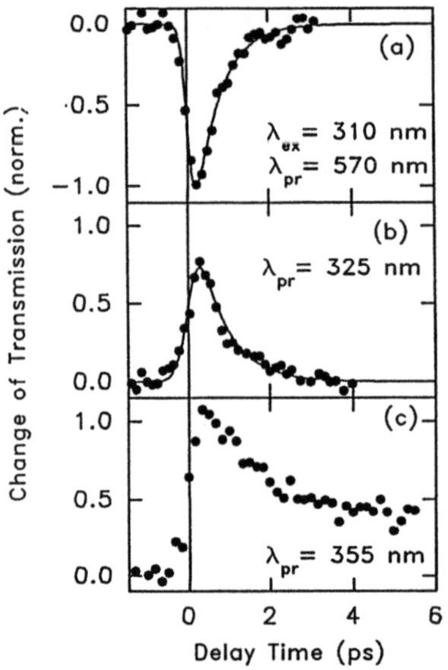

Figure 1 Time dependent transmission changes observed after femtosecond excitation at λ_{ex}=310nm for probe wavelengths λ_{pr} of (a) 570nm, (b) 325nm, and (c) 355nm. The normalized change of transmission $(T - T_0)/T_0$ is plotted up to delay times of 6 ps (points; T, T_0: transmission with and without excitation). The induced absorption at 570nm is due to the ground state of keto-TIN, whereas the bleaching of the enol ground state is observed at 325 and 355nm. The solid lines are calculated from a numerical simulation.

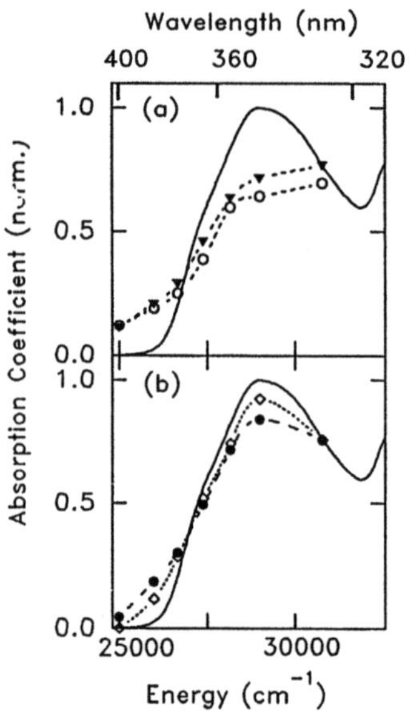

Figure 2 Transient absorption spectra of enol-TIN after femtosecond excitation (λ_{ex} = 310 nm). (a) The normalized absorption coefficient is plotted vs photon energy for delay times of 2 ps (open circles), and 5 ps (triangles). The solid line gives the steady state $S_0 \rightarrow S_1$ band. (b) Transient spectra for delay times of 10 ps (points) and 15 ps (diamonds).

are depopulated completely, originates from the **keto-type** ground state. The transients in Figs. 1 b,c were recorded in the wavelength range of the **enol** $S_0 \rightarrow S_1$ absorption band which is plotted in Fig. 2 (solid lines). At 325 nm and 355 nm, we observe an increase of transmission due to the depletion of the enol ground state. At 325 nm, this bleaching rises with the excitation pulse and decays by repopulation of the ground state with a time constant of 700±100 fs (solid line), i.e. identical to the decay time of the keto ground state absorption (Fig. 1 a). Thus keto-TIN directly transforms to the enol geometry by the back-transfer of the proton on a subpicosecond time scale. The fast proton motion points to a potential with a low or even negligible barrier along the reaction coordinate. The subpicosecond time scale of the proton transfer corresponds to vibrational frequencies in the range of 50 to 200 cm^{-1}, i.e. low-frequency vibrations of large amplitude are obviously involved in the reaction step.

At a probe wavelength of 355 nm (Fig. 1 c), the fast bleaching is followed by a slower change of transmission that decays within 40 ps. At longer wavelengths λ_{pr} between 375 and 400 nm, we observe an induced absorption exhibiting both the fast and the slow kinetics. To visualize the overall behavior of TIN, we derived transient absorption spectra from the time resolved data. The results presented in Fig. 2 demonstrate a well-pronounced reshaping of the enol $S_0 \rightarrow S_1$ absorption band with a strong enhancement of the low-energy absorption tail and a reduced extinction around the maximum. This finding gives direct evidence of vibrational redistribution processes. The fast reaction cycle of TIN transfers the energy supplied by the exciting ultraviolet photon to the vibrational manifold of the molecules, finally creating enol tautomers with a highly excited vibrational system. Our data show that the intramolecular randomization of energy occurs on a subpicosecond time scale similar to that of the reactive steps. The increase of absorption at low frequencies is caused by vibrational excess populations of modes coupling to the electronic transition. The decrease of absorption at higher frequencies is due to a reduction of population density at the bottom of the vibrational distribution. On the picosecond time scale, the transient populations are governed by the time evolution of the vibrational temperature T_{vib} of the molecules. Starting from a high initial value of $T_{vib} \simeq 1200K$, the hot molecules cool down within several tens of picoseconds by collisional interaction with the solvent [4]. As a result, the transient absorption bands directly monitored in our experiments evolve towards the steady state spectrum. The ultrafast reaction cycle and the rapid vibrational cooling account for the high efficiency of TIN as a photostabilizer of polymers.

References

[1] W. Frey, F. Laermer, T. Elsaesser, J. Phys. Chem. **95**, 10391 (1991)

[2] M. Wiechmann, H. Port, W. Frey, F. Laermer, T. Elsaesser, J. Phys. Chem. **95**, 1918 (1991)

[3] W. Frey, T. Elsaesser, Chem. Phys. Lett. **189**, 565 (1992)

[4] U. Sukowski, A. Seilmeier, T. Elsaesser, S.F. Fischer, J. Chem. Phys. **93**, 4094 (1990)

Solvent Effects on the Fast Proton Transfer of 3-Hydroxyflavone

B.J. Schwarz, L.A. Peteanu, and C.B. Harris

Department of Chemistry, University of California, Berkeley, CA 94720, USA

Abstract: The fast proton transfer (PT) in 3-Hydroxyflavone has been measured in different solvents. In non-polar environments, the PT takes 240 fsec. In methanol solution, however, PT occurs in less than 125 fsec. A simple wavepacket picture is used to rationalize these results.

One of the most widely studied systems exhibiting proton transfer (PT) has been 3-hydroxyflavone (3HF). Proton transfer in 3HF was first proposed [1] to explain the observation of two distinct fluorescence bands in the blue and green spectral regions following the absorption of a UV photon. The blue, or near UV fluorescence is thought to arise from the normal form of the molecule, whereas emission from the tautomer is the more strongly Stokes shifted green fluorescence (Fig. 1). Careful studies have shown that the blue, normal emission appears only when 3HF is in hydrogen bonding solvents or in the presence

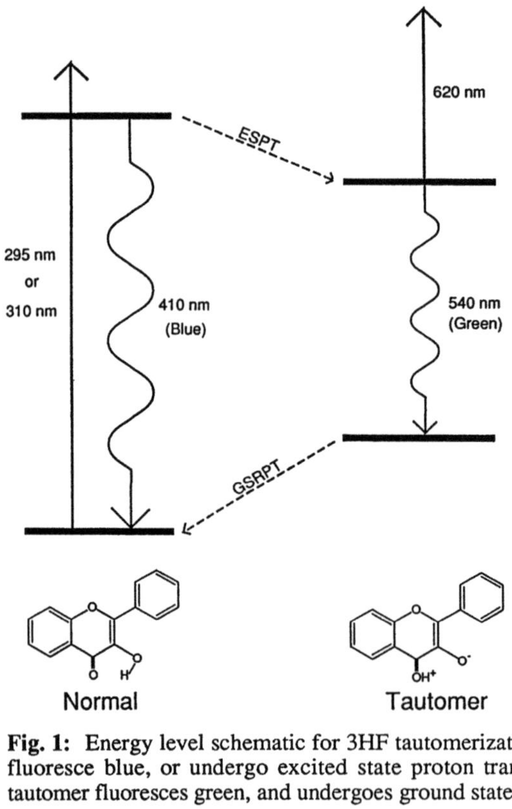

Fig. 1: Energy level schematic for 3HF tautomerization. After excitation, normal 3HF can fluoresce blue, or undergo excited state proton transfer (ESPT). After ESPT, the 3HF tautomer fluoresces green, and undergoes ground state reverse proton transfer (GSRPT).

Springer Series in Chemical Physics, Vol. 55 Ultrafast Phenomena VIII
Editors: J.-L. Martin · A. Migus · G.A. Mourou · A.H. Zewail

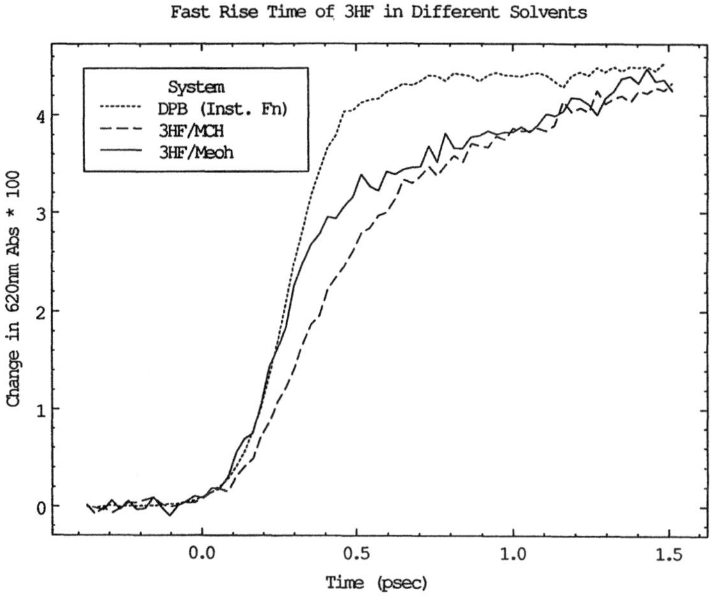

Fig. 2: Fast 620 nm absorption rise of 3HF in different solvents.

of hydrogen bonding impurities.[2] In non-polar solvents, even at low temperatures, only the green tautomer emission is observed.

Kinetic studies have observed a bi-exponential rise in the tautomer fluorescence, indicating two distinct mechanisms for PT.[3] The slower component has been well-investigated, and has been ascribed to thermally activated desolvation of 3HF-solvent hydrogen bonded complexes. The fast emission component, however, has not been previously time resolved, even with picosecond time resolution in Ar matrices at 10 K.[4] Recent picosecond work indicates that the tautomer formation can be monitored by 620 nm transient absorption (Fig. 1) as well as by the green emission.[5] *In this paper, we use femtosecond transient absorption to obtain the first measurement of the fast PT process in 3HF, and we examine the effects of the solvent environment on the fast PT.*

Femtosecond transient absorption experiments were performed on 3HF using a copper vapor laser amplified cavity-dumped CPM ring laser which produces 100 fsec 620 nm pulses with an energy of $3x10^{-6}$ J at an 8 kHz repetition rate. PT was initiated in 3HF with the doubled laser light at 310 nm, and the tautomer formation probed by 620 nm absorption with the laser fundamental. Details of the experimental apparatus and sample preparation have been published elsewhere[5].

Transient absorption data for 3HF in methanol (MeOH) and methylcyclohexane (MCH) solutions are shown in Fig. 2. The data have been scaled to have the same absorbance at a delay of 1.5 psec and superimposed on the instrument function (620 nm transient absorption of diphenylbutadiene). The difference in the fast rise of the 3HF transients is clear: the rise in MeOH is indistinguishable from the 125 fsec instrument function, while the rise in MCH can be fit to the convolution of the instrument function with a 240 fsec exponential rise. The effects of hydrogen-bonding impurities in the non-polar MCH solution were also investigating by slowly titrating the MCH solution with MeOH; the fast rise slowly decreased from 240 fsec the instrument resolution as the MeOH concentration increased, indicative of a strong 3HF-MeOH interaction.

The measured 3HF PT time of 240 fsec is in good agreement with the measurement of a 170 fsec PT rate for the similar molecule 2-(2'-hydroxyphenyl)benzothiazole (HBT).[6] In

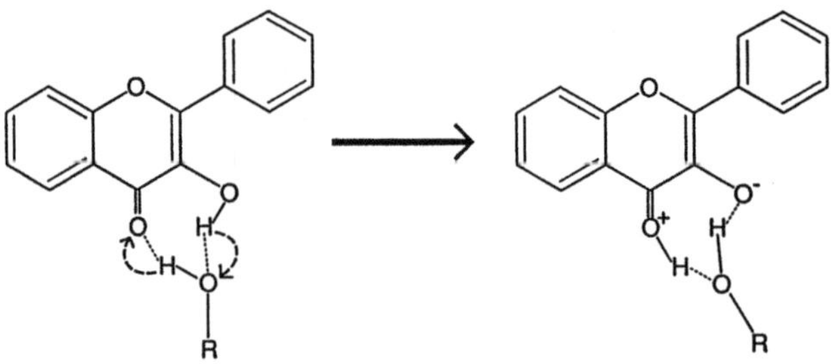

Fig. 3: Proposed dual proton transfer mechanism for 3HF-MeOH monosolvated complex.

addition, experiments on deuterated 3HF in MCH and in MeOD show no isotope effect; the proton transfer times remain unchanged upon deuterium substitution. This is also in agreement with recent work done on deuterated HBT.[7]

We picture the inherent tautomerization time in MCH as controlled by a series of coupled, intramolecular motions which lead to eventual dissociation in the O-H coordinate, bringing the wavepacket into the product region of the excited state surface after 240 fsec. The time scale of the proton transfer (240 fsec is ~10 O-H stretching periods), and the lack of deuterium isotope effect indicate that O-H stretching motion is not important to the proton transfer coordinate. This is supported by resonance raman data which shows no resonance enhancement of any modes which shift frequency upon deuterium substitution.[8] The important coordinate for tautomerization in free 3HF is probably a carbonyl bending motion--proton transfer occurs after the carbonyl group wags close enough to the hydroxy group for the proton to exchange sites. Wavepacket motion along this complex tautomerization coordinate takes 240 fsec.

In MeOH, however, 3HF can form a cyclically hydrogen-bonded complex with the solvent, leading to a new intermolecular PT pathway. (Higher order 3HF-solvent aggregates do not undergo PT[4].) This new pathway involves a dual proton transfer mechanism, as illustrated in Fig. 3. The tightly bound 3HF-MeOH complex starts the protons much closer to their final tautomeric positions than in the weakly intramolecularly H-bonded free 3HF, implying that the wavepacket has less distance to travel on the intermolecular PT pathway. In addition, the H-bonds in the complex are likely to be stiffer, and therefore, of higher frequency than the intramolecular H-bond of the unsolvated species. This implies that the wavepacket is composed of higher-frequency motions in the monosolvate than in the free molecule, leading to more rapid tautomerization. If the PT reaction coordinate is intermolecular in MeOH solution, then either or both of these ideas can be used to rationalize the more rapid evolution of the wavepacket on the excited state surface of the aggregate than of the isolated species.

In summary, we have performed the first measurements of the fast PT in 3HF. The PT rate in MCH solution is 240 fsec; that in MeOH solution was faster than the experimental resolution. The tautomerization is described by the evolution of vibrational motions coupled to the electronic excitation: the PT time corresponds roughly to the propagation time of the wavepacket comprised of the modes displaced. The faster PT rate in MeOH is explained by consideration of a new, intermolecular reaction pathway involving dual proton transfer in a 3HF-solvent complex.

References:

[1] Sengupta, P. K.; Kasha, M. *Chem. Phys. Lett.* **1979**, *68*, 382.
[2] McMorrow, D.; Kasha, M. *J. Phys. Chem.*, **1984**, *88*, 2235.

[3] Strandjord, A. J. G.; Barbara, P. F. *J. Phys. Chem.*, **1985**, *89*, 2355.
[4] Brucker, G. A.; Kelley, D. F. *J. Phys. Chem.*, **1987**, *91*, 2856.
[5] Schwartz, B. J.; Peteanu, L. A.; Harris, C. B. *J. Phys. Chem.*, **1992**, *96*, 3591.
[6] Laermer, F.; Elsaesser, T.; Kaiser, W. *Chem. Phys. Lett.*, **1988**, *148*, 119.
[7] Frey, W.; Laermer, F.; Elsaesser, T. *J. Phys. Chem.*, **1991**, *95*, 10395.
[8] Peteanu, L. A.; Mathies, R. *Unpublished Data.*

Time-Resolved Charge Separation in Acceptor-Substituted Anthrylpolyenes

H. Port[1], *G. Quapil*[1], *H.C. Wolf*[1], *F. Effenberger*[2], *C.-P. Niesert*[2], *R. Buhleier*[3,*], *Z. Gogolak*[3], *and J. Kuhl*[3]

[1]3. Physikalisches Institut Universität Stuttgart, W-7000 Stuttgart 80, Fed. Rep. of Germany
[2]Institut für Organische Chemie, Universität Stuttgart, W-7000 Stuttgart, Fed. Rep. of Germany
[3]Max-Planck-Institut für Festkörperforschung, W-7000 Stuttgart, Fed. Rep. of Germany
*On leave from JATE Univ., Dept. Opt. & Quantum Electron., Szeged, Hungary

Abstract Anthrylpolyenes of different chain lengths terminally substituted with pyridinium electron-acceptor groups have been investigated by time-resolved fluorescence and absorption techniques ($t > 100fs$).

Previously [1] we have demonstrated that in linear polyenes, with two different substituents on both ends, an intramolecular transfer of energy is possible. The present work concerns polyene systems (3) with anthracene as an electron-donor and pyridinium as an acceptor with respect to a photoinduced intramolecular electron transfer. Particular attention has been paid to the role of the polyene chain at different chain lengths. Reference is made to anthryl-polyenes with pyridine substituent (4) and to compounds without polyene chain ($n=0$) (1) and (2). Full account of the work will be given in [2]. This paper concentrates on time-resolved spectra and transients of fluorescence and absorption at room temperature measured in solution (time resolution up to 100 fs).

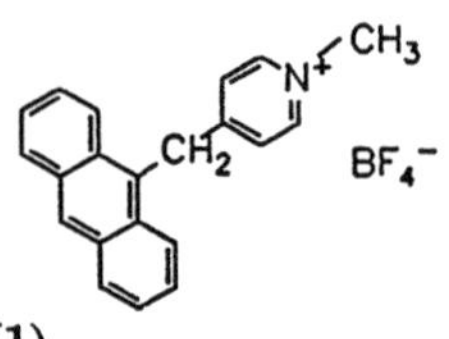

(1)

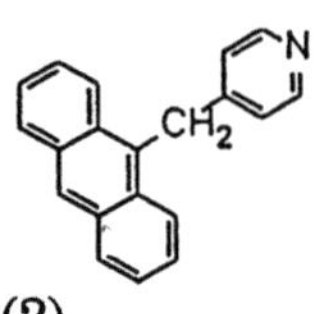

(2)

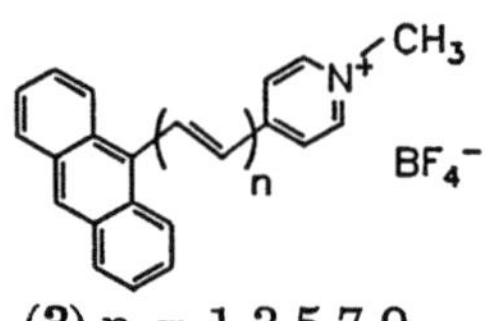

(3) n = 1,3,5,7,9

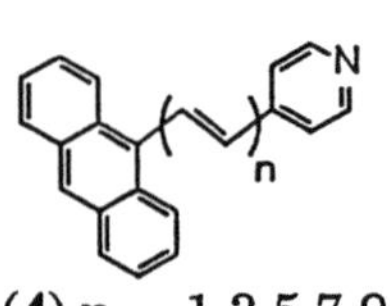

(4) n = 1,3,5,7,9

Comparison of compounds (1) and (2):

Compound (1) clearly reveals photoinduced intramolecular charge separation. Evidence is given by the transient absorption spectrum, which was found (Fig. 1) to be comparable with the absorption spectrum of the anthracene cation. It is consistent with fluorescence spectra and lifetime measurements. Upon anthracene-selective absorption no anthracene-type emission could be detected. Instead, a strongly red-shifted fluorescence component was observed. The transients of fluorescence and absorption yielded the same decay time (1.8 ns). – In contrast, for compound (2) the spectroscopic properties remain anthracene-like both in transient absorption and emission (fluorescence lifetime 7.8 ns)

Chain length dependencies of compounds (3) and (4):

Anthracene-like fluorescence emission is observed only for (4) at $n=1$. Otherwise, single broad fluorescence bands are found which are strongly red-shifted (by

Springer Series in Chemical Physics, Vol. 55 Ultrafast Phenomena VIII
Editors: J.-L. Martin · A. Migus · G.A. Mourou · A.H. Zewail

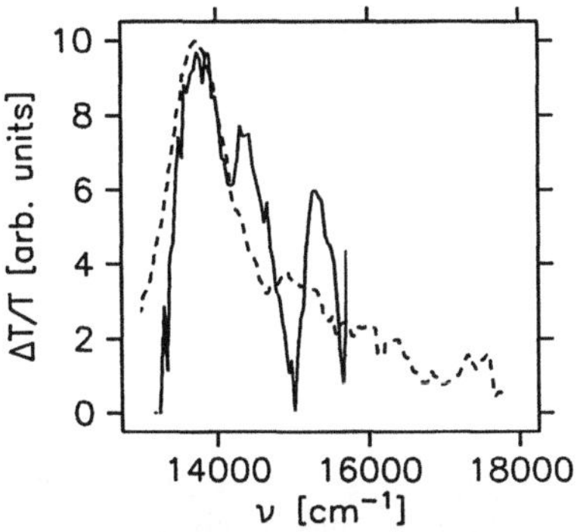

Fig. 1: Transient absorption spectra
_____ (1) in acetonitrile
_ _ _ anthracene-cation (from [3])

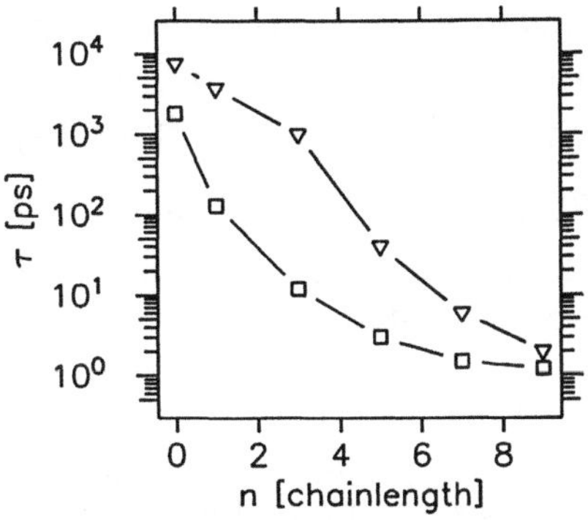

Fig. 2: Excited state decay times τ(n)
(3) in acetonitrile □
(4) in n-hexane ▽

different amounts for (3) and (4)) with increasing chain length [2]. Concomitantly, the lifetimes are drastically reduced with increasing chain length n (Fig. 2), up to n=9 by about four orders of magnitude. The τ values have been deduced from fluorescence decay and transient absorption measurements. The relative change of τ is dependent on the acceptor. For compound (3) a drastic change occurs already at small numbers n but levels off at large n. For compound (4) τ decreases slowly at small n but fast approaches the values of (3) at larger n. Both drastic red-shift and life-time shortening can be taken as indicative for charge separation in the excited state. Clear evidence, however, has been provided for compound (3) at n=5 from the transient absorption spectra. These spectra closely ressemble the absorption spectra of the polyene cation reported in the literature with maximum at about 15300 cm^{-1} very much unlike the anthracene cation in fig. 1.

From the experimental results we come to the conclusion that for short chains n of compound (4) the emitting state is anthracene-like. The more pronounced red-shift of the fluorescence of (3) is due to an effective charge separation between anthracene and pyridinium in the lowest excited state. For molecules with longer chain length the relaxation dynamics is governed by the polyene moiety for both compounds (3) and (4). From the observation of the polyene cation in transient absorption we conclude that charge separation occurs in the excited state between polyene and the acceptor substituent.

Chainlength and substitution effects for similar molecular systems have been reported in the literature [4,5]. The present results are important considering substituted polyenes as possible "molecular wires" for intramolecular electron-transfer.

This work is supported by the Deutsche Forschungsgemeinschaft (SFB 329)

References:

[1] St. Maier, H. Port, H.C. Wolf, F. Effenberger, H. Schlosser; Synth. Metals 29 (1989) E517

[2] G. Quapil, Dissertation, Universität Stuttgart (1992); paper to be published

[3] J.M. Masnovi, J.C. Huffman, J.K. Kochi, E.F. Hilinski, P.M. Rentzepis; Chem. Phys. Lett. 106 (1984) 20

[4] M.R. Wasielewski, D.G. Johnson, W.A. Svec, K.M. Kersey, D.E. Cragg, D.W. Minsek; in 'Photochemical Energy Conversion', J.R. Norris Jr., D. Meisel, Elsevier (1988) p.135ff.

[5] A. Slama-Schwok, M. Blanchard-Desce, J.M. Lehn; J.Phys.Chem. 94 (1990) 3894

Vibrationally Unrelaxed cis-Stilbene Photoproducts Examined Through Two-Color UV Pump-Probe Anti-Stokes Raman Spectroscopy

D.L. Phillips, J.-M. Rodier, and A.B. Myers

Department of Chemistry, University of Rochester, Rochester, NY 14627, USA

Abstract. We report picosecond uv antistokes resonance Raman spectra of vibrationally unrelaxed ground state trans-stilbene produced following photoexcitation of cis-stilbene, as well as solvent effects on the quantum yield for photocyclization of cis-stilbene.

The photoisomerization of cis-stilbene is of particular interest as a test for theories of solvent effects on reactions occurring on barrierless or nearly barrierless potential surfaces. Recently the dynamics of photoexcited cis-stilbene in solution have been studied by several groups through subpicosecond absorption and fluorescence methods.[1-5] While these techniques allow detailed kinetic studies to be performed, they provide relatively little information about structure. We are carrying out complementary studies which utilize pump-probe antistokes resonance Raman spectroscopy to probe the structural and vibrational relaxation of the ground-state photoproducts.

Figure 1 summarizes the experiment. A uv pump pulse excites a flowing sample of 1.3 mM cis-stilbene in cyclohexane. Due to the short excited state lifetime, the ground state photoproducts are formed on a time scale fast compared with vibrational cooling. A delayed uv probe excites antistokes Raman scattering from the ground state trans-stilbene photoproduct, and the spectra are detected with a single spectrograph and intensified diode array. The pump pulses (295 nm, 2-3 μJ) are obtained by frequency doubling the output of a modelocked Nd:YLF-pumped dye laser amplified at 500 Hz in a 3-stage dye chain pumped by a regenerative amplifier. The probe (278 nm, 0.5-0.7 μJ) is obtained by mixing the amplified dye fundamental with part of the doubled regen output. The pump-probe cross-correlation is estimated to be 10 ps based on integrating the rising edge of the photobleaching signal from trans-stilbene in cyclohexane.

Figure 2 presents the antistokes resonance Raman spectra obtained at five different delay times. For comparison, Figure 3 shows the *Stokes* spectra of both the cis and trans isomers at room temperature. Both isomers may contribute to the transient spectrum, but we expect the trans isomer to dominate based on its twofold higher extinction coefficient at the probe wavelength; the *Stokes* Raman cross section for the double bond stretching region (1500-1650 cm^{-1}) at 278 nm is a factor of five higher for room temperature trans-stilbene than for cis. That the transient features arise mainly from the trans isomer is supported by the observation that the 560 cm^{-1} cis band, present even at negative times due to thermal population, is not

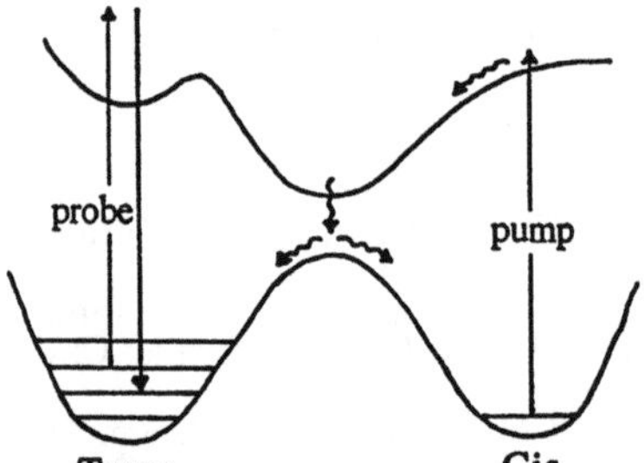

Figure 1. Potential energy surface (PES) along the reaction coordinate for the isomerization of stilbene. The straight arrows relate the experiment to the PES.

Springer Series in Chemical Physics, Vol. 55 Ultrafast Phenomena VIII
Editors: J.-L. Martin · A. Migus · G.A. Mourou · A.H. Zewail

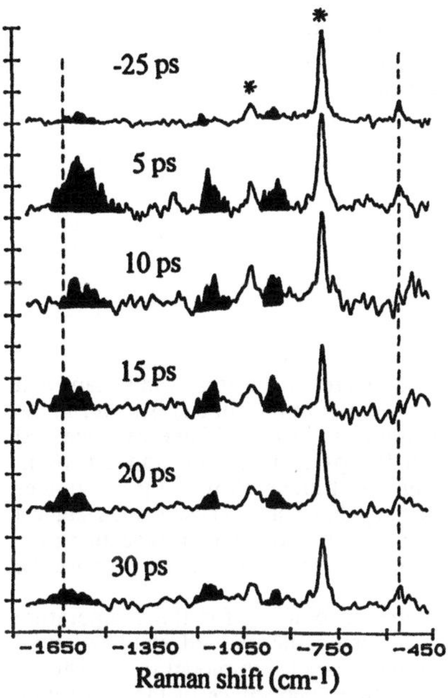

Figure 2. Antistokes spectra (pump-probe minus pump only) at the indicated delay times. Asterisks mark solvent lines. Vertical lines are at 1638 (cold trans) and 560 cm^{-1} (cis).

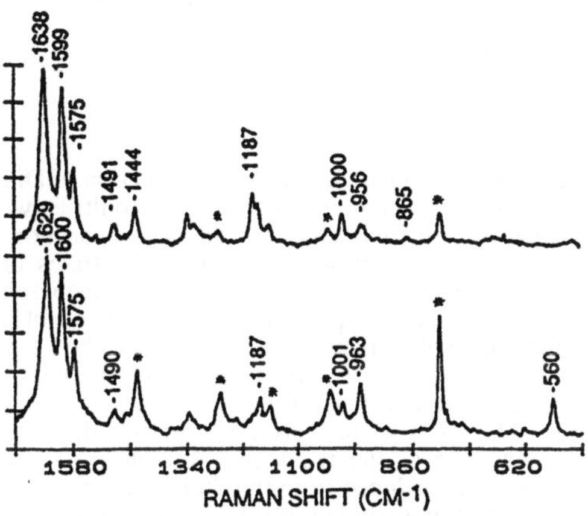

Figure 3. Stokes resonance Raman spectra of trans- (top) and cis-stilbene (bottom).

significantly enhanced in the positive time spectra. The minor ring-closure product, 4a,4b-dihydrophenanthrene (DHP), does not absorb strongly at the probe wavelength and its Raman spectrum, distinct from that of the stilbene isomers,[6] is not observed.

The C=C stretches lose intensity and shift to higher frequencies with increasing delay time, consistent with vibrational cooling. Stokes overtone spectra of the trans isomer show that the $1\rightarrow0$ transition in the central C=C stretch is 9 cm^{-1} higher than the $2\rightarrow1$ and 19 cm^{-1} higher than the $3\rightarrow2$. Based on these frequencies, we estimate that most of the antistokes intensity arises from the v=4 or 5 levels at 5 ps, v=3 at 10 ps, and v=1 at longer times. This might suggest non-Boltzmann vibrational populations, since even at the maximum temperature of 1400 K (calculated assuming all of the pump energy is distributed over the 72 vibrational modes of the electronic ground state of trans-stilbene) a Boltzmann distribution would give about four times more population in v=1 than in all higher levels of the C=C stretch. However, the frequency shifts may also have contributions from anharmonic couplings between vibrational modes and/or distortion from the planar trans geometry. The C-C single bond stretching and C-H rocking region (1100-1200 cm^{-1}) and the hydrogen out-of-plane wagging region (950-980 cm^{-1}) also exhibit significant antistokes intensity and changes as a function of delay time.

Figure 4 plots the time dependence of the antistokes intensity in the double bond stretching region. These data were obtained by accumulating signal at each delay setting for only a few minutes before moving to the next delay and then repeating the process a number of times, thereby minimizing variations in laser power and beam overlap. These very preliminary data indicate a time scale for vibrational cooling of 10-20 ps, in rough agreement with the value of 14 ps obtained by Sension *et al.* from pump-probe fluorescence measurements.[2]

We have also measured the formation (through kilosecond time scale absorption spectra) of the DHP and trans-stilbene products following excitation of cis-stilbene in several solvents.

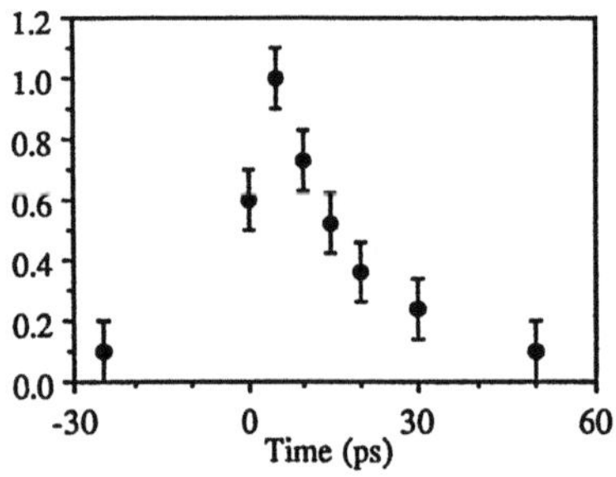

Figure 4. Antistokes intensity in 1500-1650 cm^{-1} region as a function of pump-probe delay.

The apparent quantum yield for production of DHP is two to three times greater in hexane and cyclohexane than in methanol and acetonitrile, while the yield of trans-stilbene is lower in the nonpolar solvents. Recent transient absorption measurements of cis-stilbene by Rice and Baronavski [5] show excited state lifetimes of about 0.5 ps in methanol and 1.8 ps in cyclohexane. Also, our resonance Raman spectra of cis-stilbene show slightly greater intensities in out-of-plane vibrations in methanol relative to cyclohexane. These results suggest that polar solvents increase the slope of the vertical excited state potential surface along the isomerization coordinate, resulting in faster cis-trans isomerization and a reduced yield for the competing cyclization reaction.

Quantitative interpretation of our data will require accounting for the temperature dependence of the resonance Raman cross sections by using the excited state potential parameters deduced from the Stokes resonance Raman intensities [7] to model the antistokes spectra. We also plan to measure and model the temperature dependence of the absorption spectrum. Ongoing work focuses on improving the signal-to-noise ratio of our transient spectra (presently limited by pump-induced background), obtaining reliable spectra in the region below 800 cm^{-1} where torsional vibrations may occur, and using different solvents. These results will enable us to search for non-Boltzmann vibrational populations and/or conformationally distorted transient structures and examine possible solvent effects at short times.

References

1. S. Abrash, S. Repinec and R.M. Hochstrasser, J. Chem. Phys. **93**, 1041 (1990).
2. R.J. Sension, S.T. Repinec and R.M. Hochstrasser, J. Chem. Phys. **93**, 9185 (1990).
3. S.T. Repinec, R.J. Sension, A.Z. Szarka and R.M. Hochstrasser, J. Phys. Chem. **95**, 10380 (1991).
4. D.C. Todd, J.M. Jean, S.J. Rosenthal, A.J. Ruggiero, D. Yang and G.R. Fleming, J. Chem Phys. **93**, 8658 (1990).
5. J. K. Rice and A. P. Baronavski, J. Phys. Chem. **96**, 3359 (1992).
6. J.-M. Rodier, X. Ci and A.B. Myers, Chem. Phys. Lett. **183**, 55 (1991).
7. X. Ci and A.B. Myers, Chem. Phys. Lett. **158**, 263 (1989).

Vibrational Energy Redistribution and Relaxation in the Photoisomerization of cis-Stilbene

R.J. Sension, S.T. Repinec, A.Z. Szarka, and R.M. Hochstrasser

Department of Chemistry, University of Pennsylvania,
231 South 34th Street, Philadelphia, PA 19104, USA

An important goal of condensed matter chemical dynamics is to characterize reactive coordinates and understand how interactions with the solvent influence reaction pathways and energy flow. Isomerization reactions in which a significant structural change occurs without severing chemical bonds present the opportunity to study energy reorganization directly. The cis to trans isomerization of stilbene occurs on a time scale of less than a few hundred femtoseconds [1]. This reaction cannot be slower than intramolecular vibrational energy redistribution (IVR), although the transfer of internal vibrational energy to the surrounding solvent can take much longer [2]. This combination of a fast reaction followed by slow energy transfer to the surroundings permits a spectroscopic study of energy disposal and redistribution occurring during the reaction.

Two types of experiments have been performed to investigate energy disposal in the cis to trans reaction of stilbene. In the first experiment the absorption spectrum of the photoproduct was obtained 6 ps after the excitation of cis (See Fig. 1). This delay was chosen to expose the hot product spectrum free from the cis* absorption [1]. In the second experiment, the time-evolution of the integrated fluorescence yield of photogenerated trans was measured (See Fig. 2). In order to investigate energy partitioning further, the fluorescence experiment was performed with 312 and 250 nm pump pulses and the absorption experiment was performed with 312 and 332 nm pump pulses.

The time evolution of the trans fluorescence yield has been analyzed using a model presented previously [1]. The slow rise observed in this experiment provides a sensitive measure of the temperature of the trans isomer in vibrational modes directly coupled to the trans/cis reaction coordinate owing to the dependence of the fluorescence yield on the excited state barrier crossing rate. When the data are fit to the model of Ref. 1, the temperature of trans in these modes is found to be 650 to 700 K. The result is insensitive to the extra 8000 cm^{-1} (a 25% increase) introduced into the system by using a 250 nm pump.

The 6 ps absorption spectrum has been modeled assuming a variety of population distributions for the Franck-Condon (FC) active vibrational modes. The calculations shown in Fig. 1 used thermal distributions corresponding to 1350 K (the maximum possible temperature) and 725 K, and a nonthermal distribution corresponding to an average energy of 2000 cm^{-1} in the FC active modes between 900 cm^{-1} and 1700 cm^{-1}.

Editors: J.-L. Martin · A. Migus · G.A. Mourou · A.H. Zewail

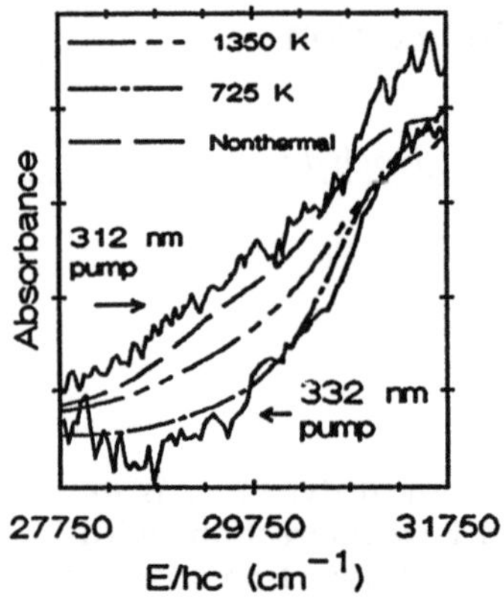

Figure 1. Comparison of 6 ps spectra with calculations.

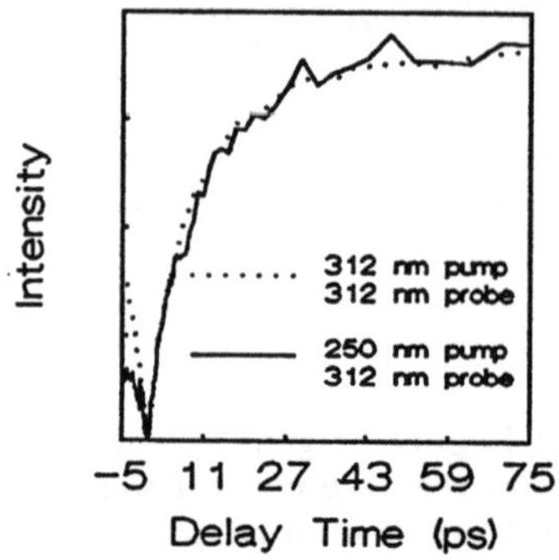

Figure 2. Time evolution of the trans fluorescence yield.

The most straightforward interpretation of the data is that the bulk of the energy placed in the Franck-Condon active modes of *cis*-stilbene in the photoexcitation process is retained in these modes during the reaction. Because the FC active mode of cis and trans are very similar, the effect of this distribution dominates the absorption spectrum.

These experiments demonstrate that IVR is not fast in stilbene. Rather, it occurs on a time scale substantially longer than 6 ps. This is in direct contrast to the fast IVR (<1 ps) generally assumed for large molecules [2]. The nascent trans isomer is also formed hot in the low frequency modes directly coupled to the reaction coordinate. The data and calculations are consistent with the following energy partitioning in the cis to trans reaction: 1.) ca. 10000 cm^{-1} is retained as vibrational excitation in the low frequency modes (this corresponds to 700 K), 2.) the bulk of the vibrational energy placed in the FC modes of *cis*-stilbene remains in these modes and has only spectator status in the reaction, 3.) ca. 20000 cm^{-1} is not accounted for and is assumed to be deposited directly into the solvent on the time scale of the reaction.

1. R.J. Sension, S.T. Repinec, and R.M. Hochstrasser, J. Chem. Phys. **93**, 9185 (1990).
2. T. Elsaesser and W. Kaiser, Annu. Rev. Phys. Chem. **42**, 83 (1991).

Photoisomerization of cis-Stilbene in Compressed Solvents

L. Nikowa[1], D. Schwarzer[1], J. Troe[1], and J. Schroeder[2]

[1]Max-Planck-Institute of Biophysical Chemistry,
Am Faßberg, W-3400 Göttingen, Fed. Rep. of Germany
[2]Institute of Physical Chemistry, University of Göttingen,
Tammannstr. 6, W-3400 Göttingen, Fed. Rep. of Germany

Abstract. The S_1-photoisomerization of cis-stilbene was investigated by femtosecond pump-probe absorption spectroscopy in compressed solvents. The viscosity dependence confirms the existence of two pathways of the reaction. One shows an inverse viscosity dependence and leads to trans-stilbene, the other one shows no viscosity dependence and leads to dihydrophenanthrene (DHP).

1. Introduction

Fast barrier crossing processes have proven to be a powerful probe to investigate various aspects of the influence of the solvent environment on reaction dynamics in solution. In the case of the photoisomerization of trans-stilbene and trans- diphenylbutadiene the solvent influence on the barrier crossing flux could be separated into a "static" interaction, i.e. a solvent-induced modification of the barrier height and a "dynamic" interaction, i.e. a viscous reduction of the flux over the effective barrier [1,2]. The Kramers-Smoluchowski description was shown to hold quantitatively for these molecules. In particular, in the high damping regime the reaction rate k_{nr} decreases as

$$k_{nr} \propto \frac{1}{\eta} \tag{1}$$

where η denotes the solvent viscosity. In order to separate the statical and dynamical effects on the reaction dynamics experiments had to be performed under conditions where the viscosity of the respective solvent was changed by pressurizing it up to several kbar. The isomerization of electronically excited cis-stilbene appeared to be a promising candidate for studies of *low-barrier* processes. After UV-excitation, the molecules disappear from the Franck-Condon region on a ps-time scale. The viscosity dependence of the decay rates of excited cis-stilbene so far was found to be only weak [3,4]. However, viscosity studies were done exclusively by changing the solvent. We went one step further and investigated the pressure and temperature dependence of the decay times in single solvents.

2. Experimental Technique

The photoisomerization of cis-stilbene was studied by femtosecond pump-probe absorption spectroscopy using laser pulses of 100 fs width. After excitation at 307 nm the transient absorption at 614 nm was monitored. High pressure experiments were performed in a stainless steel cell, that was connected to a high pressure pump which generated a constant flow through the sample cell in order to prevent accumulation of photoproducts [5].

3. Results

In agreement with the results of refs. [3] and [4] the observed decay times of the transient signals in n-alkanes at 1 bar show a weak viscosity dependence and range from τ_{nr} = 0.99 ps in n-

Springer Series in Chemical Physics, Vol. 55 Ultrafast Phenomena VIII
Editors: J.-L. Martin · A. Migus · G.A. Mourou · A.H. Zewail

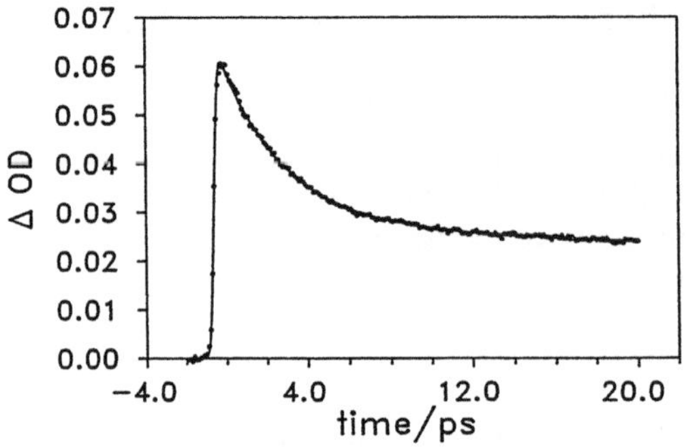

Fig.1. Transient absorption signal of cis-stilbene in n-octane at 3530 bar and 295 K (residual absorption due to incomplete removal of trans-stilbene).

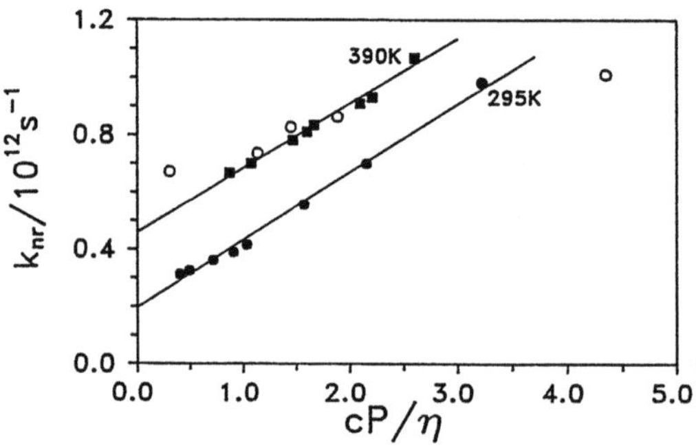

Fig.2. Inverse viscosity dependence of k_{nr} for cis-stilbene in n-hexane at 295 K (•) and 390 K (■). Rate constants obtained in n-alkanes at 1bar and 295 K (○) are plotted for comparison.

pentane to $\tau_{nr} = 1.49$ ps in n-hexadecane. Our high pressure studies, however, indicate a much stronger viscosity dependence of τ_{nr}. Fig. 1 shows the observed transient absorption signal of cis-stilbene in n-octane at 3530 bar and 295 K. The solid line represents the best fit to the data based on a convolution of a Gaussian instrument response function and a biexponential decay with the two decay times $\tau_{nr} = 3.3$ ps and $\tau_2 = 179$ ps. The fast component belongs to the cis-stilbene decay whereas τ_2 results from a trans-stilbene impurity and was used as a fixed parameter [5]. For a single solvent the decay rate constants $k_{nr} = 1/\tau_{nr}$ are found to be inversely proportional to the solvent viscosity (Fig. 2). The intercept for $1/\eta \to 0$ depends markedly on temperature. A general representation of k_{nr} in the form

$$k_{nr} = k_{DHP} + \frac{A}{\eta} \tag{2}$$

with a solvent specific parameter A is possible. A shows no temperatur dependence, whereas k_{DHP} can be represented in Arrhenius form

$$k_{DHP} = 6.3 \cdot 10^{12} \cdot exp\,(-\frac{8.5\ kJmol^{-1}}{RT})s^{-1} \tag{3}$$

4. Discussion

Eqs. (2) and (3) suggest the presence of two components of the cis-stilbene dynamics with quite different responses to the solvent viscosity. These two different chemical pathways are known:

the cis-trans isomerisation to trans-stilbene and the photocyclisation leading to dihydrophenanthrene (DHP). The observed quantum yield $\Phi_{DHP} \approx 0.3$ for the formation of electronically excited DHP [6] corresponds exactly to the result of fig. 2 if the first term of eq. (2) is attributed to DHP formation and is pressure independent over the studied pressure range. The apparent activation energy of k_{DHP} (eq. (3)) is consistent with the value found in earlier studies [7]. We attribute the second term (A/η) of eq. (2) to the pathway transforming electronically excited cis-stilbene to a 90°- conformation. As in previous work [3,4], no sign of a nonexponential decay for this channel was observed. At the same time, this term was found to be strictly inverse proportional to the viscosity. This behavior suggests the presence of a dynamical bottleneck on the multidimensional PES which could be caused by a small barrier ($E_a \leq 40cm^{-1}$) along the reaction path. Such a small barrier would be consistent with our observation that within experimental error there is no activation energy for this process. Assuming that the reaction occurs in the high-damping Smoluchowski limit a rate constant of

$$k_{nr} \approx \frac{2\omega_A}{\beta} \cdot \sqrt{\frac{k_B T}{2\pi I}} \cdot \frac{1}{\gamma_B} \tag{4}$$

can be calculated, were ω_A, β, I, and γ_B are the potential well frequency, the friction, the moment of inertia, and the position of the barrier, respectively [5]. With eq. (4) values for γ_B that range from 3.6° for n-nonane to 14° for n-pentane can be calculated.

References

[1]Ch.Gehrke, J.Schroeder, D.Schwarzer, J.Troe, and F.Voß: J.Chem.Phys. **92**, 4805(1990); Ch.Gehrke, R.Mohrschladt, J.Schroeder, J.Troe, and P.Vöhringer: Chem.Phys. **152**, 45 (1991).

[2]J.Schroeder, D.Schwarzer, J.Troe, and F.Voß: J.Chem.Phys. **93**, 2393(1990).

[3]S.Abrash, S.T.Repinec, and R.M.Hochstrasser: J.Chem.Phys. **93**, 1041(1990).

[4]D.C.Todd, J.M.Jean, S.J.Rosenthal, A.J.Ruggiero, D.Yang, and G.R.Fleming: J.Chem. Phys. **93**, 8658(1990).

[5]L.Nikowa, D.Schwarzer, J.Troe, and J.Schroeder: J.Chem.Phys.(1992), in press.

[6]H.Petek, F.Yoshihara, Y.Fujiwara, Z.Lin, J.H.Penn, and J.H.Frederick: J.Phys.Chem. **94**, 7539(1990).

[7]K.A.Muszkat and E.Fischer: J.Chem.Soc.B, **1967**, 662; Helv.Chim.Acta **54**, 1195(1971).

Ultrafast Torsional Dynamics in Adsorbates: An SSHG Study

M.J.E. Morgenthaler and S.R. Meech

Department of Chemistry, Heriot-Watt University,
Riccarton, Edinburgh EH14 4AS, UK

Abstract. The ground state recovery dynamics of malachite green on silica are studied as a function of coverage, intensity and wavelength. A fast ground state recovery indicates rapid torsional motion and excess energy dissipation in adsorbates. An intensity dependent very long lived bleaching was also observed at high coverage. This is ascribed to multiphoton induced ultrafast dissociation or desorption of dye aggregates.

1. Introduction. Primary photochemical processes in adsorbates are not well understood when compared with our understanding of the solution or gas phases. However the rapid development of surface specific nonlinear optics [1] has made possible the application of picosecond pump probe techniques to the study of ultrafast phenomena in submonolayers of adsorbates [2]. Here we report on the ultrafast ground state recovery (GSR) dynamics of malachite green (MG) adsorbed on silica. The picosecond pump probe method is employed with the ground state population being monitored by probe induced surface second harmonic generation (SSHG). Better than 2 ps time resolution is obtained by pumping a short cavity dye laser with a pulse quenched 10 Hz Nd : YAG laser. MG was chosen for study as its excited state torsional motion, leading to internal conversion and GSR by vibrational cooling, has been extensively studied in solution [3]. Questions which arise concerning the adsorbed state are (i) are the reactive potential surfaces dramatically altered by adsorption? (ii) is vibrational (torsional) cooling in the ground state efficient? (iii) are the dynamics altered by interadsorbate interactions?

2. Experimental. A conventional pump and probe geometry was employed with the probe generated SSHG signal being detected in reflection from the surface. The dynamics recorded were independent of angle of incidence and polarisation of the input beams. Other details are given elsewhere [4]. Traces with 10 ps time resolution were obtained with a short cavity dye laser and four stages of amplification pumped by a 20 ps 532 nm pulse. To provide the 2 ps time resolution required to resolve accurately the fast component a pulse quenching system for the 532 nm pulse was constructed. From the main 532 nm pulse (20 ps, 20 mJ) 400 μJ is split off and focussed via a 500 mm lens into a concentrated solution of rhodamine 6G flowing between two mirrors with >80% reflectance at 560 nm. The concentration of rhodamine was adjusted to give a transmission of ca 5 μJ. The pulse width of the transmitted 532 nm radiation was measured as 6 ps. This was used to pump the short cavity dye laser, the remainder of the 532 nm pulse pumping the amplifier stages. Under these conditions we obtain 2 ps pulses at 638 nm with energies up to 1 mJ.

3. Results and Discussion. Figures 1a and b show the ground state recovery dynamics of MG on silica at saturation coverage with two different pump intensities. Immediately noticeable is the two component character of the recovery. The amplitude of the long component is intensity and coverage dependent. This is discussed below. The fast component, resolved at 14 ± 4 ps in figure 1a, b, was found to be independent of intensity, coverage and pump wavelength. This fast initial ground state recovery is ascribed to excited state torsional motion leading to internal conversion, and its rapid reversal in the ground state. That the recovery is ultrafast suggests that the motion is barrierless in the adsorbed state, as it is in solution. That both the excited state reaction and the cooling of the ground state are rapid is

Springer Series in Chemical Physics, Vol. 55 Ultrafast Phenomena VIII
Editors: J.-L. Martin · A. Migus · G.A. Mourou · A.H. Zewail

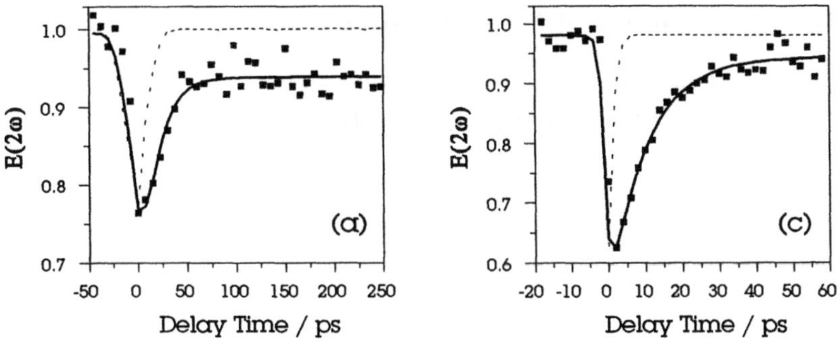

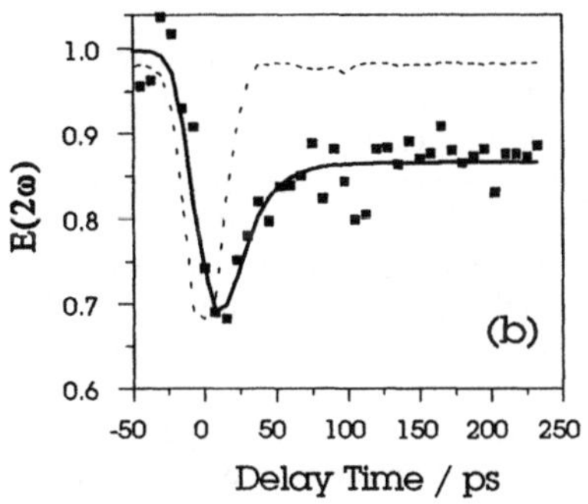

Figure 1. Malachite Green Ground State Recovery at 638nm High Coverage (a) 8mJ.cm^{-2} (b) 32mJ.cm^{-2}. Note intensity dependence. (c) As (a), but 2ps time resolution. All fitted to exponential + background.

an interesting result. This suggests that the >16000 cm^{-1} excess energy immediately after internal conversion is rapidly dissipated. In solution the low frequency tortional motion may effectively couple to solvent translational modes. These are not available in the adsorbed phase. We suggest that the low frequency modes can effectively transfer energy to the high density of states between zero and a few hundred wavenumbers in silica. In an effort to examine separately the excited state and ground state dynamics we searched for non exponential behaviour in the GSR at high coverage with higher time resolution (1c). No hint of biexponential behaviour was observed. This may imply that one of the two processes is much the fastest, or that our signal to noise ratio is insufficient to resolve them.

The long component is effectively suppressed at high coverage by reducing the pump power below a few mJ cm^{-2}. Further, the long component is very weak at any energy if the coverage is reduced to ca 0.3 monolayer. The absorption spectra show an unperturbed adsorbate, compared to solution, at low coverage, but substantial interadsorbate interaction at coverages greater than 0.6 monolayer. The long component is then ascribed to a multiphoton induced disruption of the adsorbate aggregate. Possibly IVR following the population of higher excited states leads to a break up of the aggregate.

Acknowledgement. We are grateful to Dr H. Hermann and Dr D. Klenermann of BP research for collaboration on the pulse quenching system and to SERC for an equipment grant. MJEM thanks the Carnegie Trust for a studentship.

Citations.

(1) Y.R. Shen, Annu. Rev. Phys. Chem., **40** (1989) 327.

(2) E.V. Sitzmann and K.B. Eisenthal, J. Phys. Chem., **92** (1988) 4597; S.R. Meech and K. Yoshihara, Chem. Phys. Lett., **154** (1989) 20; P.Guyot-Sionnest, Phys. Rev. Lett., **64** (1990) 2156.

(3) D. Ben-Amotz and C.B. Harris, J. Chem. Phys., **86** (1987) 4856.

(4) M.J.E. Morgenthaler and S.R. Meech, Submitted.

Barrierless Photochemical Isomerization

U. Åberg, E. Åkesson, I. Fedchenia, and V. Sundström

Department of Physical Chemistry, University of Umea,
S-901 87 Umea, Sweden

1. Introduction

Ground State Recovery experiments with 100 fs pulses have been used to further investigate the photophysics of the barrierless isomerization reaction of the cyanine molecule 1,1'-diethyl-4,4'-cyanine (1144-C). The molecule is characterized by a short excited state lifetime which depends linearly on solvent viscosity and has a negative temperature dependence of the relaxation rate [1]. Another feature which is observed for this molecule and seems to be characteristic for barrierless reactions is a wavelength dependent fluorescence kinetics [2]. As a consequence, GSR kinetics probed in the wavelength region overlapping with the fluorescence band, containing contributions from stimulated emission, result in fast non-exponential decay curves [3].

We also present Monte Carlo simulations of a modified Smulochowski equation as a description of the time evolution of the excited state population in the case of a barrierless reaction [4].

The potential surfaces are a result of the combined action of the steric interactions within the molecule together with the stabilizing double bond resonance energy which is maximum for a planar molecule. In the ground state this results in a conformation with a 40 degree twist angle between the two quinoline rings and an all downhill potential surface in the excited state [5].

The dynamics of a barrierless reaction along a harmonic excited state potential surface, followed by a rapid internal conversion was modelled by Bagchi, Fleming and Oxtoby [4], using a modified Smulochowski equation

$$\frac{\partial P(x,t)}{\partial t} = A\frac{\partial^2 P(x,t)}{\partial x^2} + B\frac{\partial}{\partial x}xP(x,t) - k_{nr}S(x)\,P(x,t)$$

where $A = k_BT/\zeta$ and $B = \omega^2\mu/\zeta$; ω is the frequency of the potential energy surface, μ is the reduced mass of the reactive motion and ζ is the friction coefficient. k_{nr} is the rate constant of the radiationless decay at the center of the sink function, where $S(x_{sink}) = 1$.

The results indicate that the long time decay of the excited state for a barrierless isomerization should be exponential and proportional to viscosity.

2. Reasults and Discussion

In pump and probe experiments with sub-picosecond pulses (100 to 150 fs) a close to linear relationship between long time decay rate and viscosity over the viscosity range 0.3 - 10 cP is observed (fig.1). A frequency $\omega = 5\star10^{12}\ S^{-1}$ was estimated from this viscosity dependence, assuming a one dimensional harmonic potential energy surface. Using this value in the simulations together with the calculated moment of inertia and a friction described by the Stoke-Einstein expression for the case of slip boundary conditions, a long time decay rate in

Springer Series in Chemical Physics, Vol. 55 Ultrafast Phenomena VIII
Editors: J.-L. Martin · A. Migus · G.A. Mourou · A.H. Zewail

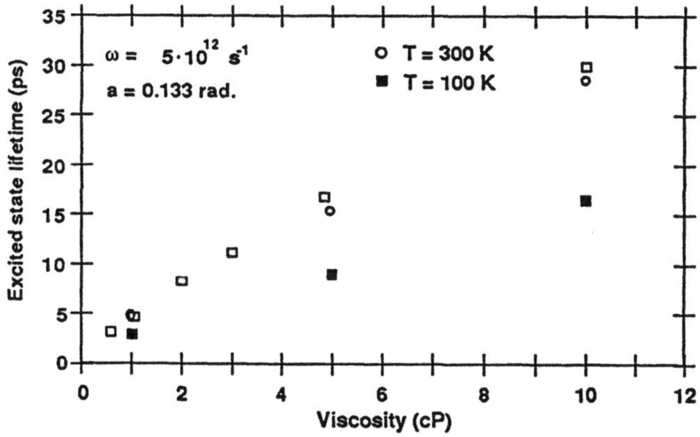

Fig.1 Excited state lifetime obtained from experiments □ and from Monte Carlo simulations ■ o For a Gaussian sink and a sink width of 0.133 rad. , there is a good agreement between experiments and simulations. There is a negative dependence of the excited state lifetime on temperature as observed in experiments.

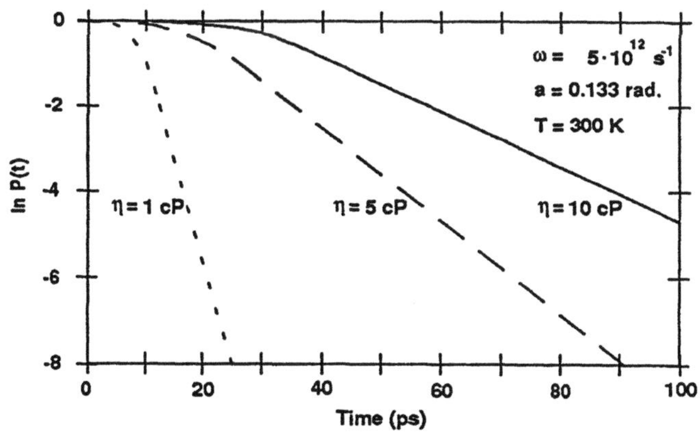

The simulated decay curves are single exponential at long times. At short times there is a substantial delay due to population relaxation down along the potential surface to the sink position. The shape and the width of the sink is determined by the long time exponential part of the decay curves ($w = 5 \cdot 10^{12}$ s^{-1} and $a = 0.133$ rad.).

good agreement with experiments is observed (fig.1). These results are obtained when a 0.133 rad. wide Gaussian sink-function was used. The negative temperature dependence of the excited state lifetime observed for different barrierless isomerizations [1,6] were also observed in the simulations.

For the short time dynamics, experiment and simulations give very different results. From the simulation, a short time delay is observed in the decay curve (fig.2). Depending on viscosity this delay varies from a couple of picoseconds up to 50 ps in the studied viscosity region. This is in sharp contrast with the short time dynamics experimentally observed where no delay is observed. This means that the delay can not be longer than a few hundred fs.

A short time delay in the decay curves is a result of fast population relaxation along the excited state surface down to the position where fast non-radiative relaxation to the ground state occur. Our results indicate that the population reaches the sink in less than a few hundred fs which raises the question wether the initial relaxation process on the excited state surface can be described by a one dimensional potential surface and how the friction in that process is related to the macroscopic viscosity.

3. References

[1] E. Åkesson, H. Bergström, V. Sundström, and T. Gillbro
Chem. Phys. Lett. 126, 385 (1986)

[2] B. Bagchi, U. Åberg and V. Sundström
Chem. Phys. Lett. 162, 227 (1989)

[3] U. Åberg and V. Sundström, Chem. Phys. Lett. 185, 461 (1991)

[4] B. Bagchi, G.R. Fleming and D.W. Oxtoby
J. Chem. Phys. 78, 7375 (1983)

[5] H. Yoshioka and K. Nakatsu, Chem. Phys. Lett. 11, 255 (1971)

[6] V. Sundström and T. Gillbro, J. Chem. Phys. 81, 3463 (1984)

Femtosecond Molecular Dynamics in Liquids

D.A. Wiersma, E.T.J. Nibbering, and K. Duppen

Department of Chemistry, Ultrafast Laser and Spectroscopy Laboratory,
University of Groningen, Nijenborgh 4,
NL-9747 AG Groningen, The Netherlands

Abstract. Ultrafast liquid phase dynamics is probed by femtosecond (stimulated) photon echo experiments. A multimode Brownian oscillator model is used to analyze the data. It is shown that in aprotic solvents a single overdamped Brownian oscillator is capable of simulating solvation; in protic solvents two additional Brownian oscillators are needed to describe the molecular dynamics in the solvation shell. A physical model for the Brownian oscillators is discussed.

1. Introduction

Grasp of liquid phase dynamics is of utmost importance to our understanding of chemistry in solution. A much debated concept is the *cage effect* which holds molecular fragments together after photo dissociation. Collisions and friction between solvent and reactant molecules determine whether or not the fragments have a chance to recombine. The direct role of nuclear solvent degrees of freedom in charge- and proton-transfer reactions is also becoming an important issue [1]. Recently ultrafast optical experiments like femtosecond photon echo [2,3], optical Kerr effect [4] and time-dependent Stokes shift [5] have provided a wealth of information concerning the molecular dynamics during the first few picoseconds after impulsive excitation. While molecular motion affects each of these observables in quite distinct ways the common denominator is the very fast dynamics that occurs upon a change of the equilibrium state. Our understanding of these motions is often based on models which are physically appealing and which have predictive power towards other experiments as well. One such model is the multimode Brownian oscillator (MBO) model that is based on the generalized Langevin equation (GLE). In the MBO model the electronic system is coupled to one or more harmonic Brownian oscillators, which may be linearly displaced upon optical excitation. These oscillators can be of intramolecular nature or may represent intermolecular motions of solvent molecules. This feature of the model assures that not only the effect of the bath on the system (probe molecule) is taken into account, but the reverse effect as well. When the random force in the GLE is assumed to have an infinitely short correlation time the Markovian limit is obtained which allows for an analytical solution of the Langevin equation. Molecular dynamics calculations on model systems can be used to test the underlying assumptions of these models and aid in the physical interpretation of the Brownian oscillators.

In this report we will present some of our recent results of femtosecond (stimulated) photon echo experiments on resorufin in dimethylsulphoxide (DMSO) and pinacyanol in DMSO and ethylene glycol (EG). We will show that the MBO model provides an excellent fit of all optical data in the case of resorufin. The first results on pinacyanol indicate that the solvent dynamics in the case of protic and aprotic solvents is very different. Prior to presenting our data we first discuss briefly the relevant equations that emerge from the MBO theory. Full details can be found, for instance, in a paper by Yan and Mukamel [6].

2. Multimode Brownian oscillator model

The central equation of the MBO theory is the GLE for the correlation function $M_j(t)$ of the jth oscillator j:

$$\ddot{M}_j(t) + \int_0^t dt' \hat{\gamma}_j(t-t')\dot{M}_j(t') + \omega_j^2 M_j(t) = 0 \ . \qquad (1)$$

Springer Series in Chemical Physics, Vol. 55 Ultrafast Phenomena VIII
Editors: J.-L. Martin · A. Migus · G.A. Mourou · A.H. Zewail

In Eq. (1) $\hat{\gamma}_j(t)$ is the non-Markovian friction kernel and ω_j represents the frequency of the oscillator. The fluctuation-dissipation relation connects the friction to the correlation function of the stochastic force:

$$[\bar{n}(\omega) + \tfrac{1}{2}]\hat{\gamma}_j(t) = \langle f_j(t) f_j(0) \rangle \quad , \tag{2}$$

where $\bar{n}(\omega)$ is the occupation number of the oscillator.

If the thermal motions of the bath, which are responsible for the stochastic force, are assumed to be fast compared to the Brownian oscillator dynamics, the Markovian limit of the MBO model is obtained. In this limit Eq. (1) can be solved analytically and one obtains:

$$M_j(t) = \tfrac{1}{2}[\alpha_+^j \exp(-i\alpha_-^j \vartheta_j t) + \alpha_-^j \exp(i\alpha_+^j \vartheta_j t)], \tag{3a}$$

where

$$\alpha_{\pm}^j = 1 \pm i\gamma_j/(2\vartheta_j) \tag{3b}$$

and

$$\vartheta_j = [\omega_j^2 - (\gamma_j/2)^2]^{1/2} \ . \tag{3c}$$

For intramolecular modes we are usually in the regime where $\omega_j > \gamma_j/2$ and the corresponding Brownian oscillator is underdamped. These modes are responsible for the vibrational structure in the optical spectrum of the solute. In the case $\gamma_j >> 2\omega_j$ the oscillator is strongly overdamped and a simple expression for $M_j(t)$ is obtained:

$$M_j(t) = \exp(-\Lambda_j \tau), \tag{4}$$

where Λ_j equals ω_j^2/γ. The strongly overdamped modes (SOMs) are mainly responsible for the breadth of the optical spectrum.

Fried, Bernstein and Mukamel have recently shown [7] that semiclassical microscopic simulations can be nicely mapped onto the MBO model, provided the dynamics of the ground and electronically excited states are similar. For a solute in water they needed six oscillators to mimic the solvent molecular dynamics simulation results. We note, however, that by using oscillators of the type given in Eq.(3a), the non-Markovian characteristics of the bath are projected into the oscillators. This fact should be recalled when a microscopic interpretation of the results is attempted. The decisive advantage of assuming Markovian bath dynamics is that our ignorance about the time-dependent friction does not prevent us from describing the molecular dynamics in terms of a physically attractive model that can be applied to many linear and nonlinear optical experiments.

The line broadening function $g_j(t)$, which determines the linear and nonlinear response functions of the polarization, is related to $M_j(t)$ in the following fashion:

$$g_j(t) = \Delta_j^2 \int_0^t dt_1 \int_0^{t_1} dt_2 M_j(t_2) + i\lambda_j \int_0^t dt_1 M_j(t_1) \ . \tag{5}$$

For a strongly overdamped oscillator, $g_j(t)$ becomes

$$g_j(t) = \frac{\Delta_i^2}{\Lambda_j^2}[\Lambda_j t - 1 + \exp(-\Lambda_j t)] + i\frac{\lambda_i}{\Lambda_j}[1 - \exp(-\Lambda_j t)] \quad , \tag{6}$$

where $\lambda_j \equiv \tfrac{1}{2} d_j^2 \omega_j$; $\Lambda_j \equiv \omega_j^2/\gamma_j$ and $\Delta_j^2 \approx 2k_B T \lambda_j/\hbar$. The last approximate equality assumes the high temperature limit to hold for the strongly overdamped modes.

When the imaginary part in Eq. (6) is ignored we recover the result obtained in the stochastic model, in which solvation is not accounted for [8]. This model is adequate to describe the results of two-pulse photon echo experiments in which only coherent superposition states of the system play a role. However, for any experiment in which populations of system states are important, solvation dynamics has to be taken into account. Examples of such experiments are the detection of time-resolved fluorescence or pump-probe spectroscopy. To study solvation dynamics at very short time scales, stimulated photon echo experiments are pre-eminently suited. Both superposition states and populations determine the outcome of an experiment, so the interplay between fluctuations and solvation in the liquid can be studied, while the time resolution is limited by the pulse duration only.

3. Liquid state optical dynamics

For the system resorufin in DMSO we showed that the two-pulse photon echo decay and optical absorption spectrum can be fitted well using the stochastic modulation model [3]. However, as pointed out above, the stochastic modulation model fails to describe the Stokes shift in emission. In the MBO model the fluorescence lineshape function can be expressed as the Laplace-Fourier transform of the line broadening function:

$$\sigma_F(\omega)=\pi^{-1}\mathrm{Re}\left[\int_0^{\infty} dt\ \exp[i(\omega-\omega_0)]\exp[\sum_j -g_j^*(t)]\right] , \quad (7)$$

where j sums over all relevant Brownian modes, and $g_j(t)$ is defined by Eq. (5). The absorption line shape is described by the same formula with $g_j^*(t)$ replaced by $g_j(t)$.

Figure 1 shows the origin region of the absorption and emission spectra of resorufin in DMSO at room temperature together with the calculated spectra using one solvent Brownian oscillator with the *same* Δ and Λ parameters that were used to fit the decay of the photon echo in the stochastic model [3]! We note that the MBO model predicts that the Stokes shift occurs on the time scale of $1/\Lambda$, which for this system is 37 fs. Such a fast time scale for solvation is in agreement with recent measurements by the Fleming group [5]. In Figure 2 the expected time dependent Stokes shift is plotted for resorufin in DMSO. In addition to the agreement between calculated and measured Stokes shift, the results from nonlinear chirped scattering experiments on the same system are also in excellent agreement with optical dynamics induced by a single overdamped oscillator [9]. The conclusion is that a single overdamped Brownian oscillator yields an excellent description of the solvent dynamics in DMSO.

We now wish to address the question of what the physical relevance is of this solvent oscillator. In order to do so we rely on the results of recent molecular dynamics (MD) calculations [10]. From these MD simulations the conclusion was drawn that the initial fast solvent dynamics, following an abrupt change in charge or dipole moment of the solute and occurring on a time scale of tens of femtoseconds, is due to inertial effects. This dynamics is due to small amplitude rotational and librational motions of molecules in their instantaneous cages. The rotational motions of different molecules in the first solvation shell were shown to be only weakly coupled during this time period. However, it is the combined effect of all cage molecules that quickly relaxes the excited state. On basis of these calculations we therefore

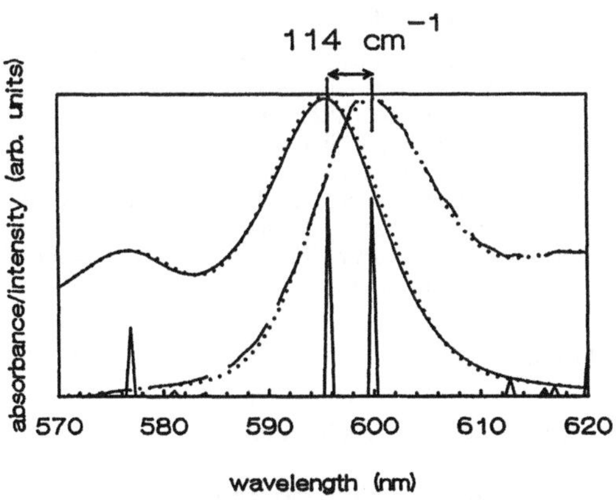

Fig. 1. Room temperature absorption (solid) and emission (dashed) spectra of resorufin in DMSO compared to simulations based on the Brownian oscillator model. Parameter values are Δ=41 THz, Λ=27 THz and λ=23 THz. This latter value is in the high temperature limit related to Δ as discussed in the text.

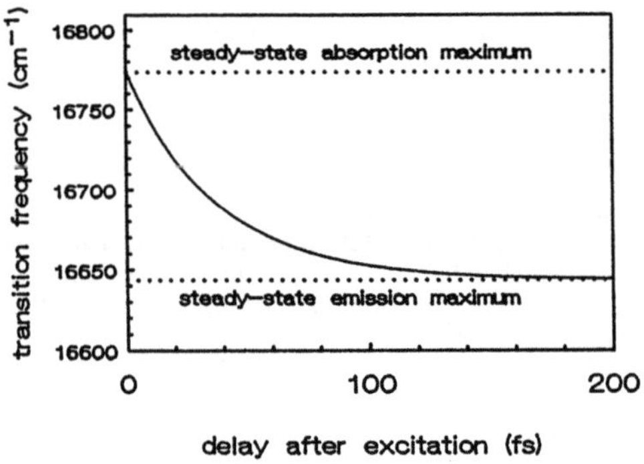

Fig. 2. Plot of the time dependent Stokes shift calculated using the Brownian oscillator model with the same parameters as in Fig. 1.

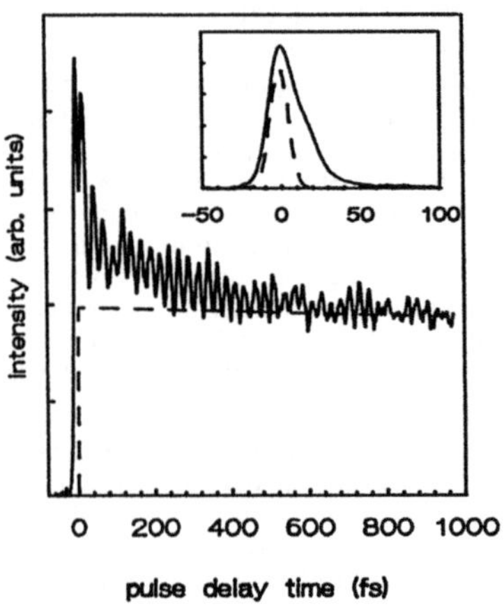

Fig. 3. Stimulated photon echo signal obtained for the system pinacyanol in ethylene glycol. The delay between the first two excitation pulses is 10 fs. The insert shows the two–pulse photon echo decay (solid line) and the autocorrelation trace (dashed curve) of the laser pulses.

suggest that our overdamped Brownian oscillator describes the *collective* effect of the inertial motions of individual DMSO molecules in the first solvation shell. While the Brownian oscillator is strongly overdamped the underlying single molecule rotational motions could be underdamped as suggested by the analysis of optical Kerr effect measurements [11].

As mentioned before, the stimulated photon echo is an excellent probe for solvation dynamics on ultrafast time scales. Figure 3 shows some very recent results obtained from stimulated photon echo experiments on the system pinacyanol in ethylene glycol. In this protic solvent the echo decay extends over a couple of hundred femtoseconds with oscillatory structure superimposed on it. The very fast beating is due to an underdamped molecular mode. The initial fast and slower decays must be due to different overdamped solvent oscillators. The observed low–frequency modulation may be due to a solvent librational mode. To analyze this complex echo decay we need an expression for the stimulated photon echo (3PSE). For the impulsively–excited 3PSE one calculates in the MBO model for the echo intensity [6]:

$$I_{3PSE}(\tau,T) \propto \int_0^\infty |[R_1(t,T,\tau)+R_2(t,T,\tau)]|^2 dt \quad . \tag{8}$$

The explicit forms of $R_1(t_3,t_2,t_1)$ and $R_2(t_3,t_2,t_1)$ in the MBO model are:

$$R_1(t_3,t_2,t_1) \equiv \exp[-g^*(t_3)-g^*(t_1)+g(t_2)-g(t_2+t_3)-g^*(t_1+t_2)+g^*(t_1+t_2+t_3)] \quad , \tag{9a}$$

$$R_2(t_3,t_2,t_1) \equiv \exp[-g(t_3)-g^*(t_1)+g^*(t_2)-g^*(t_2+t_3)-g^*(t_1+t_2)+g^*(t_1+t_2+t_3)] \quad , \tag{9b}$$

with $g(t) \equiv \sum_j g_j(t)$ given by Eq. (5).

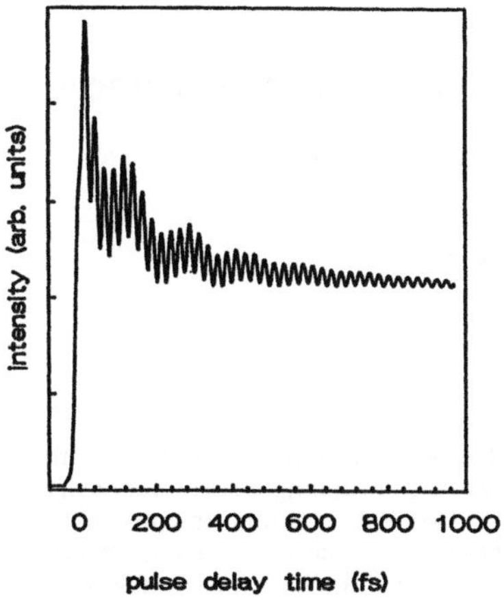

Fig. 4. Calculated decay of the stimulated photon echo (solid) in the multimode Brownian oscillator model using four different oscillators. For details of this simulation consult the text.

To simulate the 3PSE echo decay for pinacyanol in EG we first determined from the absorption spectrum an intramolecular high-frequency mode to fix the fast beating pattern; then the ultrafast (tens of fs) and the slower (few hundred fs) components in the echo decay were generated by varying the parameters Δ and Λ of two oscillators; and finally we included a third solvent mode to simulate a slow modulation in the signal. Our best result at this time is shown in Fig. 4. While the fit is, as yet, far from satisfactory, it seems that we have captured the gross features of the bath dynamics. Of the three solvent Brownian oscillators two are strongly overdamped, while the third one is an underdamped mode with a frequency of about 220 cm^{-1}. We tentatively assign these oscillators to inertial, diffusive and librational solvent molecule motion, respectively. The librational and diffusive components in the dynamics may be due to hydrogen bonding because they are not significant in aprotic solvents. Further analysis of these echo decays, however, is necessary to substantiate these conjectures. In closing we note that our results show that the 3PSE is a powerful tool for the study of solvent dynamics. "Supposing that ultimately femtosecond lasers will be as ubiquitous as 9 V batteries" [12], we expect a bright future for the fs photon echo as a probe of liquid phase dynamics.

References

1. P.F. Barbara, G.C. Walker, and T.P. Smith, Science **256**, 975 (1992).
2. J.-Y. Bigot, M.T. Portella, R.W. Schoenlein, C.J. Bardeen, A. Migus and C.V. Shank, Phys. Rev. Lett. **66**, 1138 (1991).
3. E.T.J. Nibbering, D.A. Wiersma, and K. Duppen, Phys. Rev. Lett. **66**, 2464 (1991).
4. D. McMorrow, W.T. Lotshaw, and G.A. Kenney-Wallace, IEEE J. Qu. El. **QE-24**, 443 (1988).
5. S.J. Rosenthal, X. Xie, M. Du, and G. R. Fleming, J. Chem. Phys. **95**, 4715 (1991).
6. Y.J. Yan and S. Mukamel, J. Chem. Phys. **89**, 5160 (1988).
7. L.E. Fried, N. Bernstein, and S. Mukamel, Phys. Rev. Lett. **68**, 1842 (1992)
8. R. Kubo, Adv. Chem. Phys. **15**, 101 (1969).
9. E.T.J. Nibbering, D.A. Wiersma and K. Duppen, This Conference.
10. M. Maroncelli, J. Chem. Phys. **94**, 2084 (1991).
11. M. Cho, S.J. Rosenthal, N.F. Scherer, L.D. Ziegler, and G.R. Fleming, J. Chem. Phys. **96**, 5033 (1992).
12. W.H. Knox, Optics & Photonics News **3**, 10 (1992).

Femtosecond Solvent Dynamics Studied by Time-Resolved Fluorescence and Transient Birefringence

S.J. Rosenthal, N.F. Scherer, M. Cho, X. Xie, M.E. Schmidt, and G.R. Fleming

Department of Chemistry, The University of Chicago, 5735 So. Ellis Ave., Chicago, IL 60637, USA

Femtosecond solvent dynamics in acetonitrile and methanol have been studied by both time resolved fluorescence and transient birefringence experiments. Introducing a spectral density of Brownian solvent oscillators allows comparison of the two experiments and reveals the elementary liquid motions responsible for the ultrafast dynamics. We report the first observation of a large inertial contribution to the solvent response in methanol and confirm the existence of a large inertial contribution in acetonitrile.

1. Introduction

Ultrafast dynamic solvent properties are recognized to play an important role in controlling the rate of charge transfer reactions [1] Dynamic solvation has been studied by time dependent fluorescence Stokes shift measurements of dipolar chromophores. However, only one measurement with sufficient time resolution to reveal a large contribution arising from inertial (as opposed to diffusive) solvent motion has been reported [2]. The solvent motions giving rise to the solvation response should also contribute to the collective nuclear response in transient birefringence measurements of the neat liquid [3]. To explore the connection between the two experiments and further characterize ultrafast dynamic solvation we have performed both time resolved fluorescence and transient birefringence experiments in acetonitrile and methanol.
By employing the fluorescence upconversion technique we have determined the solvation time correlation function,

$$C(t) = \frac{\nu(t) - \nu(\infty)}{\nu(0) - \nu(\infty)}$$

using coumarin 153 as a probe with ≈ 75fs time resolution. Transient birefringence measurements were carried out in an optical heterodyne configuration with ≈ 20fs resolution [4].

2. Results and Discussion

The solvent response function was constructed from the individual fluorescence decays in the usual fashion [2] and the result for acetonitrile is shown in figure 1a. This result is quite similar to that obtained using the probe molecule LDS-750 [2]. The fast initial relaxation accounts for 80% of the solvent relaxation and occurs in 200fs (≈90fs 1/e time) while the tail decays roughly exponentially with a time constant of ≈ 1ps. In order to connect this result to the transient birefringence measurements we assume a model spectral density of Brownian solvent oscillators [3]. The solvation response is used to determine the frequency distribution and oscillator damping constant representing the acetonitrile dynamics (inset 1a). The same parameters are used to simulate the nuclear contribution to the transient birefringence. The simulated and experimental result are in good agreement as seen in figure 1b.

The solvation response for methanol is shown in Figure 2a. 57% of the response arises from a 75fs decay component, 43% from a 580 fs component. While molecular dynamic simulations have predicted an ultrafast contribution to the solvent response for methanol [5] this is the first experimental observation. The nuclear response for methanol is shown in 2b and the resulting spectral density which is used to simulate the solvation response (2a dashed line) is shown in the inset of 2a. The birefringence measurement also reveals a 700 cm^{-1} component

Springer Series in Chemical Physics, Vol. 55 Ultrafast Phenomena VIII
Editors: J.-L. Martin · A. Migus · G.A. Mourou · A.H. Zewail

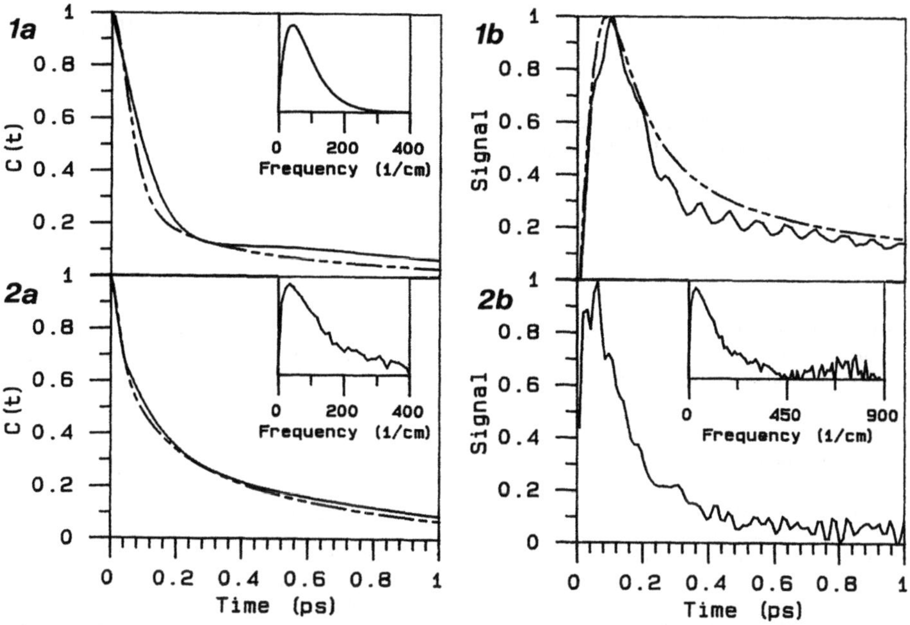

Figure 1a. Solvation response for Cu153 in acetonitrile. Dashed line: simulated response using assumed spectral density shown in the inset. Damping constant = 14 ps^{-1} Figure 1b. Nuclear response obtained from birefringence experiment. Dashed line: Simulated nuclear response using the same spectral density as in 1a. Figure 2a. Solvation response for Cu153 in methanol. Dashed line: simulation using the spectral density (inset) determined from the nuclear response. Damping constant = 75 ps^{-1}. Figure 2b. Deconvoluted nuclear response for methanol. The oscillations in the response correspond to the 700 cm^{-1} hindered O-H rotation and 225 cm^{-1} molecular rotation indicated in the spectral density directly obtained from the data inset.

corresponding to hindered O-H rotation. It is likely the solvation measurements do not have sufficient time resolution to see this component.

This work was supported by a grant from NSF. We thank Mike Armas, Colin Seaton, Ph.D., and Art Swygard of Coherent for the use of the Mira Titanium Sapphire laser used in the solvation experiments.

3. References.

1. K.Tominaga, G.C.Walker, W. Jarzeba and P.F. Barbara, J. Phys. Chem. 95 (1991) 10475; D.A. Zichi, G.Ciccotti, J.T. Hynes and M. Ferrario, J. Phys. Chem. 93 (1989) 6261
2. S.J. Rosenthal, X. Xie, M. Du and G.R. Fleming, J. Chem. Phys. 95 (1991) 4715
3. M.Cho, S. Rosenthal, N.F. Scherer, L.D. Ziegler, and G.R. Fleming, J. Chem. Phys., 96 (1992) 5033; D. McMorrow and W.T. Lotshaw, J. Phys. Chem., 95, (1991) 10395.
4. N. F. Scherer, L. D. Ziegler, and G. R. Fleming, J. Chem. Phys., 96 (1992) 5444.
5. T. Fonseca and B.M. Ladanyi, J. Phys. Chem. 95 (1991) 2116; K. Ando and S. Kato, J. Chem. Phys. 95 (1991) 5966.

Adiabatic and Nonadiabatic Effects in Solvation Dynamics

E. Neria and A. Nitzan

Sackler Faculty of Science, School of Chemistry,
Tel Aviv University, Tel Aviv 69978, Israel

Abstract. Adiabatic solvation dynamics of an ion in a polar solvent and salt solution and the nonadiabatic solvation of the hydrated electron are investigated using Molecular Dynamics simulations.

1. Introduction

In recent years much effort has been devoted to understand charge solvation dynamics. The development of short laser pulses techniques have lead to new results on solvation in polar solvents[1-5] and electrolyte solutions[6-8]. This development was accompanied by new theoretical treatments of solvation dynamics[1,9] and molecular dynamics (MD) simulations[1,10-12].

The issues considered in this work are three (1) The role of solvent motion in the solvation process. (2) The influence of added salt on the dynamics. (3) The possibility of non adiabatic transitions during the solvation process .

2. Classical Ion Solvation

In this section we address the first issue, the role of the solvent motion in the solvation process. More specifically we try to distinguish between the role of translations and rotations, to study the response of different solvation shells and to clarify the importance of the initial inertial relaxation. For this purpose we perform classical MD simulations of ion solvation. The simulations are performed using a simple model of a polar solvent (Stockmayer Liquid) characterized by spherical particles with point dipoles and Lennard Jones interaction. The parameters of the solvent were chosen to approximately simulate CH_3Cl. However the mass,M, and the moment of inertia,I, were modified in order to study the separate effects of translational and rotational degrees of freedom. A detailed description of the model can be found in [10].

The normalized solvation function defined by $S(t)=\dfrac{\Delta E_{solv}(t)-\Delta E_{solv}(\infty)}{\Delta E_{solv}(0)-\Delta E_{solv}(\infty)}$ is used to describe the solvation dynamics. $\Delta E_{solv}(t)$ is the change in the average solvation energy after step function change in the solute charge at t=0. In Fig. 1 we show the results for four model solvents. All except of the rotationless($I=\infty$) one (which relaxes exponentially) exhibit a bimodal relaxation. An initial gaussian component which is responsible for 80% of the solvation energy and a residual exponential relaxation. The initial gaussian relaxation comes from the inertial motion of the solvent which is expected to be important for short times. This is especially true for this structureless fluid with low rotational friction. Indeed this initial relaxation is slower for larger moment of inertia. The translational motion speeds up the initial inertial decay. A fit of the initial inertial relaxation to the function $\exp(-(t/\tau)^2)$ gives the values τ=170 fs for the original CH_3Cl parameters (M=50 amu) and 210 fs for the translationless case ($M=\infty$). Analysis of the contribution from three solvation shells[10] show that the

Springer Series in Chemical Physics, Vol. 55 Ultrafast Phenomena VIII
Editors: J.-L. Martin · A. Migus · G.A. Mourou · A.H. Zewail

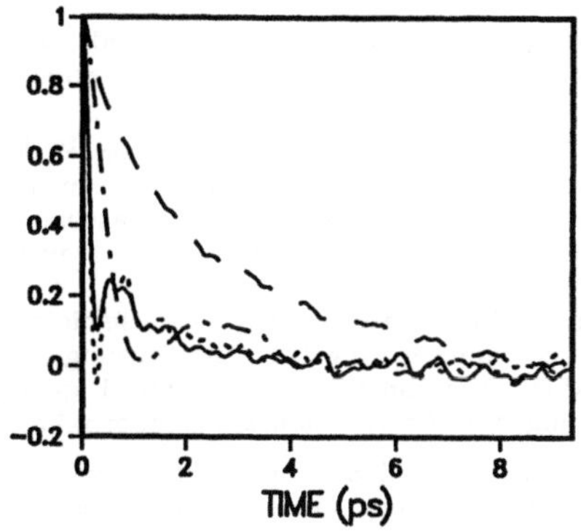

Fig.1 The solvation function for four values of M and I. Solid line: original CH_3Cl parameters; dashed-dotted line: same parameters, except that I was increased by a factor of 13 ; dotted line: frozen translations (M=∞); dashed line: frozen rotations (I=∞).

initial inertial decay exist for all three shells. However the response is faster for the inner solvation shells. The first solvation shell also exhibits a pronounced slow relaxation component which is weaker or absent in the outer shells. This may be related to the Onsager "inverse snowball" picture or to residual energy flow from the first shell. The first shell dominates the dynamics and contributes ~65% of the total solvation energy.

3. Solvation in an Electrolyte Solution

We have also studied the influence of added salt on the solvation dynamics. We have checked this influence on the initial relaxation of the solvent and have also studied the direct contribution of the salt itself. The solvent is the stockmayer fluid and the salt is composed of equal number of oppositely charged particles with Coulombic and Lennard Jones interaction. In a computer simulation we can easily distinguish between the solvent and the salt contributions to the solvation energy. Analysis of the solvent contribution at three different salt concentrations: infinite dilution, 1M and 2.5M shows that the inertial relaxation component is nearly insensitive to the salt concentration. The direct contribution of the salt to the solvation energy was studied at 1M salt concentration. This contribution, which is ~30% of the total solvation energy, displays a bimodal behaviour. An initial gaussian relaxation which is responsible for ~ 20% of the solvation energy and an exponential component. The exponential relaxation time is 28.5 ps. This should be compared with the prediction of the Debye Falkenhagen theory of ionic atmosphere relaxation [13] which predict a relaxation time of ~120ps for this solvent. The relaxation is dominated by the first solvation shell which contributes about 80% of the total ionic contribution. It should be emphasized that the simulation cannot probe the considerably longer times observed in recent experiments[6,7].

4. Hydrated Electron Solvation

Experimental studies of electron solvation in water suggest the following kinetic model for the hydrated electron formation $e_{quasifree} \xrightarrow{k_1} e_{wet} \xrightarrow{k_2} e_{solvated}$ $e_{quasifree}$ is the initially ejected electron and e_{wet} is an intermediate state. Migus et al[4a] have given an estimate of k_1^{-1}=110 fs and k_2^{-1}=240 fs. Long et al[5a] has given different estimates, k_1^{-1}=180±40 fs and k_2^{-1}=540±50 fs. Adiabatic simulations[11] of the solvated electron starting from a weakly localized state in a pre existing trap show very similar results to the solvation of a classical ion. A fast inertial relaxation (~20-30 fs) which is responsible for about 80% of the total adiabatic solvation energy. The absorption peak of the solvated electron shifts continuously during the simulated solvation process. These observations are in marked contradiction to the experimental observations.

One possible explanation for the observed dynamics is to assume e_{wet} to be an excited state of the solvated electron and k_2 to be the non adiabatic transition rate from this excited state to the ground state. It is reasonable to assume that this excited state corresponds to the three closely lying P-like states which are the lowest excited states of the hydrated electron.

We calculated this "P"⟶"S" transition rate using a semiclassical evaluation of the golden rule[12]. The calculation involves performing mixed quantum/semiclassical MD simulations in which the electron is described quantally and the water molecules semiclassically using the frozen gaussians approximation. The calculation at T=300K gives $k_2^{-1} \simeq 220$ fs for H_2O and 800 fs for D_2O (with error margins estimated at ~30%). The result for H_2O is very close to the experimental values cited above. However the isotope effect is much larger then found experimentally (35% by Long et al[5b] and very small effect of ~10% by Gauduel et al[4b]). The calculated non adiabatic transition times for this model are much larger then the adiabatic solvation times and it may therefore dominate the rate for the overall process. The disagreement between the computed and observed isotope effect makes however this assignment inconclusive at present.

5. Summary and Conclusions

Ion solvation in the Stockmayer liquid model is dominated by the initial inertial motion and by the response of the first solvation shell. Conventional dielectric response theories cannot account for the main initial component of the relaxation. The addition of salt has nearly no effect on the solvent dynamics. The direct ionic contribution is dominated by a slow exponential relaxation which comes mainly form the first solvation layer.

The proposed nonadiabatic mechanism of hydrated electron solvation was investigated by calculating the golden rule expression for the transition rate. The calculated transition time in H_2O is close to the experimental results but the calculated isotope effect is considerably larger than that observed.

References

1. M. Maroncelli ,J. Mol. Liq. (in press) and references therein.
2. S. J. Rosenthal, X. Xie, M. Du and G. R. Fleming, J. Chem. Phys. **94**, 4715(1991)
3. W. Jarzeba, G. C. Walker, A. E. Johnson and P. F. Barbara, Chem. Phys. **152**, 57 (1991)
4. (a) A. Migus, Y. Gauduel, J. L. Martin and A. Antonetti, Phys. Rev. Lett. **58**,1529 (1987) (b) Y. Gauduel, S. Pommeret, A. Migus and A. Antonetti, J. Phys. Chem. **93**, 3880 (1991)
5. (a) F. H. Long and K. B. Eisenthal, Phys. Rev. Lett. **64**, 1469 (1990) (b) F. H. Long, H. Lu and K. B. Eisenthal, Chem. Phys. Lett. **160**, 46 (1989)
6. V. Ittah and D. Huppert ,Chem. Phys. Lett. **173**, 496 (1990); in Perspectives in Photosynthesis, edited by J. Jortner and B. Pullman (Kluwer, Dordrecht, 1990)
7. C. F. Chapman and M. Maroncelli ,J. Phys. Chem. **95**, 9095 (1991)
8. Y. Gauduel, S. Pommeret, A. Migus, N. Yamada and A. Antonetti, J. Am. Chem. Soc. **112**, 2925 (1990)
9. B. Bagchi ,Ann. Rev. Phys. Chem. **40** ,115 (1989) and references therein
10. E. Neria and A. Nitzan ,J. Chem. Phys. **96**, 5433 (1992)
11. (a) R. N. Barnett, U. Landman and A. Nitzan ,J. Chem. Phys. **90**, 4413 (1989) (b) P. J. Rossky and J. Schnitker, J. Phys. Chem. **92**, 4277 (1988)
12. E. Neria, A. Nitzan,R.N. Barnett and U. Landman ,Phys. Rev. Lett. **67**, 1011 (1991)
13. P. Debye and H. Falkenhagen, Physik Z. 29, 121,401 (1928)

Excited-State Processes of 7-Azaindole

M. Négrerie[2], *F. Gai*[1], *J.-C. Lambry*[2], *J.-L. Martin*[2], *and J.W. Petrich*[1,*]

[1]Department of Chemistry, Iowa State University, Ames, Iowa 50011, USA
[2]Laboratoire d'Optique Appliquée, Ecole Polytechnique-ENSTA, INSERM U275, F-91128 Palaiseau Cedex, France
*Corresponding author

7-azaindole is the chromophoric moiety of the nonnatural amino acid 7-azatryptophan. We have proposed the use of 7-azatryptophan as an alternative to tryptophan as an optical probe of protein structure and dynamics [1]. This preference arises because 7-azatryptophan in water has a single exponential fluorescence decay over most of the pH range, when emission is collected over the entire band, whereas the naturally occurring amino acid, tryptophan, does not [2]. Furthermore, the absorption and emission spectra of 7-azatryptophan are distinguishable from those of tryptophan, and 7-azatryptophan can be incorporated into synthetic peptides and bacterial protein. We performed picosecond and femtosecond pump-probe absorption measurements to investigate in detail the non radiative decay pathways and excited-state dynamics of 7-azaindole in order to obtain a more complete characterization of the 7-azaindole chromophore.

Using 120-fs pulses, a 0.5 ps transient absorbance was detected for 7-azaindole in methanol at 400 nm (Fig. 1a). While this transient could be interpreted as an excited-state tautomerization [3], such an interpretation is unlikely since an identical transient is observed for the methylated analog, N_1-methyl-7-azaindole. The 0.5 ps component may be attributed to the dynamic solvation of an excited state. Such a time scale for solvation by methanol has been observed for other probe molecules [4]. Alternatively, this rapid component may arise from the rapid relaxation of a higher-lying excited state.

Picosecond experiments at probe wavelengths of 392 nm for 7-azaindole in water at 23°C (Fig. 1b) indicate the instantaneous appearance of a species that undergoes an ~ 70-ps decay. A transient of this duration is consistent with data obtained from time-correlated single-photon counting experiments. Measurement of the fluorescence decay at wavelengths across the emission band up to ~ 400 nm reveals a small contribution of a 40-100-ps decay component. The decay component is matched by a corresponding rise at wavelengths ≥ 400 nm. We observe similar behavior in the emission of 7-azatryptophan. The fast decay component displayed in Fig. 1b most likely arises from the absorption of the excited state singlet of a "normal" species in water, which is depleted by tautomerization. Our assignment for the rapid component in water is supported by the absence of analogous transients in the time-resolved absorption or fluorescence of 7-azaindole derivatives methylated at the N_1 or N_7 positions so that they cannot tautomerize. The temperature dependence of the short component in water yields an Arrhenius activation energy of 2.7 ± 1.8 kcal/mol, which is consistent with data for alcohols [5,6] that indicate that large-amplitude solvent motion is required for tautomerization.

The 950-ps rise time for the transient absorbance at 392 nm of 7AI in water (Fig. 1b) agrees, within experimental error, with the 915-ps fluorescence decay of 7AI in water [5]. There is evidence [7] for triplet formation in 7-azaindole, and we tentatively assign the rising absorption at 392 nm to a triplet species. Bent and Hayon observe triplet absorption for indole and tryptophan at similar wavelengths [8]. A strong transient absorbance is detected at 650 nm (Fig. 1c). This transient does not relax on a 800-ps time scale and can be detected with our picosecond apparatus from 400 to 720 nm. Such a spectrum is consistent with those for solvated electrons, e^-_{aq} [8]. The observation that the electron scavenger, KNO_3, transiently quenches this species for 7-azatryptophan confirms this assignment (lower curve in Fig. 1c).

In conclusion, excited-state tautomerization and intersystem crossing are observed for the 7-azaindole chromophore. Photoionization is a significant non-radiative decay channel in 7-azaindole and 7-azatryptophan. The solvated electron appears within 1 ps and not with a rise time corresponding to the decay of the emitting state (780 ps for 7-azatryptophan). We find, nevertheless, that ionization is monophotonic: plots of ΔA_{max} vs. the logarithm of the pump intensity yield a straight line with a slope of 1.2 ± 0.3. This result suggests that ionization arises

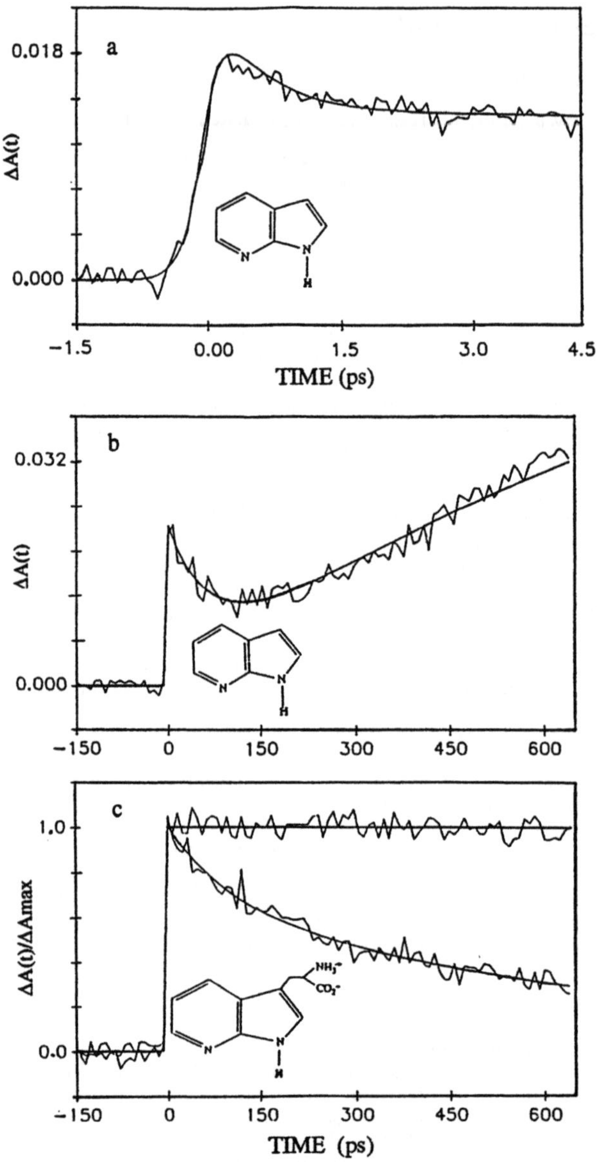

Fig. 1 : **a**, 7-azaindole in methanol, probe at 400 nm. **b**, 7-azaindole in water, probe at 392 nm. **c**, 7-azatryptophan in water, probe at 650 nm; lower curve, 7- azatryptophan + KNO_3.

from a higher-lying excited state, which is not the fluorescent state. We were able to identify this state by performing low temperature fluorescence excitation anisotropy measurements. We found as did Valeur and Weber for indole [9], that the absorption spectrum for 7-azaindole could be decomposed into ^1L_a and ^1L_b components. We propose that the upper ^1L_b state gives rise to the solvated electron in 7-azaindole. We also observe that the fluorescence quantum yield of 7-azaindole is excitation wavelength dependent. Consequently, the methods traditionally used [9] to decompose fluorescence excitation spectra into ^1L_a and ^1L_b absorption spectra are subject to modification. The existence of ^1L_a and ^1L_b states in 7-azaindole renders the assignment of the 0.5-ps component in methanol to the decay of another excited state a plausible alternative to dynamic solvation. Further work is required to distinguish the origin of this rapid component.

This work was facilitated by a NATO travel grant. J.W.P. is an Office of Naval Research Young Investigator.

1. M. Négrerie, S.M. Bellefeuille, S. Whitham, J.W. Petrich and R.W. Thornburg, J. Am. Chem. Soc. 112, 7419 (1991).
2. J.W. Petrich, M.C. Chang, D.B. McDonald and G.R. Fleming, J. Am. Chem. Soc. 105, 3824 (1983).
3. P.E. Share, M.J. Sarisky, M.A. Pereira, S.T. Repinec and R.M. Hochstrasser, J. Lumin. 48/49, 204 (1991).
4. P.F. Barbara and W. Jarzeba, Adv. photochem. 15, 1 (1990). S. J. Rosenthal, N. F. Scherer, M. Cho, X. Xie, M. E.Schmidt and G. R. Fleming. This volume.
5. M. Négrerie, F.Gai, S.M. Bellefeuille and J.W. Petrich, J. Phys. Chem. 95, 8663 (1991).
6. R.S. Moog and M. Maroncelli, J. Phys. Chem. 95, 10359 (1991).
7. H. Bulska and A. Chodkowska, J. Am. Chem. Soc. 102, 3259 (1980).
8. D.V. Bent and E. Hayon, J. Am. Chem. Soc. 97, 2612 (1975).
9. B. Valeur and G. Weber, Photochem. Photobiol. 25, 441, (1977).

Excited-State Proton Transfer and Hydrogen-Bonding Dynamics in 7-Azaindole: Time-Resolved Fluorescence and Computer Simulation

C.F. Chapman[1], *T.J. Marrone*[1], *R.S. Moog*[2], and *M. Maroncelli*[1]

[1]Department of Chemistry, The Pennsylvania State University, University Park, PA 16802, USA

[2]Franklin and Marshall College, Lancaster, PA 17604-3003, USA

7-Azaindole (7AI) is one of a class of molecules that exhibit dual fluorescence as a result of excited-state tautomer formation. The tautomer is reached through a proton transfer from the initially excited "normal" form, which is the only form present in the ground electronic state. When isolated, 7AI does not tautomerize at an observable rate. However, when labile protons are available to catalyze the reaction, tautomerization can be quite rapid. For example, recent measurements have shown that the in the 7AI dimer the reaction takes place in ~1 ps at room temperature [1].

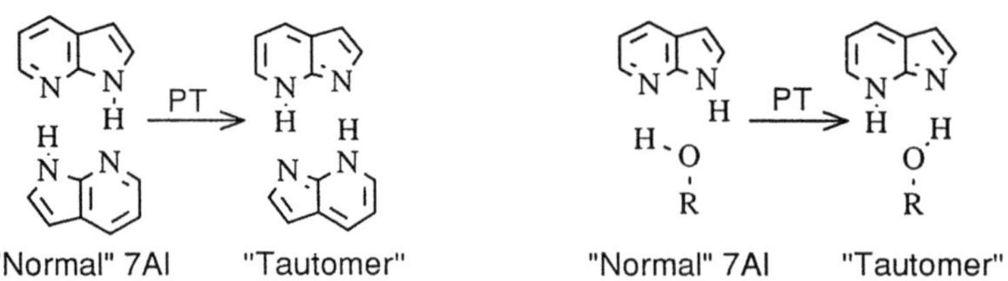

In alcohol solvents 7AI also tautomerizes, but at a much slower rate. For example, τ_{rxn}~100 ps in methanol at room temperature. Here catalysis is achieved through an alcohol molecule providing a "proton circuit", as indicated above. A number of workers have previously used time-resolved fluorescence to study the alcohol reaction [2-5] and arrived somewhat contradictory interpretations of the observed rates. We recently measured complete time- and frequency-resolved fluorescence spectra of 7AI in a number of alcohols and showed how past ambiguities could be traced to the fact that the "normal" fluorescence undergoes a time-dependent Stokes shift [6]. This Stokes shift is a result of solvation energy relaxation [7] and is not directly connected to the population kinetics of the reaction. We also have performed an extensive series of measurements of the proton transfer rate of 7AI as a function of alcohol, temperature, and solvent deuteration [6]. In order to explain the results, especially the peculiar temperature-dependent deuterium isotope effects observed, a "two"-step mechanism, similar to that originally proposed by McMorrow and Aartsma [2], is required:

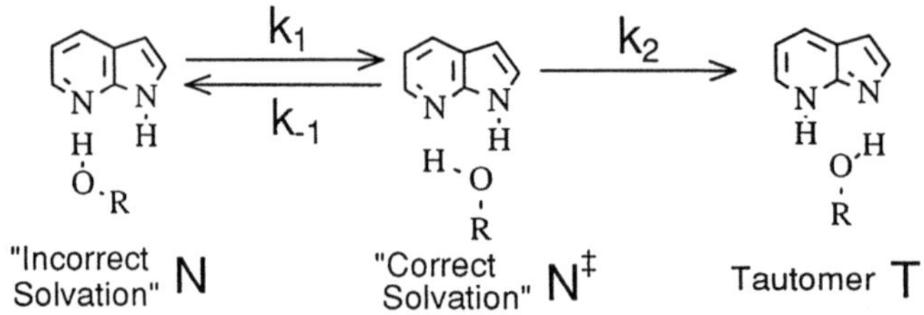

The key idea is that the proton transfer event (the k_2 step) can only occur from a select subset of "correct" solvent configurations, "N‡". Quantitative modeling shows that once the correct configuration is reached, the proton transfer step is very rapid, as it is in the 7AI dimer. What is observed as the overall reaction rate in alcohols is the much slower rate at which these "correct" configurations can be formed from "incorrect" ones (k_1) [6]. Given this explanation, the 7AI/alcohol reaction presents a unique opportunity to probe the

Springer Series in Chemical Physics, Vol. 55 Ultrafast Phenomena VIII
Editors: J.-L. Martin · A. Migus · G.A. Mourou · A.H. Zewail

dynamics of hydrogen bond formation difficult to measure in other ways. That is, the reaction rate provides a direct measure of the local solvent reorganization dynamics labelled "k_1" in the above scheme.

To further explore the nature of these hydrogen bonding dynamics we have undertaken studies of the tautomerization of 7AI in water and in dilute alcohol/alkane mixtures. Results in dilute mixtures have thus far mainly served to confirm the above mechanism. We find that isolated 1:1 7AI/alcohol complexes undergo rapid reaction, $\tau_{rxn} < 30$ ps. The difference compared to bulk alcohols is presumably due to the prevalence of cyclically hydrogen bonded 1:1 complexes (i.e. "correctly" solvated molecules). This bonding pattern is disrupted in higher 7AI:ROH_n complexes, as indicated by the fact that bulk-like solvation times are already observed in dilute solutions where only small-n complexes are expected. We have recently completed studies of the behavior of 7AI in water [8] and the results of this work are interesting in that they demonstrate that the reaction dynamics are quite sensitive to the particular hydrogen bonding patterns present in solution. From previous studies [5,9,10] it appeared that water was anomalous compared to other alcohols. In fact there was even some question as to the nature of the emitting species involved. However, time-resolved fluorescence measurements show that tautomerization takes place in water just as it does in other alcohol solvents. The only difference is that in water the reaction is almost an order of magnitude slower ($\tau \sim 850$ ps) than in the closest normal alcohol, methanol ($\tau \sim 120$ ps). This difference seems to be related to the fact that each water molecule is able to act as a multiple hydrogen-bond donor. Other polyalchols also show anomalously slow reaction rates [8].

With experimental studies alone we must be content with only a caricature of what goes on at the molecular level in these reactions. A deeper understanding can be achieved through the use of computer simulations. Using realistic intermolecular potentials generated from ab initio and semi-empirical quantum calculations, we have employed classical molecular dynamics simulations to model the solvent's participation in the tautomerization process. Differences between isolated 1:1 complexes, higher-order complexes, and bulk alcohol solvents, as well as the difference observed between water and alcohols can all be readily explained by the nature of the hydrogen bonding equilibria found in these various systems. For example, cyclically bonded structures are improbable in bulk solvents, as surmised from experiment. Because of the particularly strong network structure, cyclic structures are even rarer in water than in alcohols. Such structural differences are also accompanied by differences in the rates of interconversion between different hydrogen-bonding arrangements. We have begun to explore these differences and to try and determine reaction rates directly from simulations. Preliminary simulations in methanol show that the time scale for visits to the cyclically bonded forms assumed necessary for reaction are quite close to the experimental reaction time. We are therefore hopeful that further simulations will lead to a uniquely detailed understanding of the solute-solvent dynamics that control these solution-phase reactions.

This work was supported by the National Science Foundation.

References

1. P.E. Share, M.J. Sarisky, M.A. Pereira, S.T. Repinec, and R.M. Hochstrasser, J. Luminesc. **48/49**, 204 (1991).
2. D. McMorrow and T.J. Aartsma, Chem. Phys. Lett. **125**, 581 (1986).
3. J. Konijnenberg, A.H. Huizer, C.G.A.O. Varma **84**, 1163 (1988).
4. R.S. Moog, S.C. Bovino, and J.D. Simon, J. Phys. Chem. **92**, 6545 (1988).
5. M.Negrerie, F.Gai, S.M.Bellefeuille, and J.W.Petrich, J.Phys.Chem.**95**, 8663 (1991)
6. R.S. Moog and M. Maroncelli, J. Phys. Chem. **95**, 10359 (1991).
7. M. Maroncelli, J. Mol. Liquids **52**, 000 (1992).
8. C.F. Chapman and M. Maroncelli, J. Chem. Phys. in press.
9. P. Avouris, L.L. Yang, and M.A. El-Bayoumi, Photochem. Photobiol. **24**, 211 (1976).
10. P.T. Chou, M.L. Martinez, W.C. Cooper, S.T. Collins, D. McMorrow, and M. Kasha, J. Phys. Chem. in press.

Transient Hole Burning Studies of Electronic State Solvation: Phonon and Structural Contributions

J. Yu, J.T. Fourkas, and M. Berg

Department of Chemistry and Biochemistry, University of Texas, Austin, TX 78712, USA

Abstract. A component of the solvent/electronic state interaction that extends continuously from the glass into the liquid is identified with phonon-like motions. A second component, which becomes static below the glass transition, is linked to structural relaxation of the fluid.

Previously we showed that at room temperature the solvation of the nonpolar solute dimethyl-s-tetrazine (DMST) is largely subpicosecond, even in a variety of viscous solvents expected to show slow relaxation times.[1] Measurements of solvation of DMST in glycerol have been extended from room temperature down to the low temperature glass. The solvation is shown to decompose into two major components: one describable in terms of phonon-modulated interactions, and one connected to the structural relaxation of the solvent.[2]

Below the glass transition temperature (185 K), the only significant sources of dynamic line broadening are those driven by acoustic phonon motion. This broadening has been measured by a combination of transient and permanent hole burning and fit to a standard model of phonon-electronic state coupling in crystals. With the coupling constant established, the model can be extrapolated into the liquid phase (Fig. 1). Good agreement is found for the amount of broadening at 1.5 ps for a region up to 65 K above the glass transition. Thus at short times, the electronic-state solvation in the liquid can be described in the same manner as solvation in a disordered solid.

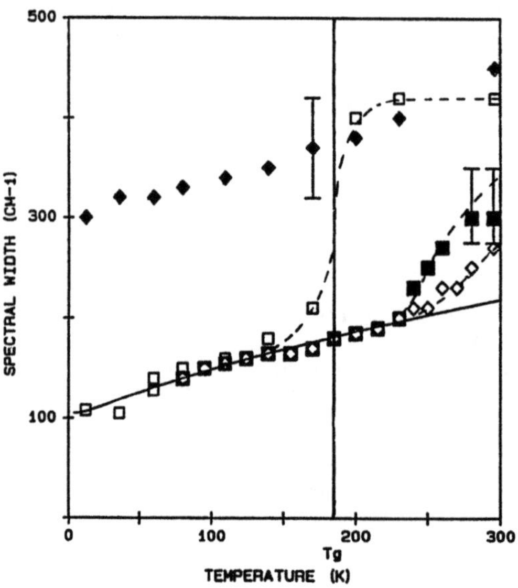

Fig. 1. Solvent-induced widths derived from absorption (◆), permanent hole burning (□), and transient hole burning at 200 ps (■) and 1.5 ps (◇). Broadening due to the vibronic structure has been removed. The solid curve is a fit to a solid-state theory of phonon-induced broadening. The glass transition temperature T_g is marked. The dashed curves are guides for the eye.

Springer Series in Chemical Physics, Vol. 55 **Ultrafast Phenomena VIII**
Editors: J.-L. Martin · A. Migus · G.A. Mourou · A.H. Zewail

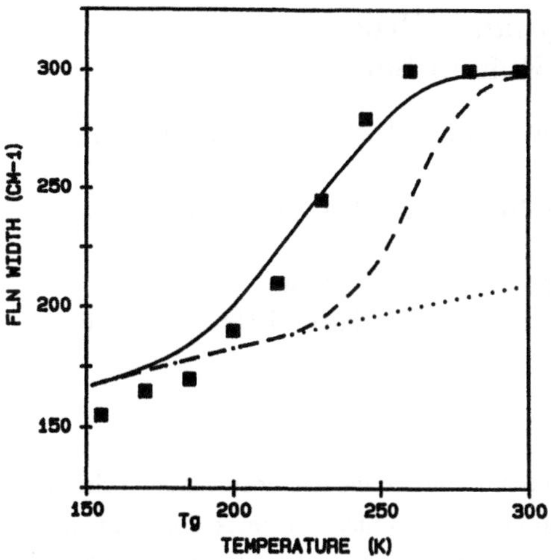

Fig. 2. Solvent-induced widths derived from fluorescence line narrowing. The dotted curve is the phonon-induced contribution. The solid curve is the width predicted from the observed Stokes' shift using a model relating spectral changes to the structural correlation function. The dashed curve is the prediction of the width based on the relaxation times found in a number of other types of experiment.[4]

However even at room temperature, the phonon-modulated interactions do not account for the full extent of solvent interaction indicated by the width of the absorption spectrum. Permanent hole burning measurements show that the additional broadening is static in the glass, but becomes dynamic at the glass transition (Fig. 1). The additional broadening must be due to interaction with structural coordinates of the solvent.

Near the glass transition temperature, the structural relaxation is too slow to affect the broadening at short times, but with increasing temperature, the relaxation rate increases. Fluorescence line narrowing (FLN), which is sensitive to solvation faster than the ~5 ns fluorescence lifetime, begins to broaden beyond the phonon contribution near 215 K (Fig. 2). Transient hole burning at 200 ps begins to have additional broadening near 240 K, and 1.5 ps measurements do not broaden until even higher temperatures (Fig. 1). By room temperature, the structural relaxation is rapid enough that both structural and phonon mechanisms contribute significantly to the subpicosecond solvation.

The structural relaxation can be described by a configuration coordinate model with Gaussian dynamics.[3] This model successfully relates the Stokes' shifts seen in ordinary fluorescence and in FLN with the widths observed in FLN (Fig. 2). It also extracts the correlation function of the structural coordinate from the observed widths. Figure 2 shows that the relaxation time found here is substantially faster than that measured by other relaxation experiments.

References

1. T.J. Kang, J. Yu and M. Berg, Chem. Phys. Lett. **174**, 476 (1990); J. Yu, T.J. Kang and M. Berg, J. Chem. Phys. **94**, 5787 (1991).
2. J. Yu and M. Berg, J. Chem. Phys., in press.
3. R.F. Loring, Y.J. Yan and S. Mukamel, J. Chem. Phys. **87**, 5840 (1987).

Subpicosecond Study of the Dynamic Processes in Push-Pull Styrenes and the Role of Solvation

P. Hébert, G. Baldacchino, T. Gustavsson, V. Kabelka, P. Baldeck, and J.-C. Mialocq*

CEA, Centre d'Etudes de Saclay, DRECAM/SCM/URA 331 CNRS, F-91191 Gif-sur-Yvette Cedex, France
*Permanent address: Laboratoire de Spectrométrie Physique, Université de Grenoble I, B.P. 87, F-38302 Saint Martin d'Hères, France

A great deal of attention has recently been devoted to the role of solvation in chemical reactions involving intramolecular charge transfer, singlet excited state relaxation and trans-cis photoisomerization. Following our previous studies of the solvatochromism of the DCM dye (4-(dicyanomethylene)-2-methyl-6-[p-(dimethylaminostyryl)-4H-pyran]) [1,2] we present in this paper a subpicosecond study of the solvation dynamics of the styryl 8 laser dye, LDS 751 (2-(4-(4-dimethylaminophenyl)-1,3-butadienyl)-3-ethylbenzothiazolium perchlorate) in various solvents. Both styryls are remarkable as regards the weak overlap between their absorption and emission spectra. However, the behaviours of the two styryls with respect to solvent polarity are quite different. On the one hand, in increasing polarity solvents, the neutral DCM presents a red shift of its absorption transition energy and a strongly dependent fluorescence spectrum indicating a large increase of the dipole moment upon electronic excitation [1,2]. On the other hand, in the cationic styryl 8 molecule, an unsymmetrical polymethine-cyanine, the large blue shift of the absorption maxima in solvents of increasing polarity indicates the enormous stabilization of the ground state with respect to the Franck-Condon excited singlet state.

The time dependent fluorescence relaxation was studied using the fluorescence upconversion technique. Styryl 8 was excited at 610nm with half of a ~1ps duration laser pulse delivered by a cavity-dumped and hybridly mode-locked dye laser (R6G/pinacyanol [3]) pumped by the second harmonic of a mode-locked cw Nd-Yag laser. The second part of the laser pulse was retarded by an optical delay line and combined with the fluorescence in a 0.5 mm urea crystal to generate the UV upconverted signal. The use of a monochromator enabled the temporal analysis of the fluorescence at different wavelengths with a spectral resolution of 1nm.

An example of styryl 8 fluorescence decays is shown in Fig. 1 for ethanol. The observed kinetics are complex and wavelength dependent. After deconvolution for the exciting pulse function and correction for the instrumental response function using steady state fluorescence spectra, the calculated decay curves show in the blue part of the spectrum, a rise time too short to be resolved and a bi-exponential function (a fast component below 10 ps and a slow component with a 90 ps lifetime). In the red part, on the other hand, the decay exhibits only the slow component and the risetime is easily extracted. The slow component, common for all wavelengths, may be assigned to the solvent-relaxed excited state, decaying mostly by a

Springer Series in Chemical Physics, Vol. 55 Ultrafast Phenomena VIII
Editors: J.-L. Martin · A. Migus · G.A. Mourou · A.H. Zewail

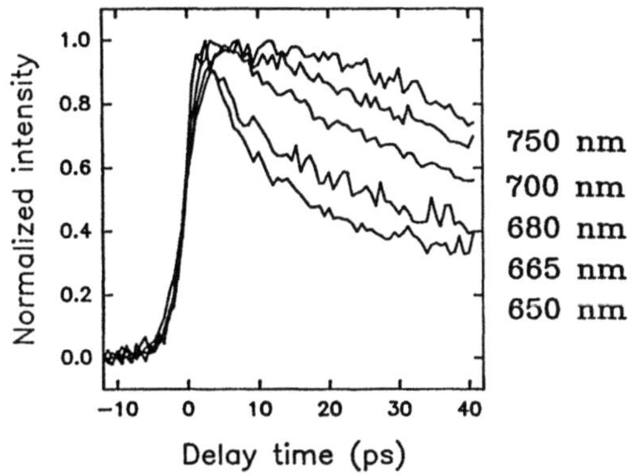

Fig. 1

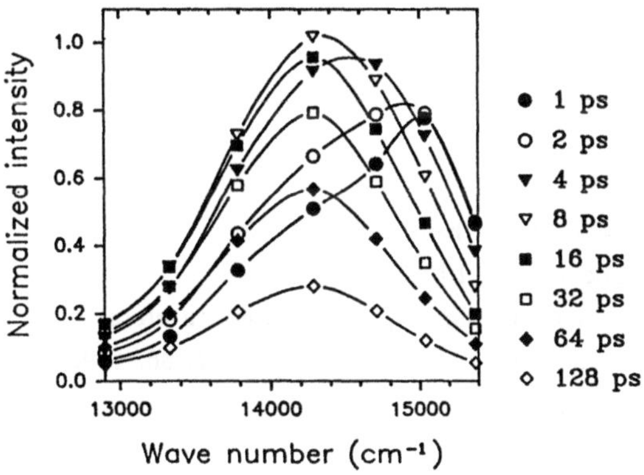

Fig. 2

rapid $S_1 \rightarrow S_0$ internal conversion process. The fast component in the blue and the rise time in the red should be seen as the result of the geometrical relaxation of the solute and its solvent cage.The calculated decay curves enabled us to plot the time dependent fluorescence spectra (Fig. 2), from which the shift of the fluorescence maximum frequency was analyzed as a function of time (Fig. 3). The shift is well described by a monoexponential decay with a 2.5 ps lifetime, which is shorter than the solvent longitudinal relaxation time. This was already observed by Castner et al. in the case of styryl 7 in methanol [4] but our spectra do not present the complex features found by Blanchard [5]. The total shift (≤ 750 cm^{-1}) consistent with the relatively small displacement of the steady state fluorescence maximum in solvents of different polarity indicates that the singlet excited state is less polar than the ground state.

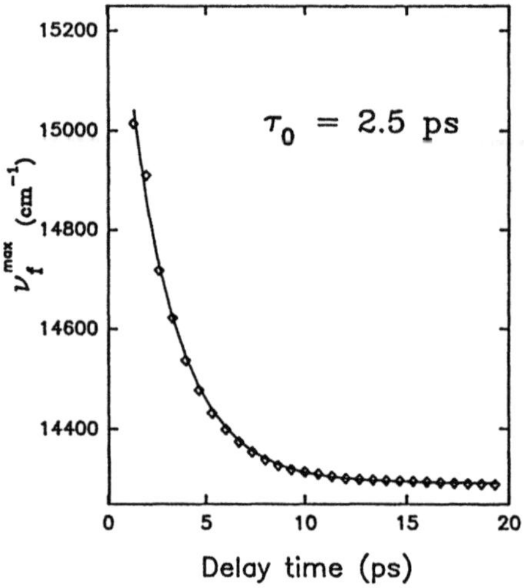

Fig. 3

[1]. M. Meyer, J.-C. Mialocq and B. Perly, J. Phys. Chem. 1990, 94, 98
[2]. J.-C. Mialocq and M. Meyer, Laser Chem. 1990, 10, 277.
[3]. P. Hébert, S. Marguet, T. Gustavsson and J. C. Mialocq, Opt. Comm (in press)
[4]. E. W. Castner, M. Maroncelli and G. R. Fleming, J. Chem. Phys. 1987, 86, 1090.
[5]. G. J. Blanchard, J. Chem. Phys. 1991, 95, 6317.

Picosecond Studies of Charge Transfer States in "Push-Pull" Linear Diphenyl Polyenes: Experimental Evidence for TICT and Bicimer States

J.M. Viallet[1], *F. Dupuy*[1], *R. Lapouyade*[2], *W.Q. Zheng*[1], *and C. Rullière*[1]

[1]Centre de Physique Moléculaire Optique et Hertzienne, Unité Associée, CNRS No. 283, 351 Cours de la Libération, F-33405 Talence Cedex, France

[2]Laboratoire de Photophysique et Photochimie Moléculaire, U.A. CNRS No. 348, 351 Cours de la Libération, F-33405 Talence Cedex, France

Abstract: Two push-pull, linear diphenyl-polyenes, 4-dimethylamino-4'-cyano-1,6-diphenylhexatriene (DCDPH) and 4-dimethylamino-4'-cyano-1,8-diphenyloctatetraene (DCDPO) compounds with large charge transfer character, have been studied. Up to three time-resolved emission bands of these compounds were seen, depending on temperature, solvant polarity, concentration and intensity of excitation. These emission bands are attributed to the locally excited state (LE), a Twisted Intramolecular Charge Transfer (TICT) excited state and a "bicimer" state.

Introduction: In a recent paper (1) we have shown that in "push-pull" stilbene derivatives a new emission band appears at high concentration and high intensity excitation. This new emission band was attributed to the formation of a new excited species (bicimer) formed by the association of two electronically excited molecules. The stabilisation of the bicimer was related to the presence of an intermediate TICT state. However we were unable to directly observe the TICT state emission. The presence of such TICT states was deduced indirectly from measurements of lifetime and fluorescence quantum yield (2). It was therefore interesting to study longer polyenic chain derivatives for two reasons: a) to observe the formation of bicimer species as for stilbene derivatives. b) to try to observe directly the formation of possible TICT states as intermediates on the pathway to the bicimer.

Experimental results: We have performed time-resolved emission and absorption experiments (on a picosecond time scale), using an experimental set-up described elsewhere (3) on DCDPH and DCDPO.

In non-polar solvents, in any experimental conditions, a single emission band was observed, corresponding to the normal "locally excited" (LE) state. In polar solvents, at all concentrations and with weak excitation (see fig. 1) ,we have observed a second emission band strongly shifted to the red. Its intensity increases as the intensity of the first band decreases and a "precursor-successor" relation-ship occurs between the two bands. We attributed this second red-shifted band to a TICT excited state.

On strong excitation the behaviour of these two compounds is different . At high concentration ($c > 10^{-4}$ M/l) a third band appears to the red of the TICT band. It disappears at lower concentrations than 10^{-4} M/l, or when the intensity of excitation is reduced (see fig.2). Further, we have observed a "precursor-successor" relationship between the intensities of the TICT band and of this third band, showing that the bicimer is formed from the TICT state. All these results were also confirmed in time resolved absorption experiments.

Springer Series in Chemical Physics, Vol. 55 Ultrafast Phenomena VIII
Editors: J.-L. Martin · A. Migus · G.A. Mourou · A.H. Zewail

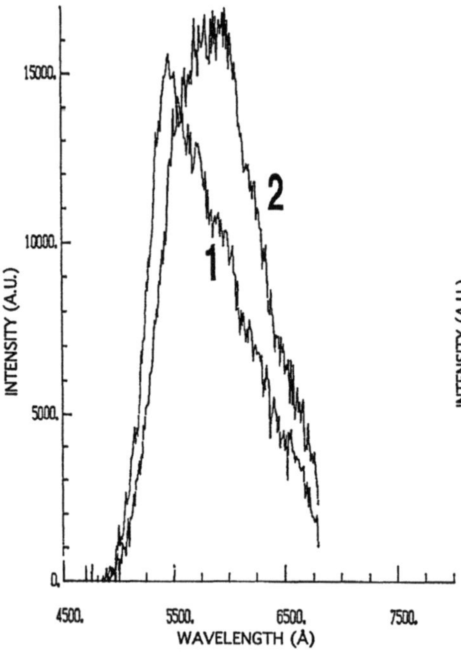

Fig.1: Time-resolved emission spectra of DCDPH
$c=2*10^{-4}$ M/l in BuCN
T=-100°C (1)t=80 ps (2) t=160 ps

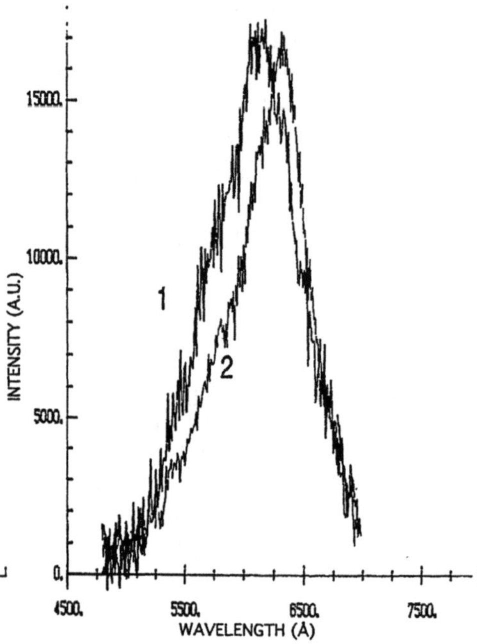

Fig. 2: Role of excitation intensity
(1) weak excitation
(2) strong excitation
$c=2*10^{-4}$ M/l in PC
t=160 ps.T=20°C

Discussion: We attribute the third band to the following mechanism. The large dipolar character and the presence of donor and acceptor groups create "loose" dimers or aggregates in the ground state. After excitation the molecules form TICT states in polar solvents. If excitation intensity is high enough, the two molecules of the "loose" dimer are excited and interact strongly due to strong dipolar interaction, and form a new excited species: a "bicimer". Bicimers emit the third band, after their formation from the TICT states. At very high excitation intensity, we have also observed sharp, directive emission in the third band which corresponds to stimulated or laser emission from the bicimer species.

Conclusion: In these two derivatives we have observed unambigously the presence of TICT states and bicimer states. Furthermore we have shown that the bicimer species is formed only when TICT states are present, the TICT state being a prerequisite for the formation of bicimers in these compounds. Bicimers are formed from the TICT state as we indicated for stilbene derivatives. These results confirm our previously proposed model (1).
We must note also that stimulated emission (laser effect) has been observed in the bicimer emission band during these experiments. We think that this property of bicimer should encourage revisiting experiments the properties of emission of those lasers dyes showing, anomalous red-shifted laser emission. Laser emission at anomalous frequencies may be due to the presence of bicimer.

(1) E. Gilabert, R. Lapouyade, C. Rullière **Chem. Phys. Lett.** 185(1991)82
(2) R. Lapouyade, K. Czeschka, W. Mayenz, W. Rettig, E Gilabert, C. Rullière (submitted to **J. Phys. Chem.**)
(3) E. Gilabert, A. Declémy, C. Rullière **Rev. Sci. Instruments** 58(1987)2049

Features of the Dual Fluorescence of 4-N,N-dialkylaminoalkylbenzoates in Alkanes

M.C.C. de Lange, D.T. Leeson, A.H. Huizer, and C.A.G.O. Varma

Leiden University, Gorlaeus Laboratories, P.O. Box 9502,
NL-2300 RA Leiden, The Netherlands

Abstract. An interpretation is given for the isoviscous temperature dependence and for the isothermal viscosity dependence of the dual fluorescence of the title compounds studied on a picosecond time scale.

1. Introduction

The appearance of a normal band F_N and an anomalous band F_X, to the red of the former, in the fluorescence spectrum of solutions of 4-N,N-diethylaminoethylbenzoate in alkanes has been considered as evidence for the existence of a low energy twisted intramolecular charge transfer (TICT) excited state in electron acceptor substituted aniline derivatives [1]. Objections have been raised against this interpretation of the dual fluorescence [2]. Nevertheless, the TICT state seems to be widely accepted as a reasonable explanation of the phenomenon. Here we will present a different explanation.

2. Results and discussion.

The photochemistry and photophysics of various 4-N,N-dialkylaminoethylbenzoates, mentioned in Table 1, have been studied. These compounds are found not to be photostable in solutions in alkanes, whereas their solutions in water or in alkanols are photostable. This photo-reaction in alkanes is attributed to a reaction from the singlet state S_1, which has strong $n\pi^*$-character. The fluorescence is thought to arise from the higher singlet

Table 1. Acronyms of the 4-N,N-dialkylaminoethylbenzoates described in this paper.

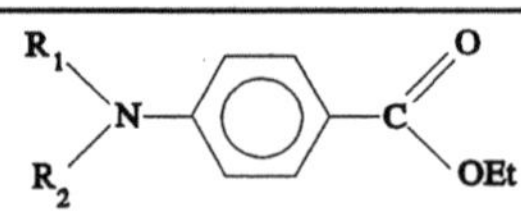

R_1	R_2	Acronym
CH_3	CH_3	DMAEB
CH_3	n-$C_{10}H_{21}$	DeMeAEB
C_2H_5	C_2H_5	DEAEB
n-C_3H_7	n-C_3H_7	DPrAEB
n-C_5H_{11}	n-C_5H_{11}	DPAEB
n-C_8H_{17}	n-C_8H_{17}	DOAEB

Springer Series in Chemical Physics, Vol. 55 Ultrafast Phenomena VIII
Editors: J.-L. Martin · A. Migus · G.A. Mourou · A.H. Zewail

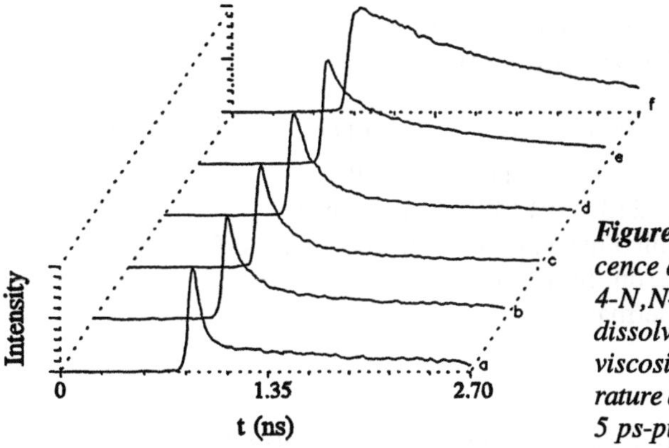

Figure 1: Decay of the fluorescence at 340 ± 5 nm of the various 4-N,N-dialkylaminoethylbenzoates dissolved in paraffin oil with a viscosity of 73.3 cP at room temperature after excitation with a 300 nm 5 ps-pulse.

state S_2, from which state the rates of electronic relaxation and emission are comparable because the electronic energy gap between S_2 and S_1 is so small that the density of accepting states is low.

A quantum chemical calculation of the CNDO/S type indicates a near degeneracy of the first three excited singlet states, which consist of one $^1n\pi^*$ and two $^1\pi\pi^*$ states. In the state S_2 the benzene ring accumulates electron density, coming from both the amino and the ester group. It is suggested that the charge on the benzene ring gets stabilized when large alkyl substituents on the amino group approach the benzene ring. This becomes feasible upon pyramidalization around the N-atom, whereby the torsion angle of the amino group does not change. The resulting folded species is supposed to be the origin of the anomalous fluorescence.

The time dependence of the dual fluorescence of the alkylaminobenzoates has been investigated, on a picosecond time scale, as a function of temperature at constant viscosity in a homologous series of linear alkanes. All the compounds reveal a interconversion between species N and X, leading in most cases to a biexponential time dependence of the fluorescence as shown in figure 1. This time dependence involves a fast process with a rate constant τ_s^{-1} and a slow process with a rate constant τ_l^{-1}. Both τ_s^{-1} and τ_l^{-1} are found to decrease as the size of the amino substituents increases. In the case of DOAEB the fast process is dominated by the rate constant, k_+, for the forward reaction N → X. Its isothermal viscosity dependence can not be satisfactorily described by Kramers's equation for barrier crossing [3]. In the present case the defect of Kramers's theory seems to be that a quantum mechanical description of the energy exchange between solute and solvent rather than a classical one is required.

This work has been submitted for publication to the Journal of Physical Chemistry.

3. References.

1. Lippert, E.; Rettig, W.; Bonacic-Koutecky, V. *Adv. Chem. Phys.* **1987**, *LXVIII* and references therein.
2. Weisenborn, P.C.M.; Huizer, A.H.; Varma, C.A.G.O. *Chem. Phys.*, **1988**, *126*, 425.
3. Kramers, H.A. *Physica* **1940**, *7*, 284.

Investigation of Fast Relaxation Processes in Non-Fluorescent Rhodamine Dyes

P. Plaza[1], *N.D. Hung*[1], *M.M. Martin*[1], *Y.H. Meyer*[1], and *W. Rettig*[2]

[1]Laboratoire de Photophysique Moléculaire du CNRS,
Université Paris-Sud, Bât. 213, F-91405 Orsay, France

[2]Iwan-N. Stranski-Institut für Physikalische und Theoretische Chemie,
Technische Universität Berlin, Strasse des 17. Juni 112,
W-Berlin, Fed. Rep. of Germany

Abstract. Evidence is given for a short-lived form of amino-Rhodamine by time-resolved absorption spectroscopy with subpicosecond resolution. This transient form is interpreted in terms of internal charge transfer with conformational change.

1. Introduction

Non-radiative relaxation in various N-substituted ethyl-ester-Rhodamine dyes (A) was interpreted by the formation of a non-emissive low lying TICT-like state through the rotation of the dialkylamino groups [1].

An other possibility for Rhodamine dyes to form a TICT-like state appears if the o-carboxy group is exchanged for a p-dimethylamino electron donor group. The fluorescence yield of this amino-Rhodamine (B) is very low in neutral ethanol but increases by more than two orders of magnitude in acidic ethanol solutions, where the electron donating property of the anilino group is blocked by protonation [1, 2].

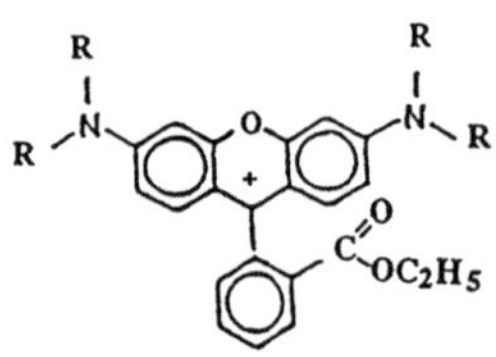

A

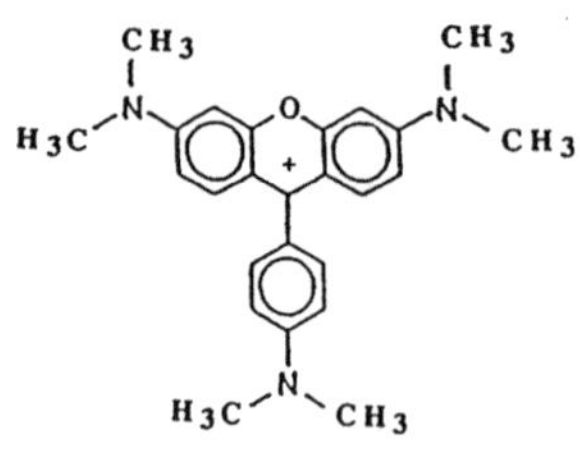

B

In order to characterize the process responsible for the low fluorescence yield of amino-Rhodamine (ARh), we studied its transient spectroscopy after subps excitation at 550 nm.

2. Experimental

Time resolved absorption and gain spectra of ARh in neutral and acidic ethanol were measured by the pump-probe technique with a non mode-locked subps dye laser system [3] pumped by a single seeded 6 ns Nd:YAG laser. Pulses of 0.5 ps at 550 nm were used for excitation and to generate a continuum of white light probe. Transient spectra were recorded in the 380 - 640 nm range through a polychromator by a double diode array detector, in a two-beam set-up [5].

3. Results and Discussion

The time resolved changes in optical density (ΔD) measured for ARh in neutral and acidic ethanol, are respectively shown in figures 1A and 1B. In neutral ethanol (fig.1A) a fast change

Springer Series in Chemical Physics, Vol. 55 **Ultrafast Phenomena VIII**
Editors: J.-L. Martin · A. Migus · G.A. Mourou · A.H. Zewail

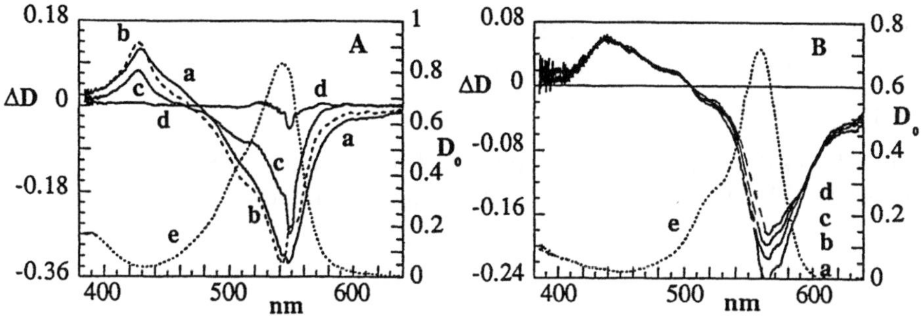

Fig.1 **(A)** Change in optical density ΔD of ARh in neutral ethanol, after 0.5 ps laser excitation at 550 nm, for: **a**: 1 ps, **b**: 5 ps, **c**: 15 ps, **d**: 50 ps pump-probe delays (left scale). **(B)**: transient spectra of ARh in acidic ethanol (~2 M HCl). **a:** 5 ps, **b:** 10 ps, **c:** 30 ps, **d:** 100 ps. **(A)** and **(B)** **e:** ground state absorption spectrum D_0 (right scale)

in shape of the 380 - 470 nm transient absorption band (ΔD > 0) occurs within 5 ps after excitation. The transient absorption decreases around 460 nm, whereas a new absorption peak grows near 430 nm. Simultaneously, a fast decrease in the gain signal (ΔD < 0) is observed above 580 nm. For a 15 ps probe delay, most of the gain signal has vanished but the transient absorption at 430 nm remains and the ground state repopulation is not complete.

In fig.1B, the transient spectra recorded for the fluorescent protonated-ARh do not show the fast changes observed in neutral ethanol. An isosbestic point between the S_1 and S_0 absorption bands is seen at 505 nm, indicating that only two states contribute to the transient spectra. The reason for the change in ΔD between 560 and 580 nm is not well understood; it might be due to solvation dynamics since gain is expected to contribute to the transmission signal in this wavelength range.

The transient spectra reported here evidence the rapid formation of a transient state from ARh S_1 state in neutral ethanol. This process is hindered in acidic ethanol where the anilino-substituent is protonated, that is when the anilino-group electron donating property is decreased. On the other hand, from experiments in decanol and dioxan, we found that solvent viscosity and polarity determine the transient state formation and decay rates. The present results are consistent with the formation of a non-emissive TICT-like state [1, 2] involving internal charge transfer from the rotating anilino-group to the xanthenic chromophore. As ARh can be regarded as an oxygen bridged TPM dye [1, 2] and shows similar fluorescence characteristics as a methylene-bridged crystal violet derivative [4], it is interesting to note that the spectral changes observed here are similar to what was previously reported for the TPM dye ethyl violet and interpreted in terms of a rapid formation of an excited state conformer [5].

References

[1] M. Vogel, W. Rettig, R. Sens, K. H. Drexhage: Chem. Phys. Lett., **147**, 452, (1988)
[2] W. Rettig: Phys. Chem. Trenn.- Messmeth. (Opt. Spectrosc. Chem. Biol) **22** 119 (1990)
[3] N. Dai Hung, P. Plaza, M.M. Martin, Y.H. Meyer: Appl. Opt. (Special Issue on Dye Lasers), *in press*
[4] M. Vogel, W. Rettig: Ber. Bunsenges. Phys. Chem., **89**, 962, (1985)
[5] M. M. Martin, P. Plaza, Y. H. Meyer: Chem. Phys., **153**, 297, (1991), J. Phys. Chem., **95**, 9310, (1991)

Femtosecond Photodissociation of Aromatic Disulfides Followed by Solvent Relaxation

N.P. Ernsting

Max-Planck-Institut für biophysikalische Chemie, Abt. Laserphysik, Postfach 2841, W-3400 Göttingen, Fed. Rep. of Germany

Abstract. Bis-(p-aminophenyl)-disulfide in solution was photodissociated at 310 nm. The solvent-induced shift of the visible absorption band of the polar radical products was followed with 100 fs resolution.

1. Introduction

The dynamical properties of liquids may be probed in the time domain in two ways: either in the neat liquid by nonlinear scattering experiments [1], or else indirectly by studying the solvation dynamics of molecular probes [2]. In the latter case, optical excitation of a suitable dye molecule leads to a change in molecular dipole moment. As the surrounding solvent adjusts to a new equlibrium, solvent-solute interaction stabilizes the excited molecule, and this results in a red-shift of the fluorescence spectrum. Typical compounds for this purpose are coumarin [2] or styryl [3] dyes. However their use for ultrafast solvation dynamics is limited by a relatively large spectral width of fluorescence, or by the possibility of geometrical isomerisation. Furthermore, measurement requires fluorescence upconversion and spectral reconstruction which may be quite tediuos.

As an alternative experimental approach to solvation dynamics, we suggest ultrafast generation of polar probe molecules in their electronic ground state, and then monitoring a visible absorption band of the products. For example, UV photodissociation of bis-(p-aminophenyl)-disulfide yields two solvatochromic [4] p-aminopheny-thiyl radicals:

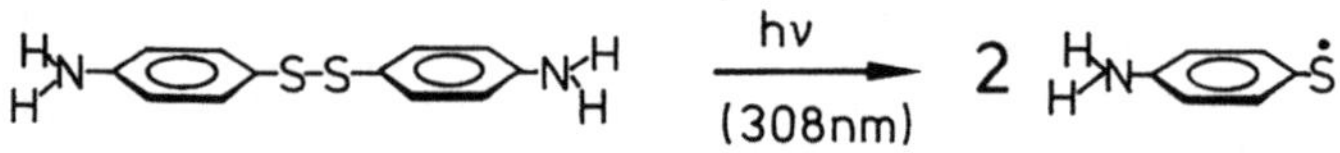

100 fs pulses at 310 nm were used to excite the disulfide parent in acetonitrile over 0.25 mm path length. The absorption spectrum was then recorded with a white-light continuum generated by 90 fs pulses at 620 nm. The rise of absorption at the band centre (near 575 nm) is given in fig. 1. Comparison with the instrument response function (dashed line) gives a deconvoluted risetime of 70 $\pm$ 10 fs. Fig. 2 shows the temporal evolution of the radical absorption band. The spectrum recorded at 150 fs shows a red shoulder which is attributed to the interacting radical pair during photodissociation. Each radical should have an excess vibrational energy of ca. 5000 cm^{-1}. The secondary processes are then vibrational relaxation - evidenced by a steepening of the red band edge - and dielectric relaxation resulting in a shift of the absorption band.

Fig. 3 gives the spectral position of the absorption maximum (from Gaussian fits to the spectra) as function of delay time. A fit to biexponential frequency shift gives characteristic times of 150 fs and 770 fs with nearly equal amplitudes. This result is in excellent

Springer Series in Chemical Physics, Vol. 55 Ultrafast Phenomena VIII
Editors: J.-L. Martin · A. Migus · G.A. Mourou · A.H. Zewail

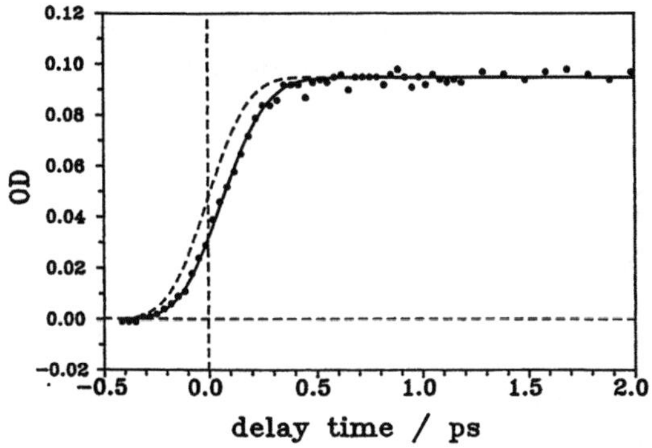

Fig 1. Rise of thiyl absorption near 575 nm.

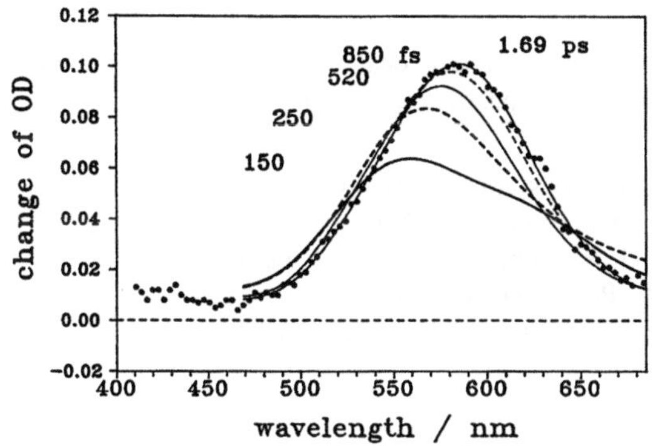

Fig 2. Spectral evolution in acetonitrile

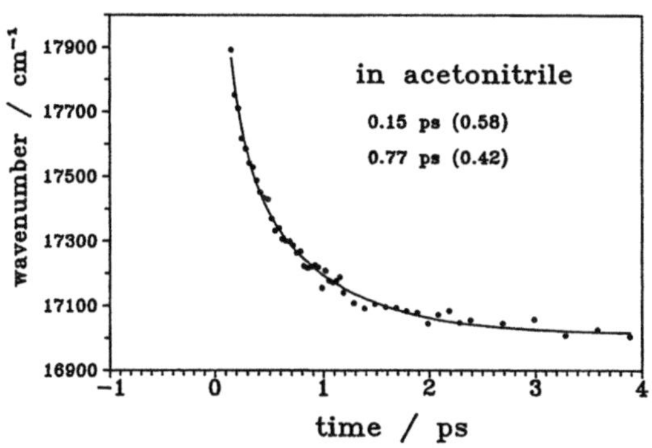

Fig 3. Spectral position of absorption maximum

agreement with solvatochromic data which have been obtained using coumarin fluorescence [2]. However the primary, dissociative part in the process studied here precludes the observation of the initial Gaussian solvent response which has been extracted from the time-resolved dynamic Stokes'shift of a styryl dye [3]. Molecular dynamics simulations are needed for an asessment of the solvent contribution to the emergence of the radical absorption band.

On a slower time scale, vibrational relaxation and geminate recombination may be studied in viscous solvents. For example, fig. 4 shows the spectral evolution of radical

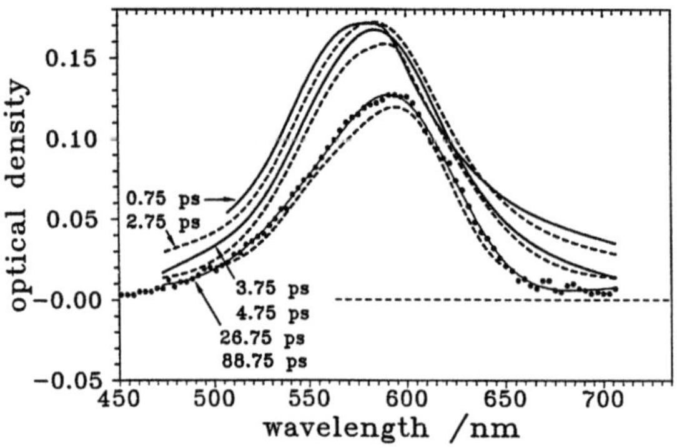

Fig. 4. Spectral evolution in propylene carbonate

absorption in propylene carbonate. Finally, coupled rotational motion of the two dipolar radicals [5] could be examined through absorption polarization anisotropy.

[1] M. Fickenscher, A. Lauberau, J. Raman Spectrosc. **21** (1990) 857.

[2] P. F. Barbara, W. Jarzeba, Adv. Photochem **15** (1990) 1.

[3] S. J. Rosenthal, X. Xie, M. Du, G. R. Fleming, J. Chem. Phys. **95** (1991) 4715.

[4] G.H. Morine, R.R. Kuntz, Chem. Phys. Lett. **67** (1979) 552.

[5] G. van der Zwan, J. T. Hynes, J. Phys. Chem. **89** (1985) 4181.

Femtosecond Dynamics of C–O Bond Cleavage of a Spirooxazine Photochromic Reaction

N. Tamai[1] *and H. Masuhara*[2]

[1]Microphotoconversion Project, ERATO, Research Development Corporation of Japan, 15 Morimoto-cho, Shimogamo, Sakyo-ku, Kyoto 606, Japan

[2]Department of Applied Physics, Faculty of Engineering, Osaka University, Osaka 565, Japan

ABSTRACT The dynamics of C-O bond cleavage followed by the structual change to merocyanine in the photochromic reaction of a spirooxazine derivative was studied by femtosecond transient absorption spectroscopy. The very broad spectra following the S_n - S_1 absorption spectrum of spirooxazine were observed within 2 ps after excitation, and assigned to the absorption from the transition state. The rate constants of C-O bond cleavage and the relaxation from the transition state to a metastable merocyanine were estimated to be approximately $(700\ fs)^{-1}$ and $(470\ fs)^{-1}$, respectively.

1. INTRODUCTION

Spirooxazines are known as typical organic compounds exhibiting photochromism [1]. Upon ultraviolet irradiation, the bond between the spiro carbon and oxygen undergoes fission, facilitating the structural change to form merocyanine as illustrated in scheme I.

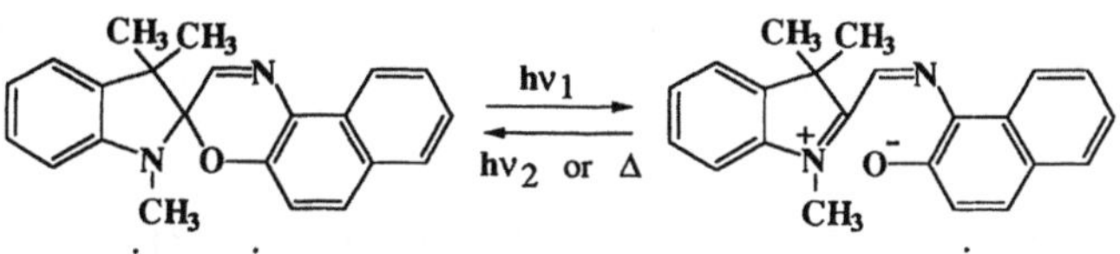

Scheme I

In the present study, transient absorption spectra of a spirooxazine were measured by a femtosecond transient absorption spectroscopic system.

2. EXPERIMENTAL

The laser system consisted of a synchronously pumped, hybridly mode-locked, dispersion compensated dye laser pumped by a mode-locked Nd:YAG laser, and a dye amplifier. The excitation beam had an average pulse energy of 40 μJ at 355 nm, a pulse width of 150 fs (FWHM), and a repetition rate of 10 Hz. All spectra were corrected with respect to the temporal dispersion of the white-light continuum. The sample solution was flowed in order to avoid photodegradation.

3. RESULTS AND DISCUSSION

Figure 1 illustrates the transient absorption spectra of spirooxazine in 1-butanol. Just after excitation, the spectrum with a peak at around 490 nm appears. This spectrum was assigned to the S_n - S_1 absorption of spirooxazine due to spectral similarity with that of phenanthrene derivatives [2] and a photochromic compound analogue [3]. As time progresses, this peak decays and is replaced by a broad transient absorption band spanning

Springer Series in Chemical Physics, Vol. 55 Ultrafast Phenomena VIII
Editors: J.-L. Martin · A. Migus · G.A. Mourou · A.H. Zewail

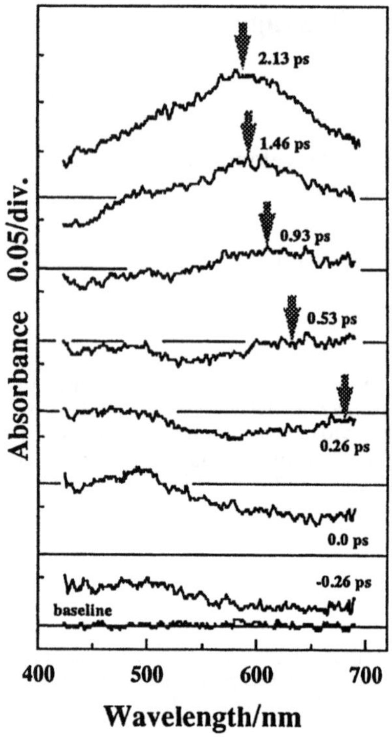

Fig. 1 Transient absorption spectra of spiro-oxazine in 1-butanol.

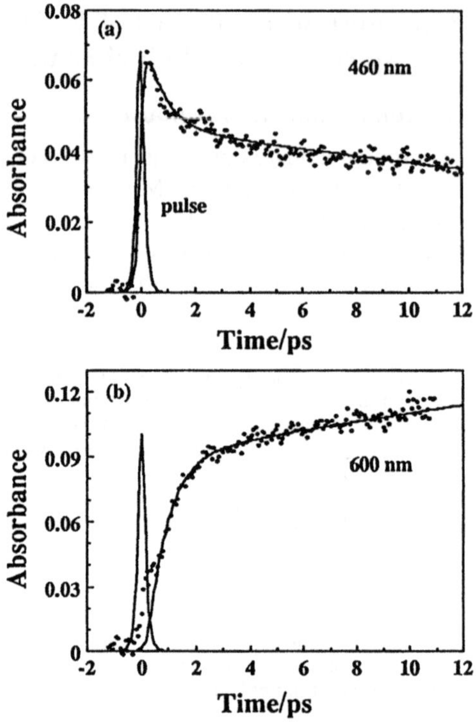

Fig. 2 Rise and decay curves of transient absorption at 460 nm (a) and 600 nm (b).

from 550 nm to 700 nm or more longer wavelength. A further point of interest is that the peak of this band appears to shift to the short wavelength region more than 3000 cm^{-1} with a time constant of approximately 470 fs, as indicated by the arrows. After 2 ps had elapsed, the transient absorption spectra peaked at 580 nm, which are similar to that of merocyanine in nonpolar solvents such as cyclohexane. Further, the bandwidth of the spectra was found to decrease with time. The rise and decay curves observed at 460 nm and 600 nm are represented by Fig. 2. The time dependence at 600 nm could be analyzed with two rise components: ~ 700 fs and 40 ~ 50 ps. Within experimental error, the lifetime of the fast decay component recorded at 460 nm coincides with the short risetime (~ 700 fs). The short lifetime of 700 fs at 460 nm represents to the relaxation of the S_1 state of spirooxazine, and should correspond to the rate of C-O bond cleavage.

These results suggest that the broad transient absorption spectrum ranging from 550 nm to 700 nm or more longer wavelength is probably due to the absorption from the nonplanar primary photoproduct X, which corresponds to the transition state of the photochromic reaction, to the upper electronic state. The rapid spectral shift with a time constant of 470 fs may represent the sliding down of the energy along the potential energy surface, from the transition state to one of the merocyanine isomers or merocyanine before solvation.

The narrowing of the bandwidth at 580 nm on a time scale of several 10 ps appears to correlate with the long risetime (40 ~ 50 ps). The long risetime, however, was found to be little dependent on the kinds of solvent, and had no correlation with the longitudinal relaxation time (τ_L) of the solvent. Therefore, the narrowing of the bandwidth is probably interpreted in terms of the thermal relaxation of vibrationally hot merocyanine. The temperature of hot merocyanine was estimated to be higher than 1200 K for several ps after excitation by the analyses of the red edge of the transient absorption spectra utilizing the Urbach relation [4].

REFERENCES

1. R. C. Bertelson, in :Techniques of chemistry, Vol III, Photochromism, ed. by G. H. Brown (Wiley-Interscience, New York, 1971) p. 45.
2. N. Tamai, H. Masuhara, and N. Mataga, J. Phys. Chem., 87 (1983) 5333.
3. N. P. Ernsting and T. Arthen-Engeland, J. Phys. Chem., 95 (1991) 5502.
4. F. Urbach, Phys. Rev. 92 (1953) 1324.

Dynamics of Molecular Rotation at the Air/Water Interface by Time-Resolved Second-Harmonic Generation

A. Castro, D. Zhang, and K.B. Eisenthal

Department of Chemistry, Columbia University, NY 10027, USA

Abstract. Picosecond time-resolved second harmonic generation is applied to the study of rotational dynamics of free Rhodamine 6G molecules at the air/water interface. Rotational motions are observed following excitation along different molecular axes. The rotational relaxation time is about 400 ~ 500 picoseconds, which is significantly slower than that in the bulk water. The data also suggests that the out-of-plane motions at the interface are dominant and that the ground- and excited-state molecules have different equilibrium orientational distributions. A theoretical model is proposed to quantify our data.

1. Introduction

Molecular motions in liquids provide a powerful handle in understanding intermolecular interactions and thereby the chemical and physical properties of the liquid state [1-3]. The asymmetry in forces at an interface differentiates it from the time-averaged isotropic environment of the bulk and is the basis for the unique properties of interfaces. Studies of orientational relaxation in the bulk liquids by fluoresce depolarization or transient absorption methods cannot be used to study interfacial molecular rotations unless the molecules are restricted to the interface, as for example tethered to an insoluble surfactant molecule. If the molecules of interest are present in the bulk liquid, their contributions will

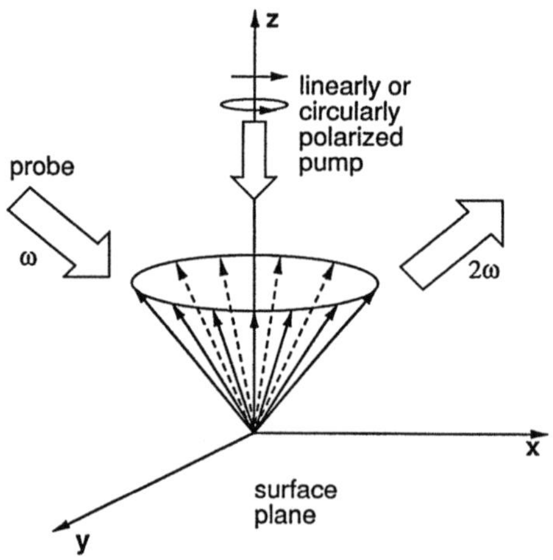

Fig. 1. Schematic experimental set-up. The arrows represent the molecular transition dipoles arranged to a common origin. The linearly or circularly polarized pump disturbs the equilibrium orientational distribution at the interface. The relaxation of the perturbed distribution is monitored by observing the SH signal (2ω) generated by a time-delayed probe (ω).

Springer Series in Chemical Physics, Vol. 55 Ultrafast Phenomena VIII
Editors: J.-L. Martin · A. Migus · G.A. Mourou · A.H. Zewail

overwhelm the far smaller interfacial population. Using the interface selective technique of second harmonic generation [4-8] as the probe in a pump/probe experiment we have succeeded in carrying out the first measurement of rotations of free (untethered) molecules at an air/liquid interface. The idea of the experiment is to use a polarized picosecond pump pulse to preferentially excite molecules of a given orientation at the interface and thereby disturb the equilibrium orientational distribution. The relaxation of the perturbed distribution is monitored by the second harmonic signal generated by a time-delayed probe laser pulse, (Fig. 1). The SH signal changes as both the ground and excited state molecules undergo rotation towards their equilibrium distributions and as the excited molecules relax back to the ground state.

2. Experimental

A 20-picosecond Nd:YAG laser running at 10 Hz was used to generate the probe pulses at 532 nm, and the pump pulses at 355 nm and 532 nm. The 532 nm probe generated the interfacial second harmonic light at 266 nm, which was detected using a polarizer, a monochromator and a photomultiplier tube and processed with a boxcar and a computer. The system was Rhodamine 6G (20 and 30 μM) adsorbed to the air/water interface.

3. Results and discussion

In our SHG experiments, we measured the time dependence of $\chi^{(2)}{}_{zxx}$ and $\chi^{(2)}{}_{xzx}$ second-order nonlinear susceptibility elements, where Z is normal to the interface. They can be expressed as a sum of ground state and excited state contributions:

$$\chi^{(2)}{}_{ijk}(t) = N_g(t)\, \Sigma < T^{i'j'k'}_{ijk}(\theta,\phi,\psi,t) >_g \alpha^{(2)}{}_{i'j'k'}(g) + N^*(t)\, \Sigma < T^{i'j'k'}_{ijk}(\theta,\phi,\psi,t) >^* \alpha^{(2)}{}_{i'j'k'}{}^* \quad (1)$$

where the subscript g and the star (*) refer to ground- and excited-state molecules respectively, and $< T^{i'j'k'}_{ijk}(\theta,\phi,\psi,t) >$ is the time-dependent direction cosine matrix that transforms the laboratory axes (i,j,k) into the molecular axes (i',j',k'). As seen in Eq. (1), the time evolution of the SH signal may be due to the change in the relative population of ground and excited state molecules N(t), and to orientational changes in the ground and excited state distributions ($< T^{i'j'k'}_{ijk}(\theta,\phi,\psi,t) >$).

3.1. Excitation with linearly polarized light at 355 nm (the short molecular axis) and 532 nm (the long molecular axis)

Figure 2 shows the time dependence of $\chi^{(2)}{}_{zxx}$ and $\chi^{(2)}{}_{xzx}$ using linearly polarized excitation at 355 nm. Our data clearly shows a different dependence of the dynamics for the two elements. The polarization dependent kinetics did not change when the interface population was increased, which shows that we have observed the rotational motions of Rhodamine 6G molecules at the air/water interface and not interfacial energy transfer. The observation that the $\chi^{(2)}{}_{zxx}$ first decreases and then increases and exceeds the negative time value suggests that the equilibrium orientational distribution of the photoexcited molecules differs from that of the ground state molecules. The faster component for both elements is in the same range, 400 ~ 500 ps, and is assigned to the rotational relaxation. The slower return to baseline component is dominated by the decay of the excited molecules on the time scale of 3 ~ 5 ns.

Further evidences of rotational motions at the air/water interface were obtained from another set of experiments involving excitation along the molecular long

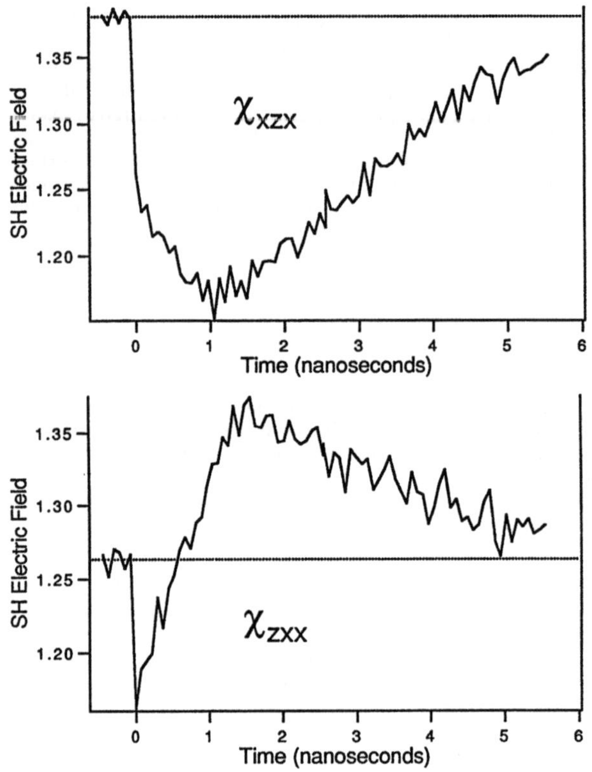

Fig. 2. Time dependence of $\chi^{(2)}{}_{xzx}$ and $\chi^{(2)}{}_{zxx}$ using linearly polarized excitation at 355 nm from the surface of a 20 μM Rhodamine 6G aqueous solution.

axis at 532 nm (Fig. 3). Once again, both elements show very different dynamics. The signal corresponding to $\chi^{(2)}{}_{xzx}$ has an initial decrease at zero-time, and then returns back to the ground state value. From a bi-exponential fitting, we find that the fast component has a relaxation time of 400 ~ 500 ps, which is the same as the initial dynamics observed with 355 nm excitation. The signal for $\chi^{(2)}{}_{zxx}$, however, remains constant within experimental error. The insensitivity of the $\chi^{(2)}{}_{zxx}$ element to the 532 nm excitation can be described in terms of the initial orientation of the ground state Rhodamine 6G molecules.

3.2. Excitation with circularly polarized light at 355 nm and 532 nm

In order to differentiate in-plane and out-of-plane molecular motions, we performed experiments using circularly polarized light at both 355 nm and 532 nm. Unlike linearly polarized excitation, the circularly polarized light excites the molecules isotropically in the surface plane. We find that the dynamic curves for $\chi^{(2)}{}_{zxx}$ and $\chi^{(2)}{}_{xzx}$ for circularly polarized excitation are qualitatively the same as those obtained using linearly polarized excitation for both 355 nm and 532 nm pumps respectively (Figs. 2, 3). This indicates that the observed rotational relaxation is due chiefly to the out-of-plane motions of the molecules at the surface. Possible explanations of this finding are: the in-plane motion is faster

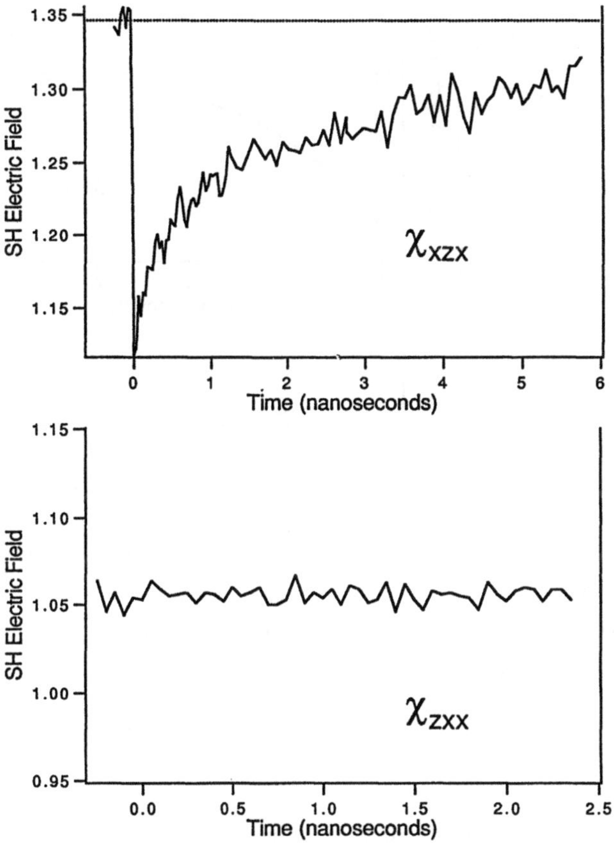

Fig. 3. Time dependence of $\chi^{(2)}{}_{xzx}$ and $\chi^{(2)}{}_{zxx}$ using linearly polarized excitation at 532 nm from the surface of a 20 μM Rhodamine 6G aqueous solution.

than our 20 ps time resolution, or such motion does not contribute significantly to the SH signal.

In order to explain the main features of our data, a simple model is proposed. It is assumed that the excited-state nonlinear polarizability, $\alpha^{(2)}$ is smaller than that of the ground state molecules; and the excited- and ground-state molecules have different equilibrium orientational distributions. We find that the out-of-plane motions of Rh 6G molecules, described by the Euler angle θ, can be used to qualitatively explain the dynamics observed. The Euler angle ϕ is random for the case of the circularly polarized excitation. Based on this model, our theoretical simulations qualitatively describe the main features of our four sets of data, (Figs. 4, 5). Theoretical fits give $\theta \approx 70° \pm 10°$ for ground state molecules and $40° \pm 10°$ for the excited state molecules. The Euler angle ψ corresponding to rotation about the molecular axis is $90° \pm 10°$.

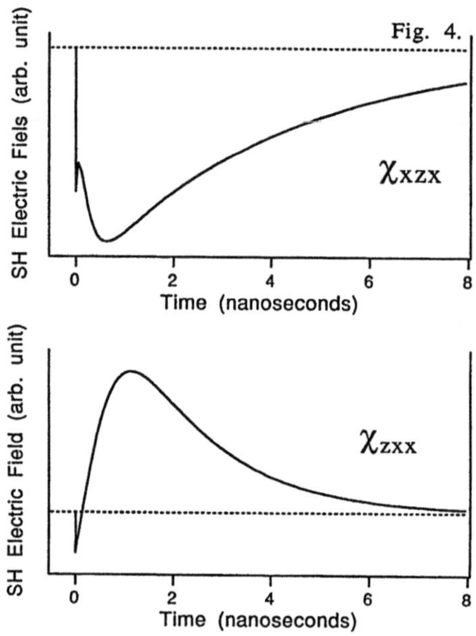

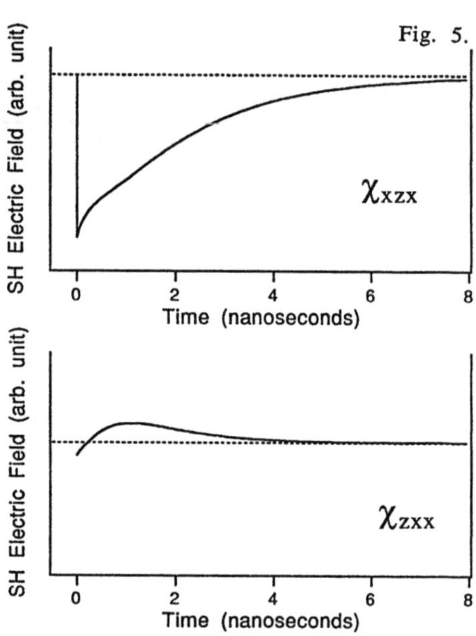

Fig. 4. Theoretical simulations of $\chi^{(2)}{}_{xzx}$ and $\chi^{(2)}{}_{zxx}$ for 355 nm excitation.

Fig. 5. Theoretical simulations of $\chi^{(2)}{}_{xzx}$ and $\chi^{(2)}{}_{zxx}$ for 532 nm excitation.

4. Conclusions

In conclusion, we have observed for the first time the rotational relaxation of free molecules at the air/water interface. Using time-resolved second harmonic generation, we found that the rotational relaxation time of Rhodamine 6G is about 400 ~ 500 ps which is significantly slower than the 200 ps value observed in the bulk water [9]. Our data also suggests that the equilibrium orientational distribution of the photoexcited molecules differs from that of the ground-state molecules. Since the circularly polarized excitation gives similar results to those of the linearly polarized excitation, it is inferred that the rotation relaxation observed by our SH experiments is chiefly due to the out-of-plane motions of the molecules at the surface. A theoretical model is proposed which qualitatively describes the basic features of our data.

We wish to acknowledge the National Science Foundation, the Air Force Office of Scientific Research and the Petroleum Research Fund for financial support of this research. We are also grateful to Dr. T. F. Heinz for valuable discussions.

References:

1. T. J. Chuang, K. B. Eisenthal, Chem. Phys. Lett. **11**, 368 (1971).
2. A. J. Cross, D. H. Waldeck, G. R. Fleming, J. Chem. Phys. **78**, 6455 (1983).
3. P. K. Wolber, B. S. Hudson, Biochemistry **20**, 2800 (1981).
4. N. Bloembergen, P. S. Pershan, Phys. Rev. **128**, 606 (1962).
5. C. K. Chen, T. F. Heinz, D. Ricard, Y. R. Shen, Phys. Rev Lett. **46**, 1010 (1981); Chem. Phys. Lett. **83**, 455 (1981); Phys. Rev. **B27**, 1965 (1983).

6. J. M. Hicks, K. Kemnitz, K. B. Eisenthal, T. F. Heinz, J. Phys. Chem. **90**, 560 (1986); K. Kemnitz, K. Bhattacharyya, G. R. Pinto, K. B. Eisenthal, T. F. Heinz, Chem. Phys. Lett. **131**, 285 (1986); K. Bhattacharyya, E. V. Sitzmann, K. B. Eisenthal, J. Chem. Phys. **87**, 1442 (1987).
7. A. L. Harris, C. E. D. Chidsey, N. J. Levinos, D. N. Loiacone, Chem. Phys. Lett. **141**, 350 (1987).
8. G. Marowsky, . Gierulski, B. Dick, Opt. Commun. **52**, 339 (1985).
9. K. Berndt, H. Durr, D. Palme, Opt. Commun. **42**, 419 (1982).

Energy Relaxation and Redistribution in Large Molecules in Solution on Ultrafast Time Scales

C.B. Harris, J.C. King, K.E. Schultz, B.J. Schwartz, and J.Z. Zhang

Department of Chemistry, University of California, Berkeley, CA 94720, USA

Abstract: Fundamental studies of the ultrafast dynamics of a series of large complex molecules in solution have yielded new information concerning their photodissociation and recombination processes, and the disposal of excess energy into the dissociated photofragments and the surrounding solvent bath.

Liquid phase energy transfer has been studied in a range of molecules of varying complexity, providing a unique opportunity to compare their dynamics in solution. The issues studied include the effect of solvent on the vibrational cooling and, where there is photodissociation, the partitioning of energy in the fragments.

The first study investigated the vibrational relaxation dynamics of azulene in simple liquids. The relaxation of vibrationally excited, ground state azulene, previously studied by Kaiser and coworkers in polyatomic liquids [1], was measured in compressed noble gases, such as xenon and krypton, whose density and temperature could be controlled independently over a wide range of values. The vibrational relaxation rate is strongly affected by the solvent mass and weakly affected by the solvent temperature. The most important result was the surprising independence of the vibrational relaxation rate from the surrounding solvent density (Fig. 1).

An analysis of the results using Isolated Binary Collision (IBC) theories accurately predicts the mass effect and temperature effect observed experimentally. However, the anomalous density effect cannot be described by conventional IBC theories. Instead large scale molecular dynamics simulations have provided the structural information crucial to understanding these results. These simulations show that the density of xenon atoms above and below the plane of the azulene rings does not change as function of solvent density. Since low frequency, out of plane modes are important to relaxation and apparently see no effective change in solvent density, a density independent relaxation rate is expected [2].

In order to study vibrational relaxation and energy redistribution in large molecules that undergo chemical reactions, a systematic study has been carried out recently on a series of transition metal carbonyl complexes, including $Mn_2(CO)_{10}$, $Re_2(CO)_{10}$, $Cr(CO)_6$, and $W(CO)_6$, several of which have long served as model systems for studying photochemical reaction dynamics in solutions [3-10]. The experiments were performed based on a pump-probe scheme using either a picosecond or a femtosecond laser system. In the femtosecond experiments (310 nm pump-620 nm probe) [11], the transient decay kinetics typically feature a pulse-width limited rise (100 fs) and a fast decay (200-500 fs), similar to what was previously observed by Nelson *et al.* [8]. These features are tentatively explained as follows: the pulse-width limited rise indicates fast dissociation (< 100 fs), and the fast decay is attributed to geminate recombination. This latter argument is supported by a study on CH_2I_2 in various halogenated methane solvents, where a similar fast decay feature is observed. The magnitude of this decay component is found to be solvent sensitive and correlates well with the expected trend in quantum yield of the photoproducts (CH_2I + I), consistent with the interpretation of geminate recombination. These results suggest that geminate recombination could be the rule of thumb rather than exception in the photodissociation of polyatomic molecules in solutions.

After photodissociation at 295 nm, internally hot photofragments are generated in solution. The vibrational energy relaxation of the photoproducts is monitored with a picosecond laser system (1 ps pulse width) in various solvents at a variety of probe wavelength. The vibrational cooling of $Mn_2(CO)_9$ in cyclohexane occurs through two different channels, a slow one on the

Springer Series in Chemical Physics, Vol. 55 **Ultrafast Phenomena VIII**
Editors: J.-L. Martin · A. Migus · G.A. Mourou · A.H. Zewail

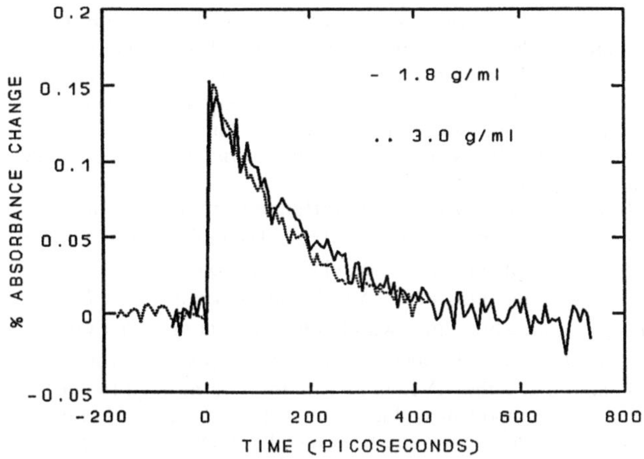

Fig.1 Effect of solvent density on ground state azulene relaxation rate in liquid xenon at 280 K.

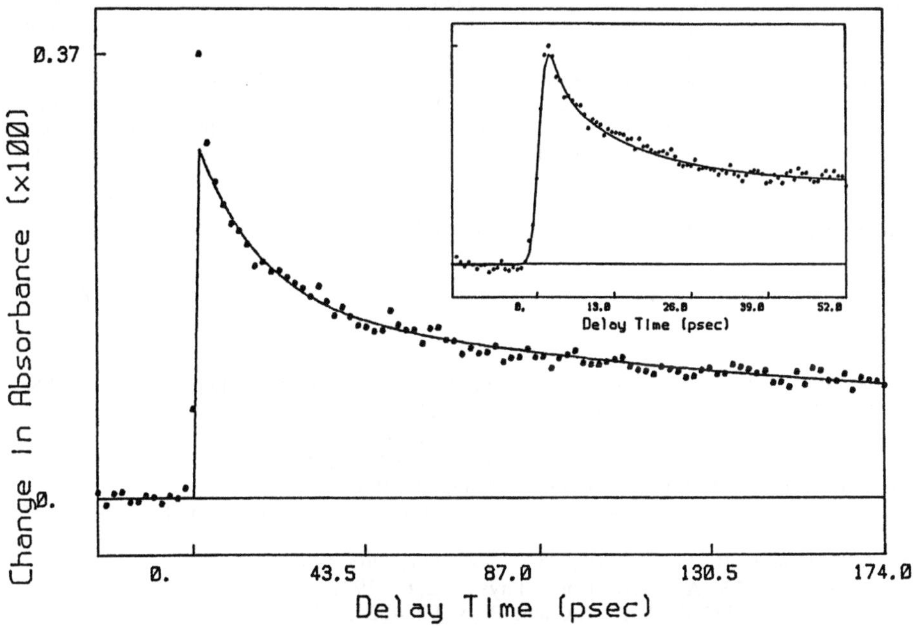

Fig.2 Transient decay kinetics of $Mn_2(CO)_9$ probed at 580 nm following dissociation of $Mn_2(CO)_{10}$ in cyclohexane (10 mM) at 295 nm, fitted with 15 ps and 170 ps decay time constants plus an offset.

time scale of 170 ps and a fast one with a time constant of 15 ps (Fig.2). Both decay components are faster in 2-propanol at the same probe wavelength. This double decay is similar to that observed in $Cr(CO)_6$ [7]. Based on resonance Raman studies [10], the slow component is assigned to relaxation through the CO mode in the photofragment while the fast component is attributed to relaxation through low frequency modes. This indicates that the Intramolecular Vibrational Redistribution (IVR) is slow in these species *and* there is significant CO excitation in the fragment following photodissociation.

The cooling behavior of $Re_2(CO)_9$ and $W(CO)_5$ are very different from that of $Mn_2(CO)_9$ and $Cr(CO)_5$ in that they show only one fast decay feature (< 20 ps). The radicals $Re(CO)_5$ and $Mn(CO)_5$ also exhibit only a fast decay component. A careful examination of the data reveals the following interesting trend: species containing heavy metal atoms (such as $Re_2(CO)_9$ and $W(CO)_5$) show a single fast decay feature while those with lighter metal atoms ($Cr(CO)_5$ and $Mn_2(CO)_9$) show double decay features. The absence of the slow component (>100 ps) in these heavy metal species indicates that either IVR is fast or there is no significant CO excitation. If the absence of the slow component is due to the lack of CO excitation in the photofragments, then the difference between heavy and light metal species can be explained qualitatively by the different kinematic coupling strengths between the CO mode in the fragment and the M-CO dissociation coordinate. This coupling strength is weaker for a heavier metal atom, and thus less CO excitation is expected. On the other hand, if IVR is primarily responsible for the difference, it indicates a sensitive dependence of energy redistribution on the vibrational modes of the molecule, i.e., IVR is faster in the heavy metal species. This could be due to the higher density of states present in the heavier species, which would provide better coupling between the high frequency modes (e.g. CO) and low frequency modes and thus more efficient energy redistribution.

These studies indicate that the rate of vibrational relaxation and IVR depend on both the distribution of solvent molecules and the modes of the solvent that can accept energy. Further work using time-resolved IR techniques should help to better reveal the particular modes responsible for the energy transfer and thus to better understand the detailed mechanism for energy redistribution/dissipation.

Acknowledgements: This research was supported by grants from NSF.

References:

1. W. Wild, A. Seilmeier, N. H. Gottfried and W. Kaiser, Chem. Phys. Lett., 119, 259, 1985.
2. K. E. Schultz, D. J. Russell and C. B. Harris, submitted to J. Chem. Phys., 1992.
3. X. Xie and J. D. Simon, J. Am. Chem. Soc. 112, 1130, 1990.
4. L. Wang, X. Zhu and K. G. Spears, J. Phys. Chem. 93, 2, 1989.
5. T. J. Meyer and J. V. Caspar, Chem. Rev. 85, 187, 1985.
6. L. J. Rothberg, N. J. Cooper, K. S. Peters and V. Vaida, J. Am. Chem. Soc., 104, 3536, 1982.
7. M. Lee and C. B. Harris, J. Am. Chem. Soc., 111, 8963, 1989.
8. A. G. Joly and K. A. Nelson, Chem. Phys., 152, 69, 1991.
9. J. Z. Zhang and C. B. Harris, J. Chem. Phys., 95, 4024, 1991.
10. S. C. Yu, X. Xu, R. Lingle, Jr. and J. Hopkins, J. Am. Chem. Soc., 112, 3668, 1990.
11. J. C. King, J. Z. Zhang, B. J. Schwartz and C. B. Harris, submitted to J. Chem. Phys.

Photodissociation and Recombination Dynamics of I_2^- in Solution

J.C. Alfano, D.A.V. Kliner, A.E. Johnson, N.E. Levinger, and P.F. Barbara

Department of Chemistry, University of Minnesota,
Minneapolis, MN 55455, USA

We have made the first measurement on the photodissociation and recombination/cage escape of I_2^- in solution using a three laser pulse sequence (synthesis-pump-probe) as follows:

$$I_3^- \xrightarrow{\text{photolysis}} I + I_2^-$$

$$I_2^-(^2\Sigma_u^+) \underset{\text{recombination/relaxation}}{\overset{\text{pump/probe}}{\rightleftarrows}} I_2^-(^2\Pi_{g,3/2}) \xrightarrow{\text{cage escape}} I(^2P_{3/2}) + I^-$$

The photolysis pulse (355nm or 380nm) excites a dissociative band of I_3^- which leads to prompt formation of I_2^-.[1] After photolysis we wait several nanoseconds in order to allow for vibrational relaxation, solvation, and diffusion apart of the I and I_2^- fragments. The pump/probe pair of pulses study the dynamics of the various electronic and nuclear processes involved in the photodissociation, recombination and cage escape of the I_2^- fragment.

In contrast to the prototypical example of I_2 photodissociation in nonpolar solvents, where short range repulsive forces dominate the solute/solvent interactions[2], the I_2^- system has strong long-range electrostatic interactions, making it a promising prototype for a variety of polar reactions in solution. For I_2^-, the gas-phase bond strength (0.9 eV)[3] is comparable to the change in solvation energy as I_2^- dissociates into $I + I^-$, thus creating an extraordinary coupling between the bond length and polar solvent coordinate during the reaction. In the gas-phase the lowest excited state of I_2^- is repulsive[4], leading to a direct dissociation into the ground state dissociation channel, I^- and $I(^2P_{3/2})$. The absorption spectrum in solution is similar to the gas-phase; therefore, it is likely that photodissociation in solution is also direct.

Figure 1 shows transient pump-probe signal for I_2^- in water, methanol, ethanol, and *n*-propanol. The rate of recovery of the spectrum is orders of magnitude faster than that observed for I_2 in nonpolar solvents. This suggests that recombination and vibrational relaxation are much faster for I_2^- than I_2, presumably resulting from much stronger solute/solvent interactions in the former case. In fact, in water, the bleach recovers entirely on the timescale of about one picosecond, except for a small residual bleach due to cage escape (vide infra). Assuming that bleach recovery dynamics in water is a semi-quantitative measure of the overall photodissociation/recombination process (i.e. photoexcitation, electronic internal conversion, and vibrational relaxation) then the overall process takes less than 10 vibrational periods of I_2^-!

Figure 2 shows pump-probe transients on a longer timescale allowing for observation of the residual bleach due to cage escape of the free radicals. The measured cage escape yields are 3±1% for water, 16±2% for methanol, 12±2% for ethanol, and 0±1% for *n*-propanol. In the alcoholic solvents, the yield decreases monotonically with increasing solvent mass in accord with conventional expectations for a cage escape process. Interestingly, while water is much less massive than methanol or ethanol, it exhibits significantly less cage escape, perhaps as a result of a hydrogen bonding network. Alternatively, this low cage escape yield may result from a much greater dielectric friction on the dissociation reaction in this solvent. Both mechanism would be novel for a dissociation reaction of a diatomic.

The transmittance changes of I_2^- is small (<1 %) due to the small concentration of I_2^- produced by the photolysis and to the small I_2^- extinction coefficient. Using a conventional amplified dye-laser spectrometer it was necessary to use identical pump and probe wavelengths at the laser fundamental, rather than the picosecond continuum, in order to achieve adequate signal-to-noise levels. Measurements at identical pump and prove wavelengths are often affect-

Springer Series in Chemical Physics, Vol. 55 Ultrafast Phenomena VIII
Editors: J.-L. Martin · A. Migus · G.A. Mourou · A.H. Zewail

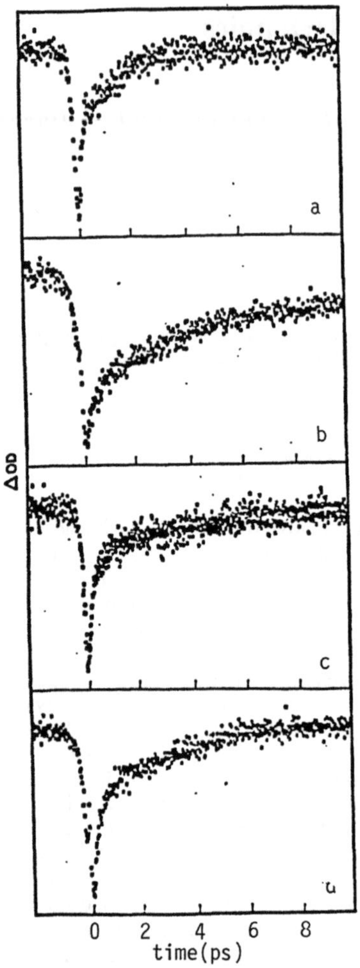

Figure 1. Pump-probe transients of I_2^- in a)water, b) methanol, c) ethanol, and d) n-propanol with $\lambda_{photolysis}$ =355nm, and λ_{pump} and λ_{probe}=790nm. Transients were taken using a YAG-regen amplified dye laser apparatus.

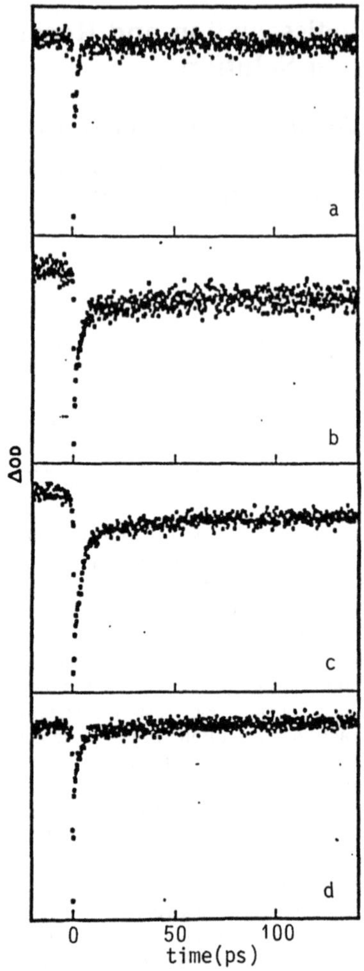

Figure 2. Pump-probe transients of I_2^- in a)water, b) methanol, c) ethanol, and d) n-propanol with $\lambda_{photolysis}$ =355nm and λ_{pump} and λ_{probe} =790nm with a longer timescale than figure 1.

ed by coherent interactions complicating the interpretation. Very recently, we have obtained high signal-to-noise variable wavelength pump-probe data (Figure 3a) on I_2^- using a newly constructed spectrometer based on an amplified femtosecond Ti:sapphire laser and continuum generation for the probe light, based on a recently reported design.[5] The results confirm the behavior we observed with the identical wavelength pump-probe measurements. We are now in the process of recording the full evolution of the transient absorption spectra of I_2^-.

The new apparatus is many times more sensitive than the dye-laser-based systems we have previously constructed, and can easily record high signal-to-noise transient pump-probe signals at various wavelengths throughout the near infrared, visible, and ultraviolet in a few minutes of averaging. For example, figure 3b shows a pump-probe measurement of betaine-30 in acetone. The maximum of the signal is about a 5% change in transmittance at the maximum.

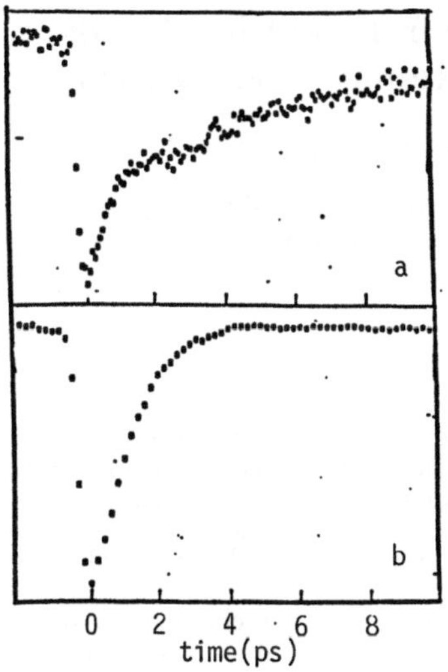

Figure 3. Pump-probe transients taken with Ti:sapphire spectrometer. a) I_2^- in ethanol, with $\lambda_{photolysis}$ =380nm, λ_{pump}=760nm, and λ_{probe}=720nm b) Betaine-30 in acetone, with λ_{pump}=760nm and λ_{probe}=720nm.

It is interesting that our measured timescale for vibrational relaxation of I_2^- in solution is comparable to that of I_2^- in CO_2 clusters[6], and other ions in solution such as N_3^- in polar solvents.[7] Similarly, the timescale we observe for the dynamics of photoexcited I_2^- in ethanol is similar to that seen in relaxation of "hot" I_2^- photofragments in photolysis of I_3^-.[1]

Acknowledgments

Support of this research by the U.S. National Science Foundation is gratefully acknowledged. NEL and DAVK were supported by National Science Foundation Postdoctoral Fellowships.

References

1. U. Banin, A. Waldman, S. Ruhman; *J. Chem. Phys.* **96**, 2416(1992).
2. a) A. L. Harris, J. K. Brown, C. B. Harris, *Ann. Rev. Phys. Chem.* **39**, 341(1988). b) N. F. Scherer, L .D. Ziegler, G. R. Fleming; *J. Chem. Phys.* **96**, 5544(1992). c) T J. Chuang, G. W. Hoffman, K. B. Eisenthal; *Chem. Phys. Lett.* **25**, 201 (1974). d) R. M. Bowman, M. Dantus, A. H. Zewail; *Chem. Phys. Lett.* **161**, 297 (1989).
3. G. N. R. Tripathi, R. H. Schuler, R. W. Fessenden; *Chem Phys. Lett.* **113**, 563(1985).
4. E. C. M. Chen, W. E.Wentworth; *J. Phys. Chem.* **89**, 4099 (1985).
5. F. Salin, J. Squier, G. Mourou, G Vaillancourt; *Opt. Letters* **16**, 1964(1991).
6. J. M. Papanikolas, J. R. Gord, N. E. Levinger, D. Ray, V. Vorsa, W. C. Lineberger; *J. Phys. Chem.* **95**, 8028(1991).
7. J. C. Owrutsky, Y. R. Kim, M. Li, M. J. Sarisky, R. M. Hochstrasser; *Chem. Phys. Lett.* **184**, 368(1991).

Probing the Microscopic Molecular Environment in Liquids with Femtosecond Fourier-Transform Raman Spectroscopy

D. McMorrow[1], *S.K. Kim*[1], *J.S. Melinger*[1], *and W.T. Lotshaw*[2]

[1]Naval Research Laboratory, Code 4613, Washington, DC 20375, USA
[2]G.E. Research and Development, P.O. Box 8, Rm. KWD-270, Schenectady, NY 12301, USA

Intermolecular dynamics in liquids that occur on the sub-500 fs timescale are determined in part by the details of the intermolecular potentials. Such dynamics traditionally have been investigated with frequency-domain techniques, such as spontaneous light scattering (LS) spectroscopy. Frequency-domain techniques, however, have provided only limited information on the intermolecular vibrational dynamics of liquid systems.

In recent work[1,2] we have demonstrated that nonlinear-optical techniques involving optical heterodyne detection (OHD) are among the more sensitive methods for measuring intermolecular dynamics, possessing tremendous potential for advancing our knowledge of liquid dynamics. We describe here results from an extensive investigation into the OHD optical Kerr effect (OKE) dynamics of numerous molecular liquids. We take advantage of the recently developed Fourier-transform/deconvolution techniques[2] to separate the nuclear and electronic contributions to the measured transients, and present the data in the physically intuitive frequency domain (spectral density) representation.[2] These results provide a particularly clear picture of how the femtosecond dynamics of a probe molecule are shaped by its microscopic environment.

The role of the local environment in determining the intermolecular dynamics of liquids is revealed in the most controlled fashion through dilution studies. In this manner, the molecules of a neat liquid are systematically replaced by solvent molecules, giving rise to an effectively continuous variation in the intermolecular potential seen by the original species. We have investigated the OKE transients of CS_2 in several solvents; in figs. 1-3 we present data for dilutions of CS_2 in the weakly interacting alkane *n*-pentane. Preliminary measurements for some alkanes have been presented.[3] Here we present new results with an analysis in terms of the Fourier-transform methods noted above.

Fig. 1 shows the nuclear part of the OHD OKE transients (generated from the measured data using the Fourier-transform methods noted above[2]). Figure 2 shows the corresponding frequency-domain representation, which contains all of the information on the nuclear response of the liquid to a δ-function optical pulse (for the frequency region shown). Changes in the time-domain transients on dilution are quite subtle, with some broadening and redistribution of intensity being evident. These changes are more clearly revealed in the spectral data of fig. 2. For *neat* CS_2, two distinct contributions are evident: a sharp, low-frequency feature associated with the 1.6 ps diffusive relaxation of the orientational anisotropy; and a broad, diffuse band that is associated with the *non-diffusive intermolecular dynamics.* On dilution with the weakly interacting pentane solvent, the diffusive peak broadens (consistent with the lower viscosity of the solution), while the spectral distribution of the non-diffusive band shifts to lower frequency. At the lowest dilution investigated (10%), only a single, structureless band is discernable (although evidence for the slowly relaxing ~1.2 ps diffusive reorientational contribution is clearly evident in the lower curve of fig. 1).

The non-diffusive contributions are of particular interest since they directly reflect the local intermolecular potential seen by the individual CS_2 molecules. Assuming that the diffusive relaxation can be described by a diffusion equation (with inertia), that contribution can be removed from the data in a simple fashion. The resulting spectra, which contain only the non-diffusive (vibrational) contributions, are shown in fig. 3. Plotted in this fashion, the changes in the vibrational spectral density are particularly clear: *the intermolecular vibrational distribution narrows and shifts to lower frequency on increasing alkane concentration.* The changes are quite significant: the band maximum

Springer Series in Chemical Physics, Vol. 55 Ultrafast Phenomena VIII
Editors: J.-L. Martin · A. Migus · G.A. Mourou · A.H. Zewail

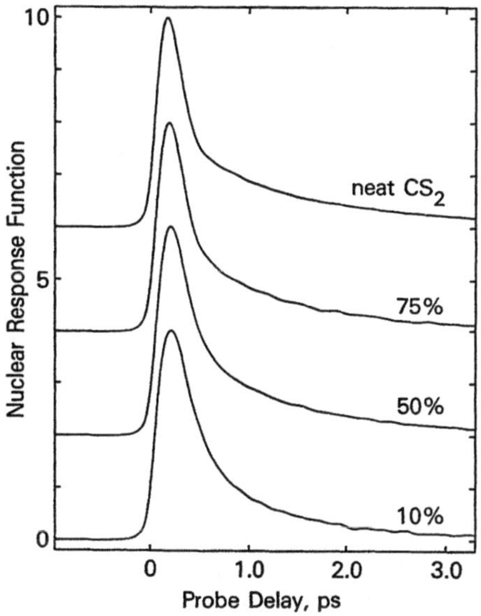

Figure 1. Nuclear part of OHD OKE response functions for CS_2 diluted in n-pentane.

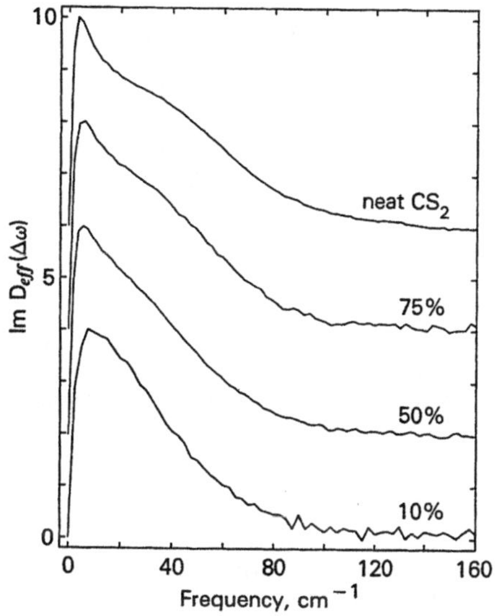

Figure 2. Impulse frequency response functions for CS_2 diluted in n-pentane.

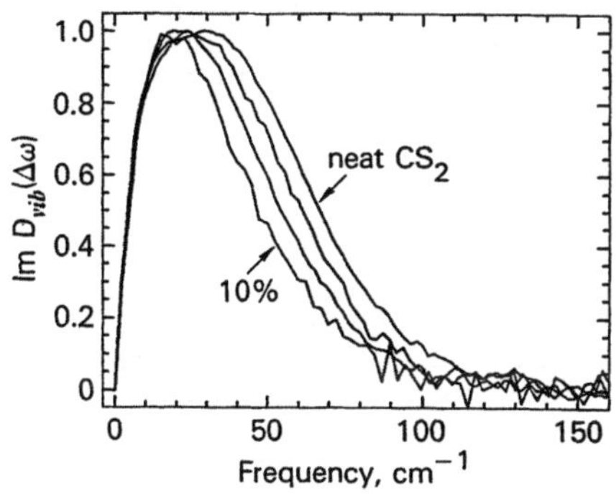

Figure 3. Non-diffusive (vibrational) part of the frequency response functions for the CS_2 dilutions shown in fig. 2.

shifts from 30 cm^{-1} in neat CS_2 to 17 cm^{-1} in the 10% solution, while the bandwidth decreases from 64 cm^{-1} to 42 cm^{-1}. The observed shifts of the spectral distribution with increasing dilution are consistent with those expected for an intermolecular vibrational mode as the effective intermolecular potential changes when CS_2 molecules are replaced by the weakly interacting pentane molecules.

In this report we have presented data for a single dilution series. These data, especially when presented in the spectral density representation, reveal the sensitive role of the local environment in shaping the non-diffusive femtosecond dynamics of molecular liquids. Dilutions of CS_2 in a range of alkanes and some more strongly interacting solvents have been investigated and will be presented elsewhere.

1. D. McMorrow, W.T. Lotshaw and G.A. Kenney-Wallace, IEEE J. Quant. Elect. QE-24, 443 (1988).
2. D. McMorrow and W.T. Lotshaw, Chem. Phys. Lett. 174, 85 (1990).
3. C. Kalpouzos, D. McMorrow, W.T. Lotshaw, and G.A. Kenney-Wallace, Chem. Phys. Lett. 150, 138 (1988).

The Homogeneity of Liquid Phase Vibrational Line Broadening from Raman Echo Experiments

L.J. Muller, D. Vanden Bout, and M. Berg

Department of Chemistry and Biochemistry, University of Texas, Austin, TX 78712, USA

Abstract. The first successful Raman echo experiment in a liquid has been performed. The results show that the sym-CH_3 stretch of acetonitrile is homogeneously broadened. The dominant interaction with the solvent is modulated faster than 0.5 ps.

1. Introduction

The dynamic interaction of a vibration with a solvent is reflected in the vibrational dephasing time. However, essential questions about the correlation time and physical origin of the interaction can only be answered if the homogeneity of the dephasing is known. Numerous indirect arguments, both experimental and theoretical, have been advanced for the occurance of either homogeneous or inhomogeneous dephasing without reaching an overall consensus.

Both isotropic Raman linewidths and time-resolved coherent Raman experiments measure the vibrational dephasing time, but neither can directly determine the homogeneity of the dephasing. Only an experiment exploiting a higher-order nonlinear response of the material, such as the Raman echo, can determine definitively whether the dephasing is homogeneous or inhomogeneous.[1]

2. The Raman Echo Experiment

The Raman echo applies to vibrational transitions the same principles of coherence spectroscopy used in the more familiar spin or photon echo.[2] However in the Raman echo, the vibration is excited by stimulated Raman scattering instead of by direct absorption, and the resulting coherence is detected by coherent Raman scattering rather than by coherent emission. The Raman echo first appears in seventh-order perturbation theory and is an eight-wave mixing or $\chi^{(7)}$ process. Because of the high order of this process and the weakness of Raman transitions, at least one group had concluded that the Raman echo effect is orders of magnitude too small to be observed under experimentally achievable conditions.[3]

A total of five, synchronized, subpicosecond pulses at two visible wavelengths is required. Initially, the vibration is coherently excited by stimulated Raman scattering between a pair of pulses separated in frequency by the vibrational frequency. After a delay time t_1, a second pair of pulses re-excites the vibration and begins a rephasing of any slowly modulated perturbations. The coherence remaining at a time t_2 after the second excitation is measured by coherent anti-Stokes' scattering from a fifth pulse.

Note that unlike the photon echo, where only the integrated intensity of the coherence is detected, the complete time dependence of the coherence is measured in the Raman echo. The interpretation is based on qualitative features of the shape. For fast solvent perturbations, the rephasing will be ineffective, and the coherence decay will be identical to the free induction decay. If slowly modulated perturbations are present, rephasing

Springer Series in Chemical Physics, Vol. 55 **Ultrafast Phenomena VIII**
Editors: J.-L. Martin · A. Migus · G.A. Mourou · A.H. Zewail

will cause a delayed coherence decay. The effects of this rephasing will increase as t_1 increases. If the slow perturbations are sufficiently strong, the coherence forms a delayed maximum, as in standard treatments of echo experiments.

3. Results and Discussion

The first Raman echo results on a liquid sample are presented here for the symmetric methyl stretch of acetonitrile.[4] Figure 1 shows the decay of the signal with t_2, the time between the second excitation pulse pair and the probe pulse, for various values of t_1, the time between the first and second excitation pulse pairs. For homogeneous dephasing, the decay in t_2 should be independent of t_1 and should match the Fourier transform of the isotropic Raman spectrum. Experimentally this is true; therefore, the dephasing is entirely homogeneous.

Figure 2 shows the sensitivity of the experiment to any inhomogeneity which might have been present. An inhomogeneous component of 5 cm^{-1} FWHM as suggested by time-resolved coherent Raman experiments and theoretical calculations [5] would have caused a substantially different decay. Note that a sharp, well defined "echo" is not anticipated even in this case.

Figure 2A also shows that an inhomogeneous component of as little as 2.2 cm^{-1} would have been detected. This coupling is small compared to the

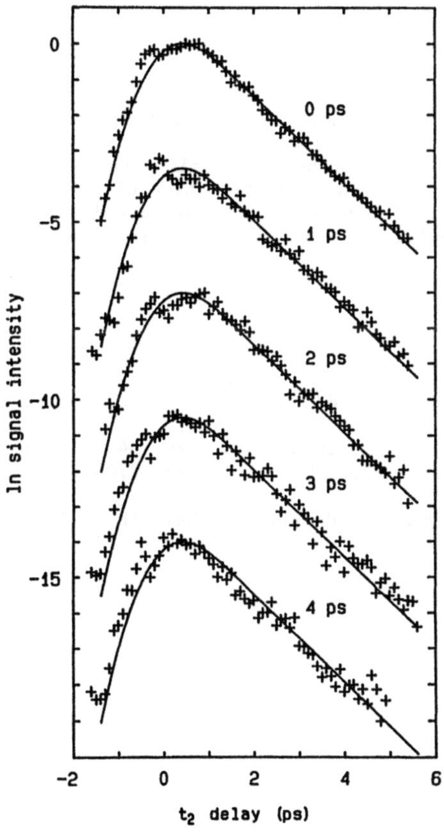

Fig. 1. Natural log of the Raman echo intensity (+) versus t_2, for increasing values of t_1. The decay shape does not change as t_1 is increased, indicating that the vibration is homogeneous. As predicted for a homogeneous vibration, the decay is always the Fourier transform of the isotropic Raman line shape (solid curves).

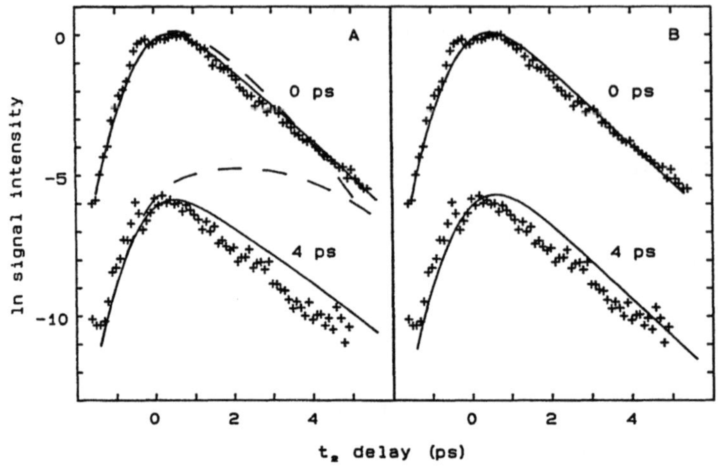

Fig. 2. Demonstration of the sensitivity of the Raman echo experiment by comparison of the experimental data for $t_1 = 0$ and 4 ps (+) with predicted decays for nonhomogeneous dephasing models. A) Dashed curve: Decay for an inhomogeneous component of 5 cm^{-1} FWHM as previously suggested.[4] Solid curve: Decay for an inhomogeneous component of 2.2 cm^{-1} FWHM (total line width of 6.5 cm^{-1}). B) Solid curve: Decay for nonhomogeneity due to a finite solvent correlation time of 0.5 ps.

30 cm^{-1} coupling to the fast, homogeneous component (estimated from the homogeneous dephasing time and Enskog collision time).

Alternately, a single modulation process can be hypothesized and its minimum rate estimated. Figure 2B shows that a modulation time of 0.5 ps or longer would have been experimentally detectable. The fast modulation actually observed is consistent with collisional models for vibrational dephasing.

References

1. R.F. Loring and S. Mukamel, J. Chem. Phys. **83**, 2116 (1985).
2. S.R. Hartmann, IEEE J. Quantum Electron. **QE-4**, 802 (1968)
3. M. Muller, K. Wynne and J.D.W. van Voorst, Chem. Phys. **128**, 549 (1988).
4. D. Vanden Bout, L. Muller and M. Berg, Phys. Rev. Lett. **67**, 3700 (1991).
5. S.M. George, A.L. Harris, M. Berg and C.B. Harris, J. Chem. Phys. **80**, 83 (1984).

Excited State Photoreactions of Chlorine Dioxide in Solution

R.C. Dunn and J.D. Simon

Department of Chemistry, University of California at San Diego,
9500 Gilman Drive, La Jolla, CA 92093-0341, USA

Abstract. The primary photochemistry of OClO in room temperature water and sulfuric acid solutions has been studied using picosecond absorption spectroscopy. In water, excitation of OClO at 355 nm results in partitioning between two excited state reaction channels: (1) bond breakage to form ClO and O and (2) isomerization to form ClOO. Following isomerization ClOO dissociates thermally into Cl and molecular oxygen. Analysis of the transient absorption data indicate that the quantum efficiency for isomerization in water solution is 0.10 ± 0.01.

1. Introduction

Chlorine oxides have been invoked in various reaction mechanisms for stratospheric ozone depletion [1-5]. Determination of the products and rate constants for reactions involving chlorine oxides has been an active area of research. Recently, there has been much interest in the photodynamics of chlorine dioxide (OClO). The photochemistry of OClO following excitation of the $^2A_2 \leftarrow {}^2B_1$ absorption band has been studied in the gas phase [6-11], rare gas and acid matrices [12-15] and various polar solutions [16-18]. These studies show that the photoreactivity of OClO involves competition between the two processes given in equation (1).

$$\begin{aligned} OClO &\rightarrow ClO + O \\ &\rightarrow ClOO \rightarrow Cl + O_2 \end{aligned} \tag{1}$$

This complex photochemistry is thought to be controlled by three closely spaced excited state surfaces. Excitation populates the bound $^2A_2$ state. Nearby in energy to the excited $^2A_2$ state are two additional excited states ($^2B_2$ and $^2A_1$). Ground state transitions to the $^2B_2$ surface are dipole forbidden. Absorption to the $^2A_1$ state is symmetry allowed, however no experimental observations of the $^2A_1 \leftarrow {}^2B_1$ absorption have been reported. These states are believed to give rise to the two photoreactions shown in Eq. 1 [19].

The formation of $Cl + O_2$ is believed to result from the thermal dissociation of the isomerized molecule, ClOO. This competition between reaction channels is important as the generation of Cl provides a pathway by which OClO can contribute a new ozone loss mechanism. Several gas phase studies have been reported which quantify the quantum efficiency of this channel. The quantum yields reported range from 10^{-5} to 0.15 [7-9]. As different experimental techniques have been used for each reported value, there is at present no consensus on the quantum yield for the formation of Cl.

Our studies have focussed on the solution photochemistry of OClO. In this paper, the photochemistry of OClO in water and sulfuric acid solution are summarized.

2. Experimental

Time resolved absorption kinetics were recorded using a high repetition rate picosecond laser system [20]. For all experiments reported, the pump and probe laser beams were depolarized to remove coherence artifacts and any other signals which might arise from polarization dependent properties of the sample. All samples were excited at 355 nm. Transient absorption dynamics were measured at several probe wavelengths between 260 nm and 420 nm. The instrument response was approximately 80 picoseconds.

OClO was generated by reaction of oxalic acid and potassium chlorate. The gaseous products, OClO and CO_2, were collected and bubbled through the solvent. CO_2 does not absorb in the spectral range studied. A one directional flow cell was used to prevent the build-up of photoproducts. Each wavelength probed required approximately 4 liters of solution.

3. Results and Discussion

Figure 1 shows the spectrum of OClO in water and sulfuric acid solutions. A complete study of the solution phase spectroscopy of OClO has been reported [17]. As seen in Figure 1, some of the vibronic structure observed in the gas phase is still present in the solution spectra. This vibronic structure varies significantly with solvent [17]. In addition, the maximum of the absorption band shifts with changing dielectric properties of the solvent.

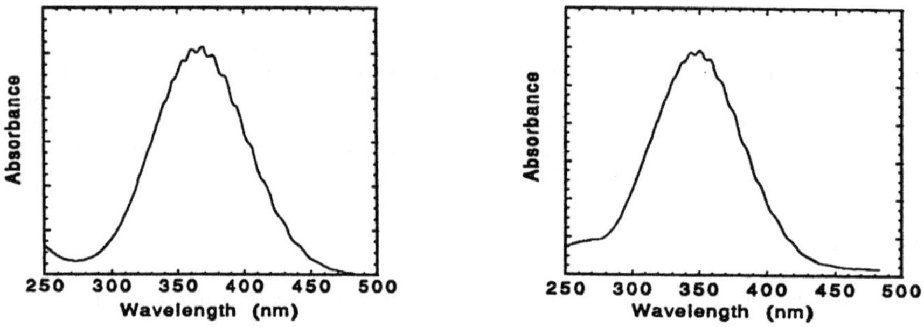

Figure 1: Steady state absorption spectra of OClO in room temperature water (left) and sulfuric acid (right) solution.

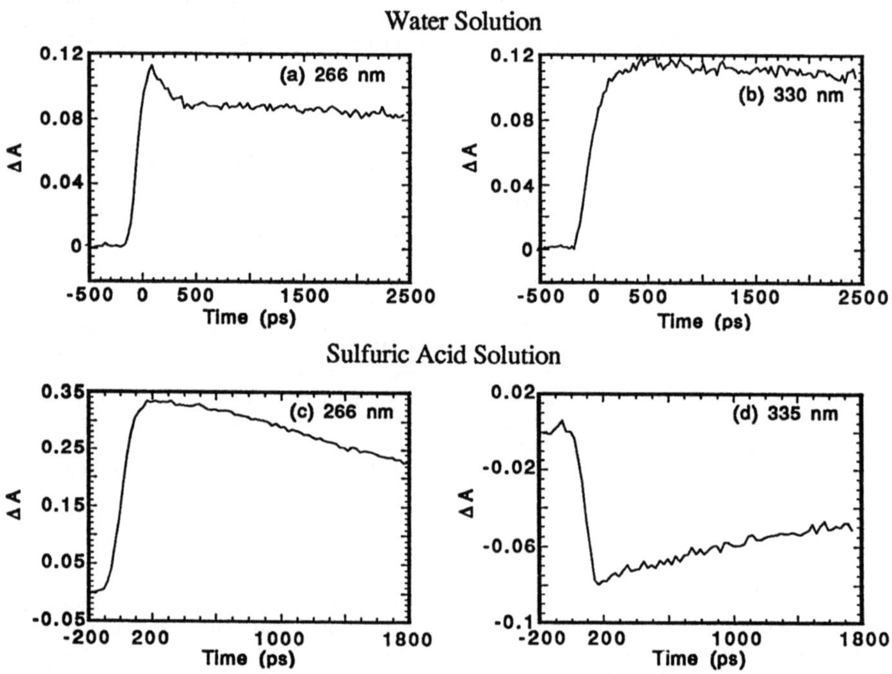

Figure 2: Time dependent absorption signals recorded at various probe wavelengths following excitation of OClO at 355 nm in room temperature water (top) and sulfuric acid (bottom) solutions. A detailed analysis of the water data has been reported [18]. A detailed account of the photodynamics of OClO in acid solution will be reported at a later date [23].

Figure 2 shows examples of the time dependent absorption signals following excitation of OClO at 355 nm in room temperature water and sulfuric acid solution.

For photolysis of OClO in water, the transient dynamics can be quantitatively modeled. This is possible as the absolute absorption cross sections and lifetimes of the reaction products Cl and ClO have been determined [21]. Solution phase spectra of ClOO have not been reported due to the thermal instability of this molecule. Gas phase spectra, however, have been reported [22]. Analysis of the time dependent absorption signals observed throughout the near-UV and visible spectral region revealed that 10% of the photoexcited OClO molecules gave rise to chlorine atom production [18].

However, the exact mechanism by which ClOO is formed in the solution photolysis of OClO is still uncertain. Within the current experimental time resolution (80 ps), the absorption data clearly show that photolysis leads to formation of ClOO and ClO. However, the detailed mechanism for formation of the ClOO isomer is unclear. Two possible pathways exist.

(1) ClOO forms via a surface crossing from the $^2B_2$ excited state of OClO and the ground electronic state of ClOO. This pathway is consistent with recent theoretical calculations [19].

(2) ClOO is formed from subpicosecond geminate cage recombination of the photodissociation products ClO and O, following rotation of the ClO fragment. Theoretical treatments also show that the reaction between ground state ClO($^2\Pi$) and O(^{3}P) correlates with the (A',A") states of ClOO.

Experiments are underway to resolve this issue.

The data shown in Figure 2 also show marked differences between water and sulfuric acid solutions. Over the wavelength range probed (266 - 420 nm), the sulfuric acid data reveal one kinetic process, consistent with dissociation of OClO into ClO and O. Unfortunately, the absorption spectrum of Cl in sulfuric acid has not been reported. Thus, the experimental data suggest one of the following two scenarios.

(1) The reaction channel leading to formation of Cl and O_2 does not occur in sulfuric acid.

(2) The Cl absorption spectrum in sulfuric acid occurs at wavelengths shorter than 260 nm.

We are presently carrying out studies in the wavelength range from 205 nm to 260 nm to address this question.

Acknowledgements. This work was supported by the NSF, and the MMFEL Program administered by ONR. We thank Professor V. Vaida for helpful discussions.

4. References

[1] M. J. Molina, F. S. Roland, *Nature*, **249**, 819, (1974).
[2] S. Solomon, G. H. Mount, R. W. Sanders, A. L. Schmeltekopf, *J. Geophys. Res.* **92**, 8329, (1987).
[3] S. Solomon, *Rev. Geophys.* **26**, 131 (1988).
[4] J. G. Anderson, W. R. Brune, R. J. Chan, *J. Geophys. Res.* **94**, 11480 (1989).
[5] V. Vaida, S. Solomon, E. C. Richard, E. Ruhl, A. Jefferson, *Nature* **342**, 405 (1989).
[6] E. C. Richard, V. Vaida, *J. Chem. Phys.* **94**, 153 (1991); **94**, 163 (1991).
[7] E. Bishenden, J. Hancock, D. J. Donaldson, *J. Phys. Chem.* **95**, 2113 (1991).
[8] H. F. Davis, Y. T. Lee, *J. Phys. Chem.*, submitted.
[9] W. G. Lawrence, K. C. Clemitshaw, V. A. Apkarian, *J. Geophys. Res.* **95**, 18591 (1990).
[10] E. Ruhl, A. Jefferson, V. Vaida, *J. Phys. Chem.* **94**, 2440 (1990).
[11] A. J. Colussi, *J. Chem. Phys.* **94**, 8922 (1990).
[12] J. L. Gole, *J. Phys. Chem.* **84**, 1333 (1980).
[13] A. Arkell, F. Schwager, *J. Amer. Chem. Soc.* **89**, 5999 (1967).
[14] M. M. Rochkind, G. C. Pimentel, *J. Chem. Phys.* **46**, 4481 (1967).
[15] F. J. Adrian, J. Bohandy, B. F. Kim, *J. Chem. Phys.* **85**, 2692 (1986).
[16] R. C. Dunn, E. Richards, V. Vaida, J. D. Simon, *J. Phys. Chem.* **95**, 6060 (1991).
[17] R. C. Dunn, B. N. Flanders, V. Vaida, J. D. Simon, *Spectrochim. Acta*, in press.
[18] R. C. Dunn, J. D. Simon, *J. Amer. Chem. Soc.*, in press.
[19] K. Peterson, H.-J. Werner, *J. Chem. Phys.*, in press.
[20] X. Xie, J. D. Simon, *Opt. Comm.* **69**, 303 (1989).
[21] U. K. Klaning, T. Wolff, *Ber. Bunsenges. Phys. Chem.* **89**, 243 (1985).
[22] R. L. Mauldin III, J. B. Burkholder, A. R. Ravishankara, *J. Phys. Chem.* **96**, 2582 (1992).
[23] R. C. Dunn, J. D. Simon, unpublished work.

Bimolecular Reactions are Power-Full

A. Masad[1], *S.Y. Goldberg*[1], *D. Huppert*[1], *and N. Agmon*[2]

[1]School of Chemistry, Sackler School of Sciences, Tel Aviv University, Tel Aviv 69978, Israel

[2]Department of Physical Chemistry, The Hebrew University, Jerusalem 91904, Israel

Abstract. A new asymptotic behavior is detected in reversible binary chemical reactions.

Bimolecular chemical reactions in solution are essentially a many-body problem. The conventional rate equations, which neglect spatial distributions, are satisfactory for intrinsically slow reactions. When reaction is diffusion controlled, competition between reacting particles introduces correlations in their diffusive motion. This should be manifested in the long-time dependence of the concentrations. Physicists [1] have focused on irreversible recombination reactions, A+B→AB, treating primarily the equal concentration and diffusion coefficient case, [A]=[B] and $D_A = D_B$. The situation is more complex when the reaction is reversible [2], since a recently bound AB molecule may subsequently dissociate to reproduce the A...B partners, thus altering their spatial distribution. After a few cycles of recombination/dissociation the initial distribution becomes irrelevant. One expects the densities to approach equilibrium, independently of the initial condition.

In the pseudo first-order case $[A] \ll [B] \equiv c$, the binding probability, $1 - \mathcal{S}(t) \propto [AB]$, approaches

$$1 - \mathcal{S}(\infty) = cK_{eq}/(1 + cK_{eq}) \tag{1}$$

as $t \to \infty$, where K_{eq} is the recombination equilibrium coefficient. The chemical rate-equations predict an exponential approach to this equilibrium limit. Mean-field approximations [3] suggest that for reactions in three dimensions the long-time behavior is [4]

$$[\mathcal{S}(\infty) - \mathcal{S}(t)]/[\mathcal{S}(\infty) - \mathcal{S}(0)] \sim K_{eq}/(1 + cK_{eq})\,(4\pi Dt)^{-3/2} \,, \tag{2}$$

where $D \equiv D_A + D_B$ is the relative diffusion coefficient. This generalizes the asymptotic behavior derived earlier [5,2] for a reversible geminate pair ($c = 0$). We have previously studied isolated pair kinetics experimentally [5], for an ROH dye molecule (8-hydroxypyrene 1,3,6-trisulfonate, HPTS) which dissociates in the excited state to produce a geminate proton–excited anion pair. Thus $\mathcal{S}(0) = 0$. The solvated proton is attracted by the quadruply charged excited-anion and recombines with it reversibly, without its quenching. We now report an extension of the measurements to low pH values [6], aimed at checking whether proton competition for reversible binding results in a power-law approach to equilibrium, Eq. (2).

Using the time-correlated single photon counting (TCSPC) technique, the blue ROH fluorescence signal in aqueous solutions is followed to 50ns with good signal/noise ratio. Careful background subtraction has been applied. This is important when probing over many decades in intensity. To obtain the binding probability, $1 - S(t)$, we multiply the corrected fluorescence signal by $\exp(t/\tau_f)$, where τ_f=5.15ns is the ROH fluorescence lifetime. In addition, we have applied a 25-point numerical data smoothing routine, effective in white-noise filtering. The maximal count (>300,000) has been normalized to the peak in the numerical propagation of the appropriate time-dependent Smoluchowski equation

Springer Series in Chemical Physics, Vol. 55 Ultrafast Phenomena VIII
Editors: J.-L. Martin · A. Migus · G.A. Mourou · A.H. Zewail

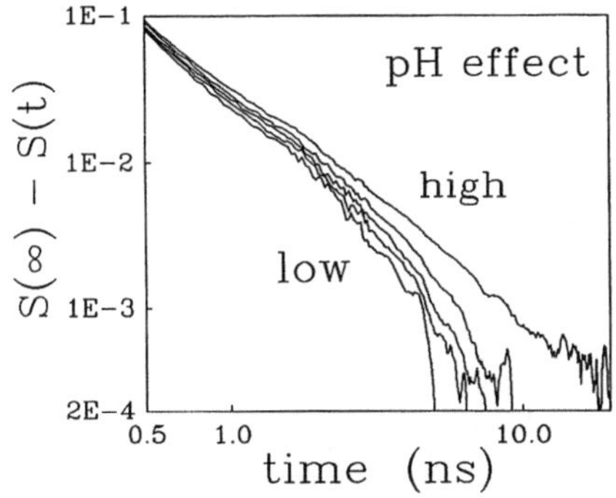

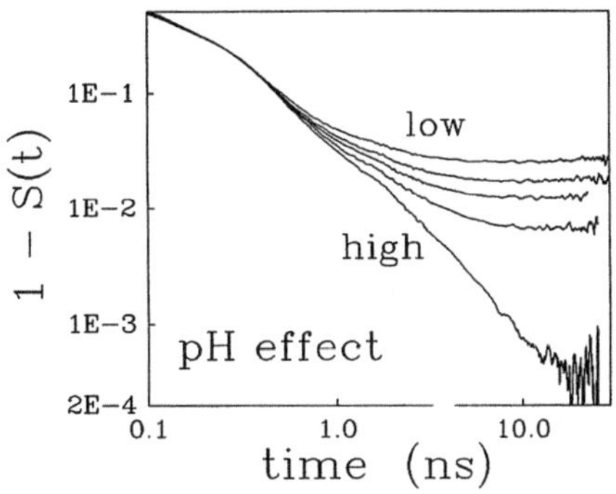

Figure 1 (left). The pH effect on HPTS fluorescence data. Proton concentrations are (bottom to top): 0, 0.3, 0.6, 0.8 and 1.4mM. Data shown after background subtraction, lifetime correction and numerical smoothing as described in the text.

Figure 2 (right). Long time approach to equilibrium in excited-state HPTS dissociation in acidic aqueous solutions. Obtained from the data in Fig. 1 by subtracting the asymptotic equilibrium values.

[5], convoluted with the experimental instrument response function (55ps FWHM). As seen from the log-log plots in Fig. 1, the data cover more than three orders of magnitude.

The lowest curve, representing a measurement in pure water (pH≈6), exhibits a power law decay with an asymptotic slope ≈3/2. As acid ($HClO_4$) is added to increase the concentration, c, of homogeneous protons, an equilibrium plateau develops. The approach to equilibrium is obtained by subtracting the asymptotic plateau, $1 - S(\infty)$, as determined numerically by averaging the data between 10 and 15ns. Results shown in Fig. 2 seem indeed to exhibit a power-law behavior

$$[\mathcal{S}(\infty) - \mathcal{S}(t)]/[\mathcal{S}(\infty) - \mathcal{S}(0)] \sim A/t^{\alpha}. \tag{3}$$

The surprise is that the asymptotic power, α, is not independent of concentration: It is 3/2 at infinite dilution and apparently increases with c. A quantitative determination of α from the noisy data of Fig. 2 depends on the range chosen for performing the linear regression. Figure 3 represents results obtained in the range 0.3-3.0ns (0.3-5.0ns for pure water), with a correlation coefficient 0.990-0.998. Present data (triangles) fit well with our previous measurements [6].

Since an increase of α with c has not been predicted by any of the approximate theories, we have tried to detect it in one dimensional Brownian simulations [3]. In these simulations, n non-interacting random walkers on a linear lattice of l sites may be trapped reversibly in a static, single-particle binding site. When the concentration, $c = n/l$, is increased by decreasing l we find that α increases above its infinite-dilution value of 1/2, but as n is varied for fixed l there are no detectable changes in α. This suggests that in the thermodynamic limit of an infinitely large lattice, α is independent of concentration. Whether the observed experimental variations should be attributed to proton-proton interactions in three dimensions or other sources remains under investigation.

The present data provides a nice proof that chemical reactions can reach true (quasi) equilibrium in the excited-state, limited only by the finite radiative lifetime, τ_f: The excited-state equilibrium coefficient shows an equilibrium salt effect, previously observed directly only in ground-state reactions. To verify this, K_{eq} is calculated from the experimental values of $S(\infty)$ via Eq. (1). It decreases with increasing c due to increased

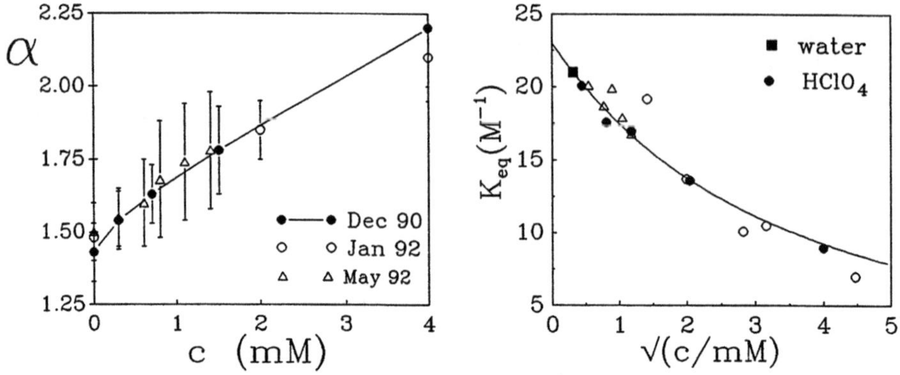

Figure 3 (left). Dependence of the asymptotic power, Eq. (3), on proton concentration. Data from the present measurements are shown by the open triangles. Previous data are from ref. [6].
Figure 4 (right). Dependence of the recombination equilibrium coefficient on proton concentration. Data sets correspond to those in Fig. 3. c includes 0.1mM Na^+ produced when dissolving the HPTS salt.

screening (by the protons) of the Coulombic proton-HPTS attraction (Fig. 4). For small c, this self-screening is well described by the Debye-Hückel theory

$$K_{eq}(c) = K_{eq}(0)\,\exp[-(R_D/a)\,B\sqrt{c}/(1+B\sqrt{c})] \quad . \tag{4}$$

The Debye radius, $R_D = 28.3$Å, corresponds to a quadruply charged ion (the HPTS anion) while $B = 2.3\text{M}^{-\frac{1}{2}}$ represents screening by singly charged ions (the protons). The contact distance, $a = 7$Å, is the same as in our Smoluchowski propagations [5]. By fitting an earlier data set (full circles) we get $K_{eq}(0) = 23.0\text{M}^{-1}$. This agrees with

$$K_{eq}(0) = 4\pi a^2\,\kappa_r\,\exp[R_D/a]\,/\kappa_d \quad , \tag{5}$$

where the intrinsic rate parameters, $\kappa_r = 7.4 \cdot 10^{-3}$Å/ps and $\kappa_d = 8 \cdot 10^{-3}\text{ps}^{-1}$, are obtained from fitting the pH≈6 kinetics to the reversible geminate recombination theory [5]. The Debye-Hückel theory is much less successful in describing *kinetic* salt effects in this reaction [7].

In conclusion, a new asymptotic behavior of reversible binary chemical reactions in solution has been revealed. While the short time behavior is affected by the initial distribution and the interaction potential, the long-time behavior carries the imprint of the equation of motion governing the time evolution, which is the fundamental reason justifying its study.

References

[1] V. Kuzovkov and E. Kotomin, *Rep. Prog. Phys.* **51**, 1479 (1988), and ref. therein.
[2] N. Agmon and A. Szabo, *J. Chem. Phys.* **92**, 5270 (1990).
[3] N. Agmon, H. Schnörer and A. Blumen, *J. Phys. Chem.* **95**, 7326 (1991).
[4] A. Szabo and R. Zwanzig, *J. Stat. Phys.* **65**, 1057 (1991);
A. Szabo, *J. Chem. Phys.* **95**, 2481 (1991).
[5] E. Pines, D. Huppert and N. Agmon, *J. Chem. Phys.* **88**, 5620, 5631 (1988);
D. Huppert, E. Pines and N. Agmon, *J. Opt. Soc. Amer.* **B7**, 1545 (1990).
[6] D. Huppert, S. Y. Goldberg, A. Masad and N. Agmon, *Phys. Rev. Lett.*, submitted.
[7] E. Pines, D. Huppert and N. Agmon, *J. Phys. Chem.* **95**, 666 (1991).

Dynamics and Mechanism of Cu-Porphyrin Triplet Quenching Through Liganding by Oxygen-Containing Solvents

V.S. Chirvony[1] *and R. Gadonas*[2]

B.I. Stepanov Inst. of Physics, Acad. Sci. Rep. Belarus, Scoryna prosp. 70, 220602 Minsk, Republic Belarus
[2]Laser Research Center, Vilnius University, Sauletekio all. 9, Build. 3, 232054 Vilnius, Lithuania

Our experiments were performed with tunable picosecond absorption spectrometer (20 ps, 380...1200 nm) for nondegassed Cu-porphyrin (CuP) solutions (10^{-4}M) at room temperature.

We have revealed that in a series of oxygen-containing solvents (O-solvents) CuP triplet lifetime is strongly shortened as compared with benzene or toluene solutions (see Table 1). A decay of the CuP triplet state absorption (lifetime τ_1) was found to be accompanied with the appearance and following decay (lifetime τ_2) of a new transient absorption different from both T_1 and S_0 states absorptions.The lifetimes τ_2 of the new transient species obtained in different solvents are listed in Table 1 for Cu(II)-octaethylporphyrin (CuOEP) and Cu(II)-tetraphenylporphyrin (CuTPP) solutions.

From experiments with mixtures of solvents (THF/benzene etc.) the lifetime τ_1 was found to be nearly inversely to the concentration of O-solvents. This proves the dynamical character of the CuP T_1 state quenching due to diffusional encounters with O-molecules.

To elucidate the nature of the observed transient species and ascertain the mechanism of the observed CuP T_1 state quenching consider the problem of CuP liganding by O-molecules (so called organic oxygen bases - O-bases). At room temperature, ground state CuP molecules are four-coordinated and not liganded by O-molecules even in net O-solvents. Only at 77K about 50% complexation is observed with the Soret band splitting into two separated bands, red-shifted band belonging to the CuP liganded by O-solvent. No luminescence T_1-S_0 was detected under excitation to the red-shifted Soret band in contrast to high intensity luminescence observed under excitation to the shortwavelength Soret band belonging to the four-coordinated CuP. Therefore, the triplet state of the CuP molecules liganded by O-molecules is strongly quenched at 77K. On the bases of this observation we believe

Table 1. Lifetimes of the CuP excited states

Solvent	CuOEP		CuTPP	
	τ_1, ps	τ_2, ps	τ_1, ps	τ_2, ps
Toluene	120x10^3	-	35x10^3	-
Pyridine	80±10	-	45±15	-
Piperidine	150±20	-	115±20	-
Tetrahydrofuran	140±10	360±40	35±10	190±10
1,4-dioxane	65±10	345±40	100±10	395±40
CH/toluene(1:1)	450±150	(*)	400±40	560±40
DMSO/toluene(1:1)	215±20	-	200±20	-
DMFA/toluene(1:1)	175±20	-	130±10	-

(*) - The two kinetics have not been correctly distinguished,
CH-cyclohexanone, DMSO-dimethylsulfoxide, DMFA-dimethylformamide.

Springer Series in Chemical Physics, Vol. 55 **Ultrafast Phenomena VIII**
Editors: J.-L. Martin · A. Migus · G.A. Mourou · A.H. Zewail

CuP triplet state quenching at room temperature to be due to its liganding by O-molecules through O atom with the following T_1 state quenching.

Similar mechanism was proposed for CuP T_1 state quenching by nitrogen-containing organic bases (N-bases) /1/. Liganding CuP triplets by N-bases was assumed results in CuP CT state lowering below T_1 state with simultaneous quenching of the last and fast dissociation of the exciplex CuP-N-base /1/. CT state population was not observed by authors /1/, and only T_1 state lifetime shortening was found experimentally in contrast to our observation of the new transient species appearance upon CuP T_1 state quenching by O-bases.

To elucidate the nature of the transient species carefull analysis of its absorption spectrum was performed. The analysis has shown the spectrum to be not similar to the structureless absorption spectra belonging to the porphyrin S_1 or T_1 states and to have distinct absorption bands in the same region as for the ground state with the transient species absorption bands being red-shifted and broadened as compared with the ground state bands.

Similar behavour of transient absorption spectrum was observed earlier /2,3/ for Ni(II)porphyrins upon (d,d) state population. It is known, however, that for four-coordinated Cu-porphyrins the (d,d) state lies much above primerely excited S_1 state and cannot be populated after excitation of the last. The picture becomes different when liganding by O-bases occurs. After this, theoretical calculations predict pronounced (d,d) state lowering together with CT state lowering /1/.We suggest, this situation is realized in our experiments where liganding CuP molecules in there T_1 state by O- bases lowers (d,d) state below T_1 with the following population of the former.

Additional data were obtained with the transient CARS and coherent Raman methods /4/ in favour of above proposed explanation of the new transient species appearance. At sufficiently high excitation energies, a revercible appearance of additional vibrational lines was found, the lines being 5-45 cm^{-1} red-shifted as compared with the corresponding ground state lines that is in full agreement with the CARS vibrational data obtained earlier upon Ni(II)porphyrins (d,d) state population /5/.

References:

1. D.Kim, D.Holten, M.Gouterman. J.Am.Chem.Soc. 106 (1984) 2793
2. V.S.Chirvony, B.M.Dzhagarov, Yu.V.Timinskyi, G.P.Gurinovich. Chem.Phys.Lett.70 (1980) 79
3. D.Kim, C.Kirmaier, D.Holten. Chem.Phys.75 (1983) 305
4. P.A.Apanasevich, R.Gadonas, V.V.Kvach, V.Krasauskas, V.A.Orlovich, V.S.Chirvony. Dokl.Akad.Nauk SSSR. 297 (1987) 1395
5. P.A.Apanasevich, V.V.Kvach, V.A.Orlovich. J.Ram.Spectr.20 (1989) 125

Fast Processes in Liquid Alkane Photolysis Above the Ionization Threshold

M. Sander, U. Brummund, K. Luther, and J. Troe

Institut für Physikalische Chemie, Universität Göttingen, Tammannstr. 6, W-3400 Göttingen, Fed. Rep. of Germany

Abstract. *Transient absorption of neat liquid alkanes after two-photon excitation at 248.5 nm is measured and the kinetics of excited state relaxation and geminate ion recombination deduced.*

Fast processes occurring in liquid alkanes upon radiolytical or optical excitation have been under investigation for many years. However, pulse radiolytical experiments have so far been limited to time scales larger than 10 ps and the only two-photon laser-experiments with subpicosecond resolution [1] have been performed at excitation energies below the ionization threshold of alkanes. The attempted interpretation of the transient absorption signals in [1] as due to geminate electron-hole recombination therefore has to be rejected.

In this work, 0.4 ps, 1 mJ - pulses at 248.5 nm with intensities of ca. 10 GW/cm² were obtained from a hybrid excimer - dye laser system similar to that described in [2] and employed to excite neat liquid alkanes 10 eV (2 · hc / 248.5 nm) above the ground state and thereby also well above the ionization onset known to lie at 8.0 - 8.9 eV [3, 4]. Transient absorption was monitored by probing with time-delayed pulses at 497 nm. Typical profiles of absorbance vs. time delay for n-pentane and n-dodecane are shown in Fig. 1.

The transient absorption is ascribed to excited neutral alkane molecules together with alkane radical cations. The kinetic behaviour is found to consist of a sum of three exponentials for all alkanes studied. It is the aim of this paper to identify the various processes giving rise to the observed kinetic components. A more detailed discussion can be found elsewhere [5].

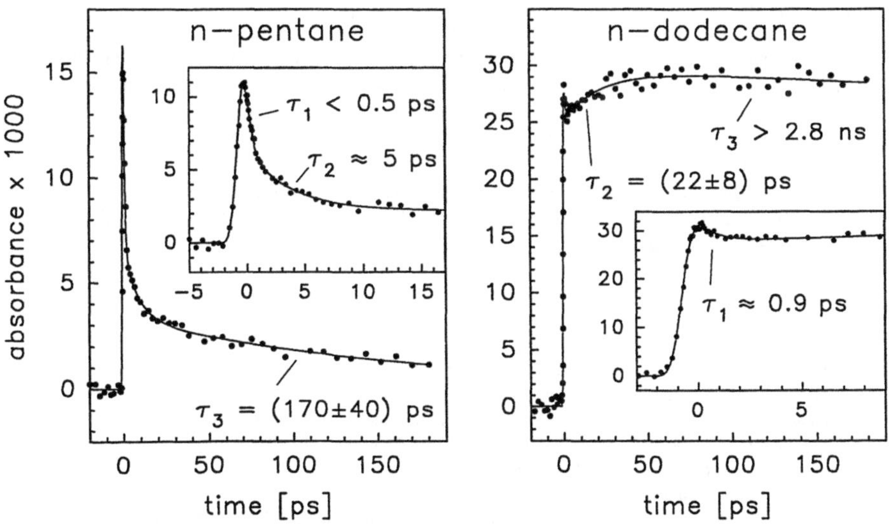

Fig. 1. Absorbance vs. time profiles of n-pentane and n-dodecane at 497 nm upon two-photon excitation at 248.5 nm. Solid line: fit to tri-exponential kinetics. Insets: measurements with higher data point density around zero delay time.

The slowest component exhibiting decay times τ_3 between (170±40) ps for n-pentane, (280±100) ps for n-hexane and several ns for the higher members of the series is for the latter known to equal the fluorescence lifetime of the respective substances. So, absorbance occurring at late times has to be ascribed predominantly - if not exclusively - to $S_n \leftarrow S_1$- absorption of electronically excited molecules in the S_1 state known to decay mainly by the molecular elimination reaction yielding H_2 and 1-alkene.

Measurements for 2,2,4-trimethylpentane (isooctane) and cyclopentane yielded shorter S_1 lifetimes than found for the studied members of the n-series: (38±12) ps and (100±50) ps respectively, the very short lifetime of isooctane(S_1) explaining the fact that attempts to detect fluorescence from this molecule have failed so far.

Assuming a nearly quantitative yield of molecules in the S_1 state at times of ca. 50 ps, the $S_n \leftarrow S_1$ absorption coefficient for the longer-chain n-alkanes can be estimated from the end absorbance and the energy absorbed: One gets a very high value of about $2 \cdot 10^4\ M^{-1}cm^{-1}$, which seems to arise from the great number of different transitions contributing to the broad absorption band.

The second component of the kinetics is identified as geminate electron-hole recombination. This component is barely recognizable in n-pentane through n-heptane, but becomes apparent when it changes its direction from a decay into a re-rise of absorption from n-octane on. It is not visible in the data of [1]. The change of direction and the increase of the amplitude of this component above C_8 is explained by the relative positions of the absorption maxima of alkane radical cations and molecules in the S_1 state as found by Tagawa et al. [6] in a pulse radiolysis study. Both maxima starting near the probe wavelength in the region of 500 nm at C_6, the maximum of cation absorption is far stronger red-shifted with increasing chain-length than that of $S_n \leftarrow S_1$- absorption. Hence, at the probe wavelength cation absorption is more and more diminished relative to excited state absorption. The mean geminate ion pair lifetime τ_2 is found for the n-alkanes from C_6 on always between 12 and 22 ps.

The amplitude of the re-rise component in n-dodecane of about 12 % of total absorption together with the observed 50 % decrease of absorption at 500 nm upon addition of CCl_4 (which nearly quantitatively converts excited molecules to radical cations) in the pulse radiolysis of dodecane [6] allow to estimate the fraction of ions among all initially excited molecules to be only about 25%.

Dissociation into radical fragments is strongly excitation-energy dependent [4], occurs mainly from higher excited singlet states, and therefore must take place prior to Internal Conversion into the S_1 state. Thus the initial decay which is almost unresolvably fast in n-pentane but slows down to a nearly constant value of τ_1 = (1.1±0.4) ps for the higher n-alkanes can now be identified as excited state fragmentation into radicals competing with Internal Conversion. The much greater proneness to radical fragmentation of lower n-alkanes than of higher ones corresponds to the observed amplitude of the initial rapid decay component decreasing with chain-length.

Further studies are required with subpicosecond time resolution and probe wavelengths below 300 nm in order to establish the exact relationship between product radical formation and excited state disappearance.

References

[1] R.M. Bowman, H.Lu, K.B. Eisenthal, *J.Chem.Phys.* **1988**, *89*, 606

[2] S. Szatmári, F.P. Schäfer, *Optics Comm.* **1988**, *68*, 196

[3] J. Casanovas, R. Grob, D. Delacroix, J.P. Guelfucci, D. Blanc, *J.Chem.Phys.* **1981**, *75*, 4661

[4] F.P. Schwarz, D. Smith, S.G. Lias, P. Ausloos, *J.Chem.Phys.* **1981**, *75*, 3800

[5] M. Sander, *Dissertation* Göttingen **1992**, and
M. Sander, U. Brummund, K. Luther, J. Troe, *in preparation*

[6] S. Tagawa, N. Hayashi, Y. Yoshida, M. Washio, Y. Tabata, *Radiat.Phys.Chem.* **1989**, *34*, 503

Index of Contributors

Springer Series in Chemical Physics

Editors: Vitalii I. Goldanskii   Fritz P. Schäfer   J. Peter Toennies

1 **Atomic Spectra and Radiative Transitions** By I. I. Sobelman
2 **Surface Crystallography by LEED** Theory, Computation and Structural Results. By M. A. Van Hove, S. Y. Tong
3 **Advances in Laser Chemistry** Editor: A. H. Zewail
4 **Picosecond Phenomena** Editors: C. V. Shank, E. P. Ippen, S. L. Shapiro
5 **Laser Spectroscopy** Basic Concepts and Instrumentation By W. Demtröder 3rd Printing
6 **Laser-Induced Processes in Molecules** Physics and Chemistry Editors: K. L. Kompa, S. D. Smith
7 **Excitation of Atoms and Broadening of Spectral Lines** By I. I. Sobelman, L. A. Vainshtein, E. A. Yukov
8 **Spin Exchange** Principles and Applications in Chemistry and Biology By Yu. N. Molin, K. M. Salikhov, K. I. Zamaraev
9 **Secondary Ion Mass Spectrometry SIMS II** Editors: A. Benninghoven, C. A. Evans, Jr., R. A. Powell, R. Shimizu, H. A. Storms
10 **Lasers and Chemical Change** By A. Ben-Shaul, Y. Haas, K. L. Kompa, R. D. Levine
11 **Liquid Crystals of One- and Two-Dimensional Order** Editors: W. Helfrich, G. Heppke
12 **Gasdynamic Laser** By S. A. Losev
13 **Atomic Many-Body Theory** By I. Lindgren, J. Morrison
14 **Picosecond Phenomena II** Editors: R. M. Hochstrasser, W. Kaiser, C. V. Shank
15 **Vibrational Spectroscopy of Adsorbates** Editor: R. F. Willis
16 **Spectroscopy of Molecular Excitions** By V. L. Broude, E. I. Rashba, E. F. Sheka
17 **Inelastic Particle-Surface Collisions** Editors: E. Taglauer, W. Heiland
18 **Modelling of Chemical Reaction Systems** Editors: K. H. Ebert, P. Deuflhard, W. Jäger
19 **Secondary Ion Mass Spectrometry SIMS III** Editors: A. Benninghoven, J. Giber, J. László, M. Riedel, H. W. Werner
20 **Chemistry and Physics of Solid Surfaces IV** Editors: R. Vanselow, R. Howe
21 **Dynamics of Gas-Surface Interaction** Editors: G. Benedek, U. Valbusa
22 **Nonlinear Laser Chemistry** Multiple-Photon Excitation By V. S. Letokhov
23 **Picosecond Phenomena III** Editors: K. B. Eisenthal, R. M. Hochstrasser, W. Kaiser, A. Laubereau
24 **Desorption Induced by Electronic Transitions DIET I** Editors: N. H. Tolk, M. M. Traum, J. C. Tully, T. E. Madey
25 **Ion Formation from Organic Solids** Editor: A. Benninghoven
26 **Semiclassical Theories of Molecular Scattering** By B. C. Eu
27 **EXAFS and Near Edge Structures** Editors: A. Bianconi, L. Incoccia, S. Stipcich
28 **Atoms in Strong Light Fields** By N. B. Delone, V. P. Krainov
29 **Gas Flow in Nozzles** By U. G. Pirumov, G. S. Roslyakov
30 **Theory of Slow Atomic Collisions** By E. E. Nikitin, S. Ya. Umanskii
31 **Reference Data on Atoms, Molecules, and Ions** By A. A. Radzig, B. M. Smirnov
32 **Adsorption Processes on Semiconductor and Dielectric Surfaces I** By V. F. Kiselev, O. V. Krylov
33 **Surface Studies with Lasers** Editors: F.R. Aussenegg, A. Leitner, M.E. Lippitsch
34 **Inert Gases** Potentials, Dynamics, and Energy Transfer in Doped Crystals. Editor: M. L. Klein
35 **Chemistry and Physics of Solid Surfaces V** Editors: R. Vanselow, R. Howe
36 **Secondary Ion Mass Spectrometry, SIMS IV** Editors: A. Benninghoven, J. Okano, R. Shimizu, H. W. Werner
37 **X-Ray Spectra and Chemical Binding** By A. Meisel, G. Leonhardt, R. Szargan
38 **Ultrafast Phenomena IV** By D. H. Auston, K. B. Eisenthal
39 **Laser Processing and Diagnostics** Editor: D. Bäuerle

MIX
Papier aus verantwortungsvollen Quellen
Paper from responsible sources
FSC® C105338

If you have any concerns about our products,
you can contact us on
ProductSafety@springernature.com

In case Publisher is established outside the EU,
the EU authorized representative is:
Springer Nature Customer Service Center GmbH
Europaplatz 3, 69115 Heidelberg, Germany

Printed by Libri Plureos GmbH
in Hamburg, Germany